Biofilms in Medicine, Industry and Environmental Biotechnology

Integrated Environmental Technology Series

The *Integrated Environmental Technology Series* addresses key themes and issues in the field of environmental technology from a multidisciplinary and integrated perspective.

An integrated approach is potentially the most viable solution to the major pollution issues that face the globe in the 21st century.

World experts are brought together to contribute to each volume, presenting a comprehensive blend of fundamental principles and applied technologies for each topic. Current practices and the state-of-the-art are reviewed, new developments in analytics, science and biotechnology are presented and, crucially, the theme of each volume is presented in relation to adjacent scientific, social and economic fields to provide solutions from a truly integrated perspective.

The *Integrated Environmental Technology Series* will form an invaluable and definitive resource in this rapidly evolving discipline.

Series Editor

Dr Ir Piet Lens, Sub-department of Environmental Technology, The University of Wageningen, P.O. Box 8129, 6700 EV Wageningen, The Netherlands. (piet.lens@algemeen.mt.wag-ur.nl)

Published titles

Biofilms in Medicine, Industry and Environmental Biotechnology:
   Characteristics, analysis and control
Decentralised Sanitation and Reuse: *Concepts, systems and implementation*
Environmental Technologies to Treat Sulfur Pollution: *Principles and engineering*
Water Recycling and Resource Recovery in Industries: *Analysis, technologies and implementation*

Forthcoming titles

Phosphorus in Environmental Technologies: *Recovery, reuse and applications*
Pond Treatment Technology

www.iwapublishing.com

Biofilms in Medicine, Industry and Environmental Biotechnology

Characteristics, Analysis and Control

Edited by
Piet Lens, Anthony P. Moran, Therese Mahony,
Paul Stoodley and Vincent O'Flaherty

Published by IWA Publishing, Alliance House, 12 Caxton Street, London SW1H 0QS, UK

Telephone: +44 (0) 20 7654 5500; Fax: +44 (0) 20 7654 5555; Email: publications@iwap.co.uk
Web: **www.iwapublishing.com**

First published 2003
© 2003 IWA Publishing

Printed by TJ International (Ltd), Padstow, Cornwall, UK
Typeset by Gray Publishing, Tunbridge Wells, Kent, UK

British Library Cataloguing in Publication Data
A CIP catalogue record for this book is available from the British Library

Library of Congress Cataloging-in-Publication Data
A catalog record for this book is available from the Library of Congress

ISBN: 1 84339 019 1

Contents

PART ONE   BIOFILM CHARACTERISTICS   **1**

Section 1   Biofilm formation   **3**

1   Macroscopic and microscopic adhesive properties of
microbial cell surfaces   5
V. Vadillo-Rodríguez, H.J. Busscher and H.C. van der Mei

2   The role of hydrophobicity and exopolymers in initial
adhesion and biofilm formation   16
J. Azeredo and R. Oliveira

3   The role of coaggregation in oral biofilm formation   32
*P.E. Kolenbrander, R.F. Lerud, D.S. Blehert,
P.G. Egland, J.S. Foster and R.J. Palmer Jr*

4   Genetics of biofilm formation   47
M. Espinosa-Urgel and J.-L. Ramos

5   The role of cell signalling in biofilm development   63
B.L. Purevdorj and P. Stoodley

Section 2   Biofilm composition   **79**

Section 2(A)   Chemical   **79**

6   Molecular architecture of the biofilm matrix   81
D.G. Allison

7   Physico-chemical properties of extracellular
polymeric substances   91
A.P. Moran and Å. Ljungh

Section 2(B)   Biological   **113**

8   Biofilms on corroding materials   115
     I.B. Beech and C.M.L.M. Coutinho

9   Biofilms in wastewater treatment systems   132
     V. O'Flaherty and P. Lens

10   Bioaerosols and biofilms   160
     S.G. Jennings, A.P. Moran and C.V. Carroll

11   Biofilms and protozoa: a ubiquitous health hazard   179
     A.W. Smith and M.R.W. Brown

PART TWO   ANALYTICAL TECHNIQUES   **193**

Section 3   Biofilm cultivation apparatus   **195**

12   Use of flow cells and annular reactors to study biofilms   197
     P. Stoodley and B.K. Warwood

13   Experimental systems for studying biofilm growth in
     drinking water   214
     R. Boe-Hansen, H.-J. Albrechtsen and E. Arvin

14   Efficacy testing of disinfectants using microbes
     grown in biofilm constructs   230
     G. Wirtanen, S. Salo and P. Gilbert

15   Steady-state heterogeneous model systems in
     microbial ecology   236
     J. Wimpenny

Section 4   Analytical techniques for biofilm properties   **257**

Section 4(A)   Physico-chemical properties   **257**

16   Use of X-ray photoelectron spectroscopy and
     Atomic force microscopy for studying interfaces
     in biofilms   259
     *P.G. Rouxhet, C.C. Dupont-Gillain, M.J. Genet
     and Y.F. Dufrêne*

17   Use of ^{1}H NMR to study transport processes in biofilms   285
     P. Lens and H. Van As

18   Screening of lectins for staining lectin-specific
     glycoconjugates in the EPS of biofilms   308
     C. Staudt, H. Horn, D.C. Hempel and T.R. Neu

Section 4(B)   Biotic properties   **329**

19   Environmental electron microscopy applied to biofilms   331
     R. Ray and B. Little

20   Use of molecular probes to study biofilms   352
     B. Zhang, B. Mariñas and L. Raskin

21   Use of microsensors to study biofilms   375
     Z. Lewandowski and H. Beyenal

22   Use of mathematical modelling to study biofilm
     development and morphology   413
     C. Picioreanu and M.C.M. van Loosdrecht

PART THREE   CONTROL OF BIOFILMS   **439**

Section 5   Biofilm monitoring   **441**

23   Biofilm monitoring by photoacoustic spectroscopy   443
     T. Schmid, U. Panne, C. Haisch and R. Niessner

24   Quartz crystal microbalance with dissipation monitoring:
     a new tool for studying biofilm formation in real time   450
     M. Rudh

25   Monitoring biofouling using infrared absorbance   461
     T.R. Bott

Section 6   Biofilm disinfection   **471**

26   Factors that affect disinfection of pathogenic biofilms   473
     S.B.I. Luppens, M.W. Reij, F.M. Rombouts and T. Abee

27   Device-associated infection: the biofilm-related problem
     in health care   503
     M. Cormican

28   Bacterial resistance to biocides: current knowledge and
     future problems   512
     A.D. Russell

Section 7   Biofilm control   **535**

29   Resistance of medical biofilms   537
     M.R.W. Brown and A.W. Smith

30   Control of biofilm in the food industry: a
     microbiological survey of high-risk processing facilities   554
     A. Peters

31   Industrial biofilms: formation, problems and control   568
     J.W. Patching and G.T.A. Fleming

32   Microbial fouling control for industrial systems   591
     W.F. McCoy

     Index   607

*Dedicated to the memory of Dr Richie Powell – esteemed
fellow scientist, colleague and friend*

Preface

Biofilms are sessile communities of microbial cells that develop on surfaces in, virtually, all aquatic ecosystems. Microbial cells within these highly structured adherent populations are phenotypically very different from their planktonic counterparts. However, for more than a century the study of cells has emphasised pure and homogenous cultures. The crude analytical/measurement techniques and data-handling capability available determined this approach. The relatively recent discovery that biofilms are the predominant form of microbial life has resulted in considerable attention being devoted to the topic. In some contexts, biofilms represent an under-utilised resource, while in others they are a poorly understood problem. In all situations, they represent a challenging area for interdisciplinary research. This book sets out to discuss the structure and function of biofilms at a fundamental level. A major priority for this volume is to bring readers information on the application of new techniques to analyse biofilms and to illustrate the use of these methodologies in practical situations. The contributions by leading experts from research institutes, industry and academia clearly illustrate the interdisciplinary nature of biofilm research and the breath of applications to which it applies.

The formation of biofilms is a complex series of events, involving interactions between physical and biological processes. Part one of this volume introduces readers to the key factors involved, namely biological, physical and chemical. During the complex process of adhesion, bacterial cells alter their phenotype in response to the proximity of a surface. At the earliest stages of biofilm formation, sessile bacteria are located in groups to form single- or multi-species microcolonies. These cellular juxtapositions, and the exuberant production of extracellular polysaccharide matrix within the developing biofilm, determine the microenvironment of each biofilm microorganism. Different biofilm microbes respond to their differing microenvironmental conditions with different growth patterns, and a structurally complex mature biofilm develops. Physiological co-operactivity is a major factor in shaping the structure and functions of biofilms, and ensures that they become very efficient microbial communities.

Part II of the book introduces the analytical approaches that have revolutionised our understanding of the structure and function of biofilms. Advances in microscopy, particularly the advent of confocal scanning laser microscopy (CSLM), and sophisticated computer image analysis tools have permitted visualisation of biofilm structure to an extent that was not possible previously. The use of fluorescently labelled molecular probes specific for individual microbial species that are present in biofilms has greatly aided this work. CSLM/fluorescent probe investigations of biofilms have shown that many biofilms possess a heterogenous structure. The accepted hypothesis suggests that, at the substratum, the biofilm cells can be arranged in a thin, dense layer of cells, to which are bound some dense, roundly shaped microcolonies filled with extracellular polymers and separated by interstitial voids. The space within the voids is filled with water or with a low-concentration solution of extracellular polymers. Other analytical advances, including the use of NMR spectroscopy and microelectrodes, have revealed information on the physico-chemical characteristics of biofilms, which support the above hypothesis and these areas are included also in Part II.

The particular properties of biofilms are beneficially utilised for environmental protection in bioreactor technology, applied to industrial and municipal wastewater treatment, and several chapters in this book introduce this field. The characteristics of biofilms are of direct practical relevance to key areas of medicine and industry. Aquatic biofilms, which are widespread in medical and dental devices, can be the source of serious nosocomial infections. In addition to their ubiquity, biofilm formation also affords the embedded microorganisms increased protection against conventional control strategies, such as antibiotics, biocides and high temperatures. Thus, not only do biofilms provide a physical environment for harbouring pathogens (such as *Pseudomonas aeruginosa*, *Escherichia coli* and *Streptococcus aureus*) but also makes them harder to kill. Issues relating to the control of biofilms in medicine and industry are dealt with in Part III of this book.

We, as editors, wish to thank all the contributors for their enthusiastic support and timely submission of their manuscripts. This book is based on the Euro Summer School, 'Biofilms in Industry, Medicine and Environmental Biotechnology – The Science', 24–29 August 2002, Galway, Ireland. This Summer School was financially supported by the 'Improving the Human Potential' Program of the EU (Grant No. HPCFCT-2001-00016). In addition to many of the oral presentations from the school, a number of invited contributions are included in this volume. Also, we thank Paul Stoodley and Peg Dirckx, Montana State University, for providing photographs for the cover of this volume. Furthermore, we acknowledge and thank John Curtin, NUI, Galway, Ireland for providing excellent technical assistance during the Euro Summer School and throughout the preparation of this work. We are grateful to Alan Click and Alan Peterson of IWA Publishing for their good-humoured help and editorial support, as well as that of their production team, in realising this book.

March, 2003   Piet Lens, *Wageningen, The Netherlands*
                                               Anthony P. Moran, *Galway, Ireland*
                                               Therese Mahony, *Galway, Ireland*
                                               Paul Stoodley, *Bozeman, Montana, USA*
                                               Vincent O'Flaherty, *Galway, Ireland*

List of contributors

Tjakko Abee
Food Hygiene and Microbiology Group,
Department of Agrotechnology and Food
Sciences, Wageningen University and
Research Center, 6703 HD Wageningen,
The Netherlands

Hans-Jørgen Albrechtsen
Environment and Resources, DTU,
Building 115, Technical University of
Denmark, 2800 Kgs. Lyngby, Denmark

David G. Allison
School of Pharmacy and Pharmaceutical
Sciences,
University of Manchester,
Manchester M13 9PL UK

Joanne Arezedo
Centro de Engenharia Biologica, IBQF,
University de Minho,
P-47100 57 Braga, Portugal

Erik Arvin
Environment and Resources, DTU,
Building 115, Technical University of
Denmark, 2800 Kgs. Lyngby, Denmark

Iwona B. Beech
School of Pharmacy and Biomedical
Sciences, University of Portsmouth, St.
Michael's Building,
Portsmouth PO1 2DT, UK

Haluk Beyenal
Centre for Biofilm Engineering,
366 EPS Building, Montana State
University, Bozeman, MT 59717, USA

David S. Blehert
Oral Infection and Immunity Branch,
National Institute of Dental and
Craniofacial Research, National Institute
of Health, Bethesda,
MD 20892-4350, USA

Rasmus Boe-Hansen
Environment and Resources, DTU,
Building 115, Technical University of
Denmark, 2800 Kgs. Lyngby, Denmark

T. Reg Bott
Department of Chemical Engineering, The
School of Engineering,
University of Birmingham,
Edgbaston, Birmingham B15 2TT, UK

Michael R.W. Brown
Department of Pharmacy and
Pharmacology, University of Bath,
BA 27AY, UK

Henk J. Busscher
Department of Biomedical Engineering,
University of Groningen, 9713 AV
Groningen,
The Netherlands

Cyril V. Carroll
Department of Microbiology,
National University of Ireland, Galway,
Ireland

Martin Cormican
Department of Bacteriology, University
College Hospital, Galway, Ireland

Claudia M.L.M. Coutinho
School of Pharmacy and Biomedical
Sciences, University of Portsmouth, St.
Michael's Building,
Portsmouth PO1 2DT, UK

Yves F. Dufrêne
Université Catholique de Louvain, Unité
de Chimie des Interfaces,
B-1348-Louvain-La-Neuve,
Belgium

Christine C. Dupont-Gillain
Université Catholique de Louvain, Unité
de Chimie des Interfaces,
B-1348-Louvain-La-Neuve,
Belgium

Paul G. Egland
Oral Infection and Immunity Branch,
National Institute of Dental and
Craniofacial Research, National Institute
of Health, Bethesda, MD 20892-4350,
USA

Manuel Espinosa-Urgel
Department Plant Biochemistry, and
Molecular and Cellular Biology, Estacion
Experimental del Zaidin, CSIC, Granada
18008, Spain

Gerard T.A. Fleming
Department of Microbiology,
National University of Ireland, Galway,
Ireland

Jamie S. Foster
Oral Infection and Immunity Branch,
National Institute of Dental and
Craniofacial Research, National Institute
of Health, Bethesda,
MD 20892-4350, USA

Michel J. Genet
Université Catholique de Louvain, Unité
de Chimie des Interfaces,
B-1348-Louvain-La-Neuve, Belgium

Peter Gilbert
School of Pharmacy and Pharmaceutical
Sciences,
University of Manchester,
Manchester M13 9PL, UK

Christoph Haisch
Institute of Hydrochemistry, Technical
University of Munich, Marchioninistrasse.
17, D-81377 Munich, Germany

Dietmar C. Hempel
Institute of Biochemical Engineering,
Technical University of Braunschweig,
38106 Braunschweig, Germany

Harald Horn
Hochschule Magdeburg-Stendal (FH),
Hydrochemistry, 39114 Magdeburg,
Germany

S. Gerard Jennings
Department of Physics,
National University of Ireland, Galway,
Ireland

Paul E. Kolenbrander
Oral Infection and Immunity Branch,
National Institute of Dental and
Craniofacial Research,
National Institute of Health,
Bethesda, MD 20892-4350, USA

Piet Lens
Environmental Technology, Wageningen
University,
PO Box 8129, 6700 EV Wageningen, The
Netherlands

Brenda Little
USN, Naval Research Laboratory, Code
7330 Stennis Space Centre,
MS 39529, USA

Suzanne B.I. Luppens
Cariology Endodontology Pedodontology,
Academic Center for Dentistry
Amsterdam, 1066 EA, Amsterdam,
The Netherlands

Rebecca F. Lerud
Southern Connecticut State University,
New Haven, CT 06515-1355, USA

Zbigniew Lewandowski
Centre for Biofilm Engineering,
366 EPS Building, Montana State
University, Bozeman, MT 59717, USA

Åsa Ljungh
Department of Medical Microbiology,
Dermatology and Infection,
Lund University,
S-223 62 Lund, Sweden

Benito Mariñas
Department of Civil and Environmental
Engineering,
1105 Newmark Civil Engineering
Laboratory, University of Illinois, Urbana,
IL 61801, USA

William F. McCoy
Aquazur Hygiene Services, Ondeo Nalco,
A Subsidiary of SUEZ, Naperville, IL,
USA

Anthony P. Moran
Department of Microbiology,
National University of Ireland, Galway,
Ireland

Reinhard Niessner
Institute of Hydrochemistry,
Technical University of Munich,
Marchioninistrasse 17,
D-81377 Munich, Germany

Thomas R. Neu
Department of Inland Water Research,
UFZ Centre for Environmental Research,
Leipzig-Halle, 39114 Magdeburg,
Germany

Vincent O'Flaherty
Department of Microbiology,
National University of Ireland, Galway,
Ireland

Rosario Oliveira
Centro de Engenharia Biologica, IBQF,
University de Minho,
P-47100 57 Braga, Portugal

Robert J. Palmer Jr
Oral Infection and Immunity Branch,
National Institute of Dental and
Craniofacial Research, National Institute
of Health, Bethesda,
MD 20892-4350, USA

Ulrich Panne
Institute of Hydrochemistry,
Technical University of Munich,
D-81377 Munich, Germany

John W. Patching
Department of Microbiology,
National University of Ireland, Galway,
Ireland

Adrian Peters
School of Applied Science, UWIC,
Llandaff Campus, Western Avenue,
Cardiff, CF5 2YB, UK

Christian Picioreanu
Department of Biochemical Engineering,
Delft University of Technology,
Julianalann 67, 2628 BC Delft, The
Netherlands

B. Laura Purevodorj
Centre for Biofilm Engineering,
366 EPS Building, Montana State
University, Bozeman, MT 59717, USA

Juan-Luis Ramos
Department Plant Biochemistry,
and Molecular and Cellular Biology,
Estacion Experimental del Zaidin, CSIC,
Granada 18008, Spain

Lutgarde Raskin
Department of Civil and Environmental
Engineering,
1105 Newmark Civil Engineering
Laboratory, University of Illinois, Urbana,
IL 61801, USA

Richard Ray
USN, Naval Research Laboratory, Code
7330 Stennis Space Centre,
MS 39529, USA

Martine W. Reij
Food Hygiene and Microbiology Group,
Department of Agrotechnology and Food
Sciences, Wageningen University and
Research Center,
6703 HD Wageningen,
The Netherlands

Frans M. Rombouts
Food Hygiene and Microbiology Group,
Department of Agrotechnology and Food
Sciences, Wageningen University and
Research Center,
6703 HD Wageningen,
The Netherlands

Paul G. Rouxhet
Université Catholique de Louvain, Unité
de Chimie des Interfaces,
B-1348-Louvain-La-Neuve,
Belgium

Mattias Rudh
Q-Sense AB, Stena Center 1B,
SE-412 92 Göteborg, Sweden

A. Denver Russell
Welsh School of Pharmacy,
Cardiff University,
Redwood Building,
Cardiff CF1 3XF, UK

Sato Salo
VTT Biotechnology, Tietotie 2, Espoo, PO
Box 1500, FIN-02044, Finland

Thomas Schmid
Institute of Hydrochemistry,
Technical University of Munich,
D-81377 Munich,
Germany

Anthony W. Smith
Department of Pharmacy and
Pharmacology, University of Bath,
BA2 7AY, UK

Christian Staudt
Department of Inland Water Research,
UFZ Centre for Environmental Research,
Leipzig-Halle,
39114 Magdeburg, Germany

Paul Stoodley
Centre for Biofilm Engineering,
366 EPS Building, Montana State
University, Bozeman, MT 59717, USA

Henk Van As
Department of Molecular Physics,
Wageningen University, PO Box 8129,
6700 EV Wageningen, The Netherlands

Virginia Vadillo-Rodríguez
Department of Biomedical Engineering,
University of Groningen, 9713 AV
Groningen, The Netherlands

Henny C. van der Mei
Department of Biomedical Engineering,
University of Groningen, 9713 AV
Groningen,
The Netherlands

Mark C.M. van Loosdrecht
Department of Biochemical Engineering,
Delft University of Technology, 2628 BC
Delft,
The Netherlands

Bryan K. Warwood
BioSurface Technologies Corporation,
Bozeman, MT 59718, USA

Julian Wimpenny
Cardiff School of Biosciences,
Cardiff University,
Cardiff CF10 3XF, UK

Gun Wirtanen
VTT Biotechnology, Tietotie 2,
Espoo, PO Box 1500,
FIN-02044, Finland

Bo Zhang
Department of Civil and Environmental
Engineering,
1105 Newmark Civil Engineering
Laboratory, University of Illinois, Urbana,
IL 61801, USA

PART ONE

Biofilm characteristics

Section 1   *Biofilm formation*   3

Section 2   *Biofilm composition*

            (A) Chemical   79
            (B) Biological   113

1

Section 1

Biofilm formation

1   Macroscopic and microscopic adhesive properties
    of microbial cell surfaces   5

2   The role of hydrofobicity and exopolymers in
    initial adhesion and biofilm formation   16

3   The role of coaggregation in oral biofilm
    formation   32

4   Genetics of biofilm formation   47

5   The role of cell signalling in biofilm development   63

1

Macroscopic and microscopic adhesive properties of microbial cell surfaces

V. Vadillo-Rodríguez, H.J. Busscher and H.C. van der Mei

1.1 INTRODUCTION

The study of microbial adhesion encompasses a broad range of scientific disciplines, ranging from medicine, dentistry and microbiology to colloid and surface science. Initially, the involvement of colloid and surface scientists originated from the simple realisation that microorganisms are, with respect to their dimensions, colloidal particles and that their adhesion should be predictable by surface thermodynamics (Absolom *et al.*, 1983) or Derjaguin–Landau–Verwey–Overbeek (DLVO) theory-like approaches (Bos *et al.*, 1999). Considering the ubiquitous nature of microbial adhesion, such generalised predictive models would be extremely valuable.

Microorganisms have, from a physico-chemical point of view, generally been considered to be similar to inert polystyrene particles. However, microorganisms are not smooth particles and, in contrast to polystyrene particles, carry long, usually very thin surface structures protruding from the cell surface and radiating outwards into the surrounding liquid. These structures are responsible for adhesion to a variety of surfaces. Bacteria, in particular, carry a wide range of surface structures that have been described on the basis of their ultrastructure and distribution on the cell surface. There are many morphologically distinct types of surface structures and almost every bacterial strain or species carries its own type of surface appendage. Yet, the function of these surface structures, in specific adhesion processes to different substratum surfaces as well as their influence on overall physico-chemical cell-surface characteristics, remains to be identified for most bacterial strains.

Application of physico-chemical models to explain microbial adhesion to solid substratum has been successful for a limited number of strains and species (Van Loosdrecht et al., 1989), despite the macroscopic nature of the input data (i.e. Hamaker constants, acid–base properties, zeta potentials and contact angles). Therefore, relevant physico-chemical measurements on microbial cell surfaces require a microscopic resolution that cannot be accomplished with most currently employed methods.

The aim of this chapter is, on the one hand, to point out the merits of a macroscopic physico-chemical approach towards microbial adhesion, while on the other hand, emphasizing the need of a microscopic physico-chemical surface characterisation.

1.2   MACROSCOPIC PHYSICO-CHEMICAL CHARACTERISTICS OF MICROBIAL CELL SURFACES

Bacterial cell-surface hydrophobicity and charge are commonly accepted as influential on bacterial interactions with their environment. Therefore, contact angles and zeta potentials of microbial cell surfaces are frequently measured for use as input data for predictive, physico-chemical models of their adhesion to surfaces. Furthermore, at a similar overall level as cell-surface hydrophobicities by water contact angles and zeta potentials, chemical composition data of cell surfaces are being obtained by unexpected techniques, like X-ray photoelectron spectroscopy (XPS) (Amory et al., 1988) or secondary-ion mass spectroscopy (SIMS) (Tyler, 1997).

1.2.1   Cell-surface hydrophobicity

Hydrophobicity is a general term utilised to describe the relative affinity of a surface for water. Microbial cell-surface hydrophobicity can only be measured by placing water droplets on carefully prepared and dried microbial lawns (Busscher et al., 1984). If water molecules have a greater preference to surround each other than to contact a microbial cell surface, the surface appears as hydrophobic and water droplets do not spread. In contrast, if water molecules favour a microbial cell surface rather than each other, a water droplet spreads and the surface appears hydrophilic.

Spreading is determined in part by the intermolecular forces, such as Lifshitz–Van der Waals and acid–base forces (Van Oss, 1995), as can be calculated amongst others, from the Hamaker constants and measured contact angles with liquids. The only non-trivial step in contact angle measurements on microbial lawns involves the degree of drying of the lawns. It must be realised that drying of the lawns determines the degree of collapse of microbial surface appendages and therewith the contact angle measured (Busscher *et al.*, 1984). Usually, contact angles measured on microbial lawns increase as a function of drying time until a so-called 'plateau' is reached. As long as the lawns are dried to the plateau for water contact angles, the cell surface is assumed to be in a physiologically relevant state. However, theoretically, cell-surface hydrophobicity cannot be measured solely by water contact angles as this does not allow to account for acid–base interactions as occurring between water molecules and between water molecules and the microbial cell surface. Indeed, whereas water contact angles have been designated as the intrinsic surface hydrophobicity of a microbial cell surface, the thermodynamic cell-surface hydrophobicity, strictly speaking, reflects its surface free energy, including Lifshitz– Van der Waals and acid–base components. Calculation of the thermodynamic cell-surface hydrophobicity requires contact angle measurements with at least three different liquids with varying acid–base properties, as water, formamide, methyleneiodine and/or α-bromonaphthalene (Van Oss *et al.*, 1987).

The cell-surface hydrophobicity describes a macroscopic surface property. The hydrophobicity of surface appendages, like fibrils or fimbriae may be substantially different from the overall cell-surface hydrophobicity. However, despite being an overall cell-surface property, microbial contact angles vary greatly between different strains and even the presence of tufts of fibrils as on *Streptococcus sanguis* strains is reflected in the measured values of contact angles (Busscher *et al.*, 1991). Both *Streptococcus salivarius* HBC12 and *S. sanguis* CR311 VAR3 have bald cell surfaces and are hydrophilic (water contact angles 21° and 31°, respectively) compared with their peritrichously fibrillated (*S. salivarius* HB) and tufted (*S. sanguis* CR311) parent strains. Hence, hydrophobicity is conveyed to the cell surface by fibrils and fimbriae and unlikely by the bald cell surface. Handley *et al.* (1991), for instance, demonstrated that hydrophobic, colloidal gold particles only adsorbed to the tip of fibrils on *S. sanguis* PSH 1b and concluded that the cell-surface hydrophobicity was confined to the ends of the long fibrils (see Figure 1.1).

Therefore, contact angles measured with liquid droplets on a microbial lawn are essentially representative of a fuzzy coat of cellular-surface material, collapsed into a lawn. Therewith results are useful to interpret the long-range interactions between an organism and a substratum surface, but not necessarily for the interpretation of short-range interactions, which may be dominated by structural and chemical cell-surface heterogeneities (Van der Mei *et al.*, 1998a).

1.2.2   Zeta potentials

Particulate electrophoresis is the most common method to determine bacterial zeta potentials and surface charge densities (James, 1991). The measurement of microbial

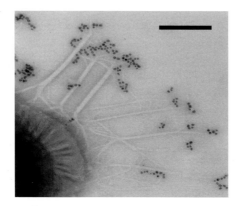

Figure 1.1   Electron micrograph of *S. sanguis* PSH 1b labelled with colloidal gold at pH 10.2. Gold is attached only to the ends of the long fibrils. Cells were stained with 1% methylamine tungstate. Bar marker is 200 nm (adapted from Van der Mei *et al.* (1998a) with permission).

electrophoretic mobilities and the derivation of zeta potentials, thereof by particulate micro-electrophoresis, proceed according to the standard methodologies in physico-chemistry (James, 1991), although the calculation of zeta potentials from measured electrophoretic mobilities is not always straightforward (Van der Wal *et al.*, 1997). Furthermore, it is important to realise that the plane of shear may be removed far away from the bacterial cell wall if long fibrils or fimbriae are present (e.g. as on *Streptococcus mitis* strains) and consequently, in these cases, the zeta potentials measured are not representative for the one of the true cell surface (Figure 1.2).

Moreover, long appendages may collapse onto the cell surface upon increasing the ionic strength, as demonstrated, e.g. by dynamic light scattering (Van der Mei *et al.*, 1994). Depending on ionic strength of the liquid, also electro-osmotic flow of fluid may occur within these polyelectrolyte layers.

Fibrils, fimbriae and even extracellular surface polymers around bacterial cells may be considered as a polyelectrolyte layer, possessing a number of fixed charges, i.e. ionic groups that are covalently linked to the polymer and thus have a strong impact on the electrostatic interactions of bacteria with surfaces. Hayashi *et al.* (2001) and Poortinga *et al.* (2001) have both described that this electrostatic repulsion is often overestimated due to the neglect of bacterial cell-surface softness, i.e. the ease with which electro-osmotic fluid flow develops in the surface layer. Soft, ion-penetrable cell surfaces experience less electrostatic repulsion than similarly charged, hard, ion-impenetrable surfaces, since their diffuse layer charges are driven into the ion-penetrable cell walls causing an effective decrease in surface potential (see also Figure 1.2) and, hence, electrostatic repulsion. Recently, Morisaki *et al.* (1999) explained adhesion of a negatively charged marine bacterium, *Vibrio alginolyticus*, onto a negatively charged substratum by considering the softness of the strains.

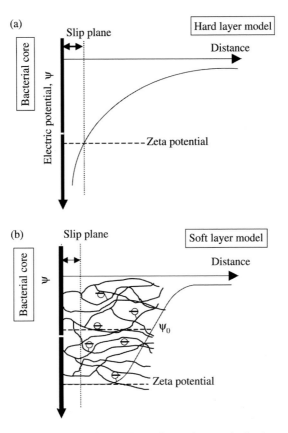

Figure 1.2   Schematic representation of the surfaces of a negatively charged ion-impenetrable (a) and ion-penetrable bacterium (b). The electrokinetic potential decreases exponentially from the ion-impenetrable bacterial core surface and the slip plane, determining its zeta potential close to the surface. The ion-penetrable bacterium is covered by a polyelectrolyte layer with fixed negative charges, through which electrophoretic fluid flow is possible. The slip plane is assumed to remain at approximately the same position as for the ion-impenetrable bacterium inside the soft layer. Therefore, the zeta potential is more negative than the potential Ψ_0 at the outside of the soft layer (taken from Kiers *et al.* (2001) with permission).

When analysing microbial adhesion data in terms of electrostatic interactions, it should not be *a priori* assumed that the electrostatic interaction is repulsive, simply because the zeta potentials of the interacting surfaces are both negative. Indeed, nearly all biological surfaces carry a net negative charge. However, at a more microscopic level than the macroscopic level of particulate micro-electrophoresis, microbial cell surfaces may have positively charged domains mediating adhesion through local electrostatic attraction despite overall repulsion, like described, e.g. the interaction between *Treponema denticola* and human erythrocytes (Cowan *et al.*, 1994). Obviously, a minor number of positively charged sites, while instrumental for adhesion, does hardly affect the macroscopic cell-surface charge density.

1.2.3   Chemical composition of microbial cell surfaces

XPS provides a mean to obtain the chemical composition of the outermost microbial cell surfaces (Amory *et al.*, 1988). XPS spectra of microbial cell surfaces are fairly similar, with carbon (C), nitrogen (N), oxygen (O) and phosphorous (P) being the main elements detected, albeit in different amounts on different isolates. Decomposition of C_{ls} and O_{ls} electron-binding energies has furthermore indicated the presence of lipids, proteins and polysaccharides. As XPS is a high vacuum technique, an extensive sample preparation, including washing, centrifuging and freeze drying, is involved before microbial cell surfaces can be studied by XPS (Rouxhet *et al.*, 1994). These steps obviously bring the cell surface in a state, i.e. far remote from its physiological one. Some authors believe that the integrity of the vulnerable cell surface of especially Gram-negative bacteria, as compared with Gram-positive bacteria, is disrupted by this extensive preparation (Marshall *et al.*, 1994) with a potential impact on the results.

However, the relevance of XPS for the analysis of microbial cell surfaces is supported by the relationships with other physico-chemical cell-surface properties, preferably measured on cells in a more physiologically relevant state than in their dehydrated, freeze-dried state as for XPS. For instance, combinations of contact angle data on a collection of widely different streptococcal strains and XPS have demonstrated that hydrophobicity is conveyed to the cell surface by nitrogen-rich groups, concurrent with the possession of a high isoelectric point (the pH at which the zeta potential is zero) (Van der Mei *et al.*, 1988c). Furthermore, the presence of tufts of fibrils on *S. sanguis* CR311 increased the N/C from 0.066 of a bald variant to 0.085 for the parent strain, indicating a nitrogen (protein)-rich composition of the tufts (Busscher *et al.*, 1991). In the same way, the progressive removal of fibrils from the cell surface of *S. salivarius* HB was accompanied by a significant decrease in the N/C elemental surface composition ratio (Van der Mei *et al.*, 1988b). However, like almost all physico-chemical methods for the study of microbial cell-surfaces properties, the spatial resolution of XPS is inadequate to deal with chemical and structural heterogeneities, like sparsely or unevenly distributed fibrils or fimbriae on cell surfaces.

SIMS is a surface-sensitive technique that also probes the chemical composition of a surface, but through a different principle as XPS. XPS involves the bombardment of a surface with X-rays and the subsequent measurement of the photo-emitted electrons. Since these photo-emitted electrons have discrete kinetic energies that are characteristic of the emitting atoms and their bonding states, they can be applied for chemical analysis. In SIMS, the surface is bombarded with a focused beam of primary ions, the impact of which produces secondary ions that are collected and focused in a mass spectrometer where they are separated according to their mass. Tyler (1997) has shown that SIMS can also be applied to probe microbial cell-surface chemistry. SIMS spectra of four freeze-dried strains, *S. salivarius* HB and three mutants, indicated the presence of proteins, hydrocarbons and carbohydrates on the bacterial cell surfaces, as well as of proteins and teichoic acid on the cell wall. The correlation between SIMS spectra and previous

XPS analysis on those strains was excellent. Taking into account that SIMS is not only capable of providing accurate analysis of surface chemistry, but is also sensitive to the composition and orientation of biomolecules, the potential of this technique to characterise bacterial cell surfaces seems promising but needs to be further explored.

1.3   MICROSCOPIC PHYSICO-CHEMICAL PROPERTIES OF MICROBIAL CELL SURFACES BY ATOMIC FORCE MICROSCOPY

Even though structural and chemical heterogeneities on microbial cell surfaces have an impact on the overall cell surfaces, methods to obtain detailed knowledge on cell-surface heterogeneities are still lacking. Indeed, considering the importance of structural and chemical heterogeneities in microbial adhesion, the development of a generalised model for microbial adhesion to surfaces seems beyond reach. However, the introduction of the atomic force microscope (AFM) and its application to biological surfaces (Dufrêne, 2001) has offered new possibilities to obtain microscopic physico-chemical properties of bacterial cell surfaces.

AFM provides exciting possibilities for probing the structural and physical properties of living microbial cells. Using topographic imaging, cell-surface nanostructures (e.g. appendages and flagella) can be directly visualised (see for instance Figure 1.3) and the changes of cell-surface morphology occurring during physiological processes can be determined).

Force–distance curves, shown in Figure 1.4, can provide complementary information on surface forces and adhesion mechanisms at a square nano-metre scale, yielding new insight into the mechanisms of biological events, such as microbial adhesion and aggregation.

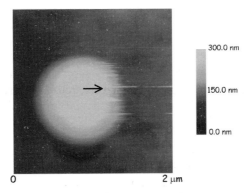

Figure 1.3   AFM contact mode topographic image of *S. mitis* T9 immersed in water. The image reveals characteristic topographic features on the right-hand side of the cell surface, i.e. lines oriented in the scanning direction (marked by an arrow), attributable to fibrils.

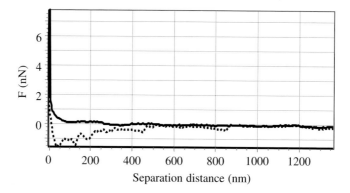

Figure 1.4   Force–distance curve of *S. mitis* T9 in water. Solid line represents the approach curve, while the dashed line indicates the retraction curve. Upon approach, a long-range repulsion, starting at a separation of ∼100 nm, was detected, while no jump-to-contact was observed. Upon retraction, multiple adhesion forces were found.

Initial AFM studies on bacterial cell surfaces have focused on probing surface morphology and surface forces. For instance, Razatos *et al.* (1998) showed that the adhesion force between a silicon nitride AFM tip and *Escherichia coli* was affected by the length of lipopolysaccharide molecules on the cell surface and by the production of a capsular polysaccharide. Furthermore, it was discovered using AFM that *E. coli* JM109 and K12J62 have different surface morphologies dependent on environmental conditions, while lysozyme treatment led to the loss of surface rigidity and eventually to dramatic changes of bacterial shape (Bolshakova *et al.*, 2001). Camesano and Logan (2000) concluded that the interaction between negatively charged bacteria and the silicon nitride tip of an AFM was dominated by electrostatic repulsion. Only a limited number of studies have focused on the characterisation of local properties of bacterial cell surfaces. Recently, for instance, the turgor pressure of a spherical bacterium, *Enterococcus hirae*, in deionised water was derived from the indentation depth caused by an AFM tip and found to be between 4 and 6 $\times$ 10^5 Pa (Yao *et al.*, 2002).

However, in order to develop a ubiquitously valid physico-chemical model for microbial adhesive interactions, microscopic characterisation of properties as hydrophobicity, surface charge density and chemical composition on microbial cell surfaces is required. Due to the structural and chemical heterogeneities that the cell surfaces present, it would be of interest to collect an array of force curves over the entire cell surface. Such an array would produce information about the distribution of different surface properties. For example, using charged or chemically functionalised AFM tips to probe the surface, would allow localizing more specific interactions at a microscopic level. At present, the only charge maps for biological samples have been made for bacteriorhodopsin membrane patches (Butt, 1992) and phospholipid bilayer patches (Heinz and Hoh, 1999) on hard substrates. From the known surface charge density of the substratum, it was possible to calculate a reasonable value for the surface charge density of the membrane

(Butt, 1992). Hydrophobicity at a microscopic level has been probed on spore surfaces of the fungus *Phanerochaete chrysosoporium* (Dufrêne, 2000) by using chemically modified AFM probes, terminated with OH (hydrophilic) and CH_3 (hydrophobic) groups.

1.4   TOWARDS RELATIONS BETWEEN MICROSCOPIC AND MACROSCOPIC PROPERTIES

Overall properties are a macroscopic expression of interactions taking place at a microscopic level. Extension of a microscopic property of microbial cell surface to the entire surface should theoretically lead to the corresponding macroscopic property. However, it will be a delicate task to amalgamate microscopic properties derived from AFM measurements into the macroscopic cell-surface properties. The different conditions, under which the macroscopic and microscopic properties are measured, should be taken into account. Properties, such as hydrophobicity and surface charge, from a macroscopic point of view, are determined in a two-component system, i.e. bacterium and liquid medium. At the microscopic level, the AFM tip interacts with the bacterium in a medium, and consequently, bacterium and medium would respond to this third component, i.e. the tip. Therefore, hydrophobicity and charge mappings derived from AFM measurements depend on the properties of the AFM tip as well. Macroscopic and microscopic properties estimated in both systems are related, but to find out how, constitutes an enormous challenge. It involves an accurate knowledge of geometry and physico-chemical characteristic of the AFM tip as well as of a theory describing long- and short-range interactions in such a system.

1.5   CONCLUSIONS

Physico-chemical properties of microorganisms can vary widely and generalisations at the species or even strain level are virtually impossible. The degree of success of physico-chemical models to explain microbial adhesion frequently decreases as the complexity of cell surface appendages on the organisms under consideration increases. Understanding how microscopic properties can be amalgamated into the macroscopic properties previously determined by many different research groups all over the world for a large variety of different strains and species is an imperative next step in the characterisation of microbial cell surfaces. Subsequently, a generalised physico-chemical theory to account for bacterial adhesion to substratum surfaces will become in reach. Recently, the introduction of the AFM and its application to biological surfaces has opened a new avenue to obtain microscopic physico-chemical properties of bacterial cell surfaces. It is a challenge for the future to develop models, based on these improved methodologies, that will allow predicting bacterial adhesion from the initial adhesion events.

REFERENCES

Absolom, D.R., Lamberti, F.V., Policova, Z., Zingg, W., Van Oss, C.J. and Neumann, A.W. (1983) Surface thermodynamics of bacterial adhesion. *Appl. Environ. Microbiol.* **46**, 90–97.

Amory, D.E., Genet, M.J. and Rouxhet, P.G. (1988) Application of XPS to the surface analysis of yeast cells. *Surf. Interf. Anal.* **11**, 478–486.

Bolshakova, A.V., Kiselyova, O.I., Filonov, A.S., Frolova, O.Y., Lyubchenkoc, Y.L. and Yaminsky, I.V. (2001) Comparative studies of bacteria with an atomic force microscopy operating in different modes. *Ultramicroscopy* **86**, 121–128.

Bos, R., Van der Mei, H.C. and Busscher, H.J. (1999) Physico-chemistry of initial microbial adhesive interactions – its mechanisms and methods for study. *FEMS Microbiol. Rev.* **23**, 179–230.

Busscher, H.J., Weerkamp, A.H., Van der Mei, H.C., Van Pelt, A.W.J., De Jong, H.P. and Arends, J. (1984) Measurement of the surface free energy of bacterial cell surfaces and its relevance for adhesion. *Appl. Environ. Microbiol.* **48**, 980–983.

Busscher, H.J., Handley, P.S., Rouxhet, P.G., Hesketh, L.M. and Van der Mei, H.C. (1991) The relationship between structural and physico-chemical surface properties of tufted *Streptococcus sanguis* strains. In *Microbial Surface Analysis: Structural and Physico-chemical Methods* (ed. N. Mozes, P.S. Handley, H.J. Busscher and P.G. Rouxhet), pp. 317–338, VCH Publishers Inc., New York.

Butt, H.-J. (1992) Measuring local surface charge densities in electrolyte solutions with a scanning force microscope. *Biophys. J.* **63**, 578–582.

Camesano, T.A. and Logan, B.E. (2000) Probing bacterial electrosteric interactions using atomic force microscopy. *Environ. Sci. Technol.* **34**, 3354–3362.

Cowan, M.M., Mikx, F.H.M. and Busscher, H.J. (1994) Electrophoretic mobility and hemagglutination of *Treponema denticola* ATCC 33520. *Coll. Surf. B: Biointerf.* **2**, 407–410.

Dufrêne, Y.F. (2000) Direct characterization of the physicochemical properties of fungal spores using functionalized AFM probes. *Biophys. J.* **78**, 3286–3291.

Dufrêne, Y.F. (2001) Application of atomic force microscopy to microbial surfaces: from reconstituted cell surface layers to living cells. *Micron* **32**, 153–165.

Handley, P.S., Hesketh, L.M. and Moumena, R.A. (1991) Charged and hydrophobic groups are localised in the short and long tuft fibrils on *Streptococcus sanguis* strains. *Biofouling* **4**, 105–111.

Hayashi, H., Tsuneda, S., Hirata, A. and Sakasi, H. (2001) Soft particle analysis of bacterial cells and its interpretation of cell adhesion behaviors in terms of DLVO theory. *Coll. Surf. B: Biointerf.* **22**, 149–157.

Heinz, W.F. and Hoh, J.H. (1999) Relative surface charge mapping with the atomic force microscope. *Biophys. J.* **76**, 528–538.

James, A.M. (1991) Charge properties of microbial cell surfaces. In *Microbial Surface Analysis: Structural and Physico-chemical Methods* (ed. N. Mozes, P.S. Handley, H.J. Busscher, and P.G. Rouxhet), pp. 221–262, VCH Publishers Inc., New York.

Kiers, P.J.M., Bos, R., Van der Mei, H.C. and Busscher, H.J. (2001) The electrophoretic softness of the surface of *Staphylococcus epidermidis* cells grown in a liquid medium and on a solid agar. *Microbiology* **147**, 757–762.

Marshall, K.C., Pembrey, R. and Schneider, R.P. (1994) The relevance of X-ray photoelectron spectrosocpy for analysis of microbial cell surfaces: a critical review. *Coll. Surf. B: Biointerf.* **2**, 371–369.

Morisaki, H., Nagai, S., Ohshima, H., Ikemoto, E. and Kogure, K. (1999) The effect of motility and cell-surface polymers on bacterial attachment. *Microbiology* **145**, 2797–2802.

Poortinga, A.T., Bos, R. and Busscher, H.J. (2001) Electrostatic interactions in the adhesion of an ion-penetrable and ion-impenetrable bacterial strain to glass. *Coll. Surf. B: Biointerf.* **20**, 105–117.

Razatos, A., Ong, Y.-L., Sharma, M.M. and Georgiou, G. (1998) Molecular determinants of bacterial adhesion monitored by atomic force microscopy. *Proc. Natl. Acad. Sci. USA* **95**, 11059–11064.

Rouxhet, P.G., Mozes, N., Dengis, P.B., Dufrêne, Y.F., Gerin, P.A. and Genet, M.J. (1994) Application of X-ray photoelectron spectroscopy to microorganisms. *Coll. Surf. B: Biointerf.* **2**, 347–369.

Tyler, B.J. (1997) XPS and SIMS studies of surfaces important in biofilm formation. Three case studies. *Ann. NY Acad. Sci.* **831**, 114–126.

Van Loosdrecht, M.C.M., Lyklema, J., Norde, W. and Zehnder, A.J.B. (1989) Bacterial adhesion: a physicochemical approach. *Microb. Ecol.* **17**, 1–15.

Van der Mei, H.C., Rosenberg, M. and Busscher, H.J. (1991) Assessment of microbial cell surface hydrophobicity. In *Microbial Cell Surface Analysis* (ed. N. Mozes, P.S. Handley, H.J. Busscher and P.G. Rouxhet), pp. 263–287, VCH Publishers Inc., New York.

Van der Mei, H.C., Meinders, J.M. and Busscher, H.J. (1994) The influence of ionic strength and pH on diffusion of micro-organisms with different structural surface features. *Microbiology* **140**, 3413–3419.

Van der Mei, H.C., Handley, P.S., Bos, R. and Busscher, H.J. (1998a) Structural and physico-chemical factors in oral microbial adhesive mechanisms. In *Oral Biofilms and Plaque Control* (ed. H.J. Busscher and L.V. Evans) pp. 19–42, Harwood Academic Publishers, Amsterdam, The Netherlands.

Van der Mei, H.C., Léonard, A.J., Weerkamp, A.H., Rouxhet, P.G. and Busscher, H.J. (1988b) Surface properties of *Streptococcus salivarius* HB and nonfibrillar mutants: measurement of zeta potential and elemental composition with X-ray photoelectron spectroscopy. *J. Bacteriol.* **170**, 2462–2466.

Van der Mei, H.C., Léonard, A.J., Weerkamp, A.H., Rouxhet, P.G. and Busscher, H.J. (1988c) Properties of oral streptococci relevant for adherence: zeta potential, surface free energy and elemental composition. *Coll. Surf.* **32**, 297–305.

Van Oss, C.J. (1995) Hydrophobicity of biosurfaces – origin, quantitative determination and interaction energies. *Coll. Surf. B: Biointerf.* **5**, 91–110.

Van Oss, C.J., Chaudhury, M.K. and Good, R.J. (1987) Monopolar surfaces. *Adv. Coll. Interf. Sci.* **28**, 35–64.

Van der Wal, A., Minor, M., Norde, W., Zehnder, A.J.B. and Lyklema, J. (1997) The electrokinetic potential of bacterial cells. *Langmuir* **13**, 165–171.

Yao, X., Walter, S., Burke, S., Steward, S., Jericho, M.H., Pink, D., Hunter, R. and Beveridge, T.J. (2002) Atomic force microscopy and theoretical considerations of surface properties and turgor pressures of bacteria. *Coll. Surf. B: Biointerf.* **23**, 213–230.

2

The role of hydrophobicity and exopolymers in initial adhesion and biofilm formation

J. Azeredo and R. Oliveira

2.1 INTRODUCTION

The formation of a biofilm includes several steps but a prerequisite is the adhesion of microbial cells to a solid surface. Studies of bacterial adhesive properties have indicated that a number of cell surface physico-chemical factors contribute to the process of adhesion. Such factors include cell surface hydrophobicity (Busscher *et al.*, 1990; Oliveira *et al.*, 2001), the presence of extracellular polymers (Allison and Sutherland, 1987; Azeredo and Oliveira, 2000a) and cell surface charge. The latter determines the electrostatic interaction between the cell and the substratum (van Loosdrecht *et al.*, 1990). However, in most common situations, i.e. aqueous media with pH near neutrality, the microbial cells and solid substrata are negatively charged. This means that surface charge normally has a repulsive effect and acts contrary to adhesiveness. The extracellular polymers are not only important for the adhesion process but also determine the structure of the biofilm. Due to the importance of hydrophobicity in the first stage of biofilm formation and the role of extracellular polymers in the final establishment of a biofilm, they will be the subjects of the following overview.

2.2   THE IMPORTANCE OF HYDROPHOBICITY IN INITIAL MICROBIAL ADHESION

In the last two decades, many studies have been referring to the effect of hydrophobicity in microbial adhesion. Most of these studies have focused on the effect of substratum on bacterial adhesion, but there is now evidence that microorganisms and viral particles have evolved ways to use the hydrophobic effect to adhere to solid substrata (Doyle, 2000). Albeit the recognised importance of hydrophobicity in the adhesion process, it has been very difficult to have an acceptable definition of hydrophobicity, especially because different experimental methodologies have been used for its measurement. In the case of solid substrata, the most common methods are contact angle measurement for flat surfaces and thin-layer wicking for particulate materials (Teixeira *et al.*, 1998). For microbial cells, there is a higher diversity of methods (van der Mei *et al.*, 1987; Doyle, 2000), with microbial adhesion to hydrocarbons (MATH), hydrophobic interaction chromatography (HIC), salt aggregation test (SAT) and microsphere adhesion being the most commonly used. All these methods have some intrinsic drawbacks, especially those based on the adhesion of cells to some liquid or solid material because they are dependent on factors as temperature, time, pH, ionic strength and relative concentration of interacting species, which can all combine to influence the adhesive event (Ofek and Doyle, 1994).

Presently, it is almost generally accepted that the contact angle method is probably the most reliable way to determine cell surface hydrophobicity (Doyle, 2000). According to this method, hydrophobicity is usually expressed in terms of the contact angle formed by a sessile drop of pure water – a contact angle higher than 50° means a hydrophobic surface. However, it is not very easy to obtain a flat non-porous cell layer enabling reproducible measurements. Thus, there are still many situations where other techniques are used and each one leads to a specific way to express hydrophobicity. On account of this, it is not correct to define the surface 'hydrophobicity' of a microbial strain other than on a comparative level with closely related strains (van der Mei *et al.*, 1987). Moreover, these approaches do not enable the calculation of the hydrophobic interaction established between the microbial cells and the supporting substratum.

2.2.1   Quantification of hydrophobicity

The above-mentioned limitations are circumvented in the approach proposed by van Oss (1995), which is based on the definition of hydrophobicity of a given entity (i) (macromolecule, microbial cell or solid surface) as the free energy of interaction between two entities (i) when immersed in water (w) – ΔG_{iwi}. If $\Delta G_{iwi} < 0$, there is a preferential interaction between entities (i) rather than between an entity (i) and water and the substance (i) is considered hydrophobic. By the same reasoning, if $\Delta G_{iwi} > 0$, the substance (i) is hydrophilic. It must be stressed that ΔG_{iwi} can be expressed in SI units, and thus, considered in calculations involving the contribution of other forms of energy of interaction.

ΔG_{iwi} is simply related to the interfacial tension between i and water, γ_{iw}, as:

$$\Delta G_{iwi} = -2\gamma_{iw} \tag{2.1}$$

However, the determination of the surface free energy (i.e. surface tension) of solids can only be obtained by indirect measurements. Therefore, γ_{iw} can be determined by contact angle measurements (van Oss *et al.*, 1988) or thin-layer wicking (Teixeira *et al.*, 1998). The latter is appropriate when the solid material is in particulate form (Teixeira *et al.*, 1998).

Considering the approach of van Oss *et al.* (1988) and van Oss (1991), the surface free energy of a solid or a liquid, γ_i^{TOT} is the sum of apolar Lifshitz–van der Waals (LW) γ_i^{LW}, and polar acid–base (AB) interactions, γ_i^{AB}:

$$\gamma_i^{TOT} = \gamma_i^{LW} + \gamma_i^{AB} = \gamma_i^{LW} + 2(\gamma_i^- \gamma_i^+)^{1/2} \tag{2.2}$$

The polar interactions are mainly due to London dispersion interactions, but induction (Debye) and orientation (Keesom) interactions may also be involved (van Oss *et al.*, 1988). In many situations, the polar AB interactions consist entirely in hydrogen bonding; and in the most general sense, they are electron donor, γ_i^-, and electron acceptor, γ_i^+, interactions. Thus, the interfacial free energy between entity (i) and water (w) can be expressed as:

$$\begin{aligned}
\gamma_{iw} = {} & \gamma_i^{LW} + \gamma_w^{LW} - 2(\gamma_i^{LW}\gamma_w^{LW})^{1/2} \\
& + 2[(\gamma_i^+\gamma_i^-)^{1/2} + (\gamma_w^+\gamma_w^-)^{1/2} - (\gamma_i^+\gamma_w^-)^{1/2} - (\gamma_i^-\gamma_w^+)^{1/2}
\end{aligned} \tag{2.3}$$

The surface free energy components of water are known, but the corresponding values for the entity (i) have to be determined. For a solid substratum or microbial cells, the surface tension components can be determined by measuring the contact angles (θ) formed by three different liquids (for which apolar, γ_i^{LW} and polar components γ_i^-, γ_i^+ are known) on its surface. Thereafter, three forms of the following equation, resulting from Young's equation, are obtained and solved simultaneously to calculate, γ_i^{LW}, γ_i^+ and γ_i^-:

$$\gamma_l(1 + \cos\theta) = 2\sqrt{\gamma_i^{LW}\gamma_l^{LW}} + 2\sqrt{\gamma_i^- \gamma_l^+} + 2\sqrt{\gamma_i^+ \gamma_l^-} \tag{2.4}$$

In the set of liquids used (Table 2.1), one has to be non-polar for the determination of γ_i^{LW}.

According to van Oss (1997), in biological systems, hydrophobic interactions are usually the strongest of all long-range non-covalent interactions. Its sole driving force is the hydrogen bonding (also designated AB forces or Lewis AB) energy of cohesion between the surrounding water molecules. This means that the AB forces, if strongly asymmetrical or monopolar, are responsible for the orientation of water molecules adsorbed on the surfaces. Moreover, water molecules oriented on the surface of one particle will repel water molecules oriented in the same manner on the surface of an adjacent particle (Parsegian *et al.*, 1985; van

Table 2.1   Surface tension parameters (mJ/m^2) of the liquids commonly used in contact angle measurements for the determination of solids surface tension.

Liquid	γ^{TOT}	γ^{LW}	γ^+	γ^-
Water	72.8	21.8	25.5	25.5
Glycerol	64.0	34.0	3.9	57.4
Formamide	58.0	39.0	2.3	39.6
Di-iodomethane	50.8	50.8	0	0
n-Decane	23.8	23.8	0	0
α-Bromonaphthalene	44.4	44.4	0	0

Oss, 1994). If the orientation of the water molecules is sufficiently strong, the two particles will not approach each other. On the other hand, if the surface is more weakly apolar, its capacity for orienting the most closely adsorbed water molecules is less pronounced and the particles will approach each other under the influence of their net LW attraction. Thus, 'hydrophobic' compounds or surfaces do not repel water: they attract water with rather substantial binding energies, albeit not quite strongly as very hydrophilic ones (van Oss, 1995). It should be stressed that hydrophobic attractions can prevail between one hydrophobic and one hydrophilic site immersed in water, as well as between two hydrophobic entities. Summarising, the so-called hydrophobic interaction is a Lewis AB interaction and for two entities (*i*) immersed in water, it can be expressed by (see Equation (2.1)):

$$\Delta G_{iwi}^{AB} = -4\left[\sqrt{\gamma_i^+\gamma_i^-} + \sqrt{\gamma_w^+\gamma_w^-} - \sqrt{\gamma_i^+\gamma_w^-} - \sqrt{\gamma_i^-\gamma_w^+}\right] \quad (2.5)$$

with γ_i^+ and γ_i^- being calculated from a system of Equations (2.4). It is also possible to calculate the AB free energy of interaction between an entity 1 and an entity 2 when immersed in water:

$$\Delta G_{1w2}^{AB} = +2\left[\sqrt{\gamma_w^+}\left(\sqrt{\gamma_1^-} + \sqrt{\gamma_2^-} - \sqrt{\gamma_w^-}\right) \right. $$
$$\left. + \sqrt{\gamma_w^-}\left(\sqrt{\gamma_1^+} + \sqrt{\gamma_2^+} - \sqrt{\gamma_w^+}\right) - (\gamma_1^+\gamma_2^-)^{1/2} - (\gamma_1^-\gamma_2^+)^{1/2}\right] \quad (2.6)$$

This means that it is possible to extend the DLVO theory (named after Derjaguin, Landau, Verwey and Overbeek (Oliveira, 1997)) to account for 'hydrophobic interactions'. The DLVO theory (Oliveira, 1997), that has been applied to explain microbial adhesion, considers that the total energy of interaction is the balance between the energy of interaction due to LW forces (ΔG_{1w2}^{LW} – usually attractive) and the electrostatic energy of interaction arising from the interpenetration of the electrical double layers of the two interacting surfaces (ΔG_{1w2}^{EL} – normally repulsive). In its extended version (XDLVO), the total energy of interaction (ΔG_{1w2}^{TOT}) is given by:

$$\Delta G_{1w2}^{TOT} = \Delta G_{1w2}^{LW} + \Delta G_{1w2}^{EL} + \Delta G_{1w2}^{AB} \qquad (2.7)$$

2.2.2   Hydrophobic effect in initial adhesion

In a study on the attachment of *Staphyloccocus epidermidis* to cellulose diacetate (Oliveira *et al.*, 2001; Fonseca *et al.*, 2001) three different strains were used: *S. epidermidis* ATCC, 35984 (RP62A) and the strains M187 and M187-Sn3 kindly offered by Gerald B. Pier (Channing Laboratory, Harvard Medical School, Boston, MA, USA). The strains RP62A and M187 have a capsule and are polysaccharide/adhesin positive (PS/A+), while M187-Sn3 is an isogenic mutant of M187 and is polysaccharide/adhesin negative (PS/A−). The polymeric material and the bacterial cells were characterised in terms of their surface tension (hydrophobicity) by contact angle measurements (Table 2.2).

As ΔG_{iwi} is positive, the three strains are hydrophilic, but a higher degree of hydrophilicity means a lower degree of hydrophobicity and directly correlates with the number of attached cells. It is interesting to note that the strain devoid of the PS/A is the one with less ability to attach, as could be expected, but it is also the less hydrophobic.

In the sequence of the same study, four polymeric materials, commonly used in indwelling devices, were assayed to test their ability to be colonised by the strain RP62A: polyethylene (PE), silicone (SI), expanded polytetrafluoroethylene (ePTFE) and cellulose diacetate (CDA). The results are presented in Figure 2.1.

In this case, the materials can all be considered hydrophobic ($\Delta G_{iwi} < 0$) and an increase in the degree of hydrophobicity linearly correlates with the number of attached cells. A similar behaviour of linear correlation between the hydrophobicity of the substrata and the number of cells attached was encountered for an anaerobic consortium (Alves *et al.*, 1999) and in the attachment of *Alcaligenes denitrificans* to polymeric supports (Teixeira and Oliveira, 1999).

The hydrophobic effect is also an important factor in yeast cells attachment. An interesting example is the comparison between the binding ability of two

Table 2.2   Number of *S. epidermidis* cells expressed as colony-forming units (CFUs) attached to cellulose diacetate after 1h of incubation in phosphate-buffered saline (PBS), for each phenotype assayed and the respective ΔG_{iwi} (mJ/m²).

Strain	Number of cells adhered (CFU/mm² × 10³)	ΔG_{iwi} (mJ/m²)
RP62A	3.31 ± 0.17	17.5
M187	3.33 ± 0.39	17.4
M187-Sn3	2.08 ± 0.40	31.9

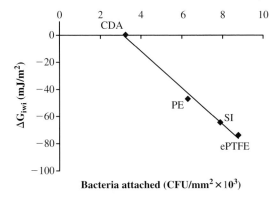

Figure 2.1   Relation between the degree of hydrophobicity (ΔG_{iwi}) of four polymeric materials and the number of *S. epidermidis* RP62A cells attached (Oliveira *et al.*, 2001).

Saccharomyces cerevisiae strains to different types of materials (Nakari-Setälä *et al.*, 2002). One strain was a transformant able to express an hydrophobin of the filamentous fungus *Trichoderma resei* (HFBI) on the yeast cell surface, the other was the parent strain. The materials used were siran, siliconised siran and immobasil. The first two are hydrophilic ($\Delta G_{iwi} = 120\,\text{mJ/m}^2$ and $\Delta G_{iwi} = 21\,\text{mJ/m}^2$, respectively) and immobasil is hydrophobic ($\Delta G_{iwi} = -29\,\text{mJ/m}^2$). The strain expressing the hydrophobin displayed a higher hydrophobicity and also had a higher-binding ability to all materials, which increased with the increasing degree of material hydrophobicity. A point to note is that the immobasil surface became less hydrophobic when coated with pure HFBI protein and a decrease in yeast cells attachment was concomitantly detected. This alteration of the coated surface is supposed to be due to the exposure of the hydrophilic sites of the macromolecule on account of the hydrophobic ones being in direct contact with the solid surface. However, for longer periods of contact with the hydrophobin, the hydrophobicity of the immobasil surface raised again. Such a phenomenon can be explained by the self-assembly of the protein on the surface to form bilayered structures (van der Vegt *et al.*, 1995). This is one of the possible effects of extracellular polymers in the process of microbial adhesion as will be outlined in the next section.

2.3   THE IMPORTANCE OF EXOPOLYMERS IN INITIAL MICROBIAL ADHESION AND BIOFILM FORMATION

2.3.1   Exopolymers in initial adhesion

The involvement of exopolymers in the advents of initial adhesion has been recognised for a long time. Marshall *et al.* (1971) suggested that exopolymers were involved in a time-dependent irreversible phase of adhesion. This hypothesis was later supported by Fletcher and Floodgate (1973), who demonstrated that the production of a secondary acidic polysaccharide-mediating irreversible adhesion of a marine bacterium to a solid was a time-dependent process.

The observation of the complementary microbial cell surface pattern on a surface (footprint) in which adhering cells were removed by shear force (Marshal et al., 1971), enzymes (Paul and Jefry, 1985) or sonication (Neu and Marshall, 1991) was a clear evidence that extracellular polymers are somehow involved in the interaction between microorganisms and surfaces. The term footprint was first used to designate polymeric materials that are left onto a surface after the bacterial cells were removed by a shear force (Marshall et al., 1971). Later, it was suggested to extend the term footprint to all molecules by which bacteria are able to label an interface (Neu, 1992). Thus, adsorption footprints (true adhesive polymers obtained when removing bacteria artificially from interfaces) or desorption footprints (molecules released by bacteria to detach themselves from interfaces) were distinguished. Nevertheless, these molecules are extracellular polymers (polysaccharides, biosurfactants, proteins, glycoproteins, lipids, etc.) that have an important role in the interaction between microorganisms and surfaces. Another important feature of microbial footprints is that they may influence the adhesion of microorganisms to a surface. In a recent survey, it was shown that footprints of *Pseudomonas aeruginosa* strains detached by passing air bubbles had a negative influence on the adhesion of newly redepositing cells (Gómez-Suárez et al., 2002).

Studies of microbial footprints have given an important contribution to the knowledge of the role of exopolymers in microbial adhesion. Other studies based on the utilisation of mutants with different capabilities to secrete exopolymers have also enlightened the contribution of exopolymers in adhesion (Flemming et al., 1998; Azeredo and Oliveira, 2000a; Gómez-Suárez et al., 2002). The adhesion to glass of three mutants of *Sphingomonas paucimobilis* having different gellan production abilities was studied in the presence and absence of solutions of their exopolymers (Azeredo and Oliveira, 2000a). The results revealed that the extent of cell adhesion in the absence of the solution of exopolymers was very similar for the three mutants, whereas in the presence of exopolymers the highest-producing mutant was able to adhere in a larger extent followed by the intermediate producer. These results have pointed out to an important feature of the involvement of exopolymers in microbial adhesion. The exopolymers released by the suspended cells can easily coat a surface, altering its surface tension and making adhesion favourable (Azeredo and Oliveira, 2001) or unfavourable (Leriche and Carpentier, 2000). This aspect is commonly found when the exopolymers have surface-active properties (Neu, 1996). Surface-active compounds (SACs) are molecules formed by a hydrophilic part and a hydrophobic one, which tend to interact with interfaces. Synthetic SACs have been used to reduce the adherence of cells to hydrophobic surfaces (Paul and Jeffrey, 1985; Stelmack et al., 1999). The assembly of microbial SACs onto surfaces can either inhibit adhesion to hydrophobic surfaces (Velraeds et al., 1996), or enhance attachment to hydrophilic ones (Neu, 1996). The physico-chemical properties of the attachment surface play an important role on the way microbial polymers assemble onto the surface (linked by the hydrophilic part, exposing the hydrophobic one or the opposite) and on the amount assembled. Capsular exopolymers of *Pseudomonas* sp. were able to cover in larger extent hydrophilic surfaces than hydrophobic ones (Kalaji and Neal, 2000).

Attempts to interpret microbial adhesion through colloidal theories, such as DLVO and more recently the XDLVO, have also revealed another valuable aspect of the contribution of exopolymers to microbial adhesion. The adhesion of *Lactococcus lactis* to glass and polystyrene could not be explained by DLVO theory, neither by the influence of surface hydrophobicity, meaning that interactions beyond physico-chemical ones have played an important role in this study (Boonaert *et al.*, 2001). The authors hypothesised that the release of macromolecular compounds, such as peptides and polysaccharides, may have a bridging or repelling action between the cells and the adhesion surface (Boonaert *et al.*, 2001). The adhesion of a high gellan-producing mutant of *Sphingomonas paucimobilis* to glass could not be explained by XDLVO theory. The explanation can be given considering that polymeric bridging between the exopolymeric layer that surrounds the cell wall and the exopolymers coating the glass surface was necessary to overcome the energy barrier between the adhesion surface and the microbial cell (Azeredo *et al.*, 1999a).

The examples presented so far only stressed the non-specific involvement of exopolymers in adhesion, either by coating the adhesion surface (by assemble or through footprints) or by bridging the cells to the substratum. The establishment of specific interactions (of adhesin-receptor type) between cells and adhesion substratum can be seen as a specific contribution of exopolymers to adhesion. Bacterial lipopolysaccharides (LPSs), capsules and polymeric slime layers (slime) undertake specific interactions with biotic and abiotic surfaces.

LPSs molecules have been suggested to function as adhesins that mediate the binding of *Campylobacter jejuni* to epithelial cells (McSweegan and Walker, 1986). LPS have also been shown to mediate the interaction of bacteria with phagocytic cells (Perry and Ofek, 1984; Wright *et al.*, 1989).

Capsules are usually acidic polysaccharides secreted by bacteria that remain cell bound following their secretion. They differ from the slime polymeric layer, which, after secretion, remains loosely associated with the cell surface. The term glycocalyx has also been used to describe both capsule and slime polymers (Ofek and Doyle, 1994). An example of the involvement of polymeric capsules in the adhesion event is the interaction of *Klebsiella pneumoniae* with macrophages. This bacterium undergoes phagocytosis mediated by capsular polysaccharides recognised by the mannose/*N*-acetylglucosamine-specific lectin of macrophages (Athamna *et al.*, 1991). In the case of *Staphylococcus epidermidis*, the expression of a capsular polysaccharide adhesin termed PS/A has been shown to be determinant for initial attachment to abiotic surfaces (Shiro *et al.*, 1994). Conversely, capsules may act as anti-adhesive structures due to their uniformly high density of negative charges (Gotschlich, 1983). Furthermore, capsules may mask potential adhesins located at the cell surface (Favre-Bonte *et al.*, 1999). The dual role of capsules (stimulation and inhibition of bacterial adhesion) can be well exemplified by *Klebsiella pneumoniae*. The initial adhesion of these bacteria to epithelium is mediated by the specific interaction between their capsules and the mucous layer. Following initial interaction, capsule expression is downregulated to facilitate interactions between bacterial cell surface adhesins and the underlying epithelial cell (Taylor and Roberts, 2002).

The ability of slime to promote adhesion through the establishment of specific interactions with the substratum is now well established. Ofek and Doyle (1994) proposed the word 'slimectin' to denominate slime-like materials that are directly involved in microbial adhesive events. Slimectin is, e.g. the polymeric slime, produced by some oral streptococci (α-1,3- and α-1,6-glucans) that participate as receptors for glucan-binding lectins.

Exopolymers are also responsible for the detachment of bacteria from surfaces. Enzymes may be able to break down different types of bonding in proteins and polysaccharides (Sutherland, 1999). The secretion of polymers at the interface is another mechanism with which bacteria can detach from a surface. This is the case of bacteria degrading hydrocarbons that can release themselves from oil droplets by producing an emulsifying substance (Rosenberg, 1986). Another example is the case of hydrophobic bacteria that can detach from hydrophilic surfaces by excreting hydrophilic polymers (Fattom and Shilo, 1985).

2.3.2   Biofilm exopolymers

Exopolymers are constituents of the polymeric matrix of biofilms and are understood as all the polymeric material, i.e. produced and secreted by microbial cells, which comprise polysaccharides, proteins, glycoproteins, LPS, lipids, among others. Apart from exopolymers, the biofilm matrix also comprises cellular debris and the products of extracellular hydrolytic activity as well as adsorbed chemicals and particles. These materials are commonly designated by the acronym EPS meaning extracellular polymeric substances (Wingender *et al.*, 1999).

Polysaccharides are considered the major constituents of the exopolymeric fraction of EPS. There are many publications referring that the exopolysaccharide fraction of the exopolymers found in the biofilm matrix is of different composition than that produced by planktonic cells. Allison *et al.* (1998) when studying the activity of a polysaccharide lyase of *P. fluorescens*, found that while lyases recovered from planktonic supernatants were only active against planktonic-derived exopolysaccharides, lyases recovered from biofilm were active against both planktonic- and biofilm-derived exopolysaccharides. This suggests that the composition of biofilm exopolymers is different from that of planktonic exopolymers. In another study, the characterisation of the exopolymers recovered from a planktonic culture of a marine *Pseudomonas* sp. and the ones obtained from the biofilm matrix of the same strain by Fourier-transform IR spectroscopy demonstrated that clear differences between these two exopolymers – namely, *O*- and *N*-acetylations – were greater in the biofilm exopolymers (Beech *et al.*, 1999).

Despite this and other evidences, one cannot straightforwardly withdraw the con-clusion that biofilm-derived exopolysaccharides are different from planktonic-derived polysaccharides. In fact, a major problem is to obtain sufficient amount of exopolysaccharides having the certainty that they are really from the exopoly-meric fraction of the EPS biofilm matrix. The method used to extract the EPS fraction of the biofilm clearly influences the amount and composition recovered. Aggressive methods, such as NaOH, vapour extraction or prolonged sonication, lead to contamination of the fraction recovered with intracellular material released during the extraction procedure. On the other hand, smooth methods (like EDTA, heating at 70°C and centrifugation) extract only a small portion of the EPS matrix (Azeredo *et al.*, 1999b; Nielsen and Jahn, 1999). Sutherland (2001) considers that the exopolysaccharides present in biofilms resemble closely the corresponding polymers synthesised by planktonic cells. This has been demonstrated through the utilisation of non-destructive and *in situ* techniques, such as the use of antibodies or specific lectins (Sutherland, 2001; Leriche *et al.*, 2000).

2.3.3   Exopolymers in biofilm formation

2.3.3.1   Cell–substratum binding

Once attached to the surface, bacteria must maintain contact with the adhesion surface and grow in order to develop a mature biofilm. The hallmark between bacteria in a biofilm is the exopolymers produced by the cells. The presence of exopolymers helps maintaining the integrity of the biofilm, allowing large num-bers of bacteria to coexist even under turbulent flow conditions (Melo and Vieira, 1999). It is clear from a large number of studies, that mutants unable to synthesise EPS are unable to form biofilms (Allison and Sutherland, 1987; Watnick and Kolter, 1999), which definitely enlightens the importance of EPS in biofilm for-mation. This was evident in a study performed with mutants of *Sphingomonas paucimobilis* having different capabilities to secrete exopolymers. The greatest exopolymer-producing mutant was able to form low dense and thick biofilms (1 mm in average); the intermediate producer gave rise to biofilms denser and ten times thinner and the lowest-producing mutant was biofilm negative (Azeredo and Oliveira, 2000b). In this case, the exopolymers were both responsible for cell attachment and biofilm growth. However, there are cases in which exopolymers are secreted after cell attachment, thus, only influencing biofilm growth. Allison and Sutherland (1987) demonstrated that two strains of freshwater bacteria only synthesised significant amounts of exopolymers after attachment. This is also the case for alginate-producing *P. aeruginosa* strains. A study performed with *P. aeruginosa* 8830 (a stable mucoid derivative of a clinical isolate) showed that the activation of alginate production occurs after cell attachment by a mechanism in which bacteria are capable of sensing the presence of the surface. Adhered bacteria are upregulated for the production of alginate. If and when there is a change in the environment (osmolarity and ethanol presence), bacteria respond by downregulating alginate production and ultimately detach from the surface (Davies *et al.*, 1993; Davies and Geesey, 1995; Davies, 1999). Alginate presumably enhances the ability of *P. aeruginosa* to remain attached to the surface. It has been reported that a non-alginate-producing mutant sticks better to glass surfaces than a producing one. However, the non-producing mutant is not able to form a thick biofilm and only develops as a monolayer of cells. This indicates that alginate production leads to the formation of cell clusters with bacteria embedded in the alginate matrix (Davies, 1999). In another study, it was demonstrated that alginate plays an important role in the biofilm structure of *P. aeruginosa*, required for the formation of a thicker three-dimensional biofilm (Nivens *et al.*, 2001).

2.3.3.2   Cell-to-cell binding

Exopolymers are also directly responsible for cell-to-cell binding, which is a *sine qua non* condition for biofilm development. Several works have revealed that the formation of *S. epidermidis* biofilms is dependent on the presence of a polysaccharide intercellular adhesin (PIA). PIA consists of two polysaccharide species that mediate cell-to-cell adhesion of the proliferating cells (Mack *et al.*, 1994; 2000). Cell-to-cell adhesion of *S. aureus* is also dependent of a PIA being determinant in biofilm formation (Cramton *et al.*, 1999).

In addition to the involvement of exopolymers in biofilm formation, exopolymers are also determinant for biofilm architecture. In a recent review of the basic structure of microbial biofilms, Wimpenny and Colasanti (1997) postulated a unifying hypothesis in which the three conceptual models revealed that three types of biofilm structure exist, namely, a heterogeneous mosaic biofilm, a penetrated water-channel biofilm and a dense confluent biofilm. According to these authors, biofilm architecture is a consequence of nutrients availability. van Loosdrecht *et al.* (1997) pointed out to the importance of biomass detachment in biofilm structure. Biofilm architecture is, thus, very dependent of the balance between biomass accumulation and detachment, and this is undoubtedly influenced by biofilm exopolymers. Because, besides being important for biofilm formation (as already seen), exopolymers determine the mechanical stability of biofilms, mediated by non-covalent interactions (Mayer *et al.*, 1999). The influence of exopolymers in biofilm architecture can be exemplified in a study in which a mutant of *E. coli* defective in colanic acid production showed a different biofilm structure from the original strain, whereas initial adhesion was not affected in this mutant, suggesting that colanic acid is not acting as an adhesin during the early attachment events but influences biofilm structure (Danese *et al.*, 2000). The biofilm produced by a high gellan-secreting mutant of *Sphingomonas paucimobilis* presented a different structure from the one formed by an intermediate gellan-secreting mutant (Azeredo and Oliveira, 2000b). In both studies, biofilms formed by high polysaccharide-producing strains were less dense and packed. Kreft and Wimpenny (2001) simulated a dual species nitrifying biofilm, elucidating the effect of exopolymer production in biofilm architecture. In this study, exopolymers lowered the density of the biofilm and the roughness of biofilm surface.

Several other reports have also emphasised the influence of exopolymers in biofilm structure. However, their specific role is still not clear, partly due to analytical limitations. For instance, there are uncertainties of binding specificities to exopolymers if antibodies or lectins are used (Neu and Lawrence, 1999).

2.4   CONCLUSIONS

In most situations, there is a strong correlation between the number of microbial cells adhered to solid surfaces and the degree of hydrophobicity of either the solid substratum and/or the cells surface. Concerning the effect of exopolymers, conversely to hydrophobicity, they can enhance or inhibit the process of cell adhesion.

The importance of exopolymers in microbial adhesion, biofilm formation and structure has been mostly elucidated by studies based on pure and well-defined cultures. However, in multispecies biofilms the interaction between the exopolymers of the different strains can strongly influence biofilm formation and structure (Skillman *et al.*, 1999; Gideon *et al.*, 1999). This can be seen as a challenging field for future work.

REFERENCES

Allison, D.G. and Sutherland, I.W. (1987) Role of exopolysaccharides in adhesion of freshwater bacteria. *J. Gen. Microbiol.* **133**, 1319–1327.

Allison, D.G., Ruiz, B., San Jose, C., Jaspe, A. and Gilbert, P. (1998) Extracellular products as mediators of the formation and detachment of *Pseudomonas fluorescens* biofilms. *FEMS Microbiol. Lett.* **167**(2), 179–184.

Alves, M.M., Pereira, M.A., Novais, J.M., Polanco, F.F. and Mota, M. (1999) A new device to select microcarriers for biomass immobilization: application to an anaerobic consortium. *Water Environ. Res.* **1**, 209–217.

Athamna, A., Ofek, I., Keisari, Y., Markowitz, S., Dutton, G.D.S. and Sharon, N. (1991) Lectinophagocytosis of encapsulated *Klebsiella pneumoniae* mediated by surface lectins of guinea pig macrophages and human monocyte-derived macrophages. *Infect. Immun.* **59**, 1673–1682.

Azeredo, J. and Oliveira, R. (2000a) The role of exopolymers in the attachment of *Sphingomonas paucimobilis*. *Biofouling* **16**, 59–67.

Azeredo, J. and Oliveira, R. (2000b) The role of exopolymers produced by *Sphingomonas paucimobilis* in biofilm formation and composition. *Biofouling* **16**, 17–27.

Azeredo, J. and Oliveira, R. (2001) The role of exopolymers in *Sphingomonas paucimobilis* attachment and biofilm formation. In *Biofilm Community Interactions: Chances or Necessity* (ed. P. Gilbert, D. Allison, M. Branding, J. Verran and J. Walker), pp. 221–230, BioLine, Cardiff, UK.

Azeredo, J., Visser, J. and Oliveira, R. (1999a) Exopolymers in bacterial adhesion: interpretation in terms of DLVO and xDLVO theories. *Colloids Surf. B: Bioint.* **14**, 141–148.

Azeredo, J., Lazarova, V. and Oliveira, R. (1999b) Methods to extract the exopolymeric matrix from biofilms: a comparative study. *Water Sci. Technol.* **39**, 243–250.

Beech, I., Hanjagsit, L., Kalaji, M., Neal, A.L. and Zinkevich, V. (1999) Chemical and structural characterisation of exopolymers produced by *Pseudomonas* sp. NCIMB 2001 in continuous culture. *Microbiology* **145**, 1491–1497.

Boonaert, C.J., Dufrêne, Y., Derclaye, S.R. and Rouxhet, P.G. (2001) Adhesion of *Lactococcus lactis* to model substrata: direct study of the interface. *Colloids Surf. B: Bioint.* **22**, 171–182.

Busscher, H.J., Sjollema J. and van der Mei H. (1990) Relative importance of surface free energy as a measure of hydrophobicity in bacterial adhesion to solid surfaces. In *Microbial Cell Surface Hydrophobicity* (ed. R.J. Doyle and M. Rosenberg), pp. 335–359, American Society for Microbiology, Washington DC.

Cramton, S., Gerke, C., Schnell, N.F., Nichols, W.W. and Götz, F. (1999) The intercellular adhesion (ica) locus is present in *Staphylococcus aureus* and is required for biofilm formation. *Infect. Immun.* **67**, 5427–5433.

Danese, P.N., Pratt, L.A. and Kolter, R. (2000) Exopolysaccharide production is required for development of *Escherichia coli* K-12 biofilm architecture. *J. Bacteriol.* **182**, 3593–3596.

Davies, D. (1999) Regulation of matrix polymer in biofilm formation and dispersion. In *Microbial Extracellular Polymeric Substances* (ed. J. Wingender, T.R. Neu and H.-C. Flemming), pp. 93–117, Springer Verlag, Berlin, Germany.

Davies, D. and Geesey, G.G. (1995) Regulation of the alginate biosynthesis gene *algC* in *Pseudomonas aeruginosa* during biofilm development in continuous culture. *Appl. Environ. Microbiol.* **59**, 1181–1186.

Davies, D., Chakrabarty, A.M. and Geesey, G.G. (1993) Exopolysaccharide production in biofilms: substratum activation of alginate gene expression by *Pseudomonas aeruginosa*. *Appl. Environ. Microbiol.* **59**, 1181–1186.

Doyle, R.J. (2000) Contribution of the hydrophobic effect to microbial infection. *Microbes and Infection* **2**, 391–400.

Fattom, A. and Shilo, M. (1985) Production of emulcyan by Phormidium J-1: its activity and function. *FEMS Microbiol. Ecol.* **31**, 3–9.

Favre-Bonte, S., Joyle, B. and Forestier, C. (1999) Consequences of reduction of *Klebsiella pneumoniae* capsule expression on interaction of this bacterium with epithelial cells. *Infect. Immun.* **67**, 554–561.

Flemming, C.A., Palmer, R.J., Arrage Jr., A.A., van der Mei, H.C. and White, D.C. (1998) Cell surface physicochemistry alters biofilm development of *Pseudomonas aeruginosa* lipopolysaccharide mutants. *Biofouling* **13**, 213–231.

Fletcher, M. and Floodgate, G.D. (1973) An electron-microscopic demonstration of an acidic polysaccharide involved in adhesion of a marine bacterium to solid surfaces. *J. General Microbiol.* **74**, 325–334.

Fonseca, A.P., Granja, P.L., Nogueira, J.A, Oliveira, R. and Barbosa, M.A. (2001) Adhesion of *Staphylococcus epidermidis* to chemically modified derivatives. *J. Mater. Sci., Mater. Med.* **12**, 543–548.

Gideon, M.W., Lawrence, J.R. and Korbell, D.R. (1999) Function of EPS. In *Microbial Extracellular Polymeric Substances* (ed. J. Wingender, T.R. Neu and H.-C. Flemming), pp. 171–200, Springer Verlag, Berlin, Germany.

Gómez-Suárez, C., Pasma, J., van der Borden, A.J., Wingender, J., Flemming, H.C., Busscher, H.J. and van der Mei, H.C. (2002) Influence of extracellular polymeric substances on deposition and redeposition of *Pseudomonas aeruginosa* to surfaces. *Microbiol.–SGM* **148**, 1161–1169.

Gotschlich, E.C. (1983) Thoughts on the evolution of strategies used by bacteria for evasion of host defenses. *Rev. Infect. Dis.* **5**(Suppl. 4), S778–S783.

Kalaji, M. and Neal, A.L. (2000) IR study of self-assembly capsular exopolymers from *Pseudomonas fluorescens* sp. NCIMB 2021 on hydrophilic and hydrophobic surfaces. *Biopolymers* **57**(1), 43–50.

Kreft, J.-U. and Wimpenny, J.W.T. (2001) Effect of EPS on biofilm structure and function as revealed by an individual-based model of biofilm structure. *Water Sci. Technol.* **43**(6), 135–141.

Leriche, V. and Carpentier, B. (2000) Limitation of adhesion and growth of *Listeria monocytogenes* on stainless steel surfaces by *Staphylococcus sciuri* biofilms. *J. Appl. Microbiol.* **88**, 594–605.

Leriche, V., Sibille, P. and Carpentier, B. (2000) Use of enzyme-linked lectinsorbent assay to monitor the shift in polysaccharide composition in bacterial biofilms. *Appl. Environ. Microbiol.* **66**, 1851–1856.

Mack, D., Nedelmann, M., Krokotsch, A., Schwarzkopf, A., Heesemann, J. and Laufs, R. (1994) Characterization of transposon mutants of biofilm producing *Staphylococcus epidermidis* impaired in the accumulative phase of biofilm production: genetic identification of a hexoxamine-containing polysaccharide intercellular adhesin. *Infect. Immun.* **62**, 3244–3253.

Mack, D., Rohde, H., Dobinski, S., Riedewald, J., Nedelmann, M., Knobloch, J.K.M., Elsner, H.-A. and Feucht, H.H. (2000) Identification of three essential regulatory gene loci governing expression of *Staphylococcus epidermidis* polysaccharide intercellular adhesin and biofilm formation. *Infect. Immun.* **68**, 3799–3807.

Marshall, K.C., Stout, R. and Mitchell, R. (1971) Mechanism of the initial events in the sorption of marine bacteria to surfaces. *J. General Microbiol.* **68**, 337–348.

Mayer, C., Moritz, R., Kirschner, C., Borchard, W., Mailbaum, R., Wingender, J. and Flemming, H.C. (1999) The role of intermolecular interactions: studies on model systems for bacterial biofilms. *Int. J. Biol. Macromol.* **26**, 3–16.

McSweegan, E. and Walker, R.I. (1986) Identification and characterisation of two *Campylobacter jejuni* adhesins for cellular and mucous substrates. *Infect. Immun.* **53**, 141–148.

Melo, L.F. and Vieira, M.J. (1999) Physical stability and biological activity of biofilms under turbulent flow and low substrate concentration. *Biop. Eng.* **20**, 363–368.

Nakari-Setälä, T., Azeredo, J., Henriques, M., Oliveira, R. Teixeira, J. and Penttilä, M. (2002) Expression of a fungal hydrophobin in *Saccharomyces cerevisiae* cell wall: effect on cell surface properties and immobilization. *Appl. Environ. Microbiol.* **68**, 3385–3391.

Neu, T.R. (1992) Microbial 'footprints' and the general ability of microorganisms to label interfaces. *Can. J. Microbiol.* **38**, 1005–1008.

Neu, T.R. (1996) Significance of bacterial surface active compounds in interaction of bacteria with interfaces. *Microbiol. Rev.* **60**, 151–166.

Neu, T.R. and Marshall, K.C. (1991) Microbial 'footprints' – a new approach to adhesive polymers. *Biofouling* **3**, 101–112.

Neu, T.R. and Lawrence, J.R. (1999) In situ characterization of extracellular polymeric substances (EPS) in biofilm systems. In *Microbial Extracellular Polymeric Substances* (ed. J. Wingender, T.R. Neu and H.-C. Flemming), pp. 21–48, Springer Verlag, Berlin, Germany.

Nielsen, P.H. and Jahn, A. (1999) Extraction of EPS. In *Microbial Extracellular Polymeric Substances* (ed. J. Wingender, T.R. Neu and H.-C. Flemming), pp. 73–92, Springer Verlag, Berlin, Germany.

Nivens, D.E., Ohman, D.E., Williams, J. and Franklin, M.J. (2001) Role of alginate and its O-acetylation in formation of *Pseudomonas aeruginosa* microcolonies and biofilms. *J. Bacteriol.* **183**(3), 1047–1057.

Ofek, I. and Doyle, R.J. (1994) *Bacterial Adhesion to Cells and Tissues.* Chapman & Hall, New York.

Oliveira, R. (1997) Understanding adhesion: a mean to prevent fouling. *Exp. Ther. Fluid Sci.* **14**, 316–322.

Oliveira, R., Azeredo, J., Teixeira, P. and Fonseca, A.P. (2001) In *Biofilm Community Interactions: Chances or Necessity* (ed. P. Gilbert, D. Allison, M. Branding, J. Verran and J. Walker), pp. 11–22, BioLine, Cardiff, UK.

Parsegian, V.A., Rand, R.P. and Rau, D.C. (1985) Hydration forces: what next? *Chem. Scripta* **25**, 28–31.

Paul, J.H. and Jeffrey, W.H. (1985) Evidences for separate adhesion mechanisms for hydrophilic and hydrophobic surfaces in *Vibrio proteolytica*. *Appl. Environ. Microbiol.* **50**, 431–437.

Perry, A. and Ofek, I. (1984) Inhibition of blood clearance and hepatic tissue binding of *Escherichia coli* by liver lectin-specific sugars and glycoproteins. *Infect. Immun.* **43**, 257–262.

Rosenberg, E. (1986) Microbial surfactants. *CRC Crit. Rev. Microbiol.* **3**, 109–132.

Shiro, H., Muller, E., Gutierrez, N., Boisot, S., Grout, M., Tosterton, T.D., Goldman, D.A. and Gier, G.B. (1994) Transposons mutants of *Staphylococcus epidermidis* deficient in elaboration of capsular polysaccharide adhesin and slime are virulent in a rabbit model of endocarditis. *J. Infect. Dis.* **169**, 1042–1049.

Skillman, L.C., Sutherland, I.W. and Jones, M.V. (1999) The role of exopolysaccharides in dual species biofilm development. *J. Appl. Microbiol. Symp. Suppl.* **85**, 12S–18S.

Stelmack, P.L., Gray, M.R. and Pickard, M.A. (1999) Bacterial adhesion to soil contaminants in the presence of surfactants. *Appl. Environ. Microbiol.* **65**, 163–168.

Sutherland, I. (2001) Biofilm exopolysaccharides: a strong and sticky framework. *Microbiology* **147**, 3–9.

Sutherland, I.W. (1999) Polysaccharases in biofilms – sources – action – consequences! In *Microbial Extracellular Polymeric Substances* (ed. J. Wingender, T.R. Neu and H.-C. Flemming), pp. 201–216, Springer Verlag, Berlin, Germany.

Taylor, C. and Roberts, I.S. (2002) The regulation of capsule expression. In *Bacterial Adhesion to Host Tissues Mechanisms and Consequences* (ed. M. Wilson), pp. 115–138, Advances in Molecular and Cellular Microbiology (AMCM), Cambridge University Press, Cambridge.

Teixeira, P. and Oliveira, R. (1999) Influence of surface characteristics on the adhesion of *Alcaligenes denitrificans* to polymeric substrates. *J. Adhes. Sci. Technol.* **13**, 1287–1294.

Teixeira, P., Azeredo, J., Oliveira, R. and Chibowski, E. (1998) Interfacial interactions between nitrifying bacteria and mineral carriers in aqueous media determined by contact angle measurements and thin layer wicking. *Coll. Surf. B: Bioint.* **12**, 69–75.

van der Mei, H.C., Weerkamp, A.H. and Busscher, H.J. (1987) A comparison of various methods to determine hydrophobic properties of streptococcal cell surfaces. *J. Microbiol. Meth.* **6**, 277–287.

van der Vegt, W., van der Mei, H.C., Wösten, H.A.B., Wessels, J.G.H. and Busscher, H.J. (1995) A comparison of the surface activity of the fungal hydrophobin SC3p with those of other proteins. *Biophys. Chem.* **57**, 253–260.

van Loosdrecht, M.C.M., Norde, W., Lyklema, L. and Zehnder, J. (1990) Hydrophobic and electrostatic parameters in bacterial adhesion. *Aquat. Sci.* **51**, 103–114.

van Loosdrecht, M.C.M., Picioreanu, C. and Heijnen, J.J. (1997) A more unifying hypothesis for biofilm structures. *FEMS Microbiol. Ecol.* **24**, 181–183.

van Oss, C.J. (1991) The forces involved in bioadhesion to flat surfaces and particles – their determination and relative roles. *Biofouling* **4**, 25–35.

van Oss, C.J. (1994) *Interfacial Forces in Aqueous Media*, Marcel Dekker, New York.

van Oss, C.J. (1995) Hydrophobicity of biosurfaces – origin, quantitative determination and interaction energies. *Coll. Surf. B. Bioint.* **5**, 91–110.

van Oss, C.J. (1997) Hydrophobicity and hydrophilicity of biosurfaces. *Curr. Opin. Coll. Inter. Sci.* **2**, 503–512.

van Oss, C.J., Good, R.J. and Chaudhury, M.K. (1988) Additive and nonadditive surface tension components and the interpretation of contact angles. *Langmuir* **4**, 884–891.

Velraeds, M., van der Mei, H.C., Reid, G. and Busscher, H.J. (1996) Inhibition of initial adhesion of urogenic *Enterococcus faecalis* by biosurfactants from *Lactobacillus* isolates. *Appl. Environ. Microbiol.* **62**, 1958–1963.

Watnick, P.I. and Kolter, R. (1999) Steps in the development of a *Vibrio cholerae* El Tor biofilm. *Mol. Microbiol.* **34**, 586–595.

Wimmpeny, J.W.T. and Colasanti, R. (1997) A unifying hypothesis for the structure of microbial biofilms based on cellular automaton models. *FEMS Microbiol. Rev.* **22**, 1–16.

Wingender, J., Neu, T.R. and Flemming, H.C. (1999) What are bacterial extracellular polymeric substances? In *Microbial Extracellular Polymeric Substances* (ed. J. Wingender, T.R. Neu and H.-C. Flemming), pp. 1–19, Springer Verlag, Berlin, Germany.

Wright, S.D., Levine, S.M., Jong, M.C.T., Chad, Z. and Kabbash, L.G. (1989) CR3 (CD11b/CD18) expresses one binding site for Arg–Gly–Asp-containing peptides and a second site for bacterial lipopolysaccharide. *J. Exp. Med.* **169**, 175–183.

3

The role of coaggregation in oral biofilm formation

P.E. Kolenbrander, R.F. Lerud, D.S. Blehert,
P.G. Egland, J.S. Foster and R.J. Palmer, Jr.

3.1 INTRODUCTION

As a group, human oral bacteria adhere to all oral surfaces (Whittaker *et al.*, 1996). They are capable of colonizing hard surfaces, such as enamel (Gibbons and Hay, 1988) and soft surfaces, such as epithelial cells (Hallberg *et al.*, 1998). Oral bacteria also bind soluble molecules, such as salivary agglutinin (Demuth *et al.*, 1996) that may cover oral surfaces, and they bind to other oral bacteria (Gibbons and Nygaard, 1970; Kolenbrander, 1988).

This chapter will focus on the latter binding process of cell–cell adherence among genetically distinct oral bacteria. When these interactions are among suspended cells, they are called coaggregations (Cisar *et al.*, 1979; Kolenbrander, 1988), and when these coaggregations occur between suspended cells and those already attached to a surface, they are termed coadhesion (Bos *et al.*, 1994). Coaggregation is different from agglutination, which is the aggregation of genetically identical cells. All of the oral bacteria so far tested exhibit coaggregation with at least one partner cell type (Kolenbrander, 1988; Whittaker *et al.*, 1996;

Andersen *et al.*, 1998). Until recently, it was thought that the oral bacteria were unique in this kind of adherence, but bacteria isolated from freshwater also exhibit extensive coaggregation partnerships (Rickard *et al.*, 2000, 2002). Distinct from oral bacteria, which coaggregate in all growth phases (Cisar *et al.*, 1979), freshwater bacteria often express their coaggregation ability optimally in the stationary phase of growth (Rickard *et al.*, 2000, 2002). This suggests significant environmental regulation of coaggregation-relevant genes and is likely to contribute to biofilm formation in freshwater systems.

While only 19 strains of freshwater bacteria have been characterized regarding their coaggregation properties, all show extensive coaggregations among themselves (Rickard *et al.*, 2002). Over 1000 strains of oral bacteria have been examined for coaggregation, and the bacteria most frequently isolated from oral sites are all coaggregation partners of several other genera of oral bacteria (Table 3.1). The prevalence of coaggregations in two distinctly different habitats suggests that these interactions play an important role in freshwater biofilms and in oral biofilms, such as dental plaque.

Table 3.1   Genera of oral bacteria and their coaggregation partners.

Genus	Coaggregation partners
Actinobacillus	*Fusobacterium*
Actinomyces	*Actinomyces, Capnocytophaga, Eikenella, Fusobacterium, Prevotella, Selenomonas, Streptococcus, Veillonella*
Capnocytophaga	*Actinomyces, Fusobacterium, Rothia, Streptococcus*
Corynebacterium	*Fusobacterium, Streptococcus*
Eikenella	*Fusobacterium, Porphyromonas, Streptococcus, Actinomyces*
Eubacterium	*Fusobacterium, Veillonella*
Fusobacterium	*Actinobacillus, Actinomyces, Capnocytophaga, Corynebacterium, Eikenella, Eubacterium, Fusobacterium, Gemella, Haemophilus, Peptostreptococcus, Porphyromonas, Prevotella, Propionibacterium, Rothia, Selenomonas, Streptococcus, Treponema, Veillonella, Wolinella*
Gemella	*Fusobacterium, Prevotella*
Haemophilus	*Fusobacterium, Streptococcus*
Peptostreptococcus	*Fusobacterium*
Porphyromonas	*Eikenella, Fusobacterium, Streptococcus*
Prevotella	*Actinomyces, Fusobacterium, Gemella, Rothia, Streptococcus*
Propionibacterium	*Fusobacterium, Streptococcus, Veillonella*
Rothia	*Capnocytophaga, Fusobacterium, Prevotella, Veillonella*
Selenomonas	*Fusobacterium, Actinomyces*
Streptococcus	*Actinomyces, Capnocytophaga, Eikenella, Fusobacterium, Haemophilus, Porphyromonas, Prevotella, Propionibacterium, Streptococcus, Veillonella, Corynebacterium*
Treponema	*Fusobacterium*
Veillonella	*Actinomyces, Eubacterium, Fusobacterium, Propionibacterium, Rothia, Streptococcus*
Wolinella	*Fusobacterium*

3.2   COAGGREGATION

Human oral bacteria exhibit multiple mechanisms of coaggregation (Whittaker *et al.*, 1996; Kolenbrander *et al.*, 2002). Coaggregations involve cellular surface components that recognize their cognate adhesins or receptors on the coaggregation partner cell surface. Most coaggregations are mediated by a heat- and protease-inactivated adhesin protein on one cell type that recognizes a heat- and protease-stable component termed a receptor on the partner cell type. Those receptors that have been characterized are polysaccharides. Cisar and colleagues have purified the receptor polysaccharide from the cell wall polysaccharide of 21 viridans streptococci (McIntire *et al.*, 1987; Cisar *et al.*, 1995; 1997). The structures were determined, and they are repeating hexasaccharide or heptasaccharide molecules that possess either of two host-like motifs, GalNAcβ1 → 3Gal or Galβ1 → 3GalNAc, and constitute six structurally related groups (Cisar *et al.*, 1997).

Receptors and adhesins are depicted as complementary sets of symbols (Figure 3.1, upper right inset). Two coaggregation partner cell types labeled 'A' and 'B' are depicted in Figure 3.1. The cells in this example can represent any genus or species of oral bacteria. The coaggregations depicted are bimodal in that each partner cell interacts by two mechanisms of coaggregation illustrated as triangle-cognate pairs and rectangle-cognate pairs, as depicted by the center circular cell that interacts with several oblong cells. Throughout this chapter, rectangular symbols of any color represent lactose-inhibitable coaggregations. Coaggregations between many oral bacteria are lactose inhibitable presumably because the galactose in beta linkage in lactose [Galβ1 → 4Glc] mimics the receptor polysaccharide linkage, e.g. Galβ1 → 3GalNAc. The third kind of adhesin shown on cell type A is semi-circular and represents an adhesin, i.e. not involved in coaggregation with partner B.

Coaggregation among oral bacteria often appears as a dramatic visual phenomenon. At the bottom of Figure 3.1, the inset containing five tubes shows the visual assay for coaggregation. An evenly turbid suspension of one cell type (tube #1) and an evenly turbid suspension of a genetically distinct second cell type (tube #2) are mixed together to test if they are coaggregation partners. If they are partners, they will clump together and form a network of mixed-species coaggregates (tube #3).

Coaggregation usually happens immediately after vortexing a suspension containing the two cell types. If the coaggregation is strong, the clumps settle to the bottom of the tube within seconds leaving a clear supernatant (tube #4). Addition of lactose at a final concentration of about 100 mM to a coaggregated suspension, such as seen in tube #4, followed by brief vortexing, can result in equally dramatic reversal of coaggregation to an evenly turbid mixed-cell type suspension (tube #5). Lactose is a convenient sugar to use to test for reversal of coaggregation, but other sugars, such as D-galactose-β(1 → 3)-*N*-acetyl-D-galactosamine glycosides (McIntire *et al.*, 1983), L-rhamnose (Weiss *et al.*, 1987), or *N*-acetyl-D-galactosamine (Kolenbrander *et al.*, 1990) are effective at 10-fold lower concentrations with some coaggregations. An example of mixed-species coaggregates

Coaggregation: receptor–adhesin interactions

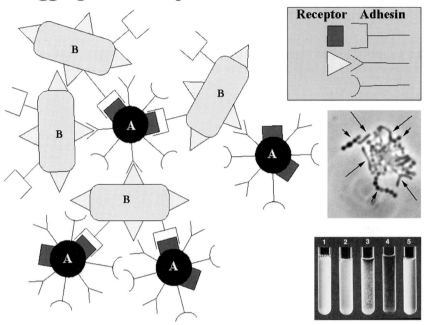

Figure 3.1   Different views of coaggregation. Diagrammatic model of coaggregation between cell type A (circular) and cell type B (oblong). Cell type A expresses three potential coaggregation-mediating surface components two of which are recognized by respective cognate components on cell type B. The components are called receptors and adhesins (inset at upper right). Adhesins are depicted as components with a stem structure and their cognate receptor has the same shape but is lacking the stem. The cognate rectangular symbols represent mediators of lactose-inhibitable coaggregations. Two of the surface components on cell type A are adhesins; the triangle-shaped adhesin recognizes its cognate triangular receptor on cell type B, but the semi-circular-shaped adhesin has no cognate on the partner. Cell type A also expresses a receptor (rectangle), whose cognate adhesin is present on cell type B. Different shaped symbols represent different kinds of adhesins or receptors and illustrate the multi-functional capabilities of oral bacteria.

Bottom right inset, visual coaggregation assay. Tube 1 represents a dense suspension of cell type A. Tube 2 represents a dense suspension of cell type B. Tube 3 shows strong coaggregation that would occur between partner cell types. Tube 4 shows the strongest kind of coaggregation in that coaggregates settle to the bottom of the tube within a few seconds after vortexing the mixed-species suspension. Tube 5 illustrates the dramatic reversal of coaggregation to an evenly turbid suspension after adding an inhibitor of coaggregation. Many coaggregations are reversed by addition of ethylenediaminetetraacetic acid or lactose.

Middle right inset, phase-contrast microscopic view of coaggregation between streptococci (short arrows) and actinomyces (long arrows). [This figure is also reproduced in colour in the plate section after page 326.]

observed by phase-contrast microscopy in a sample taken from a mixed-species suspension with the appearance seen in tube #3 is shown in the inset on the middle right (Figure 3.1). It is clearly evident that this coaggregate is composed of a network of interacting partner cell types.

3.3   COAGGREGATION VERSUS COADHESION

Large coaggregates detectable as visible clumps in a test tube are only one mani-
festation of coaggregation of oral bacteria. Small coaggregates that are not visible
by eye may also form between species. These coaggregates can be observed
either by standard phase-contrast microscopy or confocal laser microscopy
(CLM) (Figure 3.2). This distinction between visible coaggregates by eye and

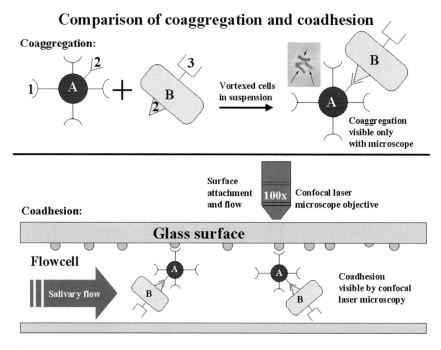

Figure 3.2   Coaggregation and coadhesion. In this illustration the number of adhesin #2
and its cognate receptor (yellow triangle on cell type B) is intentionally low to emphasize
the distinction between coaggregation and coadhesion assays. Circular cell type A and
oblong cell type B can be mixed together by vortexing. If the adhesin or receptor numbers
per cell are very low, approaching one or two per cell, then large mixed-cell-type
coaggregates visible by eye cannot form. Instead, small coaggregates occur (inset, upper
right). These can be seen using a microscope showing streptococci (short arrow)
coaggregating with actinomyces (long arrows).
   Coadhesion, on the other hand, occurs when cells have a chance to attach to a saliva-
conditioned substratum and colonize the substratum as indicated in the flowcell system
(bottom). Salivary conditioning film provides receptors (green circles) available for
binding by initial colonizing bacteria. In this example, cell type A attaches to a salivary
receptor through adhesin #1 that has no importance in the coaggregation with cell
type B. Low numbers of adhesin #2 or cognate receptor #2 have minimal effect on
retention of cell type B in the biofilm. The coadherent cells can be detected and
differentiated from the initial cell layer by immunofluorescence or FISH and using a
confocal laser microscope.[This figure is also reproduced in colour in the plate section
after page 326.]

detectable coadhesion by CLM becomes important when investigators are screening bacteria for properties that contribute to adherence-relevant phenotypes in biofilms.

In Figure 3.2, a comparison of properties of coaggregation (Figure 3.2, top panel) and coadhesion (Figure 3.2, bottom panel) is presented. Two kinds of cells are depicted. Cell type A exhibits a unimodal coaggregation function, as represented by the triangular adhesin (#2). Cell type B exhibits the cognate receptor (yellow triangle #2) as well as a rectangular adhesin (#3). In this illustration, adhesin #3 on cell type B has no importance in either coadhesion or coaggregation. It is however available for coaggregation with a partner cell with an appropriate cognate receptor #3 in which case cell type B would act as a coaggregation bridge between that potential partner cell and cell type A. The actual number of such adhesins and cognate receptors on living cells may be more than one, but Figure 3.2 diagrams interactions when the numbers of surface molecules mediating coaggregation or coadhesion are very few. In an experimental analysis of two different adhesins on *Prevotella loescheii* PK1295, maxima of 300–400 adhesins per cell were calculated by using radioactively labeled monoclonal antibodies against the two adhesins (Weiss *et al.*, 1988). Other estimates with different oral bacteria give numbers of fewer than 10 adhesins per cell (Kolenbrander, unpublished data). Thus, Figure 3.2 illustrates the distinction between results of assays that measure visible coaggregations in suspension and those that measure coadherence of a cell to an already attached cell in a biofilm.

The upper panel of the figure depicts coaggregations, which are not visible by eye but are apparent with a microscope. Contrast these small coaggregates with the large visible coaggregate shown in Figure 3.1. The microscopic nature of such coaggregations may be a result of only a few adhesin or receptor pairs available for coaggregation on the respective cell surfaces of the partners. If the number of adhesin or receptor pairs is one per respective cell type, then coaggregates can only form as two-cell units, which are visible only by using a microscope as shown in the inset at the upper right corner (Figure 3.2). This kind of unimodal coaggregation between cells expressing low numbers of adhesin or receptor cognates would be scored as no coaggregation by visual inspection. Although coaggregates are formed, they remain dispersed in the suspension and undetectable by eye.

It is possible for cells that are negative by visual inspection for coaggregation to be positive for coadhesion and, thus, biofilm colonization. This is illustrated at the bottom of Figure 3.2, where a flowcell is portrayed. Cell type A bears adhesin #1 that recognizes and binds to a salivary receptor in the conditioning film coating the glass surface. Coadhesion of cell type B to cell type A is detectable because cell type A is immobilized and thus the events of coadhesion are cumulative on the surface. Coadhesion is easily detected by CLM using fluorescently labeled antibodies against cell-surface components or fluorescently labeled oligonucleotides that hybridize to ribosomal RNA molecules [fluorescence in situ hybridization (FISH)]. Thus, the same coaggregates would be given a negative score by visual inspection of a suspension of the two partner cell types but would be positive in a biofilm formed by first adding cell type A followed by cell type B. This distinction becomes significant in investigations of the adhesins contributing to colonization of enamel surfaces of teeth.

The same results for coadhesion described for the flowcell assay (Figure 3.2, lower panel) can be obtained by coadhesion assays using filter paper supports (Lamont and Rosan, 1990) or microtiter-plate plastic surfaces (Jenkinson *et al.*, 1993). For example, radioactively labeled cell type B is easily detected as surface bound radioactivity (Lamont and Rosan, 1990) in weakly formed coaggregations that may not be detectable by visual inspection of the same partner cell-types in suspension. Likewise, measuring the coadhesion of cell type B to cell type A already attached to a microtiter-plate well surface by using an antibody to a surface component of cell type B would be scored as positive for coadhesion (Jenkinson *et al.*, 1993). These examples point out differences in the various assays for coaggregation and coadhesion. Results from the visual coaggregation assay as well as from microscopic-based assays, radioactivity measurements for coadhesion, and antibody-based detection of coadhesion define adherence phenotypes relevant to biofilm formation.

3.4   ROLE OF COAGGREGATION IN BIOFILMS

In 1970, Gibbons and Nygaard first reported coaggregation among human oral bacteria (Gibbons and Nygaard, 1970), and, in 1978, McIntire *et al.* showed that one of these coaggregations between a streptococcus and an actinomyces was mediated by lectin–carbohydrate (adhesin–receptor) interactions (McIntire *et al.*, 1978). The first review of the extensive nature of coaggregations between streptococci and actinomyces appeared in 1988 (Kolenbrander, 1988). A modification of the original diagrammatic model (Kolenbrander, 1988) is shown in Figure 3.3 and includes the coaggregations between three of the six groups of streptococci (groups 1–3) and all six groups of actinomyces (groups A–F). These coaggregation groupings are based on (1) the ability of a pair of streptococcal and actinomyces strains to coaggregate (Figure 3.1, inset with five test tubes), (2) the effect of a simple sugar, such as lactose to inhibit a coaggregating pair, (3) the ability of heat (85°C/ 30 min) or protease pre-treatment of cells to prevent coaggregation, and (4) the loss of the ability of spontaneous coaggregation-defective mutants of streptococci or actinomyces to coaggregate with partners of the parent cell. For example, heating or protease treatment of group 1 streptococcal cells (Figure 3.3, left side) completely abolished their ability to coaggregate with actinomyces group A, C, D, and E cells. A second example is that the coaggregation between streptococcal group 1 and actinomyces group C is lactose inhibitable, whereas, none of the other coaggregations with group 1 are affected by lactose.

Each coaggregation group exhibits distinct coaggregation properties (Figure 3.3). Streptococcal group 2, like streptococcal group 1, is completely inactivated by heat or protease treatment, but it is different in that it coaggregates with actinomyces

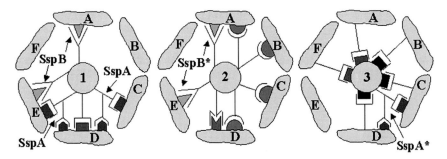

Figure 3.3   Diagrammatic representation of coaggregations between three groups of oral streptococci and six groups of oral actinomyces showing varied kinds of interactions characterizing each group. Current model of interactions emphasizing the involvement of streptococcal SspA and SspB adhesins in coaggregations with actinomyces. SspB of *S. gordonii* DL1, representative of streptococcal coaggregation group 1, mediates coaggregation with actinomyces group A and E. An SspB homolog, SspB* of streptococcal group 2, mediates the same coaggregations with actinomyces groups A and E. SspA of *S. gordonii* DL1 mediates coaggregation with actinomyces groups C, D and E. An SspA homolog, SspA* of streptococcal group 3, mediates coaggregation with actinomyces groups C and D. Possibly, the cognate receptor on actinomyces group E is distinct from the receptor on actinomyces group C and does not recognize the SspA* adhesin. Alternatively, the specificity of SspA* is slightly different than the specificity of SspA, and SspA* is unable to recognize the cognate receptor on actinomyces group E. [This figure is also reproduced in colour in the plate section after page 326.]

group B and none of its coaggregations are lactose inhibitable. Streptococcal group 3, on the other hand, is generally unaffected by heat and protease treatment. Instead, it bears polysaccharide receptors for all of its partners. In addition, it exhibits adhesins for actinomyces groups C and D. This adhesin endows strepto-coccal group 3 with bimodal capability in that it coaggregates with actinomyces groups C and D by two different mechanisms, one of which is lactose inhibitable. Streptococcal group 2 exhibits bimodal characteristics with partners in actino-myces groups A and D. Although the coaggregations exhibited by each strepto-coccal group are different, some similarities to the coaggregations mediated by the adhesins SspA and SspB in coaggregation group 1 are evident in the coaggre-gations mediated by homologs SspB* and SspA*.

Given that several of the coaggregations between streptococcal groups 2 and 3 with their partners exhibit coaggregations like those between streptococcal group 1 and its partners, it suggests that interactions between streptococci and actino-myces may frequently be mediated by SspA and SspB and their homologs. Jenkinson *et al.* (1993) reported that the coaggregations of *Streptococcus gor-donii* DL1, a representative of streptococcal group 1, with actinomyces were mediated partly by SspA on the surface of *S. gordonii* DL1 (Jenkinson *et al.*, 1993). Insertion mutants in *sspA* and *sspB* of the streptococcal group 1 represen-tative *S. gordonii* DL1 allowed the first analysis of the independent functions of SspA and SspB with respect to coaggregation with actinomyces strains and allowed detection of a previously unrecognized lactose-inhibitable interaction between *S. gordonii* and members of actinomyces coaggregation group D (Egland *et al.*, 2001). Their results indicated that the SspA and SspB adhesins were critical for coaggregation of *S. gordonii* DL1 with actinomyces coaggregation groups A, C, D, and E (Figure 3.3). SspA exhibited two coaggregation-specific functions; it participated in lactose-inhibitable and lactose-noninhibitable interactions (Figure 3.3, cognate symbols complementary to rectangle and obelisk, respectively). Mutation of *sspA* resulted in changes in coaggregation with three of the four actinomyces coaggregation groups (see left side of Figure 3.3). For example, the *sspA* mutant was unable to coaggregate with actinomyces coaggregation group C. SspB mediated only lactose-noninhibitable coaggregations (Figure 3.3, cognate symbol complementary to triangle). The *sspB* mutant was unable to coaggregate with actinomyces coaggregation group A, and its coaggregation with actinomyces coaggregation group E was changed from lactose-noninhibitable to lactose-inhibitable. An *sspAB* double mutant was generated and the resulting effects were additive; however, it retained a single previously unrecognized lactose-inhibitable coaggregation with an actinomyces coaggregation group D representative (Figure 3.3, cognate symbol complementary to gray rectangle). Thus, SspA and SspB appear to mediate all coaggregations between *S. gordonii* DL1 and the actinomyces coaggregation groups except the lactose-inhibitable coaggregation with actinomyces coaggregation group D.

Considering that oral streptococci and actinomyces exhibit distinct coaggregation partnerships, it should be possible to design experiments to test the participation of partnerships in biofilm formation. In other words, the ability of cells to coaggregate and form visible clumps in suspension (coaggregation) and the ability of the same cells to form a mixed-species biofilm in a saliva-coated flowcell (coadhesion) can be compared (Palmer and Caldwell, 1995; Kolenbrander *et al.*, 1999). To make such a comparison, wild type and an *sspAB* mutant strain of *S. gordonii* DL1 were chosen to explore the relationship between coaggregation and coadhesion. The actinomyces partner was *Actinomyces naeslundii* T14V, a representative of actinomyces coaggregation group A (see Figure 3.3). Wild type and mutant *S. gordonii* DL1 were grown overnight in brain-heart infusion broth, washed by centrifugation and resuspension of the pellet in 25% sterile saliva, and adjusted to a cell density of 3×10^9 cells/ml. Wild type or mutant cells were allowed to bind to a saliva-conditioned flowcell surface for 15 min. Biofilms of the wild type or the mutant were allowed to form for 4 h with sucrose (10 μM)- and peptone (0.025%)-supplemented saliva as the nutrient. Flow was at 0.2 ml/min. Fresh nutrient was passed through the flowcell continuously and without re-circulation, so that cells had one 15 min window, prior to initiation of flow, to bind to the substratum of the flowcell. *A. naeslundii* T14V cells, grown and washed in the same manner as the streptococci, were then introduced into the flowcell and allowed to bind to the streptococcal cell surfaces without flow for 15 min. Flow was restarted and continued for 15 min to wash out unattached actinomyces cells. The actinomyces (green) bound to the wild type *S. gordonii* biofilm twice as well as to the mutant biofilm (Figure 3.4). However, when examined by the visual coaggre-

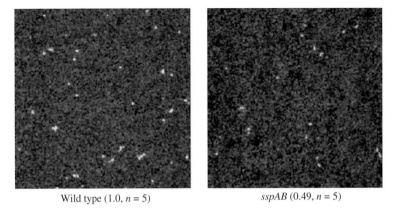

Wild type (1.0, $n = 5$) $sspAB$ (0.49, $n = 5$)

Figure 3.4   Recruitment of *A. naeslundii* T14V cells to established streptococcal biofilms. Experimental details on establishing the biofilms are given in the text. Biofilms formed for 4 h by *S. gordonii* DL1 (wild type) or an isogenic mutant (*sspAB*) were inoculated with *A. naeslundii* T14V at a density of to 3×10^9 actinomyces cells per ml. Coadherence of the actinomyces to the streptococci was measured by counting the number of coadherent actinomyces cells. The number of cells bound to the wild type was normalized to 1.0. The relative number bound to the *sspAB* mutant was 0.49. Actinomyces cells were labeled with monoclonal antibodies raised against *A. naeslundii* T14V type 1 fimbriae (Cisar *et al.*, 1988) and detected with Cy2-conjugated anti-mouse IgG (Jackson ImmunoResearch, West Grove, PA). Streptococcal cells were stained with Alexa-647 (Molecular Probes, Eugene, OR) conjugated IgG raised against whole *S. gordonii* cells. [This figure is also reproduced in colour in the plate section after page 326.]

gation assay, the effect of the *sspAB* mutation on coaggregation was even more dramatic; no coaggregation with *A. naeslundii* T14V was detected (Egland *et al.*, 2001). This demonstrates a distinction between coaggregation (Figure 3.1) and coadherence (Figure 3.4). In a separate study, the role of SspA and SspB in the coadhesion of *S. gordonii* and *Porphyromonas gingivalis*, a coadhesion partner, in biofilms was examined (Lamont *et al.*, 2002). A different *S. gordonii sspAB* null mutant was not able to support coadhesion of wild type streptococcal partner *P. gingivalis* cells, although the mutant did form a biofilm in a saliva-coated glass flowcell (Lamont *et al.*, 2002). Collectively, these results suggest that different partners and different mutations may yield different biofilm phenotypes.

The ability to detect cell–cell mixed-species interactions as coadhesion by CLM but not as coaggregation by visual inspection of mixed-species cell suspensions is relevant to investigations of the role that surface components play in forming biofilms. Coadhesion but not coaggregation may be detectable with the *sspAB* mutant because other surface components may be expressed but be sparsely distributed or may mediate very weak interactions with the actinomyces. Only a few adhesins or receptors per cell may be sufficient to detect coadhesion by CLM but not by visual inspection for coaggregates in suspension (Figure 3.2). Likewise, weak interactions may not be able to maintain coaggregate stability in suspension, but they may be sufficient to maintain coadhesion in biofilms. Another possible event that may occur and distinguish coaggregation from coadhesion is the contact-induction of a new receptor on the surface of cells attached to the substratum (Figure 3.5). Expression of the new receptor for adhesin #3 after cells are bound to the surface now permits coadhesion of a partner possessing the cognate adhesin #3. Since the receptor #3 is biofilm induced, coaggregation would be absent in suspension (Figure 3.5). Biofilm-induced changes in gene expression have been observed in *Escherichia coli* (Zhang and Normark, 1996; Prigent-Combaret *et al.*, 1999), *Pseudomonas aeruginosa* (Garrett *et al.*, 1999; Drenkard and Ausubel, 2002; Singh *et al.*, 2002), *Pseudomonas fluorescens* (O'Toole and Kolter, 1998), and *Vibrio parahaemolyticus* (McCarter and Silverman, 1990). Whereas such precedents have been reported with single species, no evidence exists that coadherence or coaggregation induces a change in gene expression. Although other explanations are possible, the difference between the results of coaggregation and those of coadherence in the *in vitro* biofilm shown in Figure 3.4 could be explained by a biofilm-induced change in expression of a streptococcal coaggregation mediator with *A. naeslundii*. This result raises the possibility that

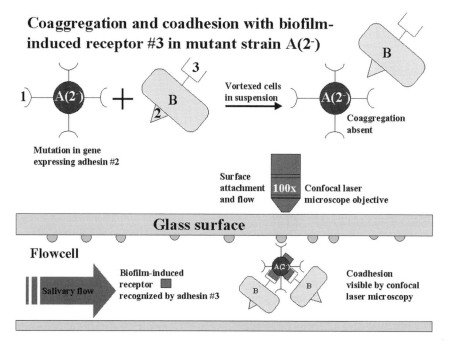

Figure 3.5   Biofilm induction of coadherence-relevant mediator. Top. Cell type A(2⁻) is a mutant lacking adhesin #2 and is unable to coaggregate with cell type B, although cell type B expresses cognate receptor #2. Cell type B expresses adhesin #3, but it is not involved in coaggregation, because cell type A does not express cognate receptor #3 in suspension. Bottom. Mutant cell type A(2⁻) binds to receptor in salivary conditioning film and is induced to synthesize receptor #3. Coadhesion occurs because cell type B expresses cognate adhesin #3. [This figure is also reproduced in colour in the plate section after page 326.]

studying cells with a mutation in a gene encoding an adhesin or a receptor may not yield the expected biofilm phenotype based on the altered coaggregation phenotype observed in screening for adherence mutants.

To test whether coaggregation contributes to successful biofilm colonization, three oral bacterial strains that were coaggregation partners of each other were examined (Palmer *et al.*, 2001). A saliva-conditioned glass flowcell as depicted in Figure 3.2 was employed, and saliva was the sole source of nutrient. Each organism, *S. gordonii* DL1, *Streptococcus oralis* 34, and *A. naeslundii* T14V, was capable of binding to the saliva-coated surface, but only *S. gordonii* formed a thick biofilm as a mono-culture. In co-culture with *S. gordonii*, neither *S. oralis* nor *A. naeslundii* grew to any significant extent, whereas *S. gordonii* grew as well in co-culture as it had in mono-culture. However, both *S. oralis* and *A. naeslundii* did exhibit coaggregation with *S. gordonii* and formed occasional mixed-species microcolonies in the biofilm. When the two species that could not form biofilms independently, *S. oralis* and *A. naeslundii*, were co-cultured, they formed a luxuriant mixed-species biofilm (Palmer *et al.*, 2001). The amount of cellular mass observed with this mutalistic growth was more than the amount of cellular mass found with the independent growth shown by *S. gordonii* in either mono-culture or co-culture with its coaggregation partners. Furthermore, these results suggested that the ability to coaggregate permits retention in the biofilm under conditions of minimal growth. Retention when conditions are unfavorable is a critical attribute of cells capable of colonizing a developing biofilm. Being retained under undesirable conditions provides opportunity to flourish when conditions change to those that encourage growth. In the above example of *S. oralis* in a mixed-species biofilm with *S. gordonii*, being retained could provide *S. oralis* with a future opportunity to coadhere with *A. naeslundii* with resultant explosive growth. This scenario is especially significant considering that in the oral cavity, many species exist and have opportunity to bind to the developing dental plaque, but perhaps the partnership is not immediately suitable for extensive growth. Thus, retention is a key element because its absence leads to removal of cells from the oral cavity by salivary flow and swallowing.

Biofilm formation *in vivo* is a primary survival mechanism for most if not all human oral bacteria. All oral bacteria tested exhibit coaggregation with at least one other group of bacteria (Table 3.1). The question remains as to the role of these coaggregations in developing dental biofilms. Preliminary studies of initial colonization of saliva-coated enamel *in vivo* (Palmer, Gordon, Cisar, and Kolenbrander, unpublished data) confirmed two earlier studies in that most of the initial colonizing bacteria appear to be streptococci (Nyvad and Kilian, 1987; 1990). Currently, the role of coaggregation in the development of *in vivo* biofilms is being investigated by using fluorescently labeled antibodies specific against actinomyces type 2 fimbriae and against the streptococcal cognate receptor polysaccharide (see Figure 3.3, interaction between actinomyces group A (type 2 fimbriae) and streptococcal group 3 (receptor polysaccharide)). A positive role for coaggregation in oral biofilms would be demonstrated by juxtaposition of cells bearing type 2 fimbriae and cells bearing cognate receptor polysaccharide on an enamel surface (Palmer, Gordon, Cisar, and Kolenbrander, unpublished data). With available molecular tools for immunofluorescence and FISH, it is now possible to establish the role of coaggregation *in vivo* in developing human oral biofilms called dental plaque.

3.5 CONCLUSIONS

The oral microbial ecosystem presents a unique opportunity to investigate interbacterial communication mechanisms. Coaggregation of oral bacteria is one form of communication, and it may play an important role in exchange of metabolites and genes. The ability of oral bacteria to use coaggregation mechanisms for retention within biofilms in the absence of cellular growth may be critical for an organism to be included in later stages of development in a flowing ecosystem, such as the oral cavity. Coaggregation and coadhesion occur by the same mechanisms. Coaggregations are detectable with the naked eye when large aggregates of mixed-species are formed. However, when small aggregates of mixed-species are formed, microscopy is required for detection. Although numerous mechanisms of cell–cell interactions between genetically distinct cells are already known, additional novel coaggregation mechanisms may be induced by contact of cells with a substratum, and this raises the possibility that other forms of communication may exist among participants of mixed-species communities, such as dental plaque.

REFERENCES

Andersen, R.N., Ganeshkumar, N. and Kolenbrander, P.E. (1998) *Helicobacter pylori* adheres selectively to *Fusobacterium* spp. *Oral Microbiol. Immunol.* **13**, 51–54.

Bos, R., van der Mei, H.C., Meinders, J.M. and Busscher, H.J. (1994) A quantitative method to study co-adhesion of microorganisms in a parallel plate flow chamber: basic principles of the analysis. *J. Microbiol. Meth.* **20**, 289–305.

Cisar, J.O., Kolenbrander, P.E. and McIntire, F.C. (1979) Specificity of coaggregation reactions between human oral streptococci and strains of *Actinomyces viscosus* or *Actinomyces naeslundii*. *Infect. Immun.* **24**, 742–752.

Cisar, J.O., Vatter, A.E., Clark, W.B., Curl, S.H., Hurst-Calderone, S. and Sandberg, A.L. (1988) Mutants of *Actinomyces viscosus* T14V lacking type 1, type 2, or both types of fimbriae. *Infect. Immun.* **56**, 2984–2989.

Cisar, J.O., Sandberg, A.L., Abeygunawardana, C., Reddy, G.P. and Bush, C.A. (1995) Lectin recognition of host-like saccharide motifs in streptococcal cell wall polysaccharides. *Glycobiology* **5**, 655–662.

Cisar, J.O., Sandberg, A.L., Reddy, G.P., Abeygunawardana, C. and Bush, C.A. (1997) Structural and antigenic types of cell wall polysaccharides from viridans group streptococci with receptors for oral actinomyces and streptococcal lectins. *Infect. Immun.* **65**, 5035–5041.

Demuth, D.R., Duan, Y., Brooks, W., Holmes, A.R., McNab, R. and Jenkinson, H.F. (1996) Tandem genes encode cell-surface polypeptides SspA and SspB which mediate adhesion of the oral bacterium *Streptococcus gordonii* to human and bacterial receptors. *Mol. Microbiol.* **20**, 403–413.

Drenkard, E. and Ausubel, F.M. (2002) *Pseudomonas* biofilm formation and antibiotic resistance are linked to phenotypic variation. *Nature* **416**, 740–743.

Egland, P.G., Dû, L.D. and Kolenbrander, P.E. (2001) Identification of independent *Streptococcus gordonii* SspA and SspB functions in coaggregation with *Actinomyces naeslundii*. *Infect. Immun.* **69**, 7512–7516.

Garrett, E.S., Perlegas, D. and Wozniak, D.J. (1999) Negative control of flagellum synthesis in *Pseudomonas aeruginosa* is modulated by the alternative sigma factor AlgT (AlgU). *J. Bacteriol.* **181**, 7401–7404.

Gibbons, R.J. and Nygaard, M. (1970) Interbacterial aggregation of plaque bacteria. *Arch. Oral Biol.* **15**, 1397–1400.

Gibbons, R.J. and Hay, D.I. (1988) Adsorbed salivary proline-rich proteins as bacterial receptors on apatitic surfaces. In *Molecular Mechanisms of Microbial Adhesion* (ed. E. Beachey), pp. 143–163, Springer-Verlag, New York.

Hallberg, K., Hammarstrom, K.J., Falsen, E., Dahlen, G., Gibbons, R.J., Hay, D.I. and Stromberg, N. (1998) *Actinomyces naeslundii* genospecies 1 and 2 express different binding specificities to *N*-acetyl-beta-D-galactosamine, whereas *Actinomyces odontolyticus* expresses a different binding specificity in colonizing the human mouth. *Oral Microbiol. Immunol.* **13**, 327–336.

Jenkinson, H.F., Terry, S.D., McNab, R. and Tannock, G.W. (1993) Inactivation of the gene encoding surface protein SspA in *Streptococcus gordonii* DL1 affects cell interactions with human salivary agglutinin and oral actinomyces. *Infect. Immun.* **61**, 3199–3208.

Kolenbrander, P.E. (1988) Intergeneric coaggregation among human oral bacteria and ecology of dental plaque. *Annu. Rev. Microbiol.* **42**, 627–656.

Kolenbrander, P.E., Andersen, R.N. and Moore, L.V. (1990) Intrageneric coaggregation among strains of human oral bacteria: potential role in primary colonization of the tooth surface. *Appl. Environ. Microbiol.* **56**, 3890–3894.

Kolenbrander, P.E., Andersen, R.N., Kazmerzak, K., Wu, R. and Palmer Jr., R.J. (1999) Spatial organization of oral bacteria in biofilms. *Meth. Enzymol.* **310**, 322–332.

Kolenbrander, P.E., Andersen, R.N., Blehert, D.S., Egland, P.G., Foster, J.S. and Palmer Jr., R.J. (2002) Communication among oral bacteria. *Microbiol. Mol. Biol. Rev.* **66**, 486–505.

Lamont, R.J. and Rosan, B. (1990) Adherence of mutans streptococci to other oral bacteria. *Infect. Immun.* **58**, 1738–1743.

Lamont, R.J., El-Sabaeny, A., Park, Y., Cook, G.S., Costerton, J.W. and Demuth, D.R. (2002) Role of the *Streptococcus gordonii* SspB protein in the development of *Porphyromonas gingivalis* biofilms on streptococcal substrates. *Microbiology* **148**, 1627–1636.

McCarter, L. and Silverman, M. (1990) Surface-induced swarmer cell differentiation of *Vibrio parahaemolyticus*. *Mol. Microbiol.* **4**, 1057–1062.

McIntire, F.C., Vatter, A.E., Baros, J. and Arnold, J. (1978) Mechanism of coaggregation between *Actinomyces viscosus* T14V and *Streptococcus sanguis* 34. *Infect. Immun.* **21**, 978–988.

McIntire, F.C., Crosby, L.K., Barlow, J.J. and Matta, K.L. (1983) Structural preferences of beta-galactoside-reactive lectins on *Actinomyces viscosus* T14V and *Actinomyces naeslundii* WVU45. *Infect. Immun.* **41**, 848–850.

McIntire, F.C., Bush, C.A., Wu, S.S., Li, S.C., Li, Y.T., McNeil, M., Tjoa, S.S. and Fennessey, P.V. (1987) Structure of a new hexasaccharide from the coaggregation polysaccharide of *Streptococcus sanguis* 34. *Carbohydr. Res.* **166**, 133–143.

Nyvad, B. and Kilian, M. (1987) Microbiology of the early colonization of human enamel and root surfaces *in vivo*. *Scand. J. Dent. Res.* **95**, 369–380.

Nyvad, B. and Kilian, M. (1990) Comparison of the initial streptococcal microflora on dental enamel in caries-active and in caries-inactive individuals. *Caries Res.* **24**, 267–272.

O'Toole, G.A. and Kolter, R. (1998) Initiation of biofilm formation in *Pseudomonas fluorescens* WCS365 proceeds via multiple, convergent signalling pathways: a genetic analysis. *Mol. Microbiol.* **28**, 449–461.

Palmer Jr., R.J. and Caldwell, D.E. (1995) A flowcell for the study of plaque removal and regrowth. *J. Microbiol. Meth.* **24**, 171–182.

Palmer, R.J., Jr., Kazmerzak, K., Hansen, M.C. and Kolenbrander, P.E. (2001) Mutualism versus independence: strategies of mixed-species oral biofilms *in vitro* using saliva as the sole nutrient source. *Infect. Immun.* **69**, 5794–5804.

Prigent-Combaret, C., Vidal, O., Dorel, C. and Lejeune, P. (1999) Abiotic surface sensing and biofilm-dependent regulation of gene expression in *Escherichia coli*. *J. Bacteriol.* **181**, 5993–6002.

Rickard, A.H., Leach, S.A., Buswell, C.M., High, N.J. and Handley, P.S. (2000) Coaggregation between aquatic bacteria is mediated by specific-growth-phase-dependent lectin-saccharide interactions. *Appl. Environ. Microbiol.* **66**, 431–434.

Rickard, A.H., Leach, S.A., Hall, L.S., Buswell, C.M., High, N.J. and Handley, P.S. (2002) Phylogenetic relationships and coaggregation ability of freshwater biofilm bacteria. *Appl. Environ. Microbiol.* **68**, 3644–3650.

Singh, P.K., Parsek, M.R., Greenberg, E.P. and Welsh, M.J. (2002) A component of innate immunity prevents bacterial biofilm development. *Nature* **417**, 552–555.

Weiss, E.I., London, J., Kolenbrander, P.E., Kagermeier, A.S. and Andersen, R.N. (1987) Characterization of lectinlike surface components on *Capnocytophaga ochracea* ATCC 33596 that mediate coaggregation with gram-positive oral bacteria. *Infect. Immun.* **55**, 1198–1202.

Weiss, E.I., London, J., Kolenbrander, P.E., Hand, A.R. and Siraganian, R. (1988) Localization and enumeration of fimbria-associated adhesins of *Bacteroides loescheii*. *J. Bacteriol.* **170**, 1123–1128.

Whittaker, C.J., Klier, C.M. and Kolenbrander, P.E. (1996) Mechanisms of adhesion by oral bacteria. *Annu. Rev. Microbiol.* **50**, 513–552.

Zhang, J.P. and Normark, S. (1996) Induction of gene expression in *Escherichia coli* after pilus-mediated adherence. *Science* **273**, 1234–1236.

4

Genetics of biofilm formation

M. Espinosa-Urgel and J.-L. Ramos

4.1 INTRODUCTION

The establishment of microbial populations as sessile communities attached to solid surfaces has long been recognised as one of the main strategies for survival of bacteria in the environment (Meadows, 1971). The capacity of microorganisms to form biofilms can be exploited not only for biotechnological applications, such as wastewater treatment (Nicolella *et al.*, 2000), but it also constitutes a significant problem in industry and medicine (see related chapters in this book), and even in space research or art preservation (Gu *et al.*, 1998; Schabereiter-Gurtner *et al.*, 2001). The study of microbial biofilms, traditionally approached from a descriptive point of view, has gained a whole new perspective in the past few years thanks to new developments in microscopic techniques allowing non-disruptive *in situ* observation, and to molecular genetics. Many research groups have geared their efforts towards unraveling the molecular mechanisms involved in bacterial settlement on solid surfaces.

Biofilms are not just aggregates of bacterial cells stuck to a surface. They have a dynamic nature, which can be viewed as a developmental process in the case of pure cultures, or as the build-up – and eventually the destruction – of a multi-species 'microbial city' (Watnick and Kolter, 2000; O'Toole *et al.*, 2000b). This dynamic nature is even more evident in the case of bacterial populations associated to biotic surfaces, such as plant roots, where the surface is not simply an inert substrate for biofilm growth, but an active player in the system.

This chapter aims to summarise (i) the methods used in the study of biofilms from a genetic point of view, and (ii) our current knowledge about genes and gene products important for the switch between the planktonic and the sessile way of life. It has become apparent that this is a complex and regulated process in which changes in the physiology, morphology and gene expression pattern of the bacterial cells take place. Rather than providing an exhaustive description of each and every element, a general view of attachment and biofilm development will be given, pinpointing those aspects that are of particular relevance in terms of microbial genetics.

4.2   OVERVIEW OF BIOFILM DEVELOPMENT

Growth of a bacterial culture in liquid medium, although a continuous process, can be differentiated in several phases, from lag to stationary phase. Similarly, several stages can be defined in the development of a pure-culture biofilm on abiotic surfaces (Figure 4.1). Although, as discussed later on, the conditions that favour biofilm formation may vary for different bacterial species, a general scheme can be drawn. Environmental conditions and/or signals promote bacterial attachment to the solid surface. Initial adhesion of individual cells, and movement along the plane gives rise to a monolayer of bacteria associated to the surface. This mono-layer will evolve towards a mature biofilm through the formation of microcolonies first, and then of macrocolonies, in a process where quorum-sensing phenomena have an important role. Eventually, when environmental conditions change and are no longer favorable for the maintenance of the biofilm, cells will start detaching and swimming away, thus, returning to the planktonic phase of the cycle.

The development of multi-species biofilms is best characterised in bacteria colonizing the oral cavity (reviewed by Rosan and Lamont, 2000). In this system, biofilm formation proceeds in a sequential way, with particular species acting as primary colonisers of teeth. In a second stage, other bacteria are able to recognise and attach to these initial colonisers and in turn be recognised by other late colonisers, in a relatively specific sequence of cell-to-cell interactions, known as coaggregation.

Therefore, biofilm formation is not simply a matter of bacteria having the ability to attach to a solid surface. Environmental signals are involved in the transitions from planktonic to sessile life and viceversa, and cell-to-cell signaling is key for the development and maturation of the biofilm, both in pure cultures and in mixed populations.

4.3   METHODS FOR GENETIC ANALYSIS OF BIOFILMS

4.3.1   Mass screening and selection of mutants

4.3.1.1   Abiotic surfaces

Random-transposon mutagenesis, combined with fast and easy methods for the selection of adhesion-deficient mutants, has been fundamental in the surge of

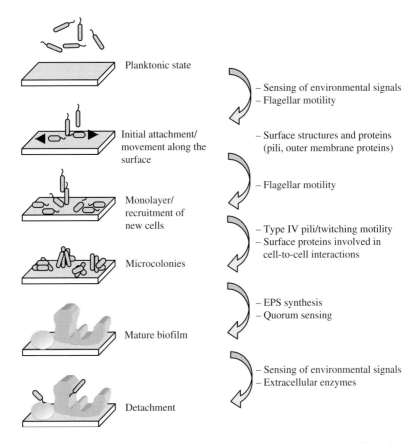

Figure 4.1   The biofilm life cycle. The functions discussed in the text are indicated in the figure.

genetic studies of biofilm formation. Transposons that integrate in the chromosome of a target strain in a fashion that, in practice, can be considered as random have been used for a long time for the purpose of insertional mutagenesis (Way *et al.*, 1984). The protocols for transposon mutagenesis are simple and well established, and a wide number of transposon derivatives carrying different antibiotic-resistance markers are available (de Lorenzo *et al.*, 1990).

A simple screening method (Figure 4.2) has been widely used to identify mutants defective in adhesion to abiotic surfaces of a variety of Gram-negative and Gram-positive bacteria, namely, *Escherichia coli*, *Pseudomonas aeruginosa*, *Pseudomonas fluorescens*, *Vibrio cholerae*, *Staphylococcus epidermidis* and *Staphylococcus aureus* (reviewed by O'Toole *et al.*, 1999). It consists of growing a collection of random-transposon mutants in liquid medium in 96-wells microtiter dishes, under conditions that promote biofilm formation (which must be previously determined for the strain of interest). Attachment of the bacteria to the plastic walls

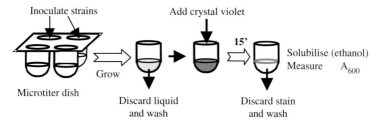

Figure 4.2   Screening and quantification method for biofilm formation on microtiter dishes.

of the dish can be monitored by discarding the liquid medium and staining with crystal violet (1% w/v). The dye will stain cells but not the plastic. Thus, after washing the wells, a purple ring or pellicle will remain, where bacteria have attached to the surface. Wells containing mutants defective in adhesion will remain clear, or significantly less stained than the wells containing the wild-type, biofilm-forming, strain. This simple method of identifying mutants potentially defective in biofilm formation has the added advantage of permitting quantification of attachment. After staining and allowing the dishes to air-dry, the stain can be solubilised by adding ethanol. Measuring absorbance at 600 nm (A_{600}) in a spectrophotometer will reflect the quantity of crystal violet retained in each well, which depends on the amount of cells attached to the surface.

4.3.1.2   Biotic surfaces

For obvious reasons, mass screening of random-transposon mutants is less straightforward in the case of biotic surfaces. Looking for mutants unable to attach to plant roots, e.g. involves sterilizing and germinating seeds, allowing the plants to grow and then placing them in individual containers with the different mutants. This is, generally, done in combination with the wild type, since what is often studied is the competitive colonisation capacity of each mutant (Figure 4.3). Analysis of attachment requires taking sections of each root and quantifying the number of bacteria associated with them, by plating dilutions after a mechanical treatment to remove the bacteria from the root. Cumbersome as this method is, it has been successfully used to identify *P. fluorescens* mutants defective in colonisation of tomato plant roots (Dekkers *et al.*, 1998a, b).

A less strenuous system has been used in studies with plant seeds (Figure 4.3). It consists of columns filled with seeds through which bacterial cultures are passed. This method, implemented by DeFlaun *et al.* (1994) to study adhesion of *P. fluorescens* mutants to seeds and sand, has been later used as a primary screen for the selection of *Pseudomonas putida* mutants deficient in adhesion to corn seeds (Espinosa-Urgel *et al.*, 2000). In this case, a pool of random-transposon

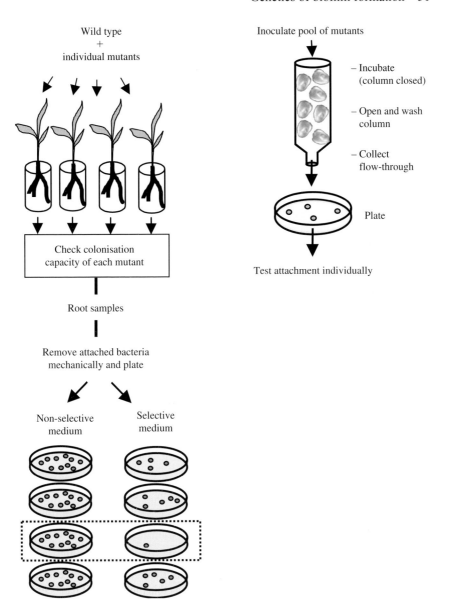

Figure 4.3   Methods to identify mutants defective in attachment to plant roots and seeds.

mutants was introduced in a closed seed column, and incubated for 1 h. The column was then opened and dilutions of the flow-through were plated on selective medium. This primary screen resulted in enrichment in adhesion-deficient mutants, reducing the number of clones that had to be then tested individually.

4.3.2   Characterisation of mutants

The methods used for phenotypic characterisation of attachment-deficient mutants involve techniques, such as analysing the architecture of the biofilm using confocal laser microscopy, or following the kinetics of biofilm formation by time-lapse microscopy. These microscopic techniques are detailed elsewhere in this volume. Quantitative methods are also typically employed, like the crystal violet staining and absorbance measurement described above. Sand columns have also been used in quantitative-attachment studies, measuring the A_{600} of the liquid flow-though over a period of time. Defects in adhesion are reflected by quick washing of the bacteria out from the column. With this system, e.g. Landini and Zehnder (2002) have recently described the negative role of the regulatory gene *hns* in adhesion of *E. coli* to sand particles. Finally, a variety of physical or mechanical methods to remove and quantify the attached cells from the substrate have been employed, from sonication to squeezing (silicone tubing), or vortexing in the presence of glass beads. The latter has proven to be a reliable method for analysing bacterial colonisation of seeds and roots (Espinosa-Urgel *et al.*, 2000; Espinosa-Urgel and Ramos, 2001). Typically, seeds or roots are introduced in tubes containing sufficient volume of minimal medium and an appropriate amount of glass beads (depending on the size of the sample to be analysed), and vortexed for 1 min at maximal speed. Dilutions are then plated to estimate the number of cells attached. The efficiency of this method is reflected by the fact that the number of cells recovered in a second vortex treatment is only 1% of those recovered in the first round.

When the strains under study come from a screening of random-transposon mutants, genetic characterisation is the next step. Identifying the location of the transposon insertion can be done by cloning the transposon and flanking chromosomal DNA, and sequencing. An alternative method that eliminates the need for cloning, known as arbitrarily-primed PCR, has been used in different systems. In brief, the method consists of PCR amplification and sequencing of the transposon–chromosome junction region. For that purpose, an oligonucleotide reading outwards from the transposon and an oligonucleotide mixture with a random portion and a fixed one are used for a first round of PCR amplification, using DNA from each mutant as template. The resulting products (a mixture in which the transposon–chromosome junction is over-represented) are used as template in a second PCR round, with another primer reading outwards from the transposon and a primer corresponding to the fixed portion of the 'arbitrary' oligonucleotide used in the first round. The products of this reaction are electrophoresed and the major band is isolated and sequenced. The great amount of data currently available from microbial genomes make it easy to find if there are similarities between even a short sequence stretch and sequences present in the databases. This method has allowed the identification of the genes or open reading frames (ORFs) affected in a wide number of attachment-deficient mutants (Pratt and Kolter, 1998; O'Toole and Kolter, 1998a, b; Espinosa-Urgel *et al.*, 2000). Not surprisingly, since the study of the genetics of biofilm formation is a relatively recent field of research, many of the identified ORFs correspond to hypothetical proteins of unknown function.

4.4   BIOFILM FORMATION ON ABIOTIC SURFACES

4.4.1   Initial attachment

As mentioned above, the transition from planktonic to sessile life is triggered by environmental conditions, which may be different for different bacteria. At least three *Pseudomonas* spp., *P. aeruginosa*, *P. fluorescens* and *P. putida*, can form biofilms on a variety of abiotic surfaces in almost any growth medium tested (O'Toole and Kolter, 1998a, b; Espinosa-Urgel *et al.*, 2000). Nutrient abundance in the medium seems to favour biofilm formation in these bacteria, since attachment begins early during growth in batch culture, reaching a maximum in 6–8 h, after which cells will start detaching and going back to the planktonic state. On the contrary, attachment of *Acinetobacter* sp. seems to be the result of low-nutrient conditions (James *et al.*, 1995). *E. coli* will form biofilms when grown in rich medium, but not in minimal medium unless it is supplemented with casamino acids (Pratt and Kolter, 1998). Another line of evidence supporting the role of environmental signals comes from the fact that the defect in biofilm formation of some *P. fluorescens* mutants can be rescued by the addition of iron (or citrate in other cases) to the medium (O'Toole and Kolter, 1998a). This suggests that in this bacterium, biofilm formation proceeds through different pathways depending on the growth medium. The role of iron in biofilm formation has been further high-lighted by a recent report indicating that human lactoferrin, an iron chelator, inhibits the development of *P. aeruginosa* biofilms (Singh *et al.*, 2002).

Flagellar motility was perhaps one of the first functions known to have a role in bacterial attachment to both biotic and abiotic surfaces (Harber *et al.*, 1983; de Weger *et al.*, 1987). Although the question remains open as to the precise role of flagella, the results obtained by Pratt and Kolter (1998) with different *E. coli* mutants suggest that flagella are not directly responsible for cell–surface inter-actions in this bacterium, since flagellated but non-motile mutants were unable to form biofilms. Rather, motility could be necessary for the cells to overcome repulsive forces present at the liquid–solid interface and reach the surface, and later on for bacterial movement and spread over the surface (Pratt and Kolter, 1998). Interestingly, the same authors reported that chemotaxis is not involved in biofilm formation on abiotic surfaces, since a $\Delta cheA$–Z mutant, deficient in chemotaxis, was indistinguishable from the wild type in its attachment ability.

Other surface structures have been associated to initial attachment processes. Glycopeptidolipids, which form part of the cell envelope of *Mycobacterium* spp., have been shown to be required for biofilm formation as well as for sliding motility (Recht *et al.*, 2000; Recht and Kolter, 2001). Lipopolysaccharides (LPS), which are relevant components of the outer membrane of Gram-negative bacteria, are also involved in initial surface attachment (Landini and Zehnder, 2002). Type I pili and the mannose-sensitive haemagglutinin pilus are essential for the early steps in biofilm formation by *E. coli* and *V. cholerae*, respectively (Pratt and Kolter, 1998; Watnick *et al.*, 1999). Type IV pili, on the other hand, seem to play a role in later stages in *P. aeruginosa* biofilm development (O'Toole *et al.*, 2000a). Some *E. coli* strains produce, under certain environmental conditions, fibre-like surface structures called curli. Production of curli has been shown to increase the efficiency of biofilm formation (Prigent-Combaret *et al.*, 2000).

Environmental conditions, such as growth in hyperosmotic medium, can also inhibit biofilm formation. In *E. coli*, the two-component regulatory system OmpR/EnvZ, which has a role in sensing medium osmolarity, participates in the regulation of initial adhesion by promoting curli production (Prigent-Combaret *et al.*, 2001).

A number of surface or outer-membrane proteins are also involved in the early biofilm formation steps. Often, these surface proteins are large polypeptides with repetitive amino acid motifs. They are best characterised in Gram-positive organisms, like the cell-wall protein Bap of *S. aureus* (Cucarella *et al.*, 2001). This protein contains 13 nearly identical repeats of 86 amino acids, corresponding to more than half of the total sequence. Some of these proteins are also involved in adhesion to biotic surfaces and cell-to-cell interactions. In the oral bacterium *Streptococcus gordonii*, two of these long repetitive surface proteins have been described, CshA and CshB, which are essential for oral cavity colonisation and participate in coaggregation with another oral microorganism, *Actinomyces naeslundii* (McNab *et al.*, 1994, 1999). CshA also contains 13 repeats of 101 amino acids. But, perhaps, the most startling example described so far is LapA, a surface protein present in *P. fluorescens* and *P. putida*. It is one of the largest bacterial proteins, 8682 amino acids, with two repetitive domains, one comprising nine almost identical repeats of 100 amino acids each, and the other with 29 imperfect repeats of 218–225 amino acids. As Figure 4.4 illustrates, this protein is essential for biofilm formation on both abiotic and biotic surfaces (Hinsa *et al.*, 2003). LapA appears to be associated to an ABC transporter whose components are also involved in biofilm formation.

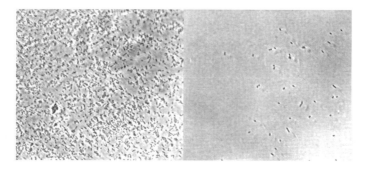

Figure 4.4   Attachment of *P. putida* to glass. Microphotographs were taken after 8 h of incubation in rich medium. Left panel: wild type. Right panel: *lapA* mutant.

4.4.2   From monolayer to mature biofilm

The development of a structured or 'mature' biofilm involves a sequence of stages after individual cells have attached to the solid surface, but relatively less is known about the elements specifically involved in each of these stages. As discussed above, flagellar motility has been proposed as a means for *E. coli* to spread throughout the surface and form a monolayer, whereas twitching motility, mediated by type IV pili, seems to be involved in further steps of biofilm development in *P. aeruginosa*. Mutants unable to synthesise type IV pili can still attach to the solid surface and form a monolayer, but do not proceed beyond that stage (O'Toole and Kolter, 1998b; O'Toole *et al.*, 2000a), and microcolonies are not formed.

Mature biofilms are typically characterised by a complex architecture, with cells embedded in an extracellular matrix, and macrocolonies often separated by the so-called water channels that might play a role in the influx of nutrients and oxygen and the efflux of waste products from the biofilm. The extracellular matrix is mainly constituted by exopolysaccharides (EPS), which, in some cases, seem to play a role in surface adhesion (McKenney *et al.*, 1998). However, in *E. coli*, the EPS colanic acid is not involved in the initial attachment, but appears to be essential for the maintenance of the complex biofilm architecture (Danese *et al.*, 2000a). Strains deficient in colanic acid production are indistinguishable from the wild type with respect to initial surface attachment, but the biofilms formed by such mutants present a 'collapsed', unstructured phenotype, with cells tightly packed in layers close to the surface, as if forming a single, flat 'microcolony'.

A similar role has been proposed for EPS produced by *V. cholerae* (Davey and O'Toole, 2000; Yildiz and Schoolnik, 1999), and for alginate, a major EPS in *P. aeruginosa*. Alginate overproduction results in a highly structured and antibiotic-resistant biofilm (Hentzer *et al.*, 2001). Changes in the levels of alginate production have been observed during biofilm development (Sauer *et al.*, 2002), and expression of *algC*, a gene involved in alginate and LPS synthesis, is induced in surface-attached cells (Davies and Geesey, 1995).

Biofilm maturation is determined not only by environmental conditions but also by intercellular quorum-sensing signals (see below) and cell-to-cell interactions. Cell-to-cell adhesion is clearly an important element in biofilm development. Whereas different elements involved in intercellular adhesion have been described in coaggregation of oral bacteria, much less is known in other types of biofilms. The major outer-membrane protein of *E. coli*, antigen 43, involved in autoaggregation, participates in both inter- and intra-specific cell-to-cell interactions in biofilms (Kjaergaard *et al.*, 2000; Danese *et al.*, 2000b). A protein involved in aggregation, AggA, has also been described in *P. putida* (Buell and Anderson, 1992), but its role in biofilm formation remains to be clarified.

4.4.3   Detachment and dispersal

Detachment is the least studied phase in the biofilm life cycle, other than in terms of methods to artificially remove attached bacteria. It could be assumed that once the environmental or nutritional conditions that are optimal for biofilm formation disappear, cells would sense this change and go back to the planktonic life. It could also be that biofilms, like cities, become overcrowded; the biofilm may exceed a growth threshold and cause cells or cell clusters to detach, or excess waste products and competition for the resources may turn the biofilm into a hostile environment for individual cells. Either way, given the complexity a biofilm can develop, dispersal may be more complicated than cells simply swimming away. Cell-to-cell interactions, through proteins or other surface components, as well as the EPS matrix, pose a barrier for the free movement of cells. Those barriers could be overcome by the production of surfactants or of extracellular enzymes. Such seems to be the case in *S. mutans*, where an endogenous surface protein-release enzyme has been described (Lee *et al.*, 1996). This enzyme removes the cell's own surface proteins, thus, allowing the bacterium to 'weigh anchor' and detach from the biofilm.

4.5   BIOFILMS ON BIOTIC SURFACES

A number of evidences suggest that bacterial attachment to biotic and abiotic surfaces takes place through different pathways, although there are some common elements that participate in both processes. Thus, flagella and motility have been shown to be important for adherence of enteropathogenic *E. coli* to cultured epithelial cells (Girón *et al.*, 2002), as well as for attachment of *P. fluorescens* to plant seeds and roots (de Weger *et al.*, 1987; DeFlaun *et al.*, 1994; Turnbull *et al.*, 2001b) and of *P. putida* to wheat roots (Turnbull *et al.*, 2001a). However, in this case the differences in attachment between motile and non-motile strains were only apparent in nutrient-poor medium, being lost in nutrient-rich medium. This suggests that nutritional signals, released by the plant, may act as attractants for the bacteria. In fact, chemotaxis has been shown to be important for initial root colonisation (Vande Broek *et al.*, 1998) in contrast with what has been observed for biofilm formation on abiotic surfaces.

Type IV pili are another example of common elements in biofilm development. They participate in attachment of pathogenic bacteria to epithelial cells, and are involved in bacterial interaction with plants and fungi (Dörr *et al.*, 1998), being crucial for the formation of microcolonies on plant roots.

A number of genes participating in adhesion of *P. putida* to seeds has been described (Espinosa-Urgel *et al.*, 2000). They encode membrane and surface proteins of which only one, LapA, has a clear role in biofilm formation on abiotic surfaces. In fact, a *lapA* mutant is unable to attach to any of the surfaces tested, either biotic or abiotic. Interestingly, a second surface protein, LapF, which shows structural similarities to LapA, seems to play a role only in adhesion to biotic surfaces (our unpublished observation). A mutant lacking a third surface protein, HlpA, with some similarity to haemolysins, shows an 'intermediate' phenotype. It is unable to adhere to some seeds, and forms poor biofilms on certain abiotic surfaces (Molina *et al.*, unpublished). The remaining genes described in that study do not seem to play a significant role in biofilm formation on abiotic surfaces (Espinosa-Urgel *et al.*, 2000).

The differences between the two processes have also been observed in *V. cholerae*, a bacterium for which biofilm formation on biotic and abiotic surfaces is a fundamental survival strategy. Whereas, in this case, the mannose-sensitive haemagglutinin takes part in biofilm formation on abiotic surfaces, it has no role in adhesion to chitin. The latter is a fundamental component of the exoskeleton of crustaceans, and one of the preferred niches for *V. cholerae* (Watnick *et al.*, 1999), since chitin is a nutrient source for this bacterium.

4.6   GENE EXPRESSION AND REGULATORY CIRCUITS IN BIOFILM DEVELOPMENT

It seems clear that all the steps in the developmental process described in the previous sections must be accompanied by regulated changes in gene expression that allow the bacteria to adapt to its life on solid substrates. An example of these regulated changes is the increase in expression of *algC* already mentioned, resulting in augmented alginate production, which has also been linked to a coordinated reduction in flagellar synthesis (reviewed by Kuchma and O'Toole, 2000). Figure 4.5 summarises the regulatory circuits and changes in gene expression discussed in this section.

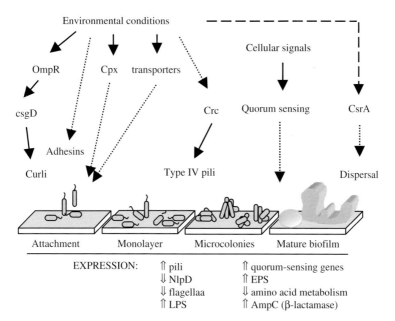

Figure 4.5   Expression changes and regulatory elements in biofilm formation. Dotted lines are pathways not fully characterised. Dashed line means environmental changes.

Whole genome approaches are beginning to be employed to investigate global gene expression changes in biofilms, either by using DNA arrays or through 2-D protein electrophoresis. One of these DNA microarray studies revealed significant differences in transcription of around 1% of the genes between planktonic and sessile populations of *P. aeruginosa* (Whiteley *et al.*, 2001), with approximately half of them being activated and the other half repressed in biofilms. These results are in contrast with earlier data obtained by random mutagenesis of *E. coli* with a mini-Mu carrying a promoterless *lacZ* gene. In that study, around 38% of the transcriptional fusions obtained were differentially expressed in biofilms (Prigent-Combaret *et al.*, 1999). Also, a recent analysis of the different stages of biofilm development in *P. aeruginosa* has revealed dramatic changes in protein expression throughout the whole process (Sauer *et al.*, 2002) with around 800 proteins differentially expressed in planktonic cells versus mature biofilms. In *P. putida*, subtractive cDNA hybridisation revealed around 40 differentially expressed genes, while 15 up-regulated and 30 down-regulated proteins were observed in attached cells, compared with their planktonic counterparts, in 2-D protein electrophoresis (Sauer and Kamper, 2001). The differentially expressed genes included those involved in biosynthesis of polysaccharide as well as components of flagella and type IV pili, which showed inverse patterns. While the flagellar protein FliC was present in planktonic cells, it did not appear in developed biofilms, whereas the opposite was true for PilA, a structural component of type IV pili.

Thus, in spite of disparities that could be due to the different methods used for biofilm growth and analysis, these global studies confirm data already obtained by previous research, but they are also offering new information about the physiology of attached cells. For example, an outer-membrane lipoprotein, NlpD, and genes involved in amino acid metabolism of *P. putida* appear to be down-regulated in biofilms (Sauer and Kamper, 2001), whereas urease, the sigma factor RpoH, and a number of hypothetical ORFs of *P. aeruginosa* seem to be induced (Whiteley *et al.*, 2001).

These changes in gene expression are induced by signals coming from the environment. We have already mentioned the role of the OmpR/EnvZ osmotic sensing system in curli production and, therefore, in attachment. Other two-component signal transduction systems appear to have a role in attachment, like Cpx (which senses physical changes in the cell envelope, such as those caused by the proximity of a solid surface) in *E. coli* (Otto and Silhavy, 2002), or the *P. aeruginosa* GacA/GacS system, which plays a role in virulence (Parkins *et al.*, 2001). Inorganic phosphate and magnesium sensing have also been implicated in biofilm formation. Mutants deficient in phosphate uptake and in a magnesium transporter of *P. aureofaciens* and *Aeromonas hydrophila*, respectively, are defective in biofilm formation (Merino *et al.*, 2001; Monds *et al.*, 2001).

Coordinated gene expression during later stages of biofilm development is achieved through cell-to-cell communication via quorum sensing, mediated by acyl-homoserine lactones in many Gram-negative microorganisms and by small peptides in Gram-positive bacteria. *P. aeruginosa* mutants affected in *lasI*, a gene involved in the synthesis of a quorum-sensing signal, are unable to proceed beyond the microcolony stage, forming flat, unstructured biofilms. Normal architecture can be restored by external addition of the acyl-homoserine lactone produced by LasI (Parsek and Greenberg, 2000). Expression of quorum-sensing signals also mediates interspecific communication in mixed biofilms (Riedel *et al.*, 2001). Further details on cell-to-cell communication in biofilms can be found elsewhere in this volume.

Global changes in gene expression require regulators that integrate and respond to the external signals. Aside from the two-component signal transduction systems mentioned above, and the previously cited *hns* gene, other general regulators are involved in biofilm formation. In *P. aeruginosa*, the catabolite repression control protein Crc, a global regulator of carbon metabolism, is required for twitching motility and formation of microcolonies, and has been proposed to integrate nutritional signals as part of the regulatory circuit for biofilm maturation (O'Toole *et al.*, 2000a). A similar link between carbon metabolism and biofilm formation is exemplified by the global regulator CsrA (carbon storage regulator) of *E. coli*, which appears to repress biofilm formation and enhance dispersal (Jackson *et al.*, 2002).

4.7 CONCLUSIONS

The identification of the two genes mentioned above as key elements in biofilm development serves as an example of how genetic studies are supporting empirical observations (in this case, the influence of the nutritional status on biofilm formation), and providing insights into the molecular mechanisms behind the sessile way of life. As a preview of the exciting discoveries still to come in this field, it is worth highlighting a recent report on the role of phenotypic variation in the appearance of *P. aeruginosa* variants with increased antibiotic resistance and enhanced biofilm formation (Drenkard and Ausubel, 2002). This phase variation process, controlled by the regulatory protein PvrR, explains the endurance of biofilms to antibiotics, and opens the way to investigate novel treatments against medically relevant biofilms.

REFERENCES

Buell, C.R. and Anderson, A.J. (1992) Genetic analysis of the *aggA* locus involved in agglutination and adherence of *Pseudomonas putida*, a beneficial fluorescent pseudomonad. *Mol. Plant–Microbe Interact.* **5**, 154–162.

Cucarella, C., Solano, C., Valle, J., Amorena, B., Lasa, I. and Penadés, J.R. (2001) Bap, a *Staphylococcus aureus* surface protein involved in biofilm formation. *J. Bacteriol.* **183**, 2888–2896.

Danese, P.N., Pratt, L.A., Dove, S.L. and Kolter, R. (2000a) The outer membrane protein, antigen 43, mediates cell-to-cell interactions within *Escherichia coli* biofilms. *Mol. Microbiol.* **37**, 424–432.

Danese, P.N., Pratt, L.A. and Kolter, R. (2000b) Exopolysaccharide production is required for development of *Escherichia coli* K-12 biofilm architecture. *J. Bacteriol.* **182**, 3593–3596.

Davey, M.E. and O'Toole, G.A. (2000) Microbial biofilms: from ecology to molecular genetics. *Microbiol. Mol. Biol. Rev.* **64**, 847–867.

Davies, D.G. and Geesey, G.G. (1995) Regulation of the alginate biosynthesis gene *algC* in *Pseudomonas aeruginosa* during biofilm development in continuous culture. *Appl. Environ. Microbiol.* **61**, 860–867.

DeFlaun, M.F., Marshall, B., Kulle, E.-P. and Levy, S.B. (1994) Tn5 insertion mutants of *Pseudomonas fluorescens* defective in adhesion to soil and seeds. *Appl. Environ. Microbiol.* **60**, 2637–2642.

Dekkers, L.C., Bloemendaal, C.J., de Weger, L.A., Wijffelman, C.A., Spaink, H.P. and Lugtenberg, B.J.J. (1998a) A two-component system plays an important role in the root-colonizing ability of *Pseudomonas fluorescens* strain WCS365. *Mol. Plant–Microbe Interact.* **11**, 45–56.

Dekkers, L.C., Phoelich, C.C., van der Fits, L. and Lugtenberg, B.J.J. (1998b) A site-specific recombinase is required for competitive root colonization by *Pseudomonas fluorescens* WCS365. *Proc. Natl. Acad. Sci. USA* **95**, 7051–7056.

de Lorenzo, V., Herrero, M., Jakubzik, U. and Timmis, K.N. (1990) Mini-Tn5 transposon derivatives for insertion mutagenesis, promoter probing, and chromosomal insertion of cloned DNA in Gram-negative eubacteria. *J. Bacteriol.* **172**, 6568–6572.

de Weger, L.A., van der Vlugt, C.I., Wijfjes, A.H., Bakker, P.A., Schippers, B. and Lugtenberg, B.J.J. (1987) Flagella of a plant-growth-stimulating *Pseudomonas fluorescens* strain are required for colonization of potato roots. *J. Bacteriol.* **169**, 2769–2773.

Dörr, J., Hurek, T. and Reinhold-Hurek, B. (1998) Type IV pili are involved in plant–microbe and fungus–microbe interactions. *Mol. Microbiol.* **30**, 7–17.

Drenkard, E. and Ausubel, F.M. (2002) *Pseudomonas* biofilm formation and antibiotic resistance are linked to phenotypic variation. *Nature* **416**, 740–743.

Espinosa-Urgel, M. and Ramos, J.L. (2001). A *Pseudomonas putida* aminotransferase involved in lysine catabolism is induced in the rhizosphere. *Appl. Environ. Microbiol.* **67**, 5219–5224.

Espinosa-Urgel, M., Salido, A. and Ramos, J.L. (2000) Genetic analysis of functions involved in adhesion of *Pseudomonas putida* to seeds. *J. Bacteriol.* **182**, 2363–2369.

Girón, J.A., Torres, A.G., Freer, E. and Kaper, J.B. (2002) The flagella of enteropathogenic *Escherichia coli* mediate adherence to epithelial cells. *Mol. Microbiol.* **44**, 361–379.

Gu, J.D., Roman, M., Esselman, T. and Mitchell, R. (1998) The role of microbial biofilms in deterioration of space station candidate materials. *Int. Biodeterior. Biodegrad.* **41**, 25–33.

Harber, M.J., Mackenzie, R. and Asscher, A.W. (1983) A rapid bioluminescence method for quantifying bacterial adhesion to polystyrene. *J. Gen. Microbiol.* **129**, 621–632.

Hentzer, M., Teitzel, G.M., Balzer, G.J., Heydorn, A., Molin, S., Givskov, M. and Parsek M.R. (2001) Alginate overproduction affects *Pseudomonas aeruginosa* biofilm structure and function. *J. Bacteriol.* **183**, 5395–5401.

Hinsa, S.M., Espinosa-Urgel, M., Ramos, J.L. and O'Toole, G.A. (2003) *Pseudomonas fluorescens* requires an ABC transporter to form biofilms on abiotic surfaces (submitted).

Jackson, D.W., Suzuki, K., Oakford, L., Simecka, J.W., Hart, M.E. and Romeo, T. (2002) Biofilm formation and dispersal under the influence of the global regulator CsrA of *Escherichia coli. J. Bacteriol.* **184**, 290–301.

James, G.A., Korber, D.R., Caldwell, D.E. and Costerton, J.W. (1995) Digital image analysis of growth and starvation responses of a surface-colonizing *Acinetobacter* sp. *J. Bacteriol.* **177**, 907–915.

Kjaergaard, K., Schembri, M.A., Ramos, C., Molin, S. and Klemm, P. (2000) Antigen 43 facilitates formation of multispecies biofilms. *Environ. Microbiol.* **2**, 695–702.

Kuchma, S.L. and O'Toole, G.A. (2000) Surface-induced and biofilm-induced changes in gene expression. *Curr. Opin. Biotechnol.* **11**, 429–433.

Landini, P. and Zehnder, A.J. (2002) The global regulatory *hns* gene negatively affects adhesion to solid surfaces by anaerobically grown *Escherichia coli* by modulating expression of flagellar genes and lipopolysaccharide production. *J. Bacteriol.* **184**, 1522–1529.

Lee, S.F., Li, Y.H. and Bowden, G.H. (1996) Detachment of *Streptococcus mutans* biofilm cells by an endogenous enzymatic activity. *Infect. Immun.* **64**, 1035–1038.

McKenney, D., Hübner, J., Muller, E., Wang, Y., Goldmann, D.A. and Pier, G.B. (1998) The *ica* locus of *Staphylococcus epidermidis* encodes production of the capsular polysaccharide/adhesin. *Infect. Immun.* **66**, 4711–4720.

McNab, R., Jenkinson, H.F., Loach, D.M. and Tannock, G.W. (1994) Cell-surface-associated polypeptides CshA and CshB of high molecular mass are colonization determinants in the oral bacterium *Streptococcus gordonii*. *Mol. Microbiol.* **14**, 743–754.

McNab, R., Forbes, H., Handley, P.S., Loach, D.M., Tannock, G.W. and Jekinson, H.F. (1999) Cell wall-anchored CshA polypeptide (259 kilodaltons) in *Streptococcus gordonii* forms surface fibrils that confer hydrophobic and adhesive properties. *J. Bacteriol.* **181**, 3087–3095.

Meadows, P.S. (1971) The attachment of bacteria to solid surfaces. *Arch. Microbiol.* **75**, 374–381.

Merino, S., Gavin, R., Altarriba, M., Izquierdo, L., Maguire, M.E. and Tomás, J.M. (2001) The MgtE Mg^{2+} transport protein is involved in *Aeromonas hydrophila* adherence. *FEMS Microbiol. Lett.* **198**, 189–195.

Monds, R.D., Silby, M.W. and Mahanty, H.K. (2001) Expression of the Pho regulon negatively regulates biofilm formation by *Pseudomonas aureofaciens* PA147-2. *Mol. Microbiol.* **42**, 415–426.

Nicolella, C., van Loosdrecht M.C. and Heijnen, J.J. (2000) Wastewater treatment with particulate biofilm reactors. *J. Biotechnol.* **80**, 1–33.

O'Toole, G.A. and Kolter, R. (1998a) Initiation of biofilm formation in *Pseudomonas fluorescens* WCS365 proceeds via multiple, convergent signaling pathways: a genetic analysis. *Mol. Microbiol.* **28**, 449–461.

O'Toole, G.A. and Kolter, R. (1998b) Flagellar and twitching motility are necessary for *Pseudomonas aeruginosa* biofilm development. *Mol. Microbiol.* **30**, 295–304.

O'Toole, G.A., Pratt, L.A., Watnick, P.I., Newman, D.K., Weaver, V.B. and Kolter, R. (1999) Genetic approaches to study of biofilms. *Meth. Enzymol.* **310**, 91–109.

O'Toole, G.A., Gibbs, K.A., Hager, P.W., Phibbs Jr., P.V. and Kolter, R. (2000a) The global carbon metabolism regulator Crc is a component of a signal transduction pathway required for biofilm development by *Pseudomonas aeruginosa*. *J. Bacteriol.* **182**, 425–431.

O'Toole, G., Kaplan, H.B. and Kolter, R. (2000b) Biofilm formation as microbial development. *Annu. Rev. Microbiol.* **54**, 49–79.

Otto, K. and Silhavy, T.J. (2002) Surface sensing and adhesion of *Escherichia coli* controlled by the Cpx-signaling pathway. *Proc. Natl. Acad. Sci. USA* **99**, 228–292.

Parkins, M.D., Ceri, H. and Storey, D.G. (2001) *Pseudomonas aeruginosa* GacA, a factor in multihost virulence, is also essential for biofilm formation. *Mol. Microbiol.* **40**, 1215–1226.

Parsek, M.R. and Greenberg, E.P. (2000) Acyl-homoserine lactone quorum sensing in Gram-negative bacteria: a signaling mechanism involved in associations with higher organisms. *Proc. Natl. Acad. Sci. USA* **97**, 8789–8793.

Pratt, L.A. and Kolter, R. (1998) Genetic analysis of *Escherichia coli* biofilm formation: roles of flagella, motility, chemotaxis and type I pili. *Mol. Microbiol.* **30**, 285–293.

Prigent-Combaret, C., Vidal, O., Dorel, C. and Lejeune, P. (1999) Abiotic surface sensing and biofilm-dependent regulation of gene expression in *Escherichia coli*. *J. Bacteriol.* **181**, 5993–6002.

Prigent-Combaret, C., Prensier, G., Le Thi, T.T., Vidal, O., Lejeune, P. and Dorel, C. (2000) Developmental pathway for biofilm formation in curli-producing *Escherichia coli* strains: role of flagella, curli and colanic acid. *Environ. Microbiol.* **2**, 450–464.

Prigent-Combaret, C., Brombacher, E., Vidal, O., Ambert, A., Lejeune, P., Landini, P. and Dorel, C. (2001) Complex regulatory network controls initial adhesion and biofilm formation in *Escherichia coli* via regulation of the *csgD* gene. *J. Bacteriol.* **183**, 7213–7223.

Recht, J. and Kolter, R. (2001) Glycopeptidolipid acetylation affects sliding motility and biofilm formation in *Mycobacterium smegmatis*. *J. Bacteriol.* **183**, 5718–5724.

Recht, J., Martínez, A., Torello, S. and Kolter, R. (2000) Genetic analysis of sliding motility in *Mycobacterium smegmatis*. *J. Bacteriol.* **182**, 4348–4351.

Riedel, K., Hentzer, M., Geisenberger, O., Huber, B., Steidle, A., Wu, H., Hoiby, N., Givskov, M., Molin, S. and Eberl, L. (2001) *N*-acylhomoserine-lactone-mediated communication between *Pseudomonas aeruginosa* and *Burkholderia cepacia* in mixed biofilms. *Microbiology* **147**, 3249–3262.

Rosan, B. and Lamont, R.J. (2000) Dental plaque formation. *Microb. Infect.* **2**, 1599–1607.

Sauer, K. and Camper, A.K. (2001) Characterization of phenotypic changes in *Pseudomonas putida* in response to surface-associated growth. *J. Bacteriol.* **183**, 6579–6589.

Sauer, K., Camper, A.K., Ehrlich, G.D., Costerton, J.W. and Davies, D.G. (2002) *Pseudomonas aeruginosa* displays multiple phenotypes during development as a biofilm. *J. Bacteriol.* **184**, 1140–1154.

Schabereiter-Gurtner, C., Piñar, G., Vybiral, D., Lubitz, W. and Rölleke, S. (2001) *Rubrobacter*-related bacteria associated with rosy discolouration of masonry and lime wall paintings. *Arch. Microbiol.* **176**, 347–354.

Singh, P.K., Parsek, M.R., Greenberg, E.P. and Welsh, M.J. (2002) A component of innate immunity prevents bacterial biofilm development. *Nature* **417**, 552–555.

Turnbull, G.A., Morgan, J.A.W., Whipps, J.M. and Saunders, J.R. (2001a) The role of motility in the *in vitro* attachment of *Pseudomonas putida* PaW8 to wheat roots. *FEMS Microbiol. Ecol.* **35**, 57–65.

Turnbull, G.A., Morgan, J.A.W., Whipps, J.M. and Saunders, J.R. (2001b) The role of bacterial motility in the survival and spread of *Pseudomonas fluorescens* in soil and in the attachment and colonisation of wheat roots. *FEMS Microbiol. Ecol.* **36**, 21–31.

Vande Broek, A., Lambrecht, M. and Vanderleyden, J. (1998) Bacterial chemotactic motility is important for the initiation of wheat root colonization by *Azospirillum brasilense*. *Microbiology* **144**, 2599–2606.

Watnick, P.I. and Kolter, R. (2000) Biofilm, city of microbes. *J. Bacteriol.* **182**, 2675–2679.

Watnick, P.I., Fullner, K.J. and Kolter, R. (1999) A role for the mannose-sensitive hemagglutinin in biofilm formation by *Vibrio cholerae* El Tor. *J. Bacteriol.* **181**, 3606–3609.

Way, J.C., Davis, M.A., Morisato, D., Roberts, D.E. and Kleckner, N. (1984) New *Tn*10 derivatives for transposon mutagenesis and for construction of *lacZ* operon fusions by transposition. *Gene* **32**, 369–379.

Whiteley, M., Bangera, M.G., Bumgarner, R.E., Parsek, M.R., Teitzel, G.M., Lory, S. and Greenberg, E.P. (2001) Gene expression in *Pseudomonas aeruginosa* biofilms. *Nature* **413**, 860–864.

Yildiz, F.H. and Schoolnik, G.K. (1999) *Vibrio cholerae* O1 El Tor: identification of a gene cluster required for the rugose colony type, exopolysaccharide production, chlorine resistance, and biofilm formation. *Proc. Natl. Acad. Sci. USA* **96**, 4028–4033.

5

The role of cell signaling in biofilm development

B.L. Purevdorj and P. Stoodley

5.1 INTRODUCTION

During the past 30–40 years with the advancement of sophisticated genetic and biochemical tools, increasing evidence of chemical communication systems in bacterial population has been provided. Requirement for a threshold of cell concentration during the process of bioluminescence production in marine symbiotic bacteria *Vibrio fischeri* (Nealson and Hastings, 1979), formation of fruiting bodies in myxobacteria (Dworkin and Kaiser, 1985) and development of competence for genetic transformation in streptococci (Dawson and Sia, 1931) are among the first evidence of cell-to-cell signaling systems in bacterial communities. At present, the term quorum sensing (QS) is defined as a cell-density dependent bacterial intercellular-signaling mechanism that enables bacteria to coordinate the expression of certain genes to coordinate the group behavior.

Although the signaling mechanisms differ, QS systems are utilized by both Gram-positive and -negative bacterial species (extensively reviewed by Miller and Bassler, 2001). As an example, a mechanism of Gram-negative *Pseudomonas aeruginosa* QS system is shown in Figure 5.1.

In general to sense their population density, Gram-negative bacteria use small chemical molecules called acylated homoserine lactones (AHSLs)

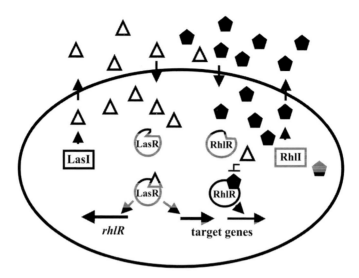

Figure 5.1   Example *P. aeruginosa* LasI/LasR–RhlI/RhlR QS system. *P. aeruginosa* uses two LuxI/LuxR-like autoinducer-sensor pairs for QS regulation of a variety of genes. The LasI protein produces the homoserine lactone signaling molecule *N*-(3-oxododecanoyl)-homoserine lactone (triangles), and the RhlI protein synthesizes *N*-(butyrl)-homoserine lactone (pentagons). The LasI autoinducer interferes with binding of the RhlI-autoinducer to RhlR. Ongoing research continues to reveal increasingly complex regulatory pathways with many interlinked positive and negative feedback loops. After Miller and Bassler, 2001.

(reviewed by Fuqua *et al.*, 2001) and Gram positives use oligo-peptides (Kleerebezem *et al.*, 1997), both collectively referred to as autoinducers. While the autoinducers are constitutively expressed, a threshold level must be achieved in the extracellular environment in order to initiate the QS system (in the closed system most often it occurs during late exponential and stationary phase). Some signal molecules are able to freely diffuse through the cell membrane but some require an active transport system (Pearson *et al.*, 1999). Inside the cell, in association with the transcriptional activators the threshold level of autoinducers is capable of inducing the expression of various genes. The target genes are involved in various physiologic activities, such as symbiosis, competence, conjugation, antibiotic production, motility, sporulation, etc. (Gutowski-Eckel *et al.*, 1994; Grossman, 1995; Ji *et al.*, 1995; Lee and Morrison, 1999; Li *et al.*, 2001a, b).

   Understanding of the QS systems in clinically relevant strains has become the focus of intense study during the last decade when many QS-regulated genes were also found to be involved in virulence (Costerton *et al.*, 1999). The idea that the biofilms are optimum sites for expression of phenotypes regulated by QS (Williams and Stewart, 1994), has led to numerous studies of QS mechanisms in the bacterial biofilms.

5.2   INTERSPECIES QUORUM SENSING INTERACTIONS

Since most signal molecules are conserved across many different species, particularly in Gram-negative bacteria, it would be reasonable to predict that some form of interspecies communication is likely in natural ecosystems where different AHSL-producing species inhabit a common habitat. Interspecies QS systems would play an important synergistic or competitive role in the dynamics of the microbial communities. In fact, recent advances in the field indicate that cell-to-cell communication via QS occurs not only within but between bacterial species (Bassler *et al.*, 1997; Surette *et al.*, 1999). *In vitro* studies of mixed species of biofilm demonstrated that *Burkholderia cepacia* was capable of perceiving the AHSL signals produced by *P. aeruginosa*, while the latter strain did not respond to the signal molecules produced by *B. cepacia* (Riedel *et al.*, 2001). Interestingly, these two species of bacteria are capable of forming mixed biofilms in the lungs of cystic fibrosis (CF) patients and perhaps the ability of these organisms to communicate with each other may facilitate successful colonization of a host lung. In the environment where two species commonly encounter each other, the ability to sense signal molecule production and interfere with the normal communication of the other species would provide an important survival advantage. A soil microorganism *Bacillus subtilis* produces an enzyme called AiiA which inactivates the HSL autoinducer of another soil habitat *Erwinia carotovora*, rendering the latter avirulent (Dong *et al.*, 2000).

In general, laboratory based findings of QS in bacterial species are helpful. However, more research is needed to understand the importance of such systems in a natural ecosystem in the presence of resident microbial communities.

5.3   QUORUM SENSING IN BIOFILM ESTABLISHMENT AND STRUCTURAL DEVELOPMENT

5.3.1   Biofilm formation

Microbial biofilms may be defined as populations of microorganisms that are concentrated at an interface (usually solid/liquid) and typically surrounded by an extracellular polymeric slime (EPS) matrix (Costerton *et al.*, 1995). Biofilms appear to be more resistant to chemicals and biocides as compared to their planktonic counterparts and also better evade host immune defense, such as phagocytes or antibodies. The resistance of biofilms confers the ability of the cells to thrive in varied and often harsh environmental conditions, such as inside host organisms, surface of the pipelines, dental unit water lines, catheters, ventilators and medical implants, causing tremendous problems in both industry and medicine. From the sessile, matrix-bound community, planktonic cells can be continuously shed from the biofilm (Bryers, 1988; Stoodley *et al.*, 2001), also the recently discovered shear-mediated downstream movement of intact biofilms along the surface (Stoodley et al., 1999a), could all lead to contamination, or in humans to a systemic and often chronic infection. It is now estimated that about 65% of all nosocomial infections are biofilm-related (Archibald and Gaynes, 1997). Life threatening infection caused by P. aeruginosa biofilms in CF patients is one of the many examples.

Since biofilm is of concern in both medical and industrial fields, enormous effort has been directed toward understanding the mechanisms of biofilm establishment and development. Biofilm establishment and development is a dynamic multifactorial process governed by both environmental and genetic control systems. Although many bacteria are capable of forming biofilms, much of what we know today about biofilms is gained through experiments largely based on pure cultures of Pseudomonas spp. The current model of biofilm establishment describes several steps starting from free-floating planktonic cells attaching on the surface, followed by growth into complex pillar like structures intervened with water channels, which eventually may disperse via detachment of individual cells into the bulk fluid. The shape and structure of the biofilm is an important aspect of the biofilm life cycle as it may determine mass transport of solutes such as antimicrobials or nutrients, biofilm detachment, and energy losses in industrial pipelines (Stoodley et al., 1997; Stewart, 1998).

Although the underlying mechanisms that define biofilm structural phenotype have not been completely characterized, previous studies have demonstrated that multiple environmental factors, such as fluid flow (Vieira et al., 1993; Stoodley et al., 1999a), carbon source (Moller et al., 1997; Wimpenny and Colasanti, 1997; Stoodley et al., 1999a), and surfaces (Dalton et al., 1994) all play important roles in biofilm processes. Also, key physiological functions, such as twitching motility mediated via type IV pili (O'Toole and Kolter, 1998), the global carbon metabolism regulator, Crc (O'Toole et al., 2000), and stationary phase s factor RpoS (Heydorn et al., 2002) were shown to be important in biofilm initial formation of microcolonies and growth. Subsequently, QS, has also been shown to be involved in differentiation of the P. aeruginosa biofilms into a mature biofilm consisting of mushroom-shaped microcolonies interspersed with water channels (Davies et al., 1998).

Since QS is a concentration dependent phenomenon, it is not likely to occur during initial stages of biofilm formation but rather in later stages when the cell density is high. This was demonstrated in a study where detailed quantitative analysis of B. cepacia biofilm structures formed by wild type (WT) and mutant strains showed that the QS was not involved in the regulation of initial cell attachment, but rather controlled the maturation of the biofilm (Huber et al., 2001). QS has also been linked to biofilm structure in oral bacteria that initiate dental plaque formation: Streptococcus gordonii (Loo et al., 2000), Streptococcus mutans (Li et al., 2002), Salmonellae spp. (Prouty et al., 2002), and in the opportunistic pathogen Aeromonas hydrophilia (Lynch et al., 2002). The potential role of QS in biofilm development has been demonstrated in in situ studies where functional autoinducers were detected in biofilms existing in river sediments (McLean et al., 1997), urinary catheters (Stickler et al., 1998) and recently in the sputum of CF patients (Singh *et al.*, 2000; Wu *et al.*, 2000; Erickson *et al.*, 2002; Middleton *et al.*, 2002). It has also been shown that QS expression in *Pseudomonas putida* biofilm coincided with marked changes in biofilm morphology which switched from consisting of microcolonies to thick distinct mushroom structures with intervening water channels (De Kievit *et al.*, 2001). The expression of QS gene was substantial at the substratum, where bacterial cell density and signal accumulation is expected to be high.

5.3.2   Biofilm structure

Although many studies pinpoint to the direct role of QS in controlling the biofilm structural development, it is increasingly evident that QS–biofilm structure relationship is more complicated. Given the many genes and pathways involved in biofilm formation, the redundancy and overlap of regulatory pathways in biofilm development, it is likely that biofilms would behave differently and may not use the same regulatory mechanisms under different environmental conditions and QS is not an exception.

To complicate the matter, bacterial biofilms often form a variable complex heterogeneous pattern even over small distances of less than a millimeter, making comparison of isogenic variants to the WT parent difficult and possibly resulting in apparently contradictory results. For example, recently, it was reported that WT *P. aeruginosa* biofilm was structurally flat (Nivens *et al.*, 2001; Heydorn *et al.*, 2002) and resembled the QS mutant biofilm described in Davies *et al.* (1998) study. Also, De Kievit *et al.* (2001) have reported that under static condition with glucose as a carbon source the biofilm developed into a thick multilayer but with citrate it formed only a sparse monolayer. This structural difference was attributed in part to *pilA* gene, responsible for twitching motility, regulated by global regulator of carbon metabolism, Crc (O'Toole *et al.*, 2000). Interestingly, in a flow through system the difference in medium composition did not significantly affect the structural development, suggesting the QS-regulating factor(s) in biofilm is differentially expressed under different external conditions (De Kievit *et al.*, 2001). Since QS is a concentration dependent phenomenon, it will be influenced by mass transfer processes or liquid flow surrounding the biomass. The flow condition may influence the concentration of signal molecules, and thus, affect QS mechanism in the biofilm. Stoodley *et al.* (1999b) and Purevdorj *et al.* (2002) have demonstrated that a mutation in the QS pathway only had a limited effect on biofilm structure but that flow dynamics had a much more pronounced effect. These findings indicate that the QS may not be the determining global system in the biofilm development, and other environmental factors, such as hydrodynamics, carbon source, mass transfer, etc., may also play a role in this dynamic process. More studies are needed to elucidate the importance of QS in the biofilm structural development and many questions are still remaining to be resolved. What QS controlled genes are involved in the biofilm development? To what extent and in what environmental conditions does QS play a significant role in the biofilm development and is there a system where the biofilm development is independent of QS?

5.4   THE INFLUENCE OF QUORUM SENSING ON BIOFILM STRUCTURE AND BEHAVIOR

Advancement in methods technology, such as improvements in fluorescent molecular techniques, design of suitable systems for biofilm cultivation, more sophisticated microscopy, and digital imaging systems, have enabled non-destructive, direct analysis of biofilms *in vitro*, which have greatly benefited our understanding of biofilm structural development as a complex community.

5.4.1   Flow cells

Naturally occurring biofilms are often found in places where it is difficult to access, making direct analysis impossible. However, biofilm study is possible through culturing in a relatively simple laboratory model system. At present, there are no standardized methods of cultivating biofilms and different laboratories utilize numerous different methods and devices for their biofilm studies. Flow cells are one of the most commonly used reactor systems. For their simplicity they are proven to be useful in growth and *in situ* visualization of biofilms. Flow cells allow control of flow velocity as well as the loading rate of nutrients or other test compounds. Multichannel flow cells exhibit efficiency by producing experimental data in replicates as well as convenience of side-by-side running of the control and test biofilms. Traditional transmission light microscopy, scanning confocal laser or epifluorescent microscopy all can be used to monitor *in situ* biofilm growing in glass flow cells. There are many types of flow cells (Hall-Stoodley *et al.*, 1999; Palmer, 1999; Zinn *et al.*, 1999) and they can be either once through or recirculating. In once through systems, nutrients or testing compounds are pumped through the flow cell into the waste container at the effluent end (Palmer, 1999). In this system, the flow rate may not be adjusted independently of residence time and the experiments are largely limited to the low-flow conditions due to the cost and time associated with media preparations. In recirculating system the flow cells are incorporated into recirculating loop attached to a mixing chamber. This system allows the flow rate to be independent of the nutrient flow rate, allowing high-flow experiments without impractical volumes of media. Stoodley *et al.* (this volume) provides a detailed description of the flow cell and associated results with the system.

5.4.2   Microorganisms

The role and function of the QS systems in autoinducer producing microorganisms can be determined via several different methods, such as utilization of reporter gene and isogenic variants of parental strain for structural comparison.

5.4.2.1   Reporter genes

The direct and temporal quantification of specific gene products within biofilm bacteria is now possible by combining flow cells with fluorescent-based reporter technology. The green fluorescent protein (GFP) originally derived from jellyfish *Aequorea victoria* (238 amino acids) has been mutated to improve the fluorescence intensity of the reporter protein and chromophore formation kinetics (Heim and Tsien, 1996; Cormack *et al.*, 1996). GFP has become an important visual marker of *in situ* gene expression because it is non-toxic and does not require a special substrate or cofactor for detection. It has been used in many different areas of protein tracking and expression analysis, such as in a quantitative marker in an insect larvae expression system (Cha *et al.*, 1999), induction of meta-pathway promoter (Moller *et al.,* 1998), to monitor chitinase gene expression (Stretton *et al.*, 1998) as well as in QS-regulated gene expression studies (De Kievit *et al.*, 2001; Heydorn *et al.*, 2002) (Figure 5.2).

One disadvantage of using GFP is that the WT and original-mutated GFP were very stable, with half-lives greater than 24 h (Tombolini *et al.*, 1997). This means that the intracellular pool of GFP may not immediately decrease after reduction of gene expression, but rather through dilution by cell proliferation. To increase temporal resolution of gene expression, mutants producing unstable GFP have been created via addition of tale-specific protease (Andersen *et al.*, 1998). This allowed analysis of downshift in gene expression. By using fusions of GFP with a half-life as short as 40 min, De Kievit *et al.* (2001) were able to analyze real-time temporal and spatial expression of the two QS – genes *lasI* and *rhlI* in *P. putida* biofilm. Another disadvantage is the dependency of fluorescence on oxygen. Care must be taken not to misinterpret heterogeneity in oxygen distribution in the biofilm with heterogeneity in gene expression. This can be tested by using a constitutively expressing mutant, which produces the same fluorophore as a control. Luciferase activity is also used to monitor gene expression in biofilms. Luciferase catalyzes the oxidation of reduced flavin ($FMNH_2$) to form intermediate peroxide, which reacts with a long-chain aldehyde to give blue-green luminescence emitting at 490 nm (Hastings, 1996). The bioluminescence activity can be meas-

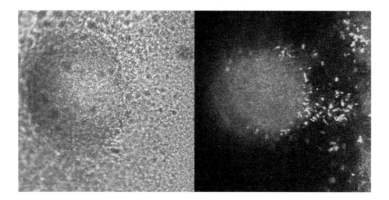

Figure 5.2   *P. aeruginosa* pMH 509 p*lasB::gfp* biofilm in the flow cell expressing the GFP (40× objective lens). Transmitted and scanning laser confocal microscopy (right) image. The strain was kindly provided by Morten Hentzer and Matthew Parsek. [This figure is also reproduced in colour in the plate section after page 326.]

ured by X-ray or photographic films, by visual or microscopic observations and by a luminometer or a scintillation counter in chemilumenescence mode. Although luciferase displays good temporal response, similar to GFP, it requires oxygen and is limited to specific hosts for expression, for example, *Escherichia coli* and *Vibrio parahaemolyticus*.

5.4.2.2   Quorum sensing mutants

To study the contribution of QS-regulated genes in biofilm development, QS mutants were generated in *P. aeruginosa* which consequently has one of the best characterized QS systems outside of the photo-luminescent marine vibrio. Null mutations in the R-proteins (transcriptional activator), the autoinducer synthases (*lasI, rhlI* and *lasIrhlI*), and mutants of QS-regulated products have been generated which has led to many important discoveries in the field (Figure 5.3).

There are few downfalls associated with the usage of gene knock out mutants. Since the mutants are created by insertionally inactivating the gene by replacing the antibiotic resistance cassette, it is possible that the cassette is excised from a chromosome, restoring the WT genotype. However, the problem can be circumvented by deleting the portion of gene of interest. Also, the chance of cross-over event can be high if there is at least 1 kb of chromosomal DNA flanking each side of the cassette (De Kievit and Iglewski, 1999). Although not usually reported in the literature, it is widely known that false positive or negative results in the mutant strains can also occur due to polar mutation. To ensure that the mutant phenotype is the direct result of the specific gene, complementation experiments by providing the copy of WT gene on a plasmid or in case of autoinducers, by exogenous addition of signal molecules to the system should be conducted.

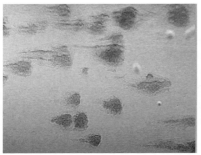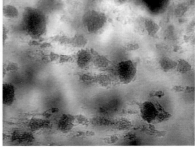

Figure 5.3   A bright field microscopic image of *P. aeruginosa* biofilm growing in the flow cell (10× objective lens, the flow direction is from left to right). Three-day-old *P. aeruginosa* Pao1 WT (left) and lasI null mutant Pao1-JP1 strain (right). In our system only subtle differences in the structure were observed between the WT and the mutant biofilm, suggesting that there is no apparent role for QS in biofilm formation and development in high-shear environment.

5.4.3   Biofilm structural analysis

Although biofilm structure has been extensively studied, at present there are few standardized methods available for quantification. Most of the previous studies were largely qualitative relying on visual interpretations of the biofilm images. However, biofilm structural parameters that can be readily measured by microscopic techniques include thickness, thickness variability (roughness), and surface area coverage (Stewart *et al.*, 1993; Murga *et al.*, 2001). Fractal dimension of activated sludge-biofilms (Hermanowics *et al.*, 1995) and density, porosity, specific surface area, and mean pore radius of waste water biofilms (Zhang and Bishop, 1994) were also measured.

Recently, more complicated software programs for systematically quantifying biofilm images have been developed. These include Image Structural Analysis (ISA) which was developed at the Center for Biofilm Engineering (www.erc.montana.edu/CBEssentials-SW/research/ImageStructureAnalyzer/default.htm), and COMSTAT which was developed at the Danish Technical University in Lyngby (http://www.im.dtu.dk/comstat).

5.4.3.1   Image structural analysis

ISA extracts information from two-dimensional biofilm images based on nine different textural and dimensional parameters for statistical comparison (Purevdorj *et al.*, 2002; Yang *et al.*, 2000). Calculated biofilm cell cluster dimensions include porosity (surface area cover), microcolony length and width, average diffusion distance (equivalent to an average diameter), maximum diffusion distance (maximum distance from the interior of the cluster to the edge), and fractal dimension (a measure of the roughness of the biofilm cell clusters). These parameters were calculated from automatically thresholded binary images to remove subjectivity from the analysis (Yang *et al.*, 2001). Briefly, each image is automatically analyzed to find the threshold region which has the most influence on porosity. In this region small changes in threshold result in large variations in dimensional parameters and a threshold value outside of this region is selected based on the same criteria for all images. ISA also calculates three textural parameters from the gray scale images, which describe the microscale heterogeneity of the image. These parameters are textural entropy (a measure of randomness between individual pixels), angular second moment (a measure of directional repeating patterns in the biofilm), and inverse difference moment (a measure of spatially repeating patterns). ISA was designed to analyze larger scale biofilm patterns in 2D gray scale images and is, therefore, useful for lower power images taken with conventional bright field or epi-fluorescence microscopy.

5.4.3.2   COMSTAT

COMSTAT was developed to analyze high-resolution 3D confocal image stacks (Heydorn *et al.*, 2000a). Prior to quantification, the image stacks are thresholded, which results in a three-dimensional matrix with a value of ONE in positions where the pixel value is above or equal to the threshold value (biomass), and ZERO when the pixel values are below the threshold value (background). Confocal images are less sensitive to thresholding and automatic thresholding is not available. There-fore, manual thresholding is required and it is advised that a fixed threshold value for all image stacks is used and that the operator is not varied (Heydorn et al., 2000a). It also features a function where noise is automatically removed from the background, by eliminating biomass pixels that are not connected to the sub-stratum. In general COMSTAT comprises ten image analysis features for biofilm structural quantification which include bio-volume (the overall volume of cells in the biofilm – EPS is not included if it is not specifically stained), the area occupied by bacteria at different heights in the biofilm, thickness and roughness, identifi-cation and distribution of microcolonies at the substratum, microcolony volume, fractal dimension, average and maximum diffusion distance, and surface to vol-ume ratio.

5.4.4   AHSL detection in the biofilm studies

Since the first discovery of AHSLs in naturally occurring biofilms (McLean et al., 1997), quantitative data for AHSLs have been reported for a number of signal pro-ducing bacteria including *Pseudomonas fluorescens* (Shaw et al., 1997), *Agrobac-terium tumefaciens* (Zhu et al., 1998) and *Pantoea stewartii* (Beck von Bodman et al., 1998). In the systems examined so far both the transcriptional activator R-proteins and the AHSL molecules are remarkably similar (Fuqua et al., 1996), all AHSLs for these organisms being identified as *N*-acylated derivatives of L-homoserine lactone, which is unique to Gram-negative organisms. Specificity of these signal molecules is conferred by the length and the nature of the substitution at carbon 3 of the acyl-side chain (Pearson et al., 1994; Passador et al., 1996). Although the transcriptional activator R-protein is specific, some infidelities, how-ever, do occur (Pearson et al., 1994). Also, it is now known that one type of bac-terium can produce more than one type of AHSL, which can also pose problems during the detection process. Potential candidates for AHSL detection are *P. aerug-inosa rhlI* and *lasI* QS systems that mediate via 4 and 12 carbon AHSLs, respec-tively; and *A. tumefaciens* system that mediates via 8 carbon AHSL system. The bacterial strains carrying genes encoding the R-proteins activated by either 4, 8, or 12 carbon AHSLs, target gene fused with lacZ (encodes β-galactosidase, an enzyme which hydrolyzes o-Nitrophenyl-β-d-Galactoside, ONPG, yielding o-nitrophenol that absorbs light at 420 nm or another reporter gene are needed for bioassay. For the indicator strain, whether *E. coli* strain harboring lasR and lasI–lacZ on a multi-ple copy plasmid. *P. aeruginosa* autoinducer synthase mutant deficient in the pro-duction of 3-oxo-C_{12}-HSL and C_4-HSL, harboring lasI–lacZ fusion plasmid can be used. Overexpression of R-proteins in the indicator strain may increase the sensi-tivity of the bioassay to non-cognate AHSLs.

Currently, there are two methods to assay autoinducer presence in the sample. First one described by McLean et al. (1997), plating both the reporter strain and the biofilm sample side by side on an agar plate. If the autoinducers are present in the biofilm sample, they will diffuse through the agar to the indicator strain resulting in the coloration (Figure 5.4).

The second method, instead of solid agar, liquid culture is used, co-incubating the sample with indicator strain. AHSL activity is detected with a standard β-galactosidase assay. So far the majority of quantitative studies have been focused on planktonic cultures and these quantitative bioassays are limited in both analyte separation and detection. Depending on the bioindicator, the detection is limited to those AHSLs to which the bioindicator responds and the signals must be present at the levels detectable by the reporter. In addition, other non-AHSL components in the extract may also interfere with the bioassay. Previous attempts to quantify the AHSL concentration in the biofilm, i.e. not based on a bioassay include high-performance liquid chromatography (HPLC) (Reimmann et al., 1997) and more sensitive the gas chromatography–mass spectrometry (GC–MS) (Charlton et al., 2000). The latter study determined the presence of diverse range of 3-oxo-AHSLs in P. aeruginosa biofilm at the levels well above those required for upregulation of QS. However, the spatial distribution of the signal molecules in the biofilm, namely whether they are localized in the biofilm extracellular matrix or limited to within the cell, still remains to be resolved.

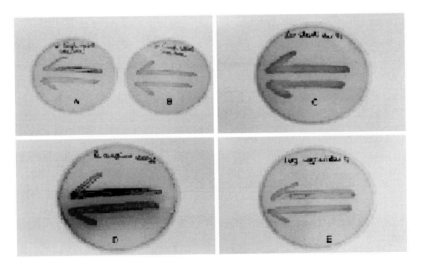

Figure 5.4   Results of cross-feeding tests. Evidence for the production of AHSLs is indicated by the expression of β-galactosidase activity in the reporter strain, A. tumefaciens A136, which in each test is streaked across the top half of the plate. Production of AHSLs by P. aeruginosa NSM35 (D) and A. tumefaciens KYC6 (positive control) (A) is visible. There is no evidence of AHSL production by Providencia stuartii NSM71 (C) or by A. tumefaciens A136 when it was incubated with itself (negative control) (B). A weakly positive response can be seen with M. morganii NSM67 (E). Stickler et al. (1998). Copyright permission from the author and Journal of Applied Environmental Microbiology.

5.5   CONCLUSION

The QS system allows both species specific and interspecies cell-to-cell communication to coordinate the behavior of a prokaryotic community. It is proven to play an important role in survival of the population by facilitating the colonization of hosts, defense against competitors, adaptation to varying physical conditions, cellular differentiation, and probably holds evolutionary importance as it directs individual cells in the population to act as a multicellular unit. At present, new pathways and mechanisms of cell-to-cell communication systems as well as their roles in physiological functions of variety different microorganisms being discovered. While many studies have demonstrated the importance of QS in biofilm structural development, there are studies that do contradict this phenomenon. It is increasingly apparent that the QS and biofilm structure relationship is far more complicated and may depend on or act together with other environmental cues, which remains to be determined. Further work is needed to elucidate the relative contribution of QS in the biofilm structural development.

Acknowledgements

This work was funded by the W.M. Keck Foundation, The National Institutes of Health RO1 grant GM60052-02 and in part by the co-operative agreement EEC-8907039 between the National Science Foundation and Montana State University, Bozeman.

REFERENCES

Andersen, B.J., Sternberg, C., Poulsen, L.K., Bjørn, S.P., Givskov, M. and Molin, S. (1998) New unstable variants of green fluorescent protein for studies of transient gene expression in bacteria. *Appl. Environ. Microbiol.* **64**(6), 2240–2246.

Archibald, L.K. and Gaynes, R.P. (1997) Hospital acquired infections in the United States: the importance of interhospital comparisons. *Nosocomial Inf.* **11**(2), 245–255.

Bassler, B.L., Greenberg, E.P. and Stevens, A.M. (1997) Cross-species induction of luminescence in the quorum sensing bacterium *Vibrio harveyi*. *J. Bacteriol.* **179**(12), 4043–4045.

Beck von Bodman, S., Majerczak, D.R. and Coplin, D.L. (1998) A negative regulator mediates quorum-sensing control of exopolysaccharide production in *Pantoea stewartii* subsp. *stewartii*. *Proc. Natl. Acad. Sci. USA* **95**, 7687–7692.

Bryers, J.D. (1988) Modeling biofilm accumulation. In *Physiological Models in Microbiology* (ed. Bazin, M.J. and Prosser J.I.), pp. 109–144, CRC Press, Inc., Boca Raton, FL.

Cha, H.J., Dalal, N.G., Pham, M.Q., Vakharia, V.N. and Bentley, W.E. (1999) Insect larval expression process is optimized by generating fusions with green fluorescent protein. *Biotechnol. Bioeng.* **65**(3), 316–324.

Charlton, T.S., de Nys, R., Netting, A., Kumar, N., Hentzer, M., Givskov, M. and Kjelleberg, S. (2000) A novel and sensitive method for the quantification of *N*-3-oxoacyl homoserine lactones using gas chromatography–mass spectrometry: application to a model bacterial biofilm. *Environ. Microbiol.* **2**(5), 530–541.

Cormack, B.P., Valdivia, R.H. and Falkow, S. (1996) FACS-optimized mutants of the green fluorescent protein (GFP). *Gene* **173**(1), 33–38.

Costerton, J.W., Lewandowski, Z., Caldwell, D.E., Korber, D.R. and Lappin-Scott, H.M. (1995) Microbial biofilms. *Ann. Rev. Microbiol.* **49**, 711–745.

Costerton, J.W., Stewart, P.S. and Greenberg, E.P. (1999) Bacterial biofilms: a common cause of persistent infections. *Science* **284**(5418), 1318–1322.

Dalton, H.M., Poulsen, L.K., Halasz, P., Angles, M.L., Goodman, A.E. and Marshall, K.C. (1994) Substratum induced morphological changes in a marine bacterium and their relevance to biofilm structure. *J. Bacteriol.* **176**(22), 6900–6906.

Davies, D.G., Parsek, M.R., Pearson, J.P., Iglewski, B.H., Costerton, J.W. and Greenberg, E.P. (1998) The involvement of cell-to-cell signals in the development of a bacterial biofilm. *Science* **280**(5361), 295–298.

Dawson, M. and Sia, R. (1931) *In vitro* transformation of pneumococcal types I, a technique for inducing transformation of pneumococcal types *in vitro*. *J. Exp. Med.* **54**, 681–699.

De Kievit, T.R. and Iglewski, B.H. (1999) Quorum sensing, gene expression and *Pseudomonas* biofilms. *Meth. Enzymol.* **310**, 117–128.

De Kievit, T.R., Gillis, R., Marx, S., Brown, C. and Iglewski, B.H. (2001) Quorum-sensing genes in *Pseudomonas aeruginosa* biofilms: their role and expression patterns. *Appl. Environ. Microbiol.* **67**(4), 1865–1873.

Dong, Y.H., Xu, J.L. and Zhang, L.H. (2000) AiiA, an enzyme that inactivates the acylhomoserine lactone quorum-sensing signal and attenuates the virulence of *Erwinia carotovora*. *Proc. Natl. Acad. Sci. USA* **97**, 3526–3531.

Dworkin, M. and Kaiser, D. (1985) Cell interactions in myxobacterial growth and development. *Science* **230**(4721), 18–24.

Erickson, D.L., Endersby, R., Kirkham, A., Stuber, K., Vollman, D.D., Rabin, R. Mitchell, H.I. and Storey, D.G. (2002) *Pseudomonas aeruginosa* quorum-sensing systems may control virulence factor expression in the lungs of patients with cystic fibrosis. *Infect. Immun.* **70**(4), 1783–1790.

Fuqua, C., Winans, S.C. and Greenberg, E.P. (1996) Census and consensus in bacterial ecosystems: the LuxR–LuxI family of quorum-sensing transcriptional regulators. *Ann. Rev. Microbiol.* **50**, 727–751.

Fuqua, C., Parsek, M.R. and Greenberg, E.P. (2001) Regulation of gene expression by cell-to-cell communication: acyl-homoserine lactone quorum sensing. *Ann. Rev. Genet.* **35**, 439–468.

Grossman, A.D. (1995) Genetic networks controlling the initiation of sporulation and the development of genetic competence in *Bacillus subtilis*. *Ann. Rev. Genet.* **29**, 477–508.

Gutowski-Eckel, Z., Klein, C., Siegers, K., Bohm, K., Hammelmann, M. and Entian, K.D. (1994) Growth phase-dependent regulation and membrane localization of SpaB, a protein involved in biosynthesis of the antibiotic subtilin. *Appl. Environ. Microbiol.* **60**(1), 1–11.

Hall-Stoodley, L., Rayner, J.C., Stoodley, P. and Lappin-Scott, H.M. (1999) Environmental monitoring of bacteria. In *Methods in Biotechnology* (ed. Edwards, C.), Vol. 12, pp. 307–319, Humana Press, Totowa, NJ.

Hastings, J.W. (1996) Chemistries and colors of bioluminescent reactions. *Gene* **173**(1), 5–11.

Heim, R. and Tsien, R.Y. (1996) Engineering green fluorescent protein for improved brightness, longer wavelengths and fluorescence resonance energy transfer. *Curr. Biol.* **6**(2), 178–182.

Hermanowicz, S.W, Schindler, U. and Wilderer, P.A. (1995) Fractal structure of biofilms: new tools for investigation of morphology. *Water Sci. Technol.* **32**(8), 99–105.

Heydorn, A., Nielsen, A.T., Hentzer, M., Sternberg, C., Givskov, M., Ersbøll, B.K. and Molin, S. (2000a) Quantification of biofilm structures by the novel computer program COMSTAT. *Microbiology* **146**(10), 2395–2407.

Heydorn, A., Ersbøll, B.K., Hentzer, M., Parsek, M.R., Givskov, M. and Molin, S. (2000b) Experimental reproducibility in flow chamber biofilms. *Microbiology* **146**(10), 2409–2415

Heydorn, A., Ersboll, B., Kato, K.J., Hentzer, M. Parsek, M.R., Nielsen, A.T., Givskov, M. and Molin, S. (2002) Statistical analysis of *Pseudomonas aeruginosa* biofilm development: impact of mutations in genes involved in twitching motility, cell-to-cell signalling, and stationary-phase sigma factor expression. *Appl. Environ. Microbiol.* **68**(4), 2008–2017.

Huber, B., Riedel, K. Hentzer, M., Heydorn, A., Gotschlich, A., Givskov, M., Molin, S. and Eberl, L. (2001) The cep quorum sensing system of *Burkholderia cepacia* H11 controls biofilm formation and swarming motility. *Microbiology* **47**(9), 2517–2528.

Ji, G., Beavis, R.C. and Novick, R.P. (1995) Cell density control of staphylococcal virulence mediated by an octapeptide pheromone. *Proc. Natl. Acad. Sci. USA* **92**, 12055–12059.

Kleerebezem, M., Quadri, L.E., Kuipers, O.P. and DeVos, W.M. (1997) Quorum sensing by peptide pheromones and two component signal-transduction systems in Gram-positive bacteria. *Mol. Microbiol.* **24**(5), 895–904.

Lee, M.S. and Morrison, D.A. (1999) Identification of a new regulator in *Streptococcus pneumoniae* linking quorum sensing to competence for genetic transformation. *J. Bacteriol.* **181**(16), 5004–5016.

Li, Y.H., Hanna, M.N., Svensäter, G., Ellen, R.P. and Cvitkovitch, D.G. (2001a) Cell density modulates acid adaptation in *Streptococcus mutans*: implications for survival in biofilms. *J. Bacteriol.* **183**(23), 6875–6884.

Li, Y.H., Lau, P.C.Y., Lee, J.H., Ellen, R.P. and Cvitkovitch, D.G. (2001b) Natural genetic transformation of *Streptococcus mutans* growing in biofilms. *J. Bacteriol.* **183**(3), 897–908.

Li, Y.H., Tang, N., Aspiras, M.B., Lau, P.C.Y., Lee, J.H., Ellen R.P. and Cvitkovitch, D.G. (2002) A quorum sensing signaling system essential for genetic competence in *Streptococcus mutans* is involved in biofilm formation. *J. Bacteriol.* **184**(10), 2699–2708.

Loo, C.Y., Corliss, D.A. and Ganeshkumar, N. (2000) *Streptococcus gordonii* biofilm formation: identification of genes that code for biofilm phenotypes. *J. Bacteriol.* **182**(5), 1374–1382.

Lynch, M.J., Swift, S., Kirke, D.F., Keevil, C.W., Dodds, C.E. and Williams, P. (2002) The regulation of biofilm development by quorum sensing in *Aeromonas hydrophila*. *Environ. Microbiol.* **4**(1), 18–28.

McLean, R.J.C., Whitely, M., Stickler, D.J. and Fuqua, W.C. (1997) Evidence of autoinducer activity in naturally occurring biofilms. *FEMS Microbiol. Lett.* **154**(2), 259–263.

Middleton, B., Rodgers, H.C., Cámara, M.M., Knox, A.J., Williams, P. and Hardman, A. (2002) Direct detection of *N*-acylhomoserine lactones in cystic fibrosis sputum. *FEMS Microbiol. Lett.* **207**(1), 1–7.

Miller, M.B. and Bassler, B.L. (2001) Quorum sensing in bacteria. *Ann. Rev. Microbiol.* **55**, 165–199.

Moller, S., Korber, D.R., Wolfaardt, G.M., Molin, S. and Caldwell, D.E. (1997) Impact of nutrient composition on a degradative biofilm community. *Appl. Environ. Microbiol.* **63**(6), 2432–2438.

Moller, S., Sternberg, C., Andersen, L.K., Christensen, B.B., Ramos, J.L., Givskov, M. and Molin, S. (1998) *In situ* gene expression in mixed-culture biofilms: evidence of metabolic interactions between community members. *Appl. Environ. Microbiol.* **64**(2), 721–732.

Murga, R.T., Foster, S., Brown, E., Pruickler, J.M., Fields, B.S. and Donlan, R.M. (2001) Role of biofilms in the survival of *Legionella pneumophila* in a model potable-water system. *Microbiology* **11**(147), 3121–3126.

Nealson, K.H. and Hastings, J.W. (1979) Bacterial bioluminescence: its control and ecological significance. *Microbiol. Rev.* **43**(4), 495–518.

Nivens, D.E., Ohman, D.E., Williams, J. and Franklin, M.J. (2001) Role of alginate and its O-acetylation in formation of *Pseudomonas aeruginosa* microcolonies and biofilms. *J. Bacteriol.* **183**(3), 1047–1057.

O'Toole, G.A. and Kolter, R. (1998) Flagellar and twitching motility are necessary for *Pseudomonas aeruginosa* biofilm development. *Mol. Microbiol.* **30**(2), 295–304.

O'Toole, G.A., Gibbs, K.A., Hager, P.W., Phibbs Jr., P.V. and Kolter, R. (2000) The global carbon metabolism regulator Crc is a component of a signal transduction pathway required for biofilm development by *Pseudomonas aeruginosa*. *J. Bacteriol.* **182**(2), 425-431.

Palmer, R.J. (1999) Microscopy flow cells: perfusion chambers for real-time study of biofilms. *Meth. Enzymol.* **310**, 160–166.

Passador, L., Tucker, K.D., Guertin, K.R., Journet, M.P., Kende, A.S. and Iglewski. B.H. (1996) Functional analysis of the *Pseudomonas aeruginosa* autoinducer PAI. *J. Bacteriol.* **178**(20), 5995–6000.

Pearson, J.P., Gray, K.M., Passador, L., Tucker, K.D., Eberhard, A., Iglewski, B.H. and Greenberg, E.P. (1994) Structure of the autoinducer required for expression of *Pseudomonas aeruginosa* virulence genes. *Proc. Natl. Acad. Sci. USA* **91**, 197–201.

Pearson, J.P., Van Delden, C. and Iglewski, B.H. (1999) Active efflux and diffusion are involved in transport of *Pseudomonas aeruginosa* cell-to-cell signals. *J. Bacteriol.* **181**(4), 1203–1210.

Prouty, A.M., Schwesinger, W.H. and Gunn, J.S. (2002) Biofilm formation and interaction with the surfaces of gallstones by *Salmonella* spp. *Infect. Immun.* **70**(5), 2640–2649.

Purevdorj, B., Costerton, J.W. and Stoodley, P. (2002) Influence of hydrodynamics and cell signalling on the structure and behavior of *Pseudomonas aeruginosa* biofilms. *Appl. Environ. Microbiol.* **68**(9), 4457–4464.

Reimmann, C., Beyeler, M., Latifi, A., Winteler, H., Fogliono, M. and Lazdunski, A. (1997) The global activator GacA of *Pseudomonas aeruginosa* PAO positively controls the production of the autoinducer *N*-butyryl-homoserine lactone and the formation of the virulence factors pyocyanin, cyanide and lipase. *Mol. Microbiol.* **24**(2), 309–319.

Riedel, K., Hentzer, M., Geisenberger, O., Huber, B., Steidle, A., Wu, H., Høiby, N., Givskov, M., Molin, S. and Eberl, L. (2001) *N*-acylhomoserine-lactone-mediated communication between *Pseudomonas aeruginosa* and *Burkholderia cepacia* in mixed biofilms. *Microbiology* **12**(147), 3249–3262.

Shaw, P.D., Ping, G., Daly, S.L., Cha, C., Cronan Jr., J.E., Rinehart, K.L. and Farrand, S.K. (1997) Detecting and characterising *N*-acyl-homoserine lactone signal molecules by thin-layer chromatography. *Proc. Natl. Acad. Sci. USA* **94**, 6036–6041.

Singh, P.K., Schaefer, A.L., Parsek, M.R., Moninger, T.O., Welsh, M.J. and Greenberg, E.P. (2000) Quorum-sensing signals indicate that cystic fibrosis lungs are infected with bacterial biofilms. *Nature* **407**(6805), 762–764.

Stickler, D.J, Morris, N.S., McLean, R.J. and Fuqua, C. (1998) Biofilms on indwelling urethral catheters produce quorum-sensing signal molecules *in situ* and *in vitro*. *Appl. Environ. Microbiol.* **64**(9), 3486–3490.

Stewart, P.S. (1998) A review of experimental measurements of effective diffusive permeabilities and effective diffusion coefficients in biofilms. *Biotechnol. Bioeng.* **59**(3), 261–272.

Stewart, P.S., Peyton, B.M., Drury, W.J. and Murga, R.J. (1993) Quantitative observations of heterogeneities in *Pseudomonas aeruginosa* biofilms. *Appl. Environ. Microbiol.* **59**(1), 327–329.

Stoodley, P., Boyle, J.D., Dodds, I. and Lappin-Scott, H.M. (1997) Consensus model of biofilm structure. In *BIOFILM: Community Interactions and Control* (ed. J.W.T. Wimpenny, P.S. Handley, P. Gilbert, H.M. Lappin-Scott, and M. Jones), pp. 1–9, Bioline, Cardiff, UK.

Stoodley, P., Lewandowski, Z., Boyle, J.D. and Lappin-Scott, H.M. (1999a) The formation of migratory ripples in a mixed species bacterial biofilm growing in turbulent flow. *Environ. Mirobiol.* **1**(5), 447–455.

Stoodley, P., Jorgensen, F., Williams, P. and Lappin-Scott, H.M. (1999b) The role of hydrodynamics and AHL signalling molecules as determinants of the structure of *Pseudomonas aeruginosa* biofilms. In *Biofilms: The Good and The Bad, and The Ugly* (ed. J. Wimpenny, P. Gilbert, J. Walker, M. Brading, and R. Bayston), pp. 223–230 Cardiff, Bioline, UK.

Stoodley, P., Wilson, S., Hall-Stoodley, L., Boyle, J.D., Lappin-Scott, H.M. and Costerton, J.W. (2001) Growth and detachment of cell clusters from mature mixed-species biofilms. *Appl. Environ. Microbiol.* **67**(12), 5608–5613.

Stretton, S., Techkarnjanaruk, S., McLennan, A.M. and Goodman, A.E. (1998) Use of green fluorescent protein to tag and investigate gene expression in marine bacteria. *Appl. Environ. Microbiol.* **64**(7), 2554–2559.

Surette, M.G., Miller, M.B. and Bassler, B.L. (1999) Quorum sensing in *Eshcerichia coli*, *Salmonella typhimurium*, and *Vibrio harveyi*: a new family of genes responsible for autoinducer production. *Proc. Natl. Acad. Sci. USA* **96**(4), 1639–1644.

Tombolini, R., Unge, A., Davey, M.E., de Bruijn, F.J. and Jansson, J.K. (1997) Flow cytometric and microscopic analysis of GFP-tagged *Pseudomonas fluorescens* bacteria. *FEMS Microbiol. Ecol.* **22**(1), 17–28.

Vieira, M.J., Melo, L.F. and Pinheiro, M.M. (1993) Biofilm formation: hydrodynamic effects on internal diffusion and structure. *Biofouling* **7**(1), 67–80.

Williams, P. and Stewart, G.S.A.B. (1994) Cell density dependent control of gene expression in bacteria – implications for biofilm development and control. In *Bacterial Biofilms and Their Control in Medicine and Industry* (ed. W.W. Nichols, J. Wimpenny, D. Stickler, and H.M. Lappin-Scott), Cardiff, Bioline, UK.

Wimpenny, J.W.T. and Colasanti, R. (1997) A unifying hypothesis for the structure of microbial biofilms based on cellular automation models. *FEMS Microbiol. Ecol.* **22**(1), 1–16.

Wu, H., Song, Z., Hentzer, M., Andersen, J.B., Heydorn, A., Mathee, K., Moser, C., Eberl, L., Molin, S., Høiby, N. and Givskov, M. (2000) Detection of *N*-acylhomoserine lactones in lung tissues of mice infected with *Pseudomonas aeruginosa*. *Microbiology* **146**(10), 2481–2493.

Yang, X., Beyenal, H., Harkin, G. and Lewandowski, Z. (2000) Quantifying biofilm structure using image analysis. *J. Microbiol. Meth.* **39**(2), 109–119.

Yang, X., Beyenal, H., Harkin, G. and Lewandowski, Z. (2001) Evaluation of biofilm image thresholding methods. *Water Res.* **35**(5), 1149–1158.

Zhang, T.C. and Bishop, P.L. (1994) Density, porosity, and pore structure of biofilm. *Water Res.* **28**(11), 2267–2277.

Zhu, J., Beaber, J.W., More, M.I., Fuqua, C., Eberhard, A. and Winans, S.C. (1998) Analogues of the autoinducer 3-oxooctanoyl-homoserine lactone strongly inhibit activity of the TraR protein of *Agrobacterium tumefaciens*. *J. Bacteriol.* **180**(20), 65398–65405.

Zinn, M.S., Kirkegaard, R.D., Palmer, R.J. and White, D.C. (1999) Laminar flow chamber for continuous monitoring of biofilm formation and sucession. *Meth. Enzymol.* **310**, 224–232.

Section 2

Biofilm composition: (A) Chemical

6   Molecular architecture of the biofilm matrix   81

7   Physico-chemical properties of   91
    extracellular polymeric substances

6

Molecular architecture of the biofilm matrix

D.G. Allison

6.1 INTRODUCTION

Characteristic to many biofilms is the production of an extracellular matrix that envelops the attached cells. This is, generally, composed of water and microbial macromolecules and provides a complex array of micro-environments surrounding the attached cells. Moreover, since matrix structure and integrity is heavily influenced by changes in the surrounding macro-environment and is constantly changing, the biofilm matrix may be considered as a dynamic environment (Sutherland, 2001). Attachment to a surface is thought to initiate a cascade of physiological changes in the cells, which leads, in part, to the overproduction of exopolymers (Allison and Sutherland, 1987; Davies and Geesey, 1995). These exopolymers not only immobilise the cells on the colonised surface, but also facilitate the spatial arrangement of different species within a biofilm (Costerton *et al.*, 1994). Such interactions give the biofilm community metabolic and physiological capabilities which are not possible for the individual, unattached cells (Gilbert *et al.*, 1997). Biofilms are, generally, considered to be problematic and can have consequences that directly affect society, the scale of which is often overlooked. Indeed, in many instances the persistent and problematic nature of biofilms is attributed to the surrounding matrix (Allison *et al.*, 2000). It is somewhat surprising, therefore, that for so ubiquitous and important a process our understanding about the properties, formation and structure of bacterial biofilms and their components is far from complete. More so, given that biofilms have been recognised for over 60 years (Henrici, 1932; Zobell and Allen, 1935).

At this juncture it is worth clarifying matrix terminology. A significant feature of the matrix is the presence of biosynthetic microbial polymers lying outside the integral cell-surface components of the resident bacteria. These extracellular biopolymers include exopolysaccharides, nucleic acids, proteins, glycoproteins and phospholipids. Since polysaccharides were identified as being a common component, the term 'glycocalyx' was introduced to describe the gelatinous mass surrounding attached cells (Costerton *et al.*, 1981). When applied to eukaryotic cells, glycocalyx suggests a defined structure. This is not appropriate for microbial systems, where not only the extracellular biopolymers, but the resident cells are constantly changing. Hence, biofilm matrix is a better descriptor, implying a multi-component, dynamic heterogeneous system. In a similar vein, the abbreviation EPS has been used interchangeably for exopolysaccharides, exopolymers and extracellular polymeric substances (Wingender *et al.*, 1999a). Although exopolysaccharides may not be always the most abundant component of the matrix, a traditional standpoint will be adopted in this chapter, whereby EPS will represent exopolysaccharides.

Very often biofilms are composed of mixed communities of microorganisms and their metabolic properties. Because of this inherent complexity, this can make isolation and characterisation of matrix components extremely difficult. Detailed analyses of the biofilm matrix have been hampered by a number of factors, not least the availability of sufficiently sensitive and specific tools for probing the matrix structure. In addition, it is also difficult to assess the composition accurately, if some components are only present in low quantities and are otherwise masked by copious quantities of more dominant macromolecules. That is not to say, however, that minority components are not important and do not play a significant role in developing and maintaining matrix architecture.

Whether matrix polymers differ from those associated with planktonically grown cells, and also whether the matrix polymers, which bind cells to other cells, differ from 'foot-print polymers', which cement the primary colonisers to the substratum is uncertain at present (Sutherland, 1995). From the limited number of studies that have attempted to characterise the matrix associated with bacterial biofilms, it is clear that they vary greatly in their composition and physical properties. However, they may not be different from the extensive range of polymers derived from planktonic cultures, which have now been characterised and probably few, if any, are biofilm specific (Sutherland, 1997). As yet, the cryptic capability of biofilm organisms to synthesise novel polymers only in the biofilm mode has not been successfully demonstrated (Sutherland, 1999a).

6.2   MATRIX COMPOSITION

Since biofilms are found in virtually every environment, both natural and artificial, in which moisture and microorganisms are present, it becomes extremely difficult to generalise about their structure and physiological activities. As summarised by Sutherland (2001), matrix composition is based upon a combination of intrinsic factors, such as the genotype of the attached cells, and extrinsic factors that include the surrounding physico-chemical environment (Wimpenny, 2000). Thus, it is probable that biofilm matrices, even those produced by identical organisms, will vary greatly in their composition and in their physical properties.

6.2.1   General composition

Water is by far the major component of the biofilm matrix, accounting for up to 97% of the mass (Table 6.1). The water may be bound within the capsules of the bacterial cells or can exist as a solvent, whose physical properties are determined by the solutes dissolved in it (Sutherland, 2001). Water binding and mobility within the biofilm matrix are integral to the diffusion processes that occur within the biofilm and, therefore, form a part of the fine structure of the biofilm (Schmitt and Flemming, 1999). Resident cells, which may include many different species, surprisingly only account for approximately 5% of the matrix. Apart from water and cells, the other components of the matrix include EPS (1–2%), proteins, which include lytic products and secreted enzymes (1–2%), nucleic acids from lysed cells (1–2%), ions and humics (trace amounts) from the surrounding environment. It is important to note that these estimates are really only snapshots of different biofilms at different points in time. Specific composition for any biofilm will vary depending upon the organism(s) present, their physiological status, the nature of the growth environment, bulk fluid-flow dynamics, the substratum and the prevailing physical conditions. Hence, it could be argued that the detailed description of a biofilm matrix is only accurate at the point at which it was measured.

Table 6.1   General composition of biofilm matrices.

Component	Matrix (%)
Cells	2–5
Water	c. 98
Biosynthetic microbial polymers	
EPS	1–2
Nucleic acids	1–2
Proteins (including enzymes and regulatory proteins)	1–2
Glycoproteins	1–2
Phospholipids	1–2
Adsorbed species	?

6.2.2   Enzymes

The composition of the matrix may be influenced by different processes, namely, active secretion of biopolymers and enzymes, shedding of cell-surface material, such as lipopolysaccharide, cell lysis and adsorption from the surrounding environment. It is worth noting that matrix material, particularly EPS, shed from biofilms can be adsorbed at places distinct from their source. Different enzymes involved in polymer degradation (hydrolases, lyases, glycosidases and other enzymes) are abundant in biofilms (Sutherland, 1999b; Wingender *et al.*,1999b), and serve to release cells from the attached community and to provide low-molecular-weight breakdown products available as carbon and energy sources for metabolism by the immobilised bacteria. It should also be remembered that in multi-species biofilms, the collective action of several different enzymes might result in the degradation or alteration of EPS, which are resistant to discrete enzymes.

6.2.3   EPS composition

Common to virtually all biofilm matrices are EPS. Although not a major component in terms of amount present, the EPS is, however, regarded as the major structural component of the matrix, providing a framework for the biofilm complex. In essence, the EPS provides the skeleton into which microbial cells and their bioactive products are inserted.

At an individual cell level, EPS occur in two basic fiorms: capsular, whereby the EPS is intimately associated with the cell surface, and as slime, which is only loosely associated with the cell. Differentiation between the two forms can often be difficult, since cells producing large quantities of capsule may 'release' some material at the periphery, giving the appearance of slime production.

Chemically, bacterial EPS are highly heterogeneous polymers containing a number of distinct monosaccharides and non-carbohydrate substituents, many of which are strain specific (Sutherland, 1985; Whitfield, 1988). As with all polysaccharides, those produced by microorganisms can be divided into homopolysaccharides and heteropolysaccharides. Most homopolysaccharides are neutral glucans, while the majority of heteropolysaccharides appear to be polyanionic. Homopolysaccharides can possess three different structures, namely linear molecules comprised of a single linkage type, linear repeat units possessing a one-sugar side chain and branched structures. Microbial heteropolysaccharides are almost all composed of repeating side units varying in size from disaccharide to octasaccharide. Structural diversity, and hence rheological properties, is increased by potential non-carbohydrate substituents (e.g. acetyl, pyruvate or sulphate groups) and linkage types. It is also worth noting that several species of bacteria are able to synthesise more than one chemically distinct EPS.

Polysaccharide chains vary in size from 10^3 to 10^8 kDa and contain subunit configurations, which may also be both functionally and species specific (Sutherland, 1985). Depending upon the components of the repeat units, polysaccharides are usually negatively charged, sometimes neutral or, rarely, positively charged.

Furthermore, polysaccharides may be hydrophilic, but can also have hydrophobic properties (Neu and Poralla, 1990). Indeed, many polymers are heterogeneous with respect to lipophilicity and hydrophobicity, let alone a fully formed, multi-component biofilm matrix (Sutherland, 1997). Polymer hydrophobicity can, therefore, play an important part in determining the behaviour of the polysaccharide at the cell surface or at an interface. Unfortunately, biochemical information of this nature has almost exclusively been provided from studies on EPS isolated from culture medium. Structure–function relationships for EPS, as reviewed by Sutherland (1997), have received little systematic investigation. Few EPS, known to be implicated in the adhesion process, have been comprehensively analysed. Moreover, EPS-producing adherent microorganisms are often cultivated in nutrient-rich growth media, irrespective of their natural habitat. This creates a nutritional environment, which for the majority of attached populations is significantly different from that of a biofilm *in situ*. It is well documented that bacterial cell-surface phenotype can change markedly in response to changes in the surrounding growth environment; particularly, those brought about by growth rate and nutrient limitation (Ellwood and Tempest, 1972; Brown and Williams, 1985; Brown and Gilbert, 1993). EPS are also subject to environmental modulation with respect to composition and molecular mass (Tait *et al.*, 1986), which in turn can affect their capacity to interact with other polymers and cations (Sutherland, 1990). Thus, in order to fully characterise any biofilm polymer, it is essential to be able to mimic *in situ* growth conditions that lead to EPS production. In this respect, previous studies on EPS synthesis and composition that have not taken these factors into consideration should be viewed with caution.

6.3   MATRIX ARCHITECTURE

The biofilm matrix provides the possibility that the resident microorganisms can form stable aggregates of different cell types, leading to the development of a functional, synergistic microconsortium. The spatial arrangement of microorganisms gives rise to nutrient and gaseous gradients as well as those of electron acceptors, products and pH. Thus, aerobic and anaerobic (anoxic) habitats can arise in close proximity and, as a consequence, the development of large variability of species takes place. Moreover, since the structure is not rigid, the organisms can move in it, thereby promoting genetic exchange.

6.3.1   Matrix structure

Understanding the mechanical and architectural properties of the matrix is very much dependent upon having suitable methods to permit accurate analysis of matrix structure and composition. Recent developments in technology, such as the use of fluorescently labelled antibodies or lectins, expression of green fluorescent protein (GFP), dissolved oxygen and pH microsensors and confocal scanning laser microscopy, especially when combined with GFP, has led to the understanding that most biofilms, irrespective of environment, comprise aggregates of cells within an EPS containing matrix, separated by interstitial voids and channels (Sutherland, 2001). Information collected by these methods along with computer simulation has been used to propose three distinct biofilm models, namely simple stalked or irregular branching structures, dense confluent structures and penetrated water channel or mushroom-shaped biofilms (Wimpenny and Colsanti, 1997). However, although these models represent three distinct forms, in reality, biofilms are a combination of all three depending on many extrinsic factors. Structure is largely dependent on resource concentration. What is basically a hetergeneous biofilm one day may be more of a dense confluent biofilm the next day following subtle changes in the bulk aqueous phase. In a study by Møller et al. (1998), biofilms grown on a poorly used substrate produced mounds of cells, but when switched to a rich medium produced a much more uniform biofilm. Similarly, Stoodley et al. (1999a) demonstrated that raising the glucose concentration changed a mixed species biofilm from a series of cell clusters with ripples to a thick biofilm without ripples. Subsequent lowering of the glucose concentration to initial levels caused a reversion of the biofilm to the initial phenotype. What have become evident is the micro-heterogeneity and the presence of water channels in all forms of biofilm. The degree of channelling does vary, but all have the same functions as that of permitting flow of nutrients, enzymes, metabolites and waste products throughout the biofilm community. The presence of relatively large channels and pores within the matrix structure might also allow entry of secondary colonisers and their establishment within the biofilm.

6.3.2   Internal factors influencing matrix architecture

The presence of polysaccharide synthesizing and degrading enzymes in the biofilm means that matrix composition will be constantly changing. New surfaces are created, while others are masked, thereby allowing new species to attach to the biofilm. Enzymes altering macromolecules within the matrix will also have a marked influence on its physical properties (Sutherland, 2001). Under conditions of nutrient starvation, EPS-degrading enzymes may be produced, which will cause local destruction of the matrix, and possible release of cells (Allison et al., 1998). This may also cause weakening of the community structure, especially if such enzymes are carried through the channels to sites remote from their synthesis.

Many microorganisms are also capable of synthesizing biosurfactants. These might clearly play a significant role in localised dissolution of matrix material. One possible role for surfactants is to enhance desorption of the microbial cells from hydrophobic surfaces, which no longer contain usable carbon sources.

At a molecular level, the physical properties of EPS are greatly influenced by the ionic content of the surrounding bulk aqueous phase and by water molecules. In most natural environments, where relatively high ionic concentrations are found, polysaccharides will invariably be found in an ordered configuration as gelled and highly hydrated polymers. EPS also vary in their water solubility, some, such as scleroglucan being highly soluble in water or dilute salt solutions; while others, such as mutan are virtually insoluble in water or form very rigid gels when in the ordered form (Sutherland, 1997). These include several polymers, which are commonly found in biofilms. In a study by Hughes (1997) on the EPS extracted from two species of *Enterobacter agglomerans* originally isolated from a biofilm contaminant of a food manufacturing surface, it was shown that both EPS possessed backbones containing a high proportion of 1,3-linked sugar residues and were poorly soluble in water. This led to the suggestion that water-soluble molecules, such as disinfectants may be effectively excluded from the interior of such biofilm matrices. Most EPS in solution undergo a change from order to disorder on heating or on removal of ions. Increasing solubility and eventual dissolution or sloughing from a biofilm (Sutherland, 1997; Willcock *et al.*, 1997) may reflect such changes. Also of great significance will be the conformation adopted by the EPS. While polymers with charged groups on the outside of the tertiary structure will bind ions, they are unlikely to form tight bonds between adjacent strands. Alternatively, polymers such as alginates, which can have block sequences of poly-L-guluronic acid forming an 'egg-box' structure can bind cations, which form salt bridges between carboxyl groups on adjacent strands, thereby forming a strong gel. By contrast, some polysaccharides lacking intra-chain hydrogen bonds will remain in a disordered conformation (Cesáro *et al.*, 1992) and, as such, will probably cause loosely attached cells to be very easily removed from the biofilm.

6.3.3   External factors influencing matrix architecture

Changes in hydrodynamic conditions can also affect matrix structure. Shear rate will influence rates of erosion of cells and regions of the matrix from the biofilm (Stoodley *et al.*, 1999b; Willcock *et al.*, 2000). Recently, it has been shown that changes in detachment from a steady-state biofilm due to a rapid change in shear force are independent of the initial shear rate and can cause temporary, elastic deformation of biofilm structures (Stoodley *et al.*, 1999b). Shear stresses applied to the biofilm will affect EPS solutions, causing flow and elastic recovery to occur. Such changes will consequently affect the shape of the matrix. Under turbulent flow, the EPS-endowed matrix will flow like a viscous fluid. Turbulent flow and increased carbon source alters the appearance of an established biofilm from patches of roughly circular cell clusters to ripples and streamers, which oscillated in the flow (Stoodley *et al.*, 1999a).

In mixed species biofilms, different polymers are likely to be produced by each component species. These are likely to blend together to generate heterogeneous regions of polymer within the biofilm matrix and, in so doing, possibly confer some structure to the biofilm (Cooksey, 1992). The physico-chemical properties of such blended EPS will differ significantly from those of purified components and will also be substantially affected by the ionic strength of the surrounding medium and the nature of the cationic species (Allison and Matthews, 1992). Furthermore, biofilm EPS from two or more species may interact synergistically, leading to the enhancement of bacterial adhesion (Skillman *et al.*, 1997). Biofilm EPS may also interact with other macromolecules, such as proteins, lipids and even nucleic acids.

6.4   CONCLUSION

Our direct knowledge of the structure and composition of the biofilm matrix is still rather limited. In addition, it would be reasonable to consider each matrix as distinct, with its own unique micro-environment. Nevertheless, it is clear that the different components of the matrix interact and provide a wide range of localised environments within the matrix, and that these micro-environments are subject to extrinsic forces and are, therefore, constantly changing. Consequently, for any biofilm in any environment, these changes will contribute to the heterogeneous nature of the matrix.

REFERENCES

Allison, D.G. and Matthews, M.J. (1992) Effect of polysaccharide interactions on anti-biotic susceptibility of *Pseudomonas aeruginosa*. *J. Appl. Bacteriol.* **73**, 484–488.

Allison, D.G. and Sutherland, I.W. (1987) The role of exopolysaccharides in adhesion of freshwater bacteria. *J. Gen. Microbiol.* **133**, 1319–1327.

Allison, D.G., Ruiz, B., San Jose, C., Jaspe, A. and Gilbert, P. (1998) Extracellular prod-ucts as mediators of the formation and detachment of *Pseudomonas fluorescens* biofilms. *FEMS Microbiol. Lett.* **167**, 179–184.

Allison, D.G., McBain, A.J. and Gilbert, P. (2000) Biofilms: problems of control. In *Community Structure and Co-operation in Biofilms* (ed. D.G. Allison, P. Gilbert, H. Lappin-Scott and M. Wilson), Vol. 59, pp. 309–327, Cambridge University Press, Cambridge, UK.

Brown, M.R.W. and Gilbert, P. (1993) Sensitivity of biofilms to antimicrobial agents. *J. Appl. Bacteriol.* **74**(Suppl.), 87S–97S.

Brown, M.R.W. and Williams, P. (1985) The influence of the environment on envelope properties affecting survival of bacteria in infections. *Ann. Rev. Microbiol.* **39**, 527–556.

Cesáro, A., Tomasi, G., Gamini, A., Vidotto, S. and Navarini, L. (1992) Solution confor-mation and properties of the galactoglucan from *Rhizobium meliloti* strain YES (S1). *Carbohydr. Res.* **231**, 117–135.

Cooksey, K.E. (1992) Extracellular polymers in biofilms. In *Biofilms: Science and Technology* (ed. L.F. Melo, T.R. Bott, M. Fletcher and B. Capdeville), pp. 137–147, Kluwer Academic Press, Dordrecht, The Netherlands.

Costerton, J.W., Irwin, R.T. and Cheng, K.J. (1981) The bacterial glycocalyx in nature and disease. *Ann. Rev. Microbiol.* **35**, 399–424.

Costerton, J.W., Lewandowski, Z., deBeer, D., Caldwell, D., Korber, D. and James, G. (1994) Biofilms, the customised microniche. *J. Bacteriol.* **176**, 2137–2142.

Davies, D.G. and Geesey, G.G. (1995) Regulation of the alginate biosynthesis gene *alg*C in *Pseudomonas aeruginosa* during biofilm development in continuous culture. *Appl. Environ. Microbiol.* **61**, 860–867.

Ellwood, D.E. and Tempest, D.W. (1972) Effects of environment on bacterial wall content and composition. *Adv. Microbial Physiol.* **7**, 83–117.

Gilbert, P., Das, J. and Foley, I. (1997) Biofilm susceptibility to antimicrobials. *Adv. Dent. Res.* **11**, 160–167.

Henrici, A.T. (1932) Studies of freshwater bacteria. *J. Bacteriol.* **25**, 277–288.

Hughes, K.A. (1997) Bacterial biofilms and their exopolysaccharides. Ph.D. thesis, University of Edinburgh, UK.

Møller, S., Sternberg, C., Andersen, J.B., Christensen, B.B., Ramos, J.L., Givskov, M. and Molin, S. (1998) *In situ* gene expression in mixed-culture biofilms: evidence of metabolic interactions between community members. *Appl. Env. Microbiol.* **64**, 721–732.

Neu, T.R. and Poralla, K. (1990) Emulsifying agents from bacteria isolated during screening for cells with hydrophobic surfaces. *Appl. Microbiol. Biot.* **32**, 521–525.

Schmitt, J. and Flemming, H.-C. (1999) Water binding in biofilms. *Water Sci. Technol.* **39**, 77–82.

Skillman, L.C., Sutherland, I.W. and Jones, M.V. (1997) Co-operative biofilm formation between two species of Enterobacteriaceae. In *Biofilms:Community Interactions and Control* (ed. J. Wimpenny, P. Handley, P. Gilbert, H. Lappin-Scott and M. Jones), pp. 119–127, Bioline Press, Cardiff, UK.

Stoodley, P., Dodds, I., Boyle, J.D. and Lappin-Scott, H. (1999a) Influence of hydrodynamics and nutrients on biofilm structure. *J. Appl. Microbiol.* **85**(Suppl.), 19S–28S.

Stoodley, P., Jørgensen, F., Williams, P. and Lappin-Scott, H. (1999b) The role of hydrodynamics and AHL signalling molecules as determinants of the structure of *Pseudomonas aeruginosa* biofilms. In *Biofilms – The Good, the Bad and the Ugly* (ed. J. Wimpenny, P. Gilbert, J. Walker, M. Brading and R. Bayston), pp. 223–230, Bioline Press, Cardiff, UK.

Sutherland, I.W. (1985) Biosynthesis and composition of Gram-negative bacterial extracellular and wall polysaccharides. *Ann. Rev. Microbiol.* **10**, 243–270.

Sutherland, I.W. (1990) Biotechnology of microbial exopolysaccharides. In *Biotechnology of Microbial Exopolysaccharides*, pp. 12–117, Cambridge University Press, Cambridge, UK.

Sutherland, I.W. (1995) Biofilm specific polysaccharides – do they exist? In *The Life and Death of Biofilm* (ed. J. Wimpenny, P. Handley, P. Gilbert and H. Lappin-Scott), pp. 103–106, Bioline Press, Cardiff, UK.

Sutherland, I.W. (1997) Microbial biofilm exopolysaccharides – superglues or velcro? In *Biofilms:Community Interactions and Control* (ed. J. Wimpenny, P. Handley, P. Gilbert, H. Lappin-Scott and M. Jones) pp. 33–39, Bioline Press, Cardiff, UK.

Sutherland, I.W. (1999a) Biofilm exopolysaccharides. In *Microbial Extracellular Polymeric Substances*, (ed. J. Wingender, T.R. Neu and H.-C. Flemming), pp. 73–92, Springer, Berlin, Germany.

Sutherland, I.W. (1999b) Polysaccharases in biofilms – source–action–consequences! In *Microbial Extracellular Polymeric Substances* (ed. J. Wingender, T.R. Neu and H.-C. Flemming), pp. 201–230, Springer, Berlin, Germany.

Sutherland, I.W. (2001) The biofilm matrix – an immobilized but dynamic environment. *Trend. Microbiol.* **9**, 222–227.

Tait, M.I., Sutherland, I.W. and Clarke-Sturman, A.J. (1986) Effect of growth conditions on the production, composition and viscosity of *Xanthomonas campestris* exopolysaccharide. *J. Gen. Microbiol.* **132**, 1483–1492.

Whitfield, C. (1988) Bacterial extracellular polysaccharides. *Can. J. Microbiol.* **34**, 415–420.

Willcock, L., Holah, J., Allison, D.G. and Gilbert, P. (1997) Population dynamics in steady-state biofilms: effects of growth environment upon dispersal. In *Biofilms: Community Interactions and Control* (ed. J. Wimpenny, P. Handley, P. Gilbert, H. Lappin-Scott and M. Jones), pp. 23–31, Bioline Press, Cardiff, UK.

Willcock, L., Gilbert, P., Holah, J., Wirtanen, G. and Allison, D.G. (2000) A new technique for the performance evaluation of clean-in-place disinfection of biofilms. *J. Ind. Microbiol.* **25**, 235–241.

Wimpenny, J.W.T. (2000) An overview of biofilms as functional communities. In *Community Structure and Co-operation in Biofilms,* (ed. D.G. Allison, P. Gilbert, H. Lappin-Scott and M. Wilson), Vol. 59, pp. 1–24, Cambridge University Press, Cambridge, UK.

Wimpenny, J.W.T. and Colsanti, R. (1997) A unifying hypothesis for the structure of microbial biofilms. *FEMS Microbiol. Ecol.* **22**, 1–16.

Wingender, J., Neu, T.R. and Flemming, H.-C. (1999a) What are bacterial extracellular polymeric substances? In *Microbial Extracellular Polymeric Substances* (ed. J. Wingender, T.R. Neu and H.-C. Flemming), pp. 1–19, Springer, Berlin, Germany.

Wingender, J., Jaeger, K.-E. and Flemming, H.-C. (1999b) Interaction between extracellular polysaccharides and enzymes. In *Microbial Extracellular Polymeric Substances* (ed. J. Wingender, T.R. Neu and H.-C. Flemming), pp. 231–251, Springer, Berlin, Germany.

Zobell, C.E. and Allen, E.C. (1935) The significance of marine bacteria in the fouling of submerged surfaces. *J. Bacteriol.* **29**, 230–251.

7

Physico-chemical properties of extracellular polymeric substances

A.P. Moran and Å. Ljungh

7.1 INTRODUCTION

7.1.1 Aggregation in biofilm formation

The majority of microorganisms occur in aggregates, of differing form and shape, in a variety of environments ranging from microcolonies adhering to soil particles, sediments, plant or animal surfaces and niches, synthetic materials, flocs and biological sludge. These phenomena have been grouped under the general term 'biofilms', which represents communities of microorganisms attached to a surface (Watnick and Kolter, 2000). Thus, biofilms develop adherent to a substratum not only at solid–water interfaces, but also at water–air and solid–air interfaces (Flemming *et al.*, 2000). Although these microbial aggregates differ profoundly, a common feature is that the microorganisms are embedded in an extracellular matrix of polymeric substances (Flemming and Wingender, 2001a,b). This matrix determines the morphology, structure, coherence, physico-chemical properties and activity of the microbial aggregates, and thus, plays an important role in biofilm development and maturation. Importantly, the biofilm matrix creates a microenvironment for sessile cells, which are influenced and conditioned by the nature and architecture of the matrix (Flemming and Wingender, 2001c). In addition to a complex association of microbial cells, extracellular products and detritus, that are secreted or released from cells, which have lysed as the biofilm ages, are trapped within the biofilm matrix (Sutherland, 2001a). Thus, polymeric substances in the biofilm matrix form a cement in which cells are immobilised and microbial products are contained.

7.1.2 Definition and abbreviations

Extracellular polymeric substances are biosynthetic polymers (biopolymers) that are highly diverse in chemical composition (Davies, 2000) and various attempts have been made to define this collective term (Flemming *et al.*, 2000; Flemming and Wingender, 2001c). Alternate definitions have been:

(1) substances of biological origin that participate in the formation of microbial aggregates (Geesey, 1982);
(2) organic polymers of microbial origin, which in biofilm systems are frequently responsible for immobilising cells and other particulate materials together (cohesion) and to the substratum (adhesion) (Characklis and Wilderer, 1989);
(3) material which can be removed from microorganisms (and in particular, bacteria) without disrupting the cell, and without which the microorganism is viable (Gehr and Henry, 1983).

Thus, a comprehensive definition would be extracellular biological polymers or oligomers and organic material of microbial origin that are not essential for microbial growth and viability, but that participate and are frequently responsible for the formation of microbial aggregates, consortia and biofilms; particularly, for microbial cell immobilisation, molecular cohesion and adhesion.

As discussed in Chapter 6, the abbreviation EPS has been used interchangeably in biofilm research to designate extracellular polymeric substances, but also exopolysaccharides and exopolymers. This has led to some confusion in the literature. Notwithstanding its usage in biofilm research, EPS is a commonly used abbreviation for exopolysaccharides in the fields of carbohydrate research, microbial glycobiology and microbial pathogenesis, and to avoid confusion for researchers in these fields and for the larger scientific community, we suggest that EPS should be retained for designation of exopolysaccharides alone. Nevertheless, for biofilm researchers, we propose to distinguish between extracellular polymeric substances and EPS by usage of the abbreviation xPS for the former. This has the advantages of allowing discrimination between these two categories of substances within biofilm research, but also comprehension of terminology for researchers in other disciplines.

7.1.3 General relevance of xPS in biofilm formation

The ability to form xPS is a widespread property among prokaryotes, i.e. bacteria and archaea, but can occur also in eukaryotic microorganisms (fungi and algae) (Kuhl *et al.*, 1996; Sutherland, 1996; Semple *et al.*, 1999; Flemming *et al.*, 2000).

The proportion of xPS in biofilms can vary between 50% and 90% of the total organic matter of biofilms (Christensen and Characklis, 1990; Nielsen *et al.*, 1997). Some of the potential roles and functions of xPS are listed in Table 7.1.

The presence of the xPS helps maintain the biofilm, allows large numbers of identical or different microbial species to co-exist within it; particularly, under flowing conditions, often in complexes that are a few tens of microns in thickness (Davies, 2000). In addition to aiding the cementing of microbial cells to one another and to the substratum, xPS molecules and the biofilm matrix are important in creating biofilm architecture, which can protect the resident microorganisms from the grazing activity of other organisms and from harmful chemicals (Davies, 2000; Flemming and Wingender, 2001a).

xPS forms a three-dimensional, gel-like, highly hydrated and locally charged matrix (Flemming *et al.*, 2000). Due to these properties, the matrix can act as an energy reserve, can increase adsorption of nutrients and metal ions (Sutherland, 1972; Flemming and Wingender, 2001b), protect against dehydration and drying (Sutherland, 1996), and act as a photon trap (Kuhl *et al.*, 1996; Flemming and Wingender, 2001a). The xPS can act as a sorption site for pollutants and heavy metal ions (Liu *et al.*, 2001; Wuertz *et al.*, 2001), which can be a burden in wastewater sludge but, on the other hand, can be exploited in water purification (Flemming and Wingender, 2001b). Moreover, the xPS molecules contribute to protection from phagocytosis (Takeoka *et al.*, 1998) and bacteriophage attacks (Sutherland, 2001a). Deposition of xPS plays an important role in the

Table 7.1 Potential functions of xPs.

Role	Function
Aggregation	Immobilisation of cells
	Concentration of cell density
Structural	Mechanical stability
	Microconsortia development
	Concentration gradients
	Retention of extracellular enzymes
	Interactions with extracellular enzymes
	Retention of lysed cell components and detritus
	Mass transport through channels
	Horizontal gene transfer
	Matrix for exchange of signalling molecules
	Light transmission
Protective against	Biocides, toxins and heavy metals
	Phagocytosis
	Enzymatic degradation
	Predation and grazing
	Dehydration and desiccation
Sorption	Nutrient accumulation
	Water retention
	Accumulation of pollutants
	Metal ions and biocides

development of dental plaque and caries (Shiroza and Kuramitsu, 1988; O'Toole *et al.*, 2000; Watnick and Kolter, 2000) and infection of medically implanted devices (Bayston, 2000).

Although xPS and the biofilm matrix do not produce a significant diffusion barrier to antimicrobial substances and biocides, xPS can restrict access to the depths of the biofilm by acting as a substrate for chemically reactive agents or through non-specific binding of highly charged antimicrobial compounds (Allison *et al.*, 2000). Biocorrosion of metals is enhanced by xPS of environmental microorganisms (Flemming and Wingender, 2001b; Chan *et al.*, 2002) and the complexing properties of xPS contribute to the dissolution of minerals in bioweathering (Flemming and Wingender, 2001b). In addition, the biofilm matrix produced by the xPS molecules not only traps ions and nutrients sequestered from the environment, but also extracellular enzymes (e.g. polysaccharases, polysaccharide lyases, proteases, β-lactamases, etc.) (Allison *et al.*, 2000; Sutherland, 2001a). However, since the biofilm matrix is not produced until after initial adhesion to a solid substratum, the xPS and matrix do not contribute to the initial adhesion events (O'Toole *et al.*, 2000). Nevertheless, xPS are considered key components in biofilm development, growth and maturation, as well as being important for the physiological functions and physico-chemical properties of biofilms (Davies, 2000; Flemming *et al.*, 2000; O'Toole *et al.*, 2000; Flemming and Wingender, 2001c). Hence, greater knowledge of EPS is of importance for understanding the occurrence, relevance and dynamics of biofilm formation.

7.2   CONSTITUENTS OF xPS

Typically, xPS consist of varying proportions of substituted and unsubstituted EPS, substituted and unsubstituted proteins, nucleic acids, phospholipids and humic acids (Wingender *et al.*, 1999; Flemming *et al.*, 2000) (Table 7.2).

Polysaccharides were considered to be the most abundant components of xPS in early biofilm research (Costerton and Irvin, 1981) and, hence, isolation and purification procedures often focused on these molecules alone. Nevertheless, more extensive analysis has shown that proteins, nucleic acids and lipids can occur in significant amounts and even predominate as constituents of xPS (Neu, 1996; Gehrke *et al.*, 1998; Nielsen and Jann, 1999; Martin-Cereceda *et al.*, 2001; Liu and Fang, 2002). Proteins have been shown to be abundant in the xPS from pure-culture biofilms and flocs of Gram-positive and Gram-negative bacteria (Wingender and Flemming, 1999). In particular, in wastewater biofilms and activated sludge, proteins have been reported to predominate over EPS (Rudd *et al.*, 1983; Nielsen *et al.*, 1997; Bura *et al.*, 1998; Dignac *et al.*, 1998; Jorand *et al.*, 1998; Martin-Cereceda *et al.*, 2001). However, this is dependent on the nature of the sludge; protein predominates in methanogenic sludges, whereas EPS predominates in acidogenic sludges (Liu and Fang, 2002). Likewise, DNA has been found regularly in the xPS of wastewater biofilms and flocs (Nielsen *et al.*, 1997;

Table 7.2   General composition of some xPs.

Macromolecule	Principal components	Main-linkage type	Polymer structure
Polysaccharides	Monosaccharides Uronic acids Amino sugars	Glycosidic bonds	Linear, branched, side chains
Proteins (polypeptides)	Amino acids (hydrophilic and hydrophobic)	Peptide bonds	Linear
Nucleic acids	Nucleotides	Phosphodiester bonds	Linear
Lipids (phospholipids)	Fatty acids Glycerol Phosphate Alcoholic compounds Isoprene	Ester bonds, ether bonds	Side chains
Humic substances	Phenolic compounds Simple sugars Amino acids	Ether bonds, C–C bonds, peptide bonds	Cross-linked

Martin-Cereceda *et al.*, 2001; Liu and Fang, 2002) as well as in microbial xPS from other environments (Platt *et al.*, 1985; Watanabe *et al.*, 1998). In some self-flocculating bacteria, nucleic acids, especially RNA, may even exceed the proportions of EPS and proteins in xPS (Watanabe *et al.*, 1998, 1999). Also, humic acids commonly accumulate in the biofilm matrix of wastewater biofilms and activated sludge (Nielsen *et al.*, 1997; Jahn and Nielsen, 1998; Liu and Fang, 2002), and various phospholipids and lipids as well as lipopolysaccharides (LPS) shed from the Gram-negative bacterial cell wall can occur as components of xPS (Gehrke *et al.*, 1998, 2001). On the other hand, the composition of xPS is influenced by nutrient availability and environmental conditions. These parameters not only influence EPS and protein composition, but also the other constituents of xPS.

7.3   PHYSICAL AND CHEMICAL PROPERTIES OF xPS

Living, hydrated biofilms contain a high proportion of water, even open water channels (DeBeer *et al.*, 1994a), and are composed of about 85% xPS and only 15% cells (by volume) (Donlan and Costerton, 2002). The xPS give the biofilm viscoelastic-like properties, and above the yield point, the gel structure fails and the system behaves like a highly viscous fluid (Korstgens *et al.*, 2001). The production of a gel-like matrix by xPS, with the lowest expense of energy and nutrients, to retain 98–99% of the required water reflects the efficient nature of the structure of the biofilm matrix (Christensen and Charackalis, 1990), which, in turn, reflects the physical and chemical properties of xPS.

7.3.1   EPS

Microbial EPS molecules vary greatly in their chemical composition which, thus, affect their physico-chemical properties. EPS are homopolysaccharides, composed of one type of monosaccharide, or heteropolysaccharides, composed of different sugar moieties (Faber *et al.*, 1998). Some are neutral, but the majority are polyanionic due to the presence of either uronic acids or ketal-linked pyruvate (Sutherland, 2001a). Some EPS contain both hexuronic acids as the anionic component with neutral sugars (Uhlinger and White, 1983; Flemming *et al.*, 2000), whereas others may completely consist of uronic acid residues, such as in the well-studied alginate of *Pseudomonas aeruginosa* (Mayer *et al.*, 1999). A few EPS may be polycationic, where cationic groups may be due to amino sugars (Hejzlar and Chudoba, 1986; Mack *et al.*, 1996; Veiga *et al.*, 1997).

In essence, the EPS are very long molecular chains with a molecular mass of the order of $0.5–2.0 \times 10^6$ Da (Sutherland, 2001a). The composition and structure of the EPS determine their primary conformation. Ordered secondary configuration frequently takes the form of aggregated helices. In some EPS polymers, a backbone sequence of $\beta(1,3)$- or $\beta(1,4)$-linkages confer considerable rigidity, whereas other linkages, as in a $\beta(1,2)$- or $\beta(1,6)$-linked EPS, may yield more flexible structures (Sutherland, 2001a). The transition in solution from random coil to ordered helical aggregates can be greatly influenced by the presence or absence of substituents on the EPS (Sutherland, 1997). Electrostatic and hydrogen bonds have been suggested to be the dominant forces between adjacent EPS molecules (Mayer *et al.*, 1999), but in the biofilm matrix interaction occurs with other classes of macromolecules, as well as involving a variety of forces (Sutherland, 2001a).

Many of the EPS are relatively soluble, and due to their high molecular mass, produce highly viscous aqueous solutions (Sutherland, 2001a,b). Some form weak gels, which may in excess solvent slough off from the exposed biofilm surface, but this is dependent on the prevalence of ions and intermolecular electrostatic forces. Nevertheless, a small number of EPS can be hydrophobic because of their composition and tertiary structure (Neu *et al.*, 1992).

EPS are often modified by substitution with acetyl, succinyl or pyruvyl groups or inorganic sulphate or phosphate groups (Sutherland, 1994, 2001a), thereby affecting the structure and physico-chemical properties including water solubility and macromolecular conformation (Chandrasekaran and Thailambal, 1990; Sutherland, 2001a). In alginates, including those of *P. aeruginosa* (Skjåk-Braek *et al.*, 1986), acetyl groups decrease the capacity and selectivity of divalent cation binding, increase solution viscosity, enhance the water retention capacity of the EPS, protect the EPS from alginate lyase degradation, and affect biofilm formation (Lange *et al.*, 1989; Skjåk-Braek *et al.*, 1989; Geddie and Sutherland, 1994; Sutherland, 1997; Flemming *et al.*, 2000). Chemical removal of the acetyl groups significantly alters EPS physical properties with resultant increases in binding of divalent ions, but reduction of solubility (Sutherland, 2001a).

Debate is ongoing concerning whether xPS molecules of the biofilm matrix, particularly EPS, differ from the extracellular polymers produced by the same microorganisms when grown planktonically (Allison *et al.*, 2000; Sutherland, 2001a,b). What is apparent is that there may be increased production of EPS as part of a stress response as has been observed with colanic acid synthesis by *Escherichia coli* (Prigent-Combaret *et al.*, 1999; Baca-DeLancey *et al.*, 1999). As suggested by Sutherland, (2001a), microorganisms that are capable of synthesising more than one type of EPS in biofilms may produce greater amounts of one form of EPS, i.e. found in lower amounts in planktonic cultures (i.e. differential synthesis). This would account for the differences in composition of planktonic and biofilm EPS that have been reported, rather than a change in the type of EPS synthesised. It is important to note that the extent and nature of polymer production are dependent on the physiological state of the microbial cells in the biofilm consortia and nutrient availability (e.g. carbon and nitrogen levels) (Sutherland, 1985). Thus, bacterial growth phase influences the EPS composition of *Pseudomonas atlantica* (Uhlinger and White, 1983), and the $C : N : P$ ratio influences the EPS composition of wastewater microbial flocs (Bura *et al.*, 1998) and in methanogenic granules (Veiga *et al.*, 1997). Therefore, it has been deduced that excess of available carbon substrate and limitations in other nutrients (nitrogen, potassium or phosphate) promotes synthesis of EPS (Sutherland, 2001a). Iron regulation may also play a role, since Deighton and Borland (1993) showed that *Staphylococcus epidermidis* increased EPS in iron-limited medium and late in the growth phase, when nutrients were potentially exhausted. Increased EPS synthesis is part of a stress response accompanied by a change in gene expression as observed in alginate synthesis in *P. aeruginosa* (Davies and Geesey, 1995), colanic acid production in *E. coli* (Prigent-Combaret *et al.*, 1999), and secretion of a galactoglucan EPS in *Vibrio cholerae* El Tor (Watnick and Kolter, 1999). On the other hand, another issue of debate is whether the molecules of xPS that bind microbial cells in the mature biofilm differ from those produced upon primary colonisation of the substratum (Allison *et al.*, 2000).

7.3.2 Proteins

Many extracellular proteins in biofilms contribute to the anionic properties of xPS, due to their possession of negative charge, because of the presence of dibasic amino acids (Higgins and Novak, 1997; Dignac *et al.*, 1998). These charges can contribute to electrostatic interactions between adjacent molecules (see below). The hydrophobic properties of xPS are enhanced by the presence of proteins with high proportions of hydrophobic amino acids. Moreover, it has been considered that the occurrence of both acidic and hydrophobic amino acids in extracellular proteins explains the electrostatic and hydrophobic interactions observed in flocs of activated sludge (Flemming *et al.*, 2000). However, it should be noted that proteins can be substituted with fatty acids to yield lipoproteins or glycosylated to produce glycoproteins (Horan and Eccles, 1986), which further influence the properties of the xPS.

7.3.3   Other molecules

Other polyanionic compounds include nucleic acids, which are anionic due to the presence of phosphate groups in the nucleotide moiety of RNA and DNA. Low-molecular-mass compounds with cationic and anionic properties (e.g. uronic acids, amino acids and nucleotides) also can occur in xPS and contribute to the electrostatic interactions in the biofilm matrix (Flemming *et al.*, 2000).

7.4   COHESIVENESS OF xPS POLYMERS

Ecologically, the xPS molecules have an important function in creating a gel-like matrix, in which the microorganisms are fixed over a time period, in communication with one another, which represents an important requirement for the formation of stable consortia (Allison *et al.*, 2000). xPS is responsible for the mechanical stability of biofilm aggregates due to intermolecular interactions between different macromolecules (Davies, 2000). For example in activated sludge, it is the surface properties and charge, hydrophobicity, and composition of xPS, rather than its quantity, that affects flocculation (Liao *et al.*, 2001). The cohesive forces between the molecules of the EPS are not covalent, but are based on weak interactions, and are reviewed extensively elsewhere (Flemming, 1999; Mayer *et al.*, 1999). The interactions include London forces, electrostatic interactions and hydrogen bonding (Table 7.3). All three types of interactions can contribute to the overall stability of biofilm matrices, to variable extents, depending on the relative hydrophilic and hydrophobic properties of EPS. Importantly, these forces result in the formation of a three-dimensional, gel-like matrix, which can adjust to changes in environmental conditions in a dynamic manner because of changes in composition, structure and properties of xPS (Mayer *et al.*, 1999). Moreover, in addition to cohesion by attractive forces, repulsive forces between xPS molecules can help to maintain an inflated, sponge-like structure of the biofilm matrix.

London forces can be weakened by surface-active substances, hence the reason for partial efficacy of surfactants on biofilms. Nevertheless, this effect is very weak and biofilms cannot be dispersed by surfactants because of the other binding forces in xPS cohesion. London forces have a binding energy of about 2.5 $kJ\,mol^{-1}$,

Table 7.3   Intermolecular interactions between different macromolecules of xPs.

Cohesive interaction	Interaction involves	Binding energy ($kJ\,mol^{-1}$)
London forces	Functional groups	2.5
Electrostatic	Ions and induced dipoles	12–29
Hydrogen bonds	Hydroxyl groups	10–30

occur mainly in hydrophobic areas, but are not localised to functional groups (Flemming *et al.*, 2000).

Electrostatic interactions are responsible for a high proportion of the overall binding energy, about $12–29\,kJ\,mol^{-1}$ for non-ionic bonds. The binding is strongly dependent on distance between the bond partners, being stronger over shorter distances, but may also act repulsively. Electrostatic interactions occur between ions and between permanent and induced dipoles. The ionic interactions involve mainly divalent cations, particularly Ca^{2+}, but various cations may compete for the same binding site (Loaëc *et al.*, 1997). For EPS, the ionic radii may be important in determining the extent of interaction between chains and the ultimate extent of aggregation of helices (Sutherland, 2001a). The interactions of xPS molecules can be influenced by ionic strength, complexing agents, and pH of the environment. For example, Mayer *et al.* (1999) studied the effect of various cations and anions on the viscosity of xPS of *P. aeruginosa*. They found that monovalent ions led to a strong decrease of viscosity indicating that electrostatic interactions are of importance in cohesiveness of the xPS. However, addition of salts led to a change in molecular conformation of xPS from stretched to coiled macromolecules, which were dispersed and coiled because of intramolecular interactions; but with increasing ionic strength, the coils aggregated and the viscosity rose. Furthermore, growth of a *P. aeruginosa* biofilm in the presence of Ca^{2+} leads to an increase in biofilm thickness and elasticity (Flemming *et al.*, 2000; Korstgens *et al.*, 2001)

Hydrogen bonds possess binding energies of about $10–30\,kJ\,mol^{-1}$. They are active between hydroxyl groups; particularly, are frequent between EPS and water molecules, and support the secondary and tertiary structure of proteins. These interactions are influenced by chaotropic agents (e.g. urea), which disturb the water-based interactions within the biofilm and between xPS molecules (Flemming, 1999; Flemming *et al.*, 2000).

Although the individual binding force of any of the interactions above is smaller by at least an order of magnitude than that of covalent bonds, accumulative binding energies of these interactions can be equivalent to those of covalent bonds (Flemming *et al.*, 2000; Sutherland, 2001a). As the composition and nature of xPS varies with the microbial species, nutrient availability and other environmental and physical parameters, thus under differing growth and environmental conditions, different interactive forces may dominate (Flemming *et al.*, 2000). Hence, when examining the role of the cohesive forces between xPS molecules in biofilms, consideration should be made of the conditions under which the biofilms are formed.

In addition, entanglements of the polymers may occur when their molecular masses exceed the critical molar mass for entanglement ($>2 \times 10^5\,mol^{-1}$) with resultant increases in viscosity (Flemming *et al.*, 2000). However, in biofilm matrices, in which the concentration of macromolecules is below a critical level ($<2\%$), entanglements are considered of lesser importance for the cohesiveness of the biofilm. In particular, entanglements act as a key factor in stable floc structure in sewage sludge (Mikkelsen and Keiding, 2002).

7.5   ROLE OF xPS IN BIOFILM FUNCTIONING

7.5.1   Chemical gradients and diffusion barriers

The nature of the molecules of the xPS determines the physical properties of the biofilm matrix, i.e. charge, hydrophobicity, elasticity, dissolution and deformation characteristics (Sutherland, 1985, 2001a). Thus, xPS, that is produced in a biofilm composed of a number of different microbial species may produce a gel-like matrix with unique characteristics compared with those of a mono-species biofilm. The diffusion properties and presence of charged reactive groups can be viewed as contributing to the establishment and maintenance of steady state conditions in the biofilm (Boult et al., 1997). Although diffusion of charged molecules across the biofilm matrix may be influenced by charges in the xPS, diffusion of uncharged solutes across the matrix is not too dissimilar to that of water (Stewart et al., 1998). Likewise, the local charge of xPS influences, profoundly, the access of ions, including protons, to the cellular membranes. Restricted diffusion of ions and charged molecules across the biofilm matrix occurs due to a combination of ionic interactions and molecular sieving events (Allison et al., 2000). Thus, xPS have been suggested to act as a protective shield and to prevent the access of antimicrobial compounds and biocides to the underlying microbial cells (Slack and Nichols, 1982; Costerton et al., 1987; Suci et al., 1994). However, ex vivo and in vivo experiments (Gristina et al., 1987; Gordon et al., 1988; Nichols et al., 1988, 1989) have not supported this hypothesis. Nevertheless, reduced susceptibilities to antibiotics have been observed in bacteria with a mucoid phenotype, e.g. P. aeruginosa (Evans et al., 1991), and therefore, although the mucoid phenotype may be associated with lesser sensitivities, reductions in diffusion coefficients alone are insufficient to explain it (Allison et al., 2000). As the biofilm matures, xPS deposition increases with the result that the magnitude of the nutrient and gaseous gradients become increased, and the net growth rate of microorganisms decreases (Allison et al., 2000). This can lead to the onset of dormancy in the more deeply lying cells, triggering the expression of stress and stringent response genes. This response also could influence the susceptibilities of biofilm-associated microorganisms to antimicrobial agents (Nickel et al., 1985). Nevertheless, biofilm resistance to toxic substances is likely to be multifactorial (Watnick and Kolter, 2000).

7.5.2   Physiological gradient properties

The close proximity of cells, along with xPS, establishes concentration gradients of gases (e.g. oxygen), ions and available nutrients within the biofilm community (Costerton et al., 1987). Thus, within the biofilm the more deeply lying microbial cells will be exposed to altered concentrations of nutrients, ions, pH and oxidation–reduction potential compared to those at the periphery, and growth rates may be reduced due to nutrient or other deficiencies (Wentland et al., 1996; Xu et al., 1998; Okabe et al., 1999; Allison et al., 2000). In contrast, at the periphery, rapid utilisation of oxygen and nutrients on the biofilm surface and interfacial region will result in depletion of availability of these in the underlying region, where anaerobic and anoxic zones occur (Marshall, 1992; Flemming and Wingender, 2001a). Overall, at any given time, a variety of microbial phenotypes will be present within the biofilm matrix. In addition, as the extent of nutrient and gaseous gradients increase as biofilms thicken and mature, reflecting changes in the xPS and biofilm matrix, the physiological effects of growth rate become greater as the biofilm ages, e.g. lesser susceptibility to biocides and antibiotics (Allison *et al.*, 2000).

In photosynthetic biofilm systems, xPS can play an important role in light transmission and may even function as photon traps (Flemming and Wingender, 2001c) and light transmitters (Flemming and Wingender, 2001a). At the biofilm surface, irradiance and spectral composition can be different to those of incident light and at the bottom of the biofilm eutrophic zone, light can be attenuated to < 5–10% of the surface irradiance (Kuhl *et al.*, 1996).

7.5.3   Reaction-sink properties

Functioning as a reaction sink, molecules of xPS may act as substrates for reactive biocides, thereby neutralising the treatment agent and, hence, reducing its availability and diffusion into the biofilm (Huang *et al.*, 1995; Allison *et al.*, 2000). Also, antibiotic-inactivating enzymes, e.g. β-lactamase, are up-regulated in biofilms (Giwercman *et al.*, 1991) and become bound to xPS. Similarly, enzymes, such as formaldehyde lyase and dehydrogenase, catalase or superoxide dismutase, which can be deposited in the biofilm matrix can aid resistance to reactive biocides (Sondossi *et al.*, 1985). Nevertheless, these barrier properties of the xPS are dependent upon the distribution of biomass and its turnover, the binding capacity of the biofilm matrix, the nature of the chemical agent and its concentration, and local hydrodynamic effects (Kumon *et al.*, 1994; DeBeer *et al.*,1994b; Allison *et al.*, 2000).

The molecules of the xPS also entrap extracellular enzymes (Sutherland, 2001a) and, thereby, prevent their washout from the biofilm and, moreover, maintain close proximity between these enzymes and microbial cells. The interaction between such enzymes and EPS can increase their heat and pH stability, as was observed between a *P. aeruginosa* lipase and alginate (Flemming and Wingender, 2001c). Although the molecules of xPS can entrap polysaccharases and polysaccharide lyases, it has been suggested that EPS are not degraded by the microorganisms that produce them (Sutherland, 2001b). On the other hand, such enzymes derived from bacteriophages are capable of degradation of biofilm EPS (Hughes *et al.*, 1998a), and other investigators have found evidence that EPS may be used as a carbon source by the producer organism (Hoffman and Decho, 1999). Furthermore, it has been hypothesised that these enzymes could play an important role in the release of microbial cells from the mature biofilm (O'Toole *et al.*, 2000) and that the biofilm matrix acts as a recycling yard for enzymes and genetic material (Flemming and Wingender, 2001a). In particular, Boyd and Chakrabarty

(1994) showed that over-expression of alginate lyase speeded detachment and cell sloughing from *P. aeruginosa* biofilms. Also, surface attachment of *Enterobacter agglomerans* and interaction in its biofilm with *Klebsiella pneumoniae* were decreased after enzymatic treatment of its EPS (Hughes *et al.*, 1998b; Skillman *et al.*, 1999). This function may not only be limited to polysaccharide-degrading enzymes, as a protease of *Streptococcus mutants* by cleaving proteins in its xPS can release the bacterium from dental plaques (Lee *et al.*, 1996). Since release of microbial cells from biofilms for attachment to uncolonised surfaces is central to the establishment of daughter biofilms, and may facilitate the search for nutrients (O'Toole *et al.*, 2000), the potential involvement of extracellular enzymes in this process is worthy of further investigation.

7.6   ROLE OF xPS IN AGGREGATION

Regardless of the nature of xPS, biosynthesis of the constituents is generally up-regulated within minutes of the irreversible attachment of the microbial cells to the substratum (Davies *et al.*, 1993). The substratum can vary in its physico-chemical properties and, as in the case of implanted medical devices, the surface can become conditioned with human plasma and human extracellular matrix proteins that change the surface properties but, nevertheless, can be recognised by micro-organisms (Ljungh and Wadström, 1995; Ljungh *et al.*, 1996). In general, initial bacterial adhesion is mediated by proteinaceous structures. For example, flagella are necessary for or facilitate initial attachment of a number of Gram-negative bacteria (O'Toole *et al.*, 2000). Deposition of xPS throughout the microcolony subsequently proceeds, and this is accompanied by changes in susceptibilities to antimicrobial agents and biocides (Das *et al.*, 1998). On the other hand, xPS may mask adhesins, thereby reducing colonisation, as has been reported for some Gram-positive bacteria (Cucarella *et al.*, 2002).

The best characterised of all xPS, alginate production by *P. aeruginosa* (Gacesa, 1998) has been shown to be de-repressed in individual bacterial cells shortly after attachment (Davies and Geesey, 1995). Alginate production is physiologically regulated, and activation of a critical alginate promoter, *algD*, occurs during nitrogen limitation, membrane perturbation induced by ethanol and exposure to high osmolarity (Berry *et al.*, 1989; DeVault *et al.*, 1989, 1990), similarly, the *algC* promoter is activated by environmental signals (e.g. high osmolarity) and is dependent on the response regulator protein AlgR1 (Zielinski *et al.*, 1991). Moreover, *P. aeruginosa* strains that produce large quantities of alginate, so-called mucoid strains, are also non-motile suggesting a link between these two phenotypes (O'Toole *et al.*, 2000). Supporting this conclusion, Garrett *et al.* (1999) showed that expression of a sigma factor (σ^{22}) required for alginate synthesis resulted in down-regulation of a key flagellar biosynthesis gene. Thus, on contacting the surface, flagellar synthesis is down-regulated, but alginate synthesis is up-regulated. Furthermore, *algC*, which encodes phosphomannomutase, is also involved in synthesis of the O-sidechain of *P. aeruginosa* LPS, and when the switch to production of alginate occurs, production of an LPS with greater adhesive properties is produced, further cementing the bacterium to the substratum (Davies, 2000). In some Gram-negative bacteria, up-regulation of xPS synthesis is partly controlled by the signal substances *N*-acyl homoserine lactones (AHL) involved in quorum sensing (Davies *et al.*, 1998). These act as global regulators of transcriptional activation and are considered potential regulators of biofilm-specific physiology. In Gram-positive bacteria, similar sensing mechanisms have been reported, but these are based on hydrophobic cyclic peptides (Kleerebezem *et al.*, 1997). However, the complexity of regulation by quorum sensing is apparent, since, in addition to alginate synthesis in *P. aeruginosa*, *algC* is necessary for the production of rhamnolipid (Olvera *et al.*, 1999), whose transcription is dependent on activation by *rhl* (Passador *et al.*, 1993), which is indicative of quorum-sensing control (Davies, 2000). Moreover, rhamnolipid renders the cell surface more hydrophobic, releasing LPS and could induce xPS transfer from one bacterial species to another (Sutherland, 2001a). Increased EPS production as part of the stress response and an accompanying major change in gene expression has been observed in other Gram-negative bacteria as well, including *E. coli* and *V. cholerae* El Tor (Prigent-Combaret *et al.*, 1999; Watnick and Kolter, 1999). For example in *E. coli*, up-regulated genes include those encoding the OmpC porin and the *wca* locus (required for colanic acid synthesis), whereas the *fliC* gene (required for flagellar biosynthesis) is down-regulated (Prigent-Combaret *et al.*, 1999; O'Toole *et al.*, 2000).

In addition to EPS, LPS has been implicated in biofilm development, since *E. coli* biofilm-defective mutants possessed mutations in LPS biosynthetic genes and a defective LPS structure (Genevaux *et al.*, 1999). Makin and Beveridge (1996) showed that loss of the B-band of *P. aeruginosa* LPS resulted in increased attachment to hydrophobic surfaces and reduced attachment to hydrophilic ones. Similar findings were reported for *Pseudomonas fluorescens,* where defects in the O-antigen of LPS could result in exposure of the lipid moiety of the membrane, thereby increasing attachment to hydrophobic surfaces (Williams and Fletcher, 1996).

As to Gram-positive bacteria, a number of investigations have examined the xPS of coagulase-negative staphylococci, especially *S. epidermidis*, and a number of molecules have been identified (Tojo *et al.*, 1988; Hussain *et al.*, 1992). Subsequent to microbial cell–substratum interactions, which are mediated by a number of factors, including proteins and a capsular polysaccharide/adhesin (PS/A), the cells enter the accumulative phase of biofilm formation, which involves cell–cell interactions and formation of cell aggregates (McKenney *et al.*, 1998; O'Toole *et al.*, 2000). A polysaccharide intracellular adhesin (PIA) has been implicated in this process (Mack *et al.*, 1994, 1996; McKenney *et al.*, 1998), and consists of a major component, PIA 1, and minor component, PIA 2 (Mack *et al.*, 1996). The genetic control of PIA production by the *ica* operon has been investigated (Heilmann *et al.*, 1996); this operon encodes an *N*-acetylglucosaminyltransferase and a helical porin peptide. Moreover, PIA production and biofilm development occur as part of the genetically controlled stress response of the organism (Mack *et al.*, 2000; Knobloch *et al.*, 2001). However, it is likely that other polymers are components of the staphylococcal xPS as well (Bayston, 2000). Interestingly, the *ica* locus may also code for PS/A (McKenney *et al.*, 1998). Expression of the *ica* locus is phase variable, the mechanism for which appears dependent on insertion elements (Ziebuhr *et al.*, 1999), and thus switching off of *ica* operon expression may be one mechanism by which staphylococci leave the biofilm. However, it is unclear whether these events are regulated in response to changes in cell physiology (O'Toole *et al.*, 2000).

The molecules of xPS are considered key components in models explaining microbial aggregation and the physico-chemical properties of the biofilm matrix (Wingender and Flemming, 1999; Kreft and Wimpenny, 2001). On the other hand, the permeability of microbial aggregates can be reduced by xPS clogging the pores within aggregates and flocs from activated sludge (Li and Yuan, 2002). In activated sludge, xPS is considered important in determining floc structure and charge, the flocculaton process, floc settleability, and dewatering properties (Flemming *et al.*, 2000; Liao *et al.*, 2001; Mikkelsen and Keiding, 2002). Flocculation is associated with xPS production and accumulation, particularly EPS production (Flemming *et al.*, 2000), but also the presence of extracellular nucleic acids released by lysis play a critical role (Watanabe *et al.*, 1998). Interestingly, cellular aggregation is dependent on the physiological state of the microorganisms; flocculation being dependent on microorganisms exhibiting restricted growth. Divalent cations, bound in the xPS, are believed to be important constituents of microbial aggregates and their removal can result in destabilisation of flocs and biofilms (Turakhia *et al.*, 1983; Bruus *et al.*, 1992; Higgins and Novak, 1997; Watanabe *et al.*, 1999). Moreover, a model of flocculation has been proposed whereby lectin-like proteins in the xPS interact with EPS, thus cross-linking to adjacent proteins, and divalent cations form ionic bridges on EPS molecules, with resultant stabilisation of the biopolymers mediating the immobilisation of microbial cells (Flemming *et al.*, 2000).

7.7   CONCLUSION

The importance of xPS for biofilm development and integrity is evident, but their study remains complex. Models aiming at prediction of biofilm structural heterogeneity are still in their infancy. Although inclusion of cell–cell communication and xPS production in the existing modelling frameworks might not prove a problem, knowledge of the inherent mechanisms involved are poorly understood. *In vitro* models mimicking the *in situ* or *in vivo* situation are becoming available, e.g. usage of hydrogels to model and examine the physico-chemical properties of xPS (Strathmann *et al.*, 2001) and perfusion flow cells to study adsorption of human proteins to different materials and adhesion of microorganisms (Lundberg *et al.*, 1997). Other promising developments include the usage of biofilm proteome-based analyses (Oosthuizen *et al.*, 2002), microscopical analyses using fluorescently labelled ligands, including two-photon laser scanning microscopy (Neu *et al.*, 2001, 2002; Strathmann *et al.*, 2002) and genetic methods using derivatives for physiological and temporal distribution studies (Sternberg *et al.*, 1999).

Moreover, there is continuing emergent evidence that biofilms, with their associated xPS, play an important role in the ecological interaction of microbial species not previously considered to be biofilm-based (e.g. Valkonen *et al.*, 1994; Stark *et al.*, 1999). As a deeper understanding of pure-culture biofilms is obtained, it is important to remember that many biofilms are multispecies consortia. Therefore, in attempting to understand the role and importance of xPS in biofilm deposition, development and maturation, greater cognizance should be made of this fact. In so doing, the knowledge gained of the molecular interactions of xPS in the biofilm matrix, be they intra- or intermolecular, promises to shed new light on the adaptation and diverse strategies employed by microorganisms for colonisation, survival and spread.

Acknowledgements

The authors acknowledge the support of grants from the Swedish Science Research Council (ÅL 6x-11229) and the Irish Higher Education Authority (PRTL-1) under the National Development Plan.

REFERENCES

Allison, D.G., McBain, A.J. and Gilbert, P. (2000) Biofilms: problems of control. In *Society for General Microbiology Symposium: Community Structure and Co-operation in Biofilms* (ed. D.G. Allison, P. Gilbert, H.M. Lappin-Scott and M. Wilson), Vol. 59, pp. 309–328, University Press, Cambridge, UK.

Baca-DeLancey, R.R., South, M.M., Ding, X. and Rather, P.N. (1999) *Escherichia coli* genes regulated by cell-to-cell signalling. *Proc. Natl. Acad. Sci. USA* **96**, 4610–4614.

Bayston, (2000) Biofilms and prosthetic devices. In *Society for General Microbiology Symposium: Community Structure and Co-operation in Biofilms* (ed. D.G. Allison, P. Gilbert, H.M. Lappin-Scott and M. Wilson), Vol. 59, pp. 295–308, University Press, Cambridge, UK.

Berry, A., DeVault, J.D. and Chakrabarty, A.M. (1989) High osmolarity is a signal for enhanced *alcD* transcription in mucoid and nonmucoid *Pseudomonas aeruginosa* strains. *J. Bacteriol.* **171**, 2312–2317.

Boult, S., Johnson, N. and Curtis, C. (1997) Recognition of a biofilm at the sediment–water interface of an acid mine drainage-contaminated stream, and its role in controlling iron flux. *Hydrol. Process* **11**, 391–393.

Boyd, A. and Chakrabarty, A.M. (1994) Role of alginate lyase in cell development of *Pseudomonas aeruginosa*. *Appl. Environ. Microbiol.* **60**, 2355–2359.

Bruus, J.H., Nielsen, P.H. and Keiding, K. (1992) On the stability of activated sludge flocs with implications to dewatering. *Water Res.* **26**, 1597–1604.

Bura, R., Cheung, M., Lioa, B., Finlayson, J., Lee, B.C., Droppo, I.G., Leppard, G.G. and Liss, S.N. (1998) Composition of extracellular polymeric substances in the activated sludge floc matrix. *Water Sci. Technol.* **37**, 325–333.

Chan, K.Y., Xu, L.C. and Fang, H.H. (2002) Anaerobic electrochemical corrosion of mild steel in the presence of extracellular polymeric substances produced by a culture enriched in sulfate-reducing bacteria. *Environ. Sci. Technol.* **36**, 1720–1727.

Chandrasekaran, R. and Thailambal, V.G. (1990) The influence of calcium ions, acetate and L-glycerate groups on the gellan double helix. *Carbohydr. Polym.* **12**, 431–442.

Characklis, W.G. and Wilderer, P.A. (1989) *Structure and Function of Biofilms*, Wiley, Chichester, UK.

Christensen, B.E. and Characklis, W.G. (1990) Physical and chemical properties of biofilms. In *Biofilms* (ed. W.G. Characklis and K.C. Marshall), pp. 93–130, Wiley, New York.

Costerton, J.W. and Irvin, R.T. (1981) The bacterial glycocalyx in nature and disease. *Annu. Rev. Microbiol.* **35**, 299–324.

Costerton, J.W., Cheng, K.J., Geesey, G.G., Ladd, T.I., Nickel, J.C., Dasgupta, M. and Marrie, T.J. (1987) Bacterial biofilms in nature and disease. *Annu. Rev. Microbiol.* **41**, 435–464.

Cucarella, C., Tormo, M.A., Knecht, E., Amorena, B., Lasa, I., Foster, T.J. and Penades, J.R. (2002) Expression of the biofilm-associated protein interferes with host protein receptors of *Staphylococcus aureus* and alters the infective process. *Infect. Immun.* **70**, 3180–3186.

Das, J.R., Bhakoo, M., Jones, M.V. and Gilbert, P. (1998) Changes in the biocide susceptibility of *Staphylococcus epidermidis* and *Escherichia coli* cells associated with rapid attachment to plastic surfaces. *J. Appl. Microbiol.* **84**, 852–858.

Davies, D.G. (2000) Physiological events in biofilm formation. In *Society for General Microbiology Symposium: Community Structure and Co-operation in Biofilms* (ed. D.G. Allison, P. Gilbert, H.M. Lappin-Scott and M. Wilson), Vol. 59, pp. 37–52, University Press, Cambridge, UK.

Davies, D.G. and Geesey, G.G. (1995) Regulation of the alginate biosynthesis gene *algC* in *Pseudomonas aeruginosa* during biofilm development in continuous culture. *Appl. Environ. Microbiol.* **61**, 860–867.

Davies, D.G., Chakrabarty, A.M. and Geesey, G.G. (1993) Exopolysaccharide production in biofilms: substratum activation of alginate gene expression by *Pseudomonas aeruginosa*. *Appl. Environ. Microbiol.* **59**, 1181–1186.

Davies, D.G., Parsek, M.R., Pearson, J.P., Iglewski, B.H., Costerton, J.W. and Greenberg, E.P. (1998) The involvement of cell-to-cell signals in the development of a bacterial biofilm. *Science* **280**, 295–298.

DeBeer, D., Stoodley, P. and Lewandowski, Z. (1994a) Liquid flow in heterogenous biofilms. *Biotechnol. Bioeng.* **44**, 636–641.

DeBeer, D., Srinivasan, R. and Stewart, P.S. (1994b) Direct measurement of chlorine penetration into biofilms during disinfection. *Appl. Environ. Microbiol.* **60**, 4339–4344.

DeVault, J.D., Berry, A., Misra, T.K. and Chakrabarty, A.M. (1989) Environmental sensory signals and microbial pathogenesis: *Pseudomonas aeruginosa* infection in cystic fibrosis. *Bio/Technology* **7**, 352–357.

DeVault, J.D., Kimbara, K. and Chakrabarty, A.M. (1990) Pulmonary dehydration and infection in cystic fibrosis: evidence that ethanol activates alginate gene expression and induction of mucoidy in *Pseudomonas aeruginosa*. *Mol. Microbiol.* **4**, 737–745.

Deighton, M. and Borland, R. (1993) Regulation of slime production in *Staphylococcus epidermidis* by iron limitation. *Infect. Immun.* **61**, 4473–4479.

Dignac, M.-F., Urbain, V., Rybacki, D., Bruchet, A., Snidaro, D. and Scribe, P. (1998) Chemical description of extracellular polymers: implication on activated sludge floc structure. *Water Sci. Technol.* **38**, 45–53.

Donlan, R.M. and Costerton, J.W. (2002) Biofilms: survival mechanisms of clinically relevant microorganisms. *Clin. Microbiol. Rev.* **15**, 167–193.

Evans, D.J., Brown, M.R.W., Allison, D.G. and Gilbert, P. (1991) Susceptibility of *Escherichia coli* and *Pseudomonas aeruginosa* biofilms to ciprofloxacin: effect of specific growth rate. *J. Antimicrob. Chemoth.* **27**, 177–184.

Faber, E.J., Zoon, P., Kamerling, J.P. and Vliegenhart, J.F.G. (1998) The exopolysaccharide produced by *Streptococcus thermophilus* Rs and Sts have the same repeating unit but differ in viscosity of their milk cultures. *Carbohydr. Res.* **310**, 269–276.

Flemming, H.-C. (1999) The forces that keep biofilms together. In *Biofilms in Aquatic Systems* (ed. W. Keevil, A.F. Godfree, D.M. Hotl and C.S. Dow), pp. 1–12, Royal Society of Chemistry, Cambridge, UK.

Flemming, H.-C. and Wingender, J. (2001a) Relevance of microbial extracellular polymeric substances (EPSs) – Part I: Structural and ecological aspects. *Water Sci Technol.* **43**, 1–8.

Flemming, H.-C. and Wingender, J. (2001b) Relevance of microbial extracellular polymeric substances (EPSs) – Part II: Technical aspects. *Water Sci Technol.* **43**, 9–16.

Flemming, H.-C. and Wingender, J. (2001c) Structural, ecological and functional aspects of EPS. In *Biofilm Community Interactions: Chance or Necessity* (ed. P. Gilbert, D. Allison, M. Brading, J. Verran and J. Walker), pp. 175–190, Bioline, Cardiff, UK.

Flemming, H.-C., Wingender, J., Mayer, C., Körstgens, V. and Borchard, W. (2000) Cohesiveness in biofilm matrix polymers. In *Society for General Microbiology Symposium: Community Structure and Co-operation in Biofilms* (ed. D.G. Allison, P. Gilbert, H.M. Lappin-Scott and M. Wilson), Vol. 59, pp. 87–106 University Press, Cambridge, UK.

Gacesa, P. (1998) Bacterial alginate biosynthesis – recent progress and future prospects. *Microbiology* **144**, 1133–1143.

Garrett, E.S., Perlegas, D. and Wozniak, D.J. (1999) Negative control of flagellum synthesis in *Pseudomonas aeruginosa* is modulated by the alternative sigma factor AlgT (AlgU). *J. Bacteriol.* **181**, 7401–7404.

Geddie, J.L. and Sutherland, I.W. (1994) The effect of acetylation on cation binding by algal and bacterial alginates. *Biotechnol. Appl. Biochem.* **20**, 117–129.

Geesey, G.G. (1982) Microbial exopolymers: ecological and economic considerations. *ASM News* **48**, 9–14.

Gehr, R. and Henry, J.G. (1983) Removal of extracellular material: techniques and pitfalls. *Water Res.* **17**, 1743–1748.

Gehrke, T., Telegdi, J., Thierry, D. and Sand, W. (1998) Importance of extracellular polymeric substances from *Thiobacillus ferrooxidans* for bioleaching. *Appl. Environ. Microbiol.* **64**, 2743–2747.

Gehrke, T., Hallmann, R., Kinzler, K. and Sand, W. (2001) The EPS of *Acidithiobacillus ferrooxidans* – a model for structure–function relationships of attached bacteria and their physiology. *Water Sci. Technol.* **43**, 159–167.

Genevaux, P., Bauda, P., DuBow, P.S., Oudega, B. (1999) Identification of Tn10 insertions in the rfaG, rfaP, and galU genes involved in lipopolysaccharide core biosynthesis that affect *Escherichia coli* adhesion. *Arch. Microbiol.* **172**, 1–8.

Giwercman, B., Jensen, E.T., Hoiby, N., Kharazmi, A. and Costerton, J.W. (1991) Induction of β-lactamase production in *Pseudomonas aeruginosa* biofilms. *Antimicrob. Agents Chemother.* **35**, 1008–1010.

Gordon, C.A., Hodges, N.A. and Marriot, C. (1988) Antibiotic interaction and diffusion through alginate and exopolysaccharide of cystic fibrosis derived *Pseudomonas aeruginosa*. *J. Antimicrob. Chemoth.* **22**, 667–674.

Gristina, A.G., Hobgood, C.D., Webb, L.X. and Myrvik, Q.N. (1987) Adhesive colonisation of biomaterials and antibiotic resistance. *Biomaterials* **8**, 423–426.

Heilmann, C., Schweitzer, O., Gerke, C., Vanittanokom, N., Mack, D. and Götz, F. (1996) Molecular basis of intracellular adhesion in biofilm-forming *Staphylococcus epidermidis*. *Mol. Microbiol.* **20**, 1083–1091.

Hejzlar, J. and Chudoba, J. (1986) Microbial polymers in the aquatic environment –. I. Production by activated sludge microorganisms under different conditions. *Water Res.* **20**, 1209–1216.

Higgins, M.J. and Novak, T. (1997) Characterization of exocellular protein and its role in bioflocculation. *J. Environ. Eng.* **123**, 479–485.

Hoffman, M. and Decho, A.W. (1999) Extracellular enzymes within microbial biofilms and the role of the extracellular polymer matrix. In *Microbial Extracellular Polymeric Substances* (ed. J. Wingender, T.R. Neu and H.-C. Flemming), pp. 217–230, Springer, Berlin, Germany.

Horan, N. and Eccles, C.R. (1986) Purification and characterization of extracellular polysaccharides from activated sludge. *Water Res.* **20**, 1427–1432.

Huang, C.T., Yu, F.P., McFeters, G.A. and Stewart, P.S. (1995) Nonuniform spatial patterns of respiratory activity within biofilms during disinfection. *Appl. Environ. Microbiol.* **61**, 2252–2256.

Hughes, K.A., Sutherland, I.W., Clark, J. and Jones, M.V. (1998a) Bacteriophage and associated polysaccharide depolymerases – novel tools for study of bacterial biofilms. *J. Appl. Microbiol.* **85**, 583–590.

Hughes, K.A., Sutherland, I.W. and Jones, M.V. (1998b) Biofilm susceptibility to bacteriophage attack: the role of phage-borne polysaccharide depolymerase. *Microbiology* **144**, 3039–3047.

Hussain, M., Hastings, J.G.M. and White, P.J. (1992) Comparison of cell-wall teichoic acid with high-molecular-weight extracellular slime material from *Staphylococcus epidermidis*. *J. Med. Microbiol.* **37**, 368–375.

Jahn, A. and Nielsen, P.-H. (1998) Cell biomass and exopolymer composition in sewer biofilms. *Water Sci. Technol.* **37**, 17–24.

Jorand, F., Boué-Bigne, F., Block, J.C. and Urbain, V. (1998) Hydrophobic/hydrophilic properties of activated sludge exopolymeric substances. *Water Sci. Technol.* **37**, 307–315.

Kleerebezem, M., Quadri, L.E.N., Kuipers, O.P. and deVos, W.M. (1997) Quorum sensing by peptide pheromones and two-component signal-transduction systems in Grampositive bacteria. *Mol. Microbiol.* **24**, 895–904.

Knobloch, J.K.-M., Bartscht, K., Sabottke, A., Rohde, H., Feucht, H.-H. and Mack, D. (2001). Biofilm formation by *Staphylococcus epidermidis* depends on functional RsbU, an activator of the *sigB* operon: differential activation mechanisms due to ethanol and salt stress. *J. Bacteriol.* **183**, 2624–2633.

Korstgens, V., Flemming, H.C., Wingender, J. and Borchard, W. (2001) Uniaxial compression measurement device for investigation of the mechanical stability of biofilms. *J. Microbiol. Meth.* **46**, 9–17.

Kreft, J.-U. and Wimpenny, J.W.T (2001) Modelling biofims with extracellular polymeric substances. In *Biofilm Community Interactions: Chance or Necessity* (ed. P. Gilbert, D. Allison, M. Brading, J. Verran and J. Walker), pp. 191–200, Bioline, Cardiff, UK.

Kuhl, M., Glud, R.N., Plough, H. and Ramsing, N.B. (1996) Microenvironmental control of photosynthesis and photosynthesis-coupled respiration in an epilithic cyanobacterial biofilm. *J. Phycol.* **32**, 799–812.

Kumon, H., Tomochika, K.-I., Matunaga, T., Ogawa, M. and Ohmori, H. (1994) A sandwich cup method for the penetration assay of antimicrobial agents through *Pseudomonas* exopolysaccharides. *Microbiol. Immunol.* **38**, 615–619.

Lange, B., Wingender, J. and Winkler, U.K. (1989) Isolation and characterization of an alginate lyase from *Klebsiella aerogenes*. *Arch. Microbiol.* **152**, 302–308.

Lee, S.F., Li, Y.H. and Bowden, G.H. (1996) Detachment of *Streptococcus mutans* biofilm cells by an endogenous enzymatic activity. *Infect. Immun.* **64**, 1035–1038.

Li, X.Y. and Yuan, Y. (2002) Collision frequencies of microbial aggregates with small particles by differential sedimentation. *Environ. Sci. Technol.* **36**, 387–393.

Liao, B.Q., Allen, D.G., Droppo, I.G., Leppard, G.G. and Liss, S.N. (2001) Surface properties of sludge and their role in bioflocculation and settleability. *Water Res.* **35**, 339–350.

Liu, H. and Fang, H.H. (2002) Extraction of extracellular polymeric substances (EPS) of sludges. *J. Biotechnol.* **95**, 249–256.

Liu, Y., Lam, M.C. and Fang, H.H. (2001) Adsorption of heavy metals by EPS of activated sludge. *Water Sci. Technol.* **43**, 59–66.

Loaëc, M., Olier, R. and Guezennec, J.G. (1997) Uptake of lead, cadmium and zinc by a novel bacterial exopolysaccharide. *Water Res.* **31**, 1171–1179.

Ljungh, Å. and Wadström, T. (1995) Binding of extracellular matrix proteins by microbes. *Methods Enzymol.* **253**, 501–514.

Ljungh, Å., Moran, A.P. and Wadström, T. (1996) Interactions of bacterial adhesins with extracellular matrix and plasma proteins: pathogenic implications and therapeutic possibilities. *FEMS Immunol. Med. Microbiol.* **16**, 117–126.

Lundberg, F., Schliamser, S. and Ljungh, Å. (1997) Vitronectin may mediate staphylococcal adhesion to polymer surfaces in perfusing human cerebrospinal fluid. *J. Med. Microbiol.* **46**, 285–296.

Mack, D., Nedelmann, M., Krorotsch, A., Schwarzkopf, A., Hessemann, J. and Laufs, R. (1994) Characterization of transposon mutants of biofilm-producing *Staphylococcus epidermidis* impaired in the accumulative phase of biofilm production: genetic identification of a hexosamine-containing polysaccharide intracellular adhesin. *Infect. Immun.* **62**, 3244–3253.

Mack, D., Fischer, W., Krorotsch, A., Leopold, K., Hartmann, R., Egge, H. and Laufs, R. (1996) The intercellualr adhesin involved in biofilm accumulation of *Staphylococcus epidemidis* is a linear β-1,6-linked glucosaminoglycan: purification and structural analysis. *J. Bacteriol.* **178**, 175–183.

Mack, D., Rohde, H., Dobinsky, S., Riedewald, J., Nedelmann, M., Knobloch, J.K.M., Elsner, H.-A. and Feucht, H.H. (2000) Identification of three essential regulatory gene loci governing expression of the *Staphylococcus epidermidis* polysaccharide intercellular adhesin and biofilm formation. *Infect. Immun.* **68**, 3799–3807.

Makin, S.A. and Beveridge, T.J. (1996) The influence of A-band and B-band lipopolysaccharide on the surface characteristics and adhesion of *Pseudomonas aeruginosa* to surfaces. *Microbiology* **142**, 299–307.

Marshall, K.C. (1992) Biofilms: an overview of bacterial adhesion, activity and control at surfaces. *ASM News* **58**, 202–207.

Martin-Cereceda, M., Jorand, F., Guinea, A. and Block, J.C. (2001) Characterization of extracellular polymeric substances in rotating biological contractors and activated sludge flocs. *Environ. Technol.* **22**, 951–959.

Mayer, C., Moritz, R., Kirschner, C., Borchard, W., Maibaum, R., Wingender, J. and Flemming, H.-C. (1999) The role of intermolecular interactions: studies on model systems for bacterial biofilms. *Int. J. Biol. Macromol.* **26**, 3–16.

McKenney, D., Hubner, J., Muller, E., Wang, Y., Goldmann, D.A. and Pier, G.B. (1998) The *ica* locus of *Staphylococcus epidermidis* encodes production of the capsular polysaccharide/ adhesin. *Infect. Immun.* **66**, 4711–4720.

Mikkelsen, L.H. and Keiding, K. (2002) Physico-chemical characteristics of full scale sewage sludges with implications to dewatering. *Water Res.* **36**, 2541–2562.

Neu, T. (1996) Significance of bacterial surface-active compounds in interaction of bacteria with interfaces. *Microbiol. Rev.* **60**, 151–166.

Neu, T.R., Dengler, T., Jann, B. and Poralla, K. (1992) Structural studies of an emulsion-stabilizing exopolysaccharide produced by an adhesive, hydrophobic *Rhodococcus* strain. *J. Gen. Microbiol.* **138**, 2531–2537.

Neu, T.R., Swerhone, G.D.W. and Lawrence, J.R. (2001) Assessment of lectin-binding analysis for *in situ* detection of glycoconjugates in biofilm systems. *Microbiology* **147**, 299–313.

Neu, T.R., Kuhlicke, U. and Lawrence, J.R. (2002) Assessment of fluorochromes for two-photon laser scanning microscopy of biofilms. *Appl. Environ. Microbiol.* **68**, 901–909.

Nielsen, P.-H. and Jann, A. (1999) Extraction of EPS. In *Microbial Extracellular Polymeric Substances* (ed. J. Wingender, T.R. Neu and H.-C. Flemming), pp. 49–72 Springer, Berlin, Germany.

Nielsen, P.-H., Jahn, A. and Palmgren, R. (1997) Conceptual model for production and composition of exopolymers in biofilms. *Water Sci. Technol.* **36**, 11–19.

Nichols, W.W., Dorrington, S.M., Slack, M.P.E. and Walmsley, H.L. (1988) Inhibition of tobramycin diffusion by binding to alginate. *Antimicrob. Agents Chemoth.* **32**, 518–523.

Nichols, W.W., Evans, M.J., Slack, M.P.E. and Walmsley, H.L. (1989) The penetration of antibiotics into aggregates of mucoid and non-mucoid *Pseudomonas aeruginosa*. *J. Gen. Microbiol.* **135**, 1291–1303.

Nickel, J., Ruseska, C.K., Wright, J.B. and Costerton, J.W. (1985) Tobramycin resistance of cells of *Pseudomonas aeruginosa* growing as a biofilm on urinary catheter material. *Antimicrob. Agents Chemoth.* **27**, 619–624.

Okabe, S., Satoh, H. and Watanabe, Y. (1999) *In situ* analysis of nitrifying biofilms as determined by *in situ* hybridization and the use of microelectrodes. *Appl. Environ. Microbiol.* **65**, 3182–3191.

Olvera, C., Goldberg, J.B., Sanchez, R. and Soberon-Chavez, G. (1999) The *Pseudomonas aeruginosa algC* gene product participates in rhamnolipid biosynthesis. *FEMS Microbiol. Lett.* **179**, 85–90.

Oosthuizen M.C., Steyn, B., Theron, J., Cosette, P., Lindsay, D., von Holy, A. and Brözel, V.S. (2002) Proteomic analysis reveals differential protein expression by *Bacillus cereus* during biofilm formation. *Appl. Environ. Microbiol.* **68**, 2770–2780.

O'Toole, G., Kaplan, H.B. and Kolter, R. (2000) Biofilm formation as microbial development. *Annu. Rev. Microbiol.* **54,** 49–79.

Passador, L., Cook, J.M., Gambello, M.J., Rust, L. and Iglewski, B.H. (1993) Expression of *Pseudomonas aeruginosa* virulence genes requires cell-to-cell communication. *Science* **260**, 1127–1130.

Platt, R.M., Geesey, G.G., Davis, J.D. and White, D.C. (1985) Isolation and partial chemical analysis of firmly bound exopolysaccharide from adherent cells of a freshwater bacterium. *Can. J. Microbiol.* **31**, 657–680.

Prigent-Combaret, C., Vidal, O., Dorel, C. and Lejeune, P. (1999) Abiotic surface sensing and biofilm-dependent regulation of gene expression in *Escherichia coli*. *J. Bacteriol.* **181**, 5993–6002.

Rudd, T., Sterritt, R.M. and Lester, J.N. (1983) Extraction of extracellular polymers from activated sludge. *Biotechnol. Lett.* **5**, 327–332.

Semple, K.T., Cain, R.B. and Schmidt, S. (1999) Biodegradation of aromatic compounds by microalgae. *FEMS Microbiol. Lett.* **170**, 291–300.

Shiroza, T. and Kuramitsu, H.K. (1988) Sequence analysis of the *Streptococcus mutans* fructosyltransferase gene and flanking regions. *J. Bacteriol.* **175**, 810–816.

Skillman, L.C., Sutherland, I.W. and Jones, M.V. (1999) The role of exopolysaccharides in dual species biofilm development. *J. Appl. Microbiol.* **85**, S13–S18.

Skjåk-Braek, G., Grasdalen, H. and Larsen, B. (1986) Monomer sequence and acetylation pattern in some bacterial alginates. *Carbohydr. Res.* **154**, 239–250.

Skjåk-Braek, G., Zanetti, F. and Paoletti, S. (1989) Effect of acetylation on some solution and gelling properties of alginates. *Carbohydr. Res.* **185**, 131–138.

Slack, M.P.E. and Nichols, W.W. (1982) Antibiotic penetration through bacterial capsules and exopolysaccharides. *J. Antimicrob. Chemoth.* **10**, 368–372.

Sondossi, M., Rossmore, H.W. and Wireman, J.W. (1985) Observation of resistance and cross-resistance to formaldehyde and a formaldehyde condensate biocide in *Pseudomonas aeruginosa*. *Int. Biodeterior.* **21**, 105–106.

Stark, R.M., Gerwig, G.J., Pitman, R.S., Potts, L.F., Williams, N.A., Greenman J., Weinzweig, I.P., Hirst, T.R. and Millar, M.R. (1999) Biofilm formation by *Helicobacter pylori*. *Lett. Appl. Microbiol.* **28**, 121–126.

Sternberg, C., Christensen, B.B., Johansen, T., Toftgaard Nielsen, A., Andersen, J.B., Givskov, M. and Molin, S. (1999) Distribution of bacterial growth activity in flow-chamber biofilms. *Appl. Environ. Microbiol.* **65**, 4108–4117.

Stewart, P.S., Grab, L. and Diemer, J.A. (1998) Analysis of biocide transport limitation in an artificial biofilm system. *J. Appl. Microbiol.* **85**, 495–500.

Strathmann, M., Griebe, T. and Flemming, H.C. (2001) Agarose hydrogels as EPS models. *Water Sci. Technol.* **43**, 169–174.

Strathmann, M., Wingender, J. and Flemming, H.-C. (2002) Application of fluorescently labelled lectins for the visualization and biochemical characterization of polysaccharides in biofilms of *Pseudomonas aeruginosa. J. Microbiol. Meth.* **50**, 237–248.

Suci, P.A., Mittelman, M.W., Yu, F.U. and Geesey, G.G. (1994) Investigation of ciprofloxacin penetration into *Pseudomonas aeruginosa* biofilms. *Antimicrob. Agents Chemoth.* **38**, 2125–2133.

Sutherland, I.W. (1972) Bacterial exopolysaccharides. *Adv. Microb. Physiol.* **8**, 143–212.

Sutherland, I.W. (1985) Biosynthesis and composition of Gram-negative bacterial extracellular and wall polysaccharides. *Annu. Rev. Microbiol.* **39**, 243–270.

Sutherland, I.W. (1994) Structure–function relationships in microbial exopolysaccharides. *Biotech. Adv.* **12**, 393–448.

Sutherland, I.W. (1996) Extracellular polysaccharides. In *Biotechnology* (ed. H.-J. Rehm and G. Reed), Vol. 6., pp. 615–657, Verlag Chemie, Weinheim, Germany.

Sutherland, I.W. (1997) Microbial exopolysaccharides – structural subtleties and their consequences. *Pure Appl. Chem.* **69**, 1911–1917.

Sutherland, I.W. (2001a) Bacterial exopolysacharides: a strong and sticky framework. *Microbiology* **147**, 3–9.

Sutherland, I.W. (2001b) The biofilm matrix – an immobilized but dynamic microbial environment. *Trends Microbiol.* **9**, 222–2227.

Takeoka, K., Ichimiya, T., Yamasaki, T. and Nasu, M. (1998) The *in vitro* effect of macrolides on the interaction of human polymorphonuclear leukocytes with *Pseudomonas aeruginosa* in biofilm. *Chemotherapy* **44**, 190–197.

Tojo, M., Yamashita, N., Goldmann, D.A. and Pier, G.B. (1988) Isolation and characterization of a polysaccharide adhesin from *Staphylococcus epidermidis. J. Infect. Dis.* **157**, 713–722.

Turakhia, M.H., Cooksey, K.E. and Characklis, W.G. (1983) Influence of a calcium-specific chelant on biofilm removal. *Appl. Environ. Microbiol.* **46**, 1236–1238.

Uhlinger, D.J. and White, D.C. (1983) Relationship between physiological status and formation of extracellular polysaccharide glycocalyx in *Pseudomonas atlantica. Appl. Environ. Microbiol.* **45**, 64–70.

Valkonen, K.H., Wadström, T. and Moran, A.P. (1994) Interaction of lipopolysaccharides of *Helicobacter pylori* with basement membrane protein laminin. *Infect. Immun.* **62**, 3640–3648.

Veiga, M.C., Jain, M.K., Wu, W.-M., Hollingsworth, R. and Zeikus, J.G. (1997) Composition and role of extracellular polymers in methanogenic granules. *Appl. Environ. Microbiol.* **63**, 403–407.

Watanabe, M., Sasaki, K., Nakashimada, Y., Kakizono, T., Noparatnaraporn, N. and Nishio, N. (1998) Growth and flocculation of a marine photosynthetic bacterium *Rhodovulum* sp. *Appl. Microbiol. Biotechnol.* **50**, 682–691.

Watanabe, M., Suzuki, Y., Sasaki, K., Nakashimada, Y. and Nishio, N. (1999) Flocculating property of extracellular polymeric substance derived from a marine photosynthetic bacterium, *Rhodovulum* sp. *J. Biosci. Bioeng.* **87**, 625–629.

Watnick, P.I. and Kolter, R. (1999) Steps in the development of a *Vibrio cholerae* El Tor biofilm. *Mol. Microbiol.* **34**, 586–595 .

Watnick, P. and Kolter, R. (2000) Biofilm, city of microbes. *J. Bacteriol.* **182**, 2675–2679.

Wentland, E.J., Stewart, P.S., Huang, C.T. and McFeters, G.A. (1996) Spatial variations in growth rate within *Klebsiella pneumoniae* colonies and biofilm. *Biotechnol. Prog.* **12**, 316–321.

Williams, V. and Fletcher, M. (1996) *Pseudomonas fluorescens* adhesion and transport through porous media are affected by lipopolysaccharide composition. *Appl. Environ. Microbiol.* **62**, 100–104.

Wingender, J. and Flemming, H.-C. (1999) Autoaggregation in flocs and biofilms. In *Biotechnology* (ed. J. Winter), Vol. 8, pp. 63–86, Verlag Chemie, Weinheim, Germany.

Wingender, J., Neu, T.R. and Flemming, H.-C. (1999) What are bacterial extracellular polymeric substances? In *Microbial Extracellular Polymeric Substances* (ed. J. Wingender, T.R. Neu and H.-C. Flemming), pp. 1–19, Springer, Berlin, Germany.

Wuertz, S., Spaeth, R., Hinderberger, A., Griebe, T., Flemming, H.C. and Wilderer, P.A. (2001) A new method for extraction of extracellular polymeric substances from biofilms and activated sludge suitable for direct quantification of sorbed metals. *Water Sci. Technol.* **43**, 25–31.

Xu, K.D., Stewart, P.S., Xia, F., Huang, C.T. and McFeters, G.A. (1998) Spatial physiological heterogeneity in *Pseudomonas aeruginosa* biofilm is determined by oxygen availability. *Appl. Environ. Microbiol.* **64**, 4035–4039.

Ziebuhr, W., Krimmer, V., Rachid, S., Lobner, I., Götz, F. and Hacker, J. (1999) A novel mechanism of phase variation of virulence in *Staphylococcus epidermidis*: evidence for control of the polysaccharide intracellular adhesion synthesis by alternating insertion and excision of the insertion sequence element IS*256*. *Mol. Microbiol.* **32**, 345–356.

Zielinski, N.A., Chakrabarty, A.M. and Berry, A. (1991) Characterization and regulation of the *Pseudomonas aeruginosa algC* gene encoding phosphomannomutase. *J. Biol. Chem.* **266**, 9754–9763.

Section 3

Biofilm composition: (B) Biological

8	Biofilms on corroding materials	115
9	Biofilms in wastewater treatment systems	132
10	Biofilms and bioaerosols	160
11	Biofilms and protozoa: a ubiquitous health hazard	179

8

Biofilms on corroding materials

I.B. Beech and C.M.L.M. Coutinho

8.1 INTRODUCTION

In both natural and man-made systems, surface-associated microbial growth (i.e.
a biofilm), can influence the physico-chemical interaction between a metallic
material (i.e. material fabricated from pure metals and/or their mixtures (alloys)),
and its environment. The interaction that leads to changes in the material properties is termed corrosion (Scully, 1990). The latter is an electrochemical process
defined as a chemical reaction involving the transfer of electrons through a series
of oxidation (anodic) reactions, which results in metal-dissolution and -reduction
(cathodic) reactions of a chemical species in contact with the metallic surface.
After a period of time, the oxidation reaction usually reaches a low rate because
the oxidation products (corrosion products) adhere to the surface and form a layer
that serves as a diffusion barrier to reactants (Figure 8.1). Such a layer can prevent
further oxidation of the underlying material. Changes in the environmental conditions can affect the stability of the protective layers and, therefore, the overall susceptibility of the material to corrosion.

8.1.1 Effects of biofilms on corrosion

A biofilm can influence corrosion by either accelerating its rate or by inhibiting
the process (Geesey *et al.*, 2000). The structure of biofilm is derived from microbial cells, their extracellular polymeric materials (EPSs), often referred to as

© IWA Publishing. *Biofilms in Medicine, Industry and Environmental Biotechnology*. Edited by Piet Lens,
Anthony P. Moran, Therese Mahony, Paul Stoodley and Vincent O'Flaherty. ISBN: 1 84339 019 1

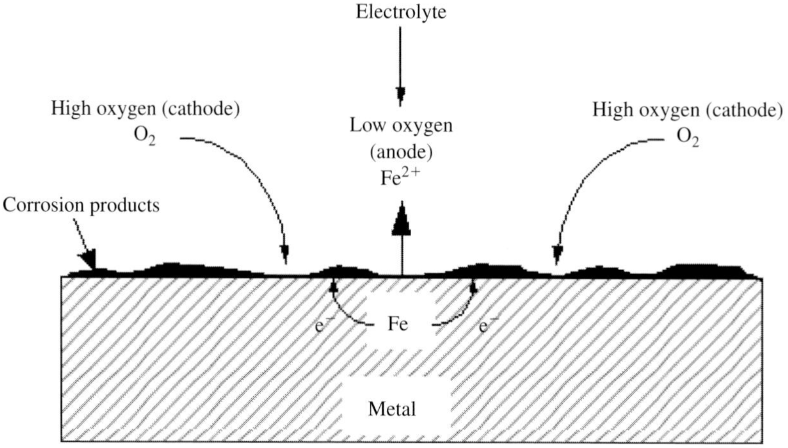

Electrolyte

High oxygen (cathode)   Low oxygen   High oxygen (cathode)
O_2 (anode) O_2
Fe^{2+}

Corrosion products

e^- Fe e^-

Metal

Figure 8.1   Basic corrosion cell under oxygenated conditions.

glycocalyx or slime, and inorganic precipitates originating from the bulk aqueous phase or present as corrosion products of the substratum. Microbial cells and/or their metabolites, e.g. enzymes, exopolysaccharides, organic and inorganic acids, as well as volatile compounds (such as ammonia or hydrogen sulphide), can effect cathodic and/or anodic reactions, thus altering electrochemical processes at the biofilm–metal interface.

The exchange of ions between the metal surface and an aqueous bulk environment can be either promoted or retarded by the presence of a biofilm. Although microorganisms and products of their metabolism, including EPS, are present in greater quantities at the metal surface compared to the bulk phase, water contributes 95% or more to the biofilm matrix. The biofilm can, therefore, act as a modified electrolyte. The accumulation of varied metabolic products, many of which are corrosive, and the high surface to volume ratio of microorganisms can alter physico-chemical conditions at metal–liquid interfaces to such a degree, that corrosion rates can be accelerated by several orders of magnitude, e.g. by factors of 10^3–10^5 (Watkins, 1998).

In contrast, certain types of biofilms may cause a barrier effect that results in a reduction of the chemical activity and, hence, a substantial decrease in the corrosion rate of the metal (Pedersen and Hermansson, 1991). For example, a biofilm formed by a bacterial consortium consisting of a thermophilic *Bacillus* sp. and *Deleya marina* was reported to reduce the rate of corrosion of carbon steel by 94% (Ford *et al.*, 1988). According to the authors, the inhibitory action of these microorganisms was due to secretion of exopolymeric substances. The latter passivated steel surface, thus, inhibiting metal-dissolution process.

Biofilm inhibition of corrosion can be accomplished through any of the following mechanisms, acting individually or in combination:

(i)   reduction in the cathodic rate by a microbial consumption of a cathodic reactant;

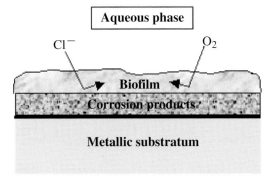

Figure 8.2   Schematic representation of a biofilm acting as a barrier protecting metallic material from corrosive species present in an aqueous phase.

(ii) stabilisation of protective films on the metal surface;

(iii) decrease in the attack of corrosive species present in the aqueous phase (Figure 8.2). These mechanisms are reviewed by Videla and Herrera (in press).

The physical–chemical properties of the metallic substratum determine the type of primary biofilms that develop on its surface. Given the same type of biofilm, different metallic materials are colonised to a varied extent, thus, indicating the specificity in cell–surface interactions (Figure 8.3 a, b).

8.1.2   Microbially influenced corrosion

Deterioration of metal under a biological influence is termed biocorrosion or microbiologically influenced corrosion (MIC) and a number of mechanisms have been identified reflecting the variety of physiological activities carried out by different types of microorganisms (Videla, 1996).

Biocorrosion in any particular system is seldom linked to a unique mechanism or to a single bacterial species. Rather, both the aggressive, as well as inhibitory effects bacterial population exerts on corrosion reactions are typically due to complex biofilm–corrosion products interactions on the material surface.

Microorganisms thriving within biofilms formed on metallic surfaces, such as, e.g. iron, copper and aluminium, and their alloys are physiologically diverse. Bacteria associated with corroding materials have, frequently, been grouped by their metabolic demand for different respiratory substrates or electron acceptors. The main types of organisms found on such materials are sulphate-reducing bacteria (SRB), sulphur-oxidising bacteria, iron-oxidising/reducing bacteria, manganese-oxidising bacteria, as well as bacteria secreting organic acids and slime (see Beech, 1999 for review). These organisms co-exist in naturally occurring biofilms often forming synergistic communities. Such consortia are able to affect electrochemical processes through co-operative metabolism in ways that a single species have difficulty to initiate and/or maintain (Dowling et al., 1991).

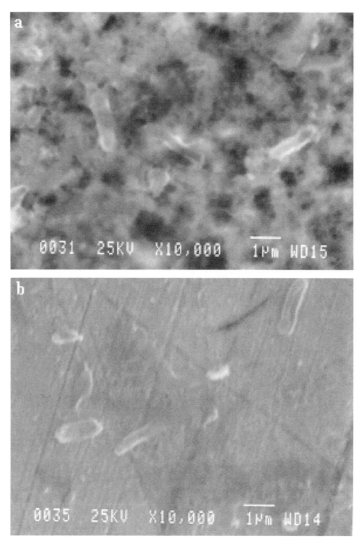

Figure 8.3   An SEM micrograph of a 7-day-old biofilm formed by marine bacteria on surfaces of (a) pure Cu foil and (b) AISI 316 stainless steel, under laboratory conditions. The biofilm was formed by a mixed bacterial population isolated from Cu–Ni alloys that had been exposed in Langstone harbour in the south of England for 5 years. The bacterial resistance to Cu was confirmed by growing cells under a different Cu ions concentration regime (Beech, unpublished).

It has to be noted that demonstrating the presence of microorganisms on a corroded metal surface, even if they are species known to produce metabolic by-products aggressive towards metals, is not a sufficient evidence for their contribution to the corrosion process (Ghassem and Adibi, 1995; Little *et al.*, 1997a). It is also generally accepted that the number of biofilm microorganisms, detected at a corroded site, does not necessarily correlate with the extent of corrosion and that the metabolic status of microbes has to be considered (Little and Wagner, 1997). To date, no clear consensus has been reached regarding the link between specific microbial metabolic rates to observed corrosion rates.

Although microorganisms detected within biofilms on corroded metallic materials include bacteria, fungi and algae, the majority of research efforts have focused on the role that bacteria play in corrosion processes. This chapter will, therefore, address solely the effect bacterial biofilms have on such materials. Furthermore, emphasis will be placed on the contribution of the biofilm matrix to the electron transfer reactions. Current trends in the biocorrosion research focus on biomineralisation processes as key corrosion mechanisms on passive alloys, such as stainless steel. The impact that bacterial metal-reduction processes have on corrosion is reviewed elsewhere (Geesey et al., 2000) and the former mechanisms are, therefore, only briefly mentioned in this chapter.

8.2   MECHANISMS OF MIC

8.2.1   Biofilm matrix

Microbial colonisation of the surface is facilitated by the production of EPS. These constitute macromolecules, such as proteins, polysaccharides, nucleic acids and lipids (Zinkevich et al., 1996). The yield, composition and properties of EPS vary temporally and spatially with bacterial species and environmental conditions. Comprehensive reviews on bacterial exopolymers are given by Allison (1998) and Wingender et al. (1999).

EPS material is associated with cells as capsules or sheaths (Costerton et al., 1992), constitutes an integral part of the biofilm matrix (Cooksey, 1992; Costerton et al., 1994; Gubner and Beech, 2000) and is released into the bulk phase of surrounding liquid as planktonic or 'free' EPS (Sutherland, 1985; Allison, 1998). Whether these three types of EPS materials contain different exopolymers is still a matter of debate, although evidence that their chemical composition is dissimilar has been presented (Beech et al., 1991, 1999a, 2000).

EPS facilitates bacterial attachment to the substratum and form the dynamic biofilm matrix (Wingender et al., 1999). Within this matrix, polysaccharides are proposed to act as fundamental structural elements responsible for the mechanical stability of a biofilm. Some studies indicate that lectin-like proteins are also likely to contribute to the matrix formation by direct or indirect cross-linking of polysaccharides through multivalent cation bridges (Higgins and Novak, 1997). However, the main function of extracellular proteins in biofilms is thought to be enzymatic.

Free exopolymers, i.e. exopolymers in the aqueous phase, can compete with bacterial cells for attachment to metallic surfaces, and can, therefore, be present on the surface (Paradies et al., 1992; Thies et al., 1995 and references therein). Atomic force microscopy (AFM) studies have revealed spatially independent

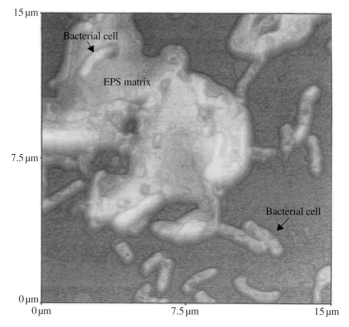

Figure 8.4   AFM micrograph of a 7-day-old biofilm formed by marine bacteria on the surface of mica. The biofilm was formed by mixed bacterial population isolated from Cu–Ni alloys that had been exposed in Langstone harbour in the south of England for 5 years.

distribution of EPS and cells (Beech *et al.*, 1996, 2002a). This is demonstrated in Figure 8.4, which shows an AFM image of a biofilm formed on mica by a mixed population of marine bacteria isolated from a corroded carbon steel structure. The question of whether the EPS material present on the surface is generated solely by sessile cells, or if it is originating from the planktonic EPS in the aqueous phase, remains unanswered.

8.2.2   Enzymes and biofilm matrix

It is documented that bacteria produce a wide range of enzymes, e.g. hydrolytic and proteolytic enzymes, as well as lyases, able to interact with substrates beyond the cell wall. Such enzymes can be broadly categorised as ectoenzymes (associated with cell, but expressed outside the cytoplasmic membrane) and extracellular enzymes, i.e. present as free forms (Chorst, 1991). The latter include polysaccharidases, proteases, lipases, esterases, peptidases, glycosidases, phosphatases and oxidoreductases (Lemmer *et al.*, 1994; Griebe *et al.*, 1997). Extracellular enzyme activities are readily observed and reported in biofilms. It has, therefore, been proposed that exoenzymes should be considered an integral part of the EPS matrix (Frolund *et al.*, 1995).

The release of enzymes by microorganisms into their external environment provides the basis for the interaction between cells and high-molecular-weight exogenous substrates. In order for such an interaction to be energy efficient, both enzymes and/or substrates and hydrolysis products should remain in close proximity (within $500\,\mu$m) of the cells (Wetzel, 1991). Biofilm environment facilitates the latter relationship. The chemical properties of biofilm matrix, i.e. the presence of different types of binding sites within exopolymers forming the matrix, promote association between EPS, extracellular enzymes and exogenous molecules, thus, enabling enzymatic hydrolysis. Subsequently, a diffusive transfer of smaller hydrolysis product to microbial cells can occur. The evidence exists that EPS matrix exhibits different retention properties depending on the size and charge of molecules interacting with the matrix and that both extracellular enzymes and hydrolysis products are localised within the matrix in a non-uniform manner (Lawrence et al., 1994).

The mechanism leading to heterogeneous spatial distribution of enzymes within biofilms is poorly understood. Two models are proposed to explain the interaction of enzymes with biofilm EPS, in particular with a polysaccharide component of the matrix. The lectin-binding model indicates that the higher-order structure of carbohydrate polymers could be involved in mediating the specific binding of certain proteins and carbohydrates (Sharon and Lis, 1993). A second theory suggests the involvement of cation bridges due to the presence of carboxyl groups common to both acidic polysaccharides and many enzymes (Stryer, 1995). Enzymes can also be bound to inorganic or organic particulate material sorbed within the EPS matrix (Vetter et al., 1998).

8.2.3   Enzymes and biocorrosion

The majority of reports on monitoring enzyme expression in biofilms focuses on activities associated with sessile cells, which is usually enhanced comparing to activities measured in planktonic population. Relatively few investigations, attempt to elucidate activity within cell-free EPS material (Frolund et al., 1995; Beech et al., 2001).

Studies of enzymatic activities in anoxic areas within biofilms are also infrequent and little is know about expression of enzymes other than hydrogenases in EPS synthesised by anaerobic microorganisms, such as, e.g. SRB. Elucidating the involvement of the hydrogenase enzyme in the SRB-influenced anaerobic corrosion of steels attracted considerable research efforts (see Beech et al., 2002b for review). Hydrogenase can play a significant role in the biocorrosion of steel not only through cathodic depolarisation mechanism (Figure 8.5), as first proposed by Von Wolzogen Kühr and Van der Vlugt (1934), but also through a direct electron transfer (DET) between the adsorbed enzyme and steel surface (Da Silva, 2002 and references therein).

Investigations addressing the role of enzymes in biocorrosion in oxygenated systems were, until recently, relatively scarce. Studies of stainless steel behaviour in natural waters, aimed to determine the mechanisms of its free corrosion

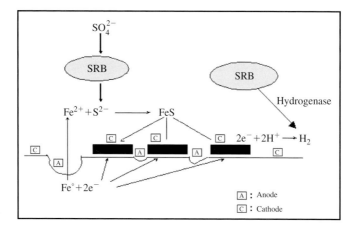

Figure 8.5   Cathodic depolarisation of a mild steel surface in the presence of SRB due to the production of iron sulphide and activity of the hydrogenase enzyme.

potential (*E*cor) increase, referred to as the ennoblement, brought enzymes into focus. The ennoblement of the free corrosion potential has been widely reported in natural waters and could not be reproduced using sterile water (Little *et al.*, 1990; Bardal *et al.,* 1993; Ismail *et al.*, 1999).

It is now accepted that biofilm development modifies the kinetics of the cathodic reaction on stainless steel (Mollica, 1992; Johnson and Bardal, 1985; Dexter and La Fontaine, 1998). Furthermore, it has been argued that the aerobic part of the biofilm plays a key role in this process. A number of studies led to the conclusion that marine biofilms induce the catalysis of oxygen reduction (Desestret, 1986; Dexter and Gao, 1988; Motoda *et al.*, 1990). Efforts were undertaken to identify the biofilm component linked to the observed electrochemical effects (Scotto *et al.*, 1993; Scotto and Lai, 1998). It has been shown that in the presence of sodium azide, which is a strong inhibitor of the enzymes involved in bacterial respiratory chain, the free corrosion potential measured in natural seawater drastically decreased (Scotto *et al.*, 1985). Thus, the involvement of biofilm enzymes, in particular oxidoreductases, in the ennoblement process has been proposed.

Lai *et al.* (1999) measured catalytic activities of superoxide dismutase, catalase and peroxidase in marine biofilms formed under field conditions. These enzymes catalyse, respectively, the following processes:

(i)   the disproportionation of the superoxide radical $O_2^{\circ-}$:

$$O_2^{\circ-} + 2H^+ \xrightarrow{\;superoxide\; dismutase\;} O_2 + H_2O_2$$

(ii)   the disproportionation of hydrogen peroxide:

$$2H_2O_2 \xrightarrow{\;catalase\;} O_2 + 2H_2C$$

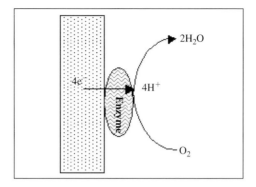

Figure 8.6   Schematic representation of the direct catalysis of oxygen reduction by oxidoreductases adsorbed to the material surface. The enzyme catalyses the DET from the material to dissolved oxygen.

(iii) the oxidation of several different substrates (substrate–H_2):

$$2\ \text{Substrate–}H_2 + H_2O_2 \xrightarrow{\textit{peroxidase}} 2\ \text{Substrate–}H^\circ + 2H_2O$$

All three enzymes were proposed to participate in the catalysis of the oxygen reduction on the stainless steel surface. Current bioelectrochemical studies confirmed that when adsorbed onto carbon and graphite catalase and horseradish peroxidase are indeed capable to catalyse the direct electrochemical reduction of dissolved oxygen, as depicted in Figure 8.6 (Lai and Bergel, 2000; Huang and Hu, 2001). The direct involvement of these enzymes in electrochemical reactions on the surface of stainless steel is yet to be demonstrated.

The effect of extracellular catalase produced by *Pseudomonas* sp. on ennoblement of Cu has also been reported (Busalmen *et al.*, 2002). The study revealed that the ability of the enzyme to reduce oxygen depended on the chemistry of the type of oxide, which was formed on the Cu surface. The authors addressed the importance of understanding the chemistry of corrosion products, when evaluating enzymically driven electrochemical reaction on surfaces of metallic materials.

The presence of oxygen gradients within the biofilm matrix can be readily demonstrated and it is accepted that anoxic niches can be found even in biofilms developed in fully oxygenated systems. Indeed, anaerobic sulphate reducers are commonly isolated from natural biofilms found in aerated environments (Gubner and Beech, 1999). Although all SRB species known to date have been described as strictly anaerobic, some SRB genera are able to survive for long periods in the presence of oxygen, indicating the existence of defence mechanisms against oxygen radicals (Hardy and Hamilton, 1981; Cypionka *et al.*, 1985; see Beech, 2002 for review). Indeed, the presence of catalase and superoxide dismutase that were constitutively expressed during anaerobic growth of *Desulfovibrio gigas* was recently confirmed (Santos *et al.*, 2000).

Current progress in microbial genomics facilitated comparative analysis of two available SRB genomes, *Desulfovibrio vulgaris* and *Desulfitobacterium hafniense*.

A particularly interesting *m*embrane-*a*ssociated *s*ensor (MASE) domain combination has been found in the unfinished genome of *D. vulgaris* demonstrating that this SRB encodes a sensor with an additional haemerythrine-like domain (Nikolskaya *et al.*, 2002). An alignment of the latter domain revealed good conservation of its iron-binding residues, suggesting involvement of this sensor in the response to iron and/or oxygen. Remarkably, *D. vulgaris* also contains a C-terminal haemerythrin domain in the recently described chemotaxis sensor DcrH, suggesting its participation in oxygen sensing (Xiong *et al.*, 2000). In addition, *D. vulgaris* encodes a haeme-containing chemotaxis sensor (Fu *et al.*, 1994). Proteins containing such domains often regulate secretion of extracellular proteins and polysaccharides (Galperin *et al.*, 2001) that are essential for the biofilm development. Delineation of the exact signal perceived by the *D. vulgaris* MASE domain would be of great interest, as it would further aid in the understanding of SRB-mediated processes within biofilms.

The genomic analysis also revealed that while no catalase encoding genes are present in Gram-negative *D. vulgaris*, Gram-positive *D. hafniense* encodes three catalases, of which at least one is secretable, i.e. extracellular (Galperin and Beech, unpublished). These catalases are HPI (CatA_ECOLI), HPII (CatE_ECOLI) and Mn-containing catalase PMID: 8939876, gi|1752756. Interestingly, a slime producing *P. aeruginosa*, encodes two catalases, CatE and Mn-dependent one, but the former exists in three copies. These findings point out that the ecology of microbial communities and, therefore, their metabolic output can be of great importance when investigating the effect of biofilms on the corrosion behaviour of metallic materials. This specificity can possibly explain why, despite identical environmental conditions, biofilms composed of different bacterial species belonging to the same genus differ in their ability to deteriorate colonised material. Whether catalases encoded in some, but not all, SRB species can be expressed in biofilms and can contribute to oxygen-reduction reactions, thus influencing corrosion processes, requires elucidating.

Several studies suggested that enzyme–metal complexes may catalyse the reduction of oxygen (Van den Brink *et al.*, 1980; Johnson and Bardal, 1985; Dexter and Gao, 1988). Metallic complexes, such as iron-porphyrins, constitute the active site of some of the oxidoreductases and these compounds are known to be efficient catalysts of the electrochemical reduction of oxygen (Jiang and Dong 1990; Arifuku *et al.*, 1992). However, the experiments using enzyme inhibitors revealed that the free corrosion potential ennoblement was drastically reduced by certain classes of inhibitors. Such a result confirms that fully-constituted enzymes, rather than the metallic complexes that form their active site, are involved in *E*cor ennoblement.

8.2.4   Reduction of metal oxides

Several investigations have shown a modification of the overall composition of the passive film of stainless steels exposed to different bacterial biofilms (Geesey *et al.*, 1996; Pendyala *et al.*, 1996; Ismail *et al.*, 1999; Beech *et al.*, 2000). The outer layer of the passive film was proven to be iron oxide rich, while the inner part contained more chromium. Iron can play the role of catalyst for the reduction of oxygen on the surface of the material. The hydrogen peroxide produced by phytoplanktons, bacteria, micro-algae, all of which can be present in mature biofilms, partially reduces the surface iron oxides, transforming trivalent iron atoms into divalent ones. It is documented that oxygen reduction is faster on divalent iron atoms (Da Silva, 2002). The reduction of iron oxides by hydrogen peroxide could, therefore, result in the catalysis of the oxygen reduction, thus leading to ennoblement.

Microorganisms are known to promote corrosion of iron and its alloys through dissimilatory reduction reactions. The consequence of these reactions is the dissolution of protective oxide/hydroxide films on the metal surface. Thus, passive layers on steel surfaces can be lost or replaced by less stable reduced metal films that allow further corrosion to occur. Obuekwe *et al.* (1981) demonstrated an increase in the corrosion of mild steel in the presence of an iron-reducing bacterium of the genus *Pseudomonas*.

Numerous types of bacteria are able to carry out manganese and/or iron oxide reduction (Arnold *et al.*, 1988; Myers and Nealson, 1988; Roden and Zachara, 1996). It has been demonstrated that in cultures of *Shewanella putrefaciens*, iron oxide surface contact was required for bacterial cells to mediate reduction of these metals. The reaction rate depended on the type of oxide film under attack (Little *et al.*, 1997b). Corrosion of carbon steel in the presence of *Shewanella putrefaciens* biofilm was monitored for up to 1300 h. The electrodes examined after this period of time were extensively colonised and pitted. The location of pits coincided with the distribution of bacterial colonies. Subsequent analysis revealed that mineral replacement reactions occurred on electrode surfaces. The biotic and abiotic manganese oxide reduction mechanisms and their importance are described elsewhere (for review see Bergel *et al.*, in press).

8.2.5   Metal binding by the biofilm matrix

The metal-binding/sorption capacity of the biofilm matrix renders it important for MIC. The sorption is both bacteria and metal species specific (Ford *et al.*, 1987, 1990; Geesey *et al.*, 1988; Beech *et al.*, 1999b). Different metal ions or/and the same type of metal ions in dissimilar oxidation states trapped within the biofilm matrix can influence electron transfer processes, which drive corrosion reactions. The importance of this mechanism in the overall corrosion process would, however, depend on the material.

Current models of metal binding by EPS emphasise the role of polysaccharides and proteins, which posses anionic properties. EPS contain ionic groups contributing to both net-negative and -positive charges on the polymers at near neutral pH values. Polysaccharides owe their negative charge either to carboxyl groups of uronic acids or to non-carbohydrate substituents, such as phosphate, sulphate, glycerate, pyruvate or succinate (Sutherland, 2001). Proteins rich in amino acids containing carboxyl groups, such as aspartic and glutamic acid can also contribute to the anionic properties of EPS (Dignac *et al.*, 1998). Nucleic acids are polyanionic due to the phosphate residues in the nucleotide moiety of the polymer molecule. Uronic acids, acidic amino acids and phosphate-containing nucleotides as negatively charged components of EPS are thought to be involved in electrostatic interactions with multivalent cations (e.g. Ca^{2+}, Cu^{2+}, Mg^{2+}, Fe^{3+}).

The possible involvement of enzymes in metal-binding processes has, as yet, not been widely considered. One of the group of enzymes, which could be potentially important for metal binding within biofilms, are phosphatases. The role of phosphatases in bioremediation processes in the presence of biofilms formed by aerobic bacteria is established and mechanism by which these enzymes mediate metal ion precipitation has been documented (Macaskie *et al.*, 1995 and references therein). Recent studies demonstarted that phosphatases can be active in either free EPS or within biofilms formed under anaerobic conditions and that phosphatase activity is bacterial species specific (Beech *et al.*, 2001).

8.3   SUMMARY AND CONCLUSIONS

Biocorrosion observed in a broad range of aquatic and terrestrial habitats differing in nutrient content, temperature, pressure and pH values, results from the presence and physiological activities of diverse microbial species thriving within biofilms. The interaction of a biofilm harbouring aerobic, facultatively anaerobic and strictly anaerobic microorganisms and the underlying substratum produces a unique physical and chemical environment at the surface. In the presence of a biofilm, conditions at the substratum surface can differ significantly from those in the bulk phase or at a biofilm-free surface, thus, influencing material deterioration. The presence of microorganisms on corroding materials promotes the establishment or maintenance of physico-chemical reactions, not normally favoured under otherwise similar conditions, i.e. in the absence of the microbes. These biofilm-driven reactions can either accelerate or inhibit the biodeterioration process.

Biofilm matrix can influence deterioration of metallic substrata. Some of the proposed mechanisms are listed below:

- accumulation of aggressive microbial metabolic products;
- enzymic activity;
- the passage of charged entities through the matrix;
- the degree of conductivity of the matrix;
- the binding/sorption of metal ions within the matrix.

The biofilm present on the metal surface can modify both morphology and chemistry of corrosion products. Frequently, these products are transferred into more aggressive species. Different mechanisms of biofilm–metal interaction, which reflect the broad range of physiological activities carried out by different types of microorganisms, have been identified. These mechanisms depend on the type of microbial species and chemistry of the colonised surface. Despite considerable research efforts, there is, however, still a need to gain a better insight into the spatial and temporal distribution of biotic and abiotic reactions, which govern electrochemical process on surfaces of corroding materials. Such knowledge would facilitate a design of improved prevention and protection measures aimed to combat material biodeterioration.

REFERENCES

Allison, D.G. (1998) Exopolysaccharide production in bacterial biofilms. *Biofilm J.* **3**, paper 2 (BF98002), Online Journals http://www.bdt.org.br/bioline/bf

Arifuku, F., Mori, K., Muratami, T. and Kurihara, H. (1992) The catalytic electroreduction of dioxygen on iron protoporphyrin IX modified glassy carbon electrodes. *Bull. Chem. Soc. Jpn* **65**, 1491–1495.

Arnold, R.G., DiChristina, T. and Hoffman, M. (1988) Reductive dissolution of Fe (III) oxides by *Pseudomonas* sp. 200, *Biotechnol. Bioeng.* **32**, 1081–1096.

Bardal E., Drugli, J.M. and Gartland, P.O. (1993) The behaviour of corrosion-resistant steels in seawater: a review. *Corr. Sci.* **35**, 257–267.

Beech, I.B. (1999) La corrosion microbienne. *Biofuture* **186**, 36–41.

Beech, I.B. (2002) Biocorrosion: role of sulfate-reducing bacteria. In *Encyclopedia of Environmental Microbiology* (ed. G. Bitton), pp. 465–475, Wiley, New York, USA.

Beech, I.B., Gaylarde, C.C., Smith, J.J. and Geesey, G.G. (1991) Extracellular polysaccharides from *Desulfovibrio desulfuricans* and *Pseudomonas fluorescens* in the presence of mild and stainless steel. *Appl. Microbiol. Biotechnol.* **35**, 65–71.

Beech, I.B., Cheung, C.W.S., Johnson, D.B. and Smith, J.R. (1996) Comparative studies of bacterial bofilms on steel surfaces using techniques of atomic microscopy and environmental scanning electron microscopy. *Biofouling* **10**(1–3), 65–77.

Beech, I.B., Hanjaknsit, L., Kalaji, M., Neal, A. and Zinkevich, V. (1999a) Exopolymer production by planktonic and biofilm *Pseudomonas* sp. NCIMB 2021 cells in continuous culture. *Microbiology* **145**, 1491–1497.

Beech, I.B., Zinkevich, V., Tapper, R. and Avci, R. (1999b) Study of the interaction of exopolymers produced by sulphate-reducing bacteria with iron using X-ray photoelectron spectroscopy and time-of-flight secondary ionisation mass spectrometry. *J. Microbiol. Meth.* **36**, 3–10.

Beech, I.B., Gubner, R., Zinkevich, V., Hanjangsit, L. and Avci, R. (2000) Characterisation of conditioning layers formed by exopolymeric substances produced by *Pseudomonas* NCIMB 2021 on surfaces of AISI 304 and 316 stainless steel. *Biofouling* **16**(1), 93–104.

Beech, I.B., Paiva, M., Caus, M. and Coutinho, C. (2001) Enzymatic activity within biofilms of sulphate-reducing bacteria. In *Biofilm Community Interactions: Chance or Necessity?* (ed. P.G. Gilbert, D. Allison, M. Brading, J. Verran and J. Walker), pp. 231–239, BioLine.

Beech, I.B., Smith, J., Steele, A., Penegar, J. and Campbell, S. (2002a) The application of AFM in investigating interactions of biofilms with inanimate surfaces. *Coll. Surf. B: Biointerf.* **23**, 231–247.

Beech, I.B., Paiva, M. and Coutinho, C. (2002b) The role of hydrogenase in biocorrosion of mild steel. In *Enzymes and Corrosion*, European Federation of Corrosion, Institute of Materials, London, in press.

Bergel, A., Mollica, A. and Feron, D. Electrochemical effects of aerobic biofilms on stainless steels: critical review of the mechanisms and models. In *Biofilms and Microbial Corrosion* (ed. I.B. Beech and H.-C. Flemming), Springer Verlag, in press.

Busalmen, J.P., Vázquez, M. and de Sánchez, S.R. (2002) New evidences on the catalase mechanism of microbial corrosion. *Eletrochem. Acta* **47**, 1857–1865.

Chrost, R.J. (1991) Environmental control of the synthesis and activity of aquatic microbial ectoenzymes. In *Microbial Enzymes in Aquatic Environments* (ed. R.J. Chorst), pp. 29–59, Springer-Verlag, New York.

Cooksey, K.E. (1992) Extracellular polymers in biofilms. In *Biofilms: Science and Technology*, pp. 137–147, Kluwer Academic Press, The Netherlands.

Costerton, J.W., Lappin-Scott, H.M. and Cheng, K.-J. (1992) Glycocalyx bacterial. In *Encyclopedia of Microbiology* (ed. J. Lederberg), Vol. 2, pp. 311–317, Academic Press, San Diego.

Costerton, J.W., Lewandowski, Z., De Beer, D., Caldwell, D., Korber, D. and James, G. (1994) Minireview: biofilms, the customized microniche. *J. Bacteriol.* **176**, 2137–2142.

Cypionka, H., Widdel, F. and Pfenning, N. (1985) Survival of sulfate-reducing bacteria after oxygen stree, and growth in sulfate-free oxygen-sulfide gradients. *FEMS Microbiol. Ecol.* **31**, 39–45.

Da Silva, S. (2002) Hydrogenases et corrosion anaerobie des aciers. Ph.D. thesis, Universite Paul sabatier-Toulose III, Toulose, France.

Desestret, A. (1986) Corrosion localisée des aciers inoxydables dans l'eau de mer. *Matériaux et Techniques* **7–8**, 317–324.

Dexter, S.C. and Gao, G.Y. (1988) Effect of seawater biofilms on corrosion potential and oxygen reduction of stainless steel. *Corrosion* **44**, 717–723.

Dexter, S.C. and La Fontaine, J.P. (1998) Effect of natural marine biofilms on galvanic corrosion. *Corrosion* **54**, 851–861.

Dignac, M.-F., Urbain, V., Rybacki, D., Bruchet, A., Snidaro, D. and Scribe, P. (1998) Chemical description of extracellular polymers: implication on activated sludge structure. *Water Sci. Technol.* **38**(8–9), 45–53.

Dowling, N.J.E., Mittelman, M.W. and White, D.C. (1991) The role of consortia in microbially influenced corrosion. In *Mixed Cultures in Biotechnology* (ed. J.G. Zeikus), pp. 341–372, McGraw Hill, New York.

Ford, T.E., Maki, J.S. and Mitchell, R. (1987) The role of metal-binding bacterial exopolymers in corrosion processes. In *Proceedings of the NACE Corrosion '87*, Paper No. 380, NACE, Houston, TX.

Ford, T., Maki, J.S. and Mitchell, R. (1988) Involvement of bacterial exopolymers in biodeterioration of metals. In *Biodeterioration* (ed. D.R. Houghton, R.N. Smith and H.O.W. Eggins), Vol. 7, pp. **378**, Elsevier Applied Science, London, UK.

Ford, T., Black, J.P. and Mitchell, R. (1990) Relationship between bacterial exopolymers and corroding metal surfaces. In *Proceedings of the NACE Corrosion '90*, Paper No. 110, NACE, Houston, TX.

Frolund, B., Griebe, T. and Nielsen, P.H. (1995) Enzymatic-activity in the activated-sludge floc matrix. *Appl. Microbiol. Biotechnol.* **43**(4), 755–761.

Fu, R., Wall, J.D. and Voordouw, G. (1994) DcrA, a c-type heme-containing methyl-accepting protein from *Desulfovibrio vulgaris* Hildenborough, senses the oxygen concentration or redox potential of the environment. *J. Bacteriol.* **176**, 344–350.

Galperin, M.Y., Nikolskaya, A.N. and Koonin, E.V. (2001) Novel domains of the prokaryotic two-component signal transduction system. *FEMS Microbiol. Lett.* **203**, 11–21.

Geesey, G.G., Jang, L., Jolley, J.G., Hankins, M.R., Iwaoka, T. and Griffiths, P.R. (1988) Binding of metal ions by extracellular polymers of biofilm bacteria. *Water Sci. Technol.* **20**, 11/12, 161–165.

Geesey, G.G., Gillis, R.J., Avci, R., Daly, D., Hamilton, M., Shope, P. and Harkin, G. (1996) The influence of surface features on bacterial colonization and subsequent substratum chemical changes of 316L stainless steel. *Corros. Sci.* **38**, 73–95.

Geesey, G.G., Beech, I.B., Bremmer, P.J., Webster, B.J. and Wells, D. (2000) Biocorrosion. In *Biofilms II: Process Analysis and Applications* (ed. J. Bryers), pp. 281–326, Wiley-Liss Inc.

Ghassem, H. and Adibi, N. (1995) Bacterial corrosion of reformer heater tubes. *Mater. Perform.* **34**(3), 47–48.

Griebe, T., Schaule, G. and Wuertz, S. (1997) Determination of microbial respiratory and redox activity in activated sludge. *J. Ind. Microbiol. Biot.* **19**(2), 118–122.

Gubner, R. and Beech, I.B. (1999) Statistical assessment of the risk of accelerated low-water corrosion in a marine environment. *Corrosion 99*, Paper 318, NACE, Houston, Texas.

Gubner, R. and Beech, I.B. (2000) The effect of extracellular polymeric substances on the attachment of *Pseudomonas* NCIMB 2021 to AISI 304 and 316 stainless steel. *Biofouling* **15**, 25–36.

Hardy, J.A. and Hamilton, W.A. (1981) The oxygen tolerance of sulfate-reducing bacteria isolated from the North Sea waters. *Curr. Microbiol.* **6**, 259–262.

Higgins, M.J. and Novak, J.T. (1997) Dewatering and settling of activated sludges: the case for using cation analysis. *Water Environ. Res.* **69**(2), 225–232.

Huang, R. and Hu, N. (2001) Direct electrochemistry and electrocatalysis with horseradish peroxidase in Eastman AQ films. *Bioelectrochemistry* **54**, 75–81.

Ismail, K.M., Jayaraman, A., Wood, T.K. and Earthman, J.C. (1999) The influence of bacteria on the passive film stability of 304 stainless steel. *Electrochim. Acta* **44**, 4685–4692.

Jiang, R. and Dong, S. (1990) Study on the electrocatalytic reduction of H_2O_2 at iron protoporphyrin modified electrode with a rapid rotation-scan method. *Electrochim. Acta* **35**, 1227–1232.

Johnson, R. and Bardal, E. (1985) Cathodic properties of different stainless steels in natural seawater. *Corrosion* **41**, 296–309.

Lai, M.E. and Bergel, A. (2000) Electrochemical reduction of oxygen on glassy carbon: catalysis by catalase. *J. Electroanal. Chem.* **494**, 30–40.

Lai, M.E., Scotto, V. and Bergel, A. (1999) Analytical characterization of natural marine biofilm: a tool for understanding biocorrosion of s.s. in seawater. In *Proceeding of the 10th International Congress on Marine Corrosion and Fouling*, Melbourne, Australia.

Lawrence, J.R., Wolfaard, G.M. and Korber, D.R. (1994) Determination of diffusion coefficients in biofilms using confocal laser microscopy. *Appl. Environ. Microbiol.* **60**, 1166–1173.

Lemmer, H., Roth, D. and Schade, M. (1994) Population-density and enzyme-activities of heterotrophic bacteria in sewer biofilms and activated-sludge. *Water Res.* **28**(6), 1341–1346.

Little, B. and Wagner, P. (1997) Myths related to microbiologically influenced corrosion. *Mater. Performance* **36**(6), 40–44.

Little, B., Ray, R., Wagner, P., Lewandowski, Z., Lee, W.C., Characklis, W.G. and Mansfeld, F. (1990) Electrochemical behavior of stainless steels in natural seawater. *Corrosion-NACE*, Paper 150.

Little, B.J., Wagner, P.A. and Lewandowski, Z. (1997a) Spatial relationships between bacteria and mineral surfaces. In *Reviews in Mineralogy, Geomicrobiology: Interactions between Microbes and Minerals* (ed. J.F. Banfield and K.H. Nealson), Vol. 35, pp. 123–159 (Series Ed., P.H. Ribbe), The Mineralogical Society of America, Washington DC, USA.

Little, B.J., Wagner, P., Hart, K., Ray, R., Lavoie, D., Nealson, K. and Aguilar, C. (1997b) The role of metal-reducing bacteria in microbiologically influenced corrosion. In *Proceedings of the NACE Corrosion '97*, Paper No. 215, NACE International, Houston, TX.

Macaskie, L.E., Hewitt, C.J., Shearer, J.A. and Kent, C.A. (1995) Biomass production for the removal of heavy metals from aqueous solutions at low pH using growth-decoupled cells of a *Citrobacter* sp. *Int. Biodeterior. Degrad.* **35**, 73–92

Mollica, A. (1992) Biofilm and corrosion on active–passive alloys in seawater. *Int. Biodeter. Biodegr.* **29**, 213–229.

Motoda, S., Suzuki, Y., Shinoara, T. and Tsujikawa, S. (1990) The effect of marine fouling on the ennoblement of electrode potential for stainless steels. *Corros. Sci.* **31**, 515–520.

Myers, C. and Nealson, K.H. (1988) Bacterial manganese reduction and growth with manganese oxide as the sole electron acceptor. *Science* **240**, 1319–1321.

Nikolskaya, A.N., Mulkidjanian, A.Y., Beech, I.B. and Galperin, M.Y. (2002) MASE1 and MASE2: two novel integral membrane sensory domains. *J. Mol. Microbiol. Biotechnol.* (in press).

Obuekwe, C.O., Westlake, D.W.S., Cook, F.D. and Costerton, J.W. (1981) Surface changes in mild steel coupons from the action of corrosion causing bacteria. *Appl. Environ. Microbiol.* **41**, 766–774.

Paradies, H.H., Fisher, W.R., Haenbel, I. and Wagner, D. (1992) Characterisation of metal biofilm interactions by extended absorption fine structure spectroscopy. In *Microbial Corrosion Publication* (ed. C.A.C. Sequeira and A.K. Tiller), European Federation of Corrosion Publication No. 8, pp. 168–189, The Institute of Materials, London.

Pedersen, A. and Hermansson, M. (1991) Inhibition of metal corrosion by bacteria. *Biofouling* **3**(1), 1.

Pendyala, J., Geesey, G.G., Stoodley, P., Hamilton, M. and Harkin, G. (1996) Chemical effects of biofilm colonization on 304 stainless steel. *J. Vac. Sci. Technol.* **A 14**, 1755–1769.

Roden, E.E. and Zachara, J.M. (1996) Microbial reduction of crystalline Fe(III) oxides: influence of oxide surface area and potential for cell growth. *Environ. Sci. Tech.* **30**, 1618–1628.

Santos, W.G., Pacheco, I., Liu, M.-Y., Teixeira, M., Xavier, A.V. and Le Gall, J. (2000) Purification and characterization of an iron superoxide dismutase and a catalase from the sulfate-reducing bacterium *Desulfovibrio gigas*. *J. Bacteriol.* **182**, 796–804.

Scotto, V. and Lai, M.E. (1998) The ennoblement of stainless steels in seawater: a likely explanation coming from the field. *Corros. Sci.* **40**, 1007–1018.

Scotto, V., Di Cintio, R. and Marcenaro, G. (1985) The influence of marine aerobic microbial film on stainless steel corrosion behaviour. *Corros. Sci.* **25**, 185–194.

Scotto, V., Beggiato, M., Marcenaro, G. and Dellepiane, R. (1993) Microbial and biochemical factors affecting the corrosion behaviour of stainless steels in seawater. European Federation of Corrosion publications No. 10, pp. 21–33, The Institute of Materials, London.

Scully, J.C. (1990) *Fundamentals of Corrosion*, 3rd edn, Pergamon Press, Oxford, UK.

Sharon, N. and Lis, H. (1993) Carbohydrates in cell recognition. *Scient. Am.* January, 82–89.

Stryer, L. (1995) *Biochemistry*, 4th edn, W.H. Freeman, New York.

Sutherland, I.W. (1985) Biosynthesis and composition of Gram-negative bacterial extracellular and wall polysaccharides. *Ann. Rev. Microbiol.* **39**, 243–262.

Sutherland, I.W. (2001) The biofilm matrix – an immobilized but dynamic microbial environment. *Trend. Microbiol.* **9**(5), 222–227.

Thies, M., Hinze, U. and Paradies, H.H. (1995) Physical behaviour of biopolymers as artificial models for biofilms in biodeterioration of copper. Solution and surface properties of biopolymers. In *Microbial Corrosion* (ed. A.K. Tiller and C.A.C. Sequeira), European Federation of Corrosion Publication No. 15, pp. 17–48, The Institute of Materials, London.

Van den Brink, F., Barendrecht, E. and Visscher, W. (1980) The cathodic reduction of oxygen. A review with emphasis on macrocyclic organic metal complexes as electrocatalysts. *J. Royal Nether. Chem. Soc.* **99**, 253–262.

Vetter, Y.A., Deming, J.W., Jumars, P.A. and Krieger-Brockett, B.B. (1998) A predictive model of bacterial foreging by means of freely released extracellular enzymes. *Microbial Ecol.* **36**, 75–92.

Videla, H.A. (1996) *Manual of Biocorrosion*. Lewis Publishers, CRC Press, Inc., Boca Raton, 273p.

Videla, H. and Herrera, L.K. Microbial inhibition of corrosion. In *Biofilms and Microbial Corrosion* (ed. I.B. Beech and H.-C. Flemming), Springer Verlag, in press.

Von Wolzogen Kuhr, C.A.V. and Van der Vlugt, S.S. (1934) Graphitisation of cast iron as an electrochemical process in anaerobic soils. *Water* (den Haag) **18**(16), 147–165.

Watkins, P.G. (1998) The corrosion of mild steel in the presence of two isolates of marine sulphate reducing bacteria. Ph.D. thesis, University of portsmouth, Portsmouth, England.

Wetzel, R.G. (1991) Extracellular enzymatic interactions: storage, redistribution and inter-specific communication. In *Microbial Enzymes in Aquatic Environments* (ed. R.J. Chorst), pp. 6–28, Springer-Verlag, New York.

Wingender, J., Neu, T.R. and Flemming, H.-C. (1999) What are bacterial extracellular polymeric substances? In *Microbial Extracellular Polymeric Substances: Characterization, Structure and Function* (ed. J. Wingender, T.R. Neu and H.-C. Flemming), pp. 1–15, Springer-Verlag, New York.

Xiong, J., Kurtz, D.M. Jr., Ai, J. and Sanders-Loehr, J. (2000) A haemerythrin-like domain in a bacterial chemotaxis protein. *Biochemistry* **39**, 5117–5125.

Zinkevich, V., Kang, H., Bogdarina, I., Hill, M.A. and Beech, I.B. (1996) The characterisation of exopolymers produced by different species of marine sulphate-reducing bacteria. *Int. Biodeter. Biodegr.* **37**(3–4), 163–172.

9

Biofilms in wastewater treatment systems

V. O'Flaherty and P. Lens

9.1 INTRODUCTION

One of the biggest environmental problems facing the world today is the quantity of liquid wastes (wastewaters) being generated and the risk these wastewaters pose to the environment. Wastewaters must be treated as they can damage the environment and threaten public health if discharged directly to a receiving water body (Metcalf and Eddy, 1991). The major components of many wastewaters, including industrial, food processing and domestic sewage, are organic compounds and nutrients. These compounds are processed and recycled by microbial decomposition as part of the biological carbon, nitrogen, phosphorus and other biogeochemical cycles. Treatment of wastewaters is often based on biofilm-based engineered systems and essentially involves the conversion of polymeric (proteins, lipids, carbohydrates, etc.) and monomeric (sugars, amino acids, fatty acids, etc.) organic compounds to CO_2, CH_4 and H_2O. Two main microbial processes are involved, aerobic and anaerobic decomposition:

$$\text{Aerobic: Organic compounds} + O_2 \rightarrow CO_2 + H_2O$$
$$\text{Anaerobic: Organic compounds} \rightarrow CO_2 + CH_4$$

© IWA Publishing. *Biofilms in Medicine, Industry and Environmental Biotechnology*. Edited by Piet Lens, Anthony P. Moran, Therese Mahony, Paul Stoodley and Vincent O'Flaherty. ISBN: 1 84339 019 1

9.1.1   Aerobic biological treatment

Bacteria, fungi and protozoa are involved in this process, which achieves complete breakdown of organic compounds to CO_2 and H_2O (stable end products with no pollution risk). Aerobic breakdown releases a large amount of energy that is used by the microbes for growth and reproduction, e.g. glucose + $6O_2$ → $6CO_2$ + $6H_2O$ ($\Delta G' = -2,840$ kJ/mol). Consequently, mineralisation (degradation) of organic compounds aerobically results in the generation of significant amounts of microbial biomass (organic matter), requiring subsequent safe disposal/re-use/recycle. The organic matter (normally measured as biochemical or chemical oxygen demand, respectively (BOD or COD)) is thus converted to CO_2, H_2O and microbial cells (typically 0.3–0.5 kg dry biomass per kg BOD converted). Another characteristic of aerobic treatment is that each substrate, e.g. a glucose molecule or a protein is metabolised completely by a single cell, such that the substrate is decomposed in a single stage and completely broken down to CO_2 and H_2O. The biofilm-based systems applied for aerobic treatment include percolating filters, fluidised bed reactors and rotating biological contactors (RBCs) (Gray, 1999).

9.1.2   Anaerobic biological treatment

Organic matter is naturally decomposed by anaerobic microbes in sediments, swamps, marshes, etc. and results in the breakdown of organics to CH_4 (methane) and CO_2. This process is also applied in engineered wastewater treatment systems. As a result of anaerobic decomposition or digestion, the bulk (70–80%) of the energy content of the waste organics is conserved in the methane product, e.g. glucose → $3CH_4$ + $3CO_2$ ($\Delta G' = -403$ kJ/mol). As a consequence of the very low amount of energy released, growth is limited and the quantity of microbial biomass produced is small (<0.1 kg dry biomass per kg BOD converted). In addition, energy recycling and renewable energy generation is accomplished through the production of a useable fuel (methane). Anaerobic biofilm-based systems used for wastewater treatment include anaerobic filters (AFs) and granular sludge-based systems (Lettinga, 1995).

9.1.3   Biofilms and wastewater treatment

The formation of biofilms and microbial aggregates are prerequisites for the successful operation of biological wastewater treatment and nutrient removal systems (Flemming, 1993; Costerton *et al.*, 1995). Practical applications of biofilm-based systems for wastewater purification have become more common and varied (Gray, 1999). There has been a significant advancement in biofilm-reactor design and operation during the past decade with the advent of sophisticated biotechnological approaches and there remains a constant state of, and need for, development, making this area one of the most active in environmental biotechnology (Gray, 1999). Aerobic, anaerobic, anoxic and combination systems are now routinely applied, but despite this success, much information remains to be elucidated on the basic biological and physico-chemical characteristics of the biofilms involved.

This chapter will outline the biological properties of the biofilms involved in two major classes of wastewater treatment systems, aerobic fixed-film systems and anaerobic granular sludge systems, used for the removal of soluble organic matter from wastewaters. Two themes are addressed: (i) the relationship between biofilm biological properties and treatment plant function in aerobic fixed-film systems and (ii) the formation of anaerobic granular biofilms. For the purposes of this chapter, the term 'wastewater' refers to settled domestic sewage (after removal of gross solids) or other industrial wastewaters with low solids content.

9.2   THE BIOLOGICAL PROPERTIES OF AEROBIC FIXED-FILM WASTEWATER TREATMENT BIOFILMS

Aerobic wastewater treatment systems can be divided into two classes: fixed-film and flocculent (activated sludge) systems (Figure 9.1). This section will discuss the fixed-film systems applied for aerobic wastewater treatment including: percolating or trickling filters, RBCs, biological aerated reactors and fluidised bed reactors. These systems are all dependent on the adhesion of microbial cells to form a biofilm on a supplied inert substratum.

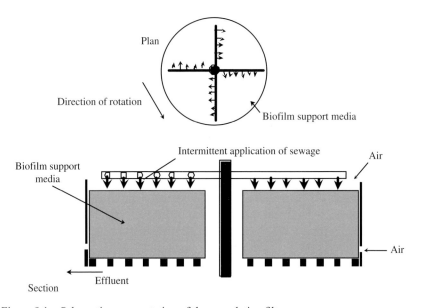

Figure 9.1   Schematic representation of the percolating filter.

9.2.1   The percolating filter

The most widely used aerobic fixed-film reactor is the percolating or trickling filter (Hawkes, 1983). In its simplest form, it comprises a bed of graded hard material, the medium, about 2 m in depth (Figure 9.1). The wastewater is spread evenly over the surface of the bed by a rotating arm distribution system, which can be used to regulate the volume and frequency of application of the wastewater. The filter has a ventilating system to ensure free access of air to the bed, which passes through the interstices or voids of the medium ensuring that all parts of the filter have sufficient oxygen. The treated effluent passes through a layer of drainage tiles that support the medium and flows away to the secondary settlement tank (Hawkes, 1983).

The loading to percolating filters depends on the physical characteristics of the medium and the degree of biofilm accumulation. Hydraulic loading is expressed in cubic metres of wastewater per cubic metre of filter medium per day, while organic loading is expressed as kg BOD per cubic metre of filter medium per day. These loadings are used to differentiate between low-rate ($<3\,m^3/m^3$/day, $0.6\,kg/m^3$/day) and high-rate ($>3\,m^3/m^3$/day $> 0.6\,kg/m^3$/day) filters. The original design of filters was for complete treatment using one filter with the wastewater being passed through only once, producing an effluent containing $<20\,mg/l$ suspended solids and $30\,mg/l$ BOD (Qasim, 1994). One of the critical aspects of filter operation is the choice of medium used for biofilm formation. Mineral (granite, gravel, slag, clinker, etc.), plastic (PVC rings, spheres, etc. often with design features to increase surface area, fins, etc.) and biological (wood chips, bamboo cuts, coconut shells, etc.) media can be employed. Also, modular- or cross-flow media can be designed to allow for maximum biofilm surface area to develop and also maximum contact between the wastewater and the biofilm microorganisms. In a percolating filter, the medium is either random packed or in modular form, leaving voids or interstitial spaces to allow the sewage to trickle through and permit circulation of air (i.e. the biofilm is never submerged).

The media must ideally be durable, low cost, preferably lightweight, have a good surface for biofilm development and have a maximum surface area (m^2/m^3) and porosity (%). The choice between cheap, locally available material such as granite (surface area: 94–$237\,m^2/m^3$; porosity: 40–55%) and more expensive but efficient PVC (surface area: 100–$300\,m^2/m^3$; porosity: 91–95%) is based on site-specific considerations (Qasim, 1994).

9.2.1.1   The formation of biofilms on percolating filters

The microbial biofilm on a percolating filter takes 3–4 weeks to become established during the summer months and up to 3 months during the winter. Only when biofilm formation has been completed, will the filter have reached fully operational capacity. If the filter is treating domestic sewage as a wastewater it is not generally necessary to supply microbial cells or inoculum to seed the biofilm, as all the necessary microbes will be present in the sewage. Seeding may well be necessary, however, for filters treating industrial wastewaters and this can be achieved using sewage sludge, biofilm from established percolating filters or waste-activated sludge. Other organisms that play important roles in the biofilm, such as macro-invertebrate grazers, tend to fly into the filter or enter via storm water drains and subsequently colonise the filter.

Analogous to the formation of many types of biofilm, the microorganisms quickly become attached to the available media in the percolating filter. However, the biofilm only develops on surfaces that receive a supply of nutrients. Thus, the effectiveness of the media in redistributing the wastewater within the filter, preventing channelling of the wastewater through a limited portion of the filter and promoting maximum wetting of the medium is an important factor affecting performance. The mature-trickling filter biofilm is an exceptionally complex community comprising bacteria, fungi, protozoa, nematodes and rotifers; plus a wide diversity of macro-invertebrates, such as enchytraeids and lumbricid worms, dipteran fly larvae and a host of other groups, which actively graze the film (including birds and rodents). The formation of the biofilm matrix is mediated by bacteria, such as *Zooglea ramigera* and similar organisms (*Pseudomonas, Chromobacter, Alcaligenes, Achromobacter, Sphaerotilus, Flavobacteriu*), which produce large amounts of extra-cellular polymeric substances (EPSs) (Thörn *et al.*, 1996; Rensink and Rulkens, 1997; Wimpenny *et al.*, 2000). The organisms in the biofilm then form the base of a food web with distinct trophic layers. The organic matter in the wastewater is degraded aerobically by the heterotrophic bacteria, which dominate the biofilm. If the filter is treating domestic sewage, intestinal organisms, such as *E. coli* and pathogens (such as *Salmonella* sp.), will also be present, although they will not generally play an active role. In the deeper biofilm zones, anaerobic bacteria, including sulphate-reducers are commonly observed using direct visualisation and molecular biological approaches (Lens *et al.*, 1995a; Wimpenny *et al.*, 2000). The biofilm has a spongy structure, which is made more porous by the feeding activities of the grazing fauna that are continually burrowing through it. The wastewater passes over the biofilm and to some extent through it, although this depends on film thickness and hydraulic loading. In filters operating under a low hydraulic load ($<3\,m^3/m^3$/day) a large proportion of the wastewater may be flowing through the biofilm matrix at any one time, and it is the physical straining effect of the matrix which allows such systems to produce extremely clear effluents (Qasim, 1994). The higher the hydraulic loading rate, such as applied to filters treating industrial wastewaters ($>3\,m^3/m^3$/day), the greater proportion of the wastewater passes over the film and a slightly inferior effluent quality is generally the result.

9.2.1.2   Relationship between biofilm structure and function in percolating filters

The function of wastewater purification in a percolating filter is mediated solely by the biofilm. The first stage of purification is the adsorption of organic nutrients onto the film. Fine particles are flocculated by EPS secreted by the microorganisms and adsorped onto the surface of the film, where, along with organic nutrients that have been physically trapped, they are broken down by extra-cellular enzymes secreted by heterotrophic bacteria and fungi. The soluble nutrients in the waste-water and those generated as a result of this extra-cellular enzymatic activity are directly absorbed by the biofilm microbes and mineralised. Oxygen diffuses from the air flowing through the filter, first into the liquid and then into the biofilm. Carbon dioxide and the end products of aerobic metabolism diffuse in the other direction. The thickness of the biofilm is critical, as the oxygen can only diffuse for a certain distance through the biofilm before being utilised leaving the deeper layers of the biofilm anoxic or anaerobic (Lens *et al.*, 1995b). The depth to which oxygen will penetrate depends on a number of factors, such as the composition of the biofilm, its density and the rate of respiration within the biofilm and has been estimated to be between 60 μm and 4 mm (Gray, 1999). The critical depth is approximately 200 μm for a predominantly bacterial percolating filter biofilm increasing to 3–4 mm for a fungal biofilm. Only the surface layer of the biofilm is efficient in terms of oxidation, so only a thin layer of film is required for efficient purification, in fact the optimum thickness in terms of performance efficiency is 150 μm. This means that in design terms it is the total surface area of active biofilm, i.e. important and not the total biomass of the biofilm, and this is a critical factor in efficient filter design (Hawkes, 1983).

The biofilm accumulates and becomes thicker over time during filter operation. This is due to both an increase in microbial biomass from use of the substrates in the wastewater and also due to accumulation of particulate material by flocculation and physical entrapment. Once the biofilm exceeds the critical thickness, anoxic and subsequently an anaerobic environment establishes below the aerobic zone. As the biofilm thickness continues to increase most of the soluble nutrients will either be metabolised before they can reach the lower layers of microbes or mass transfer limitations will limit penetration of nutrients, forcing the microbes in the deeper layers into an endogenous phase of growth. This causes instability in the biofilm as the lower layers are broken up, resulting in portions of the biofilm within the filter becoming detached and washed away in the wastewater flow (sloughing). Thick growth of biofilm generally does not reduce the efficiency of the filter, but excessive growth can reduce the void spaces leading to poor oxygen transfer and occasionally blocking them completely, preventing the movement of wastewater. Severe clogging of the void spaces is known as ponding and is normally associated with the surface of the filter. This is because higher concentrations of substrate are present at the surface and also growth of photosynthetic algal mats is promoted through exposure to sunlight. This often results in flooding of large areas of the filter (Hawkes, 1983).

Biofilm accumulation in a percolating filter follows a seasonal pattern being low in the summer due to high metabolic and grazing rates but high in the winter due to a reduced microbial metabolic rate and a reduction in the activity of the grazing fauna (Qasim, 1994). As temperature increases in the spring, a discernible sloughing of the biofilm, which has accumulated during the winter months is generally observed. This phenomenon is because at higher temperatures a greater proportion of the BOD removed by adsorption is oxidised so fewer solids accumulate

(Hawkes, 1983). The rate of oxidation decreases as temperature falls, although the rate of adsorption remains unaltered. Therefore, at lower temperatures (<10°C), there is a rapid increase in solids accumulation, which eventually results in the filters becoming clogged. In warmer months the high-microbial respiration rate may exceed the rate of adsorption, thus reducing the overall biofilm biomass. The grazing fauna also play a significant role in reducing the overall biofilm biomass by directly feeding on the film, converting it into dense faeces that are washed from the filter. The sloughing of the accumulated winter biofilm in the spring is also encouraged by the increased action of the grazers, which loosen the thick biofilm from the medium. Although the grazers suppress maximum film accumulation after sloughing, temperature is the primary factor that controls biofilm accumulation (Qasim, 1994). Hydraulic shear forces are significant in controlling biofilm accumulation only in high-rate filters such as those used for industrial wastewater treatment or those with modification, such as effluent recirculation. In these cases, the physical scouring of the biofilm becomes increasingly important. In well-designed high-rate filters employing modular plastic media the high-hydraulic loading can be used to maintain the biofilm at an optimal thickness by scouring off excess biomass (Qasim, 1994).

9.2.2 Rotating biological contactor operation and biofilm development

9.2.2.1 The rotating biological contactor

The basic design of an RBC consists of a series of disks, which can be flat or corrugated, normally made of plastic and mounted on a horizontal shaft. This shaft is driven mechanically so that the disks rotate at right angles to the incoming wastewater. The disks are spaced 20–30 mm apart and placed in a contoured tank, which fits closely to the rotating medium such that 30–40% of the disk area is immersed at any one time and they are slowly but continuously rotated (Figure 9.2). Biofilm accumulates on both sides of the disk and the organisms are, alternately, exposed to substrate and oxygen during a rotation. The principal of the operation of the RBC is that as the disks rotate, they, alternatively, adsorb organic nutrients from the wastewater and then oxygen from the atmosphere for oxidation. RBC tanks also allow for settlement of sloughed off biomass.

The flow of wastewater through the RBC tank and the action of the rotating medium produces a high-hydraulic shear on the biofilm ensuring efficient mass transfer from the liquid into the biofilm as well as preventing excessive biofilm accumulation. The disks are arranged in groups separated by baffles to reduce the effect of surges in liquid flow and to simulate plug flow (Apilánez et al., 1998). The spacing of the gaps is tapered along the shaft to ensure that the higher biofilm development in the initial stages of the process does not cause the gaps between the disks to become blocked. The rate of rotation varies between 0.75 and 1.0 rpm, although the velocity of the medium through the wastewater should not exceed 0.35 m/s. Faster rotation speeds may cause excessive sloughing of the biofilm,

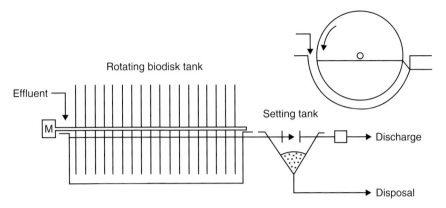

Figure 9.2   Schematic representation of a typical RBC configuration.

while too slow rotation may lead to insufficient aeration and allow settlement of biofilm in the aeration section (Apilánez *et al.*, 1998). Any solids in the treated wastewater are separated out in the settlement zone before discharge. As the disks are well balanced along the drive shaft, little energy is required to drive the motor and the units are enclosed in plastic shrouds to keep out flies, protect the motor from strain due to the wind and to provide camouflage. The main advantages of RBCs over traditional percolating filter reactors include no ponding, no oxygen limitation, low land requirement (<10% that for low-rate filtration), low-sludge production, excellent process control, ease of operation, high BOD removal, no distribution problems, no recirculation required, good settlability of sludge, low-fly nuisance, inconspicuous, low noise and low odour. Disadvantages include cost, loss of treatment during power loss, frequent motor and bearing maintenance and problem of excessive film build-up on discs after power failure resulting in damage and possible failure of the motor when restarted.

9.2.2.2   *Relationship between biofilm structure and function in RBCs*

The biological composition of the biofilm on RBC units has been studied by a number of authors (Okabe *et al.*, 1995, 1998; Neu and Lawrence, 1997; Pérez-Uz *et al.*, 1998; Nikolov *et al.*, 2002). The typical model described by these studies is of a biofilm whose matrix and architecture change with depth, and in basic terms consists of two layers: an outer aerobic layer containing *Beggiatoa* filaments and an inner anaerobic (typically black in colour) layer containing anaerobic bacteria, such as the sulphate-reducing *Desulfovibrio* sp. In outer layers, complex communities dominated by filamentous bacteria (such as *Beggiatoa*, *Sphaerotilus*, *Nocardia* and *Oscillatoria* sp.) are observed which also include protozoa, green eukaryotic algae and small metazoa. Peritrich ciliates (e.g. *Vorticella* sp.) are the most abundant group of protozoan and metazoan communities in these RBC biofilm layers. The architecture of the outer layers consists of a heterogenous distribution of microbial microcolonies or aggregates suspended in a thick network of EPS. The network of EPS has a very porous organisation, with numerous intertices or channels interconnected by fibres that represent a backbone matrix binding together biofilm components. The presence of burrowing protozoa in the biofilm channels has been reported as being essential for the maintenance of porosity in the biofilm matrix of RBC reactors (Martín-Cereceda *et al.*, 2001). A growing body of evidence also suggests, however, that channels in biofilms can originally be formed as a microbial strategy, mediated by cell–cell communication, to solve the problem of nutrient and gas transport to the inner biofilm layers (de Beer and Stoodley, 1995; Okabe *et al.*, 1998).

Inner layers of RBC biofilms contain a greatly reduced density of microorganisms (Okabe *et al.*, 1995; Neu and Lawrence, 1997), and a larger percentage of non-viable bacteria (Martín-Cereceda *et al.*, 2001). In addition, oxygen limitation in the inner biofilm layers means that anaerobic ciliates, such as *Metapus* sp., are the most common protozoa. However, viable aggregates or microcolonies of anaerobic bacteria, including sulphate-reducers, are observed in the inner layers and are likely to be important in attachment of the biofilm to the RBC disk surface. The black colour associated with the inner layers of RBC biofilms, principally, results from the precipitation of ferrous sulphide when iron reacts with the hydrogen sulphide produced by *Desulfovibrio* sp. The ecological relationship between *Desulfovibrio* and *Beggiatoa* has recently been studied. In the anaerobic zone, fermentative bacteria produce organic acids and alcohols that are used by *Desulfovibrio*. The hydrogen sulphide product of sulphate reduction diffuses to the outer aerobic zone where it is utilised by *Beggiatoa* as an electron donor. Hydrogen sulphide is oxidised to elemental sulphur, which is deposited in the cells of *Beggiatoa*. The elemental sulphur is subsequently oxidised to sulphate, which is used by *Desulfovibrio* (Martín-Cereceda *et al.*, 2001). The EPS matrix in the inner biofilms layers on RBC disks has been reported as presenting low porosity when compared to outer layers. This is probably due to the compaction of the biofilm with age, the inner being the oldest layers (Bishop *et al.*, 1995). The presence of EPS in the inner layers strengthens the hypothesis that at least some of the bacterial biomass in the inner biofilm layers is active.

The role of EPS in RBC disk biofilms is equally as important as in other biofilm systems. The composition of EPS on an RBC disk biofilms has been studied by Martín-Cereceda *et al.* (2001). They demonstrated that EPS represented approximately 28% of the volatile solids (VSs) content in the RBC biofilms studied with proteins (17% VSs) and polysaccharides (4% VSs) being the most abundant components.

9.2.3   Alternative aerobic fixed-film reactor systems

9.2.3.1   Biological aerated filters

In biological aerated filters (BAFs) the active biofilm grows over a support medium, i.e. completely submerged in wastewater with oxygen supplied by diffusers at the base of the reactor. As a result of the large surface area of the media used, a large weight of biofilm can develop, making these filters very efficient. The low shear forces exerted on the biofilm results in little natural shearing so the biofilm builds up rapidly, requiring regular backwashing to prevent the filter becoming blocked. Flow in BAFs is not dependent on gravity and so can be either in the upflow or downflow direction.

As for percolating filters, the selection of media is based on specific surface area, surface roughness, durability and cost. Modular plastic media similar to that used in high-rate percolating filters can be used in both upflow and downflow systems, as can random granular media which form a layer at the base of the reactor. This media ranges from 2 to 10 mm in diameter and include both mineral and plastic varieties. A sufficiently rough media surface ensures that enough biofilm is retained during backwashing and that the system can rapidly recover full efficiency. In upflow BAFs a floating medium can be used (plastic with a specific gravity of <1). Aeration is provided via lateral aeration pipes with sparge holes up to 2 mm in diameter. Regular backwashing of BAFs is required to remove excess biofilm that will, eventually, accumulate to the extent that the filter can become blocked. Backwashing is normally carried out by vigorously sparging the filter with air and then rinsing the loose biomass and solids with water. This process is repeated two or three times to leave only a thin layer of active biofilm on the media. After a short recovery time during which the effluent can be turbid due to solids loss, the filter will return to normal efficiency (Gray, 1999).

The high concentration of biomass in BAFs produces effluents of high quality as well as the ability to handle high-organic loading rates (0.25–2.0 kg BOD/m^3/day). As the systems are fully enclosed there is a high degree of operational control and little odour or fly nuisance. The most common application of BAFs at present is for the treatment of industrial wastewaters, although the compact design and high efficiency is also attractive for domestic wastewater treatment. They are, however, significantly more expensive than percolating filters and RBCs in terms of operation due to the costs of forced aeration.

9.2.2.2   Aerobic fluidised beds

Fluidised bed processes also use an inert medium to promote biofilm growth, but in this case the medium is very small, e.g. sand (0.2–2 mm in diameter), glass, anthracite, polyester foam pads or activated carbon granules. This choice of medium provides an extremely high specific surface area compared to other media (>3000 m^2/m^3) allowing a considerable amount of biofilm to develop, equivalent to a suspended solids concentration of up to 40,000 mg/l. Porous media, such as polyester foam pads, allow the growth of biofilm inside as well as on the surface of the media. In these systems the extremely high-biofilm (and biomass) density results in a very high oxygen demand that, in general, can only be met by the use of pure oxygen. The oxygen is injected into the influent stream as it enters the base of the reactor to give a oxygen concentration of up to 100 mg/l. The rate of wastewater input is controlled so that the media inside the filter is expanded so that extremely efficient wastewater–biofilm contact is achieved. Appropriate control of the expansion will allow treated effluent to be removed from the fluidised-bed reactor without loss of media or biomass. Any biomass, i.e. lost is, normally, recovered using a settlement tank. From time to time media is removed from the filter and washed to remove excess biofilm, which comprises the waste sludge from the system (similar to that sloughed off or backwashed from filters). This waste sludge is then treated prior to safe disposal (Gray, 1999). Fluidised beds are highly compact systems, but expensive to operate due to the requirement for pure oxygen and the use of pumps.

9.3   THE BIOLOGICAL PROPERTIES OF ANAEROBIC WASTEWATER TREATMENT SYSTEMS

9.3.1   Anaerobic wastewater treatment systems

The second major microbiological process for organic matter removal from wastewaters is anaerobic digestion. Since the early 1980s, there has been a rapid development in the application of anaerobic digestion for industrial wastewater treatment and it is now generally accepted as a proven technology (Lettinga, 1995). Together with a better understanding of the complex microbiology of the process, a major contributor to the success of anaerobic digestion for industrial wastewater treatment has been the evolution of high-rate biofilm-based reactors in which biomass retention and liquid retention is uncoupled (Lettinga, 1995). This separation allows the slow-growing anaerobic microbes to remain within the reactor, independent of wastewater flow. Biomass concentrations within the reactors are thus high, and consequently, high organic loading rates can be applied to these systems (Forster, 1994). Immobilisation within the reactors is, as for aerobic fixed-film systems, based on biofilm formation achieved either via attachment to support material or by the formation of anaerobic sludge granules, which are particulate biofilms that form spontaneously under appropriate conditions. High rate retained biomass systems currently used in industry include the upflow anaerobic sludge blanket (UASB) (Lettinga and Hulshoff Pol, 1991); the upflow staged sludge bed (USSB) (Van Lier, 1995), the anaerobic baffled reactor (ABR) (Grover et al., 1999), the expanded granular sludge bed (EGSB) (Lettinga, 1996); the AF (Young, 1991); the fluidised bed (FB) (Iza, 1991) and the anaerobic hybrid reactor (AHR) (O'Flaherty and Colleran, 1999). Figure 9.3 summarises the typical configurations employed.

Anaerobic digestion is a proven wastewater treatment solution when applied in conjunction with a post-treatment step (Lettinga et al., 1997). Previously reported difficulties, such as process instability and slow start-up, have been overcome through a better understanding of the principles of the process and the widespread availability of active anaerobic sludge from full-scale installations. In fact, anaerobic treatment offers both ecological and economical advantages over aerobic treatment whenever the wastewater composition is suitable (Svardal et al., 1993).

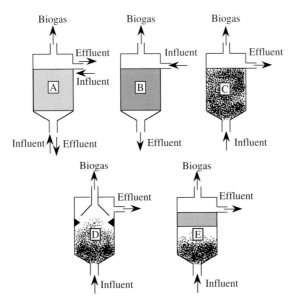

Figure 9.3   Anaerobic digester designs based on biomass retention: (a) AF;
(b) downflow stationary fixed-film reactor; (c) anaerobic fluidised bed reactor;
(d) UASB reactor; EGSB; (e) hybrid UASB/AF reactor.

The primary advantage is that it does not require expensive energy for aeration, but also that carbonaceous pollutants are converted into valuable biogas with an average methane content of approximately 60–70% (Zeikus, 1979; Zehnder *et al.*, 1982). Scammel (1975) estimated that treating a wastewater containing 1 kg glucose takes up to 0.7 kWh for oxygen transfer aerobically, while the same material could yield gas from anaerobic digestion with a calorific value of 3.7 kWh. In relation to the cost and availability of land, Verstraete and Vandevivere (1997, 1999) calculated that the total land area required (m² per inhabitant) for treatment of domestic sewage is 0.20–0.40 for activated sludge treatment whereas it is only 0.05–0.10 for a UASB reactor system.

During anaerobic digestion, up to 80% of the energy content of waste constituents is conserved in the methane product with only a small fraction of the COD being converted into biomass. Accordingly, the volume of surplus sludge generated is far less than by equivalent aerobic processes (Perez *et al.*, 1997). This is simply illustrated by the metabolism of glucose (Schink, 1988).

Aerobic: $C_6H_{12}O_6 + 6O_2 \rightarrow 6CO_2 + H_2O$ $\Delta G^{\circ\prime} = -2826\,\text{kJ/reaction}$
Anaerobic: $C_6H_{12}O_6 \rightarrow 3CH_4 + 3CO_2$ $\Delta G^{\circ\prime} = -394\,\text{kJ/reaction}$

Therefore, in the case of most wastewaters, the net sludge or biomass yield from aerobic systems is in the order of 0.20–0.30 kg VS kg/COD removed, compared with 0.05–0.15 kg VS kg/COD removed during anaerobic digestion (Van Haandel and Lettinga, 1994).

9.3.2   The microbiology of anaerobic digestion

In aerobic ecosystems, heterotrophic bacteria use oxygen as the terminal electron acceptor to carry out complete mineralisation of organic material to carbon dioxide and water. Anaerobic digestion, on the other hand, is a complex process involving the coordinated activity of a number of bacterial trophic groups (Iza *et al.*, 1991) identified on the basis of substrates used and metabolic end products generated. These microorganisms cooperate sequentially in order to achieve degradation of a variety of polymeric and monomeric substrates. Digestion is initiated by the action of facultative and obligate fermentative bacteria, whose enzymes facilitate the hydrolysis of the initial proteins and polysaccharides to monomeric sugars, amino acids, long-chain fatty acids and alcohols (Figure 9.4). Further fermentation of the monomeric products by these and other non-hydrolytic fermentative bacteria result in the generation of a wide variety of fermentation end-products including acetate, formate, methanol, H_2 and CO_2. This initial fermentative activity is known as acidogenesis.

The products of acidogenesis are further oxidised to acetate, hydrogen and carbon dioxide, which is referred to as acetogenesis and is mediated by the obligate hydrogen producing acetogens (OHPA). The action of the OHPAs is considered the link between the initial fermentation stages and the ultimate methanogenic phase (Wolin, 1976). The final step in anaerobic treatment of wastewater is methanogenesis. Methanogens are strictly anaerobic Archaea that can be subdivided into two groups: (i) hydrogenophilic species, which form methane by the reduction of H_2/CO_2 and (ii) acetoclastic methanogens, which generate methane by acetate decarboxylation. OHPA are dependent on the action of hydrogen-utilising species, such as hydrogenophilic methanogens to maintain the hydrogen partial pressure at very low levels, as the reactions carried out by OHPA are only

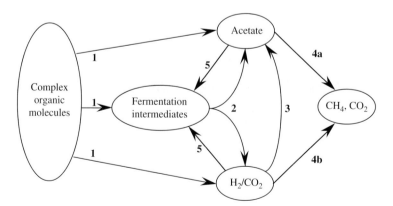

Figure 9.4   Carbon flow in anaerobic digester biofilms – 1: hydrolytic/fermentative bacteria; 2: obligate hydrogen producing bacteria; 3: homoacetogenic bacteria; 4a: acetoclastic methanogens; 4b: hydrogenophilic methanogens; 5: fatty acid synthesising bacteria.

exergonic under these conditions. For this reason the OHPA are commonly referred to as syntrophs. Acetoclastic methanogens are considered the more important methanogenic species, as 70% of the total methane generated is via this pathway (Lettinga, 1995).

9.4   ANAEROBIC GRANULATION – A UNIQUE EXAMPLE OF BIOFILM FORMATION

The UASB reactor (Figure 9.3) was the first reactor design to employ granular anaerobic sludge. Immobilisation of the bacteria into dense, well-settling particulate biofilms (Figure 9.5) is essential for reactor operation and wastewater treatment. The process has been intensively studied and the work of many authors has provided an important model in shaping our understanding of the physical, chemical and ecological factors involved in biofilm formation (Lettinga, 1995 and references therein).

Although the UASB and related systems have been intensively studied for almost two decades (Lettinga, 1995), reactor instability due to biomass loss still causes problems – a common cause of reactor instability is poor granulation. This problem arises because granules are not formed or are not retained within the reactor and biomass loss is not compensated by new growth. Alternatively, biomass is lost by flotation. In this case, the biomass in the reactor is often not present as granules, but as poor settling flocs with low volumetric activity. There have been several theories proposed on the basis for anaerobic sludge granulation, but as yet no conclusive explanation has emerged. This section will summarise the major theories of granule formation in relation to the major environmental and biological factors involved.

Figure 9.5   Anaerobic granules removed from a laboratory-scale UASB reactor. Scale bar = 5 mm.

9.4.1   Physical theories on the formation of anaerobic granules

9.4.1.1   The selection pressure theory

According to the selection pressure theory, the essence of the granulation process is believed to be the continuous selection of well-settling sludge particles that occurs in the reactor (Hulshoff Pol *et al.*, 1983). The selection pressure can be regarded as the sum of the hydraulic loading-rate and the gas loading-rate (dependent on the sludge loading rate). Both factors are important in the selection between sludge components with different settling characteristics.

Under conditions of high-selection pressure, light and dispersed sludge will be washed out while heavier components are retained in the reactor. Thus, growth of finely dispersed sludge is minimised and the bacterial growth is delegated to a limited number of growth nuclei, that can consist of inert organic and inorganic carrier materials or small bacterial aggregates present in the seed sludge (Hulshoff Pol *et al.*, 1987). These growth nuclei increase in size until a certain maximum size after which parts of granules and biofilms are detached producing a new generation of growth nuclei, and so on. The first generation consists of relatively voluminous aggregates, but gradually they become denser due to bacterial growth on the outside and inside of the aggregates. Moreover, bacterial growth is stimulated in the more voluminous aggregates as the substrate can penetrate deeper in the aggregates due to less diffusion limitation and volumetric bacterial activity inside these aggregates as compared to denser aggregates. The filamentous granules that exist in the initial stages of the granulation process become denser due to this ageing process. Under conditions of low selection pressure, growth will take place mainly as dispersed biomass, which gives rise to the formation of a bulking, or poorly settling, type of sludge. In anaerobic reactors, the predominant organism is usually the acetoclastic methanogenic archaeon *Methanosaeta* (formerly *Methanothrix* sp.) which tends to form very long filaments (200–300 μm; Figure 9.6). When these organisms grow without attachment to a solid support particle, a loosely intertwined structure of filaments will be obtained, with very low settling characteristics. Moreover, through the attachment of gas bubbles to these loosely intertwined filaments, the sludge even has a tendency to float (Hulshoff Pol *et al.*, 1987).

9.4.1.2   Growth of colonised suspended solids as the basis for granulation

This theory states that granules originate from fines formed by attrition and from colonisation of suspended solids present in the influent to the reactor (Pereboom, 1994). According to the theory, granule size increase is only due to microbial growth and, therefore, the concentric microbial layers observed on sliced granules (Yoda *et al.*, 1989) are related to small fluctuations in growth conditions rather than specific biological organisation (Pereboom, 1994). Pereboom also

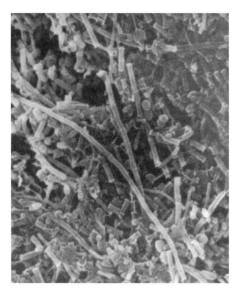

Figure 9.6   *Methanosaeta* filaments in an anaerobic granule (×1700).

reported that the most significant process limiting granule size during normal reactor operation is the regular discharge of surplus biomass, and that reactor turbulence or internal gas production appear to have no influence on the size distribution, i.e. these shear forces are not responsible for breaking or disintegration of granules and only cause attrition of small particles from the granules, which is not expected to be significant to the removal of large granules. Moreover, wastewaters with high concentration of suspended solids result in short size distributions while little or no suspended solids in the influent leads to wide size distributions (Pereboom, 1994).

9.4.2   Ecological theories for granule formation

9.4.2.1   Bridging of microflocs by *Methanosaeta* filaments

From microscopic examination and activity measurements, Dubourguier *et al.* (1987) suggest that granulation starts by a covering (or colonisation) of filamentous *Methnosaeta* sp. by colonies of cocci or rods (acidogenic bacteria), forming microflocs of 10–50 μm. Next, *Methanosaeta* filaments, due to the organisms particular morphology and surface properties, might establish bridges between several microflocs, forming larger granules (>200 μm). Further development of acidogenic bacteria and syntrophic bacteria favours the growth of the granules. Therefore, they support the idea that *Methanosaeta* plays an important role in granule strength by forming a network that stabilises the overall structure but also emphasise the role of EPS and cell walls (Dubourguier *et al.*, 1987).

9.4.2.2   The 'spaghetti theory'

Wiegant (1987) proposed this theory based on a study of granules, with *Methanosaeta* as the predominant bacteria, grown in UASB reactors treating acidified wastewaters, solutions of acetate or mixtures of volatile fatty acids. Although, sludges with *Methanosarcina* as the predominant acetoclastic methanogen species can granulate, these kinds of granules are less practically important in reactors as they bring operational problems and are unstable (Wiegant and De Man, 1986). Therefore, when the relative concentration of *Methanosaeta* bacteria is not high enough, a strong selection bias towards these bacteria must be imposed, which can be done by using low acetate concentrations during the start-up phase, as *Methanosaeta* has a high substrate affinity for acetate, in contrast to *Methanosarcina*. Wiegant divides granule formation into two phases: formation of precursors and actual growth of the granules from these precursors. The first step is considered the key part of granule formation.

During initiation of granule formation, *Methanosaeta* sp. form very small aggregates, due to the turbulence generated by the gas production, or attach to finely dispersed matter. The concentration of suspended solids should not be excessive during this phase or the increase in size of the aggregates will be slow. Selection for aggregates is achieved by imposing increases in the liquid upflow velocity within the reactor. Once the precursors are formed and a proper upflow velocity step-up routine is followed, granulation is inevitable, according to Wiegant (1987). The growth of the individual bacteria and the entrapment of non-attached bacteria lead to the expansion of the precursor particles to form granules, which due to the hydraulic shear forces of biogas and liquid moving up through the sludge bed, acquire a spherical shape. The granules in this phase still present a filamentous appearance, like a ball of spaghetti formed of very long *Methanosaeta* filaments, of which part is loose and part isin bundles. With time, rod-type granules are formed from these filamentous granules when a high biomass retention time is used, due to the increase in the density of the bacterial growth (Wiegant, 1987).

9.4.2.3   The Cape Town hypothesis

According to Sam-Soon *et al.* (1987), granulation depends on *Methanobacterium* strain AZ, an organism that utilises H_2 as its sole energy source and can produce all its amino acids, with the exception of cysteine. When this microorganism is in an environment of high H_2 partial pressure, i.e. excess substrate, cell growth and amino acid production will be stimulated. However, as *Methanobacterium* strain AZ cannot produce the essential amino acid cysteine, cell synthesis will be limited by the rate of cysteine supply. Additionally, if ammonium is available, there will be a high production of the other amino acids, which the organism secretes as extra-cellular polypeptide, binding *Methanobacterium* strain AZ and other bacteria together to form granules. However, the authors admit the possibility that other anaerobic bacteria may have characteristics similar to *Methanobacterium* strain

AZ and, thus, also contribute to granule formation. This hypothesis was proposed following the analysis along the line of flow of a UASB reactor treating a substrate mainly consisting of sugars with negligible nitrogen content and with adequate nutrients and trace elements for growth. The profiles of total soluble COD, volatile fatty acids, ammonium, organic nitrogen, total alkalinity and pH showed the existence of three zones with distinct behaviour in the sludge bed of a UASB reactor, namely a lower and upper active zone and a upper inactive zone (Figure 9.7, Sam-Soon *et al.*, 1987). In the lower active zone, the H_2 partial pressure is high, as can be deduced from the increasing propionic acid concentration. The higher boundary of this lower active zone is defined by the level at which the propionic acid concentration achieves a maximum.

Supporting observations for this hypothesis were that the net sludge production per unit mass of COD was exceptionally high in the high H_2 partial pressure zone, much higher than the yield normally expected in anaerobic systems and that the growth of sludge mass was confined to that high H_2 partial pressure zone. Furthermore, the generation of soluble organic nitrogen in the high H_2 partial pressure zone, combined with a decrease of ammonium, could not be attributed to cell growth or death. In fact, the decrease of ammonium was much more than the experimental maximum growth yield, which means that just a part of the ammonium could have been utilised for protoplasm synthesis. On the other hand, if the generation of organic nitrogen was a result of death of organisms, the death rate would have greatly exceeded the cell growth rate, which means that the death of microorganisms could not explain the observed generation of organic nitrogen in this lower active zone. Thus, the acceptable explanation given for this nitrogen behaviour was that the generation of organic nitrogen was due to the secretion of amino acids by *Methanobacterium* strain AZ, under high H_2 partial pressure and in cysteine-deficient medium and with an adequate supply of NH_4^+-

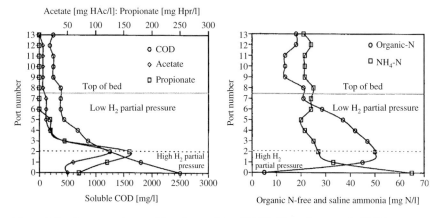

Figure 9.7   Concentration profiles observed in a UASB (Sam-Soon *et al.*, 1987).

N. According to this hypothesis, the conditions that favour granulation are the following:

(1) an environment with a high H_2 partial pressure;
(2) plug flow or semi-plug reactor (in order to achieve phase separation) with a nearly neutral pH;
(3) a non-limiting source of nitrogen, in the form of ammonium;
(4) limiting amounts of cysteine.

According to this theory, granulation is likely to occur during the conversion of carbohydrate substrates in a plug flow system. H_2 is released during the conversion of the carbohydrates to volatile fatty acids, such as propionic. Under high loading conditions, the rate of H_2 uptake by the H_2 utilising organisms is lower than the rate at which H_2 is generated and a region of high H_2 partial pressure develops, which can be maintained in a plug flow system, thus providing conditions for the development of *Methanobacterium* strain AZ. Moosbrugger *et al.* (1990) also reported that, with protein-containing substrate (casein), the granulation in a UASB reactor was easily achieved and that the system behaviour was very similar to the same system treating carbohydrate substrates.

9.4.2.4   *Granules as multi-layered biofilms with* Methanosaeta *aggregates as nucleation centres*

MacLeod *et al.* (1990), working with a UASB-filter hybrid reactor, suggested a hypothesis in which the *Methanosaeta* aggregates function as nucleation centres that initiated granule development of sucrose degrading granules (Figure 9.8). Acetate producers, including H_2-producing acetogens would then attach to this framework, providing the substrate to the *Methanosaeta* sp. and, together with H_2-consuming organisms, form a second layer around the *Methanosaeta* (the syntrophic association between H_2-producing acetogens and H_2-consuming organisms was confirmed by the degradation of propionate in the granular sludge samples). Consecutively, fermentative bacteria adhere to this mini-aggregate forming the exterior layer of the granule and being in contact with its substrates, present in the bulk solution. The products of the fermentative bacteria would then serve as substrates to the underlying acetogens. Moreover, the fact that methanogen-like organisms were found in the exterior layer lead to the idea that these H_2-consuming organisms could consume any free H_2 avoiding its diffusion into the second layer, where other H_2-consuming organisms would then be able to remove the remaining H_2 produced by the acetogens, thus guaranteeing high acetogenic activity.

Fang (2000) also states that granules probably develop as bacteria search for strategic positions for supply of substrates and for removal of products, as the layered microstructure of certain granules suggest (Sekiguchi *et al.*, 1999), and not by the random aggregation of suspended bacteria. Once a nucleus is formed, bacteria start to proliferate leading to a growth of the size of the granule that only stops when the interfacial area between bacteria and the mixed liquor decreases to a

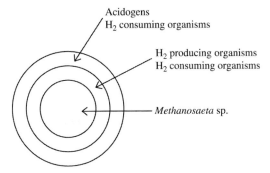

Figure 9.8   Granule as proposed by MacLeod *et al.* (1990).

Appearance	Stage	Diameter	Approximate P_{H2} condition ($\log P_{H2}$, atm)
	(A) growth of filamentous (aceticlastic) methanogens and other microorganisms in low hydrogen partial pressure condition	Filament	Low (≈ -6)
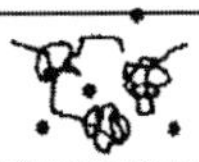	(B) bridging and rolling effects on the growth of filamentous methanogens	<100 µm	
	(C) growth of a small conglomerate as a loose core; crowded syntrophic acetogens around the surface of the core	<1 mm	
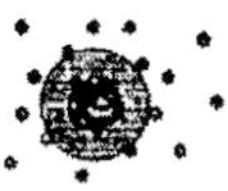	(D) growth of a small granule with a dense core; crowded syntrophic hydro-genotrophs and acidogens around the surface of a small granule	1–2 mm	
	(E) growth of a large granule with multi-layered structure, due to accumulation of extracellular polymers by hydrogenotrophs	2–5 mm	High (-2.7 to -3.7)

Figure 9.9   Ahn's proposed model (2000) for the anaerobic sludge granulation.

critical level in relation to the initial hydrolysis or fermentation that takes place at the granule surface. The microstructure observations support this theory that granules develop by evolution instead of random aggregation of suspended microbes. Vanderhaegen *et al.* (1992), although supporting the multi-layered granule structure proposed by MacLeod *et al.* (1990), states that sugar fermenting acidogens form sufficient biomass and polymers to act as 'nucleation' centres in which the rest of the methanogenic associations can develop.

Ahn (2000) proposed a similar granulation model to that presented in Figure 9.9. At the initial stage of granulation, acetoclastic methanogens (filamentous) and other organisms grow dispersed in the medium. By bridging and rolling effects, due to the hydrodynamic behaviour of the UASB, small loose conglomerates mainly composed of the filamentous methanogens are eventually formed. Subsequently, acetogens attach to this conglomerate, in syntrophic relationship with the acetoclastic methanogens, thus, forming a small granule with a dense core. Then, acidogens and hydrogenotrophs in syntrophic relationship with the acetogens crowd the small granule and due to the EPS excretion by the hydrogenotrophs the final granule structure evolves.

9.4.3   Thermodynamic theories for granule formation

9.4.3.1   Four step model for biofilm formation

Schmidt and Ahring (1996) suggest that the granulation process in UASB reactors follows the well-described four steps of biofilm formation (Costerton *et al.*, 1987; Verrier *et al.*, 1988; Gantzer *et al.*, 1989; Van Loosdrecht and Zehnder, 1990):

(1) *Transport* of cells to the surface of an uncolonised inert material or other cells (substratum);
(2) Initial reversible *adsorption* by physico-chemical forces to the substratum;
(3) Irreversible *adhesion* of the cells to the substratum by microbial appendages and/or polymers; and
(4) *Multiplication* of the cells and development of the granules.

In a UASB reactor, the cells are transported by one or a combination of the following mechanisms: diffusion (Brownian motion), advective (convective) transport by fluid flow, gas flotation or sedimentation. The initial adsorption can take place after a collision between the cells and the substratum. The substratum can either be other cells or bacterial aggregates present in the sludge or organic or inorganic materials that can function as growth nuclei (Schmidt and Ahring, 1996). The initial adsorption can be approximately described by the DLVO theory, presented by Derjaguin, Landau, Verwey and Overbeek, between 1940 and 1950, with the aim of explaining colloid stability. This theory can explain and/or predict microbial adhesion using calculations of adhesion free energy changes. By using this theory, the assumption is made that bacteria behave as inert particles and that bacterial adhesion can be understood by a physico-chemical approach. The DLVO theory postulates that the total long-range interaction over a distance of more than 1 nm is a summation of Van der Waals and Coulomb (electrostatic) interactions. According to this theory, three different situations can occur:

(1) a repulsion when electrostatic interactions dominate;
(2) a strong irreversible attraction when Van der Waals forces are dominant (primary minimum);
(3) a weak, reversible attraction when cells are located a certain distance from each other (secondary minimum).

The initial adhesion takes place predominantly in the secondary minimum of the DLVO free energy curve. The strength of adsorption depends on different physico-chemical forces like ionic, dipolar, hydrogen bonds, or hydrophobic interactions. The secondary minimum does not usually reach large negative values and particles captured in this minimum generally show reversible adhesion. In this case, there is a separation distance between the adhering bacteria and a thin water film remains present between the interacting surfaces. However, if a bacterium can reach the primary minimum, short-range interaction forces become effective and irreversible adhesion occurs. Irreversible adhesion can occur due to specific bacterial characteristics such as appendages, cell surface structures or polymers (van Loosdrecht and Zehnder, 1990; Schmidt and Ahring, 1996). However, it is not clear if bacteria first adhere reversibly and then produce EPS or if bacteria first produce EPS and then adhere irreversibly (Schmidt and Ahring, 1996). When the bacterium is adhered, colonisation has started. The immobilised cells start to divide within the EPS matrix so that the cells are trapped within the biofilm structure. This results in the formation of microcolonies of identical cells. The granulation process depends on cell division and recruitment of new bacteria from the liquid phase. The granular matrix can also contain trapped extraneous molecules, e.g. precipitates (Costerton *et al.*, 1990). The organisation of the bacteria in the granules can ease the transfer of substrates and products. The arrangement may depend on local hydrophobicity, local presence of polymers, or cell geometry (Schmidt and Ahring, 1996).

9.4.3.2   The surface tension theory

Thaveesri *et al.* (1995) related the adhesion of bacteria involved in anaerobic consortia in UASB reactors to surface thermodynamics. They found that bacteria can only obtain the maximum possible free energy of adhesion (ΔG_{adh}) when the liq-

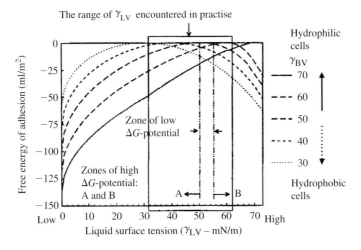

Figure 9.10   Free energies of adhesion (ΔG_{adh}) for bacteria with different γ_{BV} values as a function of γ_{LV} (Thaveesri *et al.*, 1995).

uid surface tension (γ_{LV}) is sufficiently low or high, as indicated in Figure 9.10. In the high-γ_{LV} region (zone B), low-energy surface types of bacteria (low bacterium surface tension (γ_{BV}) or hydrophobic bacteria) can adhere in order to obtain minimal energy, while in the low-γ_{LV} region (zone A), high-energy surface types of bacteria (high-γ_{BV} or hydrophilic bacteria) exhibit a greater decrease in free energy upon aggregation and thus are selected to compose aggregates. A third zone is arbitrarily defined between γ_{LV} values of 50–55 mN/m, and in this zone aggregation of neither hydrophobic nor hydrophilic cells is favoured (low ΔG_{adh} potential). Thus, operating a system at a high γ_{LV} should favour aggregation of (rather) hydrophobic bacteria, and operating a system at a low γ_{LV} should favour aggregation of (rather) hydrophilic bacteria.

9.4.3.3   The proton translocation-dehydration theory

Tay *et al.* (2000) proposed a theory for the (molecular) mechanism of sludge granulation, based on the proton translocation activity at bacterial membrane surfaces. In this theory, the sludge granulation process was considered to proceed in the four following steps (Figure 9.11):

(1) Dehydration of bacterial surfaces;
(2) Embryonic granule formation;
(3) Granule maturation; and
(4) Post-maturation.

Hydrophobic interaction between the bacterial surfaces is considered supportive for the initiation of bacterial adhesion (Mahoney *et al.*, 1987; Van Loosdrecht *et al.*, 1987). However, with decreasing surface separation distance between two bacterial cells, strong repulsive hydration interactions between the two approaching bacteria exist, due to the energy required for the removal of the tightly bonded water from the bacterial surfaces. In fact, under normal physiological conditions, a bacterial surface has a high negative charge which facilitates hydrogen bonding with water molecules, resulting in a network of water surrounding the bacterial surface (Smith and Wood, 1991), i.e. a hydration layer. However, the hydration repulsion does not normally affect the initial step of the bacterial reversible adhesion stage to a significant extent. The authors argue that acidogenic bacteria, during the acidification of substrates pump protons from the cytoplasmic side of the membrane to the exterior surface of the membrane. This proton translocation activity energises the surface and may induce breaking of the hydrogen bonds between the negatively charged groups and the water molecules and partial neutralisation of the negative charges on their surfaces, thus causing the dehydration of the cell surfaces. As a consequence of the upflow hydraulic stress, of this weakened hydration repulsion and of the hydrophobic nature of the cells, acidogens, acetogens and methanogens may adhere to each other forming embryonic granules. Moreover, due to the transfer of metabolites between cells, a further dehydration of the bacterial surfaces takes place leading to a strengthening of these initial granules.

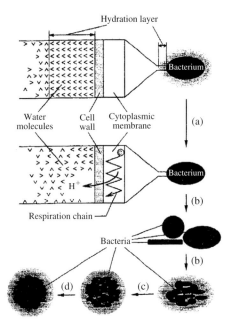

Figure 9.11   Schematic representation of proton translocation – dehydration model for sludge granulation: (a) dehydration of bacterial surfaces; (b) embryonic granule formation; (c) granule maturation; (d) post-maturation (Tay *et al.*, 2000).

The authors claim that some phenomena of sludge granulation like the advisable high-energy carbohydrate feeding during the UASB start-up period, the granular sludge washout when changing the carbon source, the existence of both uniform and layered granules, and the influence of EPS in the granulation process can be adequately explained by this translocation – dehydration theory.

9.5   CONCLUSIONS

This chapter described two wastewater treatment biofilm case studies. The role of biofilms in a wide range of wastewater treatment systems is vital. However, much research is still required to fully elucidate the environmental factors involved in biofilm formation and the ecological relationships between the microorganisms involved. Recent methodological advances, including molecular probing, microelectrodes and NMR described in this volume will clearly advance the field considerably. This information will, undoubtedly, lead to improved technological applications and better understanding of the nature of microbial communities in engineered ecosystems.

Acknowledgements

The authors would like to thank Gavin Collins and Aileen Kearney for photographs.

REFERENCES

Ahring, B.K., Schmidt, J.E., Winther-Nielsen, M., Macarop, A.J.L. and Macario, E.C. (1993) Effect of medium composition and sludge removal on the production, composition, and architecture of thermophilic (55°C) acetate-utilizing granules from and upflow anaerobic sludge blanket reactor. *Appl. Environ. Microbiol.* **59**(8), 2538–2545.

Ahn, Y.-H. (2000) Physicochemical and microbial aspects of anaerobic granular pellets. *J. Environ. Sci. Health* **A35**(9), 1617–1635.

Apilánez, I., Gutiérrez, A. and Díaz, M. (1998) Effect of surface materials on initial biofilm development. *Biores. Technol.* **66**(3), 225–230.

Bishop, P.L., Zhang Tian, C. and Yun-Chang, F. (1995) Effects of biofilm structure, microbial distributions and mass transport on biodegradation processes. *Water Sci. Technol.* **31**(1), 143–152.

Costerton, J.W., Cheng, K.J., Geesey, G.G., Ladd, T.I., Nickel, J.C., Dasgupta, M., Marrie, T.J. (1987) Bacterial biofilms in nature and disease. *Ann. Rev. Microbiol.* **41**, 435–464.

Costerton, J.W., Marrie, T.J. and Cheng, K.J. (1990) Phenomena of bacterial adhesion. In *Bacterial Adhesion, Mechanisms and Physiological Significance* (ed. D.C. Savage and M. Fletcher), pp. 3–43, Plenum Press, New York.

Costerton, J.W., Lewandowski, Z., Caldwell, D.E., Korber, D.R. and Lappin-Scott, H.M. (1995) Microbial biofilms. *Ann. Rev. Microbiol.* **49**, 1–45.

DeBeer, D. and Stoodley, P. (1995) Relationship between the structure of an aerobic biofilm and transport phenomena? *Water Sci. Technol.* **32**(8), 11–18.

De Zeeuw, W.J. (1987) Granular sludge in UASB reactors. In *Granular Anaerobic Sludge: Microbiology and Technology* (ed. G. Lettinga, A.J.B. Zehnder, J.T.C. Grotenhuis and L.W. Hulshoff Pol), pp. 132–145, Pudoc, Wageningen, The Netherlands.

Dubourgier, H.C., Prensier, G. and Albagnac, G. (1987) Structure and microbial activities of granular anaerobic sludge. In *Granular Anaerobic Sludge: Microbiology and Technology* (ed. G. Lettinga, A.J.B. Zehnder, J.T.C. Grotenhuis and L.W. Hulshoff Pol), pp. 18–33, Pudoc, Wageningen, The Netherlands.

Fang, H.H.P. (2000) Microbial distribution in UASB granules and its resulting effects. *Water Sci. Technol.* **42**(12), 201–208.

Flemming, H.C. (1993) Biofilms and environmental protection. *Water Sci. Technol.* **27**(1), 1–10.

Forster, C. (1994) Anaerobic digestion and industrial wastewater treatment. *Chem. Indus.*, 6 June, 404–406.

Gantzer, C.J., Cuunningham, A.B., Gujer, W., Gutekunst, B., Heijnen, J.J., Lightfoot, E.N., Odham, G., Rittmann, B.E., Rosenberg, E., Stolzenbach, K.D. and Zehnder, A.J.B. (1989) Exchange processes at fluid-biofilm interface. In *Structure and Function of Biofilms* (ed. W.G. Characklis and P.A. Wildered), pp. 73–90, Wiley, New York.

Gray, N.F. (1999) *Water Technology: An Introduction for Environmental Scientists and Engineers.* Arnold publishers, London.

Grover, R., Marwaha, S.S. and Kennedy, J.F. (1999) Studies on the use of an anaerobic baffled reactor for the continuous anaerobic digestion of pulp and paper mill black liquors. *Process Biochem.* **34**, 653–657.

Hawkes, H.A. (1983) The applied significance of ecological studies of aerobic processes. In *Ecological Aspects of Used Water Treatment: The Processes and Their Ecology* (ed. C.R. Curds and H.A. Hawkes), Vol. 3, pp. 163–217, Academic Press, London.

Hulshoff Pol, L.W. (1989) The phenomenon of granulation of anaerobic sludge. Ph.D. thesis, Agricultural University, Wageningen, The Netherlands.

Hulshoff Pol, L.W., De Zeeuw, W.J., Velzeboer, C.T.M. and Lettinga, G. (1983) Granulation in UASB reactors. *Water Sci. Technol.* **15**(8/9), 291–304.

Hulshoff Pol, L.W., Heijnekamp, K. and Lettinga, G. (1987) The selection pressure as a driving force behind the granulation of anaerobic sludge. In *Granular Anaerobic Sludge: Microbiology and Technology* (ed. G. Lettinga, A.J.B. Zehnder, J.T.C. Grotenhuis and L.W. Hulshoff Pol), pp. 153–161, Pudoc, Wageningen, The Netherlands.

Iza, J. (1991) Fluidised bed reactors for anaerobic wastewater treatment. *Water Sci. Technol.* **24**(1), 109–132.

Iza, J., Colleran, E., Paris, J.M. and Wu, W.M. (1991) International workshop on anaerobic treatment technology for municipal and industrial wastewaters: summary paper. *Water Sci. Technol.* **24**(1), 1–16.

Lens, P., Massone, A., Rozzi, A. and Verstraete, W. (1995a) Effect of sulfate concentration and scraping on aerobic fixed film reactors. *Water Res.* **29**(3), 857–870.

Lens P.N., De Poorter, M.-P., Cronenberg, C.C. and Verstraete, W.H. (1995b) Sulfate reducing and methane producing bacteria in aerobic wastewater treatment. *Water Res.* **29**(3), 871–880.

Lettinga, G. (1995) Anaerobic digestion and wastewater treatment systems. *Antonie van Leeuwenhoek* **67**, 3–28.

Lettinga, G. (1996) Sustainable integrated biological wastewater treatment. *Water Sci. Technol.* **33**(1), 85–98.

Lettinga, G. and Hulshoff Pol, L.W. (1991) UASB-process design for various types of wastewater. *Water Sci.Technol.* **67**(1), 3–28.

Lettinga, G., Van Velsen, A.F.M., Hobma, S.W., De Zeeuw, W. and Klapwijk, A. (1980) Use of the upflow sludge blanket (USB) reactor concept for biological waste-water treatment, specially for anaerobic treatment. *Biotechnol. Bioeng.* **22**, 699–734.

Lettinga, G., Hulshoff Pol, L.W., Zeeman, G., Field, J., Van Lier, J.B., Van Buuren, J.C.L., Jannsen, A.J.H. and Lens, P. (1997) Anaerobic treatment in sustainable environmental production concepts. *Proceedings of the 8th International Conference on Anaerobic Digestion*, Sendai International Centre, Sendai, Japan, **1**, 32–39.

Mahoney, E.M., Varangu, L.K., Cairns, W.L., Kosaric, N. and Murray, R. (1987) The effect of calcium on microbial aggregation during UASB reactor start-up. *Water Sci. Technol.* **19**(4), 249–260.

Martín-Cereceda, M., Pérez-Uz, B., Serrano, S. and Guinea, A. (2001) Dynamics of protozoan and metazoan communities in a full scale wastewater treatment plant by rotating biological contactors. *Microbiol. Res.* **156**(3), 225–238.

McLeod, F.A., Guiot, S.R. and Costerton, J.W. (1990) Layered structure of bacterial aggregates produced in an upflow anaerobic sludge bed and filter reactor. *Appl. Environ. Microbiol.* **56**(6), 1598–1607.

Metcalf and Eddy (1991) *Wastewater Engineeering: Treatment, Disposal and Reuse*, McGraw-Hill, New York.

Moosbrugger, R.E., Loewenthal, R.E. and Marais, G.R. (1990) Pelletization in a UASB system with protein (casein) as substrate. *Water SA* **16**(3), 171–178.

Neu, T.R. and Lawrence, J.R. (1997) Development and structure of microbial biofilms river water studied by confocal laser scanning microscopy. *FEMS Microbiol. Ecol.* **24**(1), 11–25.

Nikolov, L., Karamanev, D., Mamatarkova, V., Mehochev, D. and Dimitrov, D. (2002) Properties of the biofilm of *Thiobacillus ferrooxidans* formed in rotating biological contactor. *Biochem. Eng. J.* **12**, 43–48.

O'Flaherty, V. and Colleran, E. (1999) Effect of sulphate addition on volatile fatty acid and ethanol degradation in an anaerobic hybrid reactor I: process disturbance and remediation. *Biores. Technol.* **68**, 101–107.

Okabe, S., Hirata, K. and Watanabe, Y. (1995) Dynamic changes in spatial microbial distribution in mixed-population biofilms: experimental results and model simulation. *Water. Sci Technol.* **32**(1), 67–74.

Okabe, S., Kuroda, H. and Watanabe, Y. (1998) Significance of biofilm structure on transport of inert particulates into biofilms. *Water Sci. Technol.* **38**(1), 163–170.

Pereboom, J.H.F. (1994) Size distribution model for methanogenic granules from full scale UASB and IC reactors. *Water Sci. Technol.* **30**(12), 211–221.

Perez, M., Romero, L.I. and Sales, D. (1997) Immobilisation of anaerobic thermophilic biomass. *Proceedings of the 8th International Conference on Anaerobic Digestion*, Sendai International Centre, Sendai, Japan, Vol. 3, 165–168.

Pérez-Uz, B., Franco, C., Martín-Cereceda, M., Arregui, L., Campos, I., Serrano, S., Guinea, A. and Fernández-Galiano, D. (1998) Biofilm characterization of several wastewater treatment plants with rotating biological contactors in Madrid (Spain). *Water Sci. Technol.* **37**(11), 215–218.

Qasim, S.R. (1994) *Waste Water Treatment Plants: Planning, Design and Operation*, Technomic Publishing, Lancaster, PA, USA.

Rensink, J.H. and Rulkens, W.H. (1997) Using metazoa to reduce sludge production. *Water Sci. Technol.* **36**(11), 171–179.

Sam-Soon, P.A.L.N.S., Loewenthal, R.E., Dold, P.L. and Marais, G.R. (1987) Hypothesis for pelletisation in the upflow anaerobic sludge bed reactor. *Water SA* **13**(2), 69–80.

Scammel, G.W. (1975) Anaerobic treatment of industrial wastes. *Process Biochem.*, 34–36.

Schink, B. (1988) Principles and limits of anaerobic degradation: environmental and technological aspects. In *Biology of Anaerobic Microorganisms* (ed. A.J.B. Zehnder), pp. 771–846, John Wiley, New York.

Schmidt, J.E. and Ahring, B.K. (1996) Review: granular sludge formation in upflow anaerobic sludge blanket (UASB) reactors. *Biotech. Bioeng.* **49**, 229–246.

Sekiguchi, Y., Kamagata, Y., Nakamura, K., Ohashi, A. and Harada, H. (1999) Flourescence *in situ* hybridisation using 165 rRNA-targeted oligonucleotide reveals localisation of methanogens and selected uncultured bacteria in mesophilic and thermophilic sludge granules. *Appl. Env. Microbiol.* **65**, 1280–1288.

Smith, C.A. and Wood, F.J. (1991) *Molecular and Cell Biochemistry: Biological Molecules*, Chapman & Hall, London.

Svardal, K., Gotzendorfer, K., Nowak, O. and Kroiss, H. (1993) Treatment of citric acid wastewater for high quality effluent on the anaerobic–aerobic route. *Water Sci. Technol.* **28**(6), 177–186.

Tay, J.H., Xu, H.L. and Teo, K.C. (2000) Molecular mechanism of granulation. I: H^+ translocation – dehydration theory. *J. Environ. Eng.* **126**, 403–410.

Thaveesri, J., Daffonchio, D. Liessens, B, Vandermeren, P. and Verstraete, W. (1995) Granulation and sludge bed stability in upflow anaerobic sludge bed reactors in relation to surface thermodynamics. *Appl. Environ. Microbiol.* **61**(10), 3681–3686.

Thörn, M., Matteson, A. and Sorensson, F. (1996) Biofilm development in a nitrifying trickling filter. *Water Sci. Technol.* **34**(1/2), 83–89.

Vanderhaegen, B., Ysebaert, K., Favere, K., van Wambeke, M., Peeters, T., Panic, V., Vandenlangenbergh, V. and Verstraete, W. (1992) Acidogenesis in relation to in-reactor granule yield. *Water Sci. Technol.* **25**(7), 21–30.

Van Haandel, A.C. & Lettinga, G. (1994) Application of anaerobic digestion to sewage treatment. In *Anaerobic Sewage Treatment: A Practical Guide for Regions with a Hot Climate* (ed. A. Van Haandel and G. Lettinga), pp. 33–59, John Wiley, New York.

Van Lier, J.B. (1995) Thermophilic anaerobic wastewater treatment; temperature aspects and process stability. Ph.D. thesis, Wageningen Agricultural University, The Netherlands.

Van Loosdrecht, M.C.M., Lykema, J., Norde, W., Schraa, G. and Zehnder, A.J.B. (1987) The role of bacterial cell wall hydrophobicity in adhesion. *Appl. Environ. Microbiol.* **53**(8), 1893–1897.

Van Loosdrecht, M.C.M. and Zehnder, A.J.B. (1990) Energetics of bacterial adhesion. *Experientia* **46**, 817–822.

Verrier, D., Mortier, B., Dubourguier, H.C. and Albagnac, G. (1988) Adhesion of anaerobic bacteria to inert supports and development of methanogenic biofilms. In *Anaerobic Digestion* (ed. E.R. Hall and P.N. Hobson), pp. 61–70, Pergamon Press, Oxford.

Verstraete, W. and Vandevivere, P. (1997) Broader and newer applications of anaerobic digestion. *Proceedings of the 8th International Conference on Anaerobic Digestion*, Sendai International Centre, Sendai, Japan, Vol. 1, 67–74.

Verstraete, W. and Vandevivere, P. (1999) New and broader applications of anaerobic digestion. *Crit. Rev. Environ. Sci. Technol.* **28**, 151–173.

Wiegant, W.M. (1987) 'The spaghetti theory' on anaerobic sludge formation, or the inevitability of granulation. In *Granular Anaerobic Sludge: Microbiology and Technology* (ed. G. Lettinga, A.J.B. Zehnder, J.T.C. Grotenhuis and L.W. Hulshoff Pol), pp. 146–152, Pudoc, Wageningen, The Netherlands.

Wiegant, W.M. and De Man, A.W.A. (1986) Granulation of biomass in thermophilic upflow anaerobic sludge blanket reactors treating acidified wastewaters. *Biotechnol. Bioeng.* **28**, 718–727.

Wimpenny, J., Manz, W. and Szewyk, U. (2000) Heterogeneity in biofilms. *FEMS Microbiol. Rev.* **24**, 661–667.

Wolin, M.J. (1976) Microbial formation and utilisation of gases (H_2, CH_4, CO). In *Interactions between H_2-Producing and Methane-Producing Bacteria* (ed. H.G. Schlegel, G. Gottschalk and N. Pfenning), pp. 141–150, Koltze and Gottingen.

Yoda, M., Kitagawa, M. and Miyaji, Y. (1989) Granular sludge formation in the anaerobic expanded micro carrier process. *Water Sci. Technol.* **21**, 109–122.

Young, J.C. (1991) Factors affecting the design and performance of upflow anaerobic filters. *Water Sci. Technol.* **24**(7), 199–205.

Zehnder, A.J.B., Ingvorsen, K. and Marti, T. (1982) Microbiology of methane bacteria. Paper presented at the anaerobic digestion 1981, *Proceedings of the 2nd International Symposium on Anaerobic Digestion*, Travemunde, Germany, pp. 45–68.

Zeikus, J.G. (1979) Microbial populations in digesters. Paper presented at the *Proceedings of the 1st International Symposium on Anaerobic Digestion*, Cardiff, Wales, pp. 61–87.

10

Bioaerosols and biofilms

S.G. Jennings, A.P. Moran and C.V. Carroll

10.1 INTRODUCTION

Biofilms not only develop adherent to a solid surface, the so-called substratum, at solid–water interfaces and at solid–air interfaces (e.g. on soil particles), but also occur at the water– or liquid–air interface (e.g. aquatic environments) (Flemming *et al.*, 2000). Disturbance of the liquid surface by mechanical or natural means can lead to the liberation of droplets or aerosols derived from the liquid–air biofilm into the atmosphere. Likewise, soil particles and soil-derived aerosols can be liberated into the atmosphere.

The field of aerobiology is often considered the microbiology of the atmosphere. It is the study of the movement of biological particles or products of organisms within the atmosphere or an indoor environment. The interest in aerobiology has focused mainly on the occurrence of microorganisms in the air and on their dependence on environmental factors. Aerobiology is an interdisciplinary science, with meteorology and aerosol physics at its core, but with applications in the fields of microbiology, agriculture, plant pathology, allergology, public health, biochemistry, and immunology. Aerobiology often focuses on the effect of airborne agents on human health, the characteristics of indoor aerosols, and on outdoor aerosols that enter the indoor environment (Burge, 1995a, b).

An aerosol can be defined as a suspension or dispersion system of solid or liquid particles suspended in the air or in another gaseous phase. Air will often contain microorganisms, such as bacteria, fungi, protozoa, and viruses. Although these microorganisms do not live in air, they can be frequently transported in air while attached to other solid particles, e.g. soil and dust, or incorporated into liquid particles of aerosols. When aerosols contain microorganisms (bacteria, viruses, fungi, and protozoa) or their by-products (DNA, RNA, lipopolysaccharides (endotoxins), mycotoxins, proteins, lipids, and carbohydrates), the aerosol is termed a bioaerosol. The size of bioaerosols can vary in diameter from 0.5 to about 100 μm.

Despite bioaerosols being widespread in the earth's tropospheric boundary layer, they normally occur in low concentrations (Lighthart and Stetzenbach, 1994; Matthias-Maser and Jaenicke, 1994; Lighthart and Shaffer, 1995). As a result, air or surface samples will almost always contain some bacteria or fungi. Bioaerosols may be generated through natural processes or by human activity. They are found in many environments including residential houses (Burge, 1995b) and in work environments, such as sewage treatment plants and recycling or composting plants (Bünger et al., 2000).

10.2   RELEVANCE OF BIOAEROSOLS

Bioaerosols are capable of transporting microorganisms over long distances (Bovallius et al., 1978), and generally, do not pose problems when the concentration and the various types are kept within reasonable limits, but when these parameters fail they can be detrimental to health. Typically, there are two types of health effects associated with exposure to bioaerosols, namely, infectious (Frazer, 1980) and allergenic (Gravesen, 1979; Griffiths and DeCosemo, 1994).

Infections occur when living microorganisms, such as bacteria, fungi or viruses, invade and multiply in the host and cause disease. The most common route of exposure for aerosols is by the respiratory system, but infections can also affect the skin. Symptoms (such as a congested nose, sore throat, dry cough) as well as toxic reactions (such as toxic pneumonitis) have been reported (Rylander, 1994). Large, airborne particles remain in the upper respiratory tract, nose, and nasopharynx (Zeterberg, 1973), whereas particles <6 μm in diameter can be transported to the lungs. Smaller particles (<2 μm) tend to be retained in the alveoli (Salem and Gardner, 1994). It is the smallest particles (approximately 0.65 μm in diameter or smaller) that are the most harmful and most detrimental to health.

Depending on the type of bioaerosol presented, different symptoms due to exposure may occur. For example, bioaerosols containing bacteria may cause hypersensitive pneumonia, infections, or mucus membrane irritations, but if the bacteria present are those where the cell wall contains lipopolysaccharides, then they may cause headache, fever, loss of balance, muscle aches, nausea, and respiratory distress to humans (Stetzenbach, 1997).

Allergic reactions can be induced by the whole microorganism, a fragment of the organism, or its by-products. Wan and Li (1999) found a strong association between β(1,3)-glucan and lethargy or fatigue among personnel of day-care centres, homes and office buildings. Schwartz *et al.* (1995) reported that the concentration of endotoxin in the bioaerosol may be, particularly, important in the development of grain dust-induced lung disease. Moreover, many laboratory-based studies have been performed using animal model systems, and a number of models are currently used for the study of bioaerosol-induced lung responses alone (Thorne, 2000).

Several diseases transmitted between humans are acquired by inhaling bioaerosols of viral or bacterial origin from infected persons. The more important viral diseases acquired in this manner include flu, chickenpox, measles, German measles, and mumps. The more important bacterial diseases include pneumonia, tuberculosis, whooping cough, meningitis, diphtheria, and smallpox (Benenson, 1995).

Diseases that are acquired by inhaling particles and bioaerosols from environmental sources other than from an infected person include:

(1) Psittacosis which is caused by *Chlamydia psittaci*. Particles and bioaerosols containing *Ch. psittaci* can be generated from dried, powdery droppings of infected birds, e.g. pigeons.
(2) Legionnaire's disease caused by *Legionella pneumophila*, can be transmitted by droplets from air-conditioning systems, water storage tanks, etc., where the bacterium grows.
(3) Acute allergic alveolitis, which is caused by various fungal and actinomycete spores. Bioaerosols and particles containing these spores can be generated from decomposing organic matter (e.g. compost heaps, grain stores, hay, etc.). Aspergillosis which is caused by inhalation of fungal spores of a number of *Aspergillus* spp. are generated from decomposing organic matter.
(4) Histoplasmosis is caused by *Histoplasma capsulatum* spores derived from old, weathered bat or bird droppings. Coccidioidomycosis results from exposure to *Coccidioides immitis* spores in air-blown dust in the desert.

Another, more modern source of bioaerosols arises from the sorting and recycling of domestic waste (Hryhorczuk *et al.*, 2001). The causes and possible health problems associated with these sources of bioaerosols are reviewed extensively elsewhere (Poulsen *et al.*, 1995).

The development of the science of aerobiology has offered a means of investigating potentially hazardous aerosols generated from a variety of sources, e.g. domestic (Burge, 1995a), agricultural (Hameed and Khodr, 2001), and industrial and municipal sources (Hryhorczuk *et al.*, 2001; Krajewski *et al.*, 2001). To name, but a few examples, Ross *et al.* (2000) provided evidence for a relationship between asthma severity and levels of total bacteria and Gram-negative bacteria found in the home. Trout *et al.* (2001) reported the case of a worker with a respiratory illness related to bioaerosol exposure in a water-damaged building with extensive fungal contamination. Bünger *et al.* (2000) found that high exposure to bioaerosols of compost workers was significantly associated with a higher frequency of health complaints and diseases, as well as higher concentrations of specific antibodies against moulds and actinomycetes.

10.3   AEROSOLS, AEROSOL PARTICLES, DROPLETS, AND BIOAEROSOLS

The disturbance of a liquid surface can result in the generation of droplets and aerosols. As defined above, an aerosol is a suspension of solid or liquid particles in a gas medium, which is normally air. An aerosol particle consists of some amount of solid particulate matter of variable composition. A liquid particle or droplet is usually present when the medium (air) is slightly super-saturated or even slightly under-saturated with respect to water vapour, and contains particles or nuclei upon which the available water vapour condenses to create a liquid droplet. Subsequent growth of the droplet is dependent on the supply of additional water vapour. Importantly, bioaerosols contain particles of biological origin or biological activity that are (or were) derived from (dead or alive) organisms (Matthias-Maser, 1998).

10.3.1   Characteristics

The microorganism-containing liquid particles produce so-called drop nuclei which exist only for short durations. Bacterial cells when they become airborne normally rapidly die within a few seconds due to evaporation of water associated with the particle. Thus, with higher humidity, higher bioaerosol levels can prevail. Airborne fungal cells (yeasts, moulds, and their spores) can remain viable for much longer periods, even at low relative humidity and high or low temperature extremes.

For bacterial infections to occur, the characteristics which determine if exposure to the organism will cause disease include the bacterial genus and species, the microbial growth requirements, extent of microbial replication, toxicity, and concentration in the air. However, in the case of bacteria dispersed in bioaerosols, the size, shape, density, and surface characteristics of the microbial cell or the particle/droplet are also relevant to exposure (McCullough *et al.*, 1997).

The microorganisms' existence in an airborne state is of limited duration since multiplication does not occur in this environment. Their existence also depends on physical factors, such as the mass and the aerodynamic shape of the microbial aerosol particles, and the influence of the air currents. As with inert dust particles, all bioaerosols are governed by the laws of gravity, and will be affected by air movements, being transported by turbulence and diffusion. Some of their physical characteristics are discussed below.

10.3.1.1   Size

The size of aerosol particles can vary by several orders of magnitude. The upper size of aerosol particles or droplets is determined by their fall or sedimentation velocity and by updraft velocity values in the lower troposphere. These factors control the time of suspension of aerosol in the atmosphere or medium. A diameter of about 100 µm is generally accepted as the upper limit for most aerosols. A lower detectable limit of particle size is about 1 nm in diameter. Thus, aerosol

Table 10.1   Representative particle diameter
for some aerosol constituents (Schlegel, 1984).

Aerosol type	Size (diameter) range, μm
Bacteria	0.5–5
Cloud	1–100
Fungal spores	1–50
Pollen grains	5–200
Sea salt	>0.2
Soil particles	>1.0
Viruses	0.01–0.5

size can span some five orders of magnitude. The typical particle sizes of selected aerosol components are shown in Table 10.1.

Aerosol particles are generally classified into three modes:

(1) Nucleation (nuclei) mode, diameter (d) <100 nm.
(2) Accumulation mode, 100 nm $\leq d \leq$ 1 μm.
(3) Coarse mode, $d >$ 1 μm.

Particles <1 μm diameter are sometimes referred to as fine particles, and the super-micrometre (>1 μm) particles are, generally, termed coarse particles. Nevertheless, the boundaries between modes are not rigid.

10.3.1.2   Particle concentration

Aerosol particle number per unit volume of air or, alternatively, aerosol number concentration normally is expressed as number cm^{-3}. On the other hand, viable particles are frequently expressed in terms of their ability to reproduce, i.e. the number of colony-forming units per unit volume or cfu m^{-3}. It is usual to express the concentration of allergenic, endotoxic, or bacterial outer membrane material in terms of mass per unit volume of air or $\mu g\, m^{-3}$. Aerosol or bioaerosol concentration can vary over several orders of magnitude depending on the source strength, closeness to source, and ambient conditions. Representative particle concentrations of some selected aerosol constituents are shown in Table 10.2.

10.3.1.3   Size distribution

The aerosol size distribution covers about five orders of magnitude and approximately 12 orders of magnitude in number concentration. Few measurements covering all of the size range have been made because of the need to use different instrumentation for the determination of the separate size range of one order of magnitude or more. Nevertheless, models of the atmospheric aerosol size distribution have been proposed (e.g. Jaenicke, 1998; Seinfeld and Pandis, 1998).

The log-normal distribution is often used in atmospheric aerosol applications, because of the wide range of particle size and concentration, and because

Table 10.2   Typical aerosol and bioaerosol concentration for selected constituents (Lacey and Vanette, 1998; Matthias-Maser, 1998).

Aerosol type	Concentration
Bacteria	100 (rural) to 850 (city) m^{-3}
Cloud	50–1000 cm^{-3}
Fungal spores	10^4–$10^5 m^{-3}$
Pollen grains	5–300 grains m^{-3} (seasonal mean)[+]
Primary biological aerosol particles	0.3–9.5 cm^{-3}, mean = 3.11 cm^{-3}
Sea salt	1–30 cm^{-3}

it frequently matches the experimental observations. Parameters of model aerosol distribution, expressed as the sum of three log-normal modes have been presented by Jaenicke (1993) for the following aerosol types: urban, marine, rural, remote continental, free troposphere, polar, and desert.

A first attempt to characterise the primary biological aerosol particle size distribution was made by Matthias-Maser and Jaenicke (1995) over a fairly large diameter range from 0.4 up to about 100 μm. The biological particles can also be described by a log-normal size distribution (Matthias-Maser, 1998) and are lower than the rural aerosol submicron size distribution by an order of magnitude. The biological size distribution approaches the rural distribution by up to less than half an order of magnitude for the coarse mode; the coarse mode rural aerosol contains a large fraction of bioaerosol.

10.3.2   Relationship of aerosol generation to biofilms

Surface organic films or monolayers, whether on natural bodies of waters (e.g. lakes, rivers, and the sea) or on artificial waters (e.g. swimming pools, waste-treatment plants, etc.) commonly contain higher concentrations of microorganisms, including bacteria, than is found in the bulk water beneath the surfaces. Mechanical agitation of such surfaces can be caused by artificial means (e.g. aerators, used in waste-treatment plants), or by natural mechanisms (e.g. wave generation and wave disruption in the ocean). Such agitation is likely to give rise to the production of both jet and film droplets, due to the bursting of air bubbles at such surfaces. These droplets are primary sources of mechanically produced aerosol particles which generally contain enhanced concentrations of bacteria, over and above that contained in the bulk water, as well as other biological and organic material. The sea surface is frequently covered with organic films due to a combination of natural and man-made biofilm sources. Blanchard was one of the first to demonstrate that these films are transferred from the water surface of droplets ejected in the air-bubble process (Blanchard, 1963, 1989).

Surface film-active biomaterial also exists in nearly all freshwater surfaces, where it tends to be more concentrated than in the bulk. It is brought to the surface by a number of processes, such as molecular diffusion aided by organised convective motions referred to as Langmuir circulations, and by air bubbles rising through the water. The air bubbles are capable of concentrating organic and biological material by two processes:

(1) Most of the biological material is adsorbed on the bubble surface as it rises and monolayer films at the water surface have been found to contain enhanced concentrations of bacteria, dinoflagellates and other plankton.
(2) Bubbles of air reaching the air–water interface burst and eject liquid jet droplets and film droplets from the bubble film into the air. These have been found to contain enhanced concentrations of organic material.

Most previous experimental studies of the enrichment of bubble-induced droplets have been on single bubble bursting. However, multiple bubbling is more related to the natural bubbling observed in most natural biofilm systems.

10.4 BIOAEROSOLS DEPENDENCE ON ENVIRONMENTAL CONDITIONS

In order to accurately assess bioaerosols, it is necessary to understand the behaviour of microorganisms during all stages of their existence (Cox, 1987). Environmental factors can alter the water content of airborne microorganisms and as a result can lead to conditions of cellular stress (Griffiths and DeCosemo, 1994). The behaviour of bioaerosols is governed by the principles of gravitation, electricity, turbulence, and diffusion. Fluctuating levels of microorganisms found in air are attributed to

- germ carriers, such as man or animals;
- mobilisation of soil, water, or other surfaces;
- sedimentation, impaction, and adsorption on the surface of soil, water, or other surfaces.

Some environmental factors that affect bioaerosol viability and survival include temperature and relative humidity (RH), and these factors need to be considered when sampling or measuring bioaerosols (Griffiths *et al.*, 1999).

With respect to bioaerosol dependence on pH, the concentration and composition of solute material governs the acidity of droplets. A pH value of 5.6 is frequently cited as the pH of pure cloud water, which is pure water in equilibrium with the atmospheric concentration of CO_2.

At RH below 100%, sub-micrometre- and micrometre-sized aerosol particles contain water which requires the presence of dissolved, hygroscopic, inorganic, and organic compounds. If the RH of the environment is higher than that of the aerosol particles, then the particle will grow and become liquefied. An RH value between 30 and 40%, or above, is frequently used to delineate between a dry aerosol and a wet aerosol particle. On the other hand, generation of an aerosol particle from an evaporating solution leads to a more saturated solution and the

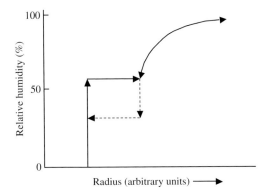

Figure 10.1   Schematic representation of variation of aerosol particle radius with relative humidity.

development of a dry particle. Phase transition occurs from solution to solid or from solid to solution at the deliquescence point which is dependent on the salt constituents. The particle size decreases at a lesser rate below the deliquescence point than when it initially grew, thus exhibiting hysteresis (Figure 10.1). The growth or evaporation of an aerosol particle is governed by the difference between the vapour pressure of the particle and the ambient vapour pressure. The increase in particle size due to water uptake is of an order $\geqslant 2$ for an increase in RH up to 90%, and by as much as 2.5 as the RH approaches 100% (Jennings, 1998).

Many biological particles, including viruses and bacteria, are hygroscopic. Some viruses without structural lipids are unstable below an RH of about 70%, whereas some studies have shown certain bacterial species are most stable at low rather than high RH (Cox, 1998).

The organic content of aerosols, which is composed of organic carbon (OC), is a complex mixture of several hundreds of organic compounds, and is usually referred to simply as OC. The OC loading or concentration is expressed as the concentration of carbon in $\mu g\ C\ m^{-3}$, and does not include the concentration of oxygen, hydrogen, and nitrogen that may be present. It has been suggested (Wolff et al., 1991) that measured OC values should be multiplied by a factor of 1.5 to yield a total organic mass. The OC loading of various aerosols is shown in Table 10.3. The mass distribution is generally bimodal with a first peak around 0.2 μm and a second peak around 1.0 μm (Seinfeld and Pandis, 1998).

Organic aerosol is emitted directly by sources (so-called primary OC) or is indirectly formed (so-called secondary organic aerosol) by the transfer to the aerosol phase of low vapour pressure and low volatile products, which result from oxidation of organic gases by chemical radical species. Primary OC particles are produced by combustion, chemical, geological, and biogenic sources. Although several hundred organic compounds have been identified as primary organic aerosol constituents, nevertheless, these only represent a relatively small fraction

Table 10.3   Organic loading of several aerosols of different origin.

Location/region	Aerosol OC $(\mu g\,(C)\,m^{-3})$		References
	Range	Average	
Urban/USA	5–20	8.8	US EPA (1996); Seinfeld and Pandis (1998)
Rural/USA		3.5	Stevens *et al.* (1984)
Remote marine/ northern hemisphere	0.2–0.8	0.35	Penner (1995)
Remote marine/ southern hemisphere	0.1–0.2	0.16	Penner (1995)
Urban/rural influenced region, Germany		6.5	Matthias-Maser and Jaenicke (1995)

(10–30%) of the total organic loading (Seinfeld and Pandis, 1998). Thus, the remaining organic content remains to be identified.

10.5   GENERATION OF NATURAL AND EXPERIMENTAL AEROSOLS

There are a number of natural mechanisms by which aerosols are generated:

(1) By condensation and chemical reactions of gases and vapours in the atmosphere. The conversion or oxidation of gaseous constituents, such as SO_2 and nitrogen oxides (NO_x) to sulphate and nitrate, respectively, in aerosol particle formation is due to photochemical processes in the presence of hydroxyl radicals. These gas to particle formation processes can also be investigated in reaction flow chambers in the laboratory.

(2) By mechanical disruption and dispersal of material from the earth's surface. The principal sources of aerosols from mechanical disruptive processes include volcanic activity, mechanical wave disruption at the ocean or freshwater surface, mechanical disruption at the surface of man-made water bodies, wind-blown dust and soil particles, and desert aerosol. A major source of primary marine aerosol is due to the bursting of air bubbles, due to entrainment of air arising from breaking waves. Air-bubble bursting produces film and jet droplets leading to the formation of marine or aquatic aerosol particles. The particles that are formed by the above mechanical processes have an overall mode of $>1\,\mu m$ diameter, and consist of sea salt or soil dust particles. Desert aerosol is present over desert or semi-arid regions but can also be transported considerable distances over oceans. Resuspension of soil particles are an additional source of mechanically generated aerosol. This group of processes is particularly important in the generation of bioaerosols.

Table 10.4   Aerosol generators with typical operational range of particle size.

Aerosol generator	Principal of operation	Aerosol diameter range (μm)
Heated wire	Voltage applied to metal wire terminals	~0.001–0.01
Condensation type aerosol generator	Condensation of a vapour on nuclei	0.2–8
Nebuliser or atomiser	Application of an air jet to a thread of aqueous suspension	0.04–20
Vibrating orifice	Disruption of a liquid jet emerging from an orifice plate under vibration	5–50
Spinning top or disk	Disruption of a film of liquid spreading from centre of air or electrically driven spinning top	1–18
Fluidised bed generator	Dispersion of the aerosol from dry material using a fluidation technique	0.5–50
Dry powder/dust generator	Suspension of particles followed by de-agglomeration	1–50
Electrostatic classifier	Classification of charged particles due to their electrical mobility	0.01–1.0

(3) Biomass burning of vegetation is both a natural and anthropogenic source of bioaerosols containing organic particulate material and black carbon. Sources of biomass burning are generally separated into five groups: boreal and temperature forest fires, savannah fires, agricultural waste, fuelwood and charcoal (Cachier, 1998). In addition, the biomass of the aerosol is influenced by the nature of the combustion; whether it occurs under smouldering or flaming conditions. Only about 10% of biomass burning events are natural occurrences (Cachier, 1998).

Laboratory-based techniques for the dispersal of microorganisms into air have been developed, but not all are adequate for microbial particles as they do not simulate natural release into air. Both wet and dry dispersal systems have been used, depending on the microbial group under examination. For bacteria- and virus-containing bioaerosols wet dispersal systems are more appropriate, whereas dry dispersal systems are more appropriate for fungal and actinomycete spores. No single laboratory-based aerosol generation technique can produce particles over the dynamic range from about 1 nm up to 100 μm in diameter. Nevertheless, a variety of aerosol generators have been used experimentally (Mitchell, 1998) and a representative number of these and their characteristics are shown in Table 10.4.

Ropenen *et al.* (1997) suggested that, when experimentally generating a bioaerosol, each microbial group may require a specific dispersal system as the characteristics of the resulting aerosols are dependent on dispersion forces, as well as the sensitivity and agglomeration of the microorganism. The natural production of particles liberated from air bubbles, entrained by breaking waves, as they burst at the air–water interface has been simulated in the laboratory by several workers. Importantly, a multiple-bubble bursting system has been devised by

Wangwongwatana *et al.* (1990) to investigate the transfer of material from the liquid to the gaseous phase using an array of capillary tube gas diffusers. The system was used to characterise the aerosol produced from a simulated activated sludge aeration tank, and could be used to simulate other processes involving gas bubbling in liquids, such as wastewater aeration tanks and fermentors.

10.6   SAMPLING OF BIOAEROSOLS

The growing concern about human exposure to bioaerosols has created a demand for advanced, more reliable, and more efficient monitoring methods to detect, identify, and enumerate airborne biological particles (Lacey and Dutkiewicz, 1994). Bioaerosol monitoring is the measurement of viable (both culturable and non-culturable) and non-viable microorganisms in both indoor and outdoor environments. This monitoring is used routinely in industry to assess indoor and outdoor environmental air quality and the air standard of clean rooms. In addition, it is applied to accessing infectious disease outbreaks, and is used in epidemiological investigations and research studies.

10.6.1   Considerations when sampling

Detection of bacteria is a time-consuming process which is not always successful (Griffiths and DeCosemo, 1994). Moreover, techniques for the monitoring of bioaerosols containing viruses, protozoa, antigenic fragments, algae, and mycoplasmas are less developed compared with those for bacteria. As most microorganisms in the air are associated with carrier particles, each of which may hold many individual organisms, the estimation of the total number of individual microorganisms present is not always possible. Nevertheless, this objective can be advanced somewhat by defining the scope of the investigation, the microorganisms sought, and the choice of sampling technologies. If one is not concerned with discriminating pathogens from each other, or from other bioaerosols, then the current technologies for aerosol sampling and detection in air streams may be adapted to this purpose (Griffiths *et al.*, 1999). For example, the number of bacteria-carrying particles per standard volume of air can be determined by sampling the air in such a way that each collected particle forms a discrete colony on an agar surface.

The collection efficiency of bioaerosols is a complex function of particle size, concentration, and free-falling velocity as well as on wind speed, direction, and turbulence (Jensen and Schafer, 1998). In many instances, it is possible to measure the volume of air sampled, but the efficiency of retention and impaction vary with sampling conditions and the type of instrument used. Also, the bacterial count and profile obtained depends on the air movement patterns in the analysis area. Under still-air conditions a considerable proportion on microorganisms will settle with their carrier particles, whereas bacterial counts increase under increased air disturbance, humidity, and precipitation (Griffiths *et al.*, 1999).

The choice of method employed for sampling (Jensen and Schafer, 1998) will depend on a combination of factors including:

- whether analysis is indoor or outdoor;
- the expected microbial concentration in the air;
- whether enumeration and/or identification of the microorganisms is required;
- whether a qualitative analysis based or morphological characteristics is sufficient;
- whether assessment of total airborne microorganisms, or a particular species, is needed;
- growth characteristics of the microorganism;
- whether a continuous temporal analysis (day and night) is required, or whether short, regular, or discontinuous sampling is required;
- what time intervals are needed to determine flora changes;
- volume size and precision of volume measurement;
- cost of apparatus and running costs;
- requirement for a power supply;
- whether the apparatus needs to be portable;
- whether apparatus efficiency varies with wind speeds.

Thus, requirements for an ideal bioaerosol sampler are diverse, but maintaining high biological efficiency is considered to be one of the main requirements for efficient performance of bioaerosol samplers (Macher, 1997).

10.6.2   Bioaerosol sampling devices

Samplers that are normally used to collect natural aerosol particles are generally unsuitable for sampling of natural bioaerosols. For physical sampling, the latter can be damaged due to impaction forces in the sampler or through dessication (Griffiths and DeCosemo, 1994; Crook, 1998). Quantitative collection of larger-sized particles is very difficult in the ambient environment (Nicholson, 1998). Physical aspects affecting sampling include the effects of sedimentation and inertia, as well as Brownian motion and likely electrical forces. Assessment of the performance of any bioaerosol sampler requires measurement of its collection efficiency over a range of wind speeds and particulate size, and can be carried out in wind tunnel studies. Collection efficiency of a number of samplers have been determined (Griffiths et al., 1993).

Many different methodologies exist for the collection and enumeration of bioaerosols but, by and large, sedimentation or gravitational, filtration, inertial, impaction, and impingement centrifugal samplers are the more common sampling techniques used (Griffiths and DeCosemo, 1994).

10.6.2.1   Gravitational sampler

The gravitational sampler, whereby a collector substrate is exposed to collect particles falling under gravity, is the simplest method of collection of bioaerosols

(Jensen and Schafer, 1998). However, being a passive sampler, it does not provide the quantitative number or mass concentration of aerosols. In the sedimentation or gravity Petri-dish method, the bacteria settle onto open agar plates due to the forces of gravity. This is a very simple technique and is used extensively in industrial and hospital environments to evaluate air quality. However, this method can only give qualitative or semi-quantitative results as no defined volume of air per time period is measured and the results are strongly influenced by air turbulence. This technique is of little value for trapping particles <30 μm in diameter.

10.6.2.2   Filtration

Filtration is the simplest method of air sample collection, whereby air can be filtered through different filter porosities at different stages to collect different sized particles. The pore sizes of the filter material can range from 0.01 to 10 μm. Deposition occurs when particles impact and are intercepted by the fibres or the surface of filter membranes (Hinds, 1982). Devices with membrane, gelatin and nucleopore filters can be employed and nucleopore filters with plastic cassette filter holders have been used to collect bioaerosols (Griffiths and DeCosemo, 1994). The efficiency of removing particles from the air depends on the rate of airflow through the pores and on the particle size. For particles <1 μm, the overall efficiency decreases with increasing face velocity (Liu *et al.*, 1983; Lippmann, 1995), whereas for particles >1 μm, the filter collection efficiency is >99%. The overall efficiency of membrane filters is approximately 100% for particles larger than the pore size (Lippmann, 1995). Employing this approach, either sieve or impaction filters can be used.

In the sieve filter method, air is drawn through a bacterial-retentive filter, consisting of a cellulose acetate or gelatin membrane. The trapped microorganisms are either cultured or examined microscopically. This method is not used routinely as it is difficult to obtain a sufficiently high rate of airflow through the filters, as filters clog upon prolonged use. In addition, viable cell recovery is low due to cell death caused by excessive drying during sampling, or cell membrane damage caused during impaction onto the dry membranes. Therefore, these drawbacks reduce somewhat the application of filtration methods (Jensen and Schafer 1998).

Impaction filters consist of deep layers of fibres or granules separated by wide air pockets. Particles are impacted onto the uppermost layers of the filter material. Different types of filter substances can be used such as nitrocellulose plugs, cotton plugs, powdered glass, or sodium alginate wool. The different filters are then washed or dissolved in a relevant solute and, subsequently, microbiological analysis is performed (Lippmann, 1995; Jensen and Schafer, 1998).

10.6.2.3   Impaction

The principle of collection by impaction utilises the tendency of a particle to deviate from an airflow streamline onto a collection surface. Impactors are devices in which the sampled air stream is directed onto a solid surface (e.g. a microscope slide with a sticky surface) or onto the surface of an agar medium. Impactors operate by use of a vacuum device which pulls air through the sampler at a known critical flow rate. This makes it possible to obtain measurements in a defined volume of air per time period. As a result quantitative analysis of the bioaerosol content of the air can be made, i.e. by performing colony counts per m^3 (Jensen and Schafer, 1998).

A wide range of impactors are available including: cascade or multiple-stage impactors, single- or two-stage impactors, and whirling or rotating arm impactors. Several different technologies exist but can be generally categorised as follows:

(1) Slit samplers, such as the Burkard Spore Trap which has a rectangular opening above a greased glass slide for particle collection. Alternatively, collection can be onto the surface of rotating agar plates. Size-dependent collection of particles is not possible.
(2) Cascade impactors, such as the six stage-Andersen sampler which collects progressively finer particles onto each Petri dish.
(3) Sieve samplers, such as the single-stage Andersen N6 Spiral System which achieves particle collection via many small holes in the sieve cover situated a few millilitres over an agar medium. Depending on the collector system, sieves with different hole sizes can be arranged. Thus, a size-dependent particle analysis is possible.

Whirling arm or rotored impactors are the most suitable samplers for collection of large airborne biological particles (e.g. pollen grains and large spore fungi). An adhesive collecting substrate is mounted on the leading edge of the rotating arms.

The main advantage of impactors is that microorganisms are collected directly onto the nutritional medium, on which the incubation and analysis takes place. This requires no further preparatory work. However, sampling with an impactor can lead to overloading of the agar whereby counting of the cfu after incubation is impossible. Moreover, microbial cells can be damaged during impaction which, thus, would affect counts.

10.6.2.4   Impingers

In impingers, air is drawn by vacuum, and the ambient aerosols impinge into a collection fluid (Jensen and Schafer, 1998). Thus, impingement differs to impaction in that a defined volume of air is drawn through a liquid collection medium or diluent, rather than onto a solid medium or glass slide. The trapped microorganisms can then be enumerated by membrane filtration or by conventional pour plate or spread plate techniques. The volume of air drawn into the device is determined by a limiting capillary orifice. These devices include bioaerosol sampling cyclone and centrifugal samplers. The collected larger-sized particles, e.g. pollen grains or spores, are usually counted and identified by conventional optical microscopy or by transmission and scanning electron microscopy. High volume samplers, equipped with a suitable collecting filter medium, are often used to determine bioaerosol concentration. These devices are cheap, efficient, and allow quantitative determination of the bioburden.

The main advantage of impingement over impaction is that the collected suspension can be diluted prior to incubation. This allows for accurate enumeration of all pollution grades. Also, it has the added advantage that the trapped microorganisms do not dehydrate during the process. Nevertheless, the primary disadvantage of impingement is that enumeration of low contaminant concentrations is difficult. In addition, cells can be damaged by mechanical forces when striking the surface of the medium, by sudden hydration or by osmotic shock. This technique is further hampered as hydrophobic particles do not stay in the liquid collector phase, but leave the collector with the waste air, and thus, a small amount of sample is lost.

10.6.2.5   Other devices

Cyclones with a wetting spray have been found to be a gentle method for collection of bioaerosols. Examples of these cyclones are the Aerojet-General liquid scrubber sampler. Griffiths *et al.* (1993) have shown that the collection efficiency of this sampler, operating at a sampling rate of 500 $l\,min^{-1}$ and at ambient wind speeds up to $4\,m\,s^{-1}$, is between 70% and 110% for particles as large as 20 µm in diameter.

In general, bioaerosol samplers have not been fully characterised in terms of inlet efficiency. Although most traditional aerosol sampling devices involve techniques that separate particles from the air stream and collect them in or on a preselected medium, more recently developed methods involve optical and electrical mobility.

Electrostatic induction can be used to attract electrically charged, artificially produced, bubble film particles to a suitably coated biological substrate (Blanchard and Syzdek, 1982). Conventional light scattering counters can be used for the physical determination of aerosol particles. Such instruments work on the principle that as an aerosol or bioaerosol particle is drawn through an illuminated (usually by a laser) volume, light scattered elastically by a single particle into a known solid angle is optically sensed, converted to an electrical signal, and electronically size-classified according to magnitude of the signal. The magnitude of the signal pulse height is related approximately to particle size. Interestingly, Pinnick *et al.* (1995) and Hairston *et al.* (1997) have shown that fluorescence particle counters that measure elastic scattering and total fluorescence can differentiate between some biological and non-biological particles. Furthermore, Pinnick *et al.* (1998) showed that a prototype aerosol fluorescence spectrum analyser can be used to classify individual airborne artificial particles into at least a few categories of bioaerosols. Thus, novel approaches are leading to the development of new sampling technologies for bioaerosols.

10.7   CONCLUSION

The generation of bioaerosols in a number of environments from biofilm-colonised surfaces and interfaces represents a novel and important mechanism for the dispersal of microorganisms into the atmosphere. Despite their importance for the transmission of infectious diseases, allergenic compounds, and pollutants, bioaerosols remain an under-studied aspect of biofilm research. Conversely, the role of biofilms in the generation of bioaerosols has not been fully considered by specialists in the field of aerobiology. Although knowledge of the physical parameters and mechanisms affecting aerosol generation are well established, knowledge of the interplay between physical and biological parameters in bioaerosol generation and dispersion from biofilms, and atmospheric survival of microorganisms remains limited. Furthermore, the stress responses of microorganisms in the unusual environment of the bioaerosol have not been considered sufficiently. In practical terms, although a number of approaches and devices exist for sampling contaminated air and bioaerosol characterisation, limitations and difficulties exist with their application. Samplers with greater collection efficiency and with wider applicability to bioaerosols of differing particle sizes are required. Nevertheless, despite these deficiencies, it is apparent that more knowledge in this area would greatly impact on healthcare, public heath, epidemiology and pollutant control.

Acknowledgements

The authors acknowledge the financial support of the Irish Higher Education Authority (PRTL-1) under the National Development Plan.

REFERENCES

Benenson, A.A. (1995) *Control of Communicable Diseases Manual*, 16th ed., American Public Health Association, Washington DC, USA.

Blanchard, D.C. (1963) The electrification of the atmosphere by particles from bubbles in jet drops from bursting bubbles. *Prog. Oceanogr.* **1**, 71–202.

Blanchard, D.C. (1989) The ejection of drops from the sea and their enrichment with bacteria and other materials: a review. *Estuaries* **12**, 127–139.

Blanchard, D.C. and Syzdek, L.D. (1982) Water-to-air transfer and enrichment of bacteria in drops from bursting bubbles. *Appl. Environ. Microbiol.* **43**, 1001–1005.

Bovallius, A., Bucht, B., Roffey, R. and Anas, P. (1978) Long range air transmission of bacteria. *Appl. Environ. Microbiol.* **35**, 1231–1232.

Bünger, J., Antlauf-Lammers, M., Schulz, T.G., Westphal, G.A., Muller, M.M., Ruhnau, P. and Hallier, E. (2000) Health complaints and immunological markers of exposure to bioaerosols among biowaste collectors and compost workers. *Occup. Environ. Med.* **57**, 458–464.

Burge, H.A. (1995a) Aerobiology of the indoor environment. *Occup. Med.* **10**, 27-40.

Burge, H. (1995b) Bioaerosols in the residential environment. In *Bioaerosols Handbook* (ed. C.S. Cox and C.M. Wathes), pp. 579–597, CRC Press, Boca Raton, FL, USA.

Cachier, H. (1998) Carbonaceous combustion aerosols. In *Atmospheric Particles* (ed. R.M. Harrison and R. van Grieken), pp. 295–348, Wiley, Chichester, UK.

Cox C.S. (1987) *The Aerobiological Pathway of Microorganisms*, Wiley, Chichester, UK.

Cox, C.S. (1998) Stability of airborne microbes and allergens. In *Bioaerosols Handbook* (ed. C.S. Cox and C.M. Wathes), pp. 77–99, CRC Press, Boca Raton, FL, USA.

Crook, B. (1998) Inertial samplers: biological perspectives. In *Bioaerosols Handbook* (ed. C.S. Cox and C.M. Wathes), pp. 247–267, CRC Press, Boca Raton, FL, USA.

Flemming, H.-C., Wingender, J., Mayer, C., Körstgens, V. and Borchard, W. (2000) Cohesiveness in biofilm matrix polymers. In *Society for General Microbiology Symposium: Community Structure and Co-operation in Biofilms* (ed. D.G. Allison, P. Gilbert, H.M. Lappin-Scott and M. Wilson), Vol. 59, pp. 87–106, University Press, Cambridge, UK.

Frazer, D.W. (1980) Legionellosis: evidence of airborne transmission. *Ann. NY Acad. Sci.* **355**, 61–66.

Gravesen, S. (1979) Fungi as a cause of allergic disease. *Allergy* **34**, 135–154.

Griffiths, W.D. and DeCosemo, G.A.L. (1994) The assessment of bioaerosols: a critical review. *J. Aerosol Sci.* **25**, 1425–1458.

Griffiths, W.D., Upton, S.L. and Mark, D. (1993) An investigation into the collection efficiency and bioefficiency of a number of aerosol samplers. *J. Aerosol Sci.* **24**, S541–S542.

Griffiths, W.D., Stewart, I.W., Clark, J.M. and Holwill, I.L. (1999) Procedures for the characterisation of bioaerosol particles–Part 1: Aerosolisation and recovery agent affects. *Aerobiologia* **15**, 267–280.

Hairston, P.P., Ho, J. and Quant, F.R. (1997) Design of an instrument for real-time detection of bioaerosols using simultaneous measurement of particle aerodynamic size and intrinsic fluorescence. *J. Aerosol Sci.* **28**, 471–482.

Hameed A.A. and Khodr M.I. (2001) Suspended particulates and bioaerosols emitted from an agricultural non-point source. *J. Environ. Monit.* **3**, 206–209.

Hinds, W.C. (1982) *Aerosol Technology*, Wiley, New York, USA.

Hryhorczuk, D., Curtis, L., Scheff, P., Chung, J., Rizzo, M., Lewis, C., Keys, N. and Moomey, M. (2001) Bioaerosol emissions from a suburban yard waste composting facility. *Ann. Agric. Environ. Med.* **8**, 177–185.

Jaenicke, R. (1993) Tropospheric aerosols. In *Aerosol–Cloud–Climate Interactions* (ed. P.V. Hobbs), pp. 1–1, Academic Press, San Diego, CA, USA.

Jaenicke, R. (1998) Atmospheric aerosol size distribution in atmospheric particles. In *Atmospheric Particles* (ed. R.M. Harrison and R. van Grieken), pp. 1–28, Wiley, Chichester, UK.

Jennings, S.G. (1998) Wet processes affecting atmospheric aerosols. In *Atmospheric Particles* (ed. R.M. Harrison and R. van Grieken), pp. 475–508, Wiley, Chichester, UK.

Jensen, P.A. and Schafer, M.P. (1998) Sampling and characterization of bioaerosols. In *National Institute for Occupational Safety and Health Manual of Analytical Methods*, Section J, pp. 82–112, National Institute for Occupational Safety and Health, Washington DC, USA.

Krajewski, J.A., Szarapinska-Kwaszewska, J., Dudkiewicz, B., Cyprowski, M., Tarkowski, S., Konczalik, J. and Stroszejn-Mrowca, G. (2001) Assessment of exposure to bioaerosols in workplace ambient air during municipal waste collection and disposal. *Med. Pract.* **52**, 417–22.

Lacey, J. and Dutkiewicz, J. (1994) Bioaerosols and occupational lung disease. *J. Aerosol Sci.* **25**, 1371–1404.

Lacey, J. and Vanette, J. (1998) Outdoor air sampling techniques. In *Bioaerosols Handbook* (ed. C.S. Cox and C.M. Wathes), pp. 407–471, CRC Press, Boca Raton, FL, USA.

Lighthart, B. and Shaffer, B. (1995) Airborne bacteria in the atmospheric surface layer: temoral distribution above a grass seed field. *Appl. Environ. Microbiol.* **61**, 1492–1496.

Lighthart, B. and Stetzenbach, L.D. (1994) Distribution of microbial bioaerosols. In *Atmospheric Microbial Aerosols, Theory and Applications* (ed. B. Lighthart and A.J. Mohr), pp. 68–98, Chapman and Hall, New York, USA.

Lippmann, M. (1995) Filters and filter holders. In *Air Sampling Instruments for Evaluation of Atmospheric Contaminants*, 8th ed. (ed. B.S. Cohen and S.V. Hering), pp. 247–279, American Conference of Governmental Industrial Hygienists Inc., Cincinnati, OH, USA.

Liu B.Y.H., Pui, D.Y.H. and Rubow, K.L. (1983) Characteristics of air sampling filter media. In *Aerosols in the Mining and Industrial Work Environments* (ed. V.A. Marple and B.Y.H. Liu), Vol. 3, pp. 989–1038, Ann Arbor Science, Ann Arbor, MI, USA.

Macher, J. (1997) Evaluation of bioaerosol sampler performance. *Appl. Occup. Environ. Hyg.* **1**, 730–736.

Matthias-Maser, S. (1998) Primary biological aerosol particles: their significance, sources, sampling methods and size distribution. In *Atmospheric Particles* (ed: R.M. Harrison and R. van Grieken), pp. 349–368, Wiley, Chichester, UK.

Matthias-Maser, S. and Jaenicke, R. (1994) Examination of atmospheric bioaerosol particles with radii greater than 0.2 micrometers. *J. Aerosol Sci.* **25**, 1605–1613.

Matthias-Maser, S. and Jaenicke, R. (1995) Size distribution of primary biological particles with radii >0.2 μm in the urban/rural influenced region. *J. Atmos. Res.* **39**, 279–286.

McCullough, N.V., Brosseau, L.M. and Vesley, D. (1997) Collection of three bacterial aerosols by respirator and surgical mask filters under varying conditions of flow and relative humidity. *Ann. Occup. Hyg.* **41**, 677–690.

Mitchell, J.P. (1998) Aerosol generation for instrument calibration. In *Bioaerosols Handbook* (ed. C.S. Cox and C.M. Wathes), pp. 101–175, CRC Press, Boca Raton, FL, USA.

Nicholson, K.W. (1998) Physical aspects of bioaerosol sampling and deposition. In *Bioaerosols Handbook* (ed. C.S. Cox and C.M. Wathes), pp. 27–53, CRC Press, Boca Raton, FL, USA.

Penner, J.E. (1995) Carbonaceous aerosols influencing atmospheric radiation: black and organic carbon. In *Aerosol Forcing of Climate* (ed. R.J. Charlson and J. Heintzenberg), pp. 91–108, Wiley, New York, USA.

Pinnick, R.G., Hill, S.C., Nachman, P. and Pendleton, J.D. (1995) Fluorescence particle counter for detecting airborne bacteria and other biological particles. *Aerosol Sci. Technol.* **23**, 653–664.

Pinnick, R.G., Hill, S.C., Nachman, P., Videen, G., Chen, G. and Cheng, R.K. (1998) Aerosol fluorescence spectrum analyzer for rapid measurement of single micrometer-sized airborne biological particles. *Aerosol Sci. Technol.* **28**, 95–104.

Poulsen, O.M., Breum, N.O., Ebbehoj, N., Hansen, A.M., Ivens, U.I., van Lilieveld, D., Malmros, P., Matthiasen, L., Nielsen, B.H., Nielsen, E.M., Schibye, B., Skov, T., Stenbaek, E.I. and Wilkins, K.C. (1995) Sorting and recycling of domestic waste. Review of occupational health problems and their possible causes. *Sci. Total Environ.* **168**, 33–56.

Ropenen, T., Willeke, K. and Ulevicius, V. (1997) Techniques for the dispersion of microorganisms into air. *Aerosol Sci. Technol.* **27**, 405–421.

Ross, M.A., Curtis, L., Scheff, P.A., Hryhorczuk, D.O., Ramakrishnan, V., Wadden, R.A. and Persky, V.W. (2000) Association of asthma symptoms and severity with indoor bioaerosols. *Allergy* **55**, 705–711.

Rylander, R. (1994) Organic dusts: from knowledge to prevention. *Scand. J. Work Environ. Health* **20**, 116–122.

Salem, H. and Gardner, D.E. (1994) Health aspects of bioaerosols. In *Atmospheric Microbial Aerosols, Theory and Applications* (ed. B. Lighthart and A.J. Mohr), pp. 304–330, Chapman and Hall, New York, USA.

Schlegel, H. (1984) *Allgemeine Mikrobiologie*, 6th ed. Thiene Verlag, Stuttgart, Germany.

Schwartz, D.A., Thorne, P.S., Yagla, S.J., Burmeister, L.F., Olenchock, S.A., Watt, J.L. and Quinn, T.J. (1995) The role of endotoxin in grain dust-induced lung disease. *Am. J. Respir. Crit. Care Med.* **152**, 603–608.

Seinfeld, J.H. and Pandis, S.N. (1998) *Atmospheric Chemistry and Physics*, Wiley, New York, USA.

Stetzenbach, L.D. (1997) Introduction to aerobiology. In *Manual of Environmental Microbiology* (ed. C.J. Hurst.), pp. 619–628, American Society for Microbiology Press, Washington DC, USA.

Stevens, R.K., Dzubay, T.G., Lewis, C.W., and Shaw Jr., R.W. (1984) Source apportionment methods applied to the determination of the origin of ambient aerosols that affect visibility in forested areas. *Atmos. Environ.* **18**, 261–272.

Thorne, P.S. (2000) Inhalation toxicology models of endotoxin- and bioaerosol-induced inflammation. *Toxicology* **152**, 13–23.

Trout, D., Bernstein, J., Martinez, K., Biagini, R. and Wallingford, K. (2001) Bioaerosol lung damage in a worker with repeated exposure to fungi in a water-damaged building. *Environ. Health Perspect.* **109**, 641–644.

U.S. Environmental Protection Agency (1996) *Air Quality Criteria for Particulate Matter.* Report EPA/600/P-95/001. Environmental Protection Agency, Washington DC, USA.

Wan, G.H. and Li, C.S. (1999) Indoor endotoxin and glucan in association with airway inflammation and systemic symptoms. *Arch. Environ. Health* **54**, 172–179.

Wangwongwatana, S., Scarpino, P.V., Willeke, K. and Baron, P.A. (1990) System for characterizing aerosols from bubbling liquids. *Aerosol Sci. Technol.* **13**, 297–307.

Wolff, G.T., Ruthkovsky, M.S., Stroup, D.P. and Korsog, P.E. (1991) A characterization of the principal PM-10 species in Claremont (summer) and Long beach (fall) during SCAQS. *Atmos. Environ.* **25A**, 2173–2186.

Zeterberg, J.M. (1973) A review of respiratory virology and the spread of virulent and possible antigenic viruses via air conditioning systems. *Ann. Allergy* **31**, 228–234.

11

Biofilms and protozoa: a ubiquitous health hazard

A.W. Smith and M.R.W. Brown

11.1 INTRODUCTION

When considering the role of protozoa and infectious diseases, thoughts naturally turn to the well-characterised pathogenic parasitic protozoa. These account for some of the most devastating diseases worldwide and include *Plasmodium falciparum* and *Plasmodium vivax* (malaria), *Toxoplasma gondii* (toxoplasmosis), *Trichomonas vaginalis* (trichomoniasis), *Leishmania* spp. (leishmaniasis), *Trypanosoma brucei* (African sleeping sickness) and *Trypanosoma cruzi* (Chagas' disease). But these are not the focus of this chapter. Rather, this chapter will focus on the role of environmental protozoa and, in particular, their interactions with bacteria in biofilms.

The word protozoa means 'first animal', reflecting the fact they are single-celled eukaryotic micro-organisms with a membrane-bound nucleus. For this reason, they are distinct from prokaryotic micro-organisms, the bacteria and blue-green algae. They are also distinct from the multicellular metazoa, such as nematodes. Recent work, beyond the scope of this chapter, indicates that there is important commonality also in the relationships between bacteria and nematodes and higher eukaryotes (Mahajan-Miklos *et al.*, 1999; Tan *et al.*, 1999a, b) and those between bacteria and protozoa discussed here. Although there are some

25,000 species of protozoa described as a phylum or subkingdom in the kingdom Protista, they are not a natural group. Only mutual relationships in their cellular organisation bring them together (Levine, 1980).

Protozoa typically range in size from 2 μm to approximately 1 mm in diameter, although some primitive forms extending into the centimetre range have been recorded (Westphal and Mühlpfordt, 1976). The four classes are the Flagellata or Mastigophora (Class I), Rhizopoda (Class II), Sporozoa (Class III) and Ciliata or Ciliophora (Class IV). This classification is based on the nature of their locomotory organelles and their ability to form cysts or spores. Thus, the flagellates and ciliates have locomotory organelles. These are long in the flagellates, typically longer than the rest of the cell, whereas in the cilia they are hair-like strands, which are shorter than the body of the cell itself. The Rhizopodia exhibit cytoplasmic streaming or pseudopodia in the vegetative trophozoite form. Under certain conditions, usually adverse environmental or nutrient conditions, members of the Rhizopodia can surround themselves with an outer capsule, and so encyst into a survival structure. Once conditions become suitable for growth, they excyst, wherein the vegetative form is released from the cyst. Finally, the Sporozoa are parasitic organisms characterised by their ability to form spores, often enclosed in a tough outer coat. Members include *Plasmodium* spp. and *T. gondii*, in which the spore form plays a major role in dissemination and pathogenesis of infection. Molecular methods now play a major role in taxonomy and genomic characterisation. Some protozoa lack mitochondria (e.g. *Giardia lamblia*, *T. vaginalis*, *Entamoeba histolytica*), whereas others possess a single mitochondrion, in contrast with higher eukaryotic cells, which often possess hundreds. Another example of the diversity of protozoa revealed by molecular studies includes RNA processing, where some species show the unusual properties of polycistronic transcription and trans-splicing as well as RNA editing. Species that are more ancient in evolutionary terms, such as *G. lamblia*, reveal their close relationship with prokaryotes by the presence of a Shine Dalgarno sequence on their ribosomal RNA (Smith and Parsons, 1996).

11.2   HABITATS OF PROTOZOA

11.2.1   The environment

The small sizes of protozoa and their cysts means that they are easily and randomly transported. However, they are not randomly distributed, but live in microhabitats within a body of water or moist environments, such as soil vegetation or the bodies of plants and animals. The types of protozoa found depend both on the chemical properties of the environment, such as water (amount and duration), salinity and levels of dissolved gases, such as O_2, CO_2 and H_2S, and the physical nature will favour protozoa of a certain body shape. Thus, mostly flagellates and some ciliates are found in open water, whereas water/surface films and bottom/submerged surfaces favour flattened ciliates, rhizopods and sessile species. Where substrates are finely branched or submerged, swimming forms take on a gliding habitat and mix with interface forms (Westphal and Mühlpfordt, 1976).

The microhabitats favoured by different protozoa are dynamic ecosystems and it is highly unlikely that they will be the only species present. It is much more likely that there will be a heterogeneous microflora composed of bacteria and possibly fungi as well as protozoa in a complex biofilm. Many protozoa feed by grazing on microbial biofilms, but they do not do so indiscriminately. A dynamic equilibrium will exist between the hunter and the hunted and the nature of the relationship changes depending on environmental conditions, namely temperature, UV light and daylight length. The overall relationship will also be determined by the flow of energy, which might come from sunlight, imported organic matter as well as bacterial or fungal decomposers. In habitats of decaying matter, there can be a succession of protozoal species beginning with colourless flagellates and small ciliates followed by large swimming and crawling ciliates, and peritrichs. As decay nears completion, large chlorophyll-bearing ciliates appear, which in turn support herbivorous species (Westphal and Mühlpfordt, 1976).

11.2.2   Biofilms in water systems

It is now well established that solid surfaces, in contact with water, will quickly become colonised by bacteria, such as *Vibrio* and *Pseudomonas* and that these will form a biofilm, often in complex association with other species. Protozoa, including amoebae, flagellates and ciliates, will also be present (Fenchel, 1987). Such protozoa will graze on the biofilm and have been proposed to be a significant factor in disrupting the biofilm by causing fragmentation and sloughing (Jackson and Jones, 1991). A number of techniques have been used to quantify this predation of bacteria, such as monitoring the release of ^{3}H label into water that had been incorporated into macromolecules of ^{3}H-thymidine-labelled bacteria. Up to 75% of bacterial species grown in model systems can be consumed (Zubkov and Sleigh, 1999).

Water-distribution networks, studies have shown that development of bacterial communities can lead to a food chain that supports the growth of protozoa and renders the water unsuitable for use. Notable examples include cases of meningococcal encephalitis caused by amoebae in swimming pools. Famously, the hot spring water from Bath spa (UK) became unsuitable for consumption for a period of time, due to contamination with thermophilic Naegleria species (Kilvington *et al.*, 1991). Relatively high levels of protozoa have been measured in water-distribution systems, sometimes in excess 5×10^5/l (Block *et al.*, 1993) and protozoa can be found throughout the system as well as in the granulated activated carbon (GAC) filters. Recent work has compared the microflora of drinking water systems, one of GAC water and one of nanofiltration (Sibille *et al.*, 1998). The nanofiltered water-supplied network contained no organisms larger than bacteria, either in the water phase (approx 5×10^7 bacterial cells/l) or the filter (7×10^6 bacterial cells/cm^2). In contrast, the GAC water-supplied network contained approximately 10-fold greater levels of bacteria in the water phase and the filter bed, as well as protozoa (10^5/l and 10^3/cm^2, respectively). Greater than 90% of the protozoa in the water phase were flagellates. Approximately, 50% of the GAC-recovered protozoa were ciliates and only these organisms had any measurable grazing activity. Interestingly, when *Escherichia coli* was deliberately added to each pilot water-distribution system, they were cleared more rapidly from the GAC system than from the nanofiltration system. Thus, it is likely that predatory protozoa can help control bacterial biomass. Equally, the converse is true, and studies using dental unit waterlines have shown that waterborne bacteria create biofilms that serve as substrata to support proliferation of a variety of free-living amoebae, including *Hartmanella*, *Vanella* and *Vahlkampfia* spp. as well as *Naegleria* and *Acanthamoeba* spp. (Barbeau and Buhler, 2001). Also, studies of factors leading to acanthamoebal keratitis in contact lens wearers have shown that in the majority of carry cases, where *Acanthamoeba* species were found, bacteria were also present (Larkin *et al.*, 1990).

11.3   THE PARADIGM OF *LEGIONELLA PNEUMOPHILA*

Much of the above discussion is based on the perspective of protozoa as hunter and bacteria as hunted, but this is an overly simplistic view. Some bacteria not only resist predation by protozoa, but also have become endosymbionts of free-living protozoa demonstrating adaptation to the normally hostile intra-cellular protozoal environment. The environmental conditions can be decisive. For example, at low temperatures, acanthamoebae may phagocytose and digest *L. pneumophila* as food, while at higher temperatures after infection with the same strain of *Legionella*, the bacteria grow to high numbers (Anand *et al.*, 1983). The concept that protozoa can be reservoirs for a number of potential pathogens is becoming more widely appreciated (Barker and Brown, 1994; Shuman *et al.*, 1998; Abu Kwaik, 1998; Brown and Barker, 1999; Cirillo, 1999; Harb *et al.*, 2000). The catalyst for much of this work came from the observation by Rowbotham that *L. pneumophila* infects and multiples within some species of free-living amoebae (Rowbotham, 1980). *L. pneumophila* is the primary cause of Legionnaires' disease, a serious form of atypical pneumonia. It has become the paradigm for a sophisticated host–parasite interaction, which is now yielding important information about the ecological and pathogenic traits of many bacteria. A number of groups have highlighted the contribution to protozoal association made by complex bacterial communities within biofilms and have gone further to propose that it is an important site for prokaryotic–eukaryotic co-evolution (Barker and Brown, 1994; Cirillo, 1999; Brown and Barker, 1999; Harb *et al.*, 2000). Most single-celled organisms present in biofilms will pre-date higher eukaryotic organisms, and thus, it is likely that the competition for growth in environmental biofilms could have selected for bacterial virulence mechanisms responsible for human disease. Typical mechanisms that can be envisaged include colonisation of surfaces, resistance to antimicrobial compounds secreted by other micro-organisms, avoidance of phagocytosis or exploitation of phagocytosis for uptake into and survival in other cells. Attention is now focused on the outcomes of interaction between bacteria and protozoa and the commonality of bacterial invasion and survival strategies between protozoa and higher eukaryotic cells.

11.3.1   Protozoa as 'Trojan horses'

L. pneumophila is unusual in that no animal reservoir has been identified, but it can infect and multiply within *Hartmanella*, *Acanthamoeba* and *Naegleria* species as well as *Tetrahymena*, a freshwater ciliate (Fields, 1996). Such protozoa have been described as 'Trojan horses' (Barker and Brown, 1994). Electron microscopy studies of *L. pneumophila* within *Acanthamoeba polyphaga* have led to estimates of up to 10^4 bacteria occupying 90% of the amoebal cell (Rowbotham, 1986). Thus, it is likely that numbers of *L. pneumophila* can be amplified significantly. Indeed, studies using a model potable water system indicated that the protozoan *Hartmanella vermiformis* was required for replication of *L. pneumophila*, although it could persist in its absence in a mixed biofilm composed of *Pseudomonas aeruginosa*, *Klebsiella pneumoniae* and *Flavobacterium* species (Murga *et al.*, 2001).

11.3.2   The intra-cellular phenotype

In addition to amplifying cell numbers, the phenotypic consequences of intra-protozoal replication should not be underestimated. A number of studies have confirmed that the phenotype of intra-protozoal-grown cells is quite distinct from their free-living counterparts cultured on complex laboratory media. Notable are their surface properties (Barker *et al.*, 1993) and their decreased susceptibility to chemical inactivation (Barker *et al.*, 1992) and antibiotics (Barker *et al.*, 1995). Moreover, bacteria trapped within amoebal cysts show enhanced resistance to inactivation by chlorine (Kilvington and Price, 1990). A further major phenotypic consequence for intra-protozoal-grown *L. pneumophila* is their enhanced invasiveness for mammalian cells compared with cells grown in conventional laboratory media (Cirillo *et al.*, 1994, 1999). Thus, bacteria emergent from protozoa are not only more resistant to chemical inactivation, but are also more invasive. But what about the mechanisms of invasion and survival, and comparisons between protozoal and mammalian cells? These will now be discussed.

11.3.3   Interaction and uptake

Much work has focused on the initial interaction between *L. pneumophila* and protozoa or higher eukaryotic cells. A Gal/GalNAc lectin on the protozoan *H. vermiformis* has been characterised as a potential receptor (Venkataraman *et al.*, 1997). Complement receptors on monocytes and macrophages may also have a role in triggering internalisation of opsonised bacteria by a pseudopod coil

(Horwitz, 1984). *L. pneumophila* expresses multiple pili of different lengths, and the type IV pilin gene (*PilE$_L$*) has a role in adherence to mammalian and protozoan cells (Stone and Abu Kwaik, 1998). Mutants in *pilE$_L$* were defective for expression of long pili and exhibited a 50% decrease in cellular attachment, but were not deficient in intra-cellular replication.

11.3.4   Intra-cellular survival

After uptake, the bacteria modulate the development of the phagocytic vacuole, such that the classical endosomal–lysosomal maturation and degradation pathways are not followed. Instead, microscopy studies have shown that replication of *L. pneumophila* both in protozoa and mammalian cells occur within ribosome-studded phagosomes. A group of 23 genes, present in two clusters, termed *dot* (defect in organelle trafficking) and *icm* (intra-cellular multiplication) has been identified. Proteins encoded by these genes control formation of the replicative vacuole, intra-cellular replication and subsequent egress/lysis. The *dot/icm* complex is thought to constitute a type IV-like secretion system capable of transferring effector molecules that perturb the normal endocytic pathway. Type II secretion processes are also involved (Hales and Shuman, 1999a). Studies with mutants in *pilD*, encoding prepilin peptidase, also indicate a role for type II protein secretion processes (Rossier and Cianciotto, 2001). Prepilin peptidases are implicated both in the formation of type IV pili and protein secretion, but since *pilE$_L$* mutants are not replication defective in protozoa and macrophages, the link with protein secretion was made. Mutations in *lspDE*, the type II outer membrane secretin and ATPase or *lspFGHIJK*, the pseudopilins showed reduced secretion of a number of *pilD*-dependent enzymatic activities, such as protease, phosphatase, lipase and phospholipase A, and were greatly impaired for growth in *H. vermiformis* and U937 macrophages. Interestingly, the *pilD* mutant was 100-fold more defective than the type II secretion mutants in U937 cells, suggesting that type II exoproteins may play a greater role in amoebae (Rossier and Cianciotto, 2001).

11.3.5   Stress responses

Adaptation to the environment of the replicative vacuole is clearly important as a prelude to replication. Mutants defective for iron acquisition and assimilation exhibited prolonged lag phases and, in some cases, replicated at slower rates in U937 cells (Pope *et al.*, 1996). Other studies have focused on the response to *L. pneumophila* to the stress imposed by the hostile microenvironment within the vacuole. *L. pneumophila* has a homologue of the *E. coli* stress and stationary phase sigma factor RpoS, which accumulates on entry into stationary phase, but paradoxically may not be required for stationary-phase-dependent resistance to stress (Hales and Shuman, 1999b). Mutants in *rpoS* could not replicate within the protozoan host *Acanthamoebae castellani*, but were able to kill macrophage-like cell lines (Hales and Shuman, 1999b).

The alarmone ppGpp, which in other organisms triggers an adaptive response including synthesis and accumulation of RpoS (Cashel *et al.*, 1996), accumulated when *L. pneumophila* was subjected to amino acid depletion. This resulted in conversion from a replicative to virulent state characterised by increased motility, cytotoxicity and infectivity characteristics of virulence (Hammer and Swanson, 1999). At least 30 proteins are expressed by *L. pneumophila* within macrophages, including GroEL/Hsp60, GroES and GspA that are induced by various stress stimuli *in vitro*. Studies with *htrA* mutants highlighted that the HtrA/DegP stress-induced protease/chaperone homologue is essential for intra-cellular replication within macrophages, but not within protozoa (Pedersen *et al.*, 2001). In macrophages, the *htrA* mutant was unable to exclude the late endosomal–lysosomal marker LAMP-2 from the phagosome and there was no association with the rough endoplasmic reticulum (ER), but rough ER recruitment did occur around protozoal vacuoles. Defective protein folding has been proposed, since some mutants, such as *dotA* that are defective in early phagosomal modulation and are trafficked into a phagolysosome, can be rescued when they reside in the same phagosome as the parent strain. The *htrA* mutant could not be rescued, suggesting an intolerance of the phagosomal microenvironment. These data suggest that the protozoal phagosome is perhaps less stressful than that of mammalian cells. However, it would be an oversimplification to consider all protozoa as being the same. Signature-tagged mutagenesis strategies aimed at identifying *L. pneumophila* genes important for infection have revealed subsets of mutants that are required for survival in *A. castellani*, but not in *H. vermiformis* (Polesky *et al.*, 2001). Some genes, such as *lphA* are completely required for replication within *A. castellani* and yet are not required for intra-cellular growth and killing of human macrophages (Segal and Shuman, 1999). Other indications of adaptation to the environment of the phagosome include identification of a *pagP* homologue, designated *rcp* (Robey *et al.*, 2001). In *Salmonella*, *pagP* increases resistance to the bactericidal effects of cationic antimicrobial peptides. *L. pneumophila rcp* mutants exhibited a 1000-fold decrease in recovery following co-culture with *H. vermiformis* and reduced recovery from and cytopathicity towards U937 macrophages (Robey *et al.*, 2001).

11.3.6   Eukaryotic cell death

Finally, what is the fate of the *L. pneumophila* and the cell in which it resides? Early studies indicated that *L. pneumophila* kills human phagocytes but not protozoan host cells by caspase 3-mediated apoptotic cell death (Gao and Abu Kwaik, 1999), induced by the Dot/Icm type IV secretion system (Zink *et al.*, 2002). Other work has addressed lysis and egress from mammalian and protozoan cells (Alli *et al.*, 2000; Gao and Abu Kwaik, 2000). Some mutants, designated *rib* (release of intra-cellular bacteria) have been identified, which are released at a later time, presumably due to apoptotic changes in the host (Alli *et al.*, 2000). Identification of the pore-forming toxin has been difficult. Type II secretion mutants are neither defective in the pore-forming toxin, nor are mutants in *rtxA* encoding the

RtxA-like protein (repeats in structural toxin), although these mutants were partially defective in causing permeability changes to the host cell membrane. Recently, it has been shown that the *icmT* gene is the defective locus in the *rib* mutants and that it is essential for egress from U937 macrophages and *A. polyphaga* (Molmeret *et al.*, 2002). The defect in *icmT* was in a stretch of poly(T) in the carboxy terminus of the gene, likely due to slipped-strand mispairing during replication. Interestingly, this causes phase variation of virulence genes in other organisms, such as *Neisseria gonorrhoea*.

11.3.7   Co-evolution of bacteria and protozoa

Clearly, there is significant conservation in the genes used to parasitise protozoa and macrophages. It has been proposed that co-evolution of bacteria and lower-order eukaryotic cells has equipped bacteria both for environmental survival and virulence towards higher-order eukaryotes (Brown and Barker, 1999; Barker *et al.*, 1999). Other authors have even described protozoa as an 'evolutionary gym' enabling *L. pneumophila* to 'train' for its role in invading higher eukaryotic organisms (Harb *et al.*, 2000).

The role played by protozoa should not be viewed only from a distant evolutionary perpective. In addition to their major role in sustaining *L. pneumophila* in the environment, emergent bacteria exhibit much decreased susceptibility to inactivation by biocides (Barker *et al.*, 1992) and antibiotics (Barker and Brown, 1995), as well as enhanced invasion and virulence in monocytes (Cirillo *et al.*, 1999). One report suggests that even supernatants from uninfected cultures of *A. castellani* can increase the intra-cellular replication of some *Legionella* species in a monocyte cell line (Neumeister *et al.*, 2000).

11.4   OTHER SPECIES COLONISING PROTOZOA

A growing list of bacteria (Table 11.1) that interact with protozoa confirm that the co-evolution between *L. pneumophila* and protozoa is not unique in nature. For example, *Mycobacterium avium* survives within environmental amoebae by the same mechanism of inhibition of lysosomal fusion as in macrophages. Compared with broth-cultured bacteria, growth in *A. castellani* enhanced invasion and intra-cellular replication in macrophages and increased virulence in a mouse model of infection (Cirillo *et al.*, 1997). *M. avium* also survives within cysts of *A. polyphaga*, which are themselves survival structures, highlighting an additional opportunity for enhanced survival and persistence in the environment (Steinert *et al.*, 1998). *L. pneumophila* and *Vibrio cholerae* also survive within amoebal cysts (Brown and Barker, 1999). It is possible that co-evolution with protozoa could have contributed to the pathogenesis of other mycobacterial infections. Indeed, some workers have speculated that it could be environmental amoebae and not solely badgers, which contribute to the current issue of bovine tuberculosis in the UK (Barker and Brown, 1999).

Table 11.1   Microbial species that associate with protozoa.

Microbial species	Protozoal host	Reference/review
Legionella spp.	*Acanthamoeba* spp., *Hartmanella, Naegleria* and others	Fields (1996); Harb *et al.* (2000); Shuman *et al.* (1998)
M. avium	*Acanthamoeba* spp.	Cirillo *et al.* (1997)
V. cholerae	*Acanthamoeba* and *Naegleria* spp.	Thom *et al.* (1992)
E. coli	*Acanthamoebae* spp.	Barker *et al.* (1999)
P. aeruginosa	*Dictyostelium discoideum*	Pukatzki *et al.* (2002); Cosson *et al.* (2002)
Simkania negevensis	*Acanthamoeba* spp.	Kahane *et al.* (2001)
Listeria monocytogenes	*Acanthamoeba* and *Tetrahymena pyriformis*	Ly and Muller (1990)
Cryptococcus neoformans	*Acanthamoeba* spp.	Steenbergen *et al.* (2001)

E. coli O157 has been shown to survive and replicate in *A. polyphaga*, a common environmental protozoan (Barker *et al.*, 1999), as has a non-O157 strain (King *et al.*, 1988). The role of protozoa in the survival of *E. coli* O157 in the natural environment has not been studied, although the organism can survive in cattle slurry for several weeks (Rodriguez-Zaragoza, 1994). There is growing concern about the survival of pathogens in sewage effluents disposed on land now that green laws prohibit sea dumping. Soil contaminated with organic matter and sewage waste contain greatly increased numbers of protozoa, such as acanthamoebae (Rodriguez-Zaragoza, 1994). Thus, it is highly likely that *E. coli* O157 will be preyed on by free-living protozoa that will be potential vectors for the spread of this pathogen.

Recent work with *P. aeruginosa* from two laboratories has shown that this bacterium utilises conserved virulence pathways to infect the social amoeba *D. discoideum* (Pukatzki *et al.*, 2002; Cosson *et al.*, 2002). Both used a simple plating assay; virulent bacteria killed the amoebae and formed lawns, whereas plaques formed in the bacterial lawn when amoebae fed on avirulent bacterial strains. In one study, mutations in *lasR*, encoding the main transcriptional regulator of the *las* quorum-sensing pathway, resulted in loss of virulence in the assay (Pukatzki *et al.*, 2002). The mediator remains unknown, although the phenazine agent pyocyanin, which can be cytotoxic via oxidative stress mechanisms, was not involved, nor was secretion of rhamnolipids. Interestingly, cell density was important. Amoebae fed even on virulent strains when the nutrient content of the medium was reduced by dilution, restricting population density. Further studies indicated that the differences between virulent and avirulent strains were not due to resistance to phagocytosis; cells from both types were digested rapidly once internalised by the amoebae. Recovery experiments indicated that virulent strains were killing the amoebae rather than simply causing starvation and the mediator was shown to be the toxin ExoU, one of a number of virulence factors injected into the host cell through a type III secretion apparatus (Pukatzki *et al.*, 2002). In the other study, isogenic mutants in the *las* quorum-sensing system were almost as inhibitory as the wild type, while *rhl* qorum-sensing mutants permitted growth of *Dictyostelium* cells. Rhamnolipids secreted by the wild-type strain could induce fast lysis of *D. discoideum* cells (Cosson *et al.*, 2002). While differences between the two studies are likely attributable to the different parent strains used, it is clear that *P. aeruginosa* may be yet another organism where pathogenic traits have developed by co-evolution with single-celled eukaryotic organisms.

Other bacteria noted to survive within amoebae include *Burkholderia cepacia* (Marolda *et al.*, 1999) and *Simkania negevensis* (Kahane *et al.*, 2001). *S. negevensis* is a chlamydia-like bacterium that has an intra-cellular development cycle comprising two different morphological entities, elementary bodies and reticulate bodies. In addition to exponential growth within *A. polyphaga* trophozoites, *S. nevegensis* was also able to survive for prolonged periods within dried amoeba cysts (Kahane *et al.*, 2001).

Finally, recent work suggests that co-evolution with protozoa is not restricted only to bacteria. The encapsulated yeast *C. neoformans* is phagocytosed by (and replicates in) *A. castellani*, which leads to death of the amoebae (Steenbergen *et al.*, 2001). A capsular strains did not survive within amoebae and a phospholipase mutant had a decreased replication rate in amoebae compared with isogenic wild-type strains. This is further evidence in support of the hypothesis that evolution of protective mechanisms against environmental predators is an explanation of virulence for mammalian cells.

11.5   CONCLUDING REMARKS: ACCIDENTAL PATHOGENS?

All human and animal pathogens must have evolved from bacteria that co-existed with lower-order eukaryotes for a billion years or so. Some commentators have proposed that *L. pneumophila* is an 'accidental' human pathogen (Shuman *et al.*, 1998) in that only a subset of the population, in this case the elderly, are susceptible. An emerging theme amongst many of the micro-organisms associating with protozoa is that they too can be regarded as accidental or not 'proper' pathogens. *B. cepacia* essentially causes infection only in the lungs of patients with cystic fibrosis (CF) or chronic granulomatous disease. Similarly, *P. aeruginosa* is essentially non-pathogenic in the greater part of the healthy population. It is typically only in patients that have CF or who are immunocompromised in some way that *P. aeruginosa* is such a devastating pathogen. *C. neoformans* also tends to cause disease in immunocompromised patients. In 'normal' hosts, the organism rarely causes clinically apparent disease, but once established it has the capacity for latency and persistence inside macrophages. Taken together, these features are suggestive of a relationship much closer to symbiosis which becomes parasitic only in special circumstances. The question remains why this is so? The answer could lie in co-evolution with lower-order eukaryotic organisms present in biofilms.

REFERENCES

Abu Kwaik, Y. (1998) Fatal attraction of mammalian cells to *Legionella pneumophila*. *Mol. Microbiol.* **30**, 689–695.

Alli, O.A.T., Gao, L.-Y., Pedersen, L.L., Zink, S., Radulic, M., Doric, M. and Abu Kwaik, Y. (2000) Temporal pore formation-mediated egress from macrophages and alveolar epithelial cells by *Legionella pneumophila*. *Infect. Immun.* **68**, 6431–6440.

Anand, C.M., Skinner, A.R., Malic, A. and Kurtz, J.B. (1983) Interaction of *Legionella pneumophila* and free living amoebae. *J. Hyg. (Camb)* **91**, 167–178.

Barbeau, J. and Buhler, T. (2001) Biofilms augment the number of free-living amoebae in dental unit waterlines. *Res. Microbiol.* **152**, 753–760.

Barker, J. and Brown, M.R.W. (1994) Trojan horses of the microbial world: protozoa and the survival of bacterial pathogens in the environment. *Microbiology* **140**, 1253–1259.

Barker, J. and Brown, M.R.W. (1995) Speculations on the influence of infecting phenotype on virulence and antibiotic susceptibility of *Legionella pneumophila*. *J. Antimicrob. Chemoth.* **36**, 7–21.

Barker, J. and Brown, M.R.W. (1999) Culling of badgers and control of bovine tuberculosis. *Lancet* **352**, 2025.

Barker, J., Brown, M.R.W., Collier, P.J., Farrell, I. and Gilbert, P. (1992) Relationship between *Legionella pneumophila* and *Acanthamoeba polyphaga*: physiological status and susceptibility to chemical inactivation. *Appl. Environ. Microbiol.* **58**, 2420–2425.

Barker, J., Lambert, P.A. and Brown, M.R.W. (1993) Influence of intra-amoebic and other growth conditions on the surface properties of *Legionella pneumophila*. *Infect. Immun.* **61**, 3503–3510.

Barker, J., Scaife, H. and Brown, M.R.W. (1995) Intraphagocytic growth induces an antibiotic-resistant phenotype of *Legionella pneumophila*. *Antimicrob. Agents Chemoth.* **39**, 2684–2688.

Barker, J., Humphrey, T.J. and Brown, M.R.W. (1999) Survival of *Escherichia coli* O157 in environmental protozoa: implications for disease. *FEMS Microbiol. Lett.* **173**, 291–295.

Block, J.C., Haudidier, K., Paquin, J.L., Miazga, J. and Lévi, Y. (1993) Biofilm accumulation in drinking water distribution systems. *Biofouling* **6**, 333–343.

Brown, M.R.W. and Barker, J. (1999) Unexplored reservoirs of pathogenic bacteria: protozoa and biofilms. *Trend. Microbiol.* **7**, 46–50.

Cashel, M., Gentry, D.R., Hernandez, V.J. and Vinella, D. (1996) The stringent response. In *Escherichia coli and Salmonella* (ed. F.C. Neidhardt, R. Curtiss III, J.L. Ingraham, E.C.C. Lin, K.B. Low, B. Magasanik, W.S. Reznikoff, M. Riley, M. Schaechter and H.E. Umbarger), pp. 1458–1496, ASM Press, Washington DC.

Cirillo, J.D. (1999) Exploring a novel perspective on pathogenic relationships. *Trend. Microbiol.* **7**, 96–98.

Cirillo, J.D., Falkow, S. and Tompkins, L.S. (1994) Growth of *Legionella pneumophila* in *Acanthamoeba castellani* enhances invasion. *Infect. Immun.* **62**, 3254–3261.

Cirillo, J.D., Falkow, S., Tompkins, L.S. and Bermudez, L.E. (1997) Interaction of *Mycobacterium avium* with environmental amoebae enhances virulence. *Infect. Immun.* **65**, 3759–3767.

Cirillo, J.D., Cirillo, S.L.G., Yan, L., Bermudez, L.E., Falkow, S. and Tompkins, L.S. (1999) Intracellular growth in *Acanthamoeba castellani* affects monocyte entry mechanisms and enhances virulence of *Legionella pneumophila*. *Infect. Immun.* **67**, 4427–4434.

Cosson, P., Zulianello, L., Join-Lambert, O., Faurisson, F., Gebbie, L., Benghezal, M., Van Delden, C., Curty, L. and Kohler, T. (2002) *Pseudomonas aeruginosa* virulence analyzed in a *Dictyostelium discoideum* host system. *J. Bacteriol.* **184**, 3027–3033.

Fenchel, T. (1987) *Ecology of the Protozoa. The Biology of Free-Living Phagotrophic Protists,* Springer-Verlag, Berlin, Germany.

Fields, B.S. (1996) The molecular ecology of legionellae. *Trend. Microbiol.* **4**, 286–290.

Gao, L.-Y. and Abu Kwaik, Y. (1999) Apoptosis in macrophages and alveolar epithelial cells during early stages of infection by *Legionella pneumophila* and its role in cytopathogenicity. *Infect. Immun.* **67**, 862–870.

Gao, L.-Y. and Abu Kwaik, Y. (2000) The mechanism of killing and exiting the protozoan host *Acanthamoeba polyphaga* by *Legionella pneumophila*. *Environ. Microbiol.* **2**, 79–90.

Hales, L.M. and Shuman, H.A. (1999a) *Legionella pneumophila* contains a type II general secretion pathway required for growth in amoebae as well as for secretion of the Msp protease. *Infect. Immun.* **67**, 3662–3666.

Hales, L.M. and Shuman, H.A. (1999b) The *Legionella pneumophila rpoS* gene is required for growth within *Acanthamoeba castellani*. *J. Bacteriol.* **181**, 4879–4889.

Hammer, B.K. and Swanson, M.S. (1999) Co-ordination of *Legionella pneumophila* virulence with entry intro stationary phase by ppGpp. *Mol. Microbiol.* **33**, 721–731.

Harb, O.S., Gao, L.-Y. and Abu Kwaik, Y. (2000) From protozoa to mammalian cells: a new paradigm in the life cycle of intracellular pathogens. *Environ. Microbiol.* **2**, 251–265.

Horwitz, M.A. (1984) Phagocytosis of the Legionnaire's disease bacterium (*Legionella pneumophila*) occurs by a novel mechanism: engulfment within a pseudopod coil. *Cell* **36**, 27–33.

Jackson, S.M. and Jones, E.B.G. (1991) Interactions within biofilms: the disruption of biofilms structure by protozoa. *Kieler Meeresforsch* **8**, 264–268.

Kahane, S., Dvoskin, B., Mathias, M. and Friedman, M.G. (2001) Infection of *Acanthamoeba polyphaga* with *Simkania negevensis* and *S. negevensis* survival within amoebal cysts. *Appl. Environ. Microbiol.* **67**, 4789–4795.

Kilvington, S. and Price, J. (1990) Survival of *Legionella pneumophila* within *Acanthamoeba polyphaga* cysts following chlorine exposure. *J. Appl. Bact.* **58**, 519–525.

Kilvington, S., Mann, P. and Warhurst, D.C. (1991) Pathogenic *Naegleria* amoebae in the waters of bath: a fatality and its consequences. In *Host Springs of Bath* (ed. G.A. Kellaway), pp. 89–96, Bath City Council, Bath, UK.

King, C.H., Shotts, E.B., Wooley, R.E. and Porter, K.G. (1988) Survival of coliforms and bacterial pathogens within protozoa during chlorination. *Appl. Environ. Microbiol.* **54**, 3023–3033.

Larkin, D.F., Kilvington, S. and Easty, D.L. (1990) Contamination of contact lens storage cases by Acanthamoeba and bacteria. *Brit. J. Ophthalmol.* **74**, 133–135.

Levine, N.D. (1980) A newly revised classification of the protozoa. *J. Protozool.* **27**, 37–58.

Ly, T.M. and Muller, H.E. (1990) Ingested *Listeria monocytogenes* survive and multiply in protozoa. *J. Med. Microbiol.* **33**, 51–54.

Mahajan-Miklos, S., Tan, M.-W., Rahme, L.G. and Ausubel, F.M. (1999) Molecular mechanisms of bacterial virulence elucidated using a *Pseudomonas aeruginosa–Caenorhabditis elegans* pathogenesis model. *Cell* **96**, 47–56.

Marolda, C.L., Hauroder, B., John, M.A., Michel, R. and Valvano, M.A. (1999) Intracellular survival and saprophytic growth of isolates from the *Burkholderia cepacia* complex in free-living amoebae. *Microbiology* **145**, 1509–1517.

Molmeret, M., Alli, O.A.T., Zink, S., Flieger, A., Cianciotto, N.P. and Abu Kwaik, Y. (2002) *icmT* Is essential for pore formation-mediated egress of *Legionella pneumophila* from mammalian and protozoan cells. *Infect. Immun.* **70**, 69–78.

Murga, R., Forster, T.S., Brown, E., Pruckler, J.M., Fields, B.S. and Donlan, R.M. (2001) Role of biofilms in the survival of *Legionella pneumophila* in a model potable-water system. *Microbiology* **147**, 3121–3126.

Neumeister, B., Reiff, G., Faigle, M., Dietz, K., Northoff, H. and Lang, F. (2000) Influence of *Acanthamoebae castellani* on intracellular growth of different *Legionella* species in human monocytes. *Appl. Environ. Microbiol.* **66**, 914–919.

Pedersen, L.L., Radulic, M., Doric, M. and Abu Kwaik, Y. (2001) HtrA homologue of *Legionella pneumophila*: an indispensable element for intracellular infection of mammalian but not protozoan cells. *Infect. Immun.* **69**, 2569–2579.

Polesky, A.H., Ross, J.T.D., Falkow, S. and Tompkins, L.S. (2001) Identification of *Legionella pneumophila* genes important for infection of amoebas by signature-tagged mutagenesis. *Infect. Immun.* **69**, 977–987.

Pope, C.D., O'Connell, W.A. and Cianciotto, N.P. (1996) *Legionella pneumophila* mutants that are defective for iron acquisition and assimilation and intracellular infection. *Infect. Immun.* **64**, 629–636.

Pukatzki, S., Kessin, R.H. and Mekalanos, J.J. (2002) The human pathogen *Pseudomonas aeruginosa* utilizes conserved virulence pathways to infect the social amoeba *Dictyostelium discoideum. Proc. Natl. Acad. Sci. USA* **99**, 3159–3164.

Robey, M., O'Connell, W. and Cianciotto, N.P. (2001) Identification of *Legionella pneumophila rcp*, a *pagP*-like gene that confers resistance to cationic antimicrobial peptides and promotes intracellular infection. *Infect. Immun.* **69**, 4276–4286.

Rodriguez-Zaragoza, S. (1994) Ecology of free-living amoebae. *Crit. Rev. Microbiol.* **20**, 225–241.

Rossier, O. and Cianciotto, N.P. (2001) Type II protein secretion is a subset of the PilD-dependent processes that facilitate intracellular infection by *Legionella pneumophila. Infect. Immun.* **69**, 2092–2098.

Rowbotham, T.J. (1980) Preliminary report on the pathogenicity of *Legionella pneumophila* for freshwater and soil amoebae. *J. Clin. Pathol.* **33**, 1179–1183.

Rowbotham, T.J. (1986) Current views on the relationships between amoebae, legionellae and man. *Israel. J. Med. Sci.* **22**, 678–689.

Segal, G. and Shuman, H.A. (1999) *Legionella pneumophila* utilizes the same genes to multiply within *Acanthamoeba castellanii* and human macrophages. *Infect. Immun.* **67**, 2117 2124.

Shuman, H.A., Purcell, M., Segal, G., Hales, L. and Wiater, L.A. (1998) Intracellular multiplication of *Legionella pneumophila*: human pathogen or accidental tourist? *Curr. Top. Microbiol. Immunol.* **225**, 99–112.

Sibille, I., Sime-Ngando, T., Mathieu, L. and Block, J.C. (1998) Protozoan bacterivory and *Escherichia coli* survival in drinking water distribution systems. *Appl. Environ. Microbiol.* **64**, 197–202.

Smith, D.F. and Parsons, M. (1996) *Molecular Biology of Parasitic Protozoa,* IRL Press, Oxford.

Steenbergen, J.N., Shuman, H.A. and Casadevall, A. (2001) *Cryptococcus neoformans* interactions with amoebae suggest an explanation for its virulence and intracellular pathogenic strategy in macrophages. *Proc. Natl. Acad. Sci. USA* **98**, 15245–15250.

Steinert, M., Birkness, K., White, E., Fields, B. and Quinn, F. (1998) *Mycobacterium avium* bacilli grow saprozoically in co-culture with *Acanthamoeba polyphaga* and survive within cyst walls. *Appl. Environ. Microbiol.* **64**, 2256–2261.

Stone, B.J. and Abu Kwaik, Y. (1998) Expression of multiple pili by *Legionella pneumophila*: identification and characterization of a type IV pilin gene and its role in adherence to mammalian and protozoan cells. *Infect. Immun.* **66**, 1768–1775.

Tan, M.-W., Mahajan-Miklos, S. and Ausubel, F.M. (1999a) Killing of *Caenorhabditis elegans* by *Pseudomonas aeruginosa* used to model mammalian bacterial pathogenesis. *Proc. Natl. Acad. Sci. USA* **96**, 715–720.

Tan, M.-W., Rahme, L.G., Sternberg, J.A., Tompkins, R.G. and Ausubel, F.M. (1999b) *Pseudomonas aeruginosa* killing of *Caeorhabditis elegans* used to identify *P. aeruginosa* virulence factors. *Proc. Natl. Acad. Sci. USA* **96**, 2408–2413.

Thom, S., Warhurst, D. and Drasar, B.S. (1992) Association of *Vibrio cholorae* with freshwater amoebae. *J. Med. Microbiol.* **36**, 303–306.

Venkataraman, C., Haack, B.J., Bondada, S. and Abu Kwaik, Y. (1997) Identification of a Gal/GalNAc lectin in the protozoan *Hartmanella vermiformis* as a potential receptor for attachment and invasion by the Legionnaire's disease bacterium. *J. Exp. Med.* **186**, 537–547.

Westphal, A. and Mühlpfordt, H. (1976) *Protozoa,* Blackie, Glasgow, UK.

Zink, S.D., Pedersen, L., Cianciotto, N.P. and Abu Kwaik, Y. (2002) The Dot/Icm type IV secretion system of *Legionella pneumophila* is essential for the induction of apoptosis in human macrophages. *Infect. Immun.* **70**, 1657–1663.

Zubkov, M.V. and Sleigh, M.A. (1999) Growth of amoebae and flagellates on bacteria deposited on filters. *Microbial Ecol.* **37**, 107–115.

PART TWO

Analytical techniques

Section 3   *Biofilm cultivation apparatus*   195

Section 4   *Analytical techniques for biofilm properties*
            A. Physico-chemical properties   257
            B. Biotic properties   329

Section 3

Biofilm cultivation apparatus

12 Use of flow cells and annular reactors
    to study biofilms   197

13 Experimental systems for studying biofilm
    growth in drinking water   214

14 Efficacy testing of disinfectants using
    microbes grown in biofilm constructs   230

15 Steady-state heterogeneous model systems
    in microbial ecology   236

12

Use of flow cells and annular reactors to study biofilms

P. Stoodley and B.K. Warwood

12.1 INTRODUCTION

In laboratory experiments biofilms are grown in a wide variety of systems or 'reactors'. Biofilm reactors generally break down into two types: those that house removable sample coupons and those which are used primarily for *in situ* microscopic observation. The latter are commonly called 'flow cells'. Flow cells allow the flow of a growth medium over a surface to which microorganisms attach and grow into a biofilm. Biofilm flow cells have been commonly used to study the attachment of single bacterial, fungal or algal cells to various surfaces (Characklis and Marshall, 1990; Cooksey and Wigglesworth-Cooksey, 1995; Dalton *et al.*, 1996). By counting individual cells over time, the surface concentration (cells/unit area) and accumulation rate can be calculated for assessing the fouling potential of different surfaces. Flow cells are also used to study the development of mature biofilms and assess how changes in the growth environment or specific genetic mutations may affect the structure of mature biofilms (Hentzer *et al.*, 2001; Nivens *et al.*, 2001; Purevdorj *et al.*, 2002), detachment from biofilms (Stoodley *et al.*, 2001a, b), spatial and temporal gene expression (De Kievit *et al.*, 2001), the distribution of extracellular polysaccharides (Wolfaardt *et al.*, 1998; Strathmann *et al.*, 2002), mechanical properties of biofilms (Stoodley *et al.*, 1999a; Klapper *et al.*, 2002) and biofilm–human leukocyte interactions (Leid *et al.*, 2002). There are many types of flow cell design but the constraints imposed by their use with commercial microscope systems has led to some common geometries and features. However, before biofilm flow cells became commercially available, individual researchers had to fabricate their own systems resulting in many variations between research groups. In part, this may explain the highly variable and sometimes conflicting data commonly reported in biofilm structures and behaviors (Heydorn *et al.*, 2000, 2002). However, biofilm flow cells are now commercially available (BioSurface Technologies [BST] Inc., www.biofilms.biz; Stovall, http://slscience.com/flowcell.html), which will allow for greater standardization in biofilm research.

There are many types of biofilm reactors which have been designed more specifically for multiple- and time-course sampling. Two of the more common reactors are the modified Robbins device (Elvers *et al.*, 2001), which has coupons fixed to sampling plugs that can be unscrewed from the walls of a square channel pipe, and annular reactors (also known as Rototorques) in which an inner drum is rotated to provide shear, and coupons are slid into and out of 'tongue and groove' slots on the stationary outside wall (Liu and Tay, 2001; Wuertz *et al.*, 2001). These reactors tend to generate high-fluid shears and are often used for industrial research and monitoring. Other types of commercially available sampling reactors include the rotating disk reactor and the 'Centers for Disease Control (CDC)' reactor (BioSurface Technologies Inc., www.biofilms.biz).

Recently, a high throughput biofilm growth system, the Calgary Biofilm Device (MBEC Biofilms Technology Ltd., Calgary, Alberta) has been designed for the use with 96 well plates in which biofilm is grown on pegs. These can then be exposed to antibiotics for rapid assessment of biofilm minimum inhibitory concentration (MIC) (Ceri *et al.*, 1999).

Although there are many types of biofilm reactors, each with its own strengths and weaknesses, this review will concentrate on the use of flow cells and the annular reactor for biofilm research and monitoring.

12.2   FLOW CELL SYSTEMS

12.2.1   Experimental setup of flow cell systems

A flow cell system consists of a number of components in addition to the flow cell itself. The simplest system is a 'recirculating batch' system in which a shake flask culture is pumped through a flow cell and then the effluent drips back into the flask. Although this type of system may demonstrate the propensity of cells to stick to surfaces, it is of limited value since the depletion of nutrients will limit biofilm growth. Also, it is difficult to quantify rates of attachment in this system since the attachment rate is a function of the suspended cell concentration, which changes as the batch cultures go through the growth cycle. More ideally for studying attachment rates of bacteria to an uncolonized surface is to use the effluent from a chemostat as an inoculum. A chemostat is a continuous culture, which

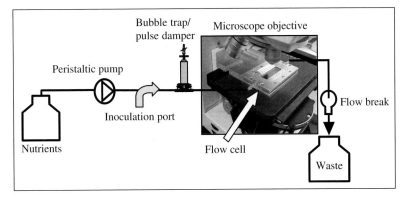

Figure 12.1   Once-through system showing basic components. A BST FC81 flat plate flow cell is shown positioned on a microscope stage.

provides a constant concentration of exponentially growing cells. The chemistry of the effluent is also constant. The flow rate through the flow cell (and therefore, the shear stress) can be extended above the flow rate of the nutrient feed into the chemostat by adding a dilution stream prior to entry into the flow cell. For a review of chemostat operating parameters see Characklis and Marshall (1990). These systems are usually set up as 'once-through' systems where the flow-cell effluent is collected into a waste container (Figure 12.1).

Initial attachment studies are usually conducted over periods of hours. When flow cells are used to study the structure of mature biofilms grown over longer time periods (days to weeks) the flow cell can be inoculated simply by injecting a small volume (i.e. 1 ml) of stock culture or shake flask culture through an inoculation port immediately upstream of the flow cell. An inoculation port, which is also useful for injecting stains or flow tracers, can be easily constructed by using a 'Y' fitting with a septum attached to one of the arms. The flow of sterile nutrients is then turned on after an attachment period of usually 1 h or less. These systems are usually restricted to low flow rates by the time and expense of media preparation. To achieve higher flow rates, a recirculating system can be used (Figure 12.2) in which medium is pumped into a mixing chamber and then recirculated through the flow cell. In this case, the flow velocity in the flow cells is independent of the nutrient supply flow rate and high shears can be obtained. However, in these systems detached cells can recirculate and possibly reattach. Also, waste products can build up in the system. The concentration of waste products will depend on the hydraulic residence time (the amount of time a volume of fluid stays in the flow-cell system until it is washed out). In essence, this system is a chemostat with an irregular geometry. For calculating the dilution rate and residence time the volume of the mixing chamber, flow cells and connective tubing need to be included in the total volume.

In recirculating systems, sterile nutrients may be added to the mixing chamber and the residence time (θ) can be controlled by the nutrient flow rate (Q_n) according to $\theta = V/Q_n$, where V is the volume of the mixing chamber *plus* the recirculation

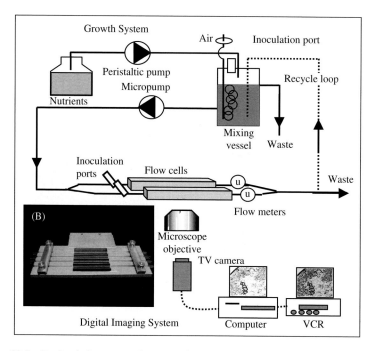

Figure 12.2   Recirculating system incorporating a square glass capillary flow cell. By redirecting the recycle stream to waste the system can easily be converted to 'once-through' mode. The mixing vessel acts as a nutrient reservoir but also functions as a bubble trap and pulse damper. The schematic shows the recirculation stream split into parallel flow cells. The insert (B) shows a four channel square (3 × 3 mm) glass capillary BST FC93 flow cell.

loop. These systems have the advantage that the flow rate in the flow cells can be adjusted independently of the nutrient flow rate so that much higher flow rates can be achieved in the flow cells without using impractical volumes of media. Also, Q_n can be adjusted so that the dilution rate ($D = Q_n/V$) is greater than the specific growth rate of the organism being investigated, so that planktonic cells are continually 'washed out' and only the attached biofilm population is retained. In this case, it can be assumed that any cells or micro-colonies in the bulk liquid must have resulted from detachment and not from planktonic growth. However, these systems have several factors to be considered, when designing experiments. First, they generally have relatively large surface areas, which are exposed to the growth nutrients in relation to the surface area in the flow cell. Possibly significant biofilm accumulation can occur in areas other than those observed in the viewing area of the flow cells (i.e. mixing chamber and connective tubing). Due to this, estimates of activity or the microbiology of the detached population in the effluent may not necessarily reflect processes occurring in the observed biofilms. Silicone tubing, which is commonly used as connective tubing, is highly oxygen permeable and biofilms growing on the tubing may have a greater effect on the chemistry and microbiology of the effluent than the biofilm in the flow cell.

Figure 12.3   BST FC81 flat plate flow cell consisting of a polycarbonate base plate, an aluminum top flange and a coverslip observation window.

A simple way of reducing biofilm accumulation in the connective tubing is to regularly squeeze along the tubing with thumb and finger. However, this will introduce a 'pulse' of detached biomass into the system. Also, care must be taken not to cause a pressure build up which will cause connections to break. Second, there will also be a range of shear stresses in different components. Thirdly, the biofilms in the flow cells are continually exposed to detached cells and waste products, which build up to steady-state concentrations in the reactor. A comprehensive mass balance analysis of various biofilm-reactor systems may be found in Characklis and Marshall (1990).

12.2.2   Types of flow cells

12.2.2.1   Flat plate flow cells

Flat plate flow cells usually consist of a polycarbonate or acrylic base plate (although polycarbonate has the advantage of being easily machined and autoclavable) and a large rectangular or square coverslip as an observation window (Figure 12.3). The channel or lumen can be made by either etching a groove into the base plate or separating the base plate from the coverslip by a gasket.

The system can be sealed by screw tightening down with a metal flange or simply by epoxy glue. The flange design has the advantage in that it can be easily disassembled for sampling or replacing the coverslip (Figure 12.4). Biofilms can be observed on the coverslip window with high-resolution oil or water microscope immersion objectives in the same way as a conventional wet mount with

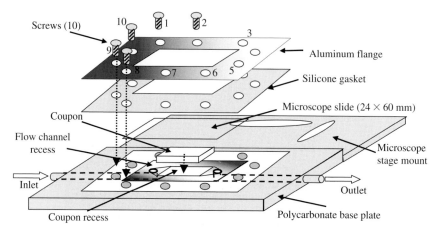

Figure 12.4   Exploded view of the components of a BST FC71 flat plate flow cell. This flow cell is shown with a recess in the base plate lumen to house a coupon which can be constructed from a wide range of materials.

any epi-fluorescence or reflected microscopy techniques, such as confocal or direct interference contrast (DIC).

Transmitted techniques can also be used but the thickness and opacity of the base plate may limit the light passage to lower power objectives. This problem can be overcome by using another coverslip or glass microscope slide on both sides. These types of flow cells are limited to studies which involve biofilm growth on glass or similarly transparent materials. However, if the material of interest is opaque then a small recess can be milled into the base plate into which a coupon of the opaque material can be inserted. The microscope is then focused through the coverslip window down onto the opaque surface. For this type of setup, epi-fluorescence or a reflected technique must be used. Additionally, the depth of the lumen may limit the objectives if the working distance is less than the depth of the lumen. Since the higher resolution, high-numerical aperture (n.a.) objectives generally have shorter working distances some may not be able to focus far enough into the flow cell (without breaking the coverslip!) to see the surface. In part, this may be overcome by using long and ultra long working distance (ULWD) objectives. However, these types of objectives may not be able to detect low levels of fluorescence, e.g., cells expressing single copy chromosomal fluorescence protein (GFP, RFP, etc.) or *lux* cassette inserts. High n.a. objectives have working distances on the order of a few hundred microns, which, therefore, limits the depth of the lumen to this depth or less. This, in turn, generally limits the flow to low Reynolds number (Re) laminar flows.

12.2.2.2   Square glass capillary flow cells

Square glass capillary flow-cell systems were designed to allow biofilms to be observed non-destructively in real time under controlled hydrodynamic conditions

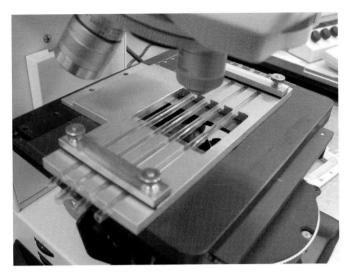

Figure 12.5   BST FC93 four channel square glass capillary flow cell positioned on the stage of an upright microscope. The capillaries are 3×3 mm square.

that cover a wide range of laminar and turbulent flows (Stoodley *et al.*, 1999a, b). The glass capillaries are commercially available in a wide variety of rectangular and square glass tubing (Friedrick & Dimmock®, Millville, NJ, USA) as are complete flow cells, which house 3×3 and 1×1 mm square glass tubes (BioSurface Technologies Inc., www.biofilms.biz) (Figure 12.5). High shear, turbulent, flow can be obtained in 3×3 mm square tubes (Stoodley *et al.*, 1999b). However, the thickness of the glass wall of these flow cells (0.5 mm) limits microscopic observations to low-n.a. long and ULWD objectives. However, 100 and 50× LWD objectives can resolve single cells with bright field and epi-fluorescent microscopy. To focus through the thickness of the entire flow cell to observe the biofilm growing on both top and bottom surfaces, objectives of 10× and lower are required. These flow cells mount directly to many types of microscope stage.

   These flow cells have the advantage of being low maintenance, and easy to use. Additionally, parallel flow cells can easily be positioned on one holder allowing the growth of replicate biofilms or quick side-by-side comparisons of biofilms growing under different flow conditions. By splitting the feed stream into parallel flow cells and adjusting the flow rate through each individual flow cell, comparisons of the influence of flow rate on biofilm growth, structure and detachment can be easily made by repositioning the stage. Since each capillary has a shared feed, the chemistry and microbiology of the influent is the same.

12.2.2.3   Capillary flow-cell hydrodynamics

The hydrodynamics in the $3 \times 3 \times 200$ mm (width × height × length) square glass capillaries have been characterized experimentally by measuring the pressure drop across a 20 cm section as a function of flow velocity (Stoodley *et al.*, 1999b).

It was found that the transition between laminar and turbulent flow occurred at a Re of approximately 1200. This was confirmed by dye tracer studies. A detailed description of the equations used to calculate the Re and shear stress in these flow cells from the flow rate, flow-cell dimensions and fluid viscosity can be found elsewhere (Stoodley *et al.*, 2001b). For turbulent flow, the flow cell should be made sufficiently long to allow for full flow development in the viewing region. The use of similar diameter connective tubing and connectors will avoid sudden expansions or contractions, thus, minimizing the entry effect (the area where flow is disturbed because of geometry changes in the lumen). A further consideration of recirculating systems is that if the flow rate in the recirculation tube is low, significant gradients may build up along the length of the recycle tube. If this occurs the flow cell cannot be assumed to be completely mixed and chemostat kinetics can no longer be used. To minimize the residence time in the recycle loop it is best to minimize the volume, i.e. use the minimum length of connective tubing required to fit the flow cell onto the microscope stage. Also, by splitting the flow into multiple flow cells the residence time in the recirculation line will be further decreased. Ideally, the residence time in the entire recycle loop and each individual flow cell (if multiple parallel flow cells are used) should not exceed a few minutes. This can be controlled by adjusting the flow rate or changing the cross section dimensions of the flow cell.

12.2.3   Microscopic resolution

The ability to resolve single cells on a surface in a flow cell is a function of the reflective or opacity of the surface, the type of microscopy, and the type, if any, of staining. Generally, with highly transparent, smooth materials, such as glass individual unstained cells attaching to a clean surface can be readily seen with transmitted light. Similarly, for cells attaching to highly reflective surfaces, such as electropolished stainless steel, single cells can be seen with reflected light. However, if the surface is rough and causes a lot of back scatter fluorescent techniques may be required. There are a wide variety of fluorescent stains that can be used for various cellular components or activities (Molecular Probes, www.molecularprobes.com). However, since many of these stains kill or compromise cell activity, they are often best used for end point sampling. An alternative is to use fluorescently tagged cells with one of a growing variety of colors of fluorescing proteins. Since sustained exposure to UV may cause detachment of bacterial cells exposure times should be minimized, which in turn limits the temporal resolution of a time course study. Use of a confocal microscope, which only exposes a very small area to the light beam per scan, significantly, reduces this problem as well as photobleaching (dimming of the fluorescent stain) caused by long-term exposure.

12.2.4   Flow-cell accessories

12.2.4.1   *Bubble traps and pulse dampers*

The most commonly used pumps in laboratory biofilm-reactor system are peristaltic pumps. These are convenient because there is no contact between the pump head and the flow stream. However, the peristaltic action causes a 'pulsing' in the flow, which induces an unsteady pulsatile flow in which the fluid velocity cycles from a static condition to a maximal value. In some cases, there is also a slight 'backwash' effect. In this case the biofilm experiences a wide range of shears, even though an 'averaged' shear is usually calculated and reported. To minimize these effects, a pulse damper can be introduced into the system between the pump and the flow cell. A pulse damper introduces a pocket of air in contact with the fluid stream. Since air is highly compressible, it acts like a shock absorber and reduces pulsing in the fluid. Another advantage of a pulse damper is that it can act as a trap for bubbles that may accumulate in the tubing. Since the lumen of flow cells is usually small, bubbles can contact both sides and effectively scour much of the biofilm off of the flow-cell walls through surface tension effects as the water–air–solid interface advances through the flow cell (Gomez-Suarez et al., 2001). Bubble traps/pulse dampers can be constructed from a glass tube or pipette held vertically on an in line tee fitting and are also commercially available (BioSurface Technologies Inc., www.biofilms.biz; Cole-Parmer, http:\www.coleparmer.com).

12.2.4.2   Flow breaks

Flow breaks are sealed drip chambers that provide a sterile air gap used to break the flow stream in the influent and effluent lines. Flow breaks are used to avoid the backgrowth of the biofilm into the medium vessel as well as the backgrowth of contamination from the effluent line. However, over long time periods (weeks) biofilm may eventually colonize around the edges of the vessel and enter the upstream tubing.

12.3   ANNULAR REACTORS

The rotating annular reactor (RAR) is a biofilm growth system designed to provide accessible sampling coupon surfaces under fluid hydrodynamic conditions representative of water distribution pipelines. The annular reactor has been used extensively in academia and industrial research settings to evaluate biological fouling (Characklis and Roe, 1984; Camper, 1996; Camper et al., 1996), biofilm structure (Gjaltema et al., 1994; Peyton, 1996; Neu and Lawrence, 1997; Lawrence et al., 2000), biocorrosion (Abernathy and Camper, 1996; Rompre et al., 1997), and biofouling and corrosion control measures relative to water distribution systems (Morin et al., 1996; Sanderson and Stewart, 1997; Gagnon and Huck, 2001; Gagnon and Slawson, 1999; Butterfield et al., 2002).

The BST RAR has 20 removable slide coupons that mate surface flush with the reactor hydrodynamic surfaces to provide a large sampling area and comparison of multiple materials, simultaneously. Studies of biofilm development within the RAR compared to a pipe distribution loop have shown a good correlation in biofilm numbers, repeatable data, and response as measured by a variety of methods and the data suggest that the annular-reactor systems may be a more sensitive monitoring device (Roe *et al.*, 1994; Camper *et al.*, 1996). It was been demonstrated that reproducible results were obtained with either reactor type (pipe loop or annular reactor) but since the annular reactor was easier to operate and the sample was less expensive and it could serve as a simple model for simple water distribution systems (Camper, 1996).

The RAR is usually operated in the same manner as a chemostat or Continuously Stirred Tank Reactor (CSTR) in which nutrients are constantly fed in at a desired flow rate and an effluent port allows waste overflow. The reactor effluent is, therefore, representative of the liquid chemistry and microbiology at any point within the reactor. Since the rotational speed can be set independently of the influent rate, the residence time of the liquid within the reactor is independent of the fluid shear at the coupon surface.

Three models of annular reactors are commercially available (BioSurface Technologies Corp., Bozeman, Montana USA): a laboratory model, a jacketed laboratory model and a pipe model.

12.3.1   Rotating annular-reactor designs

12.3.1.1   Basic rotating annular reactor

The RAR consists of a stationary outer cylinder and a rotating inner cylinder where water is circulated in the annulus between the two. A variable speed DC motor is mated to the inner cylinder through several shaft seals to provide a rotational speed that has hydrodynamic properties proportional to fluid flow in a pipe system. The reactor speed is set to a rate that will provide liquid/surface shear conditions similar to the process water system under evaluation (Figure 12.6).

12.3.1.2   Rotating annular pipe reactor

The pipe model consists of an outer pipe material either manufactured from material matching the process piping or from an actual process pipe section. Thirty stud coupons are threaded through the side of the pipe section to provide flush-mounted surfaces on the inner pipe surface (Figure 12.7). The studs are electrically grounded to the pipe by way of a stainless steel spring clip. The stud coupons are commonly manufactured from the same material as the pipe, but can be manufactured from alternate materials as well.

The top and bottom plates are manufactured from clear polycarbonate and an o-ring in a dovetail slot on each of these surfaces seals against the pipe section under compression to provide a watertight reactor vessel. Other various internal components are manufactured from 316 stainless steel, nylon/Kevlar composite, and nylon. The volume of the reactor is 1.2 l.

An advantage of the pipe design is that the rotating annular pipe reactor (RAPR) can be manufactured from an actual piece of distribution material. The disadvantage of this design is that sampling of the coupons requires partial draining of the reactor to minimize fluid loss from the sampling ports.

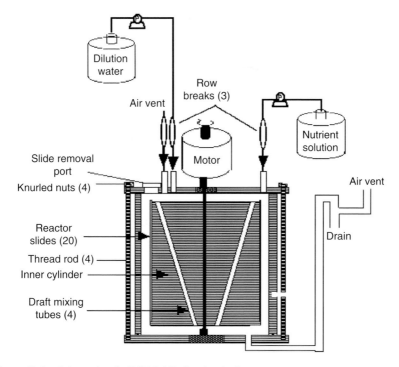

Figure 12.6   Schematic of a BST RAR showing its key components.

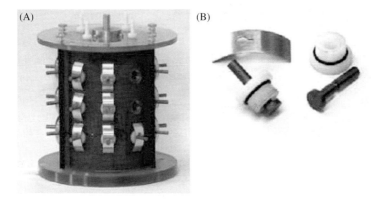

Figure 12.7   BST RAPR (A) and coupon studs (B).

12.3.1.3   Laboratory rotating annular reactors

The laboratory RAR model is manufactured using an inner, slotted polycarbonate rotor and a glass outer cylinder (Figure 12.8). Twenty slides coupons are flush mounted in the slots on the rotating inner cylinder. The slides are removed by stopping the rotation and pulling the slides out of their beveled slots through an

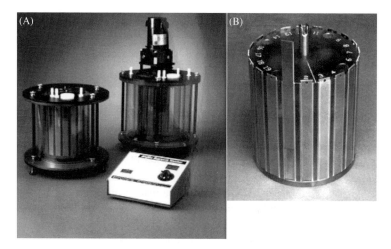

Figure 12.8   BST Laboratory annular reactor assembled with and without drive motor (A) and the inner drum with sampling coupons (B).

access port on the top of the reactor. The slides are commonly available in almost any metal or plastic and can be fabricated from most common manufacturing materials including cast iron. Base material slides can also be coated with films (epoxy or concrete, etc.). The reactor unit is fully autoclavable to 121°C. A jacketed version of the laboratory unit is also available to allow circulation of heating or cooling water in an annulus surrounding the reactor vessel. The motor drive mounts on the top of the annular reactor after autoclaving and a blade on the motor mates with a slot in the inner cylinder shaft. A variable speed motor controller provides rotational speed adjustment and displays the revolutions per minute (rpm) from 20 to 460.

The top and bottom plates are polycarbonate and seal against the glass outer cylinder (and water jacket) similar to the pipe model. Other various internal components are manufactured from 316 stainless steel, nylon/Kevlar composite, and nylon. The volume of the reactor is 1.0 l. Each of the 20 slide coupons is approximately 15 cm long by 1.5 cm wide. Many slide materials are available, and the sampling and analysis of these large flat coupons provides ample surface and size for weight loss corrosion evaluations or scraping and plating of biofilm populations. An advantage of the laboratory style reactor is the large surface area of each coupon and only one sampling port to minimize the possibility of contamination. The reactors are also readily autoclavable, while the pipe unit temperature resistance to autoclaving is dependent on the material used. The laboratory system allows easy insertion and removal of test coupons of multiple material types.

12.3.2   Reactor operations

12.3.2.1   Plumbing

Each BST RAR has three top ports, one effluent port, and one slide removal port. The reactor is designed to operate under atmospheric pressure and gravity effluent discharge. The effluent line attached at the bottom of the RAR is looped up to the top of the reactor where a vent is open to the atmosphere and the effluent line is then routed to the drain or effluent collection. The vent in the effluent line open to the atmosphere at the high point of the effluent line prevents siphoning of the reactor contents. The liquid level in the RAR can be adjusted by varying the height of the vent port in the effluent line. The liquid level in the reactor is normally operated to minimize the head space in the top of the reactor. The three top ports are used to provide influent and venting capabilities including an atmospheric vent, dilution water, concentrated growth nutrient, and/or chemical treatment for the study of biofilm or corrosion control. The dilution rate is dependent on the nutrient and chemical loading desired. Normal dilution rates are on the order of 2 h or less. A maximum flow rate through the reactor under gravity flow is nominally 50 ml/min, although increasing the effluent port diameters will increase this limit. Under drinking water distribution conditions, no gas injection is required. However, anaerobic studies may require injection of oxygen-free gas in the head space to maintain a slight positive pressure during sampling. A flow break between the influent dilution water, concentrated nutrient feed, and/or chemical feeds is often desired to prevent bacterial contamination back up the supply lines from the annular reactor.

Each additive (other than dilution water) to the reactor is pumped from a reservoir into the reactor through the flow break. Dilution water may be pumped from a clear well or reservoir vessel into the reactor or added directly from the water distribution pipeline via a pressure reducing valve. Often it is desired to remove residual disinfectant from the dilution water (if a residual is present in the distribution system) prior to introduction into the reactor. A common method for removing residual disinfectant is to pass the distribution water through a column containing granular activated carbon (GAC) just prior to injection into the reactor vessel.

A typical once-through system is shown in the schematic diagram above (Figure 12.6). Alternate configurations are possible with external recycle and operation in a number of reactors either in parallel or series. Operations in parallel can provide multiple replicates of each slide for side-by-side comparison of various biocide treatments, a combination of treatments and controls, or biofouling evaluation of different materials. Series operations can provide distribution system modeling of sequential sections of a system. External recycle may be used to increase aeration to the reactor, or heating or chilling of the contents by circulation through an outside heat or cooling source.

12.3.2.2   Autoclaving

Sterile or defined culture operations require autoclaving the reactor vessel after assembly at 121°C for 30 min. To autoclave the reactor vessel, all tubing must be in place and the tubing ends sealed or taped to prevent contamination. The reactor vessel compression screws need to be loosened to prevent stress to the seals and glass components. The reactor should be vented through either an autoclavable air vent, or autoclaving paper attached to the tubing ends to allow ready access of steam to the interior components and to vent the reactor at the end of the autoclave cycle.

12.3.2.3   Rotational speed

The rotational speed correlation for the range of operations is provided in the reactor operations manual and is based on the Hazen–Williams coefficient for the material of study, the equivalent pipe diameter and pipe fluid velocity of interest. The surface shear of the annular reactor is set to match the surface shear of the distribution system of interest (Trulear, 1980; Characklis and Roe, 1984). The rotational speed is variable from approximately 10 to 450 rpm representing a flow range of less than 0.15 m/s in a 1.22 m diameter line up to 1 m/s, or less than 0.1 m/s in a 0.01 m diameter line up to 0.9 m/s. Optional motors can provide an even broader range of rotational speeds.

12.3.3   Coupon analyses

12.3.3.1   Coupon removal

Standard protocols for biological sampling of the reactor slides is described in detail in the annular reactor operations manuals. The top access port is disinfected by spraying with 70% ethanol. The inner cylinder rotation is stopped and the port removed. The slide coupon to be analyzed is pulled from the reactor through the reactor top slide removal port using a removal hook tool and is gently rinsed by dipping the slide into sterile buffer to remove any unattached bacterial cells. The slide is then placed in a sterile glass beaker, leaning at an angle, for 1 min to drain. This treatment ensures only attached biofilm cells remain on the coupon surface for further analysis.

12.3.3.2   Biofilm analysis

The biofilm material on the slide can be imaged for various surface distributions including biofilm structure, thickness, roughness, or aerial distribution. It can also be removed by scraping for evaluation using standard microbiological techniques including cell enumeration, protein or polysaccharide quantification, and cellular differentiation (types and variety of cellular species). The biofilm on the slide can be imaged directly using microscopy and image analysis to determine the surface distribution of bulk biofilm parameters of interest. To image the slide surface transmitted light microscopy can be used on glass and clear polycarbonate slides, while reflected light or fluorescence microscopy is used on opaque materials.

To enumerate the biofilm cells on the surface, the biofilm is removed by scraping the cells from the surface into sterile suspension buffer using a spatula, or a Teflon or rubber scraper. The removed cells can then either be counted directly using a microscope and counting chamber or enumerated on growth media (media used depends on bacterial types) in petri plates or dilution series enumeration tubes. Standard methods for enumerating the removed biofilm cells are available as described elsewhere (www.standardmethods.org, and ASTM E2196-02: Standard Test Method for the Quantification of a *Pseudomonas aeruginosa* Biofilm Grown with Shear and Continuous Flow using a Rotating Disk Reactor).

12.3.3.3   Alternatives to biological evaluations

Dry weight analysis of the biofilm accumulation over time can also be monitored to provide an excellent assessment of the biofouling potential under the conditions of reactor operation. The slide coupons are weighed prior to, and after exposure and drying at 104°C for 24 h, in the RAR.

Corrosion studies using material weight loss are also possible using the annular reactor. The slide coupons are pre-weighed prior to insertion into the annular reactor. The coupons are exposed for a desired length of time and then removed from the reactor. After removal of all of the adhered corrosion products from the coupon, it is reweighed and the weight loss per time determined (Abernathy and Camper, 1997; Rompre et al., 1997).

12.4   CONCLUSIONS

The widening concept of biofilms, and the importance of surface interactions, to bacteria, archaea, protisits and fungi, in natural, medical, and industrial environments is currently in a period of exponential expansion. The embracement of these concepts by the diverse members of the microbiological research community has driven a concurrent demand for the development of research tools and techniques to study biofilms on the surfaces on which they grow. Many biofilms grow in aquatic environments, often under widely varying flow conditions. The flow environment, like the nutrient load, has a profound influence on the structure, strength and dynamic behavior of biofilms. This is not surprising since the flow will not only be a principal determinant of the mechanical forces acting on the biofilm but also the exchange of nutrients and waste metabolites. As a consequence there is not a 'one size fits all' reactor system for all biofilms, and a wide variety of systems have been designed, each with a particular research application in mind. Balanced against the need to design specific laboratory systems is the desire to standardize growing and testing methods to afford better comparability of data generated in different labs. This is, particularly, true of biofilms in which development and structure can be highly sensitive to, sometimes quite subtle, differences in experimental growth systems. In this chapter we have discussed the use of flow cells and RARs for growing, observing and sampling biofilms in controlled flowing systems. Some of these products are now commercially available which will facilitate the incorporation of biofilm research into mainstream microbiology and allow greater standardization in biofilm research techniques.

Acknowledgements

This work was funded by the National Institutes of Health RO1 grant GM60052-02, BioSurface Technologies Inc., Bozeman, MT, and a fellowship award from the Hanse Institute for Advanced Study (HWK), Delmenhorst, Germany.

REFERENCES

Abernathy, C.G. and Camper, A.K. (1997) Interactions between pipe materials, corrosion inhibitors, disinfectants, organic carbon and distribution biofilms. *AWWA Water Quality Technology Conference*, Denver, CO.

Butterfield, P.W., Camper, A.K., Ellis, B.D. and Jones, W.L. (2002) Chlorination of model drinking water biofilm: implications for growth and organic carbon removal. *Water Res.* **36**(17), 4391–4405.

Camper, A.K. (1996) Factors limiting microbial growth in distribution systems: laboratory and pilot-scale experiments. *AWWA Research Foundation and American Water Works Association, publication* #90708, Denver, CO.

Camper, A.K., Jones, W.L. and Hayes, J.T. (1996) Effect of growth conditions and substratum composition on the persistence of coliforms in mixed-population biofilms. *Appl. Environ. Microbiol.* **62**, 4014–4018.

Characklis, W.G. and Roe, F.L. (1984) Monitoring build-up of fouling deposits on surfaces of fluid handling systems, U.S. Patent 4,485,450.

Characklis, W.G. and Marshall, K.C. (1990) *Biofilms*, Wiley, New York.

Ceri, H., Olson, M.E., Stremick, C., Read, R.R., Morck, D. and Buret, A. (1999) The Calgary Biofilm Device: new technology for rapid determination of antibiotic susceptibilities of bacterial biofilms. *J. Clin. Microbiol.* **37**, 1771–1776.

Cooksey, K.E. and Wigglesworth-Cooksey, B. (1995) Adhesion of bacteria and diatoms to surfaces in the sea: a review. *Aquat. Microb. Ecol.* **9**, 87–96.

Dalton, H.M., Goodman, A.E. and Marshall, K.C. (1996) Diversity in surface colonization behavior in marine bacteria. *J. Ind. Microbiol.* **17**, 228–234.

De Kievit, T.R., Gillis, R., Marx, S., Brown, C. and Iglewski, B.H. (2001) Quorum-sensing genes in Pseudomonas aeruginosa biofilms: their role and expression patterns. *Appl. Environ. Microbiol.* **67**, 1865–1873.

Elvers, K.T., Leeming, K. and Lappin-Scott, H.M. (2001) Binary culture biofilm formation by *Stenotrophomonas maltophilia* and *Fusarium oxysporum. J. Ind. Microbiol. Biot.* **26**, 178–183.

Gagnon, G.A. and Huck, P.M. (2001) Removal of easily biodegradable organic compounds by drinking water biofilms: analysis of kinetics and mass transfer. *Water Res.* **35**(10), 2554–2564.

Gagnon, G.A. and Slawson, R.M. (1999) An efficient biofilm removal method for bacterial cells exposed to drinking water. *J. Microbiol. Meth.* **34**(3), 203–214.

Gjaltema, A., Arts, P.A.M., van Loosdrecht, M.C.M., Kuenen, J.G. and Heijnen, J.J. (1994) Heterogeneity of biofilms in rotating annular reactors: occurrence, structure, and consequences. *Biotechnol. Bioeng.* **44**, 194–204.

Gomez-Suarez, C., Busscher, H.J. and van der Mei, H.C. (2001) Analysis of bacterial detachment from substratum surfaces by the passage of air–liquid interfaces. *Appl. Environ. Microbiol.* **67**(6), 2531–2537.

Hentzer, M., Teitzel, G.M., Balzer, G.J., Heydorn, A., Molin, S., Givskov, M. and Parsek, M. (2001) Alginate overproduction affects *Pseudomonas aeruginosa* biofilm structure and function. *J. Bacteriol.* **183**, 5395–5401.

Heydorn, A., Nielsen, A.T., Hentzer, M., Sternberg, C., Givskov, M., Ersboll, B.K. and Molin, S. (2000) Quantification of biofilm structures by the novel computer program COMSTAT. *Microbiology* **146**, 2395–2407.

Heydorn, A., Ersboll, B., Kato, J., Hentzer, M., Parsek, M.R., Tolker-Nielsen, T., Givskov, M. and Molin, S. (2002) Statistical analysis of *Pseudomonas aeruginosa* biofilm development: impact of mutations in genes involved in twitching motility, cell-to-cell signaling, and stationary-phase sigma factor expression. *Appl. Environ. Microbiol.* **68**(4), 2008–2017.

Klapper, I., Rupp, C.J., Cargo, R., Purevdorj, B. and Stoodley, P. (2002) A viscoelastic fluid description of bacterial biofilm material properties. *Biotech. Bioeng.* **80**, 289–296.

Lawrence, J.R., Swerhone, G.D.W. and Neu, T.R. (2000) A simple rotating annular reactor for replicated biofilm studies. *J. Microbiol. Meth.* **42**(3), 215–224.

Leid, J.G., Shirtliff, M.E., Costerton, J.W. and Stoodley, P. (2002) Human leukocytes adhere, penetrate, and respond to Staphylococcus aureus biofilms. *Infect. Immun.* **70**(11), in press.

Liu, Y. and Tay, J.H. (2001) Metabolic response of biofilm to shear stress in fixed-film culture. *J. Appl. Microbiol.* **90**, 337–342.

Morin, P., Camper, A.K., Jones, W.L., Gatel, D. and Goldman, J.C. (1996) Colonization and disinfection of biofilms hosting coliform-colonized carbon fines. *Appl. Environ. Microbiol.* **62**, 4428–4432.

Neu, T.R. and Lawrence, J.R. (1997) Development and structure of microbial biofilms in river water studied by confocal laser scanning microscopy. *FEMS Microbiol. Ecol.* **24**(1), 11–25.

Nivens, D.E., Ohman, D.E., Williams, J. and Franklin, M.J. (2001) Role of alginate and its O acetylation in formation of *Pseudomonas aeruginosa* microcolonies and biofilms. *J. Bacteriol.* **183**, 1047–1057.

Peyton, B.M. (1996) Effects of shear stress and substrate loading rate on *Pseudomonas aeruginosa* biofilm thickness and density. *Water Res.* **30**(1), 29–36.

Purevdorj, B., Costerton, J.W. and Stoodley, P. (2002) Influence of hydrodynamics and cell signaling on the structure and behavior of *Pseudomonas aeruginosa* biofilms. *Appl. Environ. Microbiol.* **68**(9), 4457–4464.

Roe, F.L., Wentland, E., Zelver, N., Warwood, B.K., Waters, R. and Characklis, W.G. (1994) On-line side-stream monitoring of biofouling. In *Biofouling and Biocorrosion in Industrial Water Systems*, pp. 137–150, CRC Press.

Rompre, A., Prevost, M., Brisebois, P., Lavoie, J. and Lafrance, P. (1997) Comparison of corrosion control strategies efficacies and their impact on biofilm growth in a model distribution system. AWWA Water Quality Technology Conference, Denver, CO.

Sanderson, S.S. and Stewart, P.S. (1997) Evidence of bacterial adaptation to monochloramine in *Pseudomonas aeruginosa* biofilms and evaluation of biocide action model. *Biotechnol. Bioeng.* **56**(2), 201–209.

Stoodley, P., Lewandowski, Z., Boyle, J.D. and Lappin-Scott, H.M. (1999a) Structural deformation of bacterial biofilms caused by short term fluctuations in flow velocity: an *in-situ* demonstration of biofilm viscoelasticity. *Biotech. Bioeng.* **65**, 83–92.

Stoodley, P., Lewandowski, Z., Boyle, J.D. and Lappin-Scott, H.M. (1999b) The formation of migratory ripples in a mixed species bacterial biofilm growing in turbulent flow. *Environ. Microbiol.* **1**, 447–457.

Stoodley, P., Hall-Stoodley, L. and Lappin-Scott, H.M. (2001a) Detachment, surface migration, and other dynamic behavior in bacterial biofilms revealed by digital time-lapse imaging. *Method. Enzymol.* **337**, 306–319.

Stoodley, P., Wilson, S., Hall-Stoodley, L., Boyle, J.D., Lappin-Scott, H.M. and Costerton, J.W. (2001b) Growth and detachment of cell clusters from mature mixed species biofilms. *Appl. Environ. Microbiol.* **67**, 5608–5613.

Strathmann, M., Wingender, J. and Flemming, H.C. (2002) Application of fluorescently labelled lectins for the visualization and biochemical characterization of polysaccharides in biofilms of *Pseudomonas aeruginosa*. *J. Microbiol. Meth.* **50**(3), 237–248.

Trulear, M.G. (1980) Dynamics of biofilm processes in an annular reactor. M.Sc. thesis, Rice University, Houston, TX.

Wolfaardt, G.M., Lawrence, J.R., Robarts, R.D. and Caldwell, D.E. (1998) *In situ* characterization of biofilm exopolymers involved in the accumulation of chlorinated organics. *Microb. Ecol.* **35**, 213–223.

Wuertz, S., Spaeth, R., Hinderberger, A., Griebe, T., Flemming, H.C. and Wilderer, P.A. (2001) A new method for extraction of extracellular polymeric substances from biofilms and activated sludge suitable for direct quantification of sorbed metals. *Water Sci. Technol.* **43**(6), 25–31.

13

Experimental systems for studying biofilm growth in drinking water

R. Boe-Hansen, H.-J. Albrechtsen and E. Arvin

13.1 INTRODUCTION

Drinking water transported through distribution networks is subject to changes in both chemical and microbial quality. Hygienic problems may originate from contamination by external microorganisms or growth of indigenous biomass. Within the distribution system, the indigenous biomass is to a large extent located on the inner surfaces of the drinking-water pipes as a biofilm (Keevil *et al.*, 1995). Though rarely pathogenic, these microorganisms may cause numerous problems ranging from corrosion of pipes to human illness. The scale of these problems remains virtually unknown, since the monitoring frequency of the water supplies is low and the microbial techniques used are inadequate for characterisation of the indigenous microorganisms. Often, the small-to-medium size water-distribution networks rely only on a few samples a year for their quality control.

Efficient monitoring of the microbial activity in drinking-water networks is a challenge, but by gaining better understanding of the cause–effect relationships controlling the activity, and by describing the processes mathematically, predictive models may be constructed to improve overall water quality and reduce the risk of hygienic and technical problems.

© IWA Publishing. *Biofilms in Medicine, Industry and Environmental Biotechnology*. Edited by Piet Lens, Anthony P. Moran, Therese Mahony, Paul Stoodley and Vincent O'Flaherty. ISBN: 1 84339 019 1

The growth of bacteria is one of the most important processes to quantify in order to construct a dynamic model of the water quality changes in the distribution system. The bacterial growth process is complex, since it relies on a number of factors, such as concentration and composition of various substrates and nutrients. However, drinking-water engineers often rely on simplified expressions for practical purposes under specific conditions, e.g. Monod equations.

In the following, different approaches to growth rate determination in drinking water are presented and discussed. Additionally, a number of requirements are listed for studies of bacterial growth in drinking water. An overview of previously published systems is presented and discussed in order to aid future researchers in choosing a suitable approach for their studies.

13.2   DETERMINATION OF MICROBIAL GROWTH

Drinking water is characterised by a low-nutrient content (oligotrophic conditions). The bacteria living in oligotrophic systems are generally slow growing and appear to have lower growth efficiency compared to more nutrient-rich (eutrophic) systems (Del Giorgio and Cole, 1998). This makes the biofilm growth in drinking water difficult to estimate, since the level of bacterial activity often is close to, or below, the detection limit of the standard microbial methods.

On a micro-scale, the biofilm is in a constant state of change, thus the concept of microbial growth has to relate to the biofilm on a larger scale. Different approaches may be applied in order to quantify the growth rate of drinking-water biofilms:

- biofilm-formation rate;
- biomass balance at quasi-stationary phase;
- biomass detachment rate at quasi-stationary phase;
- other methods.

An overview of these methods is given in the following section.

13.2.1   Biofilm-formation rate

Perhaps the most commonly used method for determining biofilm growth is measurement of the biofilm-formation rate on clean surfaces (Pedersen, 1990; Donlan et al., 1994; Van der Kooij et al., 1995; Boe-Hansen et al., 2002b). When sterile surfaces are installed into a flow of drinking water, bacteria will immediately start forming a biofilm, i.e. the bacterial numbers on the surfaces increase until a constant level is reached.

It is noteworthy that this approach does not allow segregation between attachment, growth and decay processes, since these are pooled. However, the attachment (besides from the initial phase) and decay processes are normally neglected, which means that a simple expression can be used to describe the biofilm build-up:

$$\frac{dX_{att}}{dt} = \mu_{att} X_{att} \tag{13.1}$$

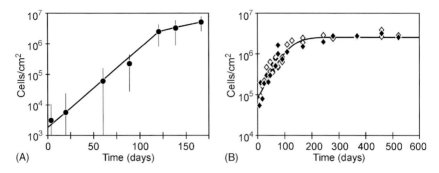

Figure 13.1   Biofilm formation in non-chlorinated drinking water. (A) Pedersen (1990) and (B) Boe-Hansen *et al.* (2002b).

where X_{att} is the concentration of attached bacteria, t the time and μ_{att} the attached bacterial growth rate.

In practice, the biofilm-formation rate can be measured by exposing clean test coupons to drinking water. The coupons are successively harvested after different periods of exposure (Figure 13.1), and the bacteria on the surfaces are quantified by a suitable microbial technique, e.g. Acridine Orange direct counts (AODC).

The approach is simple, but rather tedious, because a high number of biofilm samples and long time periods are needed for the study (i.e. the biofilm-formation rate in drinking water is low).

The main concern is that the biofilm's properties may change during maturation, which means that the net growth rate in the young biofilm may be significantly higher than the net growth rate in the mature biofilm. This has been shown in a pure-culture biofilm, most likely as a response to depletion of nutrients in the vicinity of the individual microcolony (Sternberg *et al.*, 1999). Furthermore, McBain *et al.* (2000) proposed that biofilm bacteria are able to utilise nutrients adsorbed and thereby concentrated at the surfaces where the biofilm attach, this allows the biofilm bacteria to grow faster during the early growth phase. Continued growth may deplete the nutrient pool at the surface, reducing the growth rate. In drinking water, the cell-specific ATP content during biofilm maturation has been shown to decrease considerably (Boe-Hansen *et al.*, 2002b), this strongly suggests a reduced overall growth rate.

Determination of the bacterial growth rate using biofilm-formation studies may overestimate the growth rate of the mature biofilm, but this can be acceptable for some purposes.

13.2.2   Biomass balance at quasi-stationary phase

At stationary phase, the growth of bacteria is balanced by detachment of biofilm, grazing by protozoa and cell decay. The true stationary phase can only be obtained if all extrinsic factors remain constant, which is practically impossible in biological systems. A constant selection occurs within the biofilm, which will

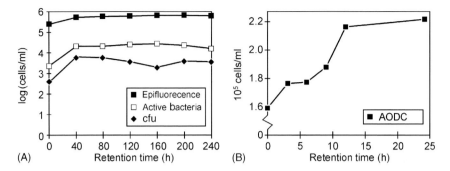

Figure 13.2   Total bacterial counts as a function of retention time in MDSs. (A) Six pipe loops in series (Haudidier *et al.*, 1988). (B) A single closed pipe loop operated as a batch (Boe-Hansen *et al.*, 2002a).

favour new organisms whenever the environmental conditions change (McBain *et al.*, 2000). The stationary biofilm is in a state of dynamic equilibrium, sometimes, referred to as a quasi- or pseudo-stationary phase.

At quasi-stationary phase the bacterial net production will be continuously transferred into the bulk water phase. By measuring the release of suspended bacteria as a function of the retention time, the net production can be estimated (Figure 13.2).

The growth rate can be calculated from the bacterial production, if the total amount of bacteria is known and decay processes are neglected. Following equation is valid for batch experiments:

$$\mu_{\text{overall}} = \frac{1}{t_{\text{batch}}} \log\left(\frac{M_{\text{total},1}}{M_{\text{total},0}}\right) \tag{13.2}$$

where μ_{overall} is the overall growth rate, t_{batch}, time of batch experiment; $M_{\text{total},0}$, $M_{\text{total},t}$, total biomass (attached + suspended) at the start (0) and after (*t*) the batch experiment.

Mass balances have been successfully used to estimate growth rates in drinking-water systems in studies by Van der Wende *et al.* (1989), Block *et al.* (1993), and Boe-Hansen *et al.* (2002a).

The mass-balance approach provides a snapshot of the situation in the distribution system at the time of measurement, which allows for studies of the dynamics in the microbial growth. The main disadvantage of the method is that it does not distinguish between biofilm growth and suspended bacterial growth. This constitutes a problem with respect to constructing predictive dynamic models of distribution networks. In some systems this poses no problems, as the suspended growth can be considered negligible compared to the biofilm growth. This is especially true when the amount of bacteria on the surface is several orders of magnitude higher than the amount of bacteria in the suspended phase, as is often the case for small pipe diameters and in chlorinated systems. However, in some systems, the suspended growth may actually be significant (Boe-Hansen *et al.*,

2002a), and when this is the case the interpretation of data from a mass-balance experiment can be complicated.

13.2.3   Biomass detachment rate at quasi-stationary phase

Another possible approach to measuring the bacterial growth rate in drinking water is by quantifying the loss rate of bacteria from a quasi-stationary biofilm. The biofilm can be labelled, e.g. by adding a ^{14}C-labelled, easily degradable carbon source like benzoic acid. The labelling of the surfaces will steadily decline due to detachment of microorganisms and respiration (endogenous):

$$\text{loss rate} = \text{detachment} + \text{respiration} \tag{13.3}$$

Although the detachment rate at the quasi-stationary phase is constant, the specific labelling of the cells being released decreases. The decrease is mainly caused by growth of the microorganisms, where every cell division halves the specific radioactivity (if endogenous respiration is neglected). The radioactivity being released from the quasi-stationary biofilm will, therefore, decrease at a rate equivalent to the net growth rate of the active bacteria in the biofilm. If ^{14}C-labelling is used, the rate of endogenous respiration may be estimated by measuring the concentration of $^{14}CO_2$ in the bulk phase. Figure 13.3 depicts the results from measurement of the detachment rate in a model drinking-water system (Boe-Hansen et al., 2002a). The reduction rate of the ^{14}C-activity in the biofilm (rate of 0.0126 day^{-1}, Figure 13.3A) is equivalent to the rate of reduction in the bulk phase (rate of 0.0130 day^{-1}, Figure 13.3B), which corresponds to the biofilm growth rate.

The major drawback of this approach is that it requires use of radioactivity, which calls for specific safety precautions during the experiments. It may be advantageous to use the stable isotope ^{13}C instead of the radioactive ^{14}C, although this will increase the cost of measurement significantly.

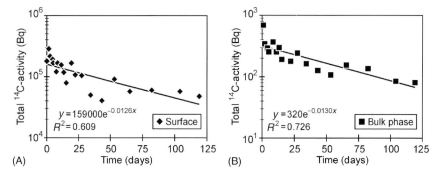

Figure 13.3   Biomass detachment rate at quasi-stationary phase in a continuously fed model distribution system (Boe-Hansen et al., 2002a). (A) ^{14}C-activity of the biofilm. (B) ^{14}C-activity of the bulk phase.

13.2.4   Other methods

The bacterial growth rate of a natural bacterial community can theoretically be determined directly from a microscopic slide by observing the fraction of cells in the later stage of cell division. The frequency of dividing cells (FDC) can be directly related to the growth rate of the community (Hagström *et al*., 1979). However, FDC is difficult to apply in the slow-growing bacterial communities in drinking water, since the number of dividing cells is very low compared to the total number of bacteria.

Measurement of incorporation rate of radioactive labelled amino acids is a measure of the bacterial activity, which is related to the bacterial growth rate (e.g. Riemann *et al*., 1990). The principle of the method is to measure the bacterial uptake of external amino acids during a period of time by addition of a known concentration of amino acid to the external environment. The measurement can be performed within a day on a single-biofilm sample, which has to be transported to and incubated in the laboratory. Several different amino acids have been applied, but thymidine and leucine are generally considered to be the best for bacterial production studies in aquatic systems (Riemann *et al*., 1990).

These methods produce a snapshot of the bacterial growth rate, which enables detailed observations of biofilm growth dynamics, e.g. studies of biofilm short-term responses to changes in substrate concentration. The amino-acid incorporation method is best used as a relative measurement within one system, since samples from different systems appear to vary significantly (Chin-Leo and Kirchman, 1988).

13.3   FUNCTIONAL DEMANDS TO TEST SYSTEMS

It is difficult to study the effect of microbial activity in distribution networks, since there is generally limited access to the biofilm, and it can be difficult to repeat the experiments when pipe sections have to be replaced after sampling. These technical hindrances have led to a wide range of drinking-water systems that facilitate routine biofilm monitoring as well as laboratory studies.

Generally, a test system for drinking-water studies should provide conditions that mimic the conditions in the distribution network. All operational and environmental parameters influencing the microbial activity should ideally be controlled or at least be monitored. The list of relevant parameters is long and disputable, however the most important can be roughly summarised as in Table 13.1.

Table 13.1   Main parameters influencing microbial activity in drinking-water distribution networks.

Operational parameters	Environmental parameters
Pre-treatment	Temperature
Disinfectants	Substrate composition and concentration
Retention time	Biomass composition
Hydraulics	Substratum (roughness, polarity, etc.)

In practice, it is rarely feasible to control/monitor such a wide range of parameters, and therefore some should be kept constant during the study and thus be eliminated. In order to study the parameters in laboratory systems, one has to consider a range of design requirements as discussed in the following.

Water and biofilm samples should be easily obtained. Sampling of a test coupon from a well-defined area is normally preferable to direct collection of biofilm from the pipes (by scraping, swabbing or a similar procedure), since all quantitative measurements require knowledge of the size of the sampled area. The sampling procedure should be rapid and aim at minimising the impact on the remaining system, i.e. sampling should not require the test system to be drained of water (different devices used for culturing and sampling biofilm are discussed in the following). Preferably, replicate biofilm samples grown under identical conditions should be available to allow for a better statistical outcome. For microscopic studies, a (near) plane surface is needed, which obviously often poses a problem because pipes are cylindrical by nature. The slow growth may also require a substantial number of biofilm-sampling points over time to ensure a steady supply of mature biofilm samples.

The test system should be constructed using materials and elements commonly used in actual distribution networks. Application of gaskets, O-rings, teflon-tape and grease to tighten the system should be minimised, since they release organic compounds to the water, which may influence microbial activity. The test coupons should be made of the same material as their immediate surroundings in order to prevent galvanic corrosion.

The temperature of the system should be kept constant, since the microbial activity is highly temperature dependent. Studies of temperature variations (e.g. seasonal changes) will often require temperature control, which will normally require a cooling system to be installed.

The mixing conditions should be well known. The continuous flow stirred tank reactor (CFSTR) is often preferable to the plug flow reactor since the biofilm in the CFSTR is uniformly exposed to the same conditions and the lack of horizontal concentration gradients reduces the number of water-sampling points needed. By combining several CFSTRs in series, one can simulate the plug flow conditions of the distribution network where the concentration of substrate may be reduced downstream from the water works. However, one should realise that the one-pass plug flow reactor bears a closer resemblance to the real-life distribution network, though it may be more difficult to employ experimentally.

The hydraulic conditions should be well defined and controllable. The hydraulic conditions in the distribution network are highly variable, ranging from turbulent to laminar flow regime. The experimental set-up should be able to simulate a variety of flow conditions. The use of bends, fittings, pumps and valves should be minimised because they introduce local turbulence. A plane test coupon inserted in a cylindrical pipe will also cause turbulence, which may affect the biofilm formation locally. The turbulence introduced by valves, pumps, bends, etc. may require insertion of extra pipe sections in order to stabilise the flow, especially in the laminar flow regime.

The retention time and flow/shear rate should be controlled independently for studies of water-quality changes. Long retention time may be needed to allow significant water-quality changes to be observed. In the one-pass plug flow system, the flow rate and the retention time are interrelated, meaning that an increase of the retention time will reduce the flow rate. This problem can be overcome by recycling the water through a pipe section, which will usually result in completely mixed conditions in the pipe section.

The liquid–air interaction should generally be minimised, mainly because the stripping of CO_2 will increase the pH value, and aeration will increase the oxygen concentration in the water. The water may also contain volatile components, like CH_4 or H_2S, which are potential substrates for the microorganisms and may easily be lost due to stripping during operation of the test system.

The operation of the system should be reliable. The biofilm formation of drinking-water bacteria is generally a slow process, and the system, therefore, normally requires long periods of operation under controlled conditions. Connecting the test system directly to the municipal drinking-water network may in some cases be advantageous. The pressure of the distribution system can be used to prohibit external contamination of the test system caused by leaks.

Equipment for on-line monitoring of different parameters should be incorporated into the experimental set-up in order to ensure reliable operation. Critical parameters are inflow rate, water velocity and temperature. However, on-line monitoring of other parameters, such as conductivity, oxygen concentration or turbidity, etc., may also prove to be useful.

Additional substrate or disinfectants may be added or removed by different pre-treatments. In some cases, a pre-treatment (such as activated carbon filters, biological filters, reverse osmosis, ion exchangers, etc.) is needed in order to comply with specific requirements to the influent water quality.

The design and operation of the system should be suitable for mathematical modelling. As an aid to the modelling, several easy-to-use modelling tools exist on the market today, e.g. AQUASIM (Reichert, 1994). Emphasis should be put on verifying system hydraulics experimentally prior to the actual experiments, e.g. by addition of a conservative tracer.

13.4   EXPERIMENTAL SYSTEMS

13.4.1   Biofilm-culturing devices

To allow for easier sampling and more controlled experimental conditions, insertions of test coupons into systems have been used in several studies. The approach is simple: test surfaces are placed within a flow of drinking water and subsequently harvested after an appropriate amount of time. In most laboratory experiments, the culturing devices are assumed to mimic a finite (small) section of the distribution system. The culturing devices can be applied in the field as well as in the laboratory.

Probably, the most popular way to culture drinking-water biofilms is to use the Robbins device (McCoy *et al.*, 1981) including several modifications (e.g. Pedersen, 1982; Manz *et al.*, 1993), where a microscopic slide or a similar surface is exposed to a constant flow of water (Figure 13.4A). The biofilm formed is very well suited for *in situ* microscopic studies. The Robbins device is cheap, easy to use and very reliable, although the hydraulic conditions under which the biofilms are formed are not well defined and practically uncontrollable. The set-up is especially unfit for studies of water-quality interaction with biofilm since the retention time is interrelated with water velocity.

A simple way to culture biofilm in the laboratory under controlled conditions is to use flow cells (Caldwell and Lawrence, 1988). A flow of water is led through removable parallel slides on which the biofilm is formed (Figure 13.4B). By changing the water flow, the hydraulic conditions can be altered. This method is unsuitable, if a large number of replicates are needed because it can be difficult to provide multiples of exactly the same conditions. The main advantage of the method is that it allows the study of the biofilm in its hydrated state because the flow-through conditions can be maintained while non-destructive direct microscopy is performed (e.g. Mueller, 1996; Kuehn *et al.*, 1998).

The annular reactor system was developed in order to provide well-defined hydraulic conditions at the test surfaces. The two most common systems are the RotoTorque and the Propella® reactor. The RotoTorque consists of a CSTR with a spinning drum (Figure 13.4C). By altering the rotational speed, the hydraulic conditions can be varied in order to mimic specific flow conditions. The hydraulic retention time of the system is controlled by the influent flow to the reactor. The reactors are completely mixed and allow for easy surface sampling. Furthermore, it is easier to perform batch experiments under constant hydraulic conditions in the annular reactors. However, the system is not well suited for studies of water-quality changes, since top and bottom surfaces are exposed to poorly defined hydraulic conditions. The biofilm formation at these surfaces will be formed under hydraulic conditions less controllable than at the test surfaces. Nevertheless, series of RotoToques have been applied in studying water-quality changes caused by drinking-water biofilm (Van der Wende *et al.*, 1989).

Another type of annular reactor is the Propella® reactor developed by Parent *et al.* (1996) (Figure 13.4D). The reactor consists of a cast-iron pipe with an inner cylinder. The water velocity in the reactor is controlled by a propeller, which pushes the water through the inner cylinder resulting in a parallel flow between the cylinder and the pipe. The system basically offers the same advantages and disadvantages as the conventional annular reactor. Other variations of the annular reactor have been proposed including an inexpensive design by Lawrence *et al.* (2000).

The biofilm monitor (Figure 13.4E), developed by Van der Kooij *et al.* (1995) enables a large number of surface samples to be taken (40 per device). Small cylinders are placed within a glass column in a flow of water. Both the inner and outer surfaces of the cylinders are in contact with the water. The main concern is that the hydraulic conditions at the surface of the cylinders and through the column are heterogeneous, which will introduce additional variability to the results.

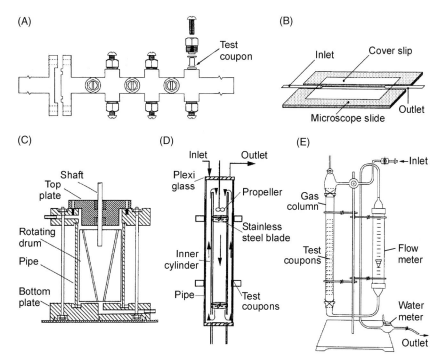

Figure 13.4   Various culturing devices. (A) Robbins device (McCoy *et al*., 1981).
(B) Flow cell (Caldwell and Lawrence, 1988). (C) RotoTorque (Van der Wende *et al*.,
1989). (D) Propella® reactor (Parent *et al*., 1996). (E) Biofilm monitor (Van der Kooij
et al., 1995).

However, the system is capable of producing a high number of surface samples,
which allows for replicate measurement in detailed studies of biofilm-formation
kinetics or tests of different materials' ability to support growth.

13.4.2   Model-distribution systems

The purpose of the model-distribution systems (MDSs) is to closely mimic the
conditions of large sections of drinking-water distribution networks. The model
systems are especially useful in determining water phase–biofilm interactions
since the hydraulics and the surface-to-volume ratio are the same as in drinking-
water networks. The MDSs generally provide closer approximation to the distribution network as a whole, in contrary to finite section approximation provided
by the culturing devices.

Sampling of surfaces without disrupting the on-going experiments can be difficult in the MDS. As a general rule, the sampling coupons should be placed flush
with the inner pipe wall in order to minimise local variations in the flow pattern.
Sampling devices (Figure 13.5), which facilitate surface sampling in a pressurised
system have been proposed in Bagh *et al*. (1999) and Boe-Hansen *et al*. (in press).

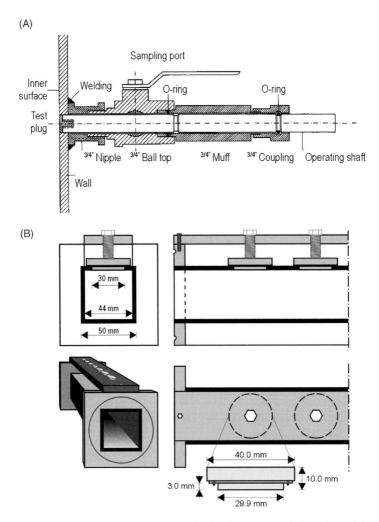

Figure 13.5   Surface-sampling devices for distribution systems. (A) Bagh *et al.* (1999) and (B) Boe-Hansen *et al.* (in press).

Overall, two types of systems exist, namely (1) one-pass systems and (2) recirculation systems. The one-pass system bears the closest resemblance of any experimental system to real-life distribution systems due to the plug flow conditions of the pipes. The main disadvantage of the one-pass system is that the water velocity determines the retention time. Thus, if realistic levels are to be attained, very long pipe lengths are needed. The largest MDS yet constructed has a length of 1.3 km (!) (McMath *et al.*, 1999), which is still only a minor fraction of the total length of a small-distribution system. The one-pass system is suitable for studies focusing exclusively on pipe-wall processes or water-quality changes. As

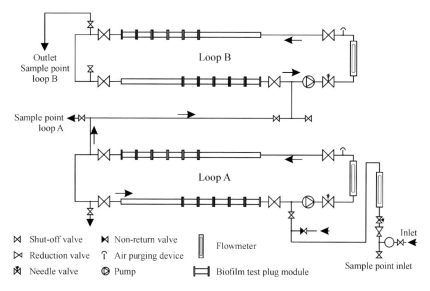

Figure 13.6   Recycled MDS (Boe-Hansen *et al.*, 2002a).

such, the one-pass system has been applied in a number of studies (LeChevallier *et al.*, 1990; McMath *et al.*, 1999).

An obvious way to by-pass the need for long pipe lengths is to recycle the water and thus enable an independent control of the water velocity and the retention time. Contrary to the one-pass system, horizontal gradients in the nutrient concentration are not present in the recycled model system. Hence, the biofilm is exposed to the same conditions throughout the loop. Different nutritional states of the biofilm can be studied simultaneously by constructing several loops in series. An example of two-loop recycled MDS is shown in Figure 13.6. It should be noted that the recycled system may be unfit for studies of degradation by-product formation and changes in substrate composition due to the ideally mixed conditions. Recycled MDS are mainly advantageous in studies of water-quality changes caused by interaction between the pipe wall and the bulk water phase (Lee *et al.*, 1980; Mathieu *et al.*, 1993; Piriou and Levi, 1994; Boe-Hansen *et al.*, 2002c).

Different types of studies have been performed using MDS ranging from microbial-induced corrosion (Lee *et al.*, 1980; LeChevallier *et al.*, 1993), coliform survival (Parent *et al.*, 1996), chlorine decay (LeChevallier *et al.*, 1990), effect from flow velocity (Percival *et al.*, 1998), etc.

13.5   CHOICE OF EXPERIMENTAL SYSTEM

In previous studies, bacterial growth rates in drinking water have been quantified by different approaches using different methods. Table 13.2 gives an overview of these studies.

Table 13.2   Different studies reporting bacterial growth rates in drinking-water systems.

Method	System[a]	Water type[b]	Chlorine present	Value (day^{-1})	References
Formation rate	mRD	SW	Yes	0.063	Pedersen (1990)
Formation rate	BM	GW	No	0.013[c]	Van der Kooij et al. (1995)
Formation rate	BM	SW	Yes	0.013[c]	Van der Kooij et al. (1995)
Formation rate	rMDS	GW	No	0.030	Boe-Hansen et al. (2002b)
Biomass balance	AR	SW	No	0.060[c]	Van der Wende et al. (1989)
Biomass balance	rMDS	SW	Yes	0.041	Block et al. (1993)
Biomass balance	rMDS	GW	No	0.049	Boe-Hansen et al. (2002a)
Detachment rate	rMDS	GW	No	0.013	Boe-Hansen et al. (2002a)
Leucine incorporation	rMDS	GW	No	0.008	Boe-Hansen et al. (2002a)

[a]AR, annular reactor (RotoTorque); mRD, modified robbins device; BM, biofilm monitor; rMDS, recycled model distribution system.
[b]SW, surface water; GW, groundwater.
[c]Our calculation (approximation).

Table 13.3   Comparison of different experimental systems for studies in drinking water.

	Robbins device	Flow cell	Annular reactor	Biofilm monitor	Model distribution system
Well-defined hydraulics	No	Yes	Yes	No	Yes
Reactor type[a]	PFR	PFR	CFSTR	PFR	CFSTR (or PFR)
Shear rate and retention time independently controlled	No	No	Yes	No	Yes
Samples suitable for direct microscopy	Yes	Yes	Yes	No	Yes
Surface samples/reactor	Varies	1	20[b]	40	Varies
Surface samples exposed to homogeneous conditions	No	–	Yes	No	Yes
Suitable for water-quality studies[c]	No	No	Partially	No	Yes
Costs	Low	Low	Medium	Medium	High

[a]PFR, plug flow reactor; CFSTR, continuous flow stirred tank reactor.
[b]RotoTorque.
[c]For water-quality studies the following requirements must be met: (i) the surface-to-volume ratio should be realistic for drinking-water networks; (ii) the entire biofilm in the system must be homogeneously exposed to the same hydraulic conditions; (iii) the retention time and shear rate should be independently controlled.

As previously mentioned, several possible design requirements should be considered, when choosing an experimental system. The previously proposed systems have different properties, which are summarised in Table 13.3.

Naturally, the choice of system should reflect the type of study to be performed, since they have different advantages and disadvantages. However, if properly constructed the MDS allows for the close simulation of the conditions in

Table 13.4   Assessment of suitability for growth studies of different experimental set-ups.

	Formation rate	Biomass balance	Detachment rate	Main concern
Robbins device	+	÷	+	Heterogeneous hydraulic conditions
Flow cell	÷	÷	÷	Too few samples
Annular reactor	++	+	++	Unrealistic surface-to-volume ratio
Biofilm monitor	+	÷	+	Heterogeneous hydraulic conditions
MDS	++	++	++	High construction cost

÷, not suitable; +, suitable; ++, very suitable.

the distribution system and a high flexibility in controlling different parameters influencing microbial activity. However, the capital and operational cost of the pipe loops systems is high, which may lead to the choice of simpler and cheaper approaches. An assessment of the various experimental systems suitability for the different types of growth-determination studies is found in Table 13.4.

13.6   CONCLUSIONS

Different approaches can be applied in order to quantify biofilm growth rate in drinking-water systems. Each approach has advantages and disadvantages, which should be considered when choosing the proper approach based on the type of study to be performed.

Most biofilm growth studies in drinking water relies on special test systems, since the 'natural' biofilm of the distribution system is generally hard to obtain and does not provide samples grown under well-defined conditions. Before constructing a test system, a number of design considerations should be made. Two types of experimental systems have been proposed in the literature, namely the biofilm-culturing devices and the MDS.

Generally, the biofilm-culturing devices only allow for detailed studies of the biofilm, while the MDS in addition allows for studies of the bulk phase–biofilm interaction. However, the cost of an MDS is generally significantly higher.

REFERENCES

Bagh, L.K., Albrechtsen, H.-J., Arvin, E. and Schmidt-Jørgensen, F. (1999) Development of equipment for *in situ* studies of biofilm in hot water systems. *Biofouling* **14**, 37– 47. Figure reprinted with permission from Taylor & Francis Group.

Block, J.C., Haudidier, K., Paquin, J.L., Miazga, J. and Levi, Y. (1993) Biofilm accumulation in drinking water distribution systems. *Biofouling* **6**, 333–343.

Boe-Hansen, R., Albrechtsen, H.-J., Arvin, E. and Jørgensen, C. (2002a) Bulk water phase and biofilm growth in drinking water at low nutrient conditions. *Water Res.* **36**, 4477–4486. Figures reprinted with permission from Elsevier Science.

Boe-Hansen, R., Albrechtsen, H.-J., Arvin, E. and Jørgensen, C. (2002b) Dynamics of biofilm formation in a model drinking water distribution system. *J. Water Suppl. Res. T. – AQUA* **51**, 399–406. Figure reprinted with permission from IWA publishing.

Boe-Hansen, R., Albrechtsen, H.-J., Arvin, E. and Jørgensen, C. (2002c) Substrate turnover at low carbon concentrations in a model drinking water distribution system. *Water Sci. Technol: Water Supply.* **2**, 89–112.

Boe-Hansen, R., Martiny, A.C., Arvin, E. and Albrechtsen, H.-J. (in press) Monitoring biofilm formation and activity in drinking water distribution networks under oligotrophic conditions. *Water Sci. Technol.*, in press. Figure reprinted with permission from IWA publishing.

Caldwell, D.E. and Lawrence, J.R. (1988) Study of attached cells in continues-flow slide culture. In *Handbook of Laboratory Model Systems for Microbial Ecosystems* (ed. J.W.T. Wimpenny), CRC Press, Boca Raton. Figure reprinted with permission from Copyright Clearance Center, Inc.

Chin-Leo, G. and Kirchman, D.L. (1988) Estimating bacterial production in marine waters from the simultaneous incorporation of thymidine and leucine. *Appl. Environ. Microbiol.* **54**, 1934–1939.

Del Giorgio, P.A. and Cole, J.J. (1998) Bacterial growth efficiency in natural aquatic systems. *Ann. Rev. Ecol. Syst.* **29**, 503–541.

Donlan, R.M., Pipes, W.O. and Yohe, T.L. (1994) Biofilm formation on cast iron substrata in water distribution systems. *Water Res.* **28**, 1497–1503.

Hagström, Å., Larsson, U., Hörstedt, P. and Normark, S. (1979) Frequency of dividing cells, a new approach to the determination of bacterial growth rates in aquatic environments. *Appl. Environ. Microbiol.* **37**, 805–812.

Haudidier, K., Paquin, J.L., Francais, T., Hartemann, P., Grapin, G., Colin, F., Jourdain, M.J., Block, J.C., Cheron, J., Pascal, O., Levi, Y., and Miazga, J. (1988) Biofilm growth in a drinking water network: a preliminary industrial pilot-plant experiment. *Water Sci. Technol.* **20**, 109–115.

Keevil, C.W., Rogers, J., and Walker, J.T. (1995) Potable-water biofilms. *Microbiol. Eur.* **3**, 10–14.

Kuehn, M., Hausner, M., Bungartz, H.J., Wagner, M., Wilderer, P.A. and Wuertz, S. (1998) Automated confocal laser scanning microscopy and semiautomated image processing for analysis of biofilms. *Appl. Environ. Microbiol.* **64**, 4115–4127.

Lawrence, J.R., Swerhone, G.D.W. and Neu, T.R. (2000) A simple rotating annular reactor for replicated biofilm studies. *J. Microbiol. Meth.* **42**, 215–224.

LeChevallier, M.W., Lowry, C.D. and Lee, R.G. (1990) Disinfecting biofilms in a model distribution system. *J. Am. Water Works Assoc.* **82**, 89–99.

LeChevallier, M.W., Lowry, C.D., Lee, R.G. and Gibbon, D.L. (1993) Examining the relationship between iron corrosion and the disinfection of biofilm bacteria. *J. Am. Water Works Assoc.* **85**, 111–123.

Lee, S.H., O'Connor, T.L. and Banerji, S.K. (1980) Biologically mediated corrosion and its effects on water quality in distribution systems. *J. Am. Water Works Assoc.* **72**, 636–645.

McBain, A.J., Allison, D.G. and Gilbert, P. (2000) Population dynamics in microbial biofilms. In *Community Structure and Co-operation in Biofilms* (eds. D.G. Allison, P. Gilbert, H.M. Lappin-Scott, and M. Wilson), SGM Symposium 59, Society for General Microbiology.

McCoy, W.F., Bryers, J.D., Robbins, J. and Costerton, J.W. (1981) Observations of fouling biofilm formation. *Can. J. Microbiol.* **27**, 910–917. Figure reprinted with permission from NRC Research Press.

McMath, S.M., Sumpter, C., Holt, D.M., Delanoue, A. and Chamberlain, A.H.L. (1999) The fate of environmental coliforms in a model water distribution system. *Lett. Appl. Microbiol.* **28**, 93–97.

Manz, W., Szewzyk, U., Ericsson, P., Amann, R., Schleifer, K.-H. and Stenström, T.-A. (1993) *In situ* identification of bacteria in drinking water and adjoining biofilms by hybridization with 16S and 23S rRNA-directed fluorescent oligonucleotide probes. *Appl. Environ. Microbiol.* **59**, 2293–2298.

Mathieu, L., Block, J.C., Dutang, M. and Reasoner, D. (1993) Control of biofilm accumulation in drinking water distribution systems. *Water Suppl.* **11**, 365–376.

Mueller, R.F. (1996) Bacterial transport and colonization in low nutrient environments. *Water Res.* **30**, 2681–2690.

Parent, A., Fass, S., Dincher, M.L., Reasoner, D., Gatel, D. and Block, J.C. (1996) Control of coliform growth in drinking water distribution systems. *J. Chart. Inst. Water E.* **10**, 442–445. Figure reprinted with permission from CIWEM.

Pedersen, K. (1982) Method for studying microbial biofilms in flowing water systems. *Appl. Environ. Microbiol.* **43**, 6–13.

Pedersen, K. (1990) Biofilm development of stainless steel and PVC surfaces in drinking water. *Water Res.* **24**, 239–243. Figure reprinted with permission from Elsevier Science.

Percival, S.L., Knapp, J.S., Edyvean, R.G.J. and Wales, D.S. (1998) Biofilms, mains water and stainless steel. *Water Res.* **32**, 2187–2201.

Piriou, P. and Levi, Y. (1994) A new tool for the study of the evolution of water quality in distribution systems: design of a network pilot. In *Proceedings of the Water Quality Technical Conference,* New York, 19th–23rd June 1994, American Water Works Association.

Reichert, P. (1994) AQUASIM – a tool for simulation and data analysis of aquatic systems. *Water Sci. Technol.* **30**, 21–30.

Riemann, B., Bell, R.T. and Jørgensen, N.G. (1990) Incorporation of thymidine, adenine and leucine into natural bacterial assemblages. *Mar. Ecol. Prog. Ser.* **65**, 87–94.

Sternberg, C., Christensen, B.B., Johansen, T., Nielsen, A.T., Andersen, J.B., Givskov, M. and Molin, S. (1999) Distribution of bacterial growth activity in flow-chamber biofilms. *Appl. Environ. Microbiol.* **65**, 4108–4117.

Van der Kooij, D., Veenendaal, H.R., Baars-Lorist, C., Van der Klift, D.W. and Drost, Y.C. (1995) Biofilm formation on surfaces of glass and teflon exposed to treated water. *Water Res.* **29**, 1655–1662. Figure reprinted with permission from Elsevier Science.

Van der Wende, E., Characklis, W.G. and Smith, D.B. (1989) Biofilms and bacterial drinking water quality. *Water Res.* **23**, 1313–1322. Figure reprinted with permission from Elsevier Science.

14

Efficacy testing of disinfectants using microbes grown in biofilm constructs

G. Wirtanen, S. Salo and P. Gilbert

14.1 INTRODUCTION

The microbes in biofilms are much more resistant to chemical disinfection measures than their planktonic counterparts (LeChevallier *et al.*, 1988; Gilbert *et al.*, 1997; Grönholm *et al.*, 1999). The efficacy of disinfectants and antimicrobial agents is usually determined in free cell suspensions, which do not mimic the growth conditions on surfaces where the agents are required to inactivate the microbes (Frank and Koffi, 1990; Gillatt, 1991). Accepted disinfection end points, for planktonic bacteria, are that the agents must reduce the microbial populations by 5-log units in order to be considered effective. A lower target for the reduction of surface-attached bacteria by disinfectants of 3-log units reflects a general acceptance that such microorganisms are more difficult to control (Mosteller and Bishop, 1993). While the standard suspension tests have proven to be sufficiently reliable because the variations of results are within the acceptable limits when replication is adequate (Bloomfield *et al.*, 1994); there can be problems associated with the repeatability and reproducibility of such suspension tests when they are performed in the presence of an organic load.

Surface tests, where bacterial inocula are bound to an inert test piece, are significantly more difficult to perform than are suspension tests because of the nature of the carrier material used, and because of the variable effects upon viability, spatial distribution and susceptibility caused by drying of cell suspensions onto a surface (Bloomfield *et al.*, 1994). Microbes growing on, or dried onto, surfaces are not susceptible to disinfectants from all sides as they are in suspension. Due to this, disinfectants are used in higher concentrations on surfaces than they are in suspension (LeChevallier *et al.*, 1988; Mattila *et al.*, 1990; Gilbert *et al.*, 1997). It is also arguable that dried films of bacterial cells do not express a characteristic biofilm phenotype.

Gilbert *et al.* (1998) describe the use of poloxamer hydrogels for the construction of model biofilms where the cell population expresses a biofilm phenotype is present locally at high cell density, and where resistance to disinfectants has increased substantially.

14.2 POLOXAMER-HYDROGEL BIOFILM CONSTRUCTS

The poloxamer hydrogels demonstrate thermo-reversible gelation properties, being liquid and fully miscible with water at temperatures <15°C but firm gels at temperatures >15°C. This means high cell densities can be cultured within the gels at 30°C and subsequently exposed to a disinfectant. After treatment a full recovery of the individual cells can be achieved simply by moving the hydrogels into neutraliser solutions/diluents at <15°C (Wirtanen *et al.*, 1998).

Poloxamer F127 is a di-block co-polymer of polyoxyethylene and polyoxypropylene. It has been investigated earlier for its potential as an agar substitute in microbiology (Gilbert *et al.*, 1998). Solutions are unaffected by autoclaving and appear to be non-toxic to all the bacterial species so far tested (Gilbert *et al.*, 1998; Wirtanen *et al.*, 1998). The poloxamer matrices in the present study not only reproduce the reaction-diffusion resistance properties of the biofilms but also simulate other aspects of the biofilm mode of growth. This chapter evaluates the possible use of such biofilm constructs of process contaminants for a comparison of commercial disinfectant formulations at normal use concentration.

14.3 DISINFECTANT TESTING USING BIOFILM CONSTRUCTS

The microbes for the disinfectant testing can be chosen amongst microbes isolated from the process or relevant microbes from a culture collection. Depending on the case studied, the panel should consist of various types of microbes, e.g. *Pseudomonas fragi*, *Enterobacter* sp., *Bacillus subtilis*, *Listeria monocytogenes*

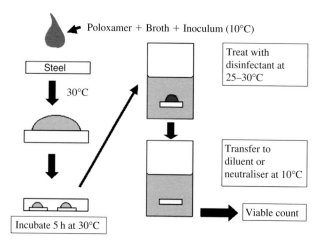

Figure 14.1   Test protocol in efficacy testing of disinfectants using biofilm constructs (Wirtanen *et al.*, 2001a).

and *Dekkera anomala*. These cultures are maintained on suitable agar slants, e.g. tryptone soy or malt extract agar in a refrigerator and broth cultures for the hydro-gel test are prepared in a similar type of broth, e.g. tryptone soya broth (TSB) or malt extract broth (50 ml). The tested disinfectants are diluted to ready-to-use concentrations or concentrations of interest. Poloxamer flakes are made up to 30% w/v in TSB and refrigerated overnight in order for hydration to take place. The dissolved poloxamer solutions are then autoclaved and returned to the refrigerator. The test procedure is described in Figure 14.1. Before use, sterile, chilled poloxamer (3 ml) is inoculated with a 1/100 dilution of an overnight (16–18 h) culture (300 μl) in fresh TSB to give 10^{3-4} colony-forming units per millilitre (cfu/ml). Drops (100 μl) of inoculated poloxamer-TSB are carefully placed onto stainless steel (AISI 304, 2B) discs placed in Petri dishes that contain a small piece of moistened cotton wool. Each Petri dish is sealed with Petrifilm and incubated at 30°C in a static incubator for 5 h. During this time period viable counts increased to circa 10^7 cfu/drop and SDS-PAGE reveals that the cells express an envelope protein profile, i.e. identical to that of submerged biofilm cultures grown in a biofilm fermenter. Viable counts can easily be performed on the incubated gels by removing them to tubes of neutraliser solution at 10°C. The gels liquify and dissolve rapidly at the lowered temperature enabling all of the cells within the constructed biofilm to be recovered. The solution may then be sterile filtered through a 0.45 μm filter and the cells recovered. Alternatively, serial dilutions may be prepared and appropriate volumes plated onto the surfaces of a suitable solid culture medium. In order to test the effectiveness of disinfectants, the steel coupons, together with the poloxamer-biofilm construct, can be submerged in disinfectant solution at 25–30°C and transferred to neutraliser solution held at 10°C at appropriate time intervals. In each instance, the effectiveness of the neutraliser has to be tested against the potential test disinfectants. The results are expressed as survival relative to appropriate controls exposed for an equal time either in sterile water, buffer or saline.

14.4   RESULTS OF THE BIOFILM-CONSTRUCT METHOD IN DISINFECTION TESTING

The reproducibility, presented as a standard deviation based on the results achieved using gels in triplicate, was an order of magnitude greater than that obtained in conventional testing systems (suspension and surface tests). The level of survival of biofilm cells within the constructs was in the order of 50–95% compared to 5-log and 3-log reductions for conventional suspension and surface tests, respectively. The survival levels obtained allowed the test to distinguish the performances of many different disinfectant formulations against a variety of test strains where this had been impossible with conventional testing methods (Table 14.1, Wirtanen et al., 1998). This test, based on hydrogel constructs, is a severe test of disinfection efficiency. While the results do not necessarily reflect the likely effects of a formulation against microbial contamination in situ, they enable discrimination between the disinfectant formulations at normal use level and choice of the most effective. Conventional suspension tests fail to discriminate between the agents in terms of their efficacy and would, therefore, not assist in the selection of agents (Gilbert et al., 1998; Wirtanen et al., 1998).

The results of testing 13 commercial disinfectants, obtained in studies by Wirtanen et al. (2001a), that Gram-negative bacteria are more resistant to

Table 14.1   Survival of *Pseudomonas aeruginosa*, *Pseudomonas fluorescens*, *Pantoea agglomerans*, *Staphyloccus epidermidis*, *Listeria innocua* and *Micrococcus luteus* grown for 5 h at 30°C within inoculated poloxamer-hydrogel contstructs and exposed to three commercial disinfectant formulations for 5 min at 30°C.

| Bacterium | Treatment (log kill [SD]) | | | | |
	Control (log cfu/gel)	IPA (100%)	HYPO (0.67%)	HPPA (0.5%)	TAAS (0.7%)
P. aeruginosa	6.62 [0.05]	0.50 [0.16]	0.40 [0.01]	0.71 [0.05]	0.06 [0.16]
P. fluorescens	6.41 [0.17]	0.69 [0.07]	0.83 [0.04]	1.03 [0.11]	0.50 [0.05]
P. agglomerans	7.15 [0.10]	No effect	0.30 [0.13]	0.37 [0.31]	0.27 [0.07]
S. epidermidis	7.51 [0.11]	0.81 [0.14]	0.84 [0.11]	1.67 [0.16]	0.86 [0.06]
L. innocua	7.52 [0.17]	1.47 [0.11]	1.40 [0.18]	0.17 [0.04]	0.41 [0.02]
M. luteus	6.80 [0.08]	1.27 [0.22]	1.15 [0.11]	1.36 [0.09]	1.16 [0.11]

TAAS, a buffered (pH 10) tertiary alkyl amine formulation in an amphoteric surfactant; HPPA, a hydrogen peroxide, peracetic acid, acetic acid (pH 1) formulation; HYPO, a sodium hypochlorite/sodium hydroxide (pH 13) formulation; IPA, an isopropyl alcohol formulation in lactic acid (pH 2.3).

disinfectant treatments than Gram-positive bacteria. This is in agreement with the general observation (Nikaido and Vaara, 1985; McKane and Kandel, 1996).

In all the gels with Gram-negative bacteria, significant levels of surviving bacteria were detected. This contrasts with the results of a conventional suspension test that gave 5-log reductions in all cases and no recoverable survivors in many cases strains (Wirtanen *et al.*, 1998). Hydrogen peroxide based disinfectants performed well against Gram negatives, e.g. *Enterobacter* sp., *Salmonella* spp. and *P. fragi*. The above-mentioned oxidising agents are also shown to be efficient against *L. monocytogenes* grown in biofilm constructs. The alcohol-based disinfectants tested proved to be more effective against vegetative cells of the Gram-positive *B. subtilis* than against the other microbes tested (Wirtanen *et al.*, 2001a). These results agree with earlier studies using bacterial cells dried on stainless steel surfaces as the inocula (Grönholm *et al.*, 1999) and with biofilm studies strains (Wirtanen *et al.*, 1998, 2001b).

14.5   CONCLUSIONS

Poloxamer-hydrogel biofilm constructs reproduce the localised high cell densities found within biofilms and are able to sustain exposure to disinfectant formulations at temperatures >20°C. The gels liquefy at temperatures <15°C enabling a full recovery of the inocula after treatment. The method reproduces the high resistance of biofilms towards disinfectants and enables highly reproducible, quantitative data to be collected for disinfection efficacy.

REFERENCES

Bloomfield, S.F., Arthur, M., Klingeren, B., van Pullen, W., Holah, J.T. and Elton, R. (1994) An evaluation of the repeatability and reproducibility of a surface test for the activity of disinfectants. *J. Appl. Bacteriol.* **76**, 86–94.

Frank, J.F. and Koffi, R.A. (1990) Surface-adherent growth of *Listeria monocytogenes* is associated with increased resistance to surfactant sanitizers and heat. *J. Food Protect.* **53**, 550–554.

Gilbert, P., Das, J. and Foley, I. (1997) Biofilm susceptibility to antimicrobials. *Adv. Dental Res.* **11**, 160–167.

Gilbert, P., Jones, M.V., Allison, D.G., Heys, S., Maira, T. and Wood, P. (1998) The use of poloxamer hydrogels for the assessment of biofilm susceptibility towards biocide treatments. *J. Appl. Microbiol.* **85**, 985–991.

Gillatt, J. (1991) Methods for the efficacy testing of industrial biocides 1. Evaluation of wet-state preservatives. *Int. Biodeterior.* **27**, 383–394.

Grönholm, L.M.O., Wirtanen, G.L., Ahlgren, K., Nordström, K. and Sjöberg, A.-M.K. (1999) Anti-microbial activities of disinfectants and cleaning agents against food spoilage microbes in the food and brewery industries. *Z. Lebensm. Unters. Forsch. A*, **208**, 289–298.

LeChevallier, M.W., Cawthon, C.D. and Lee, R.G. (1988) Inactivation of biofilm bacteria. *Appl. Environ. Microbiol.* **54**, 2492–2499.

Mattila, T., Manninen, M. and Kyläsiurola, A.-L. (1990) Effect of cleaning-in-place disin-fectants on wild bacterial strains isolated from a milking line. *J. Dairy Res.* **57**, 33–39.

McKane, L. and Kandel, J. (1996) *Microbiology: Essentials and applications.* 2nd ed. pp. 76–83, McGraw-Hill Inc., New York.

Mosteller, T.M. and Bishop, J.R. (1993) Sanitizer efficacy against attached bacteria in a milk biofilm. *J. Food Protect.* **56**, 34–41.

Nikaido, H. and Vaara, M. (1985) Molecular basis of bacterial outer membrane permeabil-ity. *Microbiol. Rev.* **49**, 1–32.

Wirtanen, G., Salo, S., Allison, D., Mattila-Sandholm, T. and Gilbert, P. (1998) Performance-evaluation of disinfectant formulations using poloxamer-hydrogel biofilm-constructs. *J. Appl. Microbiol.* **85**, 965–971.

Wirtanen, G., Aalto, M., Härkönen, P., Gilbert, P. and Mattila-Sandholm, T. (2001a) Efficacy testing of commercial disinfectants against foodborne pathogenic and spoilage microbes in biofilm-constructs. *Eur. Food Res. Technol.* **213**, 409–414.

Wirtanen, G., Salo, S., Helander, I.M. and Mattila-Sandholm, T. (2001b) Microbiological methods for testing disinfectant efficiency on *Pseudomonas* biofilm. *Coll. Surf. B: Biointerface.* **20**, 37–50.

15

Steady-state heterogeneous model systems in microbial ecology

J. Wimpenny

15.1 INTRODUCTION

15.1.1 Spatial and temporal heterogeneity in microbial ecosystems

The development of the science of microbiology has, in the main, been the story of the isolation and cultivation of pure cultures of microbes and an intensive and successful examination of their structure, physiology and genetics, a process that continues today using new terms like genomics and proteomics defining the power of recent advances in molecular biology. Although the science of microbial ecology is as old as microbiology itself the growth of microbes in natural environments presents quite different problems from those arising from pure culture studies. Natural ecosystems are almost always heterogeneous in space, in time and in genotype. Since different species proliferate in communities with a wide range of interactions between them, it is necessary to devise a methodology for investigating them not just *in situ* but in controlled systems in the laboratory.

15.1.2   The value of steady-state systems to explore microbial behaviour

One of the most intellectually satisfying approaches to studying the behaviour of pure or mixed cultures of bacteria is through the use of steady-state systems. The reason for this is that, once a system enters a steady state where no significant changes in the physico-chemical environment or in the physiological behaviour of the organisms occur, the system can be perturbed and resulting changes are unequivocally related to the imposed challenge.

15.1.3   The chemostat – an archetypal tool in microbiology

Continuous culture techniques were developed some 60 years ago by Jacques Monod (1950), also by Novick and Szillard (1950) in the same year. The most important of a variety of such flow systems is the chemostat since over a wide range of dilution rates a population of organisms, whose growth is limited for one nutrient, will enter a steady state, which can be prolonged almost indefinitely, certainly for a year or more.

The essence of a chemostat is that the environment is kept as close to perfectly mixed, and therefore as homogeneous, as possible. Once the system enters steady state, the perfect chemostat will be homogeneous both in space and in time. In addition, a pure monoculture will be apparently homogeneous in genotype. Research does indicate that there is quite stringent selection pressure in the chemostat so that mutations are continually arising and that the population undergoes a continuous process of change. There have been numerous very productive investigations into microbial behaviour using the chemostat.

Other continuous flow systems also exist. Thus, the turbidostat maintains a steady-state population based on growth rate itself by pumping more fresh medium into the system once the optical density of the culture exceeds a selected value.

15.1.4   Can heterogeneous ecosystems be explored using steady-state experimental models?

Of course the chemostat, when not in a steady state (during start up, or after perturbing the system), is temporally heterogeneous. It may also be genotypically heterogeneous where a community of different species is under investigation. It can also be used in a spatially heterogeneous fashion by linking two or more vessels together. Where the environmental optimum for growth is different to that for a process catalysed by the organisms, it can be useful to use two vessels in series. Multistage systems can be single stream where only a single flow passes through the system or multistream where different additions are made to different vessels. Multistage chemostats are interesting devices although they have one defect; they are uni-directional so possibilities of modelling bi-directional (e.g. diffusive) processes are absent.

This chapter will discuss three experimental approaches to investigating spatially heterogeneous microbial ecosystems. First, the gradostat, i.e. a bi-directional version of the multistage chemostat but incorporating source and sink flows from both ends. Second, the microstat and relatives, which incorporate two-dimensional (2-D) gradients in an agar or similar matrix on which microbes grow. Finally, the constant-depth film fermenter (CDFF) that establishes a steady-state biofilm in moving recessed pans whose surface is continually scraped to remove excess biomass. Each of these systems is discussed below.

15.2   STEADY-STATE MULTISTAGE SYSTEMS

15.2.1   Development of the gradostat

Margalef (1967) was the father of multistage systems in which some feedback in the reverse direction was mooted. He pointed out that 'Any system of chemostats affords an ideal way for mapping time series into space; going down the row of flasks is equivalent to progressing along an ecological succession'. Margalef designed systems in which 'a certain amount of diffusion and contamination upstream have been permitted as in natural systems'. His 'thought' experiments predicted multistage bi-directionally linked vessels and went much further than the systems currently employed. Thus, he considered that a system consisting of a $10 \times 10 \times 10$ array of vessels linked together by tubing and pumps could be an excellent steady-state model to examine three-dimensional (3-D) interactions in a microbial ecosystem! The logistical and operational problems of such a system put it firmly into the realms of fantasy but remain a fascinating possibility. Thus, Margalef suggested that such systems could give 'a dynamic insight on how spatial patterns develop'. Cooper and Copeland described the first truly bi-directional growth model in 1973. Their system was designed to generate a salinity gradient to study estuarine behaviour. It consisted of five stirred 40-l plastic containers linked together by diffusion couplers (open tubes!). There was a source of salt water into one end and freshwater into the other. There were also overflows from each of the end vessels.

Lovitt and Wimpenny (1981a) described the first 'gradostat'. Its ancestry was fermentation technology rather than ecology; hence, the name 'gradostat' meaning 'constant gradient' in line with 'chemostat' and 'turbidostat' whose operating principles are based on constant chemistry and constant turbidity, respectively. The gradostat consists of a series of five stirred laboratory fermentation vessels linked bi-directionally (Figure 15.1A). Flow in one direction is by pumps, in the other by gravity via overflows from each vessel. The latter are located on a stepped array to enable this to happen. The essential property of the gradostat is this bi-directional exchange. A normal configuration is that there is a source and a sink at each end of the array. The source is a reservoir and medium is fed into the system generally using a tubing pump. The sink is a pumped outlet line to a receiver, if it is the uppermost vessel; otherwise, a simple overflow from the lowest vessel.

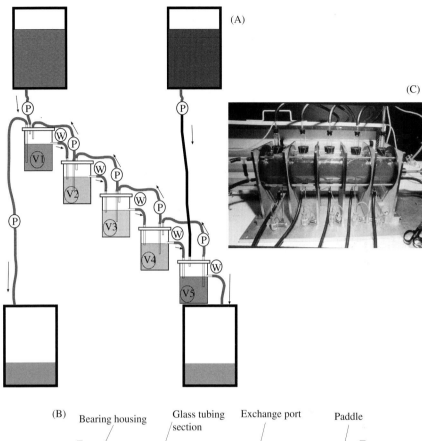

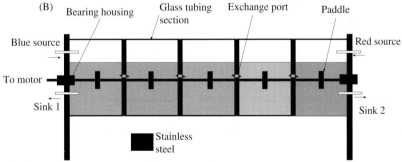

Figure 15.1   Gradostat configurations. (A) The tubing-coupled gradostat. Here material is transferred by gravity downwards over weirs (W) and upwards (also from reservoirs and into the first receiver), by tubing pumps (P). Shading is used to indicate the gradients that form under steady-state conditions. (B) A DC gradostat. Material is transferred from vessel to vessel by diffusion through an exchange port. Pumps are used only to add and remove nutrients and culture respectively into and out of each end chamber. (C) A DC gradostat in use to separate *Rhodopseudomonas marina* (salt resistant) from *Rhodobacter capsulata* (salt sensitive) in a NaCl gradient. NaCl enters the system from the right. [This figure is also reproduced in colour in the plate section after page 326.]

15.2.2   Different gradostat configurations

There are many possible configurations for the gradostat. One modification developed to provide a more compact streamlined system was the direct-coupled (DC) gradostat whose design was functionally similar to the Cooper–Copeland model (Wimpenny *et al.*, 1992).

Each vessel consists of a glass-tubing section separated from neighbouring sections by stainless steel plates. Each of these has a Teflon bearing and an accurately drilled exchange port. The complete gradostat consists of five such vessels stirred by impellors located on a drive-shaft that pass through the bearings in each section. Sources and sinks are located at each end of the array (Figure 15.1B and C). Tubing pumps supply the system from each end of the array as well as removing medium from one of the outlet tubes. The second outlet is a simple overflow that compensates for any *slight* variation in pumping rate from any of the pumps.

Gradostats can be single or double ended. Thus, *provided* that bi-directional exchange operates, a single-ended gradostat can model an ecosystem with only one surface through which solute exchange can take place (Figure 15.2A). These obviously include most biofilm as well as the majority of sedimentary system. Another possibility is to retain cells in each vessel using a membrane partition, allowing only solutes to move up and down the array (Herbert, 1988). Theoretically 2-D and 3-D arrays could be devised though as far as I am aware no one has ever tried this approach.

15.2.3   Operating the gradostat

There are no real problems in setting up a gradostat run. What discourages practical research into these devices is the relative complexity of the system. A number of fermenter vessels are needed; almost any number greater than one is possible. Most people use five, a compromise between resolution and practicality. The original form of the device was bulky since it required a stepped array of vessels to allow a gravity flow in one direction. Although theoretically all connecting lines could be pumped, in practice most pumps are not accurate enough to maintain identical flow rates so volume changes in each vessel are likely to be a problem. Each vessel is normally sterilised separately together with its flow lines. The vessels are then assembled and connected aseptically to each other and to the medium reservoirs and receivers. The vessels can all be sparged individually with whatever gas regime is needed. The overflow from one vessel to the next should enter the lower vessel well below its surface. This will prevent gas exchange down the array, provided that there is only a small gas pressure in the higher vessel.

The DC gradostat is assembled and sterilised as a unit, together with appropriate connections to both medium reservoirs and receivers.

Inoculation for both devices is similar: each vessel can be independently inoculated. Alternatively, the contents of an inoculated reservoir can be pumped in from either or both ends. The reservoirs are then replaced with sterilised medium through appropriate connectors and grow-back traps.

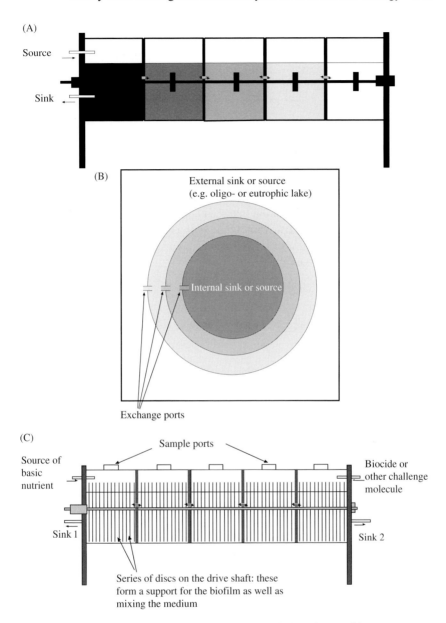

Figure 15.2   Other possible gradostats. (A) A single-ended gradostat. This system can simulate a biofilm since material enters and leaves from one surface only. Gradients form only if organisms grow acting as sinks. (B) Cylindrical DC gradostat developed as a lake mesocosm by Higashi *et al.* (1998) (C) A possible modified DC gradostat-biofilm fermenter. Our observations that extensive biofilm growth forms in a *p*-nitrophenol degrading DC gradostat suggests that the system could be modified by incorporating disks on the drive-shaft, optimising microbial attachment to surfaces.

Higashi *et al.* (1998) have described an *in situ* gradostat whose structure is quite different from the above (Figure 15.2B). It consists of three concentric cylindrical chambers, 70, 60 and 40 cm in diameter and 20 cm deep. The central chamber acts as source. Exchange, as in the DC gradostat, is via a series of calibrated holes in the walls of the chambers. The source can model a eutrophic lake, if the device is placed into an oligotrophic pond or lake. Experiments suggest that a 9.5-fold steady-state diffusion gradient can be established from inside to the outermost vessel. Incidentally, the concentration under steady-state conditions in 'conventional' gradostats is a linear function of the source versus the sink concentrations, so if these are known the value in each vessel can easily be determined. It should be noted that this only applies when no organisms using the solute are present. If they are, they become the sinks! If an organism growing in stoichiometrically equivalent concentrations of two essential nutrients, one from each source, the organism becomes a sink in the centre of the array. Concentration gradients at steady state will, therefore, also be linear (e.g. see Lovitt and Wimpenny, 1981a)

15.2.4   Examples of gradostat use

Gradostats have been used for a number of different applications. It should be noted that the gradostat is an analogue of the diffusion process, i.e. bi-directional in one dimension. Under steady-state conditions, concentrations of a solute fed from one end of a diffusion field to a sink at the other, are linear. This is also the case in a gradostat although a discretised or stepped gradient is observed. More generally, steady states between sources and sinks are the rule in the gradostat. Early applications of gradostats are reviewed by Wimpenny *et al.* (1988) and are summarised together with more recent reports in Table 15.1.

15.2.5   The gradostat and mathematics

When published, the gradostat immediately attracted the attention of mathematicians. It was regarded by one (Jager, personal communication) as the archetypal flow system of which the chemostat was just a special case! It also had close analogies to other non-biological systems, e.g. traffic flow between two cities! Its behaviour provided an excellent model system to test the competitive exclusion principle (Gause, 1934) which proposes that no two organisms can occupy an identical niche at the same time. In fact, several groups of mathematicians have shown that in a heterogeneous space (e.g. a gradostat) there exist conditions where more than one organism growing on the same nutrient, can co-exist albeit in different spatial niches (Tang, 1986; Jager *et al.*, 1987; Smith and Tang, 1989; Elowaidy and Elleithy, 1990; Smith, 1991; Smith *et al.*, 1991; Zaghrout, 1992). Most of the above articles have assumed 'pure and simple' competition employing Monod substrate relationships. Tang (1994) showed that similar outcomes to competition experiments would also apply where other uptake models were applied. The mathematicians have extended their work to examine competition for more than one substrate and/or competitor. Baltzis and Wu (1994) have

Table 15.1 Examples of gradostat use.

Type of gradostat	Operation	Comment	Reference
Tubing coupled (TC)	Solute distribution	To test a numerical model	Wimpenny and Lovitt (1984)
TC	Growth of *Paracoccus denitrificans* in counter-gradients of nitrate and succinate	Maximum cell density was reached in the central vessel of five	Lovitt and Wimpenny (1981a)
TC	*Escherichia coli* growing in counter-gradients of nitrate +oxygen versus glucose	Aerobic respiratory enzymes derepressed at the aerated end, anaerobic enzymes including nitrate reductase at the anaerobic end	Lovitt and Wimpenny (1981b)
TC	Competition in an aerobic versus anaerobic gradostat	*E. coli* survives across all vessels on own, but displaced by *Pseudomonas aeruginosa* at the aerobic end and by *Clostridium acetobutylicum* at the anaerobic end	Wimpenny (1988)
TC	Enrichment for a functional sulphur cycle	Counter-gradients of sulphate plus nitrate versus lactate anaerobically led to operational cycle	Wimpenny (1988)
TC	Mixed cultures of aerobic and anaerobic bacteria	*Paracoccus denitrificans* and *Desulfovibrio desulfuricans*, both survived in the gradostat	Wimpenny and Abdollahi (1991)
TC	Modelling nitrogen fixation in rhizosphere communities	Investigation into rhizosphere communities using the gradostat	Fritsche and Niemann (1989)
TC	Modelling rice rhizosphere systems in C, N and O_2 gradients	Growth position independent of C and N gradients but moved by oxygen due to toxicity to diazotrophic organisms	Fritsche and Niemann (1990)
TC	Influence of NH_4Cl and O_2 gradients on pure and mixed cultures of nitrogen fixing and non-fixing bacteria	Nitrogen fixers could use N_2 and NH_4Cl simultaneously. Little fixed nitrogen transferred to the non-fixers	Fritsche *et al.* (1991)

(continued)

Table 15.1 (*continued*)

Type of gradostat	Operation	Comment	Reference
Diffusion coupled using connecting tubes (DC)	Using five 40-l plastic drums to model an estuarine community in a salt gradient	Probably the first gradostat!	Cooper and Copeland (1973)
DC	Growth of a methylotroph in a methanol gradient	Linear distribution of total and viable counts away from the source vessel	Wimpenny *et al.* (1992)
DC	Separation of photosynthetic bacteria on the basis of salt tolerance	*Rhodopseudomonas marina* separated from *Rhodobacter capsulata* in a salt gradient	Wimpenny *et al.* (1992)
DC	Enrichment of *p*-nitrophenol-degrading community	A biofilm community developed on the walls of the gradostat with very high activity	Evans (1993)
In situ mesocosm gradostat (IS-MG)	Concentric cylinder system (see text) used in oligotrophic lake	Phytoplankton grew best in the mesotrophic zone (vessel 2)	Higashi *et al.* (1998)
IS-MG	Growth of typical eu- and oligotrophic phytoplankton in a eu- versus oligotrophic gradient	System placed in a eutrophic lake. Central vessel fed with distilled water to make oligotrophic conditions	Higashi and Seki (1999, 2000), Saida *et al.* (2000)
Semi-continuous gradostat	Competition between *Scenedesmus* and a pseudomonad for carbon and phosphorus	Results with a chemostat compared with those in a gradostat. The green alga excluded in the chemostat but co-existed at high C : P ratios in the gradostat	Codeco and Grover (2001)

generalised to the extent that N pure and simple competitors may co-exist in an array of 2(N − 1) bioreactors. Finally, Thomopuolos *et al.* (1998) showed that three organisms competing for two substrates could only co-exist over a very small range of parameter values in two vessels but much more easily in three! Finally, an excellent account of the theory and practice of chemostats and gradostats appeared in 1997 (Smith and Waltman).

15.2.6   Conclusions

The gradostat in its many forms is a powerful tool to examine microbial growth in a controllable steady-state system. It is a pity that it has not been followed up with further practical work. As a concept it has appealed to the mathematicians for whom experimental work is less of a problem! The latter group have demonstrated that the competitive exclusion principle might apply in a homogeneous environment but under certain conditions co-existence is possible when spatially differentiated niches are available. To our shame as microbiologists, no one has tested what would be a comparatively trivial exercise experimentally.

15.3   THE MICROSTAT AND OTHER 2-D GRADIENT PLATES

15.3.1   Gradients in agar plates

Agar, as a matrix through which solutes diffuse, has been used as an experimental tool since the pioneering work of Beijerinck in 1889. He demonstrated the responses of microorganisms to the diffusion of a range of solutes placed at positions on the agar surface and gave the name 'auxanography' to this technique. From these roots came antibiotic sensitivity testing still a standard procedure in medical microbiology today. Wedge gradient plates (see later) were developed in the fifties by Szybalski and his colleagues (Szybalski, 1951; Szybalski and Bryson, 1952; Bryson and Szybalski, 1952) to determine the response of microbes to a range of antibiotic concentrations. Sacks (1956) used such a gradient plate to examine microbial growth in pH gradients.

More interesting was the possibility of multiple dimensions examined using growth plates. A pioneer here (though not using diffusion gradients), was Halldal (1958), who investigated algal growth in 2-D light-temperature gradients. A sophisticated version of this arrangement was described in 1973 by van Baalen and Edwards.

Caldwell and Hirsch (1973) and Caldwell *et al.* (1973) were the first to generate genuine 2-D diffusion gradients in agar. These were established as steady-state systems and the broad principle of operation was that a square agar slab was fed with concentrated substrates from adjacent edges of the slab while an aqueous sink was provided at opposite edges or along one face of the slab. The presence of sources and sinks meant that at equilibrium a stable gradient would be established.

A number of variants of such a device were constructed. Caldwell *et al.* (1992) gave the name 'microstat' to this device following the precedent set for the chemostat, turbidostat and gradostat each of which could operate under steady-state conditions.

   Development of steady-state diffusion gradients in gels has not been without its problems. For example, researchers have tended to underestimate the time needed to establish a steady state, largely because diffusion is a slow process. Wolfaardt *et al.* (1993) started to develop a one-dimensional (1-D) gradient device. Here a 5-mm thick slab of agarose was fed at one end with the herbicide diclofop from a reservoir located below the slab. A sink was provided by a flow of water over the surface of the agar. Growth was observed microscopically at the agar–water interface. In a 38 mm long gel the diclofop gradient was only apparent over the first 10 mm. To extend the gradient and reduce time to steady state, the authors modified the system by investigating a range from one to five reservoirs and settled in the end, on a model with four. They suggest that this system is easy to construct, reliable and gives reproducible results. Another more versatile model was developed by Emerson *et al.* (1994). This had an 'arena' measuring 50 × 50 mm square in which is located a 15-mm deep agar slab (Figure 15.3). This is fed from each of the four sides from reservoirs, 3 ml in volume, via a 0.05-μm pore size membrane. This system does not use a flow of water as a sink rather each opposite pair of reservoirs constitutes source and sink chambers. Reservoirs are fed continuously with solutes via inlet and outlet peristaltic-tubing lines. In the cited work, the authors used only a single pair of reservoirs to investigate chemotaxis in pure and mixed bacterial cultures. Using appropriate chemo-attractors they were able to separate motile strains of *Escherichia coli* from *Pseudomonas aeruginosa*. In the

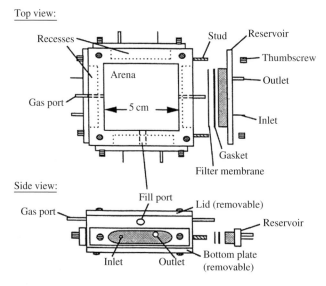

Figure 15.3 (A) 2-D gradient chamber developed by Emerson *et al.* (1994). (B) 5 × 5 cm 'arena' incorporates an agar layer in which the gradient form. Reservoirs are located along each of the four edges of the chamber but separated from it by filter membranes. See text for more details.

end the chamber was too long, and the reservoir membrane too much of a diffusion barrier, to establish a true steady-state gradient. The authors felt that the gradient that developed over a short period was good enough to be useful for a range of research projects. Emerson and Breznak (1997) used this gradient plate to investigate growth of bacteria in 2-D gradients of salt and toluene. Isolates were found that were resistant to both these and the gradient plate was used to discriminate between two of the former. Widman *et al.* (1997) also used the diffusion gradient chamber to examine chemotaxis to two agents at right angles to one another.

These have been entirely laudable attempts to develop and apply a reliable steady-state diffusion gradient. That they have not been used more is sad. The main reason, as I discussed for the gradostat, is that each gradient plate is quite complicated to set-up and requires a fair amount of equipment including four channels from a multi-channel peristaltic pump for each plate.

15.3.2   Non-steady-state gradient plates

Though not really relevant to this chapter, 2-D non-steady-state gradient plates developed first by Wimpenny and Waters (1984) have played a useful part in experimental microbiology. These plates used the wedge plate principle described by Szybalski in 1951. The Szybalski plate used two wedges opposite each other to develop a 1-D antibiotic gradient. The 2-D plates extended this principle using four wedges, two pairs at right angles to one another. After these plates had been allowed to equilibrate the result, measured at the surface, was a 2-D gradient in the first instance of pH versus NaCl concentration. The technique was extended to increase the number of possible dimensions. For example, a set of six 2-D plates can be incubated at six different temperatures generating a 3-D response surface. This is useful when trying to determine the limits to growth of an organism or a community (Thomas and Wimpenny, 1993, 1996).

I have called the object formed in Cartesian space the 'habitat domain' of the organism or community (Wimpenny, 1981). Increasing the number of dimensions is possible. Thus six sets of plates, each set containing a different concentration of a solute could be incubated at six different temperatures so that the growth domain could be mapped in four dimensions (Wimpenny and Waters, 1987). Since this structure is recursive the number of dimensions could be increased further with obvious logistical problems!

15.4   THE CONSTANT-DEPTH FILM FERMENTER

15.4.1   Can spatially heterogeneous biofilms ever be 'steady state'?

At first sight perhaps, it seems unlikely that a spatially heterogeneous structure like a biofilm could operate under steady-state conditions. We have already seen that a simple analogue of a biofilm can be investigated in a gradostat, e.g. either a single-ended version or in a double-ended gradostat with aerobic and anaerobic ends. A more accurate model is to use the biofilm itself but grown under appropriately controlled conditions. One possibility is to allow growth to a predetermined thickness and then to remove any excess growth so that the biofilm finally reaches a constant depth.

Such a model would be exposed to nutrients and have metabolic products removed from the upper surface only. After constant depth was reached, there would be nutrient and product gradients formed throughout the film. This would, in turn, lead to a gradient in metabolism and in growth rate. At some point in the array, nutrient concentration would fall to zero. Starvation, death and possible regrowth might then take place. Below, even this level, the undigested remains of dead cells would accumulate. Turnover in this area would be extremely slow. It seems probable that in time, the whole array would form a steady-state system where conditions remained constant throughout the film profile.

15.4.2   The development of steady-state biofilm fermenters

15.4.2.1   Early experiments by Atkinson and colleagues

The earliest CDFF was devised by Atkinson et al. (1968). Two models were constructed, each of which had some form of scraper bar to remove excess biofilm from the surface of the substratum. The first system consisted of a roughened glass plate, while the second consisted of a thin steel sheet template from which zones were cut to provide recessed regions in which the biofilm proliferated. This work was never followed up. The CDFF was developed by Wimpenny and colleagues (Coombe et al., 1982; Peters and Wimpenny, 1988a; Wimpenny et al., 1993).

15.4.2.2   The CDFF – development of a concept

The CDFF was developed according to the following criteria:

- For reasons already mentioned it must operate as closely as possible under steady-state conditions.
- The biofilm formed should be reproducible.
- Sufficient samples must be available at any one time to ensure statistical validity.
- The system should be able to operate under closely controlled environmental conditions.
- It should be simple to alter film substrata.
- Biofilm must be formed in a regular geometrical shape to allow both cryosectioning from the surface to the biofilm base and to enable the deployment of micro-electrodes.
- Operation, including the removal of film samples, must be aseptic.

The most recent version of the CDFF manufactured and marketed by Cardiff Consultants Ltd. (Figure 15.4A and B) is described below. In this device cells

Figure 15.4   (A) The CDFF. Biofilm forms on five recessed film plugs located in each of 15 film pans. The latter are incorporated into a stainless steel ring. This rotates at 2–3 rpm beneath two teflon or nylon spring-loaded scraper blades which remove any excess growth. The ring is irrigated with fresh medium. Spent medium and cells that have been removed fall to the base of the fermenter and are drained into a receiver. (B) Detailed view of the pans, plugs and scraper.

attach to the substratum surface and proliferate to fill a predetermined recessed space above the plug surface. Excess material is removed from the upper surface of the film using a scraper blade. After a period the film becomes 'steady state' as judged by protein content and viable count.

15.4.2.3   Operating the CDFF

The fermenter assembly is located within a borosilicate glass-tubing section (QVF plc, Stone) whose top and bottom stainless steel plates, together with their Teflon seals, are held together with long locating bolts. The film-producing assembly consists of a flat stainless steel mother ring, in which are located 15 Teflon film pans. A spring-loaded self-locating link connects the mother ring via a central axle and sterile bearing assembly to the drive unit. Drive speed is controlled by a 0–15 V DC power supply unit (Farnell, Ltd.).

Each film pan contains five removable film plugs. These can be made of any material but are normally Teflon. The film pans are held within the mother ring by neoprene 'O' rings while the plugs are a reasonably snug fit and remain in place in the pans. Each film plug is recessed a measured depth using simple stainless steel templates available in a range of sizes from 50 to 500 μm. In operation the mother ring is rotated at 2–3 rpm beneath two Acetal® or nylon scraper blades.

The latter are lightly spring loaded to maintain close contact with the rotating ring and film pans. Nutrient is fed into the system and dripped onto the scraper blades from where it is carried around the mother ring across the film pan surface. The blades remove any biofilm that grows above the surface of the film pan. Waste culture falls to the base of the CDFF and out through a tube to a receiver. Pans may be replaced during a run via a port in the top plate using a special tool inserted into a tapped hole in the centre of the film pan. The pan is then withdrawn from the mother ring. Clean sterile pans can then be reinserted in the same way and finally tamped down to the level of the mother ring with a flat-ended tool.

For use, the CDFF is autoclaved with all relevant tubing and a sterilisable air filter attached in the normal fashion. A medium reservoir plus growth medium is also sterilised as are receivers for spent medium. The system is inoculated with a starter culture. This can be done in a number of ways:

(i) by inoculating directly onto the mother ring before switching on the sterile medium feed;
(ii) (more commonly) by attaching medium feed to a small vessel or flask containing both medium and inoculum, then recirculating this through the CDFF for a time to encourage attachment before switching to the main reservoir and straight through operation;
(iii) by growing the inoculum in a chemostat and feeding the effluent from the latter directly into the CDFF.

This method though more complex, can be can be advantageous since the inoculum will itself be in a controlled steady state.

After the start-up period the CDFF is run for an appropriate time. Temperature is maintained by running the system in a temperature-controlled laboratory or in an incubator. One version of the CDFF is fitted with a glass water jacket so that it can be run on an ordinary laboratory bench with a circulating constant-temperature water bath.

15.4.2.4   Applications of the CDFF

The CDFF has been used to establish a variety of steady-state biofilm cultures. These are briefly described in Table 15.2.

15.5   DISCUSSION AND CONCLUSIONS

This chapter has outlined three separate methods by which steady states for heterogeneous ecosystems can be investigated in the laboratory. Each has its good and bad points. Gradostats possibly, have never had the popularity that they deserve as a tool for investigating tightly controlled spatial gradients. They have proved their worth, especially in their ability to establish interacting anaerobic and aerobic communities. They have as well allowed the investigation of competition where

Table 15.2   Some applications of the CDFF

Application	Reference
Growth of *Pseudomonas aeruginosa* on metal working based media	Wimpenny *et al.*, 1993
Development of a steady-state oral community	Kinniment *et al.* (1996a)
The effect of chlorhexidine on an oral community	Kinniment *et al.* (1996b)
Corrosion of intra-oral magnets by multi-species biofilms with and without sucrose	Wilson *et al.* (1997)
The effect of chlorhexidine on multi-species biofilm	Wilson *et al.* (1998)
The effects of mouthwashes on *Streptococcus sanguis* biofilms	Pratten *et al.* (1998)
Surface disinfection of thick *Pseudomonas aeruginosa* biofilms.	Wood *et al.* (1998)
Antimicrobial susceptibility and microcosm dental plaques	Pratten and Wilson (1999)
pH gradients in biofilm measured using two-photon excitation microscopy	Vroom *et al.* (1999)
Transposon transfer in a model oral biofilm	Pratten *et al.* (1999)
Use of the CDFF to study oral biofilms	Wilson (1999)
The effects of sodium hypochlorite on *Listeria monocytogenes* in a multi-species biofilm	Norwood and Gilmour (2000)
Structure of dental plaques grown under different conditions.	Pratten *et al.* (2000)
The effects of surface roughness and type of denture acrylic on biofilm formation by *Streptococcus oralis* in a CDFF	Morgan and Wilson (2001)
In vitro model for studying microbial leakage around dental restorations	Matharu *et al.* (2001)
Particulate Bioglass® and bacterial viability in an *in vitro* model	Allan *et al.* (2002)
Composition of a microcosm dental plaque before and after treating with tetracycline	Ready *et al.* (2002)

unequivocal outcomes have been noted on adding putative competitors to pure culture experiments. They have huge potential in continuously enriching organisms capable of metabolising unusual molecules since a low concentration of 'maintenance' nutrients fed in at one end of an array of vessels allows a community to survive to meet the challenge of a specific chemical fed in at the other end. They can, in addition, provide oligotrophic environments fed with limiting nutrients from one end and, perhaps, pure water from the other.

Disadvantages of the gradostat lie mainly in the perceived complexity of a multivessel array and the pumps and instruments needed to maintain the system. In addition, the gradostat only works with liquid cultures and other methods are needed to investigate dense population of the type found in microbial aggregates including biofilm.

The microstat in its steady-state form provides a good ecological tool to differentiate microbial communities as a function of both 1-D and 2-D solute gradients. Although the gradient, i.e. established can be truly steady state, it must be remembered that any growth on the surface of the matrix will lead to more complex gradients around any colony that forms. These will alter with time as the community grows.

As a research tool, the microstat has never received the attention it deserves. The problems are mostly technical. Each single-gradient plate needs a number of fluid pumps to generate and maintain gradients. Diffusion rates are very slow on the visible scale and alternative methods relying on feeding solutes at different points across the gradient field have been tried.

An alternative, the non-steady-state gradient plate has had a degree of success in a number of areas. However, this system is not strictly relevant to this chapter.

The CDFF is just one of a wide range of model systems for investigating biofilm formation. It does have the unique attribute that it can generate a biofilm, which is approximately steady state. This is clearest in the case where the biofilm formed fills the growth space completely. However, this does not always happen, although, even if the pan is not full, experience has shown that it establishes an approximate steady state due to attrition at the surface of the film. The CDFF has been used reasonably widely mainly in the area of oral microbiology. It seems to be an ideal tool for investigations in other areas, in particular to examine the effects of biocides on industrially important biofilms. As a model, the CDFF is versatile and meets the requirements of reasonably simple device capable of generating reliable and reproducible data. It has few insurmountable problems!

Finally, the three systems described here, form a powerful toolkit, well suited to tackling problems of microbial proliferation in structurally and chemically heterogeneous ecosystems. It would be interesting to see these exploited more fully by microbiologists.

REFERENCES

Allan, I., Newman, H. and Wilson, M. (2002) Particulate Bioglass® reduces the viability of bacterial biofilms formed on its surface in an *in vitro* model. *Clin. Oral Implan. Res.* **13**, 53–88.

Atkinson, B., Daoud, I.S. and Williams, D.A. (1968) A theory for the biological film reactor. *Trans. Inst. Chem. Eng. (London)* **6**, T245.

Baltzis, B.C. and Wu, M. (1994) Steady state coexistence of 3 pure and simple competitors in a 4-membered reactor network. *Math. Biosci.* **123**, 147–165.

Beijerinck, M. (1889) Auxanography, a method useful in biological research, involving diffusion in gelatine. *Arch. Neerland Sci. Exact Haarlem* **23**, 267.

Bryson, V. and Szybalski, W. (1952) Microbial selection. *Science* **116**, 45.

Caldwell, D.E. and Hirsch, P. (1973) Growth of microorganisms in two dimensional steady state diffusion gradients. *Can. J. Microbiol.* **19**, 53.

Caldwell, D.E., Lai, S. and Tiedje, J.M. (1973) A two-dimensional steady state diffusion gradient plate for ecological studies. *Bull. Ecol. Res. Comm. (Stockholm)* **17**, 151.

Caldwell, D.E., Korber, D.R. and Lawrence, J.R. (1992) Confocal laser microscopy and digital image-analysis in microbial ecology. *Adv. Microb. Ecol.* **12**, 1–67.

Codeco, C.T. and Grover, J.P. (2001) Competition along a spatial gradient of resource supply: a microbial experimental model. *Am. Nat.* **157**, 300–315.

Coombe, R.A., Tatevossian, A. and Wimpenny, J.W.T. (1982) An *in vitro* model for dental plaque. Paper given at the *30th Meeting of the International Association for Dental research,* British Division, University of Edinburgh, Scotland.

Cooper, D.C and Copeland, B.J. (1973) Responses of continuous series estuarine microsystems to point source input variations. *Ecol. Mon.* **14**, 213.

Elowaidy, H. and Elleithy, O.A. (1990) Theoretical studies on extinction in the gradostat. *Math. Biosci.* **101**, 1–2.

Emerson, D. and Breznak, J.A. (1977) The response of microbial populations from oil-brine contaminated soil to gradients of NaCl and sodium p-toluate is a diffusion gradient chamber. *FEMS Microbiol. Ecol.* **23**, 285–300.

Emerson, D., Worden, R.M. and Breznak, J.A. (1994) A diffusion gradient chamber for studying microbial behaviour and separating microorganisms. *Appl. Environ. Microbiol.* **60**, 1269–1278.

Evans, C.J. (1993) Ph.D. thesis, Cardiff University.

Fritsche, C. and Niemann, E.G. (1989) Modeling nitrogen-fixation in the rhizosphere of rice using a gradostat. *J. Gen. Appl. Microbiol.* **35**, 475–479.

Fritsche, C. and Niemann, E.G. (1990) Investigations into the growth of a nitrogen-fixing bacterium in gradients relevant to the rhizosphere of rice. *Symbiosis* **9**, 289–293.

Fritsche, C., Huckfeldt, K. and Niemann, E.G. (1991) Ecophysiology of associative nitrogen-fixation in a rhizosphere model in pure and mixed culture. *FEMS Microbiol. Ecol.* **85**, 279–292.

Gause, G.F. (1934) *The Struggle for Existence.* Williams and Wilkins, Baltimore.

Halldal, P. (1958) Pigment formation and growth of blue green algae in crossed gradients of light intensity and temperature. *Physiol. Plant* **11**, 401.

Herbert, R.A. (1988) Bidirectional compound chemostat: applications of compound diffusion-linked chemostats in microbial ecology. In *Handbook of Laboratory Model Systems for Microbial Ecosystems* (ed. J.W.T. Wimpenny), pp. 99–115, CRC press, Boca Raton, Florida.

Higashi, Y. and Seki, H. (1999) Application of an *in situ* gradostat for a natural population community in a eutrophic environment. *Environ. Pollut.* **105**, 101–109.

Higashi, Y. and Seki, H. (2000) Ecological adaptation and acclimatization of natural freshwater phytoplankters with a nutrient gradient. *Environ. Pollut.* **109**, 311–320.

Higashi, Y., Ytow, N., Saida, H. and Seki, H. (1998) *In situ* gradostat for the study of natural phytoplankton community with an experimental nutrient gradient. *Environ. Pollut.* **99**, 395–404.

Jager, W., So, J.W.H. and Waltman, P. (1987) Competition in the gradostat. *J. Math. Biol.* **25**, 23–42.

Kinniment, S.L., Wimpenny, J.W.T., Adams, D. and Marsh, P.D. (1996a) Development of a steady state oral microbial film community using the constant depth film fermenter. *Microbiology* **142**, 631–638.

Kinniment, S.L., Wimpenny, J.W.T., Adams, D. and Marsh, P.D. (1996b) The effect of chlorhexidine on defined, mixed culture oral biofilms grown in a novel model system. *J. Appl. Bacteriol.* **81**, 120–125.

Lovitt, R.W. and Wimpenny, J.W.T. (1981a) The gradostat, a bidirectional compound chemostat, and its applications in microbiological research. *J. Gen. Microbiol.* **127**, 261–268.

Lovitt, R.W. and Wimpenny, J.W.T. (1981b) Physiological behaviour of *Escherichia coli* grown in opposing gradients of glucose and oxygen plus nitrate in the gradostat. *J. Gen. Microbiol.* **127**, 269–276.

Margalef, R. (1967) Laboratory analogues of estuarine plankton systems. In *Estuaries: Ecology and Populations* (ed. G.M. Lauff), Hornshafer, Baltimore, 515.

Matharu, S., Spratt, D.A., Pratten, J., Ng, Y.L., Mordan, N., Wilson, M. and Gulabivala, K. (2001) A new *in vitro* model for the study of microbial microleakage around dental restorations: a preliminary qualitative evaluation. *Int. Endodont. J.* **34**, 547–553.

Monod, J. (1950) La technique de culture continue: théorie et applications. *Ann. Inst. Pasteur* **79**, 390–410.

Morgan, T.D. and Wilson, M. (2001) The effects of surface roughness and type of denture acrylic on biofilm formation by *Streptococcus oralis* in a constant depth film fermenter. *J. Appl. Microbiol.* **91**, 47–53.

Norwood, D.E. and Gilmour, A. (2000) The growth and resistance to sodium hypochlorite of *Listeria monocytogenes* in a steady-state multispecies biofilm. *J. Appl. Microbiol.* **88**, 512–520.

Novick, A. and Szilard, L. (1950) Experiments with the chemostat on spontaneous mutants of bacteria. *Proc. NY Acad. Sci. (USA)* **36**, 708–718.

Peters, A.C. and Wimpenny, J.W.T. (1988a) A constant depth laboratory model film fermentor. *Biotechnol. Bioeng.* **32**, 263–270.

Peters, A.C. and Wimpenny, J.W.T. (1988b) A constant depth laboratory film fermenter. In *Handbook of Laboratory Model Systems for Microbial Ecosystems* (ed. J.W.T. Wimpenny), Vol. 1, pp. 175–195, CRC Press, Boca Raton, Florida.

Pratten, J. and Wilson, M. (1999) Antimicrobial susceptibility and composition of microcosm dental plaques supplemented with sucrose. *Antimicrob. Agents Ch.* **43**, 1595–1599.

Pratten, J., Wills, K., Barnett, P. and Wilson, M. (1998) *In vitro* studies of the effect of antiseptic-containing mouthwashes on the formation and viability of *Streptococcus sanguis* biofilms. *J. Appl. Microbiol.* **84**, 1149–1155.

Pratten, J.,Wilson, M. and Mullany, P. (1999) Transfer of a conjugative transposon, Tn5397 in a model oral biofilm. *FEMS Microbiol. Lett.* **177**, 63–66.

Pratten, J., Andrews, C.S., Craig, D.Q.M. and Wilson, M. (2000) TI Structural studies of microcosm dental plaques grown under different nutritional conditions. *FEMS Microbiol. Lett.* **189**, 215–218.

Ready, D., Roberts, A.P., Pratten, J., Spratt, D.A., Wilson, M. and Mullany, P. (2002) Composition and antibiotic resistance profile of microcosm dental plaques before and after exposure to tetracycline. *J. Antimicrob. Chemoth.* **49**, 769–775.

Sacks, L.E. (1956) A pH gradient agar plate. *Nature* **178**, 269.

Saida, H., Maekawa, T., Satake, T., Higashi, Y. and Seki, H. (2000) Gram stain index of a natural bacterial community at a nutrient gradient in the fresh water environment. *Environ. Pollut.* **109**, 293–301.

Smith, H. (1991) Equilibrium distribution of species among vessels of a gradostat – a singular perturbation approach. *J. Math. Biol.* **30**, 31–48.

Smith, H. and Tang, B. (1989) Competition in the gradostat – the role of the communication rate. *J. Math. Biol.* **27**, 139–165.

Smith, H. and Waltman, P. (1997) *The Mathematical Theory of Chemostats*, Cambridge University Press.

Smith, H., Tang, B. and Waltman, P. (1991) Competition in an n-vessel gradostat. *Siam. J. Math.* **51**, 1451–1471.

Szybalski, W. (1951) Gradient plates for the study of microbial resistance to antibiotics. *Bacteriol. Proc.* p. 36.

Szybalski, W. and Bryson, V. (1952) Genetic studies on microbial cross resistance to toxic agents. I: Cross resistance of *Escherichia coli* to fifteen antibiotics. *J. Bacteriol.* **64**, 49.

Tang, B. (1986) Mathematical investigations of growth in the gradostat. *J. Math. Biol.* **23**, 319–339.

Tang, B. (1994) Competition models in the gradostat with nutrient uptake functions. *Rocky Mt. J. Math.* **24**, 335–349.

Thomas, L.V. and Wimpenny, J.W.T. (1993) Method for investigation of competition between bacteria as a function of three environmental factors varied simultaneously. *Appl. Environ. Microbiol.* **59**, 1991–1997.

Thomas, L.V. and Wimpenny, J.W.T. (1996) Investigation of the effect of combined variations in temperature, pH and NaCl concentration on nisin inhibition of *Listeria monocytogenes* and *Staphylococcus aureus*. *Appl. Environ. Microbiol.* **62**, 2006–2012.

Thomopuolos, N.A., Vayenas, D.V. and Pavlous. (1998) On the existence of three micro-bial populations competing for two complementary substrates in configurations of interconnected chemostats. *Math. Biosci.* **154**, 87–102.

van Baalen, C. and Edwards, P. (1973) Light-temperature gradient plate. In *Handbook of Phycological Methods* (ed. J.R. Stern), p. 267, Cambridge University Press.

Vroom, J.M., De Grauw, K.J., Gerritson, H.C., Bradshaw, D.J., Marsh, P.D., Watson, G.K, Birmingham, J.J. and Allison, C. (1999) Depth penetration and detection of pH gradients in biofilms by two-photon excitation microscopy. *Appl. Environ. Microbiol.* **65**, 3502–3511.

Widman, M.T., Emerson, D., Chiu, C.C. and Worden, R.M. (1997) Modeling microbial chemotaxis in a diffusion gradient chamber. *Biotechnol. Bioeng.* **55**, 191–205.

Wilson, M. (1999) Use of constant depth film fermenter in studies of biofilms of oral bac-teria. *Meth. Enzymol.* **310**, 264–279.

Wilson, M., Patel, H., Kpendema, H., Noar, J.H., Hunt, N.P. and Mordan, N.J. (1997) Corrosion of intra-oral magnets by multi-species biofilms in the presence and absence of sucrose. *Biomaterials* **18**, 53–57.

Wilson, M., Patel, H. and Noar, J.H. (1998) Effect of chlorhexidine on multi-species biofilms. *Curr. Microbiol.* **36**, 13–18.

Wimpenny, J.W.T. (1981) Spatial order in microbial ecosystems. *Biol. Rev.* **56**, 295–342.

Wimpenny, J.W.T. (1986) The DC gradostat: a novel design of fermentor for the study of heterogeneous microbial systems. *Process Biochem. Oct.*

Wimpenny, J.W.T. (1988) Bidirectionally linked continuous culture: the gradostat. In *Handbook of Laboratory Model Systems for Microbial Ecosystems* (ed. J.W.T. Wimpenny), Vol. 1, pp. 73–98, CRC Press, Boca Raton, Florida.

Wimpenny, J. (1996) Ecological determinants of biofilm formation. *Biofouling* **10**, 43–63.

Wimpenny, J.W.T. and Abdollahi, H. (1991) Growth of a mixed culture of *Paracoccus denitrificans* and *Desulfovibrio desulfuricans* in homogeneous and in heterogeneous culture systems. *Microb. Ecol.* **22**, 1–13.

Wimpenny, J.W.T. and Lovitt, R.W. (1984) The investigation and analysis of heteroge-neous environments using the gradostat. In *Microbiological Methods for Environmental Biotechnology* (ed. J.M. Grainger and J.M. Lynch), pp. 295–312, Academic Press London.

Wimpenny, J.W.T. and Waters, P. (1984) Growth of microorganisms in gel-stabilised two-dimensional gradient systems. *J. Gen. Microbiol.* **130**, 2921–6.

Wimpenny, J.W.T. and Waters, P. (1987) The use of gel-stabilized gradient plates to map the responses of microorganisms to three or four environmental factors varied simultan-eously. *FEMS Microbiol. Lett.* **40**, 263–267.

Wimpenny, J.W.T., Waters, P. and Peters, A.C. (1988) Gel-plate methods in microbiology. In *Handbook of Laboratory Model Systems for Microbial Ecosystems* (ed. J.W.T. Wimpenny), Vol.1, pp. 229–251, CRC Press, Boca raton, Florida.

Wimpenny, J.W.T., Earnshaw, R.G., Gest, H., Hayes, J.M. and Favinger, J.L. (1992) A novel directly coupled gradostat. *J. Microbiol. Meth.* **16**, 157–167.

Wimpenny, J.W.T., Kinniment, S.L. and Scourfield, M.A. (1993) The physiology and bio-chemistry of biofilm. In *Microbial Biofilms: Formation and Control* (ed. S.P. Denyer, S.P. Gorman and M. Sussman), pp. 51–94, Blackwell Scientific, Oxford.

Wolfaardt, G.M., Lawrence, J.R., Hendry, M.J., Robarts, R.D. and Caldwell, D.E. (1993) Development of steady-state diffusion gradients for the cultivation of degradative microbial consortia. *Appl. Environ. Microbiol.* **59**, 2388–2396.

Wood, P., Caldwell, D.E., Evans, E., Jones, M., Korber, D.R., Wolfhaardt, G.M., Wilson, M. and Gilbert, P. (1998) Surface catalysed disinfection of thick *Pseudomonas aeruginosa* biofilms. *J. Appl. Microbiol.* **84**, 1092–1098.

Zaghrout, A.A.S. (1992) Asymptotic behaviour of solutions of competition in a gradostat with 2 limiting complementary growth substrates. *App. Math. Comput.* **49**, 19–37.

Section 4

Analytical techniques for biofilm properties: (A) Physico-chemical properties

16 Use of X-ray photoelectron spectroscopy
and atomic force microscopy for studying
interfaces in biofilms   259

17 Use of ^{1}H NMR to study transport processes
in biofilms   285

18 Screening of lectins for staining lectin-specific
glycoconjugates in the EPS of biofilms   308

16

Use of X-ray photoelectron spectroscopy and atomic force microscopy for studying interfaces in biofilms

P.G. Rouxhet, C.C. Dupont-Gillain, M.J. Genet and Y.F. Dufrêne

16.1 INTRODUCTION

16.1.1 Peculiarities of a surface

When considering interfacial phenomena, typically adsorption (retention of ions or molecules at a surface) or aggregation and adhesion (association between two surfaces), it must be realised that a surface is characterised by particular properties.

A surface is a frontier between two phases and thus a zone of mismatch in terms of molecular interactions (broken bonds, unbalanced forces). This is responsible for an additional contribution to the free enthalpy G of the system, called surface excess free energy γ (Hiemenz and Rajagopalan, 1997; Stokes and Evans, 1997; Lyklema, 2000). In a possible transformation involving a change of interfaces i characterised by an area A, the term $\sum_i \gamma_i \, dA_i$ contributes to dG and thus to the driving force of the transformation. For a liquid–gas interface, γ is the liquid surface tension, equal to half the cohesion energy of the liquid. The nature of the forces which are unbalanced at the surface explains the trend of variation in the surface tension: hexane, 18.4 mJ/m^2 (London–van der Waals forces); water, 72.8 mJ/m^2 (hydrogen bonds); mercury, 484 mJ/m^2 (metallic bonds).

Minimisation of $\Sigma_i\ \gamma_i\ dA_i$ is the driving force of numerous processes: liquid film retraction, droplet coalescence, Ostwald ripening or powder sintering. It is the basis for understanding adsorption phenomena. These include migration of additives in polymer materials towards the surface but also surface contamination by organic compounds, which is unavoidable when a solid of 'high surface energy', such as a metal or an oxide, is exposed to air. It is also a basis for understanding adhesion phenomena.

A surface is also a zone where electrical charges often accumulate. The origin of these charges may be acid–base reactions (protonation of amine; deprotonation of carboxyl groups, presence of phosphate mono- or di-ester), selective dissolution, possibly combined with acid–base reactions (metal oxides and hydroxides, ...), structural defects (clays, ...), and specific adsorption of ions.

At the interface, there is thus a double layer of charge, one localised on the surface and the other extending into the solution. The surface is characterised by an electrical potential ψ_0 and a surface density of charge σ_0. In the model widely used now, the solution side of the electrical double layer is itself subdivided into several layers as illustrated by Figure 16.1 (Lyklema, 1995; Adamson and Gast, 1997; Hiemenz and Rajagopalan, 1997; Kleijn and van Leeuwen, 2000). In the

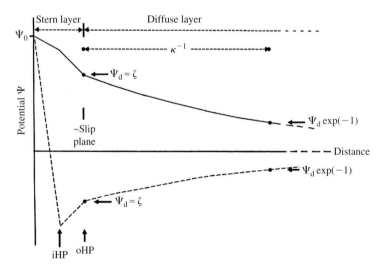

Figure 16.1   Variation of the electrical potential as a function of the distance to the surface, through the solid–water interface. The full and dashed curves arise, respectively, from a weak adsorption of counter-ions in the Stern layer and from a strong adsorption provoking charge reversal.

inner part, the Stern layer, adsorbed ions form a compact layer. The Stern layer is further subdivided to distinguish between chemically bound (chemisorbed) ions, residing in the inner Helmoltz plane (iHP), and ions retained electrostatically, residing in the outer Helmoltz plane (oHP). The latter is thus located at a distance from the surface, which is the size of the hydrated ions. Outside the compact layer is the diffuse layer or Gouy–Chapman layer, in which the ion distribution results from a balance between the electrostatic interactions with the surface side and the tendency to distribute evenly in the solution.

Neutralisation of the charge remaining on the surface side is thus not realised in a localised way. It is diffuse, being ensured by an excess of counter-ions and a depletion of co-ions over an appreciable thickness of the solution. When the electrical potential at the internal limit of the diffuse layer, ψ_d, is low, the electrical potential decays exponentially with the distance x:

$$\psi/\psi_d = \exp(-\kappa x)$$

and the distribution of ions follows the Maxwell–Boltzmann equation. The Debye length κ^{-1} has approximately the following values in water at room temperature:

- 100 nm at an ionic strength of 10^{-5} M typical of very diluted solutions,
- 10 nm at an ionic strength of 10^{-3} M typical of common water supplies,
- 1 nm at an ionic strength of 10^{-1} M typical of culture media.

The diffuse layer potential ψ_d, is considered as being close to the zeta-potential, i.e. the potential at the slip plane, which is evaluated by electrokinetic measurements (electrokinetic mobility, electro-osmosis, streaming potential measurements). Note that the point of zero charge (p.z.c.) is the pH at which ψ_0 and σ_0 are equal to zero, and the isoelectric point (i.e.p.) is the pH at which the zeta-potential is equal to zero.

16.1.2   Interactions between surfaces

Reviews and extended discussions of bacterial adhesion and aggregation viewed as the result of interactions between surfaces can be found in recent publications (Hermansson, 1999; Azeredo et al., 1999; Bos and Busscher, 1999; Rijnaarts et al., 1999; Lyklema, 2000; Wennerström, 2000; Busscher et al., 2000; Boonaert et al., 1999, 2002a).

DLVO theory, proposed by Derjaguin, Landau, Verwey and Overbeek, provides a model to evaluate the interactions between electrically charged solid surfaces by accounting for van der Waals interactions at the solid–water–solid interface and for electrostatic interactions resulting from overlap of the diffuse layers facing each other. It provides a straightforward explanation for the effect of pH, ionic strength and particular ions on adhesion and flocculation.

The often-called 'thermodynamic approach' is based on the balance of interfacial free energies associated to the adhesion process (written for 1 unit of interfacial area):

$$\Delta G_{adh} = \Sigma_i \, \gamma_i \, dA_i = \gamma_{MS} - \gamma_{ML} - \gamma_{SL}$$

where MS, ML and SL stand for the microorganism–substratum, microorganism–liquid and substratum–liquid interface, respectively. This approach explains that a hydrophobic particle will tend to attach to a hydrophobic substratum. However, it is subject to severe limitations. The interfacial free energies are deduced from measurements of liquid contact angles, whereas the theoretical frames used to perform these deductions vary according to the simplifications made and provide different results. Moreover, it is considered that the surfaces are smooth and get in molecular contact with each other, the transformation is assumed to be reversible, and electrostatic interactions between approaching surfaces are not taken into account. In practice, the relationship between interfacial free energy and bacterial adhesion is far from clear (Morra and Cassinelli, 1997).

The particular role of macromolecules at interfaces must also be pointed out. They may be the constituents of a gel covering or constituting the substratum, or be adsorbed by the substratum; they may be constituents of the cell wall or be attached to the cell surface. If the microbial adhesion is still considered as the result of the interaction between two solid surfaces, the possible influence of neutral macromolecules may be analysed as follows:

(i)   macromolecules may bridge the two surfaces, either because of their tendency to adsorb on both surfaces or because they are bound to one surface and tend to adsorb to the other one;

(ii) if macromolecules cover the two surfaces and if water is an ideal or a good solvent, steric repulsion will result as the surfaces approach each other, due to the reluctance to confining the macromolecular chains and

(iii) if macromolecules cover the two surfaces and if water is a poor solvent, attraction between segments may lead to overall attraction between the surfaces, until a balance with steric repulsion is reached.

If the macromolecules are ionised, electrostatic interactions will take place at different levels: ion–macromolecular ion, segment–segment within and between macromolecules, and macromolecule–surface. These guidelines are still valid if the two surfaces are considered as being hydrogels. They are thus helpful, both when considering macromolecules as forming a hydrogel which hosts the microorganisms (case of an established biofilm) or when considering the primary step of bacterial adhesion.

16.1.3   Need of direct surface analytical tools

It appears from the two sections above that the approaches allowing the interfacial phenomena to be understood and anticipated are subject to limitations of different kinds: assumption of smooth and homogeneous surfaces, non-deformable and non-penetrable by ions; as well as indirect and model-dependent determination of crucial physical quantities (zeta-potential, surface energy). Assumptions that are acceptable for model colloids seem unrealistic for microorganisms. In particular, questions are raised concerning the influence of surface heterogeneity (in terms of relief and chemical composition, at different scales from the nm to the μm), of surface deformability, of permeability to ions and of molecular mobility.

Information on the composition of the interfaces, which may be known *a priori* for laboratory and industrial inert systems, is not available for microbial cells and for the layer of organic compounds adsorbed on materials immersed in a fluid of environmental or biological relevance. Moreover, microbial cells are themselves time-evolving colloids; they may release macromolecules or ionic compounds which create locally a liquid environment quite different from the bulk solution.

Two approaches contribute to bring partial answers to these questions:

(i) knowing better the chemical composition of surfaces and interfaces and
(ii) probing directly the structure and properties of surfaces and/or interfacial interactions.

The aim of this contribution is to present the principle and to illustrate the application of two complementary analytical and physico-chemical methods, which play a key role in these respective approaches, i.e. X-ray photoelectron spectroscopy (XPS; also called electron spectroscopy for chemical analysis, ESCA) and atomic force microscopy (AFM).

Table 16.1 gives an overview of different chemical analysis techniques, which offer a certain surface selectivity. Among them XPS is characterised by a good balance in terms of adaptability, surface sensitivity and chemical information. It is often considered as the premier technique for surface chemical analysis and is broadly used in the field of material science.

AFM is an emerging technique, which opens new and exciting avenues for the *in situ* investigation of biosystems. Compared to electron microscopy, AFM

Table 16.1 Overview of techniques providing a chemical analysis of 'surfaces'.

Technique	Primary particle	Phenomenon – measurement	Lateral resolution	Explored depth[a]	Information[b]
ToF-SIMS[c]	Ion beam	Erosion – mass spectrum	~1 μm	Surface 1 nm	Molecular fragments[e] SQ
XPS	Photons	Ionisation – electron emission	>10 μm	Quasi-surface 5 nm	Elements, functions[e] Q
Auger[d] spectroscopy	Electron beam	Ionisation, relaxation – electron emission	~1 μm	Quasi-surface 5 nm	Elements Q
Infrared spectroscopy	Photons	Absorption, reflection	>1 mm	Sub-surface[f] 1 μm, adsorbed layer[g] 1 nm	Chemical functions[e] SQ
Electron microprobe	Electron beam	Ionisation, relaxation – X-ray emission	~1 μm	Sub-surface 1 μm	Elements Q

[a] Approximate depth.
[b] Q, possibility of quantification; SQ, quantification limited to semi-quantitative comparisons.
[c] Time-of-flight secondary-ion mass spectroscopy, or static SIMS.
[d] Requires conducting samples.
[e] Molecular compounds, with certain limitations.
[f] By attenuated total reflectance.
[g] By reflection of polarised light with a metal substratum.

offers the advantage of providing high-resolution images of surfaces directly in aqueous solutions. In addition, AFM can be used to measure local physico-chemical properties, including nanomechanics, adhesiveness, hydrophobicity and surface charges, surface forces and molecular interactions, with a lateral resolution of a few tens of nm and the possibility to map the surface.

16.2   X-RAY PHOTOELECTRON SPECTROSCOPY

16.2.1   Principle

XPS is based on recording the kinetic energy spectrum of the electrons (number of electrons versus kinetic energy E_k) emitted under irradiation by X-rays with a given energy $h\nu$. Figure 16.2 presents a scheme of the instrument with the X-ray source, the sample, the kinetic energy analyser, the detector and the recorded spectrum. The whole system is under a high vacuum (Rouxhet and Genet, 1991).

If a photoemitted electron is collected without undergoing inelastic collision, its kinetic energy is given by:

$$E_k = h\nu - E_b - W$$

where W represents the energy exchanged by the electron as it travels from the sample surface to the entrance of the analyser, which can be calibrated by an adequate procedure. The binding energy E_b is the energy required to extract the electron from the atom. As it is specific for a given energy level of a given atom, the position of an XPS peak on the kinetic energy scale is characteristic of a given element and the XPS spectrum provides an elemental analysis.

The probability that a photoemitted electron leaves a solid sample without undergoing inelastic collision is given by:

$$Q = \exp(-z/\lambda \cos \theta)$$

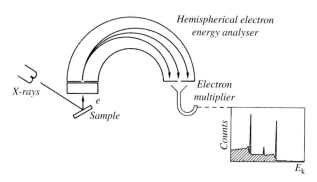

Figure 16.2   Schematic representation of an XPS spectrometer.

where z is the depth of photoemission and θ is the angle between the direction of photoelectron collection and the normal to the sample surface. λ is the inelastic mean free path which depends on the kinetic energy of the photoemitted electron and on the matrix, essentially its density. The photoelectrons, which undergo inelastic collisions, contribute to the baseline on the low kinetic energy side of the peak.

If the analysed solid is flat and smooth and has a homogeneous chemical composition, the contribution dI_A of a layer of thickness dz, located at depth z, to the peak intensity of element A, is given by:

$$dI_A = P_A C_A \, dz \exp(-z/\lambda \cos \theta)$$

where C_A is the concentration of element A and P_A is a constant for a given element and instrument. The exponential decrease of the contribution as a function of depth is illustrated by Figure 16.3; 63% of the signal originates from a layer of thickness equal to $1\lambda \cos \theta$, 86% from a layer of thickness $2\lambda \cos \theta$, 95% from a layer of thickness $3\lambda \cos \theta$. For the main elements, which are of interest for microbial cells and biochemical compounds (C, O, N), the inelastic mean free path is of the order of 3 nm. Thus XPS provides an analysis of a very thin layer at the surface of the sample.

As a matter of fact, if the layer which contributes to the information has a homogeneous chemical composition, the peak intensity is given by:

$$I_A = P_A C_A \int_0^\infty \exp\left(-z/\lambda_A \cos \theta\right) dz = P_A C_A \lambda_A \cos \theta$$

As illustrated by Figure 16.3c, everything happens as if the peak was due to a layer of thickness $\lambda_A \cos \theta$, inside which the photoelectron intensity would not be

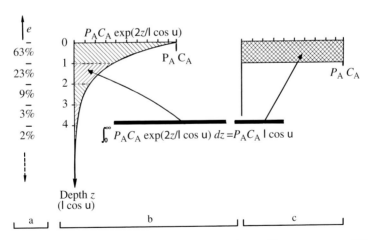

Figure 16.3   Illustration of the depth explored by XPS (case of homogeneous solid): (a) contribution to the intensity of successively deeper layers of thickness equal to $\lambda \cos \theta$; (b) variation of the contribution to the intensity as a function of depth z; (c) representation of a model situation in which the intensity would originate only from a layer of thickness equal to $\lambda \cos \theta$, in which the photoelectrons would not be subject to inelastic collisions (shaded areas are equivalent in b and c).

attenuated by inelastic scattering, the layers situated deeper giving no contribution to the signal.

For many applications, the approximation is made that the probed zone is homogeneous and concentration ratios are deduced from peak intensity ratios by calculating:

$$C_A/C_B = I_A i_B/I_B i_A$$

where i is sensitivity factor depending on the element (via λ and the probability of photoelectron emission) and on the spectrometer (transmission factor, detector efficiency). As all elements except hydrogen are detected, concentration ratios allow the elemental composition 'as seen by XPS' to be expressed in mole fraction of the different elements, excluding hydrogen.

When adsorbed layers are analysed, the approximation of a homogeneous probed zone must be considered with caution. If the adsorbed layer is thick enough ($\geqslant 10$ nm), the substratum signal is practically completely screened, the spectrum provides the chemical composition of the adsorbed layer 'as seen by XPS', which assumes that the composition of the layer itself is homogeneous.

If the approximation of a homogeneous probed zone is not applicable, information on the chemical heterogeneity of the zone analysed by XPS can be obtained by using the following strategy, illustrated in Section 16.4.2:

(i)   a hypothetical model is considered, figuring the distribution of the different constituents in space;
(ii) peak intensity ratios are computed on the basis of the considered model and adequate data, providing $(I_A/I_B)_{comp}$;
(iii) computed intensity ratios are compared with experimental ratios $(I_A/I_B)_{exp}$;
(iv) the process is repeated for different samples constituting a series within which certain experimental parameters vary (e.g. amount of adsorbed compound, sample hydrophobicity, sample handling after adsorption, ...);
(v)   alternative models are considered and the sensitivity of the computed intensity ratios to model type and parameters is explored and
(vi) the results are analysed, keeping in mind that '*a model that fits data is not necessarily a model that represents the reality.*'

16.2.2   Towards chemical functions and molecular compounds

XPS peaks present often different components. These may be due to the multiplicity of the energy level (which is not the case for 1s energy levels responsible for the C, N and O peaks) or to the influence of the chemical environment of the element (chemical shift). Peak decomposition may thus provide information on the chemical functions present in the probed zone. The variation of C, N and O peaks of microorganisms is illustrated in Figure 16.4 using different strains of *Bacillus subtilis*.

The result of peak decomposition is also shown in Figure 16.4, with the assignment of the components to chemical functions. Peak decomposition by least-square

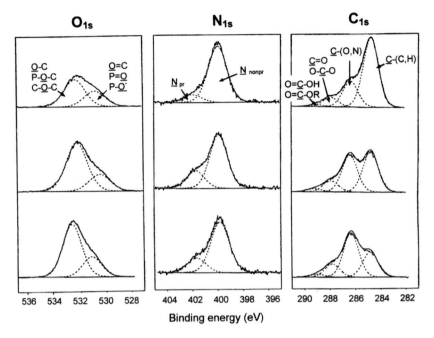

Figure 16.4   Oxygen, nitrogen and carbon XPS peaks of three strains (top, middle, bottom) of *B. subtilis*.

fitting means interpretation of spectra; it should thus be performed with care, aiming at consistency, on the one hand, and at avoiding bias, on the other hand. A procedure recommended to decompose C and O peaks of samples of biological nature may be as follows:

(i)   an appropriate shape function is selected (combination of Gauss and Lorentz functions, which may depend on the spectrometer), and a minimum number of components is selected;

(ii)   in a first fitting, the binding energy, the full width at half-maximum (FWHM), and the intensity of the components are obtained without imposing constraints and

(iii) a second fitting is performed by setting the FWHM of the components equal to the mean value of those obtained with the first fitting.

The carbon peak can satisfactorily be decomposed into four contributions:

- A component at 284.8 eV, due to carbon bound only to carbon and hydrogen [$\underline{C}$—(C,H)]; this component is used as a reference to calculate the binding energy of all the other components or peaks, i.e. to calibrate the binding energy scale.
- A component at 286.1–286.3 eV, due to carbon bound singly to oxygen or nitrogen [$\underline{C}$—(O,N)], including ether, alcohol, amine and amide.

- A component at about 288.0 eV, due to carbon making one double bond [C=O] or two single bonds with oxygen [O—C—O], including amide, carbonyl, carboxylate, acetal and hemiacetal.
- Sometimes, a weak component found near 289.0 eV and possibly due to carboxylic acid and ester.

The oxygen peak can be decomposed into two contributions:

- A component at about 531.2 eV, attributed to oxygen making a double bond with carbon [O=C] in carboxylic acid, carboxylate, ester, carbonyl, or amide, and to oxygen of phosphate [P=O].
- A component at about 532.6 eV, attributed to hydroxide [C—OH], acetal [C—O—C—O—C], hemiacetal, and phosphate ester [P—O—C].

The main nitrogen peak appearing at about 399.8 eV is due to unprotonated amine or amide functions [N_{nonpr}]. Sometimes, a weak component is found near 401.3 eV and attributed to protonated amine or ammonium [N_{pr}].

The number of components, the C and O peak decomposition procedure and the chemical attributions are supported (Rouxhet and Genet, 1991; Rouxhet *et al.*, 1994; Dufrêne *et al.*, 1996a, b; Dufrêne and Rouxhet, 1996; Boonaert and Rouxhet, 2000; Ahimou *et al.* submitted) by the consistency of results obtained over a long period with three spectrometers of different performances and by several quantitative relationships between data:

- 1 : 1 relationship between the concentration of N_{nonpr} and the concentration of C=O (involved in peptide links) as evaluated by the difference between C_{288} and $0.2 \times [C_{286.2}$–N] which reflects O—C—O of acetal.
- 1 : 2 relationship between the phosphorous concentration and the difference between $O_{531.2}$ and N_{nonpr}
- 1 : 1 relationship between the concentration of N_{nonpr} and the concentration of O=C (involved in peptide links) as evaluated by the difference between $O_{531.2}$ and twice the phosphorous concentration.

The surface concentrations of elements and chemical functions can be used to figure out roughly the surface composition in terms of molecular compounds. This requires to make assumptions about the nature and composition of the major surface constituents. For microorganisms, three classes of basic constituents have been considered: proteins (Pr), polysaccharides (PS) with a general formula $(C_6H_{10}O_5)_n$, and hydrocarbon-like compounds (HC) with a general formula $(CH_2)_n$ which refers here to lipids and other compounds that contain mainly carbon and hydrogen. In this approach, peptidoglycans are considered as a combination of proteins and polysaccharides, and (lipo)teichoic acids as a combination of hydrocarbon-like compounds and polysaccharides.

Considering the amino acid composition of a major outer membrane protein of *Pseudomonas flurorescens*, two sets of equations can be written, allowing the use of two independent sets of XPS data to evaluate the proportion of carbon associated with protein (C_{Pr}/C), polysaccharides (C_{PS}/C) and hydrocarbon-like compounds

(C_{HC}/C) (Rouxhet *et al.*, 1994). One scheme is based on the following elemental concentration ratios:

$$[N/C]_{obs} = 0.279 \ (C_{Pr}/C)$$
$$[O/C]_{obs} = 0.325 \ (C_{Pr}/C) + 0.833 \ (C_{PS}/C)$$
$$[C/C]_{obs} = (C_{Pr}/C) + (C_{PS}/C) + (C_{HC}/C) = 1$$

while the second scheme is based on the three main compounds of the carbon peak:

$$[C_{288.0}/C]_{obs} = 0.279 \ (C_{Pr}/C) + 0.167 \ (C_{PS}/C)$$
$$[C_{286.2}/C]_{obs} = 0.293 \ (C_{Pr}/C) + 0.833 \ (C_{PS}/C)$$
$$[C_{284.8}/C]_{obs} = 0.428 \ (C_{Pr}/C) + 1 \ (C_{HC}/C)$$

The proportions of carbon distributed in the different model compounds can be converted into weight fractions of the model compounds by using the carbon concentration of each of them.

Using this approach provided the following results when performing XPS on microbial surfaces:

- The surface molecular composition of *Azospirillum brasilense* varied significantly during growth. The surface protein content increased from 30% (exponential phase) to 50% (stationary phase), whereas the polysaccharides content decreased from 60% to 35% (Dufrêne and Rouxhet, 1996).
- The surface concentration in polysaccharides increased during germination of spores of *Phanerochaete chrysosporium* (Gerin *et al.*, 1993).
- The 'polysaccharides' to 'peptide' ratio at the surface of *Lactobacillus helveticus* increased between the exponential and stationary growth phases and varied appreciably from one culture to the other. In contrast, the same ratio for *Lactococcus lactis* was higher. It did not change during the culture growth phases and showed a weak variability from one culture to another (Boonaert and Rouxhet, 2000).

More sophisticated models may be used with a higher number of model constituents: chitine (for fungal spores), teichoic acids, nucleotides, ... However, the results quickly loose significance as the number of unknowns and equations increases, due to the limited precision of the experimental data and the limited accuracy of the composition selected for the model constituents.

16.2.3   Application to microorganisms

The surface analysis of microorganisms by XPS requires a sample preparation procedure which involves washing, centrifugation, quick freezing and freeze drying. A study of brewing yeast has shown that the biological material itself and the sample preparation are the major sources of variability, compared to the analysis as such (Dengis *et al.*, 1995). Freeze drying must be performed carefully: ice melting during freeze drying gives XPS data comparable to those obtained with crushed cells. The cooling rate obtained with the procedure recommended (about

1 ml of a concentrated cell suspension poured into a glass vial of about 5 cm^2 section precooled in liquid nitrogen) is expected be high enough to prevent migration of intracellular constituents to the surface. However, rearrangements of the polymer network at the surface, resulting from crystallisation and water removal cannot be excluded (Dengis and Rouxhet, 1996). Cell manipulation procedures, such as centrifugation, changes of media during washing, desiccation, were shown to affect the results of different methods of microbial surface characterisation, but XPS was not used in that context (Pembrey *et al.*, 1999).

The relevance of XPS to probe the surface composition of microbial surfaces is supported by correlations between XPS data and other data (Rouxhet *et al.*, 1994; Dufrêne *et al.*, 1999b; van der Mei *et al.*, 2000):

- fit between the peptide concentration at the surface of coryneform bacteria and the peptide concentration determined on cell walls by biochemical analyses;
- relationship between the N/C ratio determined by XPS and the amide II/CH stretching band ratio measured by transmission IR spectroscopy, or the lipoteichoic acid/teichoic acid ratio measured by biochemical analyses, for a series of *Streptococcus salivarius* strains (note that the correlation between XPS and IR was not found for *Escherichia coli*);
- relationship between the nitrogen surface concentration and structural cell surface features of streptococci;
- difference of surface composition, water contact angle, i.e.p. and flocculation behaviour between top and bottom fermentation brewing yeasts;
- relationship between culture age, increase of protein surface concentration and increase of adhesiveness of *A. brasilense*;
- relationship between swelling of spores of *P. chrysosporium*, increase of polysaccharide surface concentration and aggregation;
- relationship between phosphate concentration in the culture medium, phosphate concentration of bacteria or yeast surface and, in some cases, other properties (i.e.p., adhesion).

However, correlations must be considered with caution. While they reveal a relevance, they do not necessarily indicate a direct dependence of the variables in question (Marshall *et al.*, 1994). In particular, generalisation to other situations in terms of strains, conditions and properties should be avoided. All surface determinations raise to a certain extent the question of the representativity of the analysed surface with respect to the real problem of interest. In that respect, AFM examination under water is among the most appropriate approaches.

16.3   ATOMIC FORCE MICROSCOPY

16.3.1   Principle

AFM is a high-resolution surface imaging technique (Binnig *et al.*, 1986), which operates by scanning a sharp probe over the surface of a sample, while measuring the forces experienced by the probe (Figure 16.5). The sample is mounted on a piezoelectric scanner, which allows three-dimensional positioning with high resolution. The force is monitored by attaching the probe to a soft cantilever, typically made of silicon or silicon nitride, and measuring the bending or 'deflection' of the cantilever. The larger the cantilever deflection, the higher the force that will be experienced by the probe. To measure the cantilever deflection, a laser beam is usually focused on the free end of the cantilever and the position of the reflected beam is detected by a photodiode (position sensitive detector, PSD). The backside of the cantilever is often coated with a thin gold layer to enhance its reflectivity.

The most widely used imaging mode is the contact mode, in which sample topography can be measured in different ways. In the constant-height mode, the cantilever deflection is recorded while the sample is scanned horizontally, i.e. at constant height. However, minimising deflections, thus maintaining the applied force to small values, is often necessary to prevent sample damage, especially with biological specimens. This is achieved in the constant-deflection mode, in which the sample height is adjusted to keep the deflection of the cantilever constant using a feedback loop (Figure 16.5). The feedback output is used to display a 'height image'. Small variations of the cantilever deflection do however occur because the feedback loop is not perfect, and the resulting error signal can be used to generate a so-called 'deflection image'. It is important to note that, while the height image provides quantitative height measurements, the deflection image does not reflect true height variations but is more sensitive to fine surface details.

Several other AFM imaging modes have been developed. In contact mode, one can measure the torsional deflection of the cantilever resulting from variations in probe–sample friction while imaging surface topography. This is the principle of lateral force microscopy (LFM) which is a valuable tool to reveal chemical heterogeneities at surfaces. In tapping mode AFM (TMAFM), the cantilever is excited externally near its resonance frequency and the amplitude and phase of the probe oscillation are monitored. Although the use of TMAFM in biofilm studies has been rather limited so far, it has a great potential for imaging surfaces at minimal applied forces (for further information on the principles of AFM and on the different imaging modes, see Colton *et al.*, 1998; Morris *et al.*, 1999).

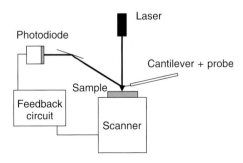

Figure 16.5   Principle of AFM.

Besides surface imaging, AFM force–distance curves can be used to measure physical properties and molecular interactions (Butt *et al.*, 1995). AFM force–distance curves are obtained by monitoring the cantilever deflection (d) as a function of the vertical displacement of the piezoelectric scanner (z). A raw force–distance curve is a plot of the PSD voltage versus scanner position. Using the slope of the retraction force curve in the region where probe and sample are in contact, the PSD voltage can be converted into a cantilever deflection. The cantilever deflection is then converted into a force (F) using the Hooke's law ($F = -kd$, where k is the cantilever spring constant). The curve can be transformed by plotting F as a function of $(z - d)$. The zero separation distance is then determined as the position of the vertical linear part of the curve in the contact region.

16.3.2   Topographic imaging

For the first time, AFM provides high-resolution images of the surface morphology of microbial cells in aqueous solution (for recent reviews, see Dufrêne, 2001, 2002). An important prerequisite for successful AFM imaging is that the sample must be well attached to a solid substrate. For imaging single cells, different methods can be used, including immobilisation in agar gels (Gad and Ikai, 1995) and in porous membranes (Dufrêne *et al.*, 1999c). These two methods offer the advantage over chemical fixation or drying that native, living cells can directly be imaged.

Figure 16.6 shows high-resolution images recorded for the surface of fungal spores, yeast and bacterial cells in aqueous solution. The surface of *Aspergillus oryzae* (van der Aa *et al.*, 2001) was covered by nanometre-scale rodlet structures, thought to play different biological functions, which is in contrast with the very smooth morphology of the *Saccharomyces cerevisiae* surface (Dufrêne, 2002). These images are consistent with previous electron microscopy images. For the bacterium *L. lactis*, however, the surface relief was strongly influenced by probe–cell interactions (Boonaert *et al.*, 2002b), emphasising the need to develop new preparation conditions and imaging modes for high-resolution imaging of bacteria by AFM.

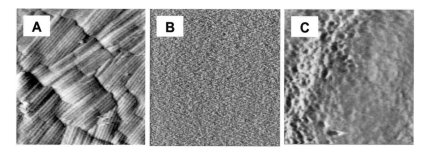

Figure 16.6   High-resolution deflection images (A, C: $500 \times 500\,nm^2$; B: $250 \times 250\,nm^2$) recorded in aqueous solution for the surface of (A) *A. oryzae*, (B) *S. cerevisiae* and (C) *L. lactis*.

AFM has also proved useful in imaging the morphology of biofilms in the hydrated state. In this case, high-resolution information cannot be obtained due to high sample softness, but direct visualisation of adhering bacteria and of hydrated extracellular polymeric substances (EPSs) is made possible. *Pseudomonas aeruginosa* cells and their associated EPS were observed in their hydrated forms in relation to the process of stainless steel corrosion (Steele *et al.*, 1994). Unsaturated biofilms of *Pseudomonas putida*, i.e. biofilms grown in humid air, were recently imaged by AFM (Auerbach *et al.*, 2000). The surface morphology, roughness and adhesion forces in the outer and basal cell layers of fresh and desiccated biofilms were determined. It was found that the EPS formed 'mesostructures' which were much larger than the discrete polymers of glycolipids and proteins that have been previously characterised on the outer surface of these Gram-negative bacteria.

16.3.3   Force measurements

AFM force–distance curves can be used to probe physico-chemical properties, i.e. surface hydrophobicity and surface charges, to measure the stiffness of the cell wall and surface appendages, to stretch cell surface macromolecules and to map cell surface adhesiveness. These measurements provide new insight into cell surface properties and biointerfacial phenomena (adhesion, aggregation and biofilm formation).

To investigate the local surface hydrophobicity and surface charges of microbial cells, AFM probes can be modified with well-defined functional groups. In this way, the surface hydrophobicity of *P. chrysosporium* spores was recently mapped by recording multiple force–distance curves using OH (hydrophilic)- and CH_3 (hydrophobic)-terminated probes (Dufrêne, 2000). As shown in Figure 16.7A, the curves obtained with the two types of probes always showed no adhesion. Comparison with experiments performed on model substrata indicated that the spore surface was uniformly hydrophilic. Presumably, the hydrophilic, nonsticky character of the fungal spore surface must play a role in determining its biological functions, namely protection and dispersion.

Using a similar approach, AFM probes functionalised with ionisable groups were employed to characterise the surface charges of *S. cerevisiae*. Figure 16.7B shows that the force–distance curves recorded between the cell surface and COOH-terminated probes were strongly influenced by pH (Ahimou *et al.*, 2002): while no adhesion was observed at neutral and alkaline pH, multiple adhesion forces were observed at pH $\leq$ 5. Theses changes can be related to a modification of the ionisation state of the cell surface functional groups. The lack of adhesion measured at pH $>$ 6 is likely to reflect the electrostatic repulsion between the negatively charged COO^--terminated probe and surface macromolecules bearing negatively charged COO^- or PO_4^- groups. In contrast, the adhesion force observed at pH $<$ 6 can be attributed to hydrogen bonding between the COOH probe and cell surface macromolecules. This interpretation is supported by the correlation obtained between the titration curve constructed by plotting the adhesion force

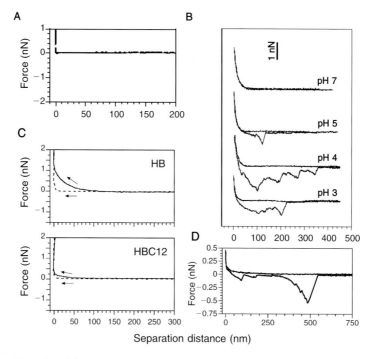

Figure 16.7   Set of force–distance curves recorded for various microorganisms: (A) Typical curve obtained in water between OH- and CH$_3$-terminated probes and the surface of dormant spores of *P. chrysosporium*. (B) Curves recorded in solutions of varying pH between a COOH-terminated probe and the surface of *S. cerevisiae*. (C) Approach force–distance curves obtained with a silicon nitride probe, in water (continuous line) or in 0.1 M KCl (dashed line), for the fibrillated strain *S. salivarius* HB (top) and for the non-fibrillated *S. salivarius* HBC12 strain (bottom). (D) Curve recorded in water on a germinating spore of *A. oryzae* using a silicon nitride probe.

versus pH and the electrophoretic mobility versus pH curve obtained by classical microelectrophoresis. In particular, the maximum of adhesion forces was observed at the i.e.p. of the cells, i.e. pH 4, indicating that AFM is capable of probing local i.e.p. Interestingly, treating the cells with Cu(II) ions caused a reversal of the cell surface charge at neutral pH and promoted the adhesion towards the negatively charged probe.

Force–distance curves may also provide information on the mechanical properties of the cell surface. Electron microscopy has revealed an array of different proteinaceous fibrillar structures on the cell surface of *S. salivarius* HB, while *S. salivarius* HBC12 appeared to have a bald cell surface. Accordingly, the nitrogen surface concentration measured by XPS was higher for the former compared to the latter (van der Mei *et al.*, 1988). As shown in Figure 16.7C, the two bacterial strains exhibited very different approach curves (van der Mei *et al.*, 2000). Upon approach of *S. salivarius* HB in water, a long-range repulsion, starting at a separation of ~100 nm, was detected while no jump-to-contact was observed. This may be attributed to the compression of the soft layer of fibrils present at the cell surface, the ~100 nm repulsion range being of the same order as the fibril length. The much shorter range observed at high ionic strength (0.1 M KCl) would reflect a stiffer cell surface due to collapse of the fibrillar mass. In contrast, the non-fibrillated strain, probed both in water (repulsion range ~20 nm) and in 0.1 M KCl (repulsion range ~10 nm), appeared much stiffer than the fibrillated strain in water. Differences in cell surface softness were correlated with previous indirect measurements by dynamic light scattering and particle microelectrophoresis. AFM is thus a valuable method for probing cell surface softness. However, it must be kept in mind that, in view of the complex and dynamic nature of microbial cell walls, distinguishing true sample deformation from repulsive surface forces (whether of electrostatic or steric nature) and defining the true 'zero separation' of force–distance curves may be a delicate task.

When recording force–distance curves on a macromolecular system, the retraction curves often show attractive elongation forces developing nonlinearly, which reflect the stretching of the macromolecules. This can be exploited to probe the molecular interactions and properties of long, flexible macromolecules in relation with biointerfacial processes. Germination of *A. oryzae* spores in liquid medium leads to spore aggregation. AFM topographic imaging revealed that, upon germination, the spore surface changed into a layer of soft granular material attributed to cell surface polysaccharides (van der Aa *et al.*, 2001). As can be seen in Figure 16.7D, force–distance curves recorded on this surface showed attractive forces of 400 ± 100 pN magnitude, along with characteristic elongation forces and rupture lengths ranging from 20 to 500 nm. Elongation forces were well fitted with an extended freely jointed chain model, using fitting parameters that were consistent with the stretching of individual poly molecules. The sticky and flexible nature of these long macromolecules may play a key role in the aggregation process by promoting bridging interactions between spores.

Finally, force mapping which consists in recording arrays (typically, 64×64) of force curves in the x, y plane, provides exciting possibilities in microbial cell surface and biofilm studies. The power of this approach is illustrated in Figure 16.8, in which differences in adhesion forces across the *P. chrysosporium* spore surface were correlated with the heterogeneous surface morphology (Dufrêne, 2001). The measured local adhesiveness is associated with an increase of the surface polysaccharide concentration, measured by XPS, and is thought to be responsible for cell aggregation observed during germination.

16.4   CASE STUDIES

16.4.1   Bacterial adhesion: *A. brasilense*

The adhesion of the soil bacterium *A. brasilense* to polystyrene has been investigated in conditions allowing cell transport to the support by sedimentation. It was found by XPS that:

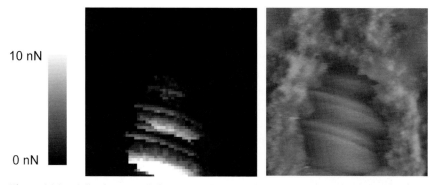

Figure 16.8   Adhesion map (left; 2 μm × 2 μm) and topographic image (right) acquired on the same area of a germinating spore of *P. chrysosporium*. The adhesion map was obtained by recording 64 × 64 force–distance curves, calculating the adhesion force for each force curve and displaying adhesion force values as grey levels (range = 0–10 nN).

(i)   proteins were the major constituent at the substratum surface after adhesion and subsequent cell detachment

(ii) changes of experimental conditions affected the protein concentration at the cell surface (growth phase, aging temperature) or at the substratum surface (adhesion tests with different contact times, at different temperatures or with addition of tetracycline) and influenced the density of adhering cells in the same direction (Dufrêne *et al.*, 1996a, b, 1999a).

Adhesion tests were performed in a parallel-plate flow chamber. Bottom and top plates were analysed by XPS after the detachment of adhering cells. Analyses revealed the presence of proteins, the concentration of which was the same on the bottom and top plates. These results demonstrated that a direct contact between the cells and the substratum was not required for the accumulation of proteins at the substratum surface and that proteins were released progressively by the cells into the solution and adsorbed at the substratum surface. The progressive release of extracellular proteins into the aqueous phase was demonstrated further by characterising the supernatant of cell suspensions by UV–visible spectrophotometry and protein assay.

More recently, the same system was re-examined by AFM (van der Aa and Dufrêne, 2002). When the adhesion test was performed under conditions favourable to adhesion (contact time of 24 h, 30°C), AFM images of substrata, obtained under water after cell detachment, clearly showed dotlike features of 8 ± 2 nm height, uniformly distributed across the surface, as illustrated by Figure 16.9B(1). These were hardly or not visible on the native substrata (Figure 16.9A(1)) or when the adhesion tests were performed under conditions unfavourable to adhesion (contact time of 2 h or temperature of 4°C). Force–distance curves recorded on bare polystyrene showed significant adhesion forces of 0.8 ± 0.2 N magnitude (Figure 16.9A(2, 3)). By contrast, the force–distance

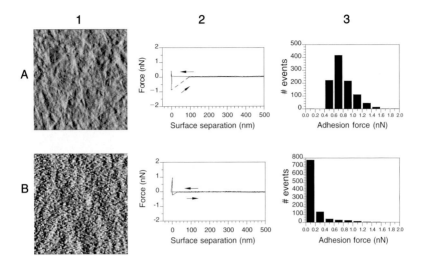

Figure 16.9   AFM examination, under water, with a silicon nitride probe, of bare polystyrene (A) and of polystyrene after adhesion of *A. brasilense* (24 h contact, 30°C) and cell detachment (B). 1, topographic-deflection images (5 μm × 5 μm); 2, typical force–distance curves; and 3, histograms of adhesion forces measured on the retraction curves.

curves recorded after adhesion and cell detachment showed either no adhesion or very small adhesion forces (0.2 ± 0.2 nN; Figure 16.9B(2, 3)). When the adhesion test was performed under conditions unfavourable to adhesion, the force–distance curves were similar to those obtained on bare polystyrene. This confirms the relationship between protein adsorption and bacterial adhesion, and indicates that protein adsorption modifies appreciably the substratum surface solvation and thereby the interactions with another surface.

16.4.2   Substrata conditioned by adsorption of macromolecules: collagen as an example

The adsorption of collagen (an extracellular matrix protein with the shape of a nanofilament of 300 nm length and 1.5 nm diameter) is taken here as an example to show

(i)   the complementarity of XPS and AFM, together with other methods, for studying the conditioning layer which is quickly formed when a material is brought in contact with a medium containing macromolecules or colloidal particles and

(ii)   the influence of the substratum on the properties of the conditioning layer and, thereby, on the specific response of living cells, in this case mammalian cells.

A   **B**   **C**

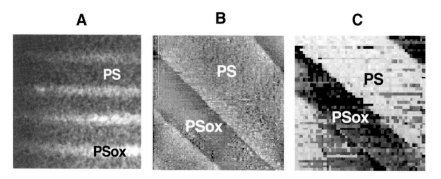

Figure 16.10   Images of a polystyrene surface with stripes of oxidised polystyrene:
(A) XPS imaging using the intensity of the O_{1s} peak (800 μm × 800 μm; stripes of
50 μm); (B and C) AFM images (silicon probe, measurements under water;
30 μm × 30 μm; stripes of 7 μm) made in lateral force mode (B) or obtained by
mapping the adhesion force (C).

16.4.2.1   Substrata

Polystyrene and polystyrene oxidised by oxygen plasma discharge are representative of a hydrophobic and a hydrophilic polymer, respectively. The XPS analysis
gave O/C concentration ratios of 0.01 to 0.03 and 0.18 to 0.26, respectively. The
shape of the carbon peak of oxidised polystyrene revealed the presence of a broad
range of oxygen-bearing functions. Figure 16.10 presents images obtained on
samples prepared by producing oxidised stripes on polystyrene. It illustrates the
possibility of chemical imaging by XPS (image A), and the generation of contrast
in AFM by friction forces (image B) or adhesion forces (image C). When using a
silicon probe in water, oxidised polystyrene gives a lower friction in lateral force
mode compared to polystyrene. On the other hand, the retraction part of force–
distance curves shows an appreciable adhesion on polystyrene, which is not the
case on oxidised polystyrene (Dupont-Gillain et al., 1999a, 2000).

The surface of polystyrene may be considered as a rigid solid. In contrast, the surface of oxidised polystyrene is constituted of mobile hydrophilic chains. These have
a polyelectrolyte character as demonstrated by a study of wetting in dynamic conditions. Figure 16.11A shows that oxidised polystyrene is more sensitive than polystyrene to probe damage in AFM examination (Dupont-Gillain and Rouxhet, 2001).

16.4.2.2   Adsorbed collagen

XPS does not provide a non-ambiguous information on the adsorbed layer
obtained after desiccation, as the peak intensities depend both on the adsorbed
amount and on the organisation of the adsorbed layer. In a first approach, the film
may be tentatively modelled as a layer characterised by a degree of surface coverage (γ) and a constant thickness (t) in the covered zones (Dufrêne et al., 1999d).

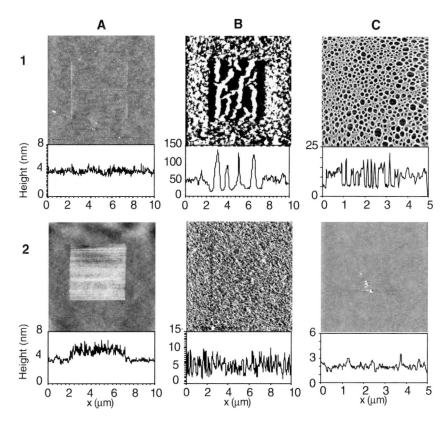

Figure 16.11   AFM height images (10 μm × 10 μm) of polystyrene (1) and oxidised polystyrene (2) with cross sections taken along a horizontal line at the centre. (A) Images recorded under water (silicon nitride probe) with a minimal loading force after scanning three times the central 5 μm × 5 μm zone at a loading force of 7.1 nN. (B) Images recorded under water after collagen adsorption (60 μg/ml) and rinsing, using the same procedure as for A. (C) Images (5 μm × 5 μm) recorded in air (silicon probe) after collagen adsorption (7 mg/ml), rinsing and slow drying.

This is illustrated by Figure 16.12 which shows that the nitrogen signal is due to the protein layer, while the carbon signal is the sum of three contributions: contribution of the uncovered substratum $(1 - \gamma)$, contribution of the substratum (γ) attenuated by the protein layer, and contribution of the protein layer (γ). The ratio of computed peak intensities I_N/I_C may be compared with experimental data, which provides an infinite set of $\gamma - t$ pairs. If the adsorbed amount Q is known independently, for instance by radiocounting, another set of $\gamma - t$ pairs may be obtained from the expression $Q = \gamma t \rho$ where ρ is the protein density. Comparison of the two plots γ versus t provides a range of γ and t values which fit the experimental data in the frame of this simple model.

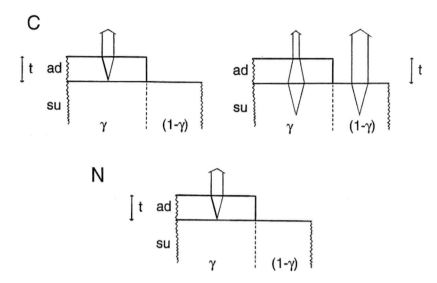

Figure 16.12   Model of a protein layer (ad) adsorbed on a polymer substratum (su) with a degree of surface coverage γ and a thickness t in the covered zones. The shapes of the arrows (width as a function of depth) illustrate the contributions to the C (top) and N (bottom) photoelectron peaks as they build up in the substratum and build up or are attenuated through the adsorbed layer.

The organisation of the adsorbed collagen layer under water was found to vary according to the hydrophilicity of the substratum and its roughness on the nanometer scale (Dufrêne *et al.*, 1999e). Figure 16.11B (Dupont-Gillain and Rouxhet, 2001) shows that the mechanical properties of the adsorbed layer also vary according to the substratum hydrophilicity; collagen adsorbed on polystyrene gather together under the action of the AFM probe while no perturbation is observed for collagen adsorbed on oxidised polystyrene.

Drying the conditioning layer may also alter the nanoscale organisation. Drying quickly the samples after adsorption had no visible effect for polystyrene or oxidised polystyrene. In contrast, Figures 16.11C(1) shows that, upon slow drying (2 days under 95% relative humidity), the collagen layer adsorbed on polystyrene reorganises to form holes, this is not observed on oxidised polystyrene (image C(2)). Depending on the substratum and conditions, the dried collagen layer may show different designs from a film pearced by holes to a net. In the case of the use of poly(methyl methacrylate) (PMMA) as substratum, XPS data and water contact angles indicated that the substratum is exposed at the outermost surface in the holes of the collagen net (Dupont-Gillain *et al.*, 1999b).

16.4.2.3   Influence on living cells

When samples with alterning polystyrene and oxidised polystyrene stripes (Figure 16.10) are exposed to a solution containing an extracellular matrix protein and a

poly(ethylene oxide)-poly(propylene oxide)-poly(ethylene oxide) co-polymer surfactant, the protein adsorbs selectively on oxidised polystyrene. Subsequent contact with mammalian cells leads to their selective adhesion on oxidised
polystyrene stripes (Dewez *et al.*, 1998).

16.5   CONCLUSION

In the context of understanding interfacial phenomena involved in biofilm initiation and development, XPS and AFM provide direct quantitative information,
which is not crucially dependent on interpretation models.

XPS gives the chemical composition (elements, chemical functions and to a
certain extent molecular compounds) of the surfaces involved (substrata, microbial cells, conditioning layers). The analysed depth is 3–10 nm, which is the size
of macromolecules or supramolecular entities. A limitation is that the samples
must be dehydrated by freeze drying. AFM offers the advantage of analysing
microbial cells and conditioning layers *in situ*, i.e. in aqueous solutions. The technique probes directly the structural and physical heterogeneity of surfaces and
interfacial phases. By playing with different modes and using modified probes,
AFM allows different local physico-chemical properties to be measured, such as
nanomechanical, adhesive and electrical properties, surface forces and molecular
interactions.

Acknowledgements

The authors thank F. Bouvy for the preparation of the manuscript. The support of
the National Foundation for Scientific Research (FNRS), the Research Department
of Communauté Française de Belgique (Concerted Research Action), the Federal
Office for Scientific, Technical, and Cultural Affairs (Interuniversity Poles of
Attraction Programme), and the European Union (Research Project No. ENV4-
CT97-0634) and the stimulation of European Union (COST Action 520 Biofouling and Materials) are gratefully acknowledged.

REFERENCES

Adamson, A.W. and Gast, A.P. (1997) *Physical Chemistry of Surfaces*, John Wiley & Sons,
    New York.
Ahimou, F., Boonaert, C.J.P., Adriaensen, Y., Jacques, P., Thonart, P., Paquot, M. and
    Rouxhet, P.G. Analysis of chemical functions at the surface of *Bacillus subtilis* strains
    (submitted).
Ahimou, F., Denis, F.A., Touhami, A. and Dufrêne, Y.F. (2002) Probing microbial cell
    surface charges by atomic force microscopy. *Langmuir*, **18**, 9937–9941.
Auerbach, I.D., Sorensen, C., Hansma, H.G. and Holden, P.A. (2000) Physical morphol-
    ogy and surface properties of unsaturated *Pseudomonas putida* biofilms. *J. Bacteriol.*
    182, 3809–3815.

Azeredo, J., Visser, J., and Oliveira, R. (1999) Exopolymers in bacterial adhesion: interpretation in terms of DLVO and XDLVO theories. *Coll. Surf. B: Biointerf.* **14**, 141–148.

Binnig, G., Quate, C.F. and Gerber, C. (1986) Atomic force microscope. *Phys. Rev. Lett.* **56**, 930–933.

Boonaert, C.J.P. and Rouxhet, P.G. (2000) Surface of lactic acid bacteria: relationships between chemical composition and physicochemical properties. *Appl. Environ. Microbiol.* **66**, 2548--2554.

Boonaert, C.J.P., Dupont-Gillain, C.C., Dengis, P.B., Dufrêne, Y.F. and Rouxhet, P.G. (1999) Cell separation, flocculation. In *Encyclopedia Bioprocess Technology: Fermentation, Biocatalysis and Bioseparation* (ed. M.C. Flickinger and S.W. Drew), pp. 531–548, Wiley, New York.

Boonaert, C.J.P., Dufrêne, Y.D. and Rouxhet, P.G. (2002a) Adhesion (primary) of microorganisms onto surfaces. In *Encyclopedia Environmental Microbiology* (ed. G. Bitton), *Biofilms* (ed. H.C. Flemming), pp. 113–132, Wiley, New York.

Boonaert, C.J.P., Toniazzo, V., Mustin, C., Dufrêne, Y.F. and Rouxhet, P.G. (2002b) Deformation of *Lactococcus lactis* surface in atomic force microscopy study. *Coll. Sur. B: Biointerf.* **23**, 201–211.

Bos, R. and Busscher, H.J. (1999) Role of acid–base interactions on the adhesion of oral streptococci and actinomyces to hexadecane and chloroform-influence of divalent cations and comparison between free energies of partitioning and free energies obtained by extended DLVO analysis. *Coll. Surf. B: Biointerf.* **14**, 169–177.

Busscher, H.J., Bos, R., van der Mei, H.C. and Handley, P.S. (2000) Physicochemistry of microbial adhesion from an overall approach to the limits. In *Physical Chemistry of Biological Interfaces* (ed. A. Baszkin and W. Norde), pp. 431–458, Marcel Dekker Inc., New York.

Butt, H.J., Jaschke, M. and Ducker, W. (1995) Measuring surface forces in aqueous electrolyte solution with the atomic force microscope. *Bioelectrochem. Bioenerg.* **38**, 191–201.

Colton, R.J., Engel, A., Frommer, J.E., Gaub, H.E., Gewirth, A.A., Guckenberger, R., Rabe, J., Heckel, W.M. and Parkinson, B. (ed.). (1998) *Procedures in Scanning Probe Microscopies*. John Wiley & Sons Ltd, Chichester, UK.

Dengis, P.B. and Rouxhet, P.G. (1996) Preparation of yeast cells for surface analysis by XPS. *J. Microbiol. Meth.* **26**, 171–183.

Dengis, P.B., Gerin, P.A. and Rouxhet, P.G. (1995) X-ray photoelectron spectroscopy analysis of biosurfaces: examination of performances with yeast cells and related model compounds. *Coll. Surf. B: Biointerf.* **4**, 199–211.

Dewez, J.L., Lhoest, J.B., Detrait, E., Berger, V., Dupont-Gillain, C.C., Vincent, L.M., Schneider, Y.J., Bertrand, P. and Rouxhet, P.G. (1998) Adhesion of mammalian cells to polymer surfaces: from physical chemistry of surfaces to selective adhesion on defined patterns. *Biomaterials* **19**, 1441–1445.

Dufrêne, Y.F. (2000) Direct characterisation of the physicochemical properties of fungal spores using functionalised AFM probes. *Biophys. J.* **78**, 3286–3291.

Dufrêne, Y.F. (2001) Application of atomic force microscopy to microbial surfaces: from reconstituted cell surface layers to living cells. *Micron* **32**, 153–165.

Dufrêne, Y.F. (2002) Atomic force microscopy, a powerful tool in microbiology. *J. Bacteriol.* **184**, 5205–5213.

Dufrêne, Y.F. and Rouxhet, P.G. (1996) Surface composition, surface properties and adhesiveness of *Azospirillum brasilense* – variation during growth. *Can. J. Microbiol.* **42**, 548–556.

Dufrêne, Y.F., Boonaert, C.J-P. and Rouxhet, P.G. (1996a) Adhesion of *Azospirillum brasilense*: role of proteins at the cell–support interface. *Coll. Surf. B: Biointerf.* **7**, 113–128.

Dufrêne, Y.J., Vermeiren, H., Vanderleyden, J. and Rouxhet, P.G. (1996b) Direct evidence for the involvement of extracellular proteins in the adhesion of *Azospirillum brasilense*. *Microbiology* **142**, 855–865.

Dufrêne, Y.F., Boonaert, C.J.P. and Rouxhet, P.G. (1999a) Role of proteins in the adhesion of *Azospirillum brasilense* to model substrata. In *Effect of Mineral–Organic–Microorganism Interactions on Soil and Freshwater Environments* (ed. J. Berthelin, P.M. Huang, J.M. Bollag and F. Andreux), pp. 261–274, Kluwer Academic/Plenum Publishers, New York.

Dufrêne, Y.F., Boonaert, C.J.P. and Rouxhet, P.G. (1999b) Surface analysis by X-ray photoelectron spectroscopy in the study of bioadhesion and biofilms. In *Methods in Enzymology*, (ed. R.J. Doyle), *Biofilms* Vol. 310, pp. 375–389, Academic Press.

Dufrêne, Y.F., Boonaert, C.J.P., Gerin, P.A., Asther, M. and Rouxhet, P.G. (1999c) Direct probing of the surface ultrastructure and molecular interactions of dormant and germinating spores of *Phanerochaete chrysosporium*. *J. Bacteriol.* **181**, 5350–5354.

Dufrêne, Y.F., Marchal, T.G. and Rouxhet, P.G. (1999d) Probing the organisation of adsorbed protein layers: complementarity of atomic force microscopy, X-ray photoelectron spectroscopy and radiolabeling. *Appl. Surf. Sci.* **144**–**145**, 638–643.

Dufrêne, Y.F., Marchal, T.G. and Rouxhet, P.G. (1999e) Influence of substratum surface properties on the organisation of adsorbed collagen films: *in situ* characterisation by atomic force microscopy. *Langmuir* **15**, 2871–2878.

Dupont-Gillain, C.C. and Rouxhet, P.G. (2001) AFM Study of the interaction of collagen with polystyrene and plasma-oxidised polystyrene. *Langmuir* **17**, 7261–7266.

Dupont-Gillain, C.C., Nysten, B., Hlady, V. and Rouxhet, P.G. (1999a) Atomic force microscopy and wettability study of oxidised patterns at the surface of polystyrene. *J. Coll. Interf. Sci.* **220**, 163–169.

Dupont-Gillain, C.C., Nysten, B. and Rouxhet, P.G. (1999b) Collagen adsorption on poly(methyl methacrylate): net-like structure formation upon drying. *Polym. Int.* **48**, 271–276.

Dupont-Gillain, C.C., Adriaensen, Y., Derclaye, S. and Rouxhet, P.G. (2000) Plasma-oxidised polystyrene: wetting properties and surface reconstruction. *Langmuir* **16**, 8194–8200.

Gad, M. and Ikai, A. (1995) Method for immobilising microbial cells on gel surface for dynamic AFM studies. *Biophys. J.* **69**, 2226–2233.

Gerin, P.A., Dufrêne, Y., Bellon-Fontaine, M.N., Asther, M. and Rouxhet, P.G. (1993) Surface properties of the conidiospores of *Phanerochaete chrysosporium* and their relevance to pellet formation. *J. Bacteriol.* **175**, 5135–5144.

Hermansson, M. (1999) The DLVO theory in microbial adhesion. *Coll. Surf. B: Biointerf.* **14**, 105–109.

Hiemenz, P.C. and Rajagopalan, R. (1997) *Principles of Colloid and Surface Chemistry*, 3rd ed., Marcel Dekker, New York.

Kleijn, J.M. and van Leeuwen, H.P. (2000) Electrostatic and electrodynamic properties of biological interfaces. In *Physical Chemistry of Biological Interfaces* (ed. A. Baszkin and W. Norde), pp. 49–83, Marcel Dekker, New York.

Lyklema, J. (1995) *Fundamentals of Interface and Colloid Science, Solid–Liquid Interfaces*, Vol. 2, Academic Press, London, UK.

Lyklema, J. (2000) Interfacial thermodynamics with special reference to biological systems. In *Physical Chemistry of Biological Interfaces* (ed. A. Baszkin and W. Norde), pp. 1–47, Marcel Dekker, New York.

Marshall, K.C., Pembrey, R. and Schneider, R.P. (1994) The relevance of X-ray photoelectron spectroscopy for analysis of microbial cell surfaces: a critical view. *Coll. Surf. B: Biointerf.* **2**, 371–376.

Morra, M. and Cassinelli, C. (1997) Bacterial adhesion to polymer surfaces: a critical review of surface thermodynamic approaches. *J. Biomater. Sci. Polym. E.* **9**, 55–77.

Morris, V.J., Kirby, A.R. and Gunning, A.P. (ed.) (1999) *Atomic Force Microscopy for Biologists*, Imperial College Press, London, UK.

Pembrey, R.S., Marhall, K.C. and Schneider, R.P. (1999) Cell surface analysis techniques: what do cell preparation protocols do to cell surface properties. *Appl. Environ. Microbiol.* **65**, 2877–2894.

Rijnaarts, H.H.M., Norde, W., Lyklema, J. and Zehnder, A.J.B. (1999) DLVO and steric contributions to bacterial deposition in media of different ionic strengths. *Coll. Surf. B: Biointerf.* **14**, 179–195.

Rouxhet, P.G. and Genet, M.J. (1991) Chemical composition of the microbial cell surface by X-ray photoelectron spectroscopy. In *Microbial Cell Surface Analysis: Structural and Physico-chemical Methods* (ed. N. Mozes, P.S. Handley, H.J. Busscher and P.G. Rouxhet), pp. 173–220, VCH Publishers, New York.

Rouxhet, P.G., Mozes, N., Dengis, P.B., Dufrêne, Y.F., Gerin P.A. and Genet, M.J. (1994) Application of X-ray photoelectron spectroscopy to microorganisms. *Coll. Surf. B: Biointerf.* **2**, 347–369.

Steele A., Goddard, D.T. and Beech, I.B. (1994) An atomic force microscopy study of the biodeterioration of stainless steel in the presence of bacterial biofilms. *Int. Biodeter. Biodegrad.* **34**, 35–46.

Stokes, R.J. and Evans, D.F. (1997) *Fundamentals of Interfacial Engineering*, Wiley-VCH, New York.

van der Aa, B. and Dufrêne, Y.F. (2002) *In situ* characterisation of bacterial extracellular polymeric substances by AFM. *Coll. Surf. B: Biointerf.* **23**, 173–-182.

van der Aa, B.C., Michel, R.M., Asther, M., Zamora, M.T., Rouxhet, P.G. and Dufrêne, Y.F. (2001) Stretching cell surface macromolecules by atomic force microscopy. *Langmuir* **17**, 3116–3119.

van der Mei, H.C., Léonard, A.J., Weerkamp, A.H., Rouxhet, P.G. and Busscher, H.J. (1988) Surface properties of *Streptococcus salivarius* HB and nonfibrillar mutants: measurement of zeta potential and elemental composition with X-ray photoelectron spectroscopy. *J. Bacteriol.* **170**, 2462–2466.

van der Mei, H.C., Busscher, H.J., Bos, R., de Vries, J., Boonaert, C.J.P., Dufrêne, Y.F. (2000). Direct probing by atomic force microscopy of the cell surface softness of a fibrillated and non-fibrillated oral streptococcal strain. *Biophys. J.* **78**, 2668–2674.

Wennerström, H. (2000) Interfacial interactions. In *Physical Chemistry of Biological Interfaces* (ed. A. Baszkin and W. Norde), pp. 85–114, Marcel Dekker, New York.

17

Use of ^{1}H NMR to study transport processes in biofilms

P.N.L. Lens and H. Van As

17.1   INTRODUCTION

Different types of transport processes, e.g. diffusion, perfusion and convection, can be distinguished in biofilms (Kühl and Revsbech, 2000; Van As and Lens, 2001). Transport processes in biofilms (Figure 17.1) strongly depend on the local amount of water, the pore dimensions/geometry (e.g. the porosity, connectivity and tortuosity) and the interaction of the transport medium with the biofilm matrix. In order to determine the rate-limiting transport process and to optimise it, there is a strong interest in the direct and non-invasive measurement of transport processes in these complex biosystems (Bear and Corapcioglu, 1984; Stewart, 1998; Wilderer *et al.*, 2002).

Advances in instrumentation technology as well as in computer control have led to spectacular progress in the development of *in situ* and/or non-invasive measurement and visualisation techniques of both diffusional and flow transport processes (Chaouki *et al.*, 1997). *In situ* measurement of transport processes can be done on biofilm slices using, e.g. microslicing techniques (Zhang *et al.*, 1995) or by microelectrodes positioned within a biofilm at a location of interest. The latter can be done using microsensors in deactivated biofilms (Lens *et al.*, 1993), by transport microsensors (Revsbech *et al.*, 1998) or by miniaturised limiting current

© IWA Publishing. *Biofilms in Medicine, Industry and Environmental Biotechnology*. Edited by Piet Lens, Anthony P. Moran, Therese Mahony, Paul Stoodley and Vincent O'Flaherty. ISBN: 1 84339 019 1

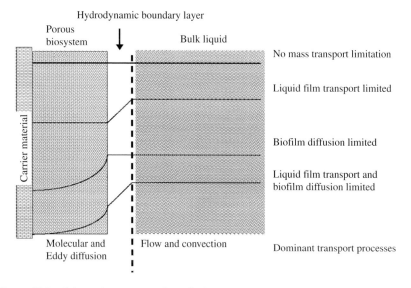

Figure 17.1   Schematic representation of substrate concentration profiles in and around a biofilm-containing microorganisms for different transport conditions. The thickness of the hydrodynamic boundary layer will depend on the velocity and profile of the bulk liquid and on the roughness of the biofilm surface. The concentration profile in the biofilm is mainly dependent on the substrate conversion rate and the substrate diffusion coefficient in the biofilm.

techniques (Yang and Lewandowski, 1995). These *in situ*, but invasive, measurements are very powerful when combined with molecular techniques (Santegoeds *et al.*, 1999).

Other tomographic and velocimetric methods allow to study biofilms and biofilm reactors non-invasively (Table 17.1). Tomography provides concentration, holdup or two-dimensional (2D) or three-dimensional (3D) density distributions of at least one component within a sample, whereas velocimetry provides the dynamic features of the phase of interest, such as the flow pattern, the velocity field or the 2D or 3D instantaneous movements. In biofilm research, most work has been done on confocal laser scanning microscopy based methods, i.e. to quantify fluorescence recovery after photobleaching (de Beer *et al.*, 1997; Bryers and Drummond, 1998).

The major advantage of nuclear magnetic resonance (NMR) for the observation of transport processes is that it is non-invasive, so that no direct contact with the fluid is necessary and that it uses natural present labels like ^{1}H, ^{13}C, ^{15}N and ^{31}P (Caprihan and Fukushima, 1990). Hence, it is well suited for studies of liquids that need to be isolated, such as those having extreme temperature, chemical reactivity or which are abrasive. NMR does not use ionising radiation, in contrast to X-ray tomography and X-ray scattering flow methods. Another special property of NMR is its potential to measure actual molecular displacement propagators (i.e. both flow, perfusion and diffusion) and that local flow patterns can be detected

Table 17.1   Non-invasive tomographic and velocimetric monitoring methods to monitor flow in multiphase systems (partially adapted from Chaouki *et al.*, 1997).

Method	Sensors	Resolution Spatial	Temporal
I. Tomography			
Ultrasonic	Ultrasonic transducers	<1 mm	I
Microwave	Microwave generators/ antennas	Fractions of the wavelength	II
Optical	Visible light source/camera	Several μm	III
Photoacoustic	Excitation: LASER in visible light range/detection: microphone	10 μm (depth)	I
Electric capacitance or resistivity imaging	Electrodes energised sequentially	About 1 mm	III
NMR imaging	Magnetic field/radiofrequency pulses	0.1 mm	II/III
X-ray diffraction	X-ray source/primary beam and diffracted X-ray detector	4 mm	I
X-ray radiography	X-ray source/X-ray detector or film	Several μm	III
II. Velocimetry			
Laser doppler anemometry	Laser source/light scattering particles or bubbles/light detector	0.1 mm inadequate for opaque systems	III
Particle image velocimetry	Laser sheet/light scattering particles or bubbles/video camera	0.2 mm	III
Fluorescent particle image velocimetry	Laser sheet/fluorescent particles/video camera	0.2 mm	III
Positron emission particle tracking	Positron emitting isotope traces, detection by angular coincidence	1 mm at 0.01 m/s 5 mm at 1 m/s	I
NMR imaging	Magnetic field gradients or contrast agents	0.1 mm	III/II
Cinematography	Coloured particle/video camera	7 mm	II

* Time needed for data acquisition: I = long, II = medium, III = fast.

in any direction within the sample, in contrast to X-ray, optical and ultra-sound scattering flow methods which only measure a net flow between the emitter and the detector. Note that, in contrast to optical methods, transparency, refraction properties and scattering in the objects to be imaged are no issue.

NMR measurements allow the combined measurement of the anatomy (via the spin (^{1}H) density and relaxation times) of a sample, simultaneously with the transport processes (Van As and Lens, 2001). NMR imaging adds another aspect to NMR, namely image formation, resulting in the spatially-resolved measurement of all information available with NMR in any selected part (volume, slice, etc.) of the system under observation. Thus, NMR imaging enables the determination of the relation between the internal structure, local water (^{1}H) density and fluid (^{1}H of water or an organic compound) transport in biofilm systems and similar media (Gladden, 1996; Van As and van Dusschoten, 1997; Baumann *et al.*, 2002).

17.2   NMR METHODS TO STUDY BIOFILM SYSTEMS

17.2.1   NMR parameters

Excitation of spins by a resonant radiofrequency (rf) pulse disturbs the equilibrium in the nuclear spin system. By applying magnetic field gradients during the NMR experiment, the rf resonance properties of the nuclei in the sample can be made spatially selective, which provides the basis for magnetic resonance imaging (MRI). As the nuclear spin system returns to equilibrium, a time-dependent rf signal is induced in the NMR measuring probe. These NMR signals are characterised by a number of different parameters. The amplitude A and the relaxation times T_1 and T_2 are the most commonly applied NMR parameters to characterise biofilm systems. More detailed descriptions of the basics of NMR spectroscopy and imaging can be found elsewhere (Callaghan, 1991; Lens and Hemminga, 1998; Blümich, 2000).

The amplitude A_0 of the NMR signal directly after excitation (time zero) is a direct measure for the amount of resonant nuclei under observation in the NMR rf coil. If the volume is known, A_0 directly relates to the density of nuclei under observation. In aqueous solutions, the amplitude of the ^{1}H NMR signal of water protons is a direct measure of the amount of water present in the sample. Thus, ^{1}H NMR can be used to determine the hydration state of solid wastes or sludges as a function of the drying process (La Heij *et al.*, 1996). Additional information can be obtained about the water status or the water-containing compartments of solid wastes when ^{1}H NMR amplitude measurements are performed spatially resolved (imaging) or combined with relaxation time and/or diffusion measurements (see below).

Two relaxation time constants describe the rate and manner at which the nuclear spin system returns to equilibrium after excitation. One time constant, the spin–lattice relaxation time T_1, describes the return to the equilibrium state in the direction of the magnetic field. The second time constant, the spin–spin relaxation time T_2, characterises the return in the plane perpendicular to the applied magnetic field. The protons in water molecules experience an intramolecular dipolar interaction between both protons within one and the same molecule, as well as an intermolecular interaction with protons of neighbouring water molecules. Both interactions fluctuate when the molecules rotate or translate. When the rotation correlation time of the molecules is short, as is the case for free water molecules $\tau(\tau_c \sim 10^{-12}\,\text{s})$, both T_1 and T_2 are equal and relatively long ($\sim 2\,\text{s}$). Water close to macromolecules or to solid surfaces generally have slower tumbling rates ($\tau_c \sim 10^{-12}$–$10^{-10}\,\text{s}$), which leads to a reduction in both relaxation times. Furthermore,

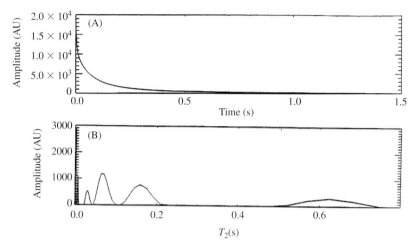

Figure 17.2 T_2 analysis of sulphidogenic granular sludge at 22°C at a 0.47 T (after Lens *et al.*, 1997): (A) Typical multi-exponential T_2-decay curve of the magnetisation as a function of time. (B) T_2 populations extracted from this multi-exponential T_2-decay curve. These T_2 populations correspond to different physical environments in the granular sludge.

exchange of protons between water and other molecules, such as sugars and proteins, also influences the relaxation times. In biofilms, these principles determine the bulk relaxation times within compartments with different chemical composition, such as the biofilm matrix, the cell walls and the cytoplasm (Figure 17.2).

The relaxation times T_1 and T_2 strongly depend on the pore size and its distribution, the type of biofilm/biocatalyst (content of minerals, organic matter, paramagnetic ions (such as Fe, Mn and Gd) and the water content. The relationship between the relaxation times of water and its physical environment has been utilised to characterise the porosity of soils, sediments and technical mineral and building materials (Blinc *et al.*, 1978; Keyon, 1992; Blümich, 2000; Song *et al.*, 2000; Valckenborg *et al.*, 2002). For a particular system, T_1 values have been successfully used to estimate the permeability (Issa and Mansfield, 1994) as well as the wettability (Howard, 1994) of the solid matrix. T_2 values have also shown to be highly correlated to the surface area and pore size distribution of soils and rocks (Kleinberg *et al.*, 1994; Song *et al.*, 2000). However, the relaxation times in porous materials are determined both by the surface area, surface material (determining the wall relaxation sink strength) as well as, if present, by the presence of paramagnetic materials on the pore walls (Bryar *et al.*, 2000; Valckenborg *et al.*, 2002). Consequently, such correlations are only valid for special classes of porous materials, and a relationship between relaxation time and porous media characteristics can only be found if paramagnetic centres are absent. In that case, the observed transverse relaxation time T_2 of water in a confined compartment as a vacuole can be described as a function of the bulk T_2 (T_2, bulk), the radii of the compartment along the x, y and z directions ($R_{x,y,z}$) and the net loss of magnetisation at the compartment boundary, the so-called magnetisation sink strength (H) (van der Weerd *et al.*, 2001):

$$1/T_{2,\text{obs}} = H(1/R_x + 1/R_y + 1/R_z) + 1/T_{2,\text{bulk}} \qquad (17.1)$$

17.2.2   Self-diffusion and flow measurements

Effects of flow and motion on the NMR signal are known for a long time (Hahn, 1950; Stejskal and Tanner, 1965). All molecules in a fluid are subject to Brownian motion. The extent of this motion depends on the temperature and the viscosity of the fluid, which are incorporated in a parameter called the bulk diffusion coefficient (D) of the fluid. When an ensemble of molecules is followed in time, the mean distance travelled increases with time as long as no boundaries are encountered. If water experiences a barrier to diffusion (e.g. a cell membrane), the cell dimensions determine the maximum displacement. As a result, the NMR diffusion experiments result in an apparent diffusion coefficient (D_{app}), which is smaller than the intrinsic D.

NMR is well capable of discriminating proton spins of flowing and stationary water on the basis of the physical properties of flow (Callaghan, 1991; Schaafsma *et al.*, 1992; Britton and Callaghan, 1997; Tallarek *et al.*, 1999). Diffusion and random dispersion do not result in a net displacement for the spin ensemble, and, therefore, do not result in a net frequency change, but in a broadening of the frequency bandwidth contributing to the signal. This results in a decrease of the signal amplitude at the original frequency. In contrast, net flow results in a frequency shift of the NMR signal originating from those moving spins, and becomes manifest as a phase shift of the NMR signal, because the NMR signal is detected with respect to a fixed reference frequency. This principle, applied in pulsed field gradient (PFG) NMR, allows to discriminate and measure flow, dispersion and diffusion. PFG NMR flow methods have been described in detail by Caprihan and Fukushima (1990), whereas PFG NMR techniques to determine self-diffusion coefficients are presented by Stilbs (1987), Le Bihan (1991) and Stallmach and Kärger (1999).

17.3   NMR AND TRANSPORT PROCESSES IN BIOFILMS

Transport processes of water (or other fluids) and of molecules or ions dissolved therein can be measured by NMR in a number of different ways. This chapter demonstrates two strategies which have been extensively used in biofilm systems:

- mapping the (spin) displacement in a well-defined time interval directly based on PFG or displacement imaging;

- mapping the distribution of contrast agents (paramagnetic molecules/ions) via single parameter spin relaxation time images or via spin relaxation-time weighted images.

17.3.1 Direct transport measurements by use of pulsed magnetic field gradients

17.3.1.1 Principle

In PFG NMR, diffusion is measured by recording the signal amplitude A as a function of strength G of the applied magnetic field gradient pulses. For free, unhindered, diffusion this results in:

$$A = A(TE) \exp(-Db) \tag{17.2}$$

with $A(TE)$, echo amplitude without gradients; D, (self-)diffusion coefficient and b, $\gamma^2\delta^2G^2$ ($\Delta - 1/3\delta$); where γ is the gyromagnetic ratio of the observed spin, δ and G are, respectively, the duration and amplitude of the magnetic field gradient pulse, and Δ is the time between two magnetic field gradient pulses.

By varying Δ, the displacement can be followed as a function of the observation time, allowing to trace the distance over which the spins can displace. In this way, the structure in which the fluid diffuses or flows (pore size and geometry) can be probed via observing effects of displacement restrictions by impermeable walls (restricted diffusion).

While the incoherent diffusive movement of the spins leads to a change in the amplitude of the available echo signal, coherent flow leads to a change in its phase. In a complex sample with a rich microstructure, such as biofilms or porous media, it is also possible that there is a superposition of flow filaments with different directions even at a small length scale. This case is called perfusion. Separating perfusion and diffusion in PFG NMR experiments requires special care (Callaghan, 1991; Le Bihan, 1991), but has been demonstrated in model systems consisting of columns packed with porous particles with an internal pore structure, i.e. chromatography columns (Tallarek et al., 2000, 2001a, 2001b).

Flow and diffusion measurements in porous biosystems by PFG can be done in a number of ways (Van As and van Dusschoten, 1997):

- Single shot at a fixed G and Δ value, either localised, but non-spatially resolved within this localisation (Snaar and Van As, 1990) or in combination with imaging (Mansfield and Issa, 1994). The advantage of this strategy is the short measurement time, allowing a high time resolution. The main disadvantage is that quantification of the measurements in terms of flow velocity requires knowledge of the flow profile, containing its distribution of velocity and direction.
- To overcome the latter problem, one can measure the signal intensity as a function of the gradient amplitude G. This type of measurements, referred to as displacement or q-space imaging, has been applied for flow and diffusion in different porous (bio-)systems (Rajanayagam et al., 1995, Lebon et al., 1996; Packer and Tessier, 1996; Seymour and Callaghan, 1996; Torres et al., 1998;

Van As *et al.*, 1998; Kuchel and Durrant, 1999; Assaf *et al.*, 2000). It should be noted that this full propagator approach requires relatively strong magnetic field gradients. If such strong gradients are not available, these measurements are limited to longer Δ values which in turn require sufficiently long spin relaxation times of in the sample.

In this context, it should be noted that the attenuation of the echo amplitudes obtained in a PFG NMR experiment on a multi-component system like a biofilm (consisting of an extracellular matrix and cells, see Figure 17.2) is due to a super-position of the relaxation-time weighted contributions of the different compo-nents of the sample, i.e.

$$A = \sum_i A_i(TE) \exp(-D_i b \qquad (17.3)$$

where $A_i(TE)$ and D_i denote the contribution of the ith component to the signal amplitude in the absence of a gradient and the diffusion coefficient in the same component, respectively. A simple exponential evaluation of these echo attenua-tion curves at small b-values leads to a relaxation-weighted mean diffusion coef-ficient (van Dusschoten *et al.*, 1995; Nestle *et al.*, 2001). At larger b-values the echo decay for such a system is typically non-exponential. Combined PFG-T_2 measurements have been used to further unravel such multi-component behaviour (van Dusschoten *et al.*, 1996; Lens *et al.*, 1997, 1999; van der Weerd *et al.*, 2002). Recently, new, sophisticated data evaluation methods involving 2D Laplace inver-sions have become available (Song *et al.*, 2002). These allow the extraction of much more detailed information from both relaxation measurements and diffu-sion measurements in complex materials.

17.3.1.2   Applications

Following the pioneering work of Stejskal and Tanner (1965), PFG NMR can be used for measuring diffusion coefficients in systems ranging from unrestricted bulk diffusion in liquids (Callaghan, 1991; Le Bihan, 1991) to the much slower motion of, e.g. sorbed molecules in zeolites (Kärger and Ruthven, 1992) or restricted diffusion in porous media or microorganisms (Le Bihan, 1991). In systems with a combination of free, unrestricted diffusive water and water con-fined in microorganisms, it is possible to discriminate both in a diffusion-weighted NMR experiment, allowing to, e.g. assay the bacteria population in porous media and soils (Potter *et al.*, 1996).

Central to the understanding of diffusion-controlled mass transfer kinetics in porous particles is the effective diffusivity (D_{eff}) of and in the non-flowing (stag-nant) fluid entrained in the intraparticle pore network of the porous biosystem. D_{eff} is defined by Fick's first law:

$$J = -D_{eff} \, dC/dx \qquad (17.4)$$

with J, flux through the biofilm and C, solute concentration in the liquid phase.

Table 17.2   Comparison of diffusion coefficients determined by PFG NMR and glucose microsensors (after Beuling et al., 1998).

Porous system	D_{rel} measured by PFG NMR	D_{rel} measured by glucose microsensor
Agar beads		
1.5% w/v	0.96	0.96
3.0% w/v	0.93	0.92
4.0% w/v	0.89	0.90
Agar beads (1.5% w/v) supplemented with polystyrene particles		
8% (homogeneous)	0.99	0.91
8% (inhomogeneous)	0.97	0.90
15% (homogeneous)	0.88	0.87
20% (homogeneous)	0.87	0.86
20% (inhomogeneous)	0.84	0.80

Data are expressed as a relative diffusion coefficient (D_{rel}), the ratio of the measured diffusion coefficient in the beads D over the diffusion coefficient in free water at the same temperature D_{aq}.

The D_{eff} is related to the bulk diffusivity in the free solution (D_{aq}) or the transient diffusivity (D_s) according to (Stewart, 1998):

$$D_{eff} = D_{aq} \left(\varepsilon_{intra}/\tau_{intra} \right) = D_s \, \varepsilon_{intra} \qquad (17.5)$$

with ε_{intra}, the internal porosity of the particle and τ_{intra}, the tortuosity factor.

In general, D_{eff} is the transport parameter relating the diffusive flux into and out of the pores to the morphology (geometry and topology) of the pores. D_{eff} includes surface characteristics (e.g. roughness or chemical modification), the pore size and its distribution, pore shape and pore interconnectivity (Wood et al., 2002). The diffusion coefficient measured by NMR correlates to D_s (Beuling et al., 1998). Diffusion coefficients determined by NMR in both well-defined biofilms and in aggregates from wastewater treatment reactors corresponded well to glucose diffusion coefficients as determined by microsensors (Table 17.2, Figure 17.3). Thus, NMR measurements provide an easy way to determine the diffusivity of biofilms (Beuling et al., 1998).

PFG NMR has been used to investigate anaerobic granular sludge from upflow anaerobic sludge bed (UASB) reactors. PFG NMR can be used to determine the main D_s of 20 ml sludge (Lens et al., 1997, 1999) or to make spatially-resolved maps of D_s with one single aggregate with a resolution up to 80 μm (Gonzalez et al., 2001; Lens et al., 2003). The latter show there is a distribution of both the T_2 and D_s within a single granule (Figure 17.4). Depending on the source of the anaerobic granules, the D_s within the granule matrix is 30–50% lower than D_{aq} (Lens et al., 1997, 2003). Also the effect of deactivation procedures on the diffusional characteristics can be easily evaluated using PFG NMR (Table 17.3).

The effective diffusivity D_{eff} can be calculated from D_s if the porosity is known. PFG NMR in combination with T_2 imaging has been used for single

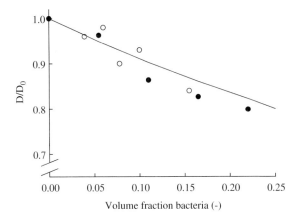

Figure 17.3   Diffusion behaviour of water (measured by PFG NMR, closed circles) and glucose (measured with a microelectrode sensor, open circles) in model biofilms containing various volume fractions of *Micrococcus luteus* (after Beuling *et al.*, 1998). The solid line represents a fit to the model of Fricke (1924).

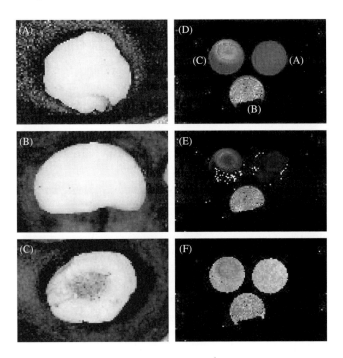

Figure 17.4   NMR images of the spatial variation of ^{1}H NMR parameters in a slice of 2 mm of an insert containing a single bead. Low magnification light microscopic images of cross-sectioned beads: (A) 1% agar, (B) 10% agar and (C) methanogenic aggregate. NMR images: (D) amplitude A, (E) R_2 ($= 1/T_2$) and (F) self-diffusion coefficient D_s. Letters in (D) indicate the different beads. [This figure is also reproduced in colour in the plate section after page 326.]

Table 17.3   Effect of deactivation on T_2 value and D_{rel} ratio (D/D_{aq}) of methanogenic aggregates grown in a full-scale anaerobic bioreactor (Eerbeek, The Netherlands) treating pulp and paper wastewater (after Lens et al., 2003).

Aggregate type and size fraction*	T_2 [ms]	D_{rel} [%]
Intact		
0.5 < d < 1.0	49.3	63
1.0 < d < 1.4	46.3	56
d > 1.4	42.0	64
Deactivated (d > 1.4)		
Glutaraldehyde	27.0	52
HgCl$_2$	45.0	64
Heat treated (70°C)	58.8	87
Heat treated (97°C)	78.9	92

* Size fractions given in mm; d = diameter.

parameter imaging of the same object: amplitude (A), the relaxation time T_2 and the diffusion coefficient (D_s) (van Dusschoten et al., 1996). The amplitude image is directly related to the proton density, i.e. the amount of protons per defined volume. By normalising this amplitude on that of pure water, it directly reflects the porosity if only extracellular water in the biofilm contributes to the observed NMR amplitude. This information has been used to derive fine scale D_{eff} images of microbial mats (Wieland et al., 2001). The apparent water diffusivity obtained in this way by NMR compared well to apparent O_2 diffusivities measured with a diffusivity microsensor (Wieland et al., 2001).

The potential of ^{1}H PFG NMR for flow characterisation, intraparticle diffusivity and mass transfer (exchange) kinetics between the stagnant mobile phase in porous beads and the interparticle void space surrounding these beads has been demonstrated by Tallarek et al. (1996, 1998, 1999). Average propagator measurements by PFG can also be combined with imaging techniques (Callaghan, 1991). The result is an image in which each pixel contains a propagator, which allows to quantify the homogeneity of the flow profile and, thus, the efficiency of hollow fibre membranes (Humbert, 2001), (capillary) chromatography columns (Tallarek et al., 1996, 2001b) and bead packings (Seymour and Callaghan, 1996; Yuen et al., 2002).

PFG NMR is not restricted to water transport, and has also been applied to study transport of oily emulsions and light oil in porous rocks (Mardon et al., 1996) and sediment sludges (Reeves and Chudek, 2001), of non-aqueous phase liquid (NAPL) in soil (Gladden, 1996) as well as water, glycerol and very immobile components in Pseudomonas aeruginosa biofilms (Vogt et al., 2000).

17.3.1.3   Time window and resolution

In general, the spatial resolution is at the expense of the signal-to-noise ratio per pixel. To date, a spatial resolution of up to 10 μm can be obtained for ¹H of water, thus, enabling NMR microscopy of individual plant and mammalian cells (Aiken *et al.*, 1995). However, for biofilm applications, a spatial resolution of about 50–100 μm is more realistic, which depends on the measurement conditions (e.g. applied magnetic field strength) and the sample (e.g. inclusion of paramagnetic ions, susceptibility artefacts induced by entrapped gas pockets). The spatial resolution is better at higher magnetic field strength than at lower fields: typically in the order of $150 \times 150 \times 1500$ μm at a magnetic field strength of 0.47 T and $20 \times 20 \times 200$ μm at 9.4 T for a small coil and a total acquisition time of about 4 min. These examples refer to 2D imaging in combination with a slice selective excitation. By use of 3D imaging, the third dimension can be reduced to the same order as the in-plane resolution in 2D imaging.

For a reliable determination of the propagator, minimal 32 different gradient values need to be measured. It is clear that this takes time. Moreover, a 2D spin echo image consisting of an $N \times N$ picture elements requires N acquisitions to be repeated. Combining this with q-space with 64 gradient steps results in $64 \times N$ acquisitions. Doing so is very time consuming (several hours). Alternatively, q-space displacement imaging has been combined with fast turbo spin–echo (TSE) imaging (Scheenen *et al.*, 2000), resulting in an N/m times faster sequence as compared to a standard $N \times N$ SE image sequence. Here m is the turbo factor, which equals to the number of echoes that can be acquired in one scan and which is defined by the actual T_2 values in the sample. Here low magnetic field NMR may be advantageous because of the longer T_2 values observed in porous (bio)systems as compared to high fields. Alternatively, an even faster line scan sequence has been presented, in which two slice selective rf pulses are used to define a line in the object or system to be measured (van Dusschoten *et al.*, 1997). Within this observation line, the normal one-dimensional (1D) spatial resolution can be obtained. This results in a N times faster sequence as compared to a 2D SE image sequence, but with reduced spatial resolution in one dimension.

Spatial resolution can be improved by increasing the data matrix (e.g. 256×256 in stead of 128×128), but this can only be done by increasing the image acquisition time and a reduction in signal-to-noise ratio of the individual pixels. By using PFG, eventually combined with T_2 measurements, multi-components or -compartments contributing to the signal in a single pixel can be discriminated (Scheenen *et al.*, 2002). Although the spatial resolution in such an image is defined by the length of the field of view over the number of matrix points in the respective direction, the information resolution is higher.

In PFG experiments the time window used to observe the transport phenomena equals the observation time Δ, which is restricted by the relaxation times (Snaar and Van As, 1990; Scheenen *et al.*, 2000). For samples with short relaxation times (i.e. materials with a low water content) or with slow transport processes, the time window available for PFG becomes too short. Then transport has to be measured by following the water intensity in time-controlled sequential images (see 17.3.2.1).

17.3.2   Transport of contrast agents or labelled molecules

17.3.2.1   Principle

A large number of paramagnetic ions exists, which are not directly observable by NMR, but which shorten the T_1 and/or T_2 of the molecules in contact with them. NMR-imaging measurements of dispersion of paramagnetic tracers in heterogeneous media via this relaxation dependence have been published (Kutchovsky *et al.*, 1996; Donker *et al.*, 1997; Chen *et al.*, 2002). The effect of such tracer ions on the relaxation times depends on the nature of the ion and on its (local) concentration. To become visible, the paramagnetic tracer has to affect the observed relaxation time. The impact of a dissolved paramagnetic tracer on the relaxation times of the protons in a liquid is described by the relaxivity r_M of the respective tracer (Banci *et al.*, 1991):

$$1/T_{obs} = 1/T_{int} + r_{M,i}c_M \qquad (17.6)$$

with T_{obs}, the observed T_1 or T_2; T_{int}, the intrinsic T_1 or T_2 value of the fluid under observation; $r_{M,i}$, the longitudinal or transverse relaxivity of the paramagnetic substance in the liquid of interest (again dependent on the magnetic field strength at which the NMR experiment is conducted, see Banci *et al.*, 1991) and c_M, the concentration of the paramagnetic substance in the fluid.

Image contrast differences in relaxation-weighted images have been used to measure transport of these ions in polysaccharide gels with and without immobilised cells and in algal frond tissues (Nestle and Kimmich, 1996a, b). Quantitative mapping of the concentration distribution of the tracer ions in the same systems was performed by susceptibility mapping MRI (Nestle and Kimmich, 1996c) using a resonance offset imaging sequence (Weis *et al.*, 1989).

Contrast differences in relaxation-weighted images can be well used for quantification of transport if the water content is constant in time. If not, a unique and real quantitative interpretation of the transport of these ions can only be obtained by quantitative single parameter T_1 or T_2 images in combination with A_0 images. Via time series of quantitative T_1 and/or T_2 and A_0 images, the presence, concentration and transport behaviour of paramagnetic tracers can be visualised. Such time series of single parameter images can also be used to characterise slow transport and conversion processes in non-stationary water content situations (Nagel *et al.*, 2002). For reliable data interpretation it is necessary to obtain quantitative spin–density images representing $A_0(r)$, which are not affected by the other parameters like the relaxation times. Quantitative spin–density images can be obtained by use of multi-echo imaging (Donker *et al.*, 1996; Edzes *et al.*, 1998), followed by an extrapolation of the amplitude images to $TE = 0$. In this way, the amplitude images are only influenced by the characteristics of the sensitivity profile of the NMR rf coil (Donker *et al.*, 1997). In samples with very short transverse relaxation times, conventional MRI techniques are not applicable anymore, and one has to switch to specialised imaging methods like single-point imaging (SPI, Szomolanyi, 2001) or strayfield imaging (STRAFI, Glover *et al.*, 1997; Carlton *et al.*, 2000).

17.3.2.2   Applications

Examples of paramagnetic ions that can be used as tracer for this type of NMR studies include: Mn^{2+}, Cu^{2+}, Fe^{2+} and Gd^{3+} (Nestle, 2002). Such 'naked' tracer ions typically are chemically very reactive and can lead to cross-linking of extra-cellular biopolymers or bind to the cellular biomass in many different ways. If one is interested in the passive diffusion of tracers, metal ions chelated by organic matter, e.g. $Mn\text{-}EDTA^{2-}$, Fe-dextran or $Gd\text{-}DTPA^{2-}$ are a better option. Such complexes have been developed for application as medical contrast agents and in their basic form are optimised for minimal long-term binding to biological material (Niendorf *et al.*, 1985; Donahue *et al.*, 1997; Thunus and Lejeune, 1999) and also exhibit not notable interaction with most environmental matrices (Möller *et al.*, 2002). Furthermore, such molecules can be manufactured in a variety of size (e.g. Fe-dextran particles can range from <0.01 to $>10\,\mu m$) or chemically functionalised as for to bind selectively to certain biological structures. Thus, one can study the behaviour of these chelated organic molecules in biofilms as a function of size, molecular weight, charge and chemical nature.

Figure 17.5 gives an example of the penetration of Fe^{2+} into a UASB aggregate. At time zero, water doped with $FeCl_2$ is added to the aggregate. By applying

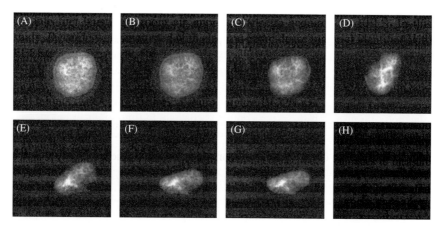

Figure 17.5   Temporal evolution of an individual aggregate from a UASB reactor treating paper mill wastewater (Eerbeek, The Netherlands) after the addition of Fe (100 mm) as manifest in fast TSE imaging; (A) prior to addition, (B) addition, $t = 0$. Next images corresponding to (C) +10 min, (D) +20 min, (E) +30 min, (F) +40 min, (G) +50 min and y (H) + 1 h. The effect of Fe is to reduce the signal intensity (ranging from black (zero), dark green, light green to white (max)) observed in these images. A single slice out of a complete 3D data set is presented (voxel size). [This figure is also reproduced in colour in the plate section after page 326.]

TSE imaging and rather short repetition times only protons with short relaxation times appear as bright signals. After some time, Fe^{2+} ions have penetrated into the aggregate, where they shorten the T_2 values of the water, resulting in darker signals. In this way, the penetration profile of paramagnetic metals in methanogenic aggregates can be followed in time (Osuna *et al.*, 2003).

17.3.2.3   *Time window and resolution*

The time resolution for this type of measurements is given by the time to acquire the data for one image, which typically range for water from some seconds to several minutes. The time window over which transport can be followed is hardly limited, because the sample under study can be measured repeatedly, over a period covering weeks or months.

17.4   NMR FOR BIOFILM REACTOR ANALYSIS

NMR procedures, used in combination with imaging and localised spectroscopy techniques, can be used to unravel the flow and transport processes in biofilm systems and enable the monitoring of growth and distribution of cells within the biofilm reactor (Table 17.4). When coupling these non-invasive ^{1}H NMR measurements to ^{13}C, ^{31}P or ^{23}Na chemical shift selective NMR, this type of NMR also allows to investigate metabolic (reaction rate) heterogeneity within a reactor and can assist in future design of biofilm reactor systems. Recent studies on the spatial distribution of conversion processes in abiotic reactor systems clearly show the potential of imaging NMR techniques in this field (Yuen *et al.*, 2002). These can than be further used in process monitoring and control (Arola *et al.*, 1997).

Until recently, 'seeing inside an object' was based on the ability to detect differential interaction of body tissues with different energy sources. The information resulting from these interactions provides 2D (e.g. X-rays) or 3D (e.g. MRI) images detailing the different densities or water content of internal tissues or structures (Figure 17.6). These data, however, are a global assessment of tissue characteristics and do not contain detailed information regarding the molecular changes underlying the image, e.g. changes in gene expression that result in altered biology. In molecular imaging, also NMR is increasingly used to image patterns of gene expression (Högemann and Basilion, 2002). A variety of magnetic resonance contrast agents have recently been developed for molecular imaging. For targets that are less abundant, strategies to NMR imaging of gene expression include static or dynamic probe accumulation, endogenous synthesis of contrast molecules or *in situ* contrast agent activation (Högemann and Basilion, 2002).

A number of parameters which play a role in a NMR-imaging experiment are temperature dependent. Magnetic resonance temperature-imaging methods, based on the temperature dependence of the molecular diffusion D (Kärger and Ruthven, 1992), the spin–spin relaxation time T_1 (Young, 1994) and the resonance

Table 17.4   Bioreactor designs used for the *in vivo* NMR of high cell densities and biofilms.

Organisms	Bioreactor type	NMR technique	Reference
I. Flat biofilms			
Pseudomonsa aeruginosa	Flow cell	^{1}H NMR	Lewandowski *et al.* (1993)
Pseudomonsa aeruginosa	Flow cell	^{31}P NMR ^{13}C NMR	Vogt *et al.* (2000); Mayer *et al.* (2001)
E. coli, Sacharomyces	Hollow fiber	^{23}Na MRI	DiBiasio *et al.* (1993)
E. coli	Planar channel	^{1}H relaxation MRI	Hoskins *et al.* (1999)
II. Spherical biofilms			
Anaerobic granular Sludge	UASB reactor	^{1}H T_2 relaxation ^{1}H MRI	Lens *et al.* (1997, 1999); Gonzalez-Gil *et al.* (2001); Osuna *et al.* (2003)
Aspergillus terreus	Airlift bioreactor	^{31}P NMR	Lyngstad and Grasdalen (1993); Melvin and Shanks (1996)
Zymomonas mobilis; *Corynebacterium glutamicum*	Membrane cyclone	^{31}P and ^{13}C NMR	Hartbrich *et al.* (1996)
Zymomonas mobilis	Continuous flow	^{31}P NMR	De Graaf *et al.* (1992)
E. coli	Glass bead supported biofilm	^{1}H relaxation MRI	Hoskins *et al.* (1999)

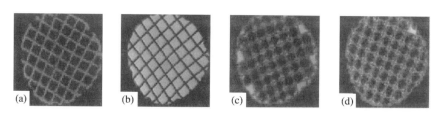

Figure 17.6   NMR imaging of water transport processes in monolith structures: (a) wetted walls, (b) water-filled channels and water film flow with (c) a bad and (d) good distribution over the channels. [This figure is also reproduced in colour in the plate section after page 326.]

frequency of water protons (Peters *et al.*, 1999) have been presented. At present, this NMR application is used in medicine to monitor temperature changes in real time during tissue heating (hyperthermia, laser surgery, focused ultra-sounds) or cooling (cryotherapy). In the context of biofilm systems, temperature mapping can assist in the design of heat-pulsed or temperature-phased bioreactors.

   Another promising evolution is the development of portable magnets, which allow *on-site* measurements of single NMR parameters A_0, T_1, T_2 (Blümich *et al.*, 1998; Meriles *et al.*, 2002) or imaging of materials (Prado *et al.*, 2000). Thus, it has become possible to measure directly subsurface geological formations *in situ* by dedicated NMR instruments (Kleinberg *et al.*, 1992; Mardon *et al.*, 1996). One can install these devices also at biofilm and membrane reactors or at composting or solid-state fermentation reactors, where they provide on-line information about transport processes occurring in biofilms and membranes or about the water content of the substrate during solid-state fermentation.

Optimisation of bioreactor performance using NMR experiments can benefit from the coupling of NMR with other analytical techniques. Optical methods are especially promising for such combined experiments as a fibre-optical system and NMR devices do not interfere with each other from the physical point of view. Fibre-optical systems can also easily be fitted into the geometrically restricted space of rf coils, gradient coils and magnet system used for NMR experiments. Conventional-optical microscopy and NMR imaging have already been realised simultaneously (Glover *et al.*, 1993). Also the combination of photoacoustics (Schmid *et al.*, 2002; see also Chapter 23 in this book) and NMR seems to be an interesting approach. Photoacoustics provide fast depth-resolved information on a certain point in a planar biofilm, while the spatial homogeneity in lateral dimensions can be monitored using MRI.

A recent approach to characterise the pore distributions in the (solid) matrix of porous (bio)media is based on the combination of cryoporometry and (NMR) relaxometry (Valckenborg *et al.*, 2002). Since both the temperature at which water in pores become frozen and the NMR relaxation times are strongly dependent on the pore sizes and frozen water has a very short T_2, relaxation time measurements as a function of sample temperature result in a direct correlation between pore size and NMR relaxation times. An alternative, which also gives insight in pore connectivity without the need to freeze the sample, is based on the measurement of diffusion through local intrinsic magnetic field gradients originating from susceptibility differences on the observed relaxation time (Song *et al.*, 2000; Chen and Song, 2002). In this way, multiple length scales, relevant for transport processes, can be determined.

17.5   CONCLUSION

NMR methods provide a range of unique possibilities for non-invasive studies of metabolic and transport processes in biofilms on various length-scales ranging from the micrometre scale to whole bioreactors. This chapter illustrated how different spectroscopic and imaging NMR methods provide valuable insights into transport processes in spherical anaerobic biofilms. Thus, they can contribute to the optimisation of bioreactor operation.

Acknowledgements

The authors thank Dr. N. Nestle for critical discussion of this manuscript.

REFERENCES

Aiken, N.R., Hsu, E.W. and Blackband, S.J. (1995) A review of NMR microimaging studies of single cells. *J. Magn. Reson. Anal.* **1**, 41–48.

Arola, D.F., Barrall, G.A., Powell, R.L., McCarthy, K.L. and McCarthy, M.J. (1997) Use of nuclear magnetic resonance imaging as a viscometer for process monitoring. *Chem. Eng. Sci.* **52**, 2049–2057.

Assaf, Y., Mayk, A., Cohen, Y. (2000) Displacement imaging of spinal cord using q-space diffusion-weighted MRI. *Magnet. Reson. Med.* **44**, 713–722.

Banci, L., Bertini, I. and Luchinat, C. (1991) *Nuclear and Electron Relaxation*, VCH-Verlag, Weinheim/Bergstraße.

Baumann, T., Petsch, R., Fesl, G. and Niessner, R. (2002) Flow and diffusion measurements in natural porous media using magnetic resonance imaging. *J. Environ. Qual.* **31**, 470–476.

Bear, J. and Corapcioglu, M.Y. (1984) *Fundamentals of Transport Phenomena in Porous Media*, Martinus Nijhoff Publ., Dordrecht.

Beuling, E.E., van Dusschoten, D., Lens, P., van den Heuvel, J.C., Van As, H. and Ottengraf, S.P.P. (1998) Characterization of the diffusive properties of biofilms using pulsed field gradient nuclear magnetic resonance. *Biotechnol. Bioeng.* **60**, 283–291.

Blinc, R., Burgar, M., Lahajnar, G., Rozmarin, M., Rutar, V., Kocuvan, I. and Ursic, J. (1978) NMR relaxation study of adsorbed water in cement and C3S pastes *J. Am. Ceram. Soc.* **61**, 35–37.

Blümich, B. (2000) *NMR Imaging of Materials*, Clarendon Press, Oxford.

Blümich, B., Blümler, P., Eidmann, G., Guthausen, A., Haken, R., Schmitz, U., Saito, K. and Zimmer, G. (1998) The NMR-MOUSE: construction, excitation, and applications. *Magn. Reson. Imaging* **16**, 479–484.

Britton, M.M. and Callaghan, P.T. (1997) Nuclear magnetic resonance visualization of anomalous flow in cone-and-plate rheometry. *J. Rheol.* **41**, 1365–1386.

Bryar, T.R., Doughney, C.J. and Knight, R.J. (2000) Paramagnetic effects of iron(III) species on nuclear magnetic relaxation of fluid protons in porous media *J. Magn. Reson.* **142**, 74–85.

Bryers, J.D. and Drummond, F. (1998) Local macromolecule diffusion coefficients in structurally non-uniform bacterial biofilms using fluorescence recovery after photobleaching (FRAP). *Biotechnol. Bioeng.* **60**, 462–473.

Callaghan, P.T. (1991) *Principles of Nuclear Magnetic Resonance Microscopy*, Clarendon Press, Oxford.

Carlton, K.J., Halse, M.R. and Strange, J.H. (2000) Diffusion-weighted imaging of bacteria colonies in the STRAFI plane. *J. Magn. Reson.* **143**, 24–29.

Caprihan, A. and Fukushima, E. (1990) Flow measurements by NMR. *Phys. Rep.* (Review Section on Physics Letters) **198**, 195–235.

Chaouki, J., Larachi, F. and Dudukovic, M.P. (1997) Noninvasive tomographic and velocimetric monitoring of multiphase flows. *Ind. Eng. Chem. Res.* **36**, 4476–4503.

Chen, Q. and Song, Y.-Q. (2002) What is the shape of pores in natural rocks? *J. Chem. Phys.* **116**, 8247–8250.

Chen, Q., Kinzelbach, W. and Oswald, S. (2002) Nuclear magnetic resonance imaging for studies of flow and transport in porous media. *J. Environ. Qual.* **31**, 477–486.

de Beer, D., Stoodley, P. and Lewandowski, Z. (1997) Measurement of local diffusion coefficients in biofilms by microinjection and confocal microscopy. *Biotechnol. Bioeng.* **53**, 151–158.

De Graaf, A.A., Wittig, R.M., Probst, U., Strohhaecker, J., Schoberth, S.M. and Sahm, H. (1992) Continuous-flow NMR bioreactor for *in vivo* studies of microbial cell suspensions with low biomass concentrations. *J. Magn. Reson.* **98**, 654–659.

DiBiasio, D., Scott, J., Harris, P. and Moore, S. (1993) The relationship between fluid flow and cell growth in hollow-fiber bioreactors: applications of magnetic resonance imaging. *BHR Group Conf. Ser. Publ.* **5**, 457–473.

Donker, H.C.W., Van As, H., Edzes, H.T. and Jans, A.H.W. (1996) NMR imaging of white button mushroom (*Agaricus bisporus*) at various magnetic fields. *Magn. Reson. Imaging* **14**, 1205–1215.

Donker, H.C.W., Van As, H., Snijder, H.J. and Edzes, H.T. (1997) Quantitative ^{1}H-NMR imaging of water in white button mushrooms. (*Agaricus bisporus*). *Magn. Reson. Imaging* **15**, 113–121.

Donahue, K.M., Weisskoff, R.M., Burstein, D. (1997) Water diffusion and exchange as they influence contrast enhancement. *J. Magn. Reson. Imag.* **7**, 102–110.

Edzes, H.T., van Dusschoten, D. and Van As, H. (1998) Quantitative T_2 imaging in plant tissues by means of multi-echo MRI microscopy. *Magn. Reson. Imaging* **16**, 185–196.

Fricke, H. (1924) A mathematical treatment of the electric conductivity and capacity of disperse systems. *Phys. Rev.* **24**, 575–587.

Gladden, L.F. (1996) Structure–transport relationships in porous media. *Magn. Reson. Imaging* **14**, 719–726.

Glover, P.M., Bowtell, R.W., Brown, G.D. and Mansfield, P. (1993) A microscope slide probe for high-resolution imaging at 11.7, T. *Magnet. Reson. Med.* **31**, 423–428.

Glover P.M., McDonald, P.J. and Newling, B. (1997) Stray-field imaging of planar films using a novel surface coil. *J. Magn. Reson.* **126**, 207–212.

Gonzalez, G., Lens, P., Van Aelst, A., Van As, H., Versprille, A.I. and Lettinga, G. (2001) Cluster structure of anaerobic aggregates of an expanded granular sludge bed reactor. *Appl. Environ. Microbiol.* **67**, 3683–3692.

Hahn, E. (1950) Spin Echoes. *Phys. Rev.* **80**, 580–594.

Hartbrich, A., Schmitz, G., Weuster-Botz, D., de Graaf, A.A. and Wandrey, C. (1996) Development and application of a membrane cyclone reactor for *in vivo* NMR spectroscopy with high microbial cell densities. *Biotechnol. Bioeng.* **51**, 624–635.

Högemann, D. and Basilion, J.P. (2002) 'Seeing inside the body': MR imaging of gene expression. *Eur. J. Nucl. Med.* **29**, 400–408.

Hoskins, B.C., Fevang, L., Majors, P.D., Sharma, M.M. and Georgious, G. (1999) Selective imaging of biofilms in porous media by NMR relaxation. *J. Magn. Reson.* **139**, 67–73.

Howard, J.J. (1994) Wettability and fluid saturations determined from NMR T_1 distributions. *Magn. Reson. Imaging* **12**, 197–200.

Humbert, F. (2001) Potentials of radio-frequency field gradient NMR microscopy in environmental science. *J. Ind. Microbiol. Biotechnol.* **26**, 53–61.

Issa, B. and Mansfield, P. (1994) Permeability estimation from T_1 mapping. *Magn. Reson. Imaging* **12**, 213–214.

Kärger, J. and Ruthven, D.M. (1992) *Diffusion in Zeolites and Other Microporous Solids,* John Wiley, New York.

Kenyon, W.E. (1992) Nuclear magnetic resonance as a petrophysical measurement. *Nucl. Geophys.* **6**, 153–171.

Kleinberg, R.L., Sezginer, A., Griffin, D.D. and Fukuhara, M. (1992) Novel NMR apparatus for investigating an external sample. *J. Magn. Reson.* **97**, 466–485.

Kleinberg, R.L., Kenyon, W.E. and Mitra, P.P. (1994) Mechanism of NMR relaxation of fluids in rocks. *J. Magn. Reson. A* **108**, 206–214.

Kuchel, P.W. and Durrant, C.J. (1999) Permeability coefficients from NMR q-space data: models with unevenly spaced semi-permeable parallel membranes. *J. Magn. Reson.* **139**, 258–272.

Kühl, M. and Revsbech, N.P. (2000) Microsensors for the study of interfacial biogeochemical processes. In *The Benthic Boundary Layer: Transport Processes and Biogeochemistry* (ed. Boudreau, B and Jørgensen, B.B.), Oxford Press, in press.

Kutchovsky, Y.E., Alvarado, V., Davis, H.T. and Scriven, L.E. (1996) Dispersion of paramagnetic tracers in bead packs by T_1 mapping: experiments and simulations. *Magn. Reson. Imaging* **14**, 833–840.

La Heij, E.J., Kerkhof, P.J.A.M., Kopinga, K. and Pel, L. (1996) Determining porosity profiles during filtration and expression of sewage sludge by NMR imaging. *AICHE J.* **42**, 953–959.

Le Bihan, D. (1991) Molecular diffusion nuclear magnetic resonance imaging. *Magn. Reson. Quart.* **7**, 1–30.

Le Bihan, D., Mangin, J.F., Poupon, C., Clark, C.A., Pappata, S., Molko, N. and Chabriat, H. (2001) Diffusion tensor imaging: concepts and applications. *J. Magn. Reson. Imaging* **13**, 534–546.

Lebon, L., Oger, L., Leblond, J., Hulin, J.P., Martys, N.S. and Schwartz, L.M. (1996) Pulsed gradient NMR measurements and numerical simulation of flow velocity distribution in sphere packings. *Phys. Fluid.* **8**, 293–301.

Lens, P.N.L. and Hemminga, M.A. (1998) Nuclear magnetic resonance in environmental engineering: principles and applications. *Biodegradation* **9**, 393–409.

Lens, P., de Beer, D., Cronenberg, C., Houwen, F., Ottengraf, S. and Verstraete, W. (1993) Inhomogenic distribution of microbial activity in UASB aggregates: pH and glucose microprofiles. *Appl. Environ. Microbiol.* **59**, 3803–3815.

Lens, P., Hulshoff Pol, L., Lettinga, G. and Van As, H. (1997) Use of ^{1}H NMR to study transport processes in sulfidogenic granular sludge. *Water Sci. Technol.* **36**(6–7), 157–163.

Lens, P., Vergeldt, F., Lettinga, G. and Van As, H. (1999) ^{1}H-NMR study of the diffusional properties of methanogenic aggregates. *Water. Sci. Technol.* **39**(7), 187–194.

Lens, P.N.L., Gastesi, R., Vergeldt, F., Pisabarro, G. and Van As, H. (2003) Diffusional properties of methanogenic granular sludge: ^{1}H-NMR characterisation. *Appl. Environ. Microbiol*, in press.

Lewandowski, Z., Altobelli, S.A. and Fukushima, E. (1993) NMR and microelectrode studies of hydrodynamics and kinetics in biofilms. *Biotechnol. Prog.* **9**, 40–45.

Lyngstad, M. and Grasdalen, H. (1993) A new NMR airlift bioreactor used in phosphorus 31 NMR studies of itaconic acid producing *Aspergillus terreus*. *J. Biochem. Biophy. Meth.* **27**, 105–116.

Mansfield, P. and Issa, B. (1994) Studies of fluid transport in porous rocks by echo-planar MRI. *Magn. Reson. Imaging* **12**, 275–278.

Mardon, D., Prammer, M.G. and Coates, G.R. (1996) Characterization of light hydrocarbon reservoirs by gradient-NMR well logging. *Magn. Reson. Imaging* **14**, 769–777.

Mayer, C., Lattner, D. and Schürks, N. (2001) ^{13}C nuclear magnetic resonance studies on selectively labeled bacterial biofilms. *J. Ind. Microbiol. Biotechnol.* **26**, 62–69.

Meriles, C.A., Sakellariou, D., Heise, H., Moule, A.J. and Pines, A. (2002) Approach to high-resolution ex situ NMR spectroscopy. *Science* **293**, 82–85.

Melvin, B.K. and Shanks, J.V. (1996) Influence of aeration on cytoplasmic pH of yeast in an NMR airlift bioreactor. *Biotechnol. Prog.* **12**, 257–265.

Möller, P., Paces, T., Dulski, P. and Morteani, G. (2002) Anthropogenic Gd in surface water, drainage system, and the water supply of the city of Prague, Czech Republic. *Environ. Sci. Technol.* **36**, 2387–2394.

Nagel, F.J., Rinzema, A., Tramper J. and Van As, H. (2002) Water and glucose gradients in the substrate measured with NMR imaging during solid-state fermentation with *Aspergillus oryzae*. *Biotechnol. Bioeng.* **79**, 653–663.

Nestle, N. (2002) NMR studies on heavy metal immobilization in biosorbents and other matrices. *Rev. Environ. Sci. Biotech.* **1**, 215–225.

Nestle, N. and Kimmich, R. (1996a) NMR imaging of heavy metal absorption in alginate, immobilized yeast cells and kombu algal biosorbents. *Biotechnol. Bioeng.* **51**, 538–543.

Nestle, N. and Kimmich, R. (1996b) NMR microscopy of heavy metal absorption in calcium. *Appl. Biochem. Biotechnol.* **56**, 9–17.

Nestle, N. and Kimmich, R. (1996c) Susceptibility NMR microimaging of heavy metal uptake in alginate biosorbents. *Magn. Reson. Imaging* **14**, 905–906.

Nestle, N., Galvosas, P., and Kärger, J. (2001) Direct measurement of water self-diffusion in hardening blast furnace slag cement pastes by means of nuclear magnetic resonance techniques. *J. Appl. Phys.* **90**, 518–520.

Niendorf, H.P., Felix, R., Laniado, M., Schörner, W., Claussen, C. and Weinmann, H.J. (1985) Gadolinium-DTPA: a new contrast agent for magnetic resonance imaging. *Radiat. Med.* **3**, 7–12.

Osuna, B., Gerkema, E., Isa J., Van As, H. and Lens, P.N.L. (2003) Heavy metal transport in methanogenic aggregates: ^{1}H-NMR characterisation. *Appl. Environ. Microbiol.*, in preparation.

Packer, K.J. and Tessier, J.J. (1996) The characterization of fluid transport in a porous solid by pulsed gradient stimulated echo NMR. *Mol. Phys.* **87**, 267–272.

Peters, R.D., Hinks, R.S. and Henkelman, R.M. (1999) Heat-source orientation and geometry dependence in proton-resonance frequency shift magnetic resonance themometry. *Magnet. Reson. Med.* **41**, 909–918.

Potter, K., Kleinberg, R.L., Brockman, F.J. and McFarland, E.W. (1996) Assay for bacteria in porous media by diffusion-weighted NMR. *J. Magn. Reson. Series B* **113**, 9–15.

Prado P.J., Blumich, B. and Schmitz, U. (2000) One-dimensional imaging with a palm-size probe. *J. Magn. Reson.* **144**, 200–206.

Rajanayagam, V., Yao, S. and Pope, J. (1995) Quantitative magnetic resonance flow and diffusion imaging in porous media. *Magn. Reson. Imaging* **13**, 729–738.

Reeves, A.D. and Chudek, J.A. (2001) Nuclear magnetic resonance imaging (MRI) of diesel oil migration in estiarine sediment samples. *J. Ind. Microbiol. Biotechnol.* **26**, 77–82

Revsbech, N.P., Nielsen, L.P. and Ramsing, N.B. (1998) A novel microsensor for the determination of apparent diffusivity in sediments. *Limnol. Oceanogr.* **43**, 986–992.

Santegoeds, C.M., Damgaard, L.R., Hesselink, G., Zopfi, J., Lens, P., Muyzer, G. and de Beer, D. (1999) Distribution of sulfate reducing and methanogenic bacteria in UASB aggregates determined by microsensors and molecular techniques. *Appl. Environ. Microbiol.* **65**, 4618–4629.

Schaafsma, T.J., Van As, H., Palstra, W.D., Snaar, J.E.M. and de Jager, P.A. (1992) Quantitative measurement and imaging of transport processes in plants and porous media by ^{1}H NMR. *Magn. Reson. Imaging* **10**, 827–836.

Scheenen, T.W.J., van Dusschoten, D., de Jager, P.A. and Van As, H. (2000) Microscopic displacement imaging with pulsed field gradient turbo spin–echo NMR. *J. Magn. Reson.* **142**, 207–215.

Scheenen, T.W.J., Heemskerk, A., de Jager, P.A., Vergeldt, F. and Van As, H. (2002) Functional imaging of plants: a NMR study of a cucumber plant. *Biophys. J.* **82**, 481–492

Schmid, T., Panne, U., Haisch, C., Hausner, M. and Niessner, R. (2002) A photoacoustic technique for depth-resolved *in situ* monitoring of biofilms. *Environ. Sci. Technol.* **36**, 4135–4141.

Seymour, J.D. and Callaghan, P.T. (1996) Flow-diffraction structural characterization and measurement of hydrodynamic dispersion in porous media by PGSE NMR. *J. Magn. Reson. Series A* **122**, 90–93.

Snaar, J.E.M. and Van As, H. (1990) Discrimination of different types of motion by modified stimulated-echo NMR. *J. Magn. Reson.* **87**, 132–140.

Song, Y.Q., Ryu, S. and Sen, B. (2000) Determining multiple length scales in rocks. *Nature* **406**, 178–181.

Song, Y.Q., Venkataramanan, L. Hürlimann, M.D., Flaum, M., Frulla, P. and Straley, C. (2002) T_1–T_2-correlation spectra obtained using a fast two-dimensional Laplace inversion. *J. Magn. Reson.* **154**, 261–268.

Stallmach, F. and Kärger, J. (1999) The potentials of pulsed field gradient NMR for investigation of porous media. *Adsorption* **5**, 117–133.

Stejskal, E.O. and Tanner, J.E. (1965) Spin diffusion measurements: spin echoes in the presence of a time dependent field gradient. *J. Chem. Phys.* **42**, 288–293.

Stewart, P.S. (1998) A review of experimental measurements of effective diffusive permeabilities and effective diffusion coefficients in biofilms. *Biotechnol. Bioeng.* **59**, 261–272.

Stilbs, P. (1987) Fourier transform pulsed-gradient spin–echo studies of molecular diffusion. *Prog. Nucl. Mag. Res. Spectrosc.* **19**, 1–45.

Szomolanyi, P., Goodyear, D., Balcom, B. and Matheson, D. (2001) SPIRAL-SPRITE: a rapid single-point MRI technique for application to porous media. *Magn. Reson. Imaging* **19**, 423–428.

Tallarek, U., Albert, K., Bayer, E. and Guiochon, G. (1996) Measurement of transverse and axial apparent dispersion coefficients in packed beds. *AICHE J.* **42**, 3041–3054.

Tallarek, U., van Dusschoten, D., Van As, H., Guiochon, G. and Bayer, E. (1998) Direct observation of fluid mass transfer resistance in porous media by NMR spectroscopy. *Angew. Chem. Int. Edit.* **37**, 1882–1885.

Tallarek, U., Vergeldt, F.J. and Van As, H. (1999) Stagnant mobile phase mass transfer in chromatographic media: intraparticle diffusion and exchange kinetics. *J. Phys. Chem. B* **103**, 7654–7664.

Tallarek, U., Rapp, E., Scheenen, T., Bayer, E. and Van As, H. (2000) Electroosmotic and pressure-driven flow in open and packed capillaries: velocity distributions and fluid dispersion. *Anal. Chem.* **72**, 2292–2301

Tallarek, U., Rapp, E., Van As, H. and Bayer, E. (2001a) On the electrokinetics in fixed beds: experimental demonstration of electro-osmotic perfusion. *Angew. Chem. Int. Edit.* **40**, 1684–1687.

Tallarek, U., Scheenen, T.W.J. and Van As, H. (2001b) Macroscopic flow heterogeneity in electro-osmotic and pressure-driven flows through fixed beds at low column-to-particle diameter ratio. *J. Phys. Chem. B.* **105**, 8591–8599.

Thunus, L. and Lejeune, R. (1999) Overview of transition metal and lanthanide complexes as diagnostic tools. *Coordin. Chem. Rev.* **184**, 125–155.

Torres, A.M., Michniewicz, R.J., Chapman, B.E., Young, G.A.R and Kuchel, P.W. (1998) Characterisation of erythrocyte shapes and sizes by NMR diffusion–diffraction of water: correlations with electron micrographs. *Magn. Reson. Imaging* **16**, 423–434.

Valckenborg, R.M.E., Pel, L. and Kopinga, K. (2002) Combined NMR cryoporometry and relaxometry. *J. Phys. D Appl. Phys.* **35**, 249–256.

Van As, H. and Lens, P.N.L. (2001) Use of ^{1}H NMR to measure transport processes in porous biosystems. *J. Ind. Microbiol. Biotechnol.* **26**, 43–-52.

Van As, H. and van Dusschoten, D. (1997) NMR methods for imaging of transport processes in microporous systems. *Geoderma* **80**, 389–403.

Van As, H., Palstra, W., Tallarek, U. and van Dusschoten, D. (1998) Flow and transport studies in (non)consolidated porous (bio)systems consisting of solid or porous beads by PFG NMR. *Magn. Reson. Imaging* **16**, 569–573.

van der Weerd, L., Melnikov, S.M., Vergeldt, F.J., Novikov, E.G. and Van As, H. (2002) Modelling of self-diffusion and relaxation time NMR in multi-compartment systems with cylindrical geometry. *J. Magn. Reson.* **156**, 213–221.

van Dusschoten, D., de Jager, P.A. and Van As, H. (1995) Flexible PFG NMR desensitezed for susceptibility artifacts using the PFG-multiple-spin–echo sequence, *J. Magn. Reson. A* **112**, 237–240.

van Dusschoten, D., Moonen, C.T.W., de Jager, P.A. and Van As, H. (1996) Unravelling diffusion constants in biological tissue by combining CPMG imaging and pulsed field gradient NMR. *Magnet. Reson. Med.* **36**, 907–913.

van Dusschoten, D., van Noort, J. and Van As, H. (1997) Displacement imaging in porous media using the linescan NMR technique. *Geoderma* **80**, 405–416.

Vogt, M., Flemming, H.-C. and Veeman, W.S. (2000) Diffusion in *Pseudomonas aeruginosa* biofilms: a pulsed field gradient NMR study. *J. Biotechnol.* **77**, 137–146.

Weis, J., Frollo, I. and Budins'y, L. (1989) Magnetic field distribution measurement by the modified flash method. *Z. Naturforsch.* **44a**, 1151–1154.

Wieland, A., de Beer, D., Damgaard, L.R., Kuhl, M., van Dusschoten, D. and Van As, H. (2001) Fine-scale measurements of diffusivity in a microbial mat with NMR imaging. *Limnol. Oceanogr.* **46**, 248–259.

Wilderer, P.A., Bungartz, H.-J., Lemmer, H., Wagner, M., Keller, J. and Wuertz, S. (2002) Modern scientific methods and their potential in wastewater science and technology. *Water Res.* **36**, 370–393.

Wood, B.D., Quintard, M. and Whitaker, S. (2002) Calculation of effective diffusivities for biofilms and tissues. *Biotechnol. Bioeng.* **77**, 495–744.

Yang, S. and Lewandowski, Z. (1995) Measurement of local mass transfer coefficient in biofilms. *Biotechnol. Bioeng.* **48**, 737–744.

Young, I.R. (1994) Modeling and observation of temperature changes *in vivo* using MRI. *Magn. Reson. Med.* **32**, 358–369.

Yuen, E.H.L., Sederman, A.J. and Gladden, L.F. (2002) *In situ* magnetic resonance visualization of the spatial variation of catalytic conversion within a fixed-bed reactor. *Appl. Catal. A* **232**, 29–38.

Zhang, T.C., Fu, Y.-C. and Bishop, P.L. (1995) Competition for substrate and space in biofilms. *Water Environ. Res.* **67**, 992–1003.

18

Screening of lectins for staining lectin-specific glycoconjugates in the EPS of biofilms

C. Staudt, H. Horn, D.C. Hempel and T.R. Neu

18.1   INTRODUCTION

Extracellular polymeric substances (EPS) as multi-functional components in biofilms remain to be a serious challenge with respect to their visualisation and characterisation (Cooksey, 1992; Neu, 1994; Wolfaardt *et al.*, 1999; Sutherland 2001). This challenge already starts with their definition as EPS are considered to be a mixture of polysaccharides, proteins, nucleic acids and amphiphilic polymeric compounds produced by procaryotic and eucaryotic cells in biofilms (Neu, 1996; Wingender *et al.*, 1999). Consequently, there is no single technique available to characterise them. Another challenge is based on the huge structural variety of the different polymers. This is, especially, true for polysaccharides (Sharon and Lis, 1989). From these facts, the basic question derived is: How many different EPS polymers do really exist? The number of described species for procaryotes is approximately 4800 strains. If this equals 0.1% of procaryotes actually present in nature, about 4,800,000 species have to be considered (Amann, 2000). From this follows another question: Does this equal the number of possible EPS compounds? If we, for example, consider that each species produces and excretes only one type of polysaccharide and one type of protein, the answer would be about 9,600,000 EPS compounds. This is a conservative calculation, as most microorganisms are capable of producing several different extracellular polymers. Furthermore, in this short calculation, the number of possible EPS compounds produced by algae and fungi are completely neglected. Presently, we are far away from analysing this enormous variety of roughly 10 million polymers, which comprise a significant part of complex microbial biofilms.

The classical approach to analyse EPS is by means of wet chemistry. This approach was mainly used to investigate the composition of activated sludge. However, already the very first step, extraction of the polymers from the biofilm is critical, especially, if environmental biofilms have to be characterised (Nielsen and Jahn, 1999). The subsequent analysis is still difficult, as different polymer classes have to be analysed by different analytical techniques. Mostly, a bulk analysis was performed to address the major components, e.g. proteins, polysaccharides and lipids (see references in Nielsen and Jahn, 1999).

Due to the problems to extract and analyse EPS by means of wet chemistry, other *in situ* approaches have been suggested. These include FT-IR spectroscopy, suitable for thin layers only, such as conditioning films (Jolley *et al.*, 1989; Schmitt *et al.*, 1995); NMR spectroscopy, a new potent technique for EPS analysis, which is just emerging (see special issues of Hemminga and Visser, 2000; Lens and Hemmings, 2001); and fluorescence-based techniques (Neu and Lawrence, 1999a, b). The latter of these approaches ideally employs laser-scanning microscopy (LSM) in combination with specific probes for EPS compounds in biofilms. LSM has been described in detail as the most suitable and versatile tool to investigate microbial biofilms (Lawrence *et al.*, 1998a, 2002; Lawrence and Neu, 1999; Neu and Lawrence, 2002).

18.2   LECTINS

Due to the fact that there is neither a general stain for EPS compounds nor a fluorescent stain for all types of polysaccharides, the approach using lectins seems to be most appropriate. Lectins are non-enzymatic and non-immunogenic proteins with a specificity for carbohydrates (see introduction in Neu and Lawrence, 1999a). As most lectins are hazardous compounds, precautions for a safe application have been provided (Neu and Lawrence, 1999a). They are usually employed in cell biology studies to characterise specific glycoconjugates on cell surfaces. Due to their specificity for carbohydrates, they can also be employed to stain the lectin-specific EPS compounds in biofilms. Lectins have been used in microbiology to stain bacterial surface glycoconjugates in pure cultures and the extracellular matrix in environmental biofilms (see Neu *et al.*, 2001 for references).

The application of lectin-binding analysis, a critical evaluation of the approach and its possible combination with other techniques have been described elsewhere (Neu and Lawrence, 1999a, b; Neu *et al.*, 2001; Böckelmann *et al.*, 2002;

Neu and Lawrence, 2002). The advantages of lectin-binding analysis can be sum-marised as follows:

(1) *in situ* applicable,
(2) non-destructive,
(3) visualisation of glycoconjugates,
(4) characterisation of glycoconjugates,
(5) quantification using digital image analysis
(6) combination with other fluorescent probes.

On the one hand, there is the advantage of simultaneous visualisation and characterisation of glycoconjugates due to the lectin specificity. On the other hand, there is the disadvantage of only staining the lectin-specific glycoconjugate fraction of the EPS. Nevertheless, this problem may be solved by applying other stains, e.g. specific for proteins and nucleic acids in combination with colocalisa-tion procedures. By this means, additional EPS compounds can be stained and detected using LSM (Neu and Lawrence, 2002).

A requirement of lectin-binding analysis is the need to test a specific biofilm against a range of lectins (Neu *et al.*, 2001). This is necessary in order to select the most appropriate lectins for a biofilm from a specific habitat consisting of a spe-cific community of glycoconjugate producers. In the following paragraph, a screening of all the lectins commercially available, which were tested with three types of biofilms, is presented.

18.3   LECTIN-BINDING ANALYSIS

In order to evaluate lectins with different specificities and to find the most suitable ones, all the commercially available lectins have been tested. For this purpose, biofilms were grown in rotating annular reactors (RAR), which were inoculated with water from the river Saale (for details see Neu and Lawrence, 1997; Lawrence *et al.*, 2000). For comparison, three reactors were fed with different substrates:

(1) a control reactor grown with river water only,
(2) a reactor grown with glucose ($mg \, l^{-1} \, d^{-1}$)
(3) a reactor grown with methanol ($\mu g \, l^{-1} \, d^{-1}$).

The reactors were run in semi-batch mode, fed from a reservoir (15 l) with a flow through of 15 l/h. The biofilms in the RAR developed at 200 rpm at room temperature over a minimum period of 21 days.

For the screening, 63 different types of lectins were purchased from the three major suppliers (Sigma, St. Louis, Missouri, USA; EY, San Mateo, California, USA; Vector, Burlingame, California, USA). The lectins used and their details are listed in Table 18.1. Their specificity and molecular weight is given according to the data sheets of the suppliers. All the lectins were either custom-labelled with

FITC or self-labelled with ALEXA-488 (kit from Molecular Probes, Eugene, Oregon, USA). The procedure of lectin staining has been described elsewhere (Neu and Lawrence, 1999a; Neu et al., 2001). Briefly, biofilm samples of about 1 cm^2 were stained with 100 μl lectin solution (100 μg/ml) and incubated for 20 min at room temperature. Then samples were carefully rinsed with tap water to remove unbound lectin. Subsequently, bacteria were stained with the nucleic acid specific fluorochromes SYTO 60 or SYTO 64 (Molecular Probes, Eugene, Oregon, USA). Biofilms were examined by LSM using a TCS SP controlled by the LCS Version 2.00 Build 0477 (Leica, Heidelberg, Germany). The LSM system was equipped with an Ar (458, 476, 488, 514 nm), Kr (568 nm) and He/Ne (633 nm) laser. Images were collected using an upright microscope and a 20 × 0.5 NA water immersible lens. Sections of biofilms up to thickness of 100 μm and up to 200 μm or more were taken every 1 or 5 μm, respectively.

For quantification of lectin signals, SCION IMAGE (Scion Corporation, Frederick, Maryland, USA) was used (see Lawrence et al., 1998 for basic description). For faster analysis of multiple images, a macro, allowing semi-automated quantification, was developed. Segmentation of images was done with a 20% threshold.

18.4 LECTIN SCREENING

For comparison and visualisation, sample images of all lectins are presented in Table 18.1, indicating the specific-binding pattern of lectins on the three different biofilm types. Lectin binding revealed a differential binding pattern, which is related to the substrates available for growth. The biofilms grown with river water only showed mostly no binding or only a weak binding to cell surface glycoconjugates. Of the 63 lectins tested on the biofilms grown with an additional carbon source, nearly all had a binding pattern useful for lectin-binding analysis.

Overall, about 40 lectins showed a good differential binding pattern, if glucose and methanol biofilms are compared. About 17 of the lectins had a very weak-binding pattern and were not suitable for staining these specific biofilms. However, they may be well suitable for other types of biofilms. The lectins from *Aleuria aurantia*, *Phaseolus coccineus*, *Solanum tuberosum*, and *Triticum vulgaris* showed a strong-binding pattern to the EPS matrix of glucose biofilms. In the methanol biofilms, the lectins of *A. aurantia* and *Hippeastrum hybrid* showed a

Table 18.1   Comparison of 63 commercially available lectins demonstrating their binding pattern in biofilms grown under different conditions.

Name of Lectin Fluorochrome Molecular weight in kDa specificity	**Glucose biofilm**	**Methanol biofilm**	**Control biofilm**
	Thickness of biofilm	Thickness of biofilm	Thickness of biofilm

Table 18.1   (continued)

Abrus precatorius FITC 28-32 65 134 β-Gal Lac	 138 µm	 440 µm	 44 µm
Aegopodium podagraria FITC 55/60 480 α-GalNAc β-GalNAc	 138 µm	 185 µm	 24 µm
Agaricus bisporus FITC 24/50-90 58,5 β-Gal Gal(β 1,3)NAc β-Gal(1-3)GalNAc	 168 µm	 530 µm	 23 µm
Aleuria aurantia Alexa-488 36 Fuc	 400 µm	 530 µm	 48 µm
Anguilla anguilla FITC 30-32 72-84 Fuc	 230 µm	 320 µm	 39 µm
Arachis hypogaea FITC 28 110 120 Fuc β-Gal Gal(β 1,3)GalNAc	 210 µm	 280 µm	 34 µm
Artocarpus integrifolia FITC 42 65 α-Gal α-GalNAc Gal(β 1,3)GalNAc	 580 µm	 192 µm	 13 µm

Table 18.1   (continued)

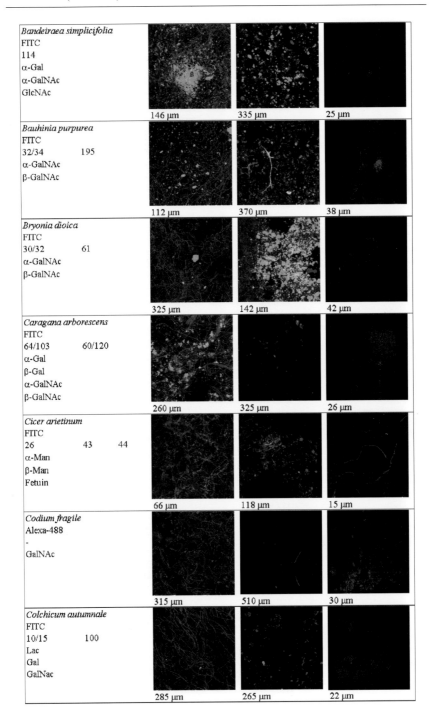

Bandeiraea simplicifolia FITC 114 α-Gal α-GalNAc GlcNAc	146 µm	335 µm	25 µm
Bauhinia purpurea FITC 32/34 195 α-GalNAc β-GalNAc	112 µm	370 µm	38 µm
Bryonia dioica FITC 30/32 61 α-GalNAc β-GalNAc	325 µm	142 µm	42 µm
Caragana arborescens FITC 64/103 60/120 α-Gal β-Gal α-GalNAc β-GalNAc	260 µm	325 µm	26 µm
Cicer arietinum FITC 26 43 44 α-Man β-Man Fetuin	66 µm	118 µm	15 µm
Codium fragile Alexa-488 - GalNAc	315 µm	510 µm	30 µm
Colchicum autumnale FITC 10/15 100 Lac Gal GalNac	285 µm	265 µm	22 µm

Table 18.1   (continued)

Canavalia ensiformis FITC 26,5 106 102 α-Man α-Glc α-GlcNAc			
	116 µm	260 µm	41 µm
Cytisus scoparius FITC 31 120 β-Gal GalNAc Lac			
	285 µm	335 µm	13 µm
Datura stramonium FITC 40/46 86 β-GlcNAc (GlcNAc)$_2$			
	255 µm	210 µm	22 µm
Dolichos biflorus FITC 26-30 110-120 140 α-GalNAc GalNAc(α1,3)GalNAc			
	240 µm	260 µm	18 µm
Erythrina corallodendron Alexea-488 60 β-Gal(1-4)GlcNAc			
	112 µm	345 µm	19 µm
Erythrina cristagalli FITC 28-30 56,8 β-Gal(1-4)GlcNAc			
	128 µm	295 µm	19 µm
Euonymus europaeus FITC 166 α-Gal(1-3)Gal Gal(α1,3)Fuc(α1,2)Gal			
	480 µm	215 µm	15 µm

Table 18.1   (continued)

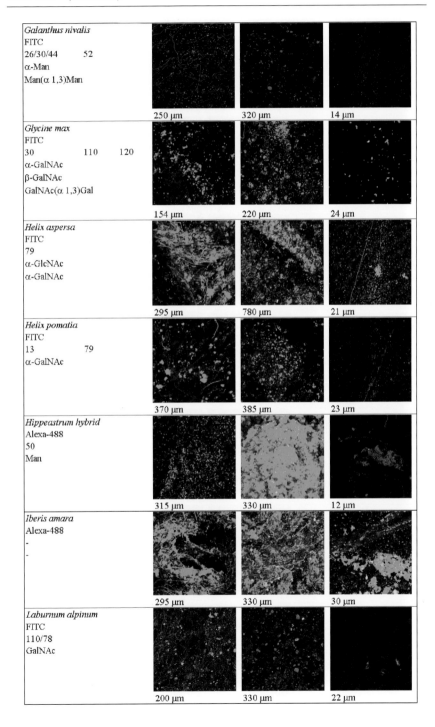

Galanthus nivalis FITC 26/30/44 52 α-Man Man(α 1,3)Man	250 µm	320 µm	14 µm
Glycine max FITC 30 110 120 α-GalNAc β-GalNAc GalNAc(α 1,3)Gal	154 µm	220 µm	24 µm
Helix aspersa FITC 79 α-GlcNAc α-GalNAc	295 µm	780 µm	21 µm
Helix pomatia FITC 13 79 α-GalNAc	370 µm	385 µm	23 µm
Hippeastrum hybrid Alexa-488 50 Man	315 µm	330 µm	12 µm
Iberis amara Alexa-488 - -	295 µm	330 µm	30 µm
Laburnum alpinum FITC 110/78 GalNAc	200 µm	330 µm	22 µm

Table 18.1   (continued)

Lathyrus odoratus FITC 40-43 α-Man	 340 μm	 804 μm	 16 μm
Lens culinaris FITC 20-28 49 α-Man α-Glc α-GlcNAc	 400 μm	 285 μm	 17 μm
Limulus polyphemus FITC 58/62/90/116 350-500 400 GalNAc GlcNAc Sialic Acids	 186 μm	 154 μm	 20 μm
Lycopersicon esculentum FITC 71 β-GlcNAc (GlcNAc)$_3$	 300 μm	 580 μm	 34 μm
Maackia amurensis FITC 130 NANA(α 2,3)Gal Sialic Acid	 280 μm	 235 μm	 13 μm
Maclura pomifera FITC 14,7/2,2 40-43 40-46 α-Gal α-GalNAc Gal(β 1,3)GalNAc	 345 μm	 240 μm	 21 μm
Momordica charantia Alexa-488 115-119 Gal GalNAc	 460 μm	 350 μm	 16 μm

Table 18.1   (continued)

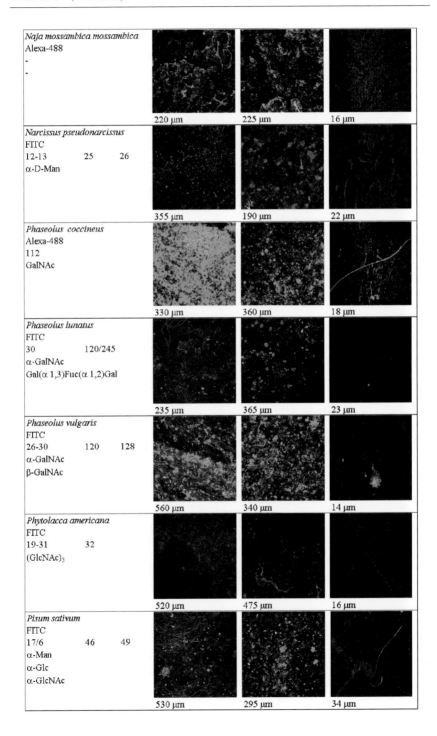

Naja mossambica mossambica Alexa-488 - -	220 µm	225 µm	16 µm
Narcissus pseudonarcissus FITC 12-13 25 26 α-D-Man	355 µm	190 µm	22 µm
Phaseolus coccineus Alexa-488 112 GalNAc	330 µm	360 µm	18 µm
Phaseolus lunatus FITC 30 120/245 α-GalNAc Gal(α 1,3)Fuc(α 1,2)Gal	235 µm	365 µm	23 µm
Phaseolus vulgaris FITC 26-30 120 128 α-GalNAc β-GalNAc	560 µm	340 µm	14 µm
Phytolacca americana FITC 19-31 32 (GlcNAc)₃	520 µm	475 µm	16 µm
Pisum sativum FITC 17/6 46 49 α-Man α-Glc α-GlcNAc	530 µm	295 µm	34 µm

Table 18.1   (continued)

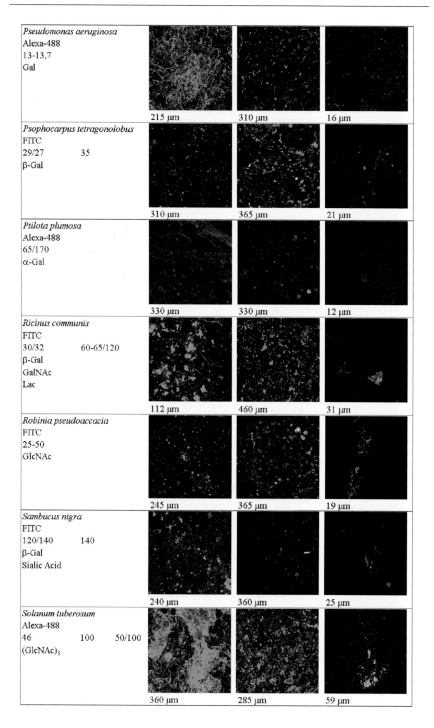

Pseudomonas aeruginosa Alexa-488 13-13,7 Gal		
215 µm	310 µm	16 µm
Psophocarpus tetragonolobus FITC 29/27 35 β-Gal		
310 µm	365 µm	21 µm
Ptilota plumosa Alexa-488 65/170 α-Gal		
330 µm	330 µm	12 µm
Ricinus communis FITC 30/32 60-65/120 β-Gal GalNAc Lac		
112 µm	460 µm	31 µm
Robinia pseudoaccacia FITC 25-50 GlcNAc		
245 µm	365 µm	19 µm
Sambucus nigra FITC 120/140 140 β-Gal Sialic Acid		
240 µm	360 µm	25 µm
Solanum tuberosum Alexa-488 46 100 50/100 (GlcNAc)₃		
360 µm	285 µm	59 µm

Table 18.1   (continued)

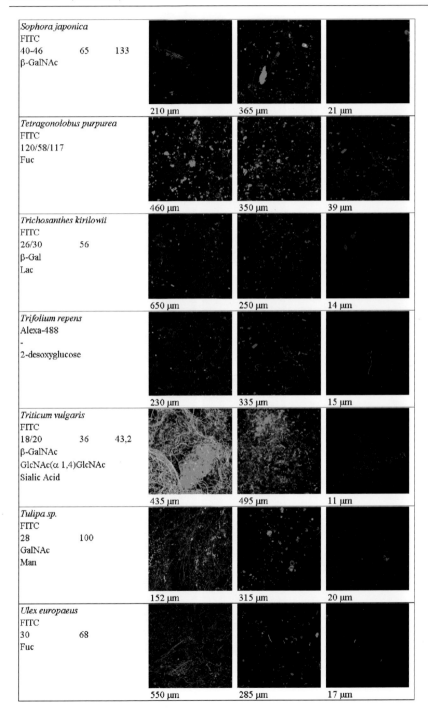

Sophora japonica FITC 40-46 65 133 β-GalNAc		
210 µm	365 µm	21 µm
Tetragonolobus purpurea FITC 120/58/117 Fuc		
460 µm	350 µm	39 µm
Trichosanthes kirilowii FITC 26/30 56 β-Gal Lac		
650 µm	250 µm	14 µm
Trifolium repens Alexa-488 - 2-desoxyglucose		
230 µm	335 µm	15 µm
Triticum vulgaris FITC 18/20 36 43,2 β-GalNAc GlcNAc(α 1,4)GlcNAc Sialic Acid		
435 µm	495 µm	11 µm
Tulipa sp. FITC 28 100 GalNAc Man		
152 µm	315 µm	20 µm
Ulex europaeus FITC 30 68 Fuc		
550 µm	285 µm	17 µm

Table 18.1 (continued)

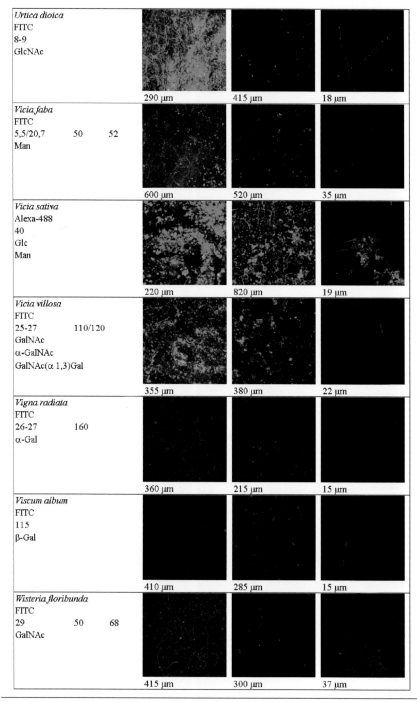

Urtica dioica FITC 8-9 GlcNAc			
	290 μm	415 μm	18 μm
Vicia faba FITC 5,5/20,7 50 52 Man			
	600 μm	520 μm	35 μm
Vicia sativa Alexa-488 40 Glc Man			
	220 μm	820 μm	19 μm
Vicia villosa FITC 25-27 110/120 GalNAc α-GalNAc GalNAc(α 1,3)Gal			
	355 μm	380 μm	22 μm
Vigna radiata FITC 26-27 160 α-Gal			
	360 μm	215 μm	15 μm
Viscum album FITC 115 β-Gal			
	410 μm	285 μm	15 μm
Wisteria floribunda FITC 29 50 68 GalNAc			
	415 μm	300 μm	37 μm

Images were presented as maximum intensity projection without any manipulation of intensity and contrast. Image size: 501 × 501 μm. In the table, lectin details and results of staining are shown according to the heading given.

strong-binding pattern to the EPS matrix. The lectins of *Canavalia ensiformis* and *Iberis amara* had a strong affinity to all types of biofilms. Some of the lectins, e.g. *Helix aspersa Lycopersicum esculentum, Momordica charantia, Pseudomonas aeruginosa* and *T. vulgaris*, preferably, bound to the cell surface of filamentous microorganisms in the glucose-grown biofilms. Other lectins did not bind at all to these filaments.

A rating of the lectins suitable for lectin-binding analysis on the biofilms examined is shown in Table 18.2. This information may be used as a guideline for biofilms of similar origin and growth conditions. For other biofilms, there is the need of screening against a panel of lectins. Figure 18.1 shows the volume per-cent of the 19 lectins with the strongest-binding pattern. The results underline the effect of biofilm origin and biofilm growth conditions on the lectin-binding pattern. Furthermore, it proves the necessity of screening a specific biofilm developed under certain conditions against a panel of lectins in order to select the most suitable ones.

For verification of the lectin-binding pattern recorded, controls are compulsory (Neu *et al.*, 2001). In Figures 18.2 and 18.3, controls with *A. aurantia* lectin after incubation with various carbohydrate concentrations are shown. It is obvious that rising carbohydrate concentrations eventually saturate the specific binding site of the lectin resulting in a lower binding to the glycoconjugates in the biofilm.

18.5   BIOFILM DEVELOPMENT

According to Characklis and Marshall 'One of the ultimate aims in studying biofilms is to evolve the means for manipulating biofilm processes for technical and ecological advantage' (Characklis and Marshall, 1990). This idea was extended by Bryers in the second edition of 'Biofilms' (Bryers, 2000). From both publications, it becomes clear that one key issue is the investigation of biofilm development not only at the cellular but also at the EPS level. The second key issue comprises the effects and interactions of physico-chemical factors onto biofilm development. Both are most important in order to fully understand the complex-ity of biofilm processes.

Biofilm models are an additional tool to understand the complexity of trans-port and reaction processes in the biofilm matrix (Horn and Hempel, 1995, 2001). For the verification of these biofilm models, more experimental data with respect to the growth and decay of EPS inside the biofilm matrix must be available (Horn *et al.*, 2001). Until now the following issues are not completely understood:

- the production of EPS in biofilms (growth related or non-growth related);
- the role of EPS with respect to biofilm structure and density (Wäsche *et al.*, 2000);
- the function of EPS with respect to transport of dissolved components into and inside the biofilm matrix;
- the function of EPS with respect to protection against, e.g. biozides.

Table 18.2   Rating of suitable lectins according to their binding pattern in biofilms grown with glucose, methanol or river water.

Lectin	Glucose	Methanol	Riverwater
Abrus	+ +	+	–
Aleuria	+ + + +	+ + + +	+
Arachis	+ +	+ +	–
Bandeiraea	+ +	+ +	–
Bauhinia	+	+	–
Bryonia	+	+ + +	–
Caragana	+ +	+	–
Canavalia	+ + +	+ + + +	+ +
Cytisus	+	+	–
Datura	+	+	–
Dolichos	+	+	–
Erythrina cor.	+ +	+ +	–
Erythrina cri.	–	+ +	+
Glycine	+ +	+ +	+ +
Helix a	+ + +	+ + +	+ +
Helix p	+ +	+ +	+
Hippeastrum	+	+ + + +	–
Iberis	+ + +	+ + +	+ + +
Lens	+	+ +	–
Limulus	+	–	–
Lycopersicum	+ + +	+ +	+
Maclura	–	+	–
Momordica	+ +	+ +	–
Naja	+	+	+
Phaseolus c.	+ + + +	+ +	+ +
Phaseolus v.	+ + +	+ + +	+ +
Pisum	+	+ +	–
Psophocarpus	+	+	+
Ricinus	+ +	+ +	+ +
Robinia	+	+	+
Sambuca	+	+	–
Solanum	+ + + +	+ + +	+ +
Sophora	–	+	–
Tetragonolobus	+ +	+ +	–
Triticum	+ + + +	+ + +	–
Tulipa	+	+	–
Ulex	–	+	–
Urtica	+ + + +	–	–
Vivia f.	+	–	–
Vicia s.	+ + +	+ + +	+ +
Vicia v.	+ +	+ +	–

+ + + +: extremely good, + + +: very good, + +: good, +: fair, –: hardly any binding.

All these issues can be investigated in biofilms by a systematic use of lectin-binding analysis in combination with microelectrode studies (Horn, 2000). Lectin-binding analysis and staining of bacteria with suitable nucleic acid stains will offer

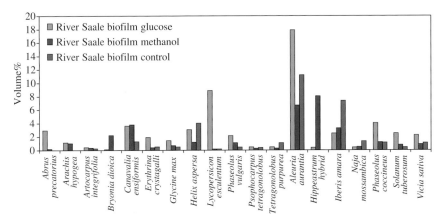

Figure 18.1   Quantitative comparison of 19 lectins showing the strongest-binding pattern. All lectins up to *Tetragonolobus purpurea* were bought pre-labelled with FITC, the others were self-labelled with Alexa-488. Volume% is the percentage of the total recorded volume showing lectin staining.

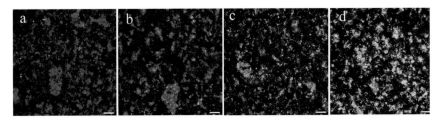

Figure 18.2   Inhibition tests of the lectin *A. aurantia* with fucose. The images show (a) highest fucose concentration (50 mg/ml), (b) dilution 1 : 100, (c) dilution 1 : 10,000 and (d) lectin without fucose. Colour allocation: lectin = green, SYTO 60 nucleic acid stain = red, yellow indicates colocalisation of bacteria and glycoconjugates. Scale bar: 50 μm.

data on the structure and distribution of the particulate components inside the biofilm matrix. Microelectrode studies will offer data on transport and reaction processes in the biofilm. The combination of both techniques promises a more detailed look on structure and function of biofilms.

In the study presented, lectin-binding analysis has been employed to investigate the development of biofilms grown under different physico-chemical growth conditions. The nucleic acid stain SYTO 60 and the Alexa-488 conjugated lectin of *A. aurantia* proved to be the best choice for this investigation. The routine staining experiments with this lectin showed a direct relation between the increase of lectin binding and the thickness, measured by two different methods (Figure 18.4). From Figure 18.4, it is obvious that there is a good match between biofilm wet weight and the calculated volume of the EPS glycoconjugate signals recorded by LSM. In this biofilm, the volume of nucleic acid stained bacteria was negligible compared to the volume of EPS glycoconjugate staining.

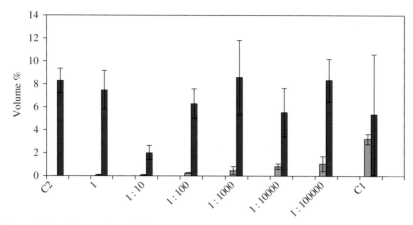

Figure 18.3   Comparison of the stained volume% of bacteria (red) and glycoconjugates (green) in the inhibition tests of *A. aurantia* lectin with fucose (Figure 18.2). C2 shows a control with 50 mg/ml fucose without lectin, C1 shows a control without fucose but with lectin. The others are dilutions of the 50 mg/ml fucose stock solution with water added to equal lectin concentrations.

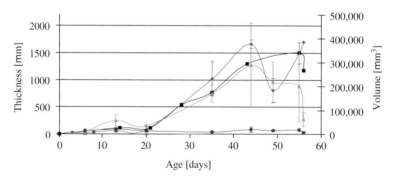

Figure 18.4   Biofilm development when grown on glucose showing: Average thickness of recorded CLSM stacks (blue), biofilm thickness calculated from the wet biomass (black), average volume of bacteria (red) and average volume of EPS glycoconjugates (green).

18.6   CONCLUSIONS

From the data shown, it is obvious that some lectins prefer binding to the EPS matrix, whereas others have a higher affinity to cell surface glycoconjugates. Depending on the type of biofilm and the type of glycoconjugates produced, lectins may be employed to stain a large fraction of the EPS matrix. However, one has to keep in mind that lectins have a specificity, and thus, will only label a specific fraction of the lectin-specific glycoconjugates. Nevertheless, after selecting the most suitable lectin(s) for a specific biofilm, it (they) may be used to follow the production of lectin-specific glycoconjugates over time. The production of glyco-conjugates during different biofilm growth stages may, however, change due to the different physiological status and the succession of microorganisms.

So far lectins have been detected by LSM with one-photon excitation, but, recently, their suitability for two-photon excitation has been demonstrated. This will allow to use and detect them in deep biofilm locations (Neu *et al.*, 2002). One of the major limitations of the approach in environmental biofilms with a thickness of 500 µm or more, may be the penetration of laser light into the biofilm. If this constitutes a serious problem, either two-photon LSM or embedding and physical sectioning has to be used in order to record the lectin signal in deep biofilm regions. Despite some limitations (Neu *et al.*, 2001), lectin-binding analysis in combination with LSM is the only *in situ* approach to study the lectin-specific EPS compounds in environmental biofilms.

Acknowledgements

The excellent technical assistance of Ute Kuhlicke is highly appreciated. For image analysis, Annett Eitner and Martin Tröger developed the semi-automated macro for SCION IMAGE. We would like to thank John R. Lawrence and George D.W. Swerhone for introduction into SCION IMAGE macro-routine. C.S. was financed by a project of the German Research Foundation, DFG (HE 1515/12-1).

REFERENCES

Amann, R. (2000) Who is out there? Microbial aspects of biodiversity. *Syst. Appl. Microbiol.* **23**, 1–8.

Böckelmann, U., Manz, W., Neu, T.R. and Szewzyk, U. (2002) A new combined technique of fluorescent *in situ* hybridisation and lectin binding analysis (FISH-LBA) for the investigation of lotic microbial aggregates. *J. Microbiol. Meth.* **49**, 75–87.

Bryers, J.D. (2000) Biofilms: an introduction. In *Biofilms II. Process Analysis and Applications* (ed. J.D. Bryers), pp. 3–11, Wiley-Liss, New York.

Characklis, W.G. and Marshall, K.C. (1990) Biofilms: a basis for an interdisciplinary approach. In *Biofilms* (ed. W.G. Characklis and K.C. Marshall), pp. 3–15, John Wiley & Sons, New York.

Cooksey, K.E. (1992) Extracellular polymers in biofilms. In *Biofilms – Science and Technology*. NATO ASI Series E: Applied Sciences (ed. L.F. Melo, T.R. Bott, M. Fletcher and B. Capdeville), Vol. 23, pp. 137–147, Kluwer, Dordrecht.

Hemminga, M.A. and Visser, J. (2000) NMR in biotechnology. *J. Biotechnol.* **77**, special issue.

Horn, H. (2000) Microelectrodes and tube reactors in biofilm research. In *Biofilms: Investigative Methods and Applications* (ed. H.-C. Flemming, U. Szewzyk and T. Griebe), pp. 125–140, Technomic, Lancaster.

Horn, H. and Hempel, D.C. (1995) Mass transfer coefficients for an autotrophic and heterotrophic biofilm system. *Water Sci. Technol.* **32**, 199–204.

Horn, H. and Hempel, D.C. (2001) Simulation of substrate conversion and mass transport in biofilm systems. *Eng. Life Sci.* **1**, 225–228.

Horn, H., Neu, T.R. and Wulkow, M. (2001) Modelling the structure and function of extracellular polymeric substances in biofilms with new numerical techniques. *Water Sci. Technol.* **43**, 121–127.

Jolley, J.G., Geesey, G.G., Hankins, M.R., Wright, R.B. and Wichlacz, P.L. (1989) *In situ*, real-time FT-IR/CIR/ATR study of the biocorrosion of copper by gum arabic, alginic acid, bacterial culture supernatant and *Pseudomonas atlantica* exopolymer. *Appl. Spectrosc.* **43**, 1062–1067.

Lawrence, J.R. and Neu, T.R. (1999) Confocal laser scanning microscopy for analysis of microbial biofilms. *Meth. Enzymol.* **310**, 131–144.

Lawrence, J.R., Wolfaardt, G. and Neu, T.R. (1998a) The study of microbial biofilms by confocal laser scanning microscopy. In *Digital Image Analysis of Microbes* (ed. M.H.F. Wilkinson and F. Shut), pp. 431–465, John Wiley & Sons, Chichester.

Lawrence, J.R., Neu, T.R. and Swerhone, G.D.W. (1998b) Application of multiple parameter imaging for the quantification of algal, bacterial, and exopolymer components of microbial biofilms. *J. Microbiol. Meth.* **32**, 253–261.

Lawrence, J.R., Swerhone, G.D.W. and Neu, T.R. (2000) A simple rotating annular reactor for replicated biofilm studies. *J. Microbiol. Meth.* **42**, 215–224.

Lawrence, J.R., Korber, D.R., Wolfaardt, G.M., Caldwell, D.E. and Neu, T.R. (2002) Analytical imaging and microscopy techniques. In *Manual of Environmental Microbiology* (ed. C.J. Hurst, R.L. Crawford, G.R. Knudsen, M.J. McInerney and L.D. Stetzenbach), pp. 39–61, American Society for Microbiology, Washington.

Lens, N.L. and Hemmings, M.A. (2001) NMR in environmental microbiology. *J. Ind. Microbiol. Biot.* **26**, special issue.

Neu, T.R. (1994) The challenge to analyse extracellular polymers in biofilms. In *Microbial Mats, Structure, Development and Environmental Significance.* NATO ASI Series (ed. L.J. Stal and P. Caumette), Vol. G 35, pp. 221–227, Springer-Verlag, Berlin.

Neu, T.R. (1996) Significance of bacterial surface-active compounds in interaction of bacteria with interfaces. *Microbiol. Rev.* **60**, 151–166.

Neu, T.R. and Lawrence, J.R. (1997) Development and structure of microbial stream biofilms as studied by confocal laser scanning microscopy. *FEMS Microbiol. Ecol.* **24**, 11–25.

Neu, T.R. and Lawrence, J.R. (1999a) Lectin-binding-analysis in biofilm systems. *Meth. Enzymol.* **310**, 145–152.

Neu, T.R. and Lawrence, J.R. (1999b) *In situ* characterization of extrazellular polymeric substances (EPS) in biofilm systems. In *Microbial Extracellular Polymeric Substances* (ed. J. Wingender, T.R. Neu and H.-C. Flemming), pp. 22–42, Springer-Verlag, Berlin.

Neu, T.R. and Lawrence, J.R. (2002) Laser scanning microscopy in combination with fluorescence techniques for biofilm study. In *The Encyclopedia of Environmental Microbiology* (ed. G. Bitton), Vol. 4, pp. 1772–1788, John Wiley & Sons, New York.

Neu, T.R., Swerhone, G.D.W. and Lawrence, J.R. (2001) Assessment of lectin-binding analysis for *in situ* detection of glycoconjugates in biofilm systems. *Microbiology* **147**, 299–313.

Neu T.R., Kuhlicke, U. and Lawrence, J.R. (2002) Assessment of fluorochromes for two-photon laser scanning microscopy of biofilms. *Appl. Environ. Microbiol.* **68**, 901–909.

Nielsen, P.H. and Jahn, A. (1999) Extraction of EPS. In *Microbial Extracellular Polymeric Substances* (ed. J. Wingender, T.R. Neu and H.-C. Flemming), pp. 50–69, Springer-Verlag, Berlin.

Schmitt, J., Nivens, D.E., White, D.C. and Flemming, H.-C. (1995) Changes of biofilm properties in response to sorbed substances – an FTIR–ATR study. *Water Sci. Technol.* **32**, 139–155.

Sharon, N. and Lis, H. (1989). Lectins as cell recognition molecules. *Science* **246**, 227–234.

Sutherland, I.W. (2001) The biofilm matrix – an immobilized but dynamic microbial environment. *Trends Microbiol.* **9**, 222–227.

Wäsche, S., Horn, H. and Hempel, D.C. (2000) Mass transfer phenomena in biofilm systems. *Water Sci. Technol.* **41**, 357–360.

Wingender, J., Flemming, H.-C. and Neu, T.R. (1999) Introduction – What are bacterial extracellular polymeric substances? In *Microbial Extracellular Polymeric Substances* (ed. J. Wingender, T.R. Neu and H.-C. Flemming), pp. 1–19, Springer-Verlag, Berlin.

Wolfaardt, G., Lawrence, J.R. and Korber, D.R. (1999) Function of EPS. In *Microbial Extracellular Polymeric Substances* (ed. J. Wingender, T.R. Neu and H.-C. Flemming), pp. 171–200, Springer-Verlag, Berlin.

Coaggregation: receptor-adhesin interactions

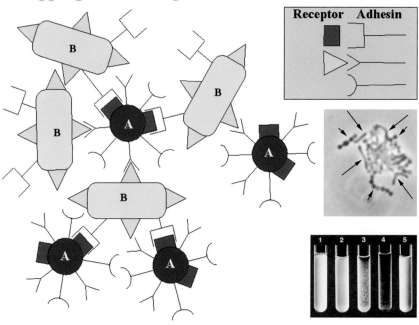

Figure 3.1   Different views of coaggregation. Diagrammatic model of coaggregation between cell type A (circular) and cell type B (oblong). Cell type A expresses three potential coaggregation-mediating surface components two of which are recognized by respective cognate components on cell type B. The components are called receptors and adhesins (inset at upper right). Adhesins are depicted as components with a stem structure and their cognate receptor has the same shape but is lacking the stem. The cognate rectangular symbols represent mediators of lactose-inhibitable coaggregations. Two of the surface components on cell type A are adhesins; the triangle-shaped adhesin recognizes its cognate triangular receptor on cell type B, but the semi-circular-shaped adhesin has no cognate on the partner. Cell type A also expresses a receptor (rectangle), whose cognate adhesin is present on cell type B. Different shaped symbols represent different kinds of adhesins or receptors and illustrate the multi-functional capabilities of oral bacteria.

Bottom right inset, visual coaggregation assay. Tube 1 represents a dense suspension of cell type A. Tube 2 represents a dense suspension of cell type B. Tube 3 shows strong coaggregation that would occur between partner cell types. Tube 4 shows the strongest kind of coaggregation in that coaggregates settle to the bottom of the tube within a few seconds after vortexing the mixed-species suspension. Tube 5 illustrates the dramatic reversal of coaggregation to an evenly turbid suspension after adding an inhibitor of coaggregation. Many coaggregations are reversed by addition of ethylenediaminetetraacetic acid or lactose.

Middle right inset, phase-contrast microscopic view of coaggregation between streptococci (short arrows) and actinomyces (long arrows).

Colour plates

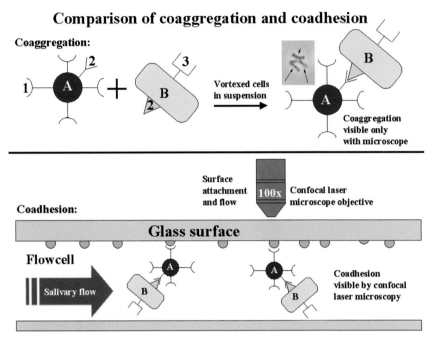

Comparison of coaggregation and coadhesion

Coaggregation:

Vortexed cells in suspension

Coaggregation visible only with microscope

Coadhesion:

Surface attachment and flow

100x Confocal laser microscope objective

Glass surface

Flowcell

Salivary flow

Coadhesion visible by confocal laser microscopy

Figure 3.2   Coaggregation and coadhesion. In this illustration the number of adhesin #2 and its cognate receptor (yellow triangle on cell type B) is intentionally low to emphasize the distinction between coaggregation and coadhesion assays. Circular cell type A and oblong cell type B can be mixed together by vortexing. If the adhesin or receptor numbers per cell are very low, approaching one or two per cell, then large mixed-cell-type coaggregates visible by eye cannot form. Instead, small coaggregates occur (inset, upper right). These can be seen using a microscope showing streptococci (short arrow) coaggregating with actinomyces (long arrows).

Coadhesion, on the other hand, occurs when cells have a chance to attach to a saliva-conditioned substratum and colonize the substratum as indicated in the flowcell system (bottom). Salivary conditioning film provides receptors (green circles) available for binding by initial colonizing bacteria. In this example, cell type A attaches to a salivary receptor through adhesin #1 that has no importance in the coaggregation with cell type B. Low numbers of adhesin #2 or cognate receptor #2 have minimal effect on retention of cell type B in the biofilm. The coadherent cells can be detected and differentiated from the initial cell layer by immunofluorescence or FISH and using a confocal laser microscope.

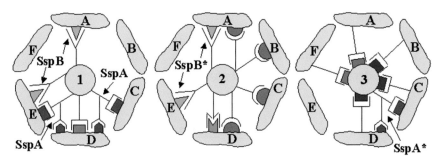

Figure 3.3 Diagrammatic representation of coaggregations between three groups of oral streptococci and six groups of oral actinomyces showing varied kinds of interactions characterizing each group. Current model of interactions emphasizing the involvement of streptococcal SspA and SspB adhesins in coaggregations with actinomyces. SspB of *S. gordonii* DL1, representative of streptococcal coaggregation group 1, mediates coaggregation with actinomyces group A and E. An SspB homolog, SspB* of streptococcal group 2, mediates the same coaggregations with actinomyces groups A and E. SspA of *S. gordonii* DL1 mediates coaggregation with actinomyces groups C, D and E. An SspA homolog, SspA* of streptococcal group 3, mediates coaggregation with actinomyces groups C and D. Possibly, the cognate receptor on actinomyces group E is distinct from the receptor on actinomyces group C and does not recognize the SspA* adhesin. Alternatively, the specificity of SspA* is slightly different than the specificity of SspA, and SspA* is unable to recognize the cognate receptor on actinomyces group E.

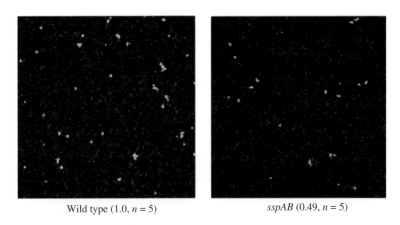

Wild type (1.0, $n = 5$)   *sspAB* (0.49, $n = 5$)

Figure 3.4 Recruitment of *A. naeslundii* T14V cells to established streptococcal biofilms. Experimental details on establishing the biofilms are given in the text. Biofilms formed for 4 h by *S. gordonii* DL1 (wild type) or an isogenic mutant (*sspAB*) were inoculated with *A. naeslundii* T14V at a density of to 3×10^9 actinomyces cells per ml. Coadherence of the actinomyces (green) to the streptococci (red) was measured by counting the number of coadherent actinomyces cells. The number of cells bound to the wild type was normalized to 1.0. The relative number bound to the *sspAB* mutant was 0.49. Actinomyces cells were labeled with monoclonal antibodies raised against *A. naeslundii* T14V type 1 fimbriae (Cisar *et al.*, 1988) and detected with Cy2-conjugated anti-mouse IgG (Jackson ImmunoResearch, West Grove, PA). Streptococcal cells were stained with Alexa-647 (Molecular Probes, Eugene, OR) conjugated IgG raised against whole *S. gordonii* cells.

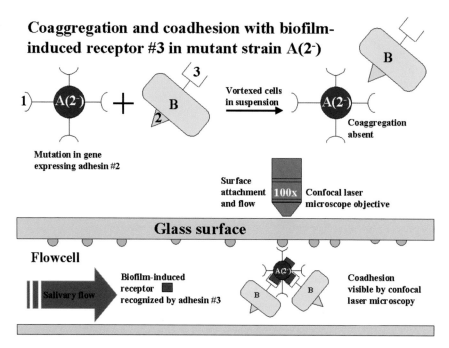

Coaggregation and coadhesion with biofilm-induced receptor #3 in mutant strain A(2⁻)

Figure 3.5 Biofilm induction of coadherence-relevant mediator. Top. Cell type A(2⁻) is a mutant lacking adhesin #2 and is unable to coaggregate with cell type B, although cell type B expresses cognate receptor #2 (yellow triangle). Cell type B expresses adhesin #3, but it is not involved in coaggregation, because cell type A does not express cognate receptor #3 in suspension. Bottom. Mutant cell type A(2⁻) binds to receptor (green circle) in salivary conditioning film and is induced to synthesize receptor #3. Coadhesion occurs because cell type B expresses cognate adhesin #3.

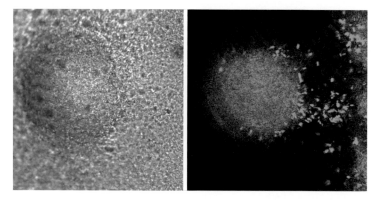

Figure 5.2 *P. aeruginosa* pMH 509 p*lasB*::*gfp* biofilm in the flow cell expressing the GFP (40× objective lens). Transmitted and scanning laser confocal microscopy (right) image. The strain was kindly provided by Morten Hentzer and Matthew Parsek.

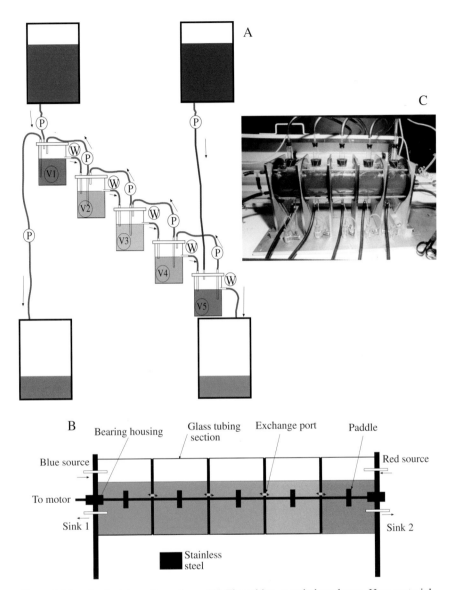

Figure 15.1 Gradostat configurations. (A) The tubing-coupled gradostat. Here material is transferred by gravity downwards over weirs (W) and upwards (also from reservoirs and into the first receiver), by tubing pumps (P). Colours are used to indicate the gradients that form under steady-state conditions. (B) A DC gradostat. Material is transferred from vessel to vessel by diffusion through an exchange port. Pumps are used only to add and remove nutrients and culture respectively into and out of each end chamber. (C) A DC gradostat in use to separate *Rhodopseudomonas marina* (salt resistant) from *Rhodobacter capsulata* (salt sensitive) in a NaCl gradient. NaCl enters the system from the right.

Colour plates

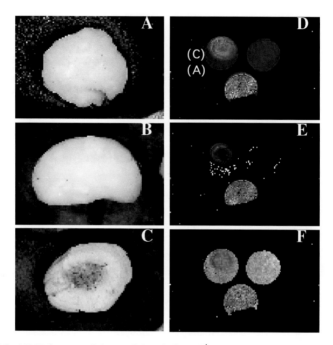

Figure 17.4   NMR images of the spatial variation of ^{1}H NMR parameters in a slice of 2 mm of an insert containing a single bead. Low magnification light microscopic images of cross-sectioned beads: (A) 1% agar, (B) 10% agar and (C) methanogenic aggregate. NMR images: (D) amplitude A, (E) R_2 ($=1/T_2$) and (F) self-diffusion coefficient D_s. Letters in (D) indicate the different beads.

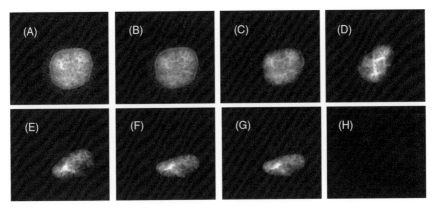

Figure 17.5   Temporal evolution of an individual aggregate from a UASB reactor treating paper mill wastewater (Eerbeek, The Netherlands) after the addition of Fe (100 mm) as manifest in fast TSE imaging; (A) prior to addition, (B) addition, $t = 0$. Next images corresponding to (C) +10 min, (D) +20 min, (E) +30 min, (F) +40 min, (G) +50 min and y (H) + 1 h. The effect of Fe is to reduce the signal intensity (ranging from black (zero), dark green, light green to white (max)) observed in these images. A single slice out of a complete 3D data set is presented (voxel size).

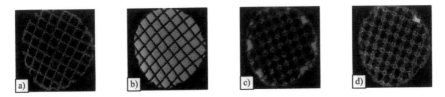

Figure 17.6   NMR imaging of water transport processes in monolith structures: (a) wetted walls, (b) water-filled channels and water film flow with (c) a bad and (d) good distribution over the channels.

Figure 20.6   Image of a biofilm sample treated with SYTOX green followed by FISH with a bacterial probe (S-D-Bact-0338-a-A-18) labelled with Cy5.

Section 4

Analytical techniques for biofilm properties: (B) Biotic properties

19 Environmental electron microscopy
    applied to biofilms   331

20 Use of molecular probes to study biofilms   352

21 Use of microsensors to study biofilms   375

22 Use of mathematical modelling to study
    biofilm development and morphology   413

329

19

Environmental electron microscopy applied to biofilms

R. Ray and B. Little

19.1 INTRODUCTION

Environmental electron microscopy includes both scanning and transmission techniques for the examination of biological materials with a minimum of manipulation, i.e. fixation and dehydration. Both techniques have been used to examine hydrated biological material and biofilms. Environmental electron microscopy provides fast, accurate images of biofilms and their spatial relationship to substrata, and when coupled with analytical tools can provide their elemental composition.

Biofilms containing bacteria and associated extracellular polymers develop on all engineering materials placed in biologically active liquids. Biofilms can be either beneficial or detrimental in industrial processes. Biofilms can remove dissolved and particulate contaminants in fixed film biological systems, such as trickling filters, rotating biological conductors, and fluidized bed wastewater treatment plants. Biofilms determine water quality by influencing dissolved oxygen content and by serving as a sink for toxic and/or hazardous materials. Biofilm reactors are used for commercial fermentation processes, including the manufacture of vinegar (Nilsson and Ohlson, 1982). Microorganisms within biofilms are used to recover minerals (Lawrence and Bruynesteyn, 1983; Geesey and Jang, 1990) and to remove sulfur from coal (Chandra *et al.*, 1979). Biofilms form undesirable deposits on industrial equipment causing reduced heat transfer, increased fluid frictional resistance (Characklis *et al.*, 1990), plugging (Moreton and Glover, 1980), corrosion (Little *et al.*, 1991), and other types of deterioration (Mansfeld *et al.*, 1990).

Biofilm formation has been evaluated as a function of substratum, liquid medium (Little *et al.*, 1986), carbon source (McEldowney and Fletcher, 1988), pH, and hydrodynamic parameters including flow rate (Characklis *et al.*, 1990). Electron microscopy enables one to visualize biofilms and the supporting substrata simultaneously, demonstrating areal coverage, relationship of cells to substrata, distribution of cells and, in some cases, biodeterioration. Many of the conclusions about biofilm development, composition, distribution, and relationship to substratum have been derived from traditional scanning electron microscopy (SEM) (Sieburth, 1975; Marszalek *et al.*, 1979; Mitchell and Kirchman, 1981) and transmission electron microscopy (TEM) (Characklis and Marshall, 1990).

19.2   ENVIRONMENTAL SCANNING ELECTRON MICROSCOPY

19.2.1   Principles

In traditional SEM non-conducting samples including biofilms must be dehydrated and coated with a conductive film of metal before the specimen can be viewed. Uncoated non-conductors build up local concentrations of electrons, referred to as 'charging', that prevent the formation of usable images. Energy dispersive X-ray (EDS) analysis can be used to determine the elemental composition of surface films in the SEM, but EDS analysis must be completed prior to deposition of the thin metal coating. EDS data are typically collected from an area. The specimen is removed from the specimen chamber, coated with a conductive layer, and then returned to the SEM. The operator attempts to relocate and image the precise area from which the EDS data were collected.

The environmental SEM (ESEM) provides a technique for imaging hydrated material at high magnification with no prior sample preparation. ESEM can be achieved by using isolated sample cells in a conventional SEM or by a system of differential pumping. ESEM images of biofilms in this chapter were produced using an ESEM with a differential pumping system and pressure limiting apertures to achieve a graduation of vacuum and an environmental secondary electron detector (ESED) capable of forming high-resolution images at pressures in the range of 0.1–20 Torr (Figure 19.1). At these relatively high pressures, specimen charging is dissipated into the gaseous environment of the specimen chamber, enabling direct observation of uncoated, non-conducting specimens. If water vapor is used as the specimen environment, and a Peltier® cooling stage is used, wet, non-conducting samples can be viewed. Wet biofilms can be examined directly, and EDS spectra can be collected immediately after morphology/topography is imaged. In the following sections comparisons will be made of images and chemistries from identical biofilms using SEM/EDS and ESEM/EDS.

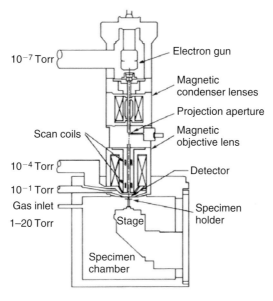

Figure 19.1   Cross section of ESEM.

19.2.2   Methods and materials for ESEM examination of biofilms

Two types of biofilms were evaluated with SEM and ESEM (Little *et al.*, 1991). In one series of experiments, naturally occurring estuarine biofilms formed over a 4-month period on types 304 and AL6X stainless steels and commercially pure copper coupons (1 cm $\times$ 1 cm) in an aquarium maintained with flowing Gulf of Mexico estuarine water. The turnover time for the aquarium water was 63 h. In a second series of experiments, a mixed culture of obligate and facultative anaerobic bacteria, including sulfate-reducing bacteria (SRB), was used to inoculate copper-containing foils (90Cu : 10Ni and commercially pure copper) in an enrichment medium using lactate as the electron donor and carbon source for growth. Table 19.1 lists the elemental composition of all substrata. The medium was supplemented with 2.5% (w/v) NaCl for enrichment of marine microbes. Cultures were grown anaerobically at room temperature in sealed bottles for 4 months.

Biofilms on metal surfaces were fixed in 4% glutaraldehyde in filtered seawater or in distilled water buffered with sodium cacodylate (0.1 M, pH 7.2) at 4°C for a minimum of 4 h rinsed to distilled water, and examined in the ESEM (Figure 19.2b). Micrographs and EDS data were collected for the wet biofilms. Specimens were removed from the ESEM, dehydrated through a series of distilled water/acetone washes to acetone and examined. Specimens were returned to acetone and rinsed through a series of acetone/xylene washes to xylene before final examination. Areal coverage was evaluated directly from the distilled water and after water removal with acetone and xylene. EDS data were collected from the biofilms at

Table 19.1   Major alloy constituents (wt %) of biofilm substrata used for corrosion studies.

	UNS[a] No.	Cu	Ni	Fe	Mn	Mo	Cr
Commercially pure copper	C11000	99.9 oxygen free					
90Cu:10Ni	C70600	86.73 to 87.02	10.47 to 10.81	1.68 to 1.69	0.58 to 0.67		
SS304	S30400		8.0 to 10.5	REM[b]	2.00		18.0 to 20.0
SSAL6X	N08366		23.5 to 25.5	REM[b]	2.00	6.0 to 7.0	20.0 to 22.0

[a]Unified numbering system.
[b]REM: remainder.

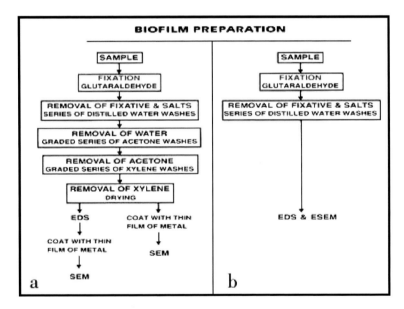

Figure 19.2   Flowchart comparing sample preparation for (a) traditional SEM and (b) ESEM.

each stage of sample manipulation. Bacteria on the copper-containing foils were imaged in the ESEM after fixation and washing in distilled water (Marszalek *et al.*, 1979).

ESEM examination was performed in an ElectroScan® (Wilmington, MA) Type II microscope. Specimens were attached to a Peltier® stage maintained at 4°C and imaged in an environment of water vapor at 2–5 Torr to maintain samples in a hydrated state. At the acetone or xylene stage, specimens were imaged using acetone or xylene vapor. The ESEM was operated at 20 kV using ESED. EDS data were obtained with a Tracor Northern® (Middleton, WI) System II X-ray micro-analyzer equipped with a diamond window light element detector. Samples were held at a 20° tilt during spectrum acquisition. A program correcting for atomic number (Z), absorption (A), and fluorescence (F) was used for semiquantitative analysis during data acquisition. A complete review of ZAF correction procedures has been prepared by Beaman and Isasi (1974).

A reference mark was etched on the edge of each metal specimen to facilitate relocation of precise positions on the specimen surface. The reference mark was centered on the display screen and the x–y-coordinates recorded. Coordinates were also recorded for each subarea examined in detail. Differences in coordinates were established for subareas in relation to the reference mark. After removal of the specimen from the ESEM, experimental treatment, and replacement in the specimen chamber, subareas could be precisely relocated using x–y-coordinates and the reference mark.

In SEM examination biofilms were removed from the growth medium and fixed in 4% glutaraldehyde. Estuarine biofilms fixed in 4% glutaraldehyde in filtered seawater at 4°C for a minimum of 4 h were dipped through a series of filtered sea-water/distilled water washes to remove residual salts. Distilled water was removed from biofilms through a graded series of distilled water/acetone washes to acetone. Acetone was removed through a series of acetone/xylene washes to xylene and air dried (Figure 19.2a). Dried specimens were sputter-coated with gold (20 nm thickness) using a Polaron® (Line Lexington, PA) Series II Sputtering System, Type E5100. SEM images were collected with an AMRAY® (Bedford, MA) 1000A SEM operated at an accelerating voltage of 30 kV and a vacuum of 10^{-5} Torr.

19.2.3 SEM versus ESEM

19.2.3.1 Estuarine biofilm

Figure 19.3a is an ESEM image of a wet estuarine biofilm on 304 stainless steel. The distribution is patchy. There is no visible corrosion and machining marks are evident on the surface. Figure 19.3b is an ESEM image of the same area after the biofilm was treated with acetone and xylene. Areal coverage by the biofilm was reduced. Table 19.2 summarizes the elemental composition of the biofilm before and after solvent extraction of water. The wet biofilm contained elevated levels of Ni, Si, Al, and Ti. After solvent removal of water and polysaccharides from the sample biofilm, only elements from the substratum were detected.

Figure 19.4a shows the gelatinous appearance of the wet biofilm on the AL6X stainless steel surface. Individual bacteria could not be distinguished within the wet biofilm, but numerous diatoms were enmeshed within the gel-like matrix. After preparation for SEM, bacteria could be observed as a single layer of cells (Figure 19.4b and c). At low magnification, the bacterial cells in Figure 19.4a appear to be partially covered with a surface film, while the extracellular polymer

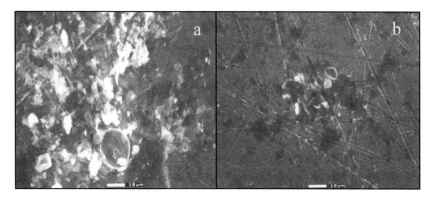

Figure 19.3 (a) ESEM image of wet estuarine biofilm on 304 stainless steel surface. (b) ESEM image of estuarine biofilm on 304 stainless steel surface after treatment with acetone/xylene.

Table 19.2 Weight percent of elements found on 304 stainless steel surfaces after exposure to estuarine water and after treatment through xylene.

Element	Base metal	After exposure to estuarine water	After xylene
Fe	72.55	11.24	72.43
Cr	18.10	4.71	18.17
Ni	8.02	34.00	7.78
Si	0.93	7.35	1.06
Al	0.40	3.33	0.57
Cl	0	0.66	0
S	0	0.99	0
K	0	1.88	0
Na	0	2.12	0
Mg	0	4.98	0
Ti	0	28.74	0

surrounding the cells in Figure 19.4b and c has been reduced to fine filaments connecting the cells.

Wet estuarine biofilms on commercially pure copper surfaces appeared to be uniform with prominent diatom frustules (Figure 19.5a). After specimen dehydration through acetone, diatoms were removed from the biofilm (Figure 19.5b). After treatment with xylene, the same specimen area exhibited extensive cracking (Figure 19.5c). The elemental composition of the wet biofilm on the copper surface (Table 19.3) reflects the presence of diatom frustules with an elevated proportion of Si. Al and Fe were also concentrated in the wet biofilm. After removal of the diatoms, Si and Al concentrations decreased. After solvent removal of the water, only chlorides remained on the copper surface.

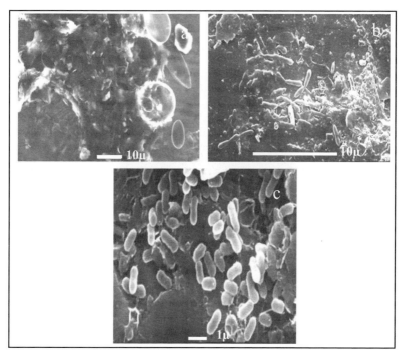

Figure 19.4   (a) ESEM image of wet estuarine biofilm on AL6X stainless steel surface showing gelatinous structure and accumulation of diatoms (marker: 10 μm). (b) SEM of bacterial cells on AL6X stainless steel surface. Cells appear to be embedded in residual polymeric matrix (marker: 10 μm). (c) SEM of bacterial cells on AL6X. Extracellular polymer appears as thin strands connecting cells (marker: 1 μm).

Table 19.3   Weight percent of elements found on commercially pure copper surfaces after exposure to estuarine water for 4 months and sequential treatment with acetone and xylene.

Element	Base metal	After exposure to estuarine water	After acetone	After xylene
Al		9.49	1.22	0.74
Si		21.38	1.89	1.27
Cl		0.93	15.9	15.93
Cu	99.9	59.62	80.99	82.06
Mg		1.96	0	0
P		0.98	0	0
S		0.95	0	0
Ca		0.49	0	0
K		0.67	0	0
Fe		3.52	0	0

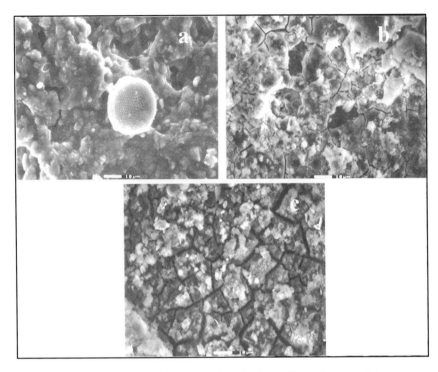

Figure 19.5   (a) ESEM image of wet estuarine biofilm on Cu surface containing embedded diatoms. (b) ESEM images of estuarine biofilm with extensive cracking after treatment with acetone. (c) ESEM images of estuarine biofilm with extensive cracking after treatment with xylene.

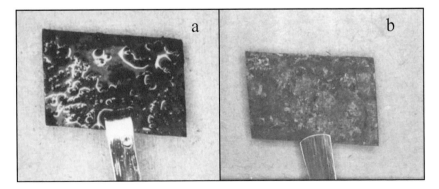

Figure 19.6   (a) 90Cu : 10Ni surface colonized by SRB immediately after removal from culture medium. (b) 90Cu : 10Ni surface colonized by SRB after preparation for SEM.

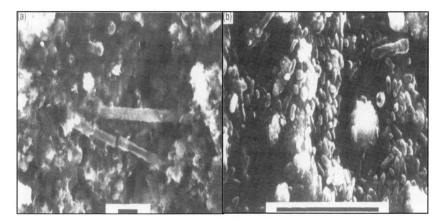

Figure 19.7   (a) ESEM image of bacteria within sulfide corrosion layers on 90C u : 10Ni foil (marker: 2 μm). (b) SEM image of bacteria on residual corrosion layers of 90C u : 10Ni after fixation, dehydration, and critical-point drying (marker: 2 μm).

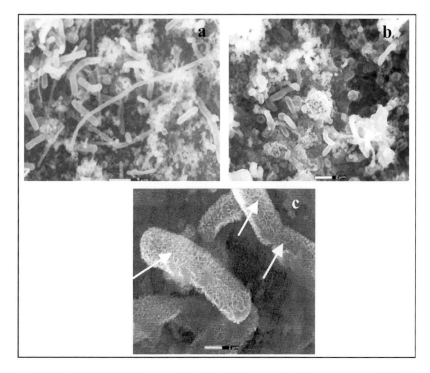

Figure 19.8   (a)–(c) ESEM images of bacteria within corrosion layers on copper foils ((a) and (b) markers: 5 μm; (c) marker: 1 μm). Arrows indicate sulfide encrusted cells in (c).

19.2.3.2   Copper foils

Surfaces of the copper-containing foils colonized by anaerobic bacteria including SRB, immediately after removal from the culture medium, were uniformly covered with a black gelatinous film and localized deposits (Figure 19.6a). After fixation, dehydration, and critical-point drying, much of the surface material was removed from the specimen (Figure 19.6b). Non-tenacious corrosion layers sloughed from the surface during fixation and drying, exposing bare metal. Adherent layers ranging in color from gray to yellowish brown were curled and easily dislodged from the surface. ESEM micrographs of the wet 90Cu:10Ni surface show bacteria within the corrosion layers, while a SEM image of the same area indicates the presence of monolayer of cells attached to a corrosion layer (Figure 19.7a and b). This was typical of all copper-containing materials colonized by the SRB. ESEM images of the wet copper-containing surfaces document several bacterial morphologies within the copper sulfide corrosion layers (Figure 19.8a and b). Some cells were encrusted with sulfides (Figure 19.8c). Bacterial cells attached to base metal under the corrosion layers could be located with both SEM and ESEM. The shape and dimension of the bacterial cells attached to the base metal were not altered by the SEM preparation.

19.3   ENVIRONMENTAL TRANSMISSION ELECTRON MICROSCOPY

19.3.1   Principles

Traditional TEM methods for imaging biofilms require fixation of biological material, embedding in a resin and thin-sectioning to achieve a section that can transmit an electron beam (Characklis and Marshall, 1990). Environmental TEM (ETEM) can be achieved by differential pumping as described for ESEM (Parsons, 1974). In contrast, the TEM used in the studies described in this chapter uses a self-contained cell that requires no modification to the transmission electron microscope. The cell is of the closed cell type (Fukami et al., 1991) and confinement of a pressurized environment within the cell is achieved with electron transparent windows (Figures 19.9–19.12). Windows are fabricated from 15 to 20 nm thick amorphous carbon (a-C) films that cover seven hexagonally arrayed, 0.15 mm apertures on a 3.5 mm diameter Cu disk (Figure 19.13). Prior to use, the windowed grids are tested to withstand a pressure differential of 250 Torr for 1 min. The a-C films can sustain a pressure differential up to 0.5 atm. The microscope is equipped with two environmental cells: a two-line gas cell and a four-line gas/liquid cell. Both cells are capable of circulating dry or water-saturated gas, the latter enables the examination of hydrated specimens (gas circulation is supported by two lines). The four-line cell has two additional lines, which can be used to independently inject several microliters of liquids. Each cell consists of a small cylindrical cell sealed by two electron transparent windows on the top and bottom. The JEOL® JEM-3010 TEM operating at 300 kV was equipped with interchangeable specimen holders, EDS, and a Gatan® imaging filter (GIF200) capa-

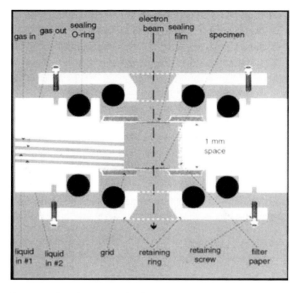

Figure 19.9   Diagram of environmental cell, gas circulation lines, and liquid injection lines.

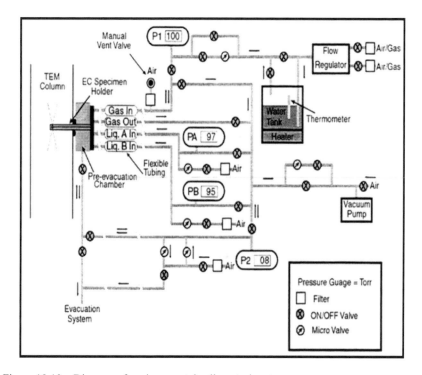

Figure 19.10   Diagram of environmental cell control system.

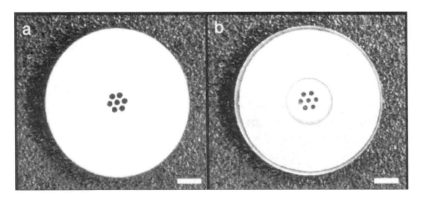

Figure 19.11   Seven hole environmental cell grids. (a) Upper specimen surface of grid. (b) Under side of grid showing concave recess. Markers: 0.5 mm.

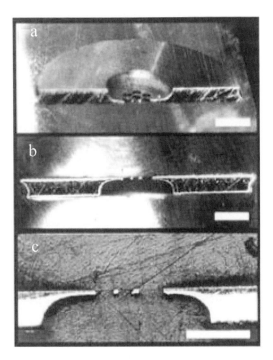

Figure 19.12   Cross section of seven hole environmental cell grid showing: (a) under side; (b) recess; (c) grid hole size. Scale bar: 0.5 mm.

ble of electron energy loss spectroscopy (EELS) measurements. EDS can be used to detect and precisely locate elements associated with cells. EELS has the advantage that the oxidation state of elements can be determined. A computer controls the environmental cell, facilitating insertion and removal of holders from the microscope column without breaching the delicate windows. The environmental cell was operated at

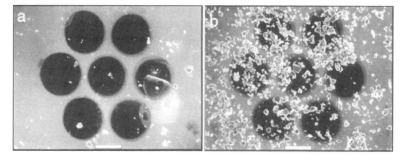

Figure 19.13   ESEM images of windowed grids following use in the environmental cell. (a) Upper grid containing material dislodged from lower grid during the course of an experiment. (b) Lower grid containing corroded iron filings and bacteria. Note that the carbon support film is intact on both the upper and lower grids. Scale bar: 100 μm.

100 Torr with room air saturated with water vapor circulating through the cell at a rate of ≈2 l/min. Specimens were supported on the lower a-C film window. Once sealed the environmental cell was inserted in the column of the TEM and examined under an atmosphere of 100 Torr of saturated water vapor.

The following conditions were used for the collection of EELS spectra under ETEM conditions: an illumination angle $2\alpha = 4$–10 mrad, a collection angle of $2\beta = 10.8$ mrad, a 2 mm entrance aperture, and an energy dispersion of 0.1 eV/channel. Low-loss spectra were acquired with an integration time of 0.128 s and core-loss spectra between 0.512 and 1.02 s. For each acquisition, 10 spectra were summed. Spectra were collected in diffraction mode of the TEM (i.e. image coupling to the EELS spectrometer) and were corrected for dark current and channel-to-channel gain variation of the charge coupled device detector.

Energy calibration of the core-loss regime and measurement of energy drift during data acquisition were performed by collecting zero-loss spectra before and following collection of core-loss spectra. The energy of the core-loss spectra was calibrated using the average position of the two zero-loss peaks.

19.3.2   Methods and materials

Two monospecies biofilms were examined using TEM and ETEM (Little *et al.*, 2001). *Oceanospirillum* sp. was maintained in marine broth at room temperature and grown on copper foil (120 mm thick). Samples for TEM were fixed in 3% glutaraldehyde at 4°C for overnight, buffered at pH of 7.3 with 0.2 M cacodylate, and postfixed in 1% osmium tetroxide for 1–2 h at room temperature. Subsamples were dehydrated through a graded seawater/deionized water and deionized water/acetone series. Dehydrated samples were trimmed to a small triangular shape and put into a BEEM® capsule filled with Spurr® embedding medium. Ultrathin cross sections were cut with a diamond knife on a Sorvall® MT-2 ultramicrotome, and stained with uranyl acetate for 20–30 min and lead citrate for 5 min to improve the contrast. Ultrathin sections were coated with a thin layer of carbon (Chiou *et al.*, 1996).

For ETEM copper filings (99% Cu) were prepared from copper foil by sanding with 600 grit sand paper, collected by rinsing into a test tube with distilled water, degreased with acetone, washed with ethanol, and placed into filter sterilized seawater (0.2 μm pore filteration pasteurization, 70°C for 30 min). Filings were allowed to corrode overnight. Fresh log phase *Oceanospirillum* cells were washed three times in sterile seawater and placed into a tube containing corroding copper filings and incubated overnight on a rotator at room temperature. Filings with biofilms were centrifuged, supernatant removed, and the precipitate rinsed with dilute seawater followed with distilled water. The process was repeated using distilled water to remove trace salts. An aliquot of solution containing suspended copper particles with biofilm was loaded into the environmental cell and examined as previously described.

Pseudomonas putida was maintained in brain heart infusion broth at room temperature and cells were transferred to fresh medium every two days. Iron filings were prepared from carbon steel (C1010) by light sanding with 600 grit abrasive paper and isolated from the abrasive grit using a Teflon® coated magnet. Iron filings were degreased with acetone, washed with ethanol, and placed into sterile distilled water. Filings exposed to log phase cultures of *P. putida* were washed three times with distilled water, and allowed to incubate over night on a rotator at room temperature. An aliquot of solution containing suspended iron particles and bacteria were placed onto the a-C film of the lower environmental cell windowed grid. Particles settled for 5 min and excess solution was wicked away with filter paper.

19.3.3   ETEM imaging of samples

19.3.3.1   Oceanospirillum *on copper foil*

ESEM images of *Oceanospirillum* on copper foil (Figure 19.14) prepared as a reference for TEM images, demonstrated the expected spirillum shape of the organism. TEM micrographs of thin sections demonstrate the distribution of the

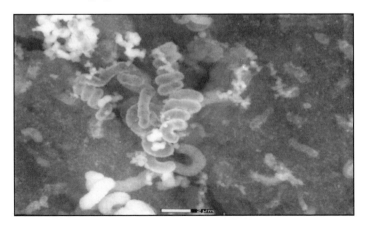

Figure 19.14   ESEM image of *Oceanospirillum* on Cu foil.

organism throughout a gel-like matrix (Figure 19.15). Dark and rounded particles were identified as copper by EDS.

When hydrated *Oceanospirillum* on corroding copper filings was examined in the environmental cell, cells and extracellular polymeric substances (EPS) were intact (Figure 19.16a). Cells remain plump/hydrated and the extracellular polymeric materials retained moisture and appeared as a continuous layer. Many of the *Oceanospirillum* were encrusted with electron dense material (Figure 19.16b) identified as copper using EDS. Using EELS one can demonstrate a change in the core-loss spectrum (Figure 19.17a and b) consistent with Cu^o at the base metal to Cu^{2+} in the corrosion products.

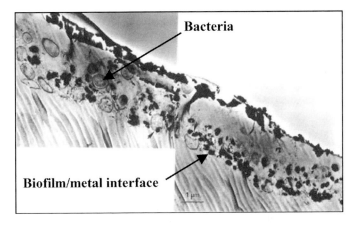

Figure 19.15   A montage of TEM images showing a biofilm of *Oceanospirillum* on Cu foil.

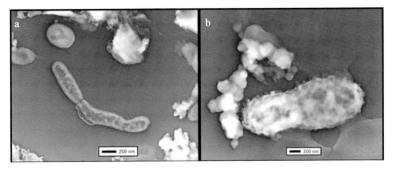

Figure 19.16   (a) Hydrated *Oceanospirillum* on copper filings in environmental cell with circulating air-saturated water vapor. (b) *Oceanospirillum* encrusted with electron dense copper corrosion products.

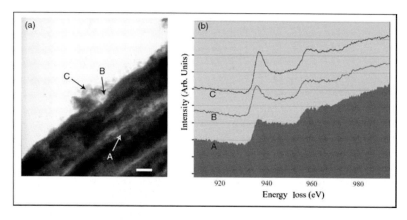

Figure 19.17   EELS of Cu corrosion. (a) TEM of Cu foil with *Oceanospirillum* and associated corrosion products. (b) EELS spectra from three regions indicated in (a). The three spectra show Cu^0 at the base metal (A), and Cu^{2+} corrosion products (B and C).

19.3.3.2   Pseudomonas *on iron fillings*

Examination of viable *P. putida* deposited on an a-C coated TEM grid placed directly into the microscope column under high vacuum demonstrated rod-shaped bacteria, however, damage was readily apparent (Figure 19.18). Internal structures, such as the nucleoid and other electron dense structures were visible, however cell membranes were ruptured by decompression under high vacuum. Furthermore, extracellular material became filamentous and web-like with exposure to the electron beam.

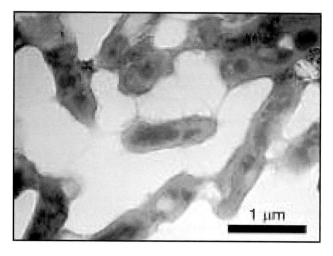

Figure 19.18   *P. putida* on an a-C coated grid in the TEM showing dehydration and collapse of cell membrane and extracellular polymers.

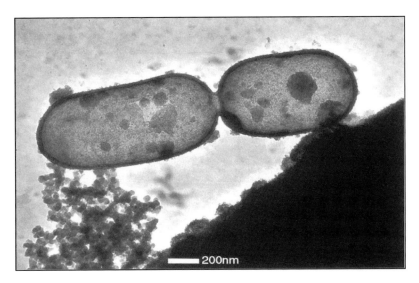

Figure 19.19   Hydrated *P. putida* after removal of excess moisture by circulation of air through the environmental cell.

In contrast, when *P. putida* on iron filings was examined in the environmental cell at 100 Torr, under a circulation of air saturated with water vapor, cells remain plump/hydrated (Figure 19.19). Iron corrosion products were located between cells and the iron substratum. At lower magnifications (Figure 19.20a and b) ETEM micrographs demonstrate that the organisms were not directly in contact with the metal. Instead, the cells were attached to the substratum with extracellular material. Under ETEM conditions the EPS retained moisture and appeared as a well-defined interface.

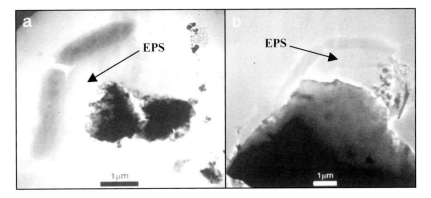

Figure 19.20   (a and b) Two images of *P. putida* on corroding iron filings in the environmental cells showing complete hydration of bacterial cell and EPS.

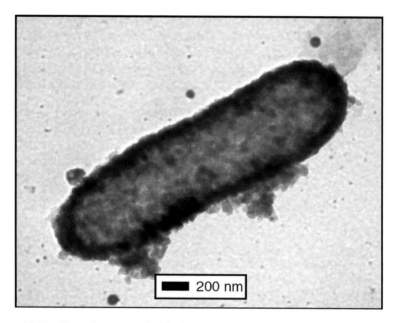

Figure 19.21   *P. putida* encrusted with electron beam dense material.

If the water-saturated gas circulating through the environmental cell was replaced with circulating room air, excess moisture in EPS was removed and image resolution improved (Figure 19.21). Dehydrating the EPS by circulating (or injecting bursts of) room air improved spectra acquired by EELS. In this study, *P. putida* were encrusted with electron beam dense material identified with EELS as iron.

Although decompression and dehydration damage was avoided in the environmental cell, electron beam damage of viable *P. putida* was observed within several minutes of electron beam exposure (Figure 19.22a–c).

19.4   CONCLUSION

The capability to image hydrated specimens using electron microscopy is not a new development (Clarke *et al.*, 1973; Parsons, 1974; Robinson, 1975; Hui, 1976; Fukushima *et al.*, 1982; Suda *et al.*, 1992). Recently, Daulton *et al.* (2002) used ETEM coupled with EELS to image *Shewanella oneidensis* and to document microbial reduction of Cu (VI) to Cr (III). However, environmental microscopy techniques have only recently been applied to biofilms. For example, ESEM has been used to examine algal biofilms on stone, painted surfaces, and metal surfaces (Beech *et al.*, 1996; Surman *et al.*, 1996).

Each of the electron microscopies described in this chapter have advantages and limitations. The SEM and ESEM provide excellent resolution with the ability

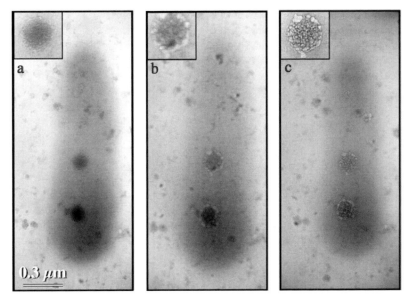

Figure 19.22   (a–c) Progressive electron beam damage in the environmental cell after several minutes.

to image complex shapes. The ESEM allows one to accurately determine aerial coverage of biofilms with cells and EPS associated with surfaces and, in combination with EDS, to determine relative concentrations of elements, heavier than sodium, localized in the biofilm (Figures 19.3–19.5, Tables 19.2 and 19.3). It is sometimes difficult to image individual calls embedded in extracellular material. Sample preparation for SEM reduces extracellular material associated with cells, removes elements bound with that material and makes imaging of individual cells easier. In all cases, preparation of biofilms for SEM results in images that make biofilms appear as monolayers of cells (Figures 19.4c and 19.7b).

TEM images of thin sections can demonstrate the distribution of cells within biofilms and the relationship between cells and corrosion products (Figure 19.15). Cross sectioning has advantages and disadvantages. One can determine spatial arrangements and cellular ultrastructure. Resolution for the TEM and ETEM are the same and are typically an order of magnitude greater than SEM/ESEM counterparts. There are no significant differences in resolution between modern ESEM and SEM. ETEM has the advantage that hydrated cells can be imaged and the relationship to substrata can be examined. The present limitation is the size of the environmental cell.

Acknowledgements

This work was supported by the Office of Naval Research, Program Element 61153N, through the NRL Defense Research Sciences Program. N7303RL

Contribution Number NRL/BC/7303/02/0002. ETEM images were collected by Dr. Tyrone Daulton, NRL, SSC, and Dr. Robert K. Pope, currently at Indiana University, South Bend, IN.

REFERENCES

Beaman, D.R. and Isasi, J.A. (1974) Electron beam microanalysis. ASTM STP 506 (Philadelphia, PA).

Beech, I.B., Chung, C.W.S., Johnson, D.B. and Smith, T.R. (1996) Comparative studies of bacterial biofilms on steel surfaces using atomic force microscopy and environmental scanning electron microscopy. *Biofouling* **10**(1–3), 65–78.

Chandra, D., Roy, P., Mishra, A.K., Chakrabarti, J.N. and Sengupta, B. (1979) Microbial removal of organic sulfur from coal. *Fuel* **58**, 549.

Characklis, W. and Marshall, K.C. (1990) Biofilms: a basis for an interdisciplinary approach. In *Biofilms* (ed. W.G. Characklis and K.C. Marshall), pp. 3–15, John Wiley & Sons, New York.

Characklis, W., Turakhia, M. and Zelver, N. (1990) Transport and interfacial transfer phenomena. In *Biofilms* (ed. W.G. Characklis and K.C. Marshall), pp. 265–340, John Wiley & Sons, New York.

Chiou, W.A., Kohyama, N., Little, B., Wagner, P. and Meshi, M. (1996) *TEM Study of a Biofilm on Copper Corrosion* (ed. G.W. Bailey, J.M. Corbett, R.V.W. Dimlich, J.R. Michael and N.J. Zaluzec), pp. 220–221, San Francisco Press, San Francisco, CA.

Clarke, J.A., Ward, P.R. and Salsbury, A.J. (1973) High voltage electron microscopy of cellular changes in the wet state. *J. Microsc.* **97**(3), 365–368.

Daulton, T., Little, B.J., Lowe, K. and Jones-Mehan, J. (2002) A technique for examination of microbial chromium reduction. *J. Microbiol. Meth.* **50**, 39–54.

Geesey, G. and Jang, L. (1990) Extracellular polymers for metal binding. In *Mineral Recovery* (ed. H.L Ehrlich and C.L. Brierley), pp. 223–247, McGraw-Hill, New York.

Fukami A., Fukushima, K. and Kohyama, N. (1991) Observation technique for wet clay minerals using film-sealed environmental cell equipment attached to high-resolution electron microscope. In *Microstructure of Fine-Grained Sediments from Mud to Shale* (ed. R.H. Bennett, W.R. Bryant and M.H. Hulbert), pp. 321–331, Springer-Verlag, New York.

Fukushima, K., Katho, M. and Fukami, A. (1982) Quantative measurements of radiation damage to hydrated specimens observed in a wet gas environment. *J. Electron Microsc.* **31**(2), 119–126.

Hui, S.W. (1976) Direct observation of membrane movement by electron microscopy. In *Membranes and Neoplasia: New Approaches and Strategies* (ed. V.T. Marchesi), pp. 159–170, Alan R. Liss, New York.

Lawrence, R.W. and Bruynesteyn, A. (1983) Biologic preoxidation to enhance gold and silver recovery from refractory pyritic ores and concentrates. *CIM Bull.* **76**, 107.

Little, B., Wagner, P. and Mansfeld, F. (1991) An overview microbiologically influenced corrosion of metals and alloys. *Int. Mater. Rev.* **36**(6), 253–272.

Little, B., Wagner, P., Ray, R. and Sheetz, R. (1991) Biofilms: an ESEM evaluation of artifacts introduced during SEM preparation. *J. Indus. Microbiol.* **8**, 213–222.

Little, B.J., Wagner, P., Maki, J.S., Walch, M. and Mitchell, R. (1986) Factors influencing the adhesion of microorganisms to surfaces. *J. Adhesion* **20**, 187.

Little, B.J., Daulton, T., Pope, R.K. and Ray, R. (2001) Application of environmental cell transmission electron microscopy to microbiologically influenced corrosion. In *Proceedings of the NACE International Conference,* Houston, TX, March.

Mansfeld, F., Postyn, A., Shih, H., Devinny, J., Islander, R. and Chen, C.L. (1990) Corrosion monitoring and control in concrete sewer pipes. In *Proceedings of the Corrosion '90,* Las Vegas, NV, April 22–27; p. 113.

Marszalek, D.S., Gerchakov, M. and Udey, L.R. (1979) Influence of substrate composition on marine microfouling. *Appl. Environ. Microbiol.* **38**, 987.

McEldowney, S. and Fletcher, M. (1988) Bacterial desorption from food container and food processing surfaces. *Microb. Ecol.* **15**, 229.

Mitchell, R. and Kirchman, D. (1981) The microbial ecology of marine surfaces. In *Marine Biodeterioration* (ed. J.D. Costlow and R.C. Tipper), pp. 49–56, Naval Institute Press, Annapolis, MD.

Moreton, B.B. and Glover, T.G. (1980) New marine industry applications for corrosion and biofouling resistant, copper nickel alloys. In *Proceedings of the 5th International Congress on Marine Corrosion and Fouling*, Biologia Marina, Barcelona, Spain, May 19–23; p. 267.

Nilsson, I. and Ohlson, S. (1982) Columnar denitrification of water by immobilized Pseudomonas denitrificans cells. *Eur. J. Appl. Microbiol. Biotechnol.* **14**, 86.

Parsons, D.F. (1974) Structure of wet specimens in electron microscopy. *Science* **186**, 407–414.

Robinson, V.N.E. (1975) A wet stage modification to a scanning electron microscope. *J. Microsc.* **103**, 71–77.

Sieburth, J.M. (1975) *A Pictorial Essay on Marine Microorganisms and Their Environments*, University Press, Baltimore, MD.

Suda H., Ishikawa, A. and Fukami, A. (1992) Evaluation of the critical electron dose on the contractile activity of hydrated muscle fibers in the film-sealed environmental cell. *J. Electron Microsc.* **41**, 223–229.

Surman, S.B., Walker J.T., Goddard, D.T., Morton, L.H.G., Keevil, C.W., Weaver, W., Skinner, A., Hanson, K. Caldwell, D. and Kurtz, J. (1996) Comparison of microscope techniques for the examination of biofilms. *J. Microbiol. Meth.* **25**, 57–70.

20

Use of molecular probes to study biofilms

B. Zhang, B. Mariñas and L. Raskin

20.1 INTRODUCTION

Up to the early 1990s, meso- (0.001–10 m) or macro- (>10 m) scale parameters, such as the concentrations of degraded organics and the levels of dissolved oxygen, were primarily measured in biofilm research. Accordingly, the techniques used were mostly physico-chemically based. Most microbiological methods employed were culture dependent. Since then, techniques have become available that make it possible to measure micro-scale (<0.001 m) parameters in biofilm research, such as the thickness of thin biofilms and the composition of biofilm communities. The analysis of biofilm communities requires the use of molecular techniques and fits in the general area of microbial ecology. In addition to providing a fundamental understanding of the biofilm communities by describing population composition and interactions, molecular microbial ecology studies help engineers to diagnose and prevent possible problems and achieve optimal operation of their systems.

A number of different nucleic-acid-based techniques have been used in biofilm research. Until now, the 16S and 23S ribosomal RNA (rRNA) and ribosomal DNA (rDNA) have been the most widely used target nucleic acids. Although techniques targeting rRNA and rDNA do not provide much information on the physiology of microorganisms, they are powerful tools to determine the identity of microorganisms, trace the presence of certain populations, and study overall community structure and diversity of biofilms. In certain cases, they can be used to obtain information on activity levels (Poulsen *et al.*, 1993; Licht *et al.*, 1999). One commonly used approach is based on the construction of clone libraries of community rDNAs. Sequencing of the target molecule allows identification of the community members represented in clone libraries through comparison with existing sequences present in databases. In addition, several nucleic acid fingerprinting techniques, such as denaturing gradient gel electrophoresis (DGGE) (Santegoeds *et al.*, 1998) and terminal restriction fragment length polymorphism (T-RFLP) (Sakano *et al.*, 2002) have been applied in biofilm studies to follow changes in community structure or to compare community diversity in various samples. None of these techniques can be used to study the three-dimensional structure of biofilms because of the need to extract DNA from biofilm samples to perform the necessary analyses.

Another approach, which relies on the identification of rRNA 'signature' sequences of specific populations to design oligonucleotide hybridisation probes, has the potential to non-destructively study biofilms and thus allows studying the three-dimensional structure of biofilms. Nevertheless, many hybridisation methods are still destructive since they rely on obtaining a pool of representative community rRNA in solution. Membrane-based hybridisations (Mobarry *et al.*, 1996; Raskin *et al.*, 1995), microarray methods (Koizumi *et al.*, 2002), and solution-based hybridisations (Xi *et al.*, 2003) fall in this category of 'destructive' hybridisations. In contrast, whole cell hybridisations (hybridisations performed with morphologically intact cells) and *in situ* hybridisations (whole cell hybridisations targeting cells in their natural environment) (Amann *et al.*, 1995) can be used to study the three-dimensional distribution of cells while maintaining the structural integrity of biofilms. In this chapter, as in most other studies, the term 'fluorescence *in situ* hybridisation' (FISH) is used to describe both whole cell and *in situ* hybridisations. As discussed in detail by Oerther *et al.* (1999), either approach (hybridisations relying on extraction of nucleic acids and FISH) has inherent advantages and drawbacks. FISH can provide discrete information regarding individual cells, in addition to providing information on the three-dimensional structure of biofilms, while extraction based assays provide 'lumped' results (since information on cell abundance and rRNA content is combined) for an entire population. FISH also has the potential to provide 'lumped' data through the quantification of probe conferred signals for large numbers of cells. However, the generation of meaningful statistical data with FISH is currently still difficult.

This chapter focuses on the use of 16S rRNA oligonucleotide probe hybridisations to study biofilm composition and structure. In Section 20.2, general procedures for probe design and characterisation are discussed and this discussion is illustrated with data on the design and characterisation of probes developed to track two opportunistic pathogens of concern in drinking water treatment, i.e. *Mycobacterium avium* and *Aeromonas hydrophila*. Subsequently, a summary of concerns related to membrane hybridisations is presented in Section 20.3. Section 20.4 provides an overview of the FISH procedure as well as a detailed discussion of challenges related to several of the experimental steps involved in FISH. Reference is made to data collected with probes for *M. avium* and *A. hydrophila*, when appropriate. Finally, Section 20.5 includes a discussion of methods that combine FISH with other techniques to broaden the applicability of FISH.

20.2   PROBE DESIGN AND CHARACTERISATION

20.2.1   Oligonucleotide probe design

The primary structure of rRNA molecules composed of sequence regions with varying degrees of conservation enables the design of oligonucleotide probes with specificities at different phylogenetic levels, ranging from the domain- to the species-level. To localise specific rRNA signature sequences that are unique for the desired phylogenetic group, aligned rRNA sequences must be obtained from appropriate databases, such as the GenBank Sequence Database (http://www.ncbi.nlm.nih.gov), the EMBL Nucleotide Sequence Database (http://www.ebi.ac.uk), and the Ribosomal Database Project (RDP) (http://www.cme.msu.edu/RDP/html/). These databases are regularly updated to incorporate newly obtained sequences. Alternatively, probe design can be performed using a software package called ARB. This package, based on a Linux platform, provides multi-functional phylogenetic analyses and can be downloaded from the World Wide Web (http://www.mpi-bremen.de/molecol/).

The specificity of a potential probe targeting a signature sequence for a defined phylogenetic group of organisms should be carefully checked during the process of probe design. When hybridisation and/or washes are performed under relatively high stringency, oligonucleotide probes can be used to discriminate between target sequences that differ in only one nucleotide (Stahl and Amann, 1991). If an intended target species has a single mismatch between the probe and the corresponding rRNA sequence, it may be possible to lower hybridisation and/or wash stringencies to include the hybridisation signal contributed by this microorganism. However, by doing this, hybridisation to non-target species with a small number of mismatches may compromise probe specificity. Consequently, database searches are essential to identify mismatches in non-target and target species. RDP, GenBank, and ARB provide such evaluations of probe specificity through functions termed Probe Match (http://rdp.cme.msu.edu/html/analyses.html), basic local alignment search tool (BLAST) network service (http://www.ncbi.nlm.nih.gov/BLAST), and Probe Check (a tool box incorporated in the ARB software package), respectively. Evaluations of probe specificity should be repeated on a regular basis in order to include the rapidly expanding collection of nucleic acid sequences in the analysis. In addition, since the position and composition of a mismatch between probe and target influence the extent of hybrid destabilisation (Stahl and Amann, 1991), and since it is still difficult to predict hybrid stability based on sequence composition, it is necessary to evaluate the specificity of newly designed probes experimentally (Section 20.2.2). This is particularly important when the probes are to be used for studies of complex microbial communities, such as those in the majority of biofilms, because they often contain novel organisms.

Another factor, which should be considered during the process of probe design, involves internal complementarity within an oligonucleotide probe: the formation of stable stem structures within probes may prevent hybrid formation with the target sequence or reduce hybridisation rates. Raskin *et al.* (1997) indicate that internal complementarity involving less than five canonical base pairs does not appreciably reduce the expected hybridisation response, based on their experience with oligonucleotide probes (15–25 nucleotides) targeting 16S rRNA. Internal complementarity involving five or more canonical base pairs generally results in low hybridisation signals. Since it is sometimes difficult to eliminate secondary structure during the process of probe design, we recommend using on-line programs for a systematic evaluation of the thermodynamic stability of stem loop structures formed by oligonucleotide probes (http://www.rna.icmb.utexas.edu/ and http://www.bioinfo.rpi.edu/applications/mfold/).

Several studies indicate that differential target accessibility should be considered during the process of probe design. Even though the higher-order structures of rRNAs have been determined at a relatively detailed level (Gutell *et al.*, 1994) and these models can be used to help evaluate inter- and intra-molecular structures (e.g. RNA–RNA or RNA–protein interactions), target accessibility is not well understood and has been poorly documented in the literature. This problem has been cited in the context of FISH (Amann *et al.*, 1995; Frischer *et al.*, 1996) and a first attempt was made to evaluate higher-order structure during the process of probe design (Frischer *et al.*, 1996). Subsequently, Fuchs *et al.* (1998) reported a systematic study on the accessibility of 16S rRNA target sites for FISH probes. The target sites were grouped into six classes of brightness according to their relative fluorescence. The fluorescence intensities obtained with more than 200 probes with fixed cells of *E. coli* were reported as a map distribution of relative fluorescence intensities of oligonucleotide probes on the 16S rRNA molecule. Thus, during the process of FISH probe design, it is important to evaluate target sites for new probes using the map of fluorescence intensities of oligonucleotide probes on the 16S rRNA molecule developed by Fushs *et al.* (1998) as well as a list of target sites for probes that have been used successfully in FISH (Amann *et al.*, 1995). If selection of probe target sites in desirable regions is not possible, unlabelled oligonucleotides, termed 'helpers', can be used to improve site accessibility (Fuchs *et al.*, 2000). These helpers are designed to bind to sites adjacent to the probe target site to help open target sites with low accessibility. Using these helpers, Fuchs *et al.* (2000) were able to substantially increase FISH signals. The use of peptide nucleic acid (PNA) oligonucleotide probes is another way to alleviate the problem of differential target accessibility. PNA is a DNA analogue with a polyamide backbone instead of a sugar phosphate backbone (Egholm *et al.*, 1993). PNA can hybridise to nucleic acids in the absence of counterions because of its neutral backbone. Thus, PNA oligonucleotide probes exhibit superior hybridisation characteristics compared to DNA probes, especially under low-salt conditions (Perry-O' Keefe *et al.*, 2001; Stender *et al.*, 2002). Even though differential target accessibility has been cited primarily as a concern in FISH, Raskin *et al.* (1997) demonstrated that it might be important in membrane hybridisations as well.

Finally, hybrid duplex stability should be considered during the process of probe design. There are several empirically derived relationships to estimate temperature of dissociation (T_d) values (Stahl and Amann, 1991). They should be used during the process of probe design. It is often possible to alter the predicted T_d value by making a probe longer or shorter even though probe design does not always allow much flexibility. Obviously, probe specificity should not be compromised during this process. When multiple probes are being used to determine the abundance/activity of multiple populations in the same sample, it is preferable to design a set of probes with T_d values that are relatively close to each other. The availability of probes with similar T_d values is mostly a matter of experimental convenience when employing membrane hybridisations. However, when multiple probes are to be used simultaneously in FISH, hybridisation conditions can only be optimised for all probes involved if their T_d values are similar. Despite the importance of evaluating duplex stability during the process of probe design, it is recommended to experimentally determine the hybridisation and wash stringencies for each newly designed probe (see Section 20.2.2), since detailed rules for predicting duplex stability are not yet well established.

Illustration

Based on comparative 16S rRNA sequence analyses using RDP and ARB, a set of nested 16S rRNA targeted oligonucleotide probes for the opportunistic pathogens *Mycobacterium avium* complex (MAC) and *A. hydrophila* were selected and designed based on the phylogeny of the genera *Mycobacterium* and *Aeromonas*. The oligonucleotide probes and their target groups are listed in Table 20.1.

Figure 20.1 shows a 16S rRNA based phylogenetic tree for the genus *Mycobacterium*. This tree indicates a clear distinction between fast-growing (above

Table 20.1   Oligonucleotide probes with target groups.

Probe	Sequence (5'–3')	E. coli numbering	Target organisms
S-G-Mycob-0829-a-A-18	AAGGAAGGAAACCCACAC	829–847	Most *Mycobacterium* spp.
S-*-Mycob-0998-a-A-23	CTATCTCTAGACGCGTCCTGTGC	998–1021	Most slow-growing mycobacteria
S-*-Mycob-0068-a-A-23	AGTACCTCCGAAGAGGCCTTTCC	68–91	*M. avium* and *M. paratuberculosis*
S-G-Aero-0066-a-A-18	CTACTTTCCCGCTGCCGC	66–83	Most *Aeromonas* spp.
S-*-Aero-0222-a-A-18	ACCTGGGCATATCCAATC	222–239	Subgroup of aeromonads
S-*-Aero-0456-a-A-22	GTTGATACGTATTAGGCATCAA	456–477	*A. hydrophila* and *A. encheleia*

the dashed line) and slow-growing mycobacteria (below the dashed line). Many of the slow-growing mycobacteria, including *M. avium*, are human and animal pathogens. *M. avium* is closely related to *M. paratuberculosis* and *M. intracellulare*. Therefore, these three organisms are often referred to together as the MAC. Based upon the phylogeny of *Mycobacterium* spp., a set of three previously published probes were selected. Urbance (1992) worked on the phylogeny of *Mycobacterium* and developed an extensive set of probes for *Mycobacterium* spp., which have not been published in the peer-reviewed literature. Probes S-S-Mycob-0068-a-A-23 and S-*-Mycob-0998-a-A-23 were selected from his Ph.D. dissertation (Urbance, 1992). Schwartz *et al.* (1998) also reported a set of nested probes for *Mycobacterium* spp. Probe S-G-Mycob-0829-a-A-18 was chosen from this study.

Since these three probes were developed a number of years ago, the specificity of each probe was checked again by using the Probe Match program provided by RDP and the BLAST network service available through GenBank. The results of the specificity analysis indicate that probe S-G-Mycob-0829-a-A-18 targets 121 of the 150 accessible 16S rRNA sequences of *Mycobacterium* spp. For the remaining 29 sequences, nine display more than one mismatches or a one-nucleotide deletion in the target site; for the other 20 sequences, no sequence information is available within the target site of the probe. All bacteria not affiliated with the genus *Mycobacterium* display at least two mismatches in the target region. Probe

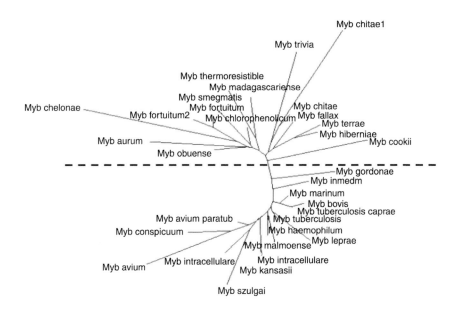

Figure 20.1   16S rRNA based phylogenetic tree for mycobacteria. The tree was derived by using a neighbour-joining method integrated in the ARB software package.

S-*-Mycob-0998-a-A-23 targets all the slow-growing mycobacteria except *M. triviale, M. simiae, M. gordonae*, and *M. leprae*. At least one mismatch was found in the corresponding sequences of *M. leprae* and *M. komossense*. Other fast-growing mycobacteria and bacteria not affiliated with the genus *Mycobacterium* have at least two mismatches with this probe. The third probe, S-*-Mycob-0068-a-A-23, targets *M. avium, M. paratuberculosis*, and *M. lentiflavum*, but has at least two mismatches with other mycobacteria and bacteria not affiliated with the genus *Mycobacterium*.

Figure 20.2 shows a 16S rRNA based phylogenetic tree for the genus *Aeromonas*. Kampfer *et al.* (1996) identified several oligonucleotide probes for this genus. Probes S-G-Aero-0066-a-A-18 and S-*-Aero-0222-a-A-18 were selected from their study. A new probe, S-*-Aero-0456-a-A-22, specific for *A. hydrophila* and *A. encheleia*, was also designed. The specificity of each probe was checked using the Probe Match program provided by RDP and the BLAST network service available through GenBank. This evaluation indicated that probe S-G-Aero-0066-a-A-18 targets all available *Aeromonas* species except for *A. shubertii* (> two mismatches) and *A. popoffii* (one mismatch in the middle of the target region). This probe has at least two mismatches with bacteria not affiliated with the genus *Aeromonas*. Probe S-*-Aero-0222-a-A-18 targets a subgroup of *Aeromonas* spp. that includes *A. hydrophila, A. media, A. caviae, A. trata*, and *A. encheleia* (Figure 20.2). This probe has at least two mismatches with other *Aeromonas* spp. and

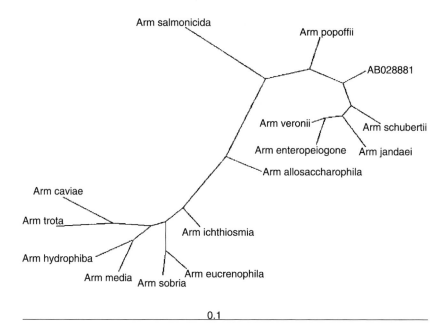

Figure 20.2   16S rRNA based phylogenetic tree for aeromonads. The tree was derived by using a neighbour-joining method integrated in the ARB software package.

bacteria not affiliated with the genus *Aeromonas*. Finally, probe S-*-Aero-0456-a-A-22 targets *A. hydrophila* and *A. encheleia*, a recently described species. This probe has at least two mismatches with other *Aeromonas* spp. and bacteria not affiliated with the genus *Aeromonas*.

20.2.2   Experimental probe characterisation

During the last decade, a large number of studies have been published that performed detailed characterisations of oligonucleotide hybridisation probes before using these probes to detect and quantify microbial populations in environmental samples. On-line databases are available that summarise probe information from these studies, such as the oligonucleotide probe database (http://www.cme.msu.edu/OPD) and oligo retrieval system (http://soul.mikro.biologie.tu-menchen.de/ORS).

Raskin *et al.* (1997) provide a detailed discussion of issues to be considered when characterising oligonucleotide probes experimentally. First, they emphasise the need to determine probe-target duplex stability experimentally, despite the availability of empirically derived relationships to estimate T_d values. Furthermore, they recommend evaluating probe specificity using a number of target and non-target species to confirm that the experimentally determined optimal hybridisation and wash stringencies result in the desired specificity. They also indicate that probe specificity can be tested further using the concept of 'probe nesting'. Oligonucleotide probes designed at different levels of specificity (e.g. family-, genus-, and species-specific probes) can be used to characterise a single environment. Assuming that each set of probes encompasses the complete diversity of target species present in a certain sample, the sum of the amounts of 16S rRNAs quantified by a set of specific probes (e.g. species-specific probes) should equal the amount determined by a more general probe (e.g. genus-specific probe). Thus, consistency between hybridisation results of general and specific probes increases the confidence in probe specificity, whereas inconsistencies may indicate that at least one probe of the nested set lacks specificity. Alternatively, the possibility remains that uncultured (novel) species related to the probe target group might not be detected by one (or more) of the nested set of probes, while being detected by another probe of different specificity.

Illustration

Optimal hybridisation wash temperatures and probe specificities for the probes listed in Table 20.1 were evaluated experimentally using a variety of pure cultures. The following bacteria were obtained from the American Type Culture Collection (ATCC, Manassas, VA): *M. avium* (ATCC 15769), *M. intracellulare* (ATCC 35767), *M. paratuberculosis* (ATCC 19698), *M. kansasii* (ATCC 12478), *M. komossense* (ATCC 33013), *A. hydrophila* (ATCC 35654), *A. media* (ATCC 33907), *A. shubertii* (ATCC 43700), *A. salmonicida* (ATCC 33658) and *N. nova* (ATCC 33727). The organisms were grown according to recommendations by ATCC and harvested

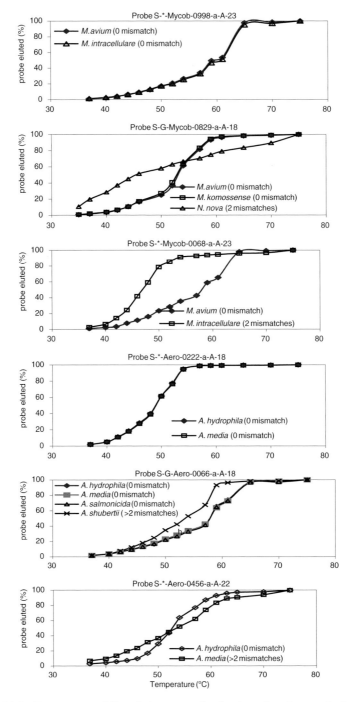

Figure 20.3   Temperature of dissociation curves for the six probes characterised in this study.

at mid-log phase. RNA was extracted from cell pellets using a low-pH hot-phenol method. The concentration of extracted RNA was quantified with a UV spectrophotometer by assuming that 1 mg of RNA per 1 ml is equal to 20 optical density units at an absorbance of 260 nm (A_{260}). The quality of extracted RNA was evaluated by polyacrylamide gel electrophoresis (Alm and Stahl, 2000).

To determine the optimal hybridisation wash temperatures, T_d studies were performed as previously described (Zheng et al., 1996). In brief, RNA was extracted from a variety of target and non-target organisms, denatured with glutaraldehyde, and 50 ng of denatured RNA samples were applied to nylon membranes (Magna Charge, Micron Separation Inc., Westborough, MA). The membranes were baked at 80°C for 2 h, pre-hybridised for 12 h, and hybridised for 16 h using oligonucleotide probes that were 5' end labelled with $[\gamma^{-32}P]ATP$. Washes were performed at increasing temperatures and the amount of probe washed off was determined with scintillation counting. The T_d of the probe was defined as the temperature at which 50% of the hybridised probes were removed from the membrane (Figure 20.3). Table 20.2 lists the experimentally determined T_d values for the six probes characterised in this study.

Using the experimentally determined T_d for perfect duplexes as the post-hybridisation wash temperature usually ensures that duplexes with one or more mismatches are dissociated (Stahl and Amann, 1991). Therefore, the T_d values were used as the post-hybridisation wash temperatures in experimental specificity studies. A universal probe S-*-Univ-1390-a-A-20 (Zheng et al., 1996), which targets the 16S rRNA of most known organisms, was also included in the specificity studies as a control. The results of the specificity study are shown in Figure 20.4. The experimental specificity results agree well with the theoretical predictions. Since probe S-*-Univ-1390-a-A-18 targets all organisms, all 10 rRNA samples provide a positive hybridisation response. The lower signal associated with some samples can be attributed to smaller amounts of rRNA applied to the membrane. Hybridisation with probe S-G-Mycob-0829-a-A-18 resulted in a positive signal for all members of the *Mycobacterium* genus and in an absence of response for *N. nova* (two mismatches). Hybridisations with probe S-*-Mycob-0998-a-A-23 gave signals for *M. avium*, *M. paratuberculosis*, *M. intracellulare*, and *M. kansasii*.

Table 20.2   Oligonucleotide probes, temperature of dissociation (T_d), and optimal formamide (FA) concentration range for FISH.

Probe	T_d	FA (%)
S-G-Mycob-0829-a-A-18	50.5	20–35
S-*-Mycob-0998-a-A-23	62.5	30–50
S-*-Mycob-0068-a-A-23	61.0	30–50
S-G-Aero-0066-a-A-18	61.5	40–50
S-*-Aero-0222-a-A-18	51.0	30–50
S-*-Aero-0456-a-A-22	53.5	40–55

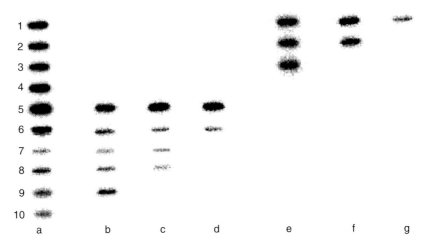

Figure 20.4   Probe specificity study. Membrane hybridisation results were analysed with an Instant Imager (Packard Instrument Co. Meriden, CT) and with Adobe Photoshop 5.0 (Adobe, Seattle, WA). The organisms from which rRNA was extracted and immobilised on the membranes were: (1) *A. hydrophila*; (2) *A. media*; (3) *A. salmonicida*; (4) *A. shubertti*; (5) *M. avium*; (6) *M. paratuberculosis*; (7) *M. intracellulare*; (8) *M. kansasii*; (9) *M. komossense*; (10) *N. nova*. The oliglonucleotide probes used were: (a) S-*-Univ-1390-a-A-18; (b) S-G-Mycob-0829-a-A-18; (c) S-*-Mycob-0998-a-A-23; (d) S-*-Mycob-0068-a-A-23; (e) S-G-Aero-0066-a-A-18; (f) S-*-Aero-0222-a-A-18; (g) S-*-Aero-0456-a-A-22.

This probe has one mismatch with *M. komossense* and two mismatches with *N. nova*. The third *Mycobacterium* probe, probe S-*-Mycob-0068-a-A-23 targets *M. avium* and *M. paratuberculosis* as anticipated. None of the *Mycobacterium* probes hybridised with any of the *Aeromonas* spp. Similarly, hybridisation results with the three *Aeromonas* probes, S-G-Aero-0066-a-A-18, S-*-Aero-0222-a-A-18, and S-*-Aero-0456-a-A-22, indicated anticipated specificities.

The optimal FISH hybridisation/wash conditions for each probe were evaluated by using different concentrations of FA in the hybridisation buffer (see Section 20.4.3 for fixation and hybridisation conditions). After hybridisation, the microscope slides were washed using wash solutions with concentrations of NaCl corresponding to the percentage of FA used during hybridisation (de los Reyes *et al.*, 1997). Increasing the FA concentration by 1% has the approximate equivalent effect of increasing the T_d by 0.7°C (Stahl and Amann, 1991). Thus, different wash temperatures in membrane hybridisation can be correlated with different FA concentrations in the hybridisation buffer and various NaCl concentrations in the wash solutions. The selected FA concentrations were: 0%, 10%, 20%, 30%, 35%, 40%, 45%, 50%, 60%, and 70%. The six probes were evaluated using target and non-target organisms for each of these conditions. Cells were visualised with a microscope fitted with the appropriate filter sets and signal intensities were quantified using the arbitrary unit of average signal intensity per pixel. Signals for target species were consistently higher than those for non-target species (data not shown).

The optimum ranges of FA concentrations were determined for each probe and are listed in Table 20.2. The ranges of FA concentrations for the six probes exhibit certain levels of overlap, which makes it possible to simultaneously apply several probes in one hybridisation experiment.

20.3   MEMBRANE HYBRIDISATION

Membrane hybridisation or slot (dot) blot hybridisation involves the application of RNA extracted from samples that need to be characterised on nylon or other types of membranes. If quantification is an objective, a dilution series of RNA obtained from a pure culture targeted by the probe used in the hybridisation (reference RNA) is also applied to the membrane. Subsequently, the membranes are pre-hybridised, hybridised with the probes, and washed. Usually, membrane hybridisations are conducted at low stringency (high salt concentrations and low temperatures) and the washing step is used to remove non-specifically bound probe (see Section 20.2.2).

Raskin *et al.* (1997) and Hofman-Bang *et al.* (2002) provide detailed discussions of issues to be considered when performing quantitative membrane hybridisations. These reviews include discussions of the sensitivity and precision of the hybridisation method, problems associated with compounds co-extracted with RNA (e.g. DNA and humic acids), saturation of membranes, non-linear hybridisation responses, use of non-native RNA (i.e. *in vitro* transcribed rRNA) as reference RNA, and interpretation of hybridisation outputs.

Membrane hybridisation has been used to determine the composition of biofilms (Mobarry *et al.*, 1996; Raskin *et al.*, 1996). As discussed in Section 20.1, membrane hybridisations cannot provide information on the three-dimensional structure of biofilms because of the destructive nature of the RNA extraction procedure, unless it is possible to slice the biofilm before RNA extractions are performed. For this reason, it has been much more common to use FISH to analyse microbial community composition and three-dimensional distribution of populations.

20.4   FLUORESCENCE *IN SITU* HYBRIDISATION

Oligonucleotide probes may be labelled chemically or enzymatically either directly with fluorochromes and enzymes (alkaline phopsphatase, horseradish peroxidase) or indirectly using haptens. The application of hapten- or enzyme-linked probes is limited by the molecular size of the antihapten–antibody or the oligonucleotide-enzyme conjugate, which may be too large to penetrate biofilm matrices and cell walls of certain target cells (Manz, 1999). Theoretically, any fluorochrome with preferred excitation and emission wavelengths can be used to label the probes. Commonly used fluorochromes are cyanine dyes (Cy5, Cy3, and Cy7), tetra-methylrhodamine isothiocyanate (TRITC), thiocyanate (FITC), and Texas Red.

FISH involves several steps, including sample preparation (cell fixation, cell permeabilisation, and cell dehydration), hybridisation, removal of unbound probes in a washing step, and fluorescence signal detection using epifluorescence microscopy, laser scanning confocal microscopy, or flow cytometry. In contrast to membrane hybridisations, the hybridisation step of FISH is usually conducted at a relatively high stringency and the washing step is used to remove excess probes. Although temperature can be used to control stringency, salt or FA concentrations are more often used (see Section 20.2.2). This is experimentally more convenient since only one temperature is needed for hybridisations at different stringencies.

Despite the success of FISH in many applications, some challenges related to several of the experimental steps remain, such as poor sensitivity if the rRNA content of target cells is low, variable target site accessibility, non-uniform cell permeability, data analysis, and background fluorescence derived from autofluorescence of samples, insufficient washing, or unspecific binding of probes to compounds present in samples that can not be removed by washing. In the following sections, some of these issues and concepts are addressed and illustrated with data from our study with *Mycobacterium* and *Aeromonas* probes.

20.4.1   Poor sensitivity

The fact that slow-growing organisms usually have a lower rRNA content than fast-growing organisms makes it sometimes difficult to use FISH to detect cells exhibiting low-growth activities. When encountering a sensitivity problem, possibly confounded with a background problem, it is important to evaluate the imaging method used. A laser scanning confocal microscope is much more powerful than a normal epifluorescence microscope. The use of commercial UV systems and multiphoton laser microscopy systems allows collection of multiple quantitative data for a single site within a biofilm (Lawrence *et al.*, 1998). Furthermore, a tyramide signal amplification system combined with FISH is able to amplify the signal 7–12-fold (Lebaron *et al.*, 1997). If possible, multiple probes with the same specificity can be used simultaneously to enhance signal intensity (Amann *et al.*, 1995).

20.4.2   Variable target site accessibility

The issue of variable target site accessibility is discussed in detail in the context of probe design (Section 20.2.1). It is best to evaluate whether a newly designed probe targets an accessible region during the process of probe design so that alternative probes with the same specificity can be selected without investment in experimental probe characterisation work.

Based on the map of fluorescence intensities of oligonucleotide probes on the 16S rRNA molecule developed by Fuchs *et al.* (1998) (Section 20.2.1), the site accessibilities of three of the six probes listed in Table 20.1 may be limited and experimental conditions will need to be optimised to ensure that the use of these probes will be possible. So far, we have not experienced problems with these probes when using them with pure cultures. Helper probes or PNA probes (Section 20.2.1) could provide a solution if the combination of poor target site accessibility and sensitivity becomes an issue.

20.4.3   Non-uniform cell permeability

Fixation is a critical step in preparing the sample for FISH. Fixation helps to maintain the integrity of cell morphology as well as to increase the permeability of the cell envelop. Thus, the ideal fixative should be able to penetrate cells or tissue rapidly as well as preserve cellular structure. Also, the fixative should not cause autofluorescence. Since poor fixation can be the reason behind unsuccessful FISH, it is recommended to evaluate the permeability of the cell wall using previously characterised and well-labelled universal or domain-specific probes (Amann *et al.*, 1995).

Chemical fixatives can be categorised into coagulating or cross-linking fixatives. Ethanol, methanol, and acetone are representatives of the first group, while formaldehyde and glutaraldehyde fall in the second group. In general, 50% ethanol offers good results for fixation of Gram-positive cells, whereas 4% paraformaldehyde (PFA) works well for Gram-negative cells (Amann *et al.*, 1995). Gram-negative cells can also be fixed with 50% ethanol, but the fixed cells were found to have low stability. Earlier studies indicated that formaldehyde is not suitable for Gram-positive cells since it can cause cross linkage of proteins within the Gram-positive cell wall to the extent that probes cannot penetrate the cells. However, excessive cross linking can be avoided by shortening the PFA fixation time (de los Reyes *et al.*, 1997). Some Gram-positive cells need treatment with lytic enzymes (mutanolysin or lysozyme), hydrophobic solvents (toluene or diethylether), or acids for proper fixation (Amann, 1995). Burggraf *et al.* (1994) demonstrated that 4% PFA was a suitable fixative for most *Archaea* tested in their study, whereas Zheng and Raskin (2000) and Sorensen *et al.* (1997) found that optimal fixation conditions were quite variable for methanogens with different cell wall structures. Since optimal fixation conditions appear to be species dependent even within a group of closely related species (de los Reyes *et al.*, 1997; Zheng and Raskin, 2000), it is difficult to fix all cells in a complex community with a single treatment. Therefore, different fixation methods should be used according to the cell wall properties of the populations that are to be detected.

Illustration

A range of fixation/permeabilisation conditions was evaluated with pure cultures of *A. hydrophila* and *M. avium* to determine the combined optimal fixation conditions in order to allow simultaneous detection of these Gram-negative and Gram-positive cells in biofilm samples. Oligonucleotide probes were synthesised with an aminolinker at the 5' end and were coupled to a fluorescent dye (Megabase, Evanston, IL). The following conditions were evaluated: 4% PFA for 1, 10, and 30 min, 1, 2, 4, 6, and 15 h; 50% ethanol for 10, and 30 min, 1, 2, 6, and 14 h;

80% ethanol for 5, 15, and 30 min, 1, and 2 h; 1-M HCl for 30 min, 50 min, and 1 h; 4% PFA (0, 30, 60, and 120 min) followed by 5000 U/ml mutanolysin treatment at 37°C (0, 20, 40, and 60 min). A general bacterial probe (probe S-D-Bact-0338-a-A-18; Stahl and Amann, 1991) labelled with FITC was used in this optimisation experiment.

Hybridisations were performed as described by de los Reyes *et al.* (1997) with slight modifications. One volume of overnight culture was added to three volumes of fresh fixative. After fixation, samples were washed with $1 \times$ phosphate-buffered saline (PBS) buffer (130 mM NaCl, 10 mM sodium phosphate, pH 7.2), re-suspended in 1 : 1 (vol : vol) PBS : ethanol, and stored at -20°C. Fixed cells were then applied to wells on a microscope slide, air-dried, and dehydrated by serial immersion in 50%, 80%, and 96% ethanol (2 min each). For some treatments, cells were treated with mutanolysin at 37°C and dehydration was repeated. Subsequently, 8 μl of hybridisation solution and 1 μl (25 ng) of FITC-labelled probe were applied to each well. The slides were incubated at 46°C in a moisture chamber for 2 h. After hybridisation, the slides were rinsed once with pre-warmed (48°C) wash solution, washed for 20 min at 48°C, dipped into ice-cold water, 4′(,6′(-diamidino-2′-phenylindolehydrochloride (DAPI) stained, air dried, and mounted in Citifluor (Citifluor Ltd., London, UK). The cells were visualised with an epifluorescence microscope (Axioskop; Carl Zeiss, Germany) fitted with filter sets (Chroma Tech. Corp., Brattleboro, Vt.) for TRITC (filter set 41002), FITC (filter set 41001), and DAPI (filter set 31000) and a Zeiss 50-W high-pressure bulb. Images were captured with a liquid-cooled charge coupled device (CCD) camera (Photometrics MXC200L, class 2 Kodak KAF 1400 CCD, 1317-by-1035 pixel array, 6.8-mm pixel size; Photometrics Ltd., Tucson, AZ) controlled by a PowerPC 7100 (Apple Computer Inc., Cupertino, CA). To provide a comparative measure of fluorescence, a constant exposure time of 2 s was used for images captured to compare fixation treatments. Image acquisition was controlled with IPLab Spectrum image analysis software (Signal Analytics, Vienna, Va). The images were analysed with NIH Scion imaging software and Adobe Photoshop 5.0 (Adobe, Seattle, WA). Background signals were quantified similarly and subtracted from the mean pixel value for each image.

For all fixation/permeabilisation treatments, the Gram-negative aeromonads exhibited strong fluorescence after FISH, except for the 1-h, 1-M HCl treatment. The best fixation conditions (based upon the average signal intensity per pixel) for the Gram-positive mycobacteria included a 1-h, 1-M HCl treatment or a 2-h PFA with 40-min mutanolysin treatment. Therefore, the 2-h PFA with 40-min mutanolysin treatment was selected as the best fixation condition for the simultaneous detection of *Aeromonas* and *Mycobacterium* species (Figure 20.5).

20.4.4   Analysis of FISH results

If FISH is used to obtain quantitative information, several issues related to the analysis of FISH images need to be considered. The first question is: How much certainty do we have that a FISH image represents the real sample? More specifically, are the cell counts obtained from analysing the FISH image the same as the cell density

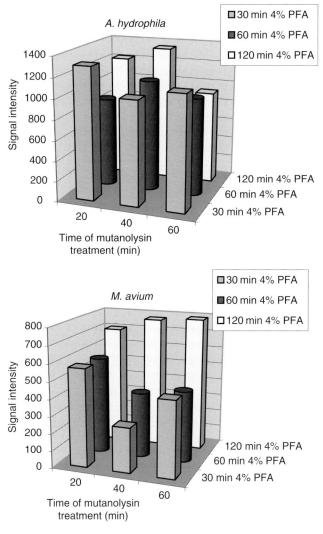

Figure 20.5   Optimisation of fixation conditions using 4% PFA followed by treatment with mutanolysin (5000 U/ml).

in the sample? Does the signal intensity obtained from the FISH image provide information on the activity of the cells in the sample? To answer these quantitative questions, statistical tools should be applied to data analysis.

Assuming that the optimal experimental conditions have been identified and FISH images have been captured, two questions need to be resolved. First, how many microscopic fields need to be analysed to obtain statistically significant cell density information to allow correlation of changes in cell density with environmental factors imposed on the microbial community? Second, how many cells need to be analysed to obtain statistically significant cell intensity information to allow correlation of changes in signal intensity with experimental conditions that resulted in activity changes?

Assuming that the experiments are completely randomised designs, the analysis starts with an analysis of variance, which evaluates if the experimental errors satisfy the assumption of homogeneous variance and normal distribution. If this assumption is met, then the analysis proceeds to the next step. Otherwise, the variances should be stabilised with an appropriate transformation (Tukey, 1977; Velleman and Hoaglin, 1981). For example, observations of cell counts on a surface per unit area may have the Poisson distribution for which the mean is equal to the variance. The square root transformation $x = \text{SQRT}(y)$ is recommended to stabilise the variances for observation from the Poisson distribution. The next step consists in constructing a linear model that equates the response variable (e.g. cell density or fluorescence intensity) to the treatment condition (e.g. concentration of FA in the hybridisation buffer). Statistical software packages such as SAS (SAS Institute Inc., Cary, NC) and Statview (SAS Institute Inc., Cary, NC) are powerful tools to facilitate analysis of the linear model to reach the final conclusion. In our analysis of different fixation/permeabilisation conditions (Section 20.4.3), it is important to determine whether the mean response for different treatments differs significantly at pre-determined significance criteria.

20.4.5   FISH in combination with other techniques

In addition to the optimisation and careful evaluation of the experimental steps of FISH, FISH can be combined with other techniques to address some of its limitations. For example, to improve the sensitivity of FISH, antibody probes can be combined with oligonucleotide probes targeting the 16S rRNA molecule. An advantage of this approach for certain applications is that cells with different metabolic activities are equally visualised due to the use of antibody probes. Oerther *et al.* (1999) and de los Reyes *et al.* (1998) combined antibody and oligonucleotide probes for the *in situ* quantification of slow-growing filamentous bacteria (*Gordonia* species) in activated sludge systems and anaerobic digesters. Some branched filaments were clearly stained by the antibodies, but gave a low signal with the FISH probe, whereas other filaments exhibited good FISH signals. Thus, individual cells within a population may differ in metabolic activity and, if so, it may be difficult to estimate cell abundance with FISH alone.

FISH has been combined with microautoradiography (MAR) to simultaneously identify a single cell and to evaluate its metabolic activity (Lee *et al.*, 1999; Ouverney and Fuhrman, 1999). MAR involves the uptake of a radio-labelled substrate by cells in an environmental sample, the exposure and development of a radiosensitive emulsion, and the evaluation using microscopy to determine which cells have taken up or metabolised the radio-labelled substrate. Cells on the developed slide can then be investigated further by FISH to correlate phylogenetic affiliation and metabolic activity.

Recently, FISH was combined with viability staining using ethidium bromide (EMA) (Oldenburg, 1999) and 5-cyano-2,3-ditolyl tetrazolium chloride (CTC) (Brehm-Stecher and Johnson, 2001). Fluorescence-based viability tests fall into two groups: (1) those based on membrane integrity and (2) those based on metabolic activity. With respect to the first group, ideal indicators of membrane integrity concentrate only in cells with compromised membranes and exhibit marked fluorescence enhancement within these cells. DNA and RNA provide large numbers of intracellular binding sites that promote fluorescence enhancement of several different stains. For example, phenanthridium nucleic acid stains such as ethidium bromide (EMA), propidium iodide (PI), ethidium homodimer 1 (EthD-1), and SYTOX Green have been used to evaluate cell membrane integrity in a variety of animal and bacterial species (Roth *et al.*, 1997). Regarding the second group, the most frequently used viability stains based on metabolic activity evaluate the presence of respiration. Common redox dyes in this group include 5-cyano-2,3-ditolyl tetrazolium chloride (CTC) and 2-(p-iodophenyl)-3-(p-nitrophenyl)-5-phenyl tetrazolium chloride (INT).

Considering the spectrum of bacterial metabolic states in the laboratory and the environment (growing on agar, actively respiring, injured, dormant, residual enzymatic activity, dead, and lysed), staining methods based on compromised membranes likely target 'lysed' and a portion of the 'dead' cells. Staining methods based on metabolic activity likely target bacteria that are 'growing on agar', 'actively respiring', 'injured', or 'dormant', and at least part of the bacteria that exhibit 'residual enzymatic activity'.

Illustration

The electron transport chain is present in mycobacteria and aeromonads since they are both respiring bacteria. Thus, in this case, it is possible to use viability stains that rely on detection of respiration. One dye from each group was evaluated, with SYTOX Green (Molecular Probes Inc., Eugene, OR) representing the first group and CTC (PolySciences, Warrington, PA) the second.

SYTOX Green is an unsymmetrical cyanine dye with three positive charges. It stains nucleic acids and is completely excluded from live cells. Binding of SYTOX Green stain to nucleic acids resulted in a >500-fold enhancement in fluorescence emission, rendering bacteria with compromised membranes brightly green fluorescent. In addition, SYTOX Green provides several important advantages over other cell-extrusion dyes such as easiness in manipulation, quicker reaction, etc. The redox dye CTC was initially introduced for enumeration of active bacteria in water samples in 1992 (Rodriguez *et al.*, 1992). Similar to INT, CTC is readily reduced to insoluble, highly fluorescent, and intracellularly accumulated CTC-formazan through bacterial respiration. Fluorescence emission of CTC-formazan is primarily in the red region. Using CTC staining, actively respiring cells can be distinguished from non-respiring cells and abiotic material that typically emit in the blue or blue-green regions. CTC staining was considered superior to INT staining by Yu *et al.* (1995).

Table 20.3   Viability assessment.

	Signal intensity		
	Live	Dead	Background
SYTOX Green			
M. avium	157 ± 34	403 ± 79	<50
A. hydrophila	343 ± 54	879 ± 101	<50
CTC			
M. avium	412 ± 60	<50	<50
A. hydrophila	680 ± 69	<50	<50

The viability stains were prepared as previously reported or according to the manufacturers' instructions. Bacterial suspensions containing around 10^8 cells/ml were stained with appropriate concentration of the dye by incubating the cell-dye mixture for 6 h at room temperature. The bacterial mixture contained both live and dead cells. Live cells were harvested from over-night cultures of *A. hydrophila* and *M. avium*. Dead cells were prepared by exposing live *A. hydrophila* and *M. avium* cells to monochloramine for 10 min and 300 min, respectively (100 ml of cell suspension was mixed with pre-formed monochloramine to obtain a final cell density of 10^6 CFU/ml and monochloramine concentration of 8 mg/l). The disinfection conditions were selected to achieve over seven logs of inactivation based on previously determined disinfection kinetics.

Both *A. hydrophila* and *M. avium* were used to evaluate the effect of staining with each dye by preparing three samples for each organism and for each dye in PBS buffer: live cells, dead cells, and the dye only. The signals for each sample were quantified (signals for the third set of samples served to evaluate the background signal). The results are summarised in Table 20.3. Background signals for both SYTOX Green and CTC were low. Signals yielded by SYTOX Green were generally higher than those obtained with CTC. However, the difference in signal intensity between live and dead cells for CTC was higher than that for SYTOX Green.

We evaluate the combined use of FISH and viability staining using SYTOX Green and CTC. It has been reported that both these dyes diffuse out of the cell once the cell is dehydrated using ethanol (Brehm-Stecher and Johnson, 2001; Molecular Probes Product Data Sheet). Therefore, FISH without an ethanolic dehydration step was evaluated using *A. hydrophila* and *M. avium* cells. Both organisms yielded good FISH signals without ethanol treatment.

Glass slides covered with biofilm were collected from a bench-scale flow-through reactor (diameter of 1.1 cm and length of 45 cm) fed with groundwater (flow rate of 1 ml/s), which had been passed through a green sand filter to remove iron and manganese and was pre-ozonated at a contact time of 2.2 min using a dosage of 1.55 mg O_3/l. To evaluate the combined use of FISH and viability staining, biofilm samples, collected after 5 weeks of reactor operation, were treated as in the viability assay, fixed, hybridised, and visualised as discussed above (Section 20.4.3). Figure 20.6 shows a representative image of the biofilm after staining with

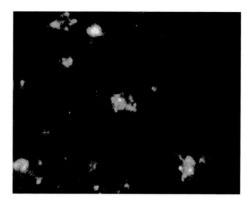

Figure 20.6   Image of a biofilm sample treated with SYTOX green followed by FISH with a bacterial probe (S-D-Bact-0338-a-A-18) labelled with Cy5. [This figure is also reproduced in colour in the plate section between pages 328 and 329.]

SYTOX Green and FISH with a general bacterial probe. The majority of cells that are red (bacteria) do not exhibit a green colour, indicating that most cells are alive. A few cells inside clusters of bacterial cells exhibit a green signal, indicating that their membranes are compromised and that these cells are dead.

20.5   CONCLUSIONS

During the last decade, molecular biology tools have been introduced in biofilm research and it has become common to identify biofilm populations and study biofilm communities using rRNA-based approaches. In this chapter, some of the techniques and procedures that have contributed to this change in biofilm research were reviewed, with particular emphasis on challenges associated with the use of these techniques. Most of the work to date in this area has focussed on identification of populations and on interactions between populations based on the characterisation of the three-dimensional structure of biofilm communities. Future techniques will need to focus on metabolic activity of different populations and determine how the activity of one cell or one population affects other cells or other populations. In this chapter, we discussed some of the modifications of FISH that have been introduced to assess metabolic activity in parallel with identification. However, additional techniques under development (e.g. see Chapters 4 and 5) will make it possible to exploit the use of molecular methods to fully understand complex ecosystems such as biofilms.

Acknowledgements

This research was supported by the US Environmental Protection Agency under Cooperative Agreement No. CR-826461010.

REFERENCES

Alm, E. and Stahl, D.A. (2000) Evaluation of key parameters for extraction of native rRNA from different environmental matrices. *J. Microbiol. Meth.* **40**, 153.

Amann, R.I., Ludwig, W. and Schleifer, K.H. (1995) Phylogenetic identification and *in situ* detection of individual microbial cells without cultivation. *Microbiol. Rev.* **59**, 143–169.

Brehm-Stecher, B.F. and Johnson, E.A. (2001) Combined fluorescence *in situ* hybridization and tetrazolium-based viability estimation of *Salmonella* and *Listeria monocytogenes*. *101th American Society for Microbiology General Meeting*, Orlando, FL.

Burggraf, S., Mayer, T., Amann, R., Schadhauser, S., Woese, C.R. and Stetter, K.O. (1994) Identifying members of the domain Archaea with rRNA-targeted oligonucleotide probes. *Appl. Environ. Microbiol.* **60**, 3112–3119.

de los Reyes, F.L., Ritter, W. and Raskin, L. (1997) Group-specific small-subunit rRNA hybridization probes to characterize filamentous foaming in activated sludge systems. *Appl. Environ. Microbiol.* **63**, 1107–1117.

de los Reyes, F.L., Oerther, D., de los Reyes, M.F., Hernandez, M. and Raskin, L. (1998) Characterization of filamentous foaming in activated sludge systems using oligonucleotide hybridization probes and antibody probes. *Water Sci. Technol.* **37**(4–5), 485–493.

Egholm, M., Buchardt, O., Christensen, L., Behrens, C., Freier, S.M., Driver, D.A., Berg, R.H., Kim, S.K., Norden, B. and Nielsen, P.E. (1993) PNA hybridizes to complementary oligonucleotides obeying the Watson–Crick hydrogen-bonding rules. *Nature* **365**, 566–568.

Frischer, M.E., Floriani, P.J. and Nierzwicki-Bauer, S.A. (1996) Differential sensitivity of 16S rRNA targeted oligonucleotide probes used for fluorescence *in situ* hybridization is a result of ribosomal higher order structure. *Can. J. Microbiol.* **42**, 1061–1071.

Fuchs, B.M., Wallner, G., Beisker, W., Schwippl, I., Ludwig, W. and Amann, R. (1998) Flow cytometric analysis of the *in situ* accessibility of *Escherichia coli* 16S rRNA for fluorescently labeled oligonucleotide probes. *Appl. Environ. Microbiol.* **64**, 4973–4982.

Fuchs, B.M., Glockner, F.O., Wolf, J. and Amann, R. (2000) Unlabeled helper oligonucleotides increase the *in situ* accessibility to 16S rRNA of fluorescently labeled oligonucleotide probes. *Appl. Environ. Microbiol.* **66**, 3603–3607.

Gutell, R.R., Larsen, N. and Woese, C.R. (1994) Lessons from an evolving rRNA: 16S and 23S rRNA structures from a comparative perspective. *Microbiol. Rev.* **58**, 10–26.

Hofman-Bang, J., Zheng, D., Westermann, P., Ahring, B.K. and Raskin, L. (2002) Molecular ecology of anaerobic reactor systems. In *Biomethanation, Advances in Biochemical Engineering/Biotechnology* (ed. B.K. Ahring), Springer-Verlag Inc. (in press).

Hristova, K.R., Mau, M., Zheng, D., Aminov, R.I., Mackie, R.I., Gaskins, H.R. and Raskin, L. (2000) *Desulfotomaculum* genus- and subgenus-specific 16S rRNA hybridization probes for environmental studies. *Environ. Microbiol.* **2**, 143–159.

Imlay, J.A. and Linn, S. (1987) Mutagenesis and stress response induced in *Escherichia coli* by hydrogen peroxide. *J. Bacteriol.* **169**, 2967–2976.

Kampfer, P., Erhart, R., Beinfohr, C., Bohringer, J., Wagner, M. and Amann, R. (1996) Characterization of bacterial communities from activated sludge: culture-dependent numerical identification versus *in situ* identification using group- and genus-specific rRNA-targeted oligonucleotide probes. *Microbiol. Ecol.* **32**, 101–121.

Koizumi, Y., Kelly, J.J., Nakagawa, T., Urakawa, H., El-Fantroussi, S., Al-Muzaini, S., Fukui, M., Urushigawa, Y. and Stahl, D.A. (2002) Parallel characterization of anaerobic toluene- and ethylbenzene-degrading microbial consortia by PCR-denaturing gradient gel electrophoresis, RNA–DNA membrane hybridization, and DNA microarray technology. *Appl. Environ. Microbiol.* **68**, 3215–3225.

Lawrence, J.R., Neu, T.R. and Swerhone, G.E.W. (1998) *Can. J. Microbiol.* **44**, 825.

Lebaron, P., Catala, P., Fajon, C., Joux, F., Baudart, J. and Bernard, L. (1997) A new sensitive, whole-cell hybridization technique for detection of bacteria involving a biotinylated oligonucleotide probe targeting rRNA and tyramide signal amplification. *Appl. Environ. Microbiol.* **63**, 3274–3278.

Lee, N., Nielsen, P.H., Andreasen, K.H., Juretschko, S., Nielsen, J.L., Schleifer, K.-H. and Wagner, M. (1999) Combination of fluorescent *in situ* hybridization and microautoradiography – a new tool for structure–function analyses in microbial ecology. *Appl. Environ. Microbiol.* **65**, 1289–1297.

Licht, T.R., Tolker-Nielsen, T., Holmstrom, K., Krogfelt, K.A. and Molin, S. (1999) Inhibition of *Escherichia coli* precursor- 16S rRNA processing by mouse intestinal contents. *Environ. Microbiol.* **1**, 23–32.

Lisle, J.T., Broadaway, S.C., Prescott, A.M., Pyle, B.H., Fricker, C. and McFeters, G.A. (1998) Effects of starvation on physiological activity and chlorine disinfection resistance in *Escherichia coli* O157:H7. *Appl. Environ. Microbiol.* **64**, 4658–4662.

Manz, W. (1999) *In situ* analysis of microbial biofilms by rRNA-targeted oligonucleotide probing. In *Methods in Enzymology* (ed. R.J. Doyle), Vol. 310, *Biofilms*, pp. 79–91. Academic Press Ltd., San Diego, CA.

Mobarry, B.K., Wagner, M., Urbain, V., Rittmann, B.E. and Stahl, D.A. (1996) Phylogenetic probes for analyzing abundance and spatial organization of nitrifying bacteria. *Appl. Environ. Microbiol.* **62**, 2156–2162.

Molin, S., Nielsen, A.T., Christensen, B.B., Andersen, J.B., Licht, T.R., Tolker-Nielsen, T., Sternberg, C., Hansen, M.C., Ramos, C. and Givskov, M. (2000) Molecular ecology of biofilms (ed. J.D. Bryers), *Biofilms II Process Analysis and Applications*, Wiley Series in Ecological and Applied Microbiology (series ed. R. Mitchell, pp. 89–120, A John Wiley & Sons Inc., New York, USA.

Oerther, D.B., de los Reyes III, F.L., Hernandez, M. and Raskin, L. (1999) Simultaneous oligonucleotide probe hybridization and immunostaining for *in situ* detection of *Gordona* species in activated sludge. *FEMS Microbiol. Ecol.* **29**, 129–136.

Oerther, D.B., de los Reyes III, F.L. and Raskin, L. (1999) Interfacing phylogenetic oligonucleotide probe hybridizations with representations of microbial populations and specific growth rates in mathematical models of activated sludge processes. *Water Sci. Technol.* **39**(1), 11–20.

Oldenburg, P.S. (1999) Ammonia-oxidizing bacteria: inactivation kinetics in chloraminated water and a method for their rapid enumeration. M.S. thesis, Department of Civil and Environmental Engineering, University of Wisconsin-Madison, p. 147.

Ouverney, C.C. and Fuhrman, J.A. (1999) Combined microautoradiography-16S rRNA probe technique for determination of radioisotope uptake by specific microbial cell types *in situ. Appl. Environ. Microbiol.* **65**, 1746–1752.

Perry-O' Keefe, H., Rigby, S., Oliveira, K., Sorensen, D., Stender, H., Coull, J. and Hyldig-Nielsen, J.J. (2001) Identification of indicator microorganisms using a standardized PNA FISH method. *J. Microbiol. Meth.* **47**, 281–292.

Poulsen, L.K., Ballard, G. and Stahl, D.A. (1993) Use of rRNA fluorescence *in situ* hybridization for measuring the activity of single cells in young and established biofilms. *Appl. Environ. Microbiol.* **59**, 1354–1360.

Raskin, L., Zheng, D., Griffin, M.E., Stroot, P.G. and Misra, P. (1995) Characterization of microbial communities in anaerobic bioreactors using molecular probes. *Anton. Leeuw.* **68**, 297–308.

Raskin, L., Capman, W.C., Sharp, R., Poulsen, L.K. and Stahl, D.A. (1997) Molecular ecology of gastrointestinal ecosystems (ed. R.I. Mackie, B.A. White and R.E. Isaacson), *Ecology and Physiology of Gastrointestinal Microbes*, Vol. 2, *Gastrointestinal Microbiology and Host Interactions*, pp. 243–298, Chapman and Hall.

Rodriguez, G.G., Phipps, D., Ishiguro, K. and Ridgway, H.F. (1992) Use of a fluorescent redox probe for direct visualization of actively respiring bacteria. *Appl. Environ. Microbiol.* **58**, 1801–1808.

Roth, B.L., Poot, M., Yue, S.T. and Millard, P.J. (1997) Bacterial viability and antibiotic susceptibility testing with SYTOX green nucleic acid stain. *Appl. Environ. Microbiol.* **63**, 2421–2431.

Sakano, Y., Pickering, K.D., Strom, P.F. and Kerkhof, L.J. (2002) Spatial distribution of total, ammonia-oxidizing, and denitrifying bacteria in biological wastewater treatment reactors for bioregenerative life support. *Appl. Environ. Microbiol.* **68**, 2285–2293.

Santegoeds, C.M., Ferdelman, T.G., Muyzer, G. and Beer, D. (1998) Structural and functional dynamics of sulfate-reducing populations in bacterial biofilms. *Appl. Environ. Microbiol.* **64**, 3731–3739.

Schwartz, T., Kalmbach, S., Hoffmann, S., Szewzyk, U. and Obst, U. (1998) PCR-based detection of *Myco*bacteria in biofilm from a drinking water distribution system. *J. Microbiol. Meth.* **34**, 113–123.

Sorensen, A.H., Torsvik, V.L., Torsvik, T., Poulsen, L.K. and Ahring, B.K. (1997) Whole-cell hybridization of *Methanosarcina* cells with two new oligonucleotide probes. *Appl. Environ. Microbiol.* **63**, 3043–3050.

Stahl, D.A. and Amann, R. (1991) Development and application of nucleic acid probes. In *Nucleic Acid Techniques in Bacterial Systematics* (ed. E. Stackebrandt and M. Goodfellow), pp. 205–248, John Wiley and Sons Ltd., New York, USA.

Stender, H., Fiandaca, M., Hyldig-Nielsen, J.J. and Coull, J. (2002) PNA for rapid microbiology. *J. Microbiol. Meth.* **48**, 1–17.

Tukey, J.W. (1977) *Exploratory Data Analysis*, Addison-Wesley, Reading, Mass.

Urbance, J.W. (1992) Phylogenetic analysis of the mycobacteria and diagnosis of paratuberculosis by 16S rRNA amplification and probing. Ph.D. dissertation, School of Veterinary Medical Science, The University of Illinois at Urbana-Champaign, p. 159.

Velleman, P.F. and Hoaglin, D.C. (1981) *Applications, Basics and Computing of Exploratory Data Analysis*, Duxbury Press, Boston, Mass.

Wagner, M., Erhart, R., Manz, W., Amann, R., Lemmer, H., Wedi, D. and Schleifer, K.H. (1994) Development of an rRNA-targeted oligonucleotide probe specific for the genus Acinetobacter and its application for *in situ* monitoring in activated sludge. *Appl. Environ. Microbiol.* **60**, 792–800.

Xi, C., Balberg, M., Boppart, S.A. and Raskin, L. (2003) Use of DNA and peptide nucleic acid (PNA) molecular beacons for the detection and quantification of rRNA in solution and in whole cells, in preparation.

Yu, W., Dodds, W.K., Banks, M.K., Skalsky, J. and Strauss, E.A. (1995) Optimal staining and sample storage time for direct microscopic enumeration of total and active bacteria in soil with two fluorescent dyes. *Appl. Environ. Microbiol.* **61**, 3367–3372.

Zheng, D., Alm, E.W., Stahl, D.A. and Raskin, L. (1996) Characterization of universal small-subunit rRNA hybridization probes for quantitative molecular microbial ecology studies. *Appl. Environ. Microbiol.* **62**, 4504–4513.

Zheng, D. and Raskin, L. (2000) Quantification of *Methanosaeta* species in anaerobic bioreactors using genus- and species-specific hybridization probes. *Microbiol. Ecol.*, **39**, 246–262.

21

Use of microsensors to study biofilms

Z. Lewandowski and H. Beyenal

21.1 INTRODUCTION

Microsensors, typically in the form of microelectrodes and microoptodes (fiber-optic microsensors), have assumed a prominent position as indispensable tools in biofilm research because they allow for the probing of local microenvironments and the quantification of local chemistries at the microscale level with high spatial resolution, providing information that is difficult to get otherwise. As the thickness of most biofilms does not exceed a few hundred micrometers and the measurements need to be done within the space occupied by the biofilm, it is easy to see why microsensors are indispensable. Biofilm researchers use them to measure the concentration profiles of dissolved substances within the space occupied by the biofilm as well as in the bulk solution near the biofilm surface. From such measurements, two types of factors are quantified: (1) those characterizing microbial activity in the biofilm (e.g. rates of nutrient consumption), and (2) those characterizing nutrient transfer from the bulk solution to the biofilm and within the biofilm (e.g. depth of nutrient penetration, mass-transport coefficient, and diffusivity).

To relate the intra-biofilm chemistry to the biofilm structure, microsensors are often used in conjunction with other techniques common in studying biofilms, such as confocal scanning laser microscopy (CSLM) and fluorescent *in situ*

© IWA Publishing. *Biofilms in Medicine, Industry and Environmental Biotechnology*. Edited by Piet Lens, Anthony P. Moran, Therese Mahony, Paul Stoodley and Vincent O'Flaherty. ISBN: 1 84339 019 1

hybridization (FISH) probes. In combination with the other measurements, microsensors provide information about local chemistries associated with the presence and distribution of various physiological groups of microorganisms, and with the physical structure of microbial aggregates in the biofilm. Microsensors alone are important tools in biofilm research, and in combination with the other techniques their importance increases further.

Even though the utility of microsensors in biofilm research is beyond question, their use is not as widespread as might be expected. One reason for this is that the commercial availability of microsensors has been limited. Microsensors have to be manufactured by the user, which practically limits their use to the laboratories where such activity can be sustained. Even in these laboratories, constructing microsensors is often a celebrated activity that develops independently from other research and addresses questions that are more relevant to sensor technology than to biofilm research. This is now changing, as some microsensors can be purchased from specialized vendors. It is expected that their use in biofilm research will soon increase.

The limited commercial availability of microsensors is not the only factor that inhibits their widespread use in biofilm research. Not all difficulties end once the microsensors are available. Biofilm researchers often expect that the microsensor measurements will help explain, or even predict, biofilm behavior at the macroscale. This expectation is not always fulfilled for two reasons:

(1) microsensor measurements provide isolated pieces of information about a specific biofilm, which need to be amended by other pieces of information to be of real value and
(2) the lack of suitable mathematical models of biofilm activity and accumulation with which to interpret the microsensor measurements prevents scaling up the data collected at the microscale to predict biofilm behavior at the macroscale.

This chapter provides an overview of the various microsensors currently available to biofilm researchers. It is not a compendium of the available literature, but a guide to the available tools and procedures, including personal opinions about individual measurements based on the authors' experience in constructing and using microsensors in the Microsensors Laboratory at the Center for Biofilm Engineering. General principles for using microsensors and interpreting the resultant measurements are discussed as well.

21.2   MICROSENSORS USED TO PROBE BIOFILMS – MICROELECTRODES AND MICROOPTODES

The term microsensor is often used when referring to small chemical sensors. However, the adjective 'micro-' may refer to almost any linear dimension. Therefore, it is better to specify the required features that make a microsensor suitable for biofilm research rather than to try to justify any arbitrarily selected linear dimensions and refer to it as 'micro-' For example, the pH microelectrode with a tip diameter of 1.2 mm described by Liermann *et al.* (2000) is too large for measuring pH profiles in biofilms, although it may be perfectly well suited to other microscale applications, such as measurements of pH in small volumes of water.

Figure 21.1 exemplifies the shape of microsensors used to measure concentration profiles in biofilms. The most important part of such devices is the sensing tip, which is small, typically several micrometers in diameter. When using microsensors to probe biofilms, we tacitly assume that their tips are small enough not to damage the biofilm structure (Santegoeds *et al.*, 1998). To justify this assumption, the microsensors used to probe biofilms have elongated shapes and are tapered to a sensing tip that is often less than 10 μm in diameter, although the tip sizes vary and some sensors may have tip diameters as large as 50 μm or even, on occasion, 100 μm. Sensors with tips larger than 100 μm are considered less useful in probing biofilms, unless they are used entirely outside of the space occupied by the biofilm; the possibility of damaging biofilm structure and thus producing artifacts increases with tip diameter.

Another reason for using sensors with small tip diameters is that the spatial resolution of the measurements is roughly equal to the tip diameter of the sensor. While measuring substrate concentrations across a biofilm, the locations of the measurements need to be vertically separated by a distance equal to or exceeding the spatial resolution of the sensor, which is approximately the tip diameter of the sensor. Consequently, a microelectrode with a 10-μm tip diameter in a 500-μm thick biofilm would allow for the collection of about 50 meaningful data, while an electrode with a 100-μm tip diameter would allow for the collection of only about five.

However, sensors with a smaller tip diameter are not always better than those with larger tips. The quest for sensors with small tip diameters is limited by their mechanical properties; smaller tips are more prone to damage, and many microsensors had been destroyed by careless handling before they produced a single result. An acceptable compromise in tip size are the microelectrodes with tip diameters

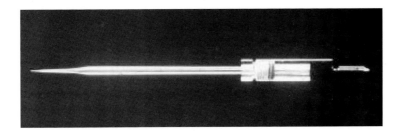

Figure 21.1   A popular microsensor in biofilm research – a pH microelectrode with an LIX membrane. If needed, it can be made with a tip diameter of less than 1 μm. The shape of the microelectrode, elongated and gradually tapered to a small tip diameter, makes it ideally suited for use in biofilms.

between 10 and 20 μm; such tips are small enough not to damage biofilm structure, yet large enough to survive moderate abuse by the operator.

Among the many devices advertised as microsensors, two types are particularly useful with biofilms, namely electrochemical microsensors (built as microelectrodes) and optical microsensors (built as fiber-optic sensors). Both types of sensors can easily be manufactured with small and elongated tips:

- Useful electrochemical microsensors include potentiometric microelectrodes, which measure potential across membranes, and amperometric microelectrodes, which measure current between the working and the reference electrode.
- Useful fiber-optic microsensors, often called microoptodes, measure light absorption, light reflection, and fluorescence.

Potentiometric sensors measure membrane potentials. Various materials are used as membranes to make potentiometric sensors, e.g. metal surfaces of the first-kind electrodes, metal salts precipitated on surfaces of the second-kind electrodes, ion-selective glasses, and liquid ion exchangers (LIXs). The principle of the measurement is to determine the potential difference of two reference electrodes, immersed on either side of the membrane, the inner and the outer side. When the chemical potential of the measured ions in the inner side of the sensor is kept constant, the Nernst equation describing the sensor's response can be simplified to the form:

$$E = \text{const} \pm \frac{RT}{nF} \ln a_{\text{b}} \qquad (21.1)$$

The sign is positive when the measured ion is a cation and negative when it is an anion, and the constant incorporates the chemical potential of the ions in the inner part of the sensor. Potentiometric sensors are calibrated by relating the measured potential to the activity of the ions in the external solution.

Amperometric sensors measure current resulting from the transfer of electrons between members of redox couples. Typically, these sensors are used in two-electrode configurations and the current is controlled by polarizing the working electrode (the sensor) to a known potential against a suitable reference (counter)-electrode. The resultant current between the working electrode and the reference electrode is equivalent to the rate of the electrode reaction, and its magnitude depends on the reactant concentration and the mass-transport rate in the vicinity of the electrode. Two major challenges that face those who construct amperometric sensors are:

(1) to make them selective, which means that only one redox couple is involved in the electrode reaction and
(2) to make them insensitive to stirring of the solution,

and these requirements not always can be satisfied.

To satisfy the first requirement, the electrodes are polarized to a properly selected potential. To satisfy the second requirement, the electrodes are covered with membranes. Membranes on amperometric electrodes serve as barriers separating the chemistries in the electrode shafts from the chemistries in the external solutions. The purpose to using membranes in amperometric devices is thus different from the purpose to using membranes in potentiometric devices. Membranes covering amperometric confine the dimensions of the mass-transport boundary layer to the thickness of the membrane, which makes the sensor insensitive to stirring. The limiting current measured by a properly designed amperometric sensor is described by the equation:

$$i = nFAC_b \frac{P_m}{\Delta}$$

(21.2)

The available optic-fiber microsensors use photosensitive reagents that are: (1) immobilized on their tips, (2) added to the solution, and (3) naturally present in biofilms. Examples of such sensors are those that measure the fluorescence of immobilized ruthenium salts sensitive to quenching by oxygen, those that measure light generated by fluorescently labeled antibodies, and those that measure the concentration of NADH.

The response of the fiber-optic microsensors is interpreted using relevant equations quantifying light intensity:

(1) If the analyte absorbs the light delivered to the sample, the sensor response is quantified using the Beer–Lambert's law:

$$I = I_0 \exp(-\alpha L)$$

(21.3)

(2) If a fluorescent reagent is immobilized at the tip of the fiber sensor and the analyte quenches the fluorescence, the response of the sensor is quantified using the Stern–Volmer equation:

$$\frac{\tau_0}{\tau} = 1 + k_q \tau_0 [C_q]$$

(21.4)

The product ($k_q \tau_0$) is referred to as the Stern–Volmer quenching constant, K_{SV}
Fluorescence lifetime in the Stern–Volmer equation can be substituted for fluorescence intensity:

$$\frac{I_0}{I} = \frac{\tau_0}{\tau}$$

(21.5)

The above equation shows that (micro)sensors based on fluorescence quenching can be operated in two modes: (1) measuring the fluorescence intensity and (2) measuring the fluorescence lifetime. Although measuring fluorescence intensity is simpler, and most sensors do just that, measuring fluorescence lifetime offers a distinctive advantage: it does not depend on the amount of the fluorophore immobilized or on the changes in that amount that may occur during the measurement due to gradual leaking or photobleaching.

21.3   POTENTIOMETRIC MICROELECTRODES

21.3.1   Sulfide microelectrode

Sulfide ion-selective electrodes use solid-state silver sulfide membranes, as described in the early works by Hseu and Rechnitz (1968) and Pungor and Toth (1970). The ion-selective sulfide microelectrodes are just miniature copies of their macroscale counterparts and have been used in microscale ecological research (Jorgensen and Revsbech, 1988), microbially influenced corrosion (MIC) studies (Lee *et al.*, 1993), and in combination with redox potential measurements in biofilms (Bishop and Yu, 1999). They are easy to make: a silver wire, appropriately cleaned and sealed in a glass microcapillary, is immersed in a solution of ammonium sulfide or another sulfide, which causes spontaneous precipitating of the silver sulfide membrane, which is in equilibrium with the sulfide ions in the solution.

There is an inherent problem with using this construction, however. The lack of precise thermodynamic data prevents the calibration of sulfide electrodes with confidence. The electrodes are calibrated in solutions of sulfide ions of various activities. However, sulfide ions in water are protonated and yield hydrogen sulfide, bisulfide ions, and sulfide ions in various proportions depending on pH. Therefore, the concentration of sulfide ions must be calculated from the total concentration of all sulfide forms [TS] present in the solution and the pH. When the pH of the solutions used to calibrate the electrodes is between pK_1 and pK_2 of hydrogen sulfide, sulfide ion concentration is calculated as

$$S^{2-} = \frac{K_2[TS]}{[H^+]} \tag{21.6}$$

To calculate the concentration of sulfide ions in standard solutions and to calibrate sulfide electrodes, the second dissociation constant, K_2, for hydrogen sulfide is needed. Most textbooks (Benjamin, 2001; Snoeyink and Jenkins, 1980) report a pK_2 for hydrogen sulfide around 14. However, a closer inspection of the reported equilibrium constants reveals that it is less certain than is usually assumed. In 1971, Giggenbach reported a pK_2 value for hydrogen sulfide of 17.1 (Giggenbach, 1971). Since Giggenbach's original work, other researchers have corroborated that the pK_2 was well beyond the generally accepted values. To make matters even worse, in 1986, Myers suggested that Giggenbach erred in his use of acidity function values when establishing the 1971 value, and suggested a more appropriate pK_2 value of 19 ± 2.0, which effectively puts the pK_2 between 17 and 21 (Myers, 1986). Consequently, the reported pK_2 for hydrogen sulfide varies in the range between 13 and 21, over eight orders of magnitude! This controversy about the magnitude of the second dissociation constant of hydrogen sulfide makes the calibration of potentiometric sulfide electrodes uncertain.

21.3.2   Redox microelectrodes

As with macroscale redox electrodes, measurements with redox microelectrodes are easy to take but the results are difficult to interpret. Redox potential depends on pH and on the rate at which the redox couples present in water equilibrate on platinum electrodes. The latter effect is difficult to quantify, therefore, making the measurement difficult to interpret. Despite this, the information provided by the profiles of concentration profiles potential in biofilms can be useful as auxiliary information when combined with redox profiles of the species that control the redox potential in natural waters, such as oxygen and sulfide (Tankere *et al.*, 2002; Bishop and Yu, 1999). Such combined oxygen profiles, sulfide profiles, and redox profiles give the researcher more confidence in the measurements. Redox microelectrodes are easy to make: a platinum wire (preferably platinized) is sealed in a glass capillary, and its potential in the solution is measured against a suitable reference electrode (Zhang and Pang, 1999; Bishop and Yu, 1999).

21.3.3   Ion-selective microelectrodes

The most popular ion-selective microelectrodes are built using LIXs (Figure 21.2) (Buhlmann *et al.*, 1998; Verschuren *et al.*, 1999; Zhang and Pang, 1999; Okabe and Watanabe, 2000; *Okabe et al.*, 2002). There are several LIX membranes available from specialized vendors. Table 21.1 shows LIX from Fluka that we use in our laboratory. A complete list of ionophores from this vendor is available at www.fluka.com. Some of the LIX membranes are sold as ready-to-use cocktails. For example, Fluka 95297 is a ready-to-use membrane for hydrogen ions; the carrier is 4-nonadecylpyridine (6%), and the solvent is 2-nitropheny-

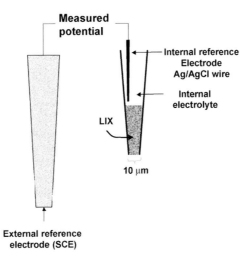

Figure 21.2   Microelectrode with an LIX membrane.

Table 21.1   LIX membranes for some selected ions and their vendors.

Measured ion	LIX	Filling solution (mmol/l)
H^+	Fluka 95297	NaCl – 100, KH_2PO_4 – 250, Na_2HPO_4 – 250
NH_4^+	Fluka 09879	NH_4Cl – 10
NO_3^-	Fluka 72549	KNO_3 – 10
NO_2^-	Fluka 72590	$NaNO_2$ – 10, NaCl – 10

loctylether (NPOE). The cocktail also contains an additive, potassium tetrakis (4-chlorophenyl) borate (1%). If needed, the membrane is solidified using high-molecular-weight PVC (Fluka # 81392) added to a concentration of 10–30% PVC dissolved in tetrahydrofuran (THF, Fluka #87369). Solidifying membranes have increased lifetimes and increase the mechanical stability of microelectrodes.

When using ion-selective microelectrodes with LIX membranes, it is advisable to test their selectivity with respect to other ions that may be present in the tested solution. Very few ion-selective electrodes are selective enough to be used in precise measurements, and ion-selective electrodes with LIX membranes are not an exception.

There is an inherent difficulty in using ion-selective microelectrodes in biofilms. By definition, these electrodes measure ion activities, which depend on ionic strength, which in turn is affected by all ions in the tested solution. Therefore, in principle, ion-selective electrodes are sensitive to all ions in the tested solution, not only the one that is being measured. As a remedy, when ion-selective electrodes are used to measure concentrations in homogeneous solutions, special buffers are added to increase the ionic strength of the solution, and to stabilize the changes in ionic strength. However, this remedy cannot be used in biofilms, which are heterogeneous, and the spatial distribution of ionic strength in the biofilms has an unknown effect on the measurements.

21.3.3.1   pH microelectrodes

Microelectrodes that measure pH are among the most popular and useful in biofilm research, and they have been constructed using various principles (Allan et al., 1999; Zhang and Pang, 1999; Dexter and Chandrasekaran, 2000; Okabe et al., 2002). We will discuss here the three most popular constructions of pH microelectrodes: (1) those made with LIX membranes, (2) those made with glass membranes, and (3) metal oxides pH microelectrodes.

Microelectrodes with LIX membranes are easy to make and easy to use. The membranes are available from specialized vendors as ready-to-use cocktails composed of the ion exchanger, organic solvent, and stabilizers (see Table 21.1). The cocktails are applied to the tip of a micropipette by capillary action. As the cocktails are hydrophobic, to ensure that the membrane will adhere the tip of the glass micropipette needs to be hydrophobic as well. This is accomplished by 'silanizing' the glass by immersing the tip of the microelectrode in a silanization

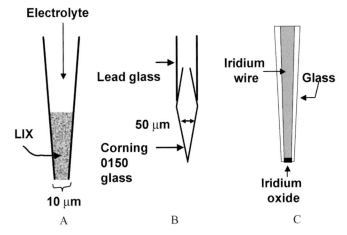

Figure 21.3   pH microelectrodes with: (A) an LIX membrane (Thomas, 1978); (B) a glass membrane (Thomas, 1978; Jorgensen and Revsbech, 1988): and (C) iridium oxide microelectrode (VanHoudt *et al.*, 1992).

solution, e.g. Fluka #85120 silanization solution, followed by baking the micropipette at 200°C for 15 min. Usually, the electrodes are prepared just before the measurements and disposed of right after the measurement. Figure 21.3A shows a pH microelectrode with an LIX membrane.

Glass pH microelectrodes are miniaturized versions of the macroscale pH electrodes, as those described by Hinke (1969). The ion-sensitive glasses used in these microelectrodes are provided by specialized vendors, e.g. the pH-sensitive Corning glass #0150. Constructing glass microelectrodes is not particularly difficult (Thomas, 1978; Pucacco *et al.*, 1986), but they have a distinct disadvantage that limits their popularity for probing biofilms: the sensing tip is of considerable length. Figure 21.3B shows the construction: the sensor is made by sealing a micropipette made of pH-sensitive glass to another micropipette (e.g. made of lead glass), which serves as the shaft of the microelectrode (Thomas, 1978). The tip of the microelectrode is small, less than 20 μm, which qualifies the device as useful for probing biofilms. However, as seen in Figure 21.3B, the sensing tip protrudes from the shaft and its length can be 100 μm or more. Somewhere along this distance, pH is measured; however, it is not clear where. Therefore, glass pH microelectrodes are particularly useful for measuring pH gradients in thick layers, as in microbial mats or bottom sediments in rivers, whose thickness exceeds the *length* of the sensing tip by many times. For use with thin biofilms, however, their usefulness is questionable.

An alternative glass pH microelectrode design was developed in our laboratory to measure pH at polarized metal surfaces in a study on MIC (Lewandowski *et al.*, 1990). In this design, the pH-sensitive membrane did not protrude from the tip, but instead it was recessed into the electrode shaft. However, the external tip diameter exceeded 100 μm, and this electrode was used to measure pH at metal surfaces in the absence of biofilms, and was not useful for probing biofilms.

A separate category among pH electrodes is occupied by metal oxide pH electrodes (Figure 21.3C). An example is the iridium oxide pH microelectrode described by VanHoudt *et al.* (1992). The oxides of many noble metals are pH sensitive and can be used to make pH microelectrodes. Such pH microelectrodes are not particularly difficult to prepare. The main difficulty is in oxidizing the noble metals, which requires using cyclic voltametry, as noble metals do not oxidize easily. Iridium oxide electrodes are prepared by potential cycling between -0.25 and $+1.25$ V vs. standard calomel electrode (SCE) in sulfuric acid. Larger electrodes can be prepared by sputter deposition or thermal oxidation. However, sputtering (Katsube *et al.*, 1982) and thermal oxidation (Ardizzone *et al.*, 1981; Augustynski *et al.*, 1984) produce predominately anhydrous iridium oxide, while electrochemical oxidation of iridium produces predominately hydrated oxides (Burke *et al.*, 1984). Anhydrous IrO_2 responds to pH changes with a slope of 59 mV/pH unit, while hydrated iridium oxides are superior and present super-Nernstian responses, explained by the mechanisms predicting 1.5 electrons transferred per H^+, with a slope of ≈ 90 mV/pH unit (VanHoudt *et al.*, 1992).

Miniaturized iridium oxide pH sensors are comparable in performance to commercial glass electrodes. The slope, d(mV)/d(pH), of iridium oxide microelectrodes is usually slightly higher than that of glass microelectrodes, and it ranges from 70 to 80 mV/pH unit for a freshly cycled sensor to ≈ 65 mV/pH, characteristic of microsensors after several uses. This compares favorably to the 59 mV/pH unit of the ideal Nernstian slope for glass electrodes. The iridium oxide microelectrode is an excellent analytical tool, but its use is restricted in many environmental studies. Iridium oxide is not stable in the presence of H_2S, and probably of other reductants. This disadvantage practically disables the microelectrode when measurements are conducted in the presence of sulfate-reducing bacteria (SRBs). However, many laboratory measurements are conducted in well-controlled conditions, and in biofilms of pure culture microorganisms, of which we know that they do not reduce sulfate ions; in these conditions, the iridium oxide pH microelectrodes are exemplary.

21.3.4   Carbon dioxide microelectrode

Carbon dioxide dissolves in water, and upon hydrolysis forms a weak diprotic acid, carbonic acid, which then dissociates and affects pH. Therefore the concentration of carbon dioxide can be evaluated from changes in pH if the experimental system is so designed that pH variations can be solely ascribed to carbon dioxide production or consumption (Zhao and Cai, 1997; Suzuki *et al*, 1999). The principle of the measurement is based on the buffer action (de Beer *et al.*, 1997; Zhao and Cai, 1997). The internal electrolyte in the microelectrode shaft is made of (sodium) bicarbonate that equilibrates with carbonic acid (carbon dioxide) (Figure 21.4). During measurements, carbon dioxide dissolved in the external solution, $CO_2(g)$, diffuses through a gas-permeable membrane and reaches the internal solution, where part of it dissolves to form $CO_2(aq)$:

$$CO_2(g) \leftrightharpoons CO_2(aq) \tag{21.7}$$

The dissolved carbon dioxide, $CO_2(aq)$, undergoes hydration and forms carbonic acid, H_2CO_3:

$$CO_2(aq) + H_2O \leftrightharpoons H_2CO_3 \tag{21.8}$$

The carbonic acid, H_2CO_3, dissociates in two steps. The first step produces bicarbonate ions:

$$H_2CO_3 \leftrightharpoons H^+ + HCO_3^- \tag{21.9}$$

The internal solution is composed of carbonic acid and bicarbonate, and its pH can be calculated from the Henderson–Hasselbach equation:

$$pH = pK_a + \log \frac{[HCO_3^-]}{[H_2CO_3]} \tag{21.10}$$

From Equation (21.10), the concentration of carbonic acid, which for all practical purposes is equal to the concentration of carbon dioxide, can be estimated:

$$\log[H_2CO_3] = pK_a + \log[HCO_3^-] - pH \tag{21.11}$$

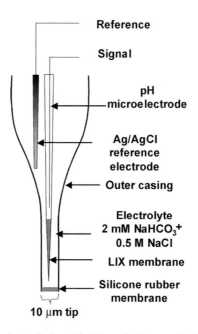

Reference

Signal

pH microelectrode

Ag/AgCl reference electrode

Outer casing

Electrolyte 2 mM NaHCO₃+ 0.5 M NaCl

LIX membrane

Silicone rubber membrane

10 µm tip

Figure 21.4 CO_2 microelectrode (modified from de Beer *et al.*, 1997).

The internal electrolyte is prepared in such a way that the concentration of bicarbonate ions is exceedingly large, which makes the *change* in bicarbonate concentration, $[HCO_3^-]$, negligible, and therefore:

$$\log[H_2CO_3] \cong \text{const} - pH \qquad (21.12)$$

In summary, by preparing the internal electrolyte with an excess of bicarbonate, the pH in the internal electrolyte is proportional to the activity of carbon dioxide in the external solution, which makes the electrode sensitive to carbon dioxide, even though it really measures pH.

In the authors' laboratory, CO_2 microelectrodes with tip diameters less than 20 μm are constructed according to procedures described by Zhao and Cai (1997) and de Beer *et al.* (1997) (Figure 21.4). In part, the construction is similar to that of the Clark-type dissolved oxygen (DO) microelectrode (Figure 21.5) in that the external micropipette is covered with a gas-permeable membrane, silicone rubber. The internal sensing device, however, is a pH microelectrode. The response time of the CO_2 microelectrodes is long, between 2 and 5 min, which limits their usefulness in some applications.

21.4   AMPEROMETRIC MICROELECTRODES

21.4.1   (DO) microelectrode

Figure 21.5 shows the flagship of microsensors used to probe biofilms, the DO microelectrode. Since many microorganisms preferentially use oxygen as the terminal electron acceptor in respiration, oxygen electrodes are often used to quantify biofilm activity (e.g. Zhu *et al.*, 2001; Okabe *et al.*, 2002). In these cases, biofilm activity is identified with the oxygen consumption rate. Incidentally, oxygen microelectrodes are among the most reliable microsensors used in biofilm research.

In an oxygen microelectrode, oxygen diffuses through the silicone rubber membrane, arrives at the cathodically polarized working electrode, and is reduced to water (Figure 21.5). The device uses an Ag/AgCl half-cell as the counter-electrode and a noble metal, such as gold or platinum, as the working electrode. The reduction of oxygen is achieved at potentials between -0.4 and -1.2 V. Applying -0.8 V typically satisfies the limiting-current conditions, and the measured current is proportional to the DO concentration in the vicinity of the sensor's tip.

Oxygen is reduced on the platinum working electrode in the reaction:

$$2e^- + \tfrac{1}{2}O_2 + H_2O \rightarrow 2OH^- \qquad (21.13)$$

by the electrons derived from the anodic reaction on the counter-Ag/AgCl electrode:

$$2Ag + 2Cl^- \rightarrow 2AgCl + 2e^- \qquad (21.14)$$

Oxygen microelectrodes are calibrated in water alternately aerated and de-aerated by purging with nitrogen and air. The measured current is typically between

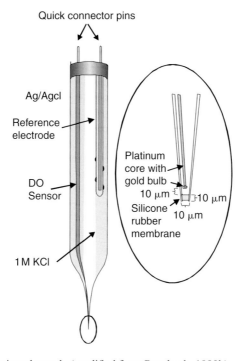

Figure 21.5   DO microelectrode (modified from Revsbech, 1989b).

10 and 150 pA for N_2-saturated water and between 100 and 700 pA for air-saturated water. Two-point calibration is sufficient because the calibration curve is linear. The response time of a good oxygen microelectrode is 1–5 s (95% of the maximum current), and it should not increase the current by more than 5% when the solution is stirred.

21.4.2   Sulfide microelectrode

The development of reliable sulfide microelectrodes was an important task in environmental research because sulfate is an important terminal electron acceptor in anaerobic microbial respiration (Santegoeds *et al.*, 1998; Okabe *et al.*, 2002). Initially, the microelectrodes to measure sulfide ion concentration were built as potentiometric devices (see Section 21.3.1) using a silver sulfide membrane (Revsbech *et al.*, 1983). Later, this construction was eclipsed by the amperometric sulfide microelectrode, developed by Jeroschewski *et al.* (1994, 1996), which measures the concentration of the hydrogen sulfide, not of the sulfide ion.

Amperometric sulfide electrodes measure the dissolved hydrogen sulfide concentration, rather than the sulfide ion concentration, and therefore avoid the controversies about the correct pK_2 for hydrogen sulfide. The microsensor is shown in

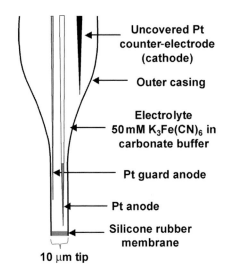

Figure 21.6   Sulfide microelectrode (modified from Jeroschewski *et al.*, 1996).

Figure 21.6. During measurements, the gas, hydrogen sulfide, penetrates across the gas-permeable membrane, silicone rubber, and dissolves in the internal electrolyte. The internal electrolyte includes a redox mediator, ferricyanide $[Fe(CN)_6]^{-3}$, which is reduced by the hydrogen sulfide to ferrocyanide, $[Fe(CN)_6]^{-4}$, and then re-oxidized on a platinum electrode. The sequence of the reactions involved in this process follows:

1. hydrogen sulfide gas penetrates the membrane and dissolves in the internal electrolyte:

$$H_2S(g) \leftrightharpoons H_2S(aq) \tag{21.15}$$

2. aqueous hydrogen sulfide dissociates to bisulfide and protons:

$$H_2S(aq) \leftrightharpoons HS^- + H^+ \tag{21.16}$$

3. bisulfide is chemically oxidized by ferricyanide to elemental sulfur, while ferricyanide is reduced to ferrocyanide:

$$HS^- + 2[Fe(CN)_6]^{-3} \rightarrow 2[Fe(CN)_6]^{-4} + S_0 + 2H^+ \tag{21.17}$$

4. ferrocyanide produced in the step 3 is re-oxidized electrochemically to ferricyanide at the platinum electrode:

$$[Fe(CN)_6]^{-4} \rightarrow [Fe(CN)_6]^{-3} + e \tag{21.18}$$

The internal platinum anode (placed $10\,\mu m$ behind the silicone membrane) and the guard anode (placed behind the internal anode) are polarized to $(+)100\,mV$ against the uncovered platinum counter-electrode. The response time of a good sulfide microelectrode is less than 1 s, and it is linear over a large range of H_2S concentrations, from 0 to $2000\,\mu M$ H_2S. Sulfide microelectrodes are calibrated in solutions of H_2S and the current generated by oxidizing ferro-to ferricyanide is proportional to the H_2S concentration near the tip of the microelectrode.

21.4.3   Microelectrodes to measure concentration of oxidants

Much biofilm research is devoted to removing unwanted biofilms. One way of accomplishing this is to kill the biofilm microorganisms using strong oxidants, such as chlorine (Stewart *et al.*, 2001) or hydrogen peroxide (Liu *et al.*, 1998). Therefore, there is strong interest among biofilm researchers in developing micro-electrodes that can take profiles of the oxidants used as antimicrobial agents. Examples of such microelectrodes are the chlorine microelectrode (de Beer *et al.*, 1994) and the hydrogen peroxide microelectrode (Liu *et al.*, 1998). We will dis-cuss the principles of making and using these devices using hydrogen peroxide microelectrodes as an example.

21.4.3.1   Chlorine and hydrogen peroxide microelectrodes

Electrodes that measure the oxidant concentration in a biofilm are amperometric devices that can either reduce the oxidant by delivering electrons from an external circuit or further oxidize the oxidant by accepting electrons from the oxidant and transferring them to an external circuit (de Beer *et al.*, 1994; Liu *et al.*, 1998; Stewart *et al.*, 2001). This principle is exemplified by the hydrogen peroxide microelectrode, which oxidizes the hydrogen peroxide to oxygen at the tip of a platinum wire (Figure 21.7).

The microsensor consists of a platinum wire inserted into a glass capillary and covered with a cellulose acetate membrane. The electrical circuit is completed by an external SCE used as the counter-electrode, and a picoammeter to measure the

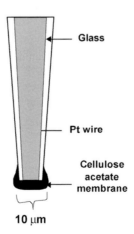

10 μm

Figure 21.7   Hydrogen peroxide microelectrode.

current. To facilitate the electrode reaction, oxidation of hydrogen peroxide, the platinum working electrode is polarized anodically at $+0.8\,\text{V}$ against the SCE:

$$H_2O_2 \rightarrow O_2(g) + 2H^+ + 2e^- \tag{21.19}$$

The electrons are used in the cathodic reaction at the counter-electrode:

$$Hg_2Cl_2(s) + 2e^- \rightarrow 2Hg(l) + 2Cl^- \tag{21.20}$$

The microelectrodes are calibrated by plotting the measured current versus the concentration of H_2O_2 in standard solutions. Calibration curves are electrode specific because the measured current depends not only on the concentration of hydrogen peroxide, but also on the active surface area of the electrode, which is unknown and not trivial to measure. Before calibrating the electrodes, the exact oxidant concentration of hydrogen peroxide in the stock solution must be determined using methods of volumetric chemical analysis. Oxidants are notoriously unstable in water solutions, and their concentration in the stock solutions changes over time. It is well known that hydrogen peroxide is unstable in water solutions, and its concentration must be determined before standard solutions can be prepared. In our laboratory, we use a 0.3% (w/w) hydrogen peroxide working solution prepared from a 30% stock solution. To prepare the working solutions, we standardize the stock solution by titrating it with $KMnO_4$.

21.4.3.2   Microelectrodes to quantify mass-transport rate

Since most biofilm reactions are mass-transport limited, it is important to quantify mass-transport rates in biofilm reactors. This task is difficult because, like most biofilm processes, mass-transport rates vary from location to location. To visualize the spatial distribution of mass-transport rates throughout a biofilm, we use microelectrodes that can quantify local mass-transport rates (Lewandowski and Beyenal, 2001). Such microelectrodes are mobile and can quantify mass-transport dynamics at selected locations, and with high spatial resolution. For that purpose, we use amperometric microelectrodes without membranes. As discussed in the section on amperometric electrodes, the electrode reaction rate, which is equivalent to the measured current, depends on the rate at which the reactant is transported to the tip of the microelectrode. Using mass-transport microelectrodes requires extensive modifications to the experimental procedure to make sure that the variables are properly isolated and that the system responds to one variable only, the one selected by the operator.

Depending on the experimental arrangement used, the mass-transport microelectrodes can measure

(1) local mass-transport coefficient,
(2) local effective diffusivity, or
(3) local flow velocity.

The measurements are conducted in flat-plate biofilm reactors, which allow for control of the hydrodynamics. Before the measurements are taken, the nutrient solution in the biofilm reactor is replaced with a solution of 25 mM potassium ferricyanide, $K_3Fe(CN)_6$, in 0.2 M KCl. This step is critical and controversial. It is critical because it provides the electroactive reactant that is used in the electrode reaction. It is controversial because it inhibits most physiological reactions in the biofilm. However, it has been demonstrated that replacing the nutrient solution with the electrolyte does not change the biofilm structure (Yang and Lewandowski, 1995).

Once the reactor is filled with the electrolyte, the microelectrodes are inserted into the biofilm and polarized cathodically. With proper experimental control, the ferricyanide is reduced to ferrocyanide, $Fe(CN)_6^{4-}$, at the surface of cathodically polarized microelectrodes:

$$Fe(CN)_6^{3-} + e^- \rightarrow Fe(CN)_6^{4-} \qquad (21.21)$$

Increasing the polarization potential applied between the microelectrode and the reference electrode increases the rate of the electrode reaction (i.e. the rate of reduction of ferricyanide to ferrocyanide) until the rate reaches its limit for the existing set of conditions. At the limiting current, the concentration of ferricyanide at the electrode surface is zero, and the concentration gradient cannot increase any further. The current is controlled by the rate of ferricyanide transport to the tip of the microelectrode.

The electrodes always measure the limiting current, which is equivalent to the rate of the electrode reaction. However, depending on the experimental arrangement, the rate of the electrode reaction is affected by various factors. Therefore, the experimental conditions, and the calibration procedures define what is actually measured. For example, to measure the local mass-transfer coefficient, the measurements are conducted in flowing electrolytes and the electrode reaction rate, the limiting current, is affected by the convective mass transport of the reactant to the microelectrode. However, to measure local effective diffusivity the electrolyte in the reactor is stagnant and the limiting current reflects only the diffusional component of mass transport.

We use several procedures to quantify various aspects of the mass-transport dynamics in biofilms. The simplest to measure is the local mass transfer coefficient (Yang and Lewandowski, 1995). More complicated are local effective diffusivity (Beyenal et al., 1998; Beyenal and Lewandowski, 2000) and local flow velocity (Xia et al., 1998). The greatest benefits of these measurements can be expected when the results are compared with the local nutrient consumption rates measured by other microelectrodes (Beyenal and Lewandowski, 2001). Such combined microelectrode measurements provide powerful images of the relations between local mass-transport dynamics and local microbial activity. Since the application of limiting-current-type sensors requires the introduction of ferricyanide, which inactivates metabolic reactions in biofilms, measurements of mass-transport dynamics need to be conducted separately from the measurement of the primary substrate utilization rate. As measurements of mass-transport rates are destructive, local substrate utilization rates are measured first, and then the nutrient solution is replaced by the electrolyte and the factors affecting mass-transport dynamics are quantified (Beyenal and Lewandowski, 2001).

Mass-transport microelectrodes are also used to measure local flow velocities in biofilms (Xia *et al.*, 1998). The principle of this measurement rests on the fact that the factor defining the rate of mass-transport in flowing electrolytes is convection. Since convective mass-transport rate can be quantified by the mass-transport microelectrodes, using an appropriate calibration procedure, it is possible to relate the rate of convection to the flow velocity in the vicinity of the microelectrode tip. The procedure works at low velocities, not exceeding several centimeters per second. The calibration is based on simultaneous measurement of the limiting current and of the local flow velocity near the tip of the electrode. Local flow velocities in the vicinity of the microelectrode tip are measured using velocimetry combined with CSLM, as described in Xia *et al.* (1998), and the local flow velocities are plotted against the limiting current measured by the mass-transport microelectrodes.

21.5   MICRO-BIOSENSORS

Micro-biosensors are a separate category of microsensors used in biofilm research. They use biological material immobilized on sensing tips to modify the chemical signal that reaches the transducer. Sensors built on this principle constitute the majority of macroscale biosensors, but their use to probe biofilms is severely inhibited by the technical difficulties caused by biosensor miniaturization. The main problem is that reducing the size of the sensor's tip to acceptable dimensions also reduces the space available for immobilizing the biological materials, which decreases the sensitivity of the biosensor. Examples of successfully miniaturized biosensors are glucose, methane, nitrate, and nitrous oxide microsensors.

21.5.1   Glucose microsensor

Even though glucose is not a particularly relevant reagent in environmental studies, many biofilm researchers use glucose as an electron donor to grow biofilms in laboratories. Since glucose sensors are one of the most intensively studied biosensors for biomedical applications (Peteu *et al.*, 1996), there are several constructions available that can be used as models for constructing glucose microsensors to probe biofilms. The one that was successfully miniaturized is an amperometric microsensor, which measures hydrogen peroxide, H_2O_2, generated by the enzymatic oxidation of glucose catalyzed by the glucose oxidase enzyme immobilized at the tip of the microelectrode (Cronenberg and van den Heuvel, 1991; Kim and Lee, 1988):

$$\text{glucose} + O_2 \rightarrow H_2O_2 + \text{D-gluconolactone} \tag{21.22}$$

$$H_2O + \text{D-gluconolactone} \rightarrow \text{D-gluconic acid} \tag{21.23}$$

A glucose microsensor is shown in Figure 21.8. The construction is simple; platinum wire is sealed in a glass capillary, and the tip of the wire is exposed and

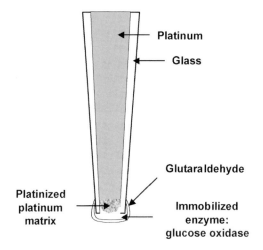

Figure 21.8   Glucose microsensor (modified from Cronenberg and van den Heuvel, 1991).

platinized to obtain a porous matrix for immobilizing the glucose oxidase enzyme. To immobilize the enzyme, the tip is dipped in the enzyme solution and then covered with a thin layer of glutaraldehyde. To measure glucose concentration, the platinum wire is polarized to $+0.8\,V$ against a SCE, in a similar manner to the sensors used to measure hydrogen peroxide (Kim and Lee, 1988; Cronenberg and van den Heuvel, 1991). The measured current is linearly correlated with glucose concentration near the microsensor's tip.

The response of the glucose microsensor depends on the oxygen concentration and pH in the external solution, because enzyme activity is affected by these factors. For that reason the microelectrode must be calibrated in solutions of known and constant oxygen concentration and pH. To evaluate the glucose concentration in a biofilm it is necessary to measure oxygen, pH, and glucose at the same location. The sensor calibrates linearly between 0 and $3\,mM$ when the oxygen concentration is kept near saturation and pH = 6.8 (Cronenberg and van den Heuvel, 1991). The response time is several seconds.

21.5.2   Methane microsensor

Methane (CH_4) is an important byproduct of many natural and engineered processes. In natural systems, methanogenic bacteria generate methane by reducing carbon dioxide. Part of the methane produced by methanogens is then utilized as an energy source by methane oxidizers. Consequently, the concentration of methane in natural systems is the net effect of methane production and consumption. To determine the rate of methanogenesis in sediments, methane concentration needs to be measured with high spatial resolution and a micro-biosensor was developed to make this possible. The sensor uses immobilized methane-oxidizing microorganisms to oxidize methane, and an oxygen electrode to measure oxygen

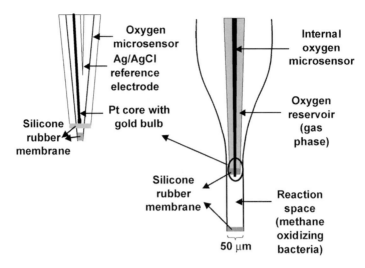

Figure 21.9   Methane microsensor (modified from Damgaard and Nielsen, 1997; Damgaard *et al.*, 1998, 2001).

consumption rate by these microorganisms (Damgaard and Nielsen, 1997; Damgaard *et al.*, 1998, 2001).

Figure 21.9 shows the tip of the micro-biosensor that is used to measure methane concentration profiles in sediments. Methane gas passes through the gas-permeable membrane to the reaction space, where it dissolves in the internal solution and is oxidized by the methane-oxidizing bacteria, *Methylosinus trichosporium* OB3b. The oxygen used by these microorganisms is delivered from the internal oxygen reservoir, through another silicone rubber membrane. The concentration of oxygen in the internal solution is affected by the rate of oxygen delivery and oxygen consumption, and is correlated with the methane concentration in the external solution by calibrating the sensor in solutions of known concentrations of methane. Depending on the size of the reaction space and the thickness of the silicone rubber, the micro-biosensor may show a linear response between 0 and 1 atm partial pressure of methane. The response time is between 30 and 60 s. As expected, oxygen and H_2S in the external solution can interfere with the measurement.

21.5.3   Nitrous oxide and nitrate microelectrodes

Nitrate microsensors are useful in studying nitrogen conversion rates in the sediments of aquifers. Initially, nitrate-ions-selective microelectrodes constructed as LIX microelectrodes were used to measure nitrate ion concentration profiles in sediments and biofilms. However, these devices suffered from poor ion selectivity, which prompted development of microsensors base on measuring nitrous oxide concentration. Since nitrous oxide (N_2O) is an intermediate product in

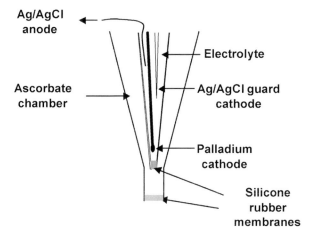

Figure 21.10   The nitrous oxide microsensor (modified from Kjaer *et al.*, 1999; Anderson *et al.*, 2001).

microbial denitrification, the activity of denitrifiers in biofilms can be estimated by measuring the concentration of nitrous oxide, which is an attractive alternative as the nitrate-ion-selective electrodes were less than perfect (Buhlmann *et al.*, 1998).

The first available nitrous oxide microsensor was a miniaturized Clark-type nitrous oxide sensor with a guard cathode, silver wire (Revsbech, 1989a, b). This construction was then improved by Anderson *et al.* (2001) by adding an oxygen scavenger, ascorbate (Figure 21.10). Nitrous oxide and oxygen penetrate the silicone rubber membrane at the tip of the microelectrode and dissolve in the electrolyte made of 1 M KCl and 0.1 NaOH. However, oxygen in the internal solution is consumed by ascorbate and only nitrous oxide reaches the internal nitrous oxide sensor consisting of a palladium-plated cathode placed behind a silicone rubber membrane. An Ag/AgCl electrode is used as the reference and counter-electrode (Kjaer *et al.*, 1999). The cathode is polarized at −0.75 V against the Ag/AgCl electrode (Andersen *et al.*, 2001). Nitrous oxide is reduced at the palladium cathode. The sensor when calibrated in saturated solutions of N_2O (Anderson *et al.*, 2001) showed a linear response between 0 and 1100 μM N_2O.

Nitrous oxide is reduced at the tip of the sensor to nitrogen gas in the reaction:

$$N_2O + H_2O + 2e^- \rightarrow N_2 + 2H^+ \tag{21.24}$$

Using the nitrous oxide microsensor, Larsen *et al.* (1996) developed an amperometric microsensor that directly measures nitrate concentration using immobilized denitrifying bacteria. The nitrate micro-biosensor was made by attaching a small capillary filled with denitrifying bacteria to the tip of a nitrous oxide microsensor. The immobilized denitrifying bacteria converted nitrate to nitrous

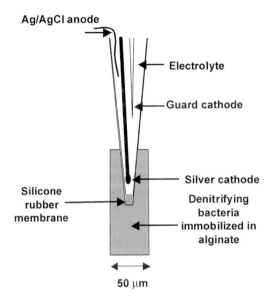

Figure 21.11   Nitrate micro-biosensor (modified from Larsen *et al.*, 1996).

oxide, and the nitrous oxide concentration was linearly correlated with the nitrate concentration at the vicinity of the microelectrode tip (Larsen *et al.*, 1996). Figure 21.11 shows the nitrate micro-biosensor based on the nitrous oxide microsensor. While the sensor in Figure 21.11 uses silver cathode, palladium is used in the nitrous oxide sensor in Figure 21.10. It is not clear whether this difference in the materials used to construct the two sensors is of any relevance.

21.6   FIBER-OPTIC MICROSENSORS

Fiber-optic sensors offer important advantages over electrochemical sensors: freedom from electromagnetic interference, internal optical reference, compactness, and geometric versatility. Some optical fibers can withstand high temperatures, aggressive chemicals, and other harsh environments. However, technical difficulties in making fiber-optic sensors with small tips, which are required to make them useful in biofilm research, inhibit both the construction of fiber-optic microsensors and their application in probing biofilms. The main problem is that when the size of the tip of the fiber-optic sensor decreases, the performance of the sensor rapidly deteriorates because of light coupling problems. Therefore, typical fiber-optic microsensors have tip diameters much larger than 20 μm, and such sensors may damage biofilm structure. Thus far, only DO fiber-optic microsensors (Holst *et al.*, 1997; Klimant *et al.*, 1995, 1997) are popular in biofilm research. It is unfortunate that the impressive advances in developing the macroscale fiber-optic sensors are

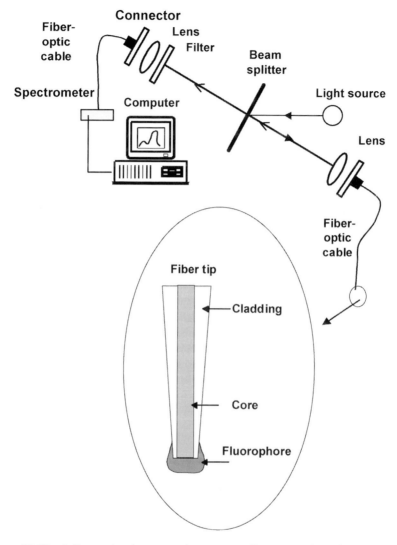

Figure 21.12   A fiber-optic microsensor that measures fluorescence intensity used in our laboratory.

not paralleled by the advances in developing the fiber-optic microsensors useful in studying biofilms.

The only popular for probing biofilms fiber-optic microsensor, the oxygen microsensor, correlates fluorescence intensity (Figure 21.12) with oxygen concentration using ruthenium(II)-tris-4,7-diphenyl-1,10 phenanthroline perchlorate (Ru[dpph]$_3$) immobilized in a polystyrene matrix as the fluorophore (Preininger *et al.*, 1994; Klimant *et al.*, 1997).

21.7   MEASURING SUBSTRATE CONCENTRATION PROFILES AND ESTIMATING BIOKINETIC PARAMETERS OF BIOFILM REACTIONS

Metabolic reactions in biofilms are limited by two factors:

(1) the rate at which the nutrients are delivered to the space occupied by the biofilm and
(2) the intrinsic rates of the metabolic reactions.

As most reactions in biofilms are controlled by the rate of mass transport, dissolved reactants and products of the metabolic reactions form concentration profiles in biofilms. The position of each data point on these profiles reflects the equilibrium between nutrient delivery to the vicinity of this location by respective mass-transport mechanisms and nutrient removal from the vicinity of that location by microbial consumption and mass transport. Using this approach, a mass balance that reflects the equilibrium between nutrient delivery and nutrient removal can be written for each data point on the substrate concentration profile. For relatively short periods of time, in relation to the time of microbial growth, these concentration profiles do not change. This fact is referred to as the pseudo-steady state and ultimately justifies using the steady-state approach to quantifying the kinetic constants of the underlying metabolic reactions.

As the shape of the concentration profile is affected by the biokinetic parameters of the underlying metabolic microbial reactions, these profiles can be used to quantify biokinetic parameters. It is important to estimate these parameters because they are used in mathematical models of biofilm growth and activity. The process of selecting the right parameters for such models is called model calibration, and it is fundamental for validating existing mathematical models of biofilms. As the shape of the concentration profiles measured by microelectrodes is also affected by factors other than the biokinetic parameters, notably by the hydrodynamics, appropriate experimental procedures are used to separate the variables before the biokinetic parameters are estimated.

21.7.1   Substrate concentration profiles across biofilms

To measure nutrient concentration profiles in biofilms, a measurement system like that in Figure 21.13 can be used. We describe the procedure of measuring oxygen concentration profiles as an example. The biofilm is grown in a flat-plate reactor (Beyenal and Lewandowski, 2001), and when the biofilm is ready the reactor is positioned on the stage of an inverted microscope. A micromanipulator is used to hold and to move the oxygen microelectrode. It is equipped with a stepper motor and remotely manipulated through a computer controller. The microelectrode, attached to the micromanipulator, is moved from the bulk liquid and down through the biofilm in 10-μm increments. The data acquisition system records the data from the voltmeter/picoammeter/DC voltage source. The concentration

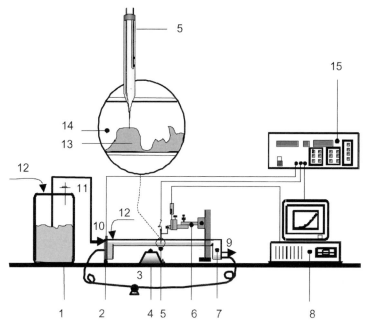

Figure 21.13   Experimental setup used to measure oxygen concentration profiles in biofilms: (1) growth medium, (2) reference electrode (used only with microelectrodes that are not equipped with an internal reference), (3) peristaltic pump, (4) inverted microscope, (5) microelectrode, (6) micromanipulator, (7) flat-plate flow cell, (8) computer, (9) outlet, (10) fresh feed, (11) vent, (12) air, (13) cell cluster, (14) interstitial void, and (15) picoammeter.

profiles are displayed in real time on the monitor, and are stored numerically in a computer's file for further analysis.

21.7.2   Computing kinetic parameters from the substrate concentration profiles

The available computational procedures for determining the biokinetic parameters of microbial metabolic reactions from the substrate concentration profiles require the assumption that biofilms are uniform, not heterogeneous. This assumption is in obvious disagreement with the general image of heterogeneous biofilms that is essential for much of our research (Yang and Lewandowski, 1995; Beyenal and Lewandowski, 2000, 2002), and we are aware of that. However, until better mathematical models of mass transport and reaction in biofilms are available, this assumption is widely used. While acknowledging this disagreement between the current conceptual models of biofilms and the mathematical models available to interpret microelectrode measurements, the researcher can minimize the effect of biofilm heterogeneity by judiciously selecting the locations of the microelectrode measurements. For example, measuring concentration profiles at the centers of large microcolonies minimizes the effect of biofilm heterogeneity on the lateral mass transport to the location of the measurement. Often the centers of large microcolonies are also less porous than their edges, and measuring substrate concentration profiles near the centers of large microcolonies minimizes the effect of non-uniform distribution of biomass at the location of the measurement as well. Such biased selection of the locations for measuring substrate concentration profiles is entirely acceptable when the purpose of the microelectrode measurements is to estimate biokinetic parameters of the biofilm microorganisms.

Thus, assuming that the biofilms are uniform and that mass transport in biofilms is one dimensional, normal to the substratum, continuity equation (21.25) can be used to model mass transport and reaction in biofilms (Rittmann and McCarty, 1980a, b; Strand and McDonell, 1985; Suidan et al., 1989; Saez and Rittmann, 1992):

$$\left(\frac{\delta C}{\delta t}\right)_f = D_f \left(\frac{\delta^2 C}{\delta x^2}\right)_f - V_{max}\left(\frac{C}{K_s + C}\right) \quad 0 \leqslant x \leqslant x_s \qquad (21.25)$$

The first term on the right side of Equation (21.25) corresponds to the mass-transport resistance within the biofilm, which is assumed to be diffusive and to follow Fickian mass transport. The second term corresponds to the substrate utilization (reaction), which is assumed to follow Monod kinetics.

Due to the non-linearity of the Monod-type reaction terms, Equation (21.25) has no analytical solutions, except for the cases where the Monod-type expressions are simplified by zero-order kinetics ($K_s \ll C$), first-order kinetics ($K_s \gg C$), or other simplifying assumptions (Golla et al., 1990). Most authors solve Equation (21.25) using numerical techniques (Rittmann and McCarty, 1980a, b; Simkins and Alexander, 1985; Suidan and Wang, 1985) or pseudo-analytical methods (Rittmann and McCarty, 1980a; Ritmann et al., 1986; Saez and Rittman, 1988; Kim and Suidan, 1989). To predict biofilm responses to different stimuli, and to calibrate models of biofilm activity and growth, it is important to evaluate the constants in the kinetic equation.

The pseudo-steady state within a biofilm $(\delta C/\delta t = 0)_f$ is achieved when the concentration profiles of the growth-limiting substrate do not change in reasonably long periods of time, longer than the time needed to measure the profiles:

$$D_f \left(\frac{d^2 C}{dx^2}\right)_f = \frac{V_{max}C}{K_s + C} \qquad (21.26)$$

Using the described computational procedure we estimate D_f, V_{max} and K_s by Taylor expansion of the function describing the concentration profile around the point $x = x_s$ positioned at the biofilm surface. Thus, substrate concentration in the vicinity of this point is described by the following equation:

$$C = C_s + \left(\frac{dC}{dx}\right)_{x_s} \times (x - x_s) + \frac{1}{2!}\left(\frac{d^2c}{dx^2}\right)_{x_s} \times (x - x_s)^2$$

$$+ \frac{1}{3!}\left(\frac{d^3C}{dx^3}\right)_{x_s} \times (x - x_s)^3 + \cdots + \frac{1}{n!}\left(\frac{d^nC}{dx^n}\right)_{x_s} \times (x - x_s)^1 \quad (21.27)$$

The first derivative (dC/dx) is estimated by noticing that the flux of the substrate across the biofilm surface has to be continuous:

$$J = J_{w,x_s} = J_{f,x_s} \quad (21.28)$$

Subscript f refers to the biofilm side, and subscript w to the water side, of the water–biofilm interface.

On the biofilm side:

$$J_{f,x_s} = D_f\left(\frac{dC}{dx}\right)_{f,x_s} \quad (21.29)$$

On the water side:

$$J_{w,x_s} = D_w\left(\frac{dC}{dx}\right)_{w,x_s} \quad (21.30)$$

Therefore, $(dC/dx)_{f,xs}$ can be estimated from the flux across the surface from the water side and the diffusion coefficient in biofilm:

$$\left(\frac{dC}{dx}\right)_{f,x_s} = \frac{J_{w,x_s}}{D_f} \quad (21.31)$$

The higher-order derivatives are estimated as

$$\frac{d^2C}{dx^2} = \frac{V_{max}}{D_f} - \frac{V_{max}K_s}{D_f}\frac{1}{K_s + C} \quad (21.32)$$

and

$$\frac{d^3C}{dx^3} = \frac{d}{dx}\left(\frac{d^2C}{dx^2}\right) = \frac{d}{dC}\left(\frac{d^2C}{dx^2}\right) \times \frac{dC}{dx}$$

$$= \frac{V_{max}K_s}{D_f}\frac{1}{(K_s + C)^2} \times \frac{dC}{dx} \quad (21.33)$$

Once we have the general form of these derivatives, they are evaluated at the biofilm surface, at the location where $x = x_s$. The first derivative is estimated using Equation (21.31). The second and the third derivatives at $x = x_s$, are:

$$\left(\frac{d^2 C}{dx^2}\right)_{x_s} = \frac{V_{max}}{D_f} \frac{C_s}{K_s + C_s} \tag{21.34}$$

$$\left(\frac{d^3 C}{dx^3}\right)_{x_s} = \left(\frac{dC}{dx}\right)_{x_s} \times \frac{V_{max} K_s}{D_f} \times \frac{1}{(K_s + C_s)^2} \tag{21.35}$$

If needed, further derivatives can also be calculated. However, for this computational procedure, calculating the three first derivatives suffices because we have three unknowns, V_{max}, K_s, and D_f, and we need three equations.

21.7.3   Example calculations

To exemplify the computational procedure we use to estimate biokinetic parameters from substrate concentration profiles, we will calculate these parameters from the oxygen concentration profile shown in Figure 21.14.

The first task in this procedure is to locate the position of the biofilm surface, which in this example was located at the inflection point of the substrate concentration profile (Figure 21.14). For the purpose of further computations, this point (x_s, C_s) segments the profile into two parts, the part below the biofilm surface and

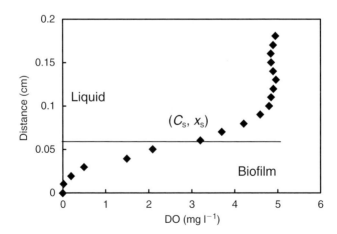

Figure 21.14   DO concentration profile across a biofilm. The biofilm surface was positioned at the inflection point of the profile. The location C_s, x_s, segments the set of data into two parts, one above and one below the biofilm surface. From the figure, $x_s = 0.06$ cm, $C_s = 3.2$ mg/l, and $C_b = 5$ mg/l.

the part above the biofilm surface, and the procedure generates two respective sets of data. The set of data from below the biofilm surface contains information about the biofilm reaction rate and substrate diffusivity through the biofilm, and this is used to calculate the biokinetic parameters of oxygen consumption. The set of data from above the biofilm surface contains information about the mass-transport rate to the biofilm, and this is used to calculate the oxygen flux to the biofilm (Equation (21.31)). The biofilm surface in Figure 21.14 was located at $x_s = 0.06$ cm, and the oxygen concentration at that location, $C_s = 3.2$ mg/l, segments the set of data into two parts.

The shape of the oxygen profile *above the biofilm surface* is described by the following empirical exponential function:

$$\frac{C - C_s}{C_b - C_s} = 1 - \exp[-B(x - x_s)] \tag{21.36}$$

Equation (21.36) can be linearized to conveniently find coefficient B from the experimental data:

$$Ln\left(1 - \frac{C - C_s}{C_b - C_s}\right) = -B(x - x_s) \tag{21.37}$$

Coefficient B, calculated as the slope of the line when presenting the data in coordinates $(x - x_s)$ versus $Ln(1 - (C - C_s)/C_b - C_s)$, equals -54 cm^{-1} (Figure 21.15A). The model adequately reflects the distribution of the experimental data (Figure 21.15B).

The slope in Figure 21.15A is $B = 54$ cm^{-1} in coordinates $(x - x_s)$ versus $Ln[1 - (C - C_s)/(C_b - C_s)]$.

Then, differentiating equation (21.36):

$$\left(\frac{dC}{dx}\right)_{w,x_s} = B(C_b - C_s) \tag{21.38}$$

From the part of the profile above the surface, flux $J_{w,xs}$ at point $x = x_s$ is calculated as 1.84×10^{-3} by substituting the oxygen diffusion coefficient in water at 21°C, $D_w = 2.0 \times 10^{-5}$ cm^2/s; $(dC/dx)w,xs$, calculated from Equation (21.31), is 91.8 mg/l cm.

To calculate oxygen concentration near the biofilm surface we use the first, second, and the third derivatives at the biofilm surface integrated to a third-order polynomial:

$$C = C_s + \frac{J_{w,x_s}}{D_f} \times (x - x_s) + \frac{1}{2!} \frac{V_{max}C_s}{D_f(K_s + C_s)} \times (x - x_s)^2$$
$$+ \frac{1}{3!} \frac{J_{w,x_s}}{D_f} \times \frac{V_{max}K_s}{D_f(K_s + C_s)^2} \times (x - x_s)^3 \tag{21.39}$$

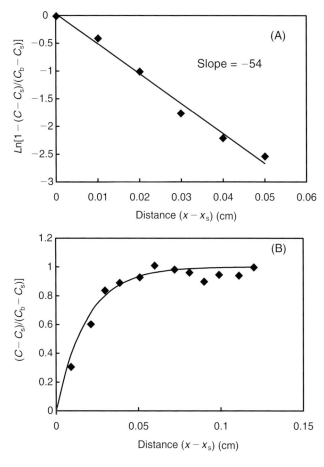

Figure 21.15   (A) $Ln[1 - (C - C_s/C_b)]$ versus $(x - x_s)$. The slope gives coefficient B in Equation (21.39). (B) The continuous line shows model data (Equation (21.38)) and the diamonds show experimental data.

A least-square-based best fit to experimental data to the third-order polynomial equation (Figure 21.16) for the points in the biofilm is done and the coefficients obtained are as follows:

$$C = 3.2 + 121.8(x - 0.06) + 1161.6(x - 0.06)^2 + 156.8(x - 0.06)^3 \quad (21.40)$$

Comparing Equations (21.39) and (21.40), we have

$$\frac{J_{w,x_s}}{D_f} = 121.8 \quad (21.41)$$

$$\frac{1}{2}\frac{V_{max}}{D_f}\frac{C_s}{K_s + C_s} = 1161.6 \quad (21.42)$$

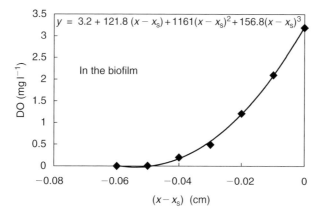

Figure 21.16   DO concentration versus $(x - x_s)$ in the biofilm. The squares are experimental data and the continuous line is a curve fit using a third-order polynomial.

$$\frac{1}{6}\frac{J_{w,x_s}}{D_f} \frac{V_{max}K_s}{D_f(K_s + C_s)^2} = 156.8 \qquad (21.43)$$

Now, these equations are solved by substituting numerical values $C_s = 3.2\,\text{mg/l}$ and $J_w = 1.84 \times 10^{-3}\,\text{mg cm l}^{-1}$:

$$D_f = 1.51 \times 10^{-5}\,(\text{cm}^2/\text{s}) \qquad (21.44)$$

$$V_{max} = 0.0354\,(\text{mg l}^{-1}\text{s}^{-1}) \qquad (21.45)$$

$$K_s = 0.0344\,(\text{mg l}^{-1}) \qquad (21.46)$$

In conclusion, from the oxygen concentration profiles, and using the Monod-growth model, the flux of oxygen, the coefficient of diffusion, and the maximum reaction rate can be estimated at a given location in a biofilm. The procedure is general and can be used for any substrate (e.g. organic compound, ion, or dissolved gas) for which a concentration profile across the biofilm can be measured. However, the calculations are based on Taylor's expansion at the biofilm–water interface, and the precision of the solution obtained is directly related to the precision with which the biofilm–water interface can be located. In addition, we assume that the biofilm is uniform and that mass transfer is one dimensional.

21.8   CONCLUSIONS

The most useful microsensors in biofilm research have elongated shafts tapered down to a tip diameter of several micrometers. As a rule, the tip diameter should be small enough not to damage biofilm structure or produce other artifacts, e.g. excessive consumption of the measured reactant by the amperometric microsensors. As microsensors with submicron-sized tip diameters are notoriously fragile, the commonly used tip diameters are between 10 and 20 μm. Sensors with tip diameters of 50 μm or more should be evaluated for the possible effects they may have on biofilm structure.

The majority of microsensors used by biofilm researchers are electrochemical sensors. Among them, the most useful are amperometric microsensors, which can be used to measure the concentration of dissolved gases, ions, and organic and inorganic molecules. The chemistries involved in their operation are straightforward, and similarly constructed microsensors can be used to measure the concentrations of various substances depending on the potential at which the working electrode is polarized. Care should be exercised when using amperometric sensors to make sure that the measurements are selective, and only one substance serves as the donor or acceptor of electrons. It is important to note that amperometric sensors actually consume the reactant whose concentration they measure.

Ion-selective microelectrodes are popular for probing biofilms because they are easy to make, but they all suffer from the disadvantages common to all ion-selective electrodes, like poor selectivity, plus additional disadvantages related to: (1) the fact that their tips have to be small and (2) the heterogeneous nature of biofilms. The small tip size makes them susceptible to electromagnetic noise. Potentiometric microelectrodes with small tip diameters are many times more susceptible to electromagnetic noise than the large-scale ion-selective electrodes (the small tip has high electrical impedance). Heterogeneous nature of biofilms defeats the basic principles of using ion-selective electrodes accepted in analytical chemistry. The signal measured by ion-selective electrodes is in part due to the activity of the measured ion and in part to the activity of all other ions present in the solution, through the ionic strength of the solution. When using macroscale ion-selective electrodes it is a common practice to increase the ionic strength of the solution, using appropriate buffers, to make the electrode response less susceptible to changes in the ionic strengths of the ions originally present in the solution. This practice is obviously not possible when probing biofilms. Therefore, the response of the ion-selective microelectrodes in biofilms is affected, to an unknown extent, by the variations in ionic strength across the biofilm.

Following the trends in the design and construction of macroscale biosensors, a few micro-biosensors have been constructed with immobilized active biological materials, e.g. glucose, methane, nitrate, and nitrous oxide microsensors. It is encouraging to see that these are amperometric sensors, and that they replace similar potentiometric sensors, such as the nitrate microsensor. However, micro-biosensors are difficult to construct, and their performance is affected by factors difficult to isolate and to quantify, e.g the effects of oxygen and pH on the activity of the biologically active materials immobilized in the sensor. Some of these problems can be traced to biofilm heterogeneity. When using macroscale biosensors to measure the concentration of an analyte in homogeneous solutions, some interfering effects can be mitigated by keeping their intensity constant; e.g. a constant and known oxygen concentration in the sample or a constant and known pH. This is obviously not possible when probing heterogeneous biofilms, and the variations in these parameters may affect the activity of the biological material immobilized in the sensor to an unknown extent.

The sensors that quantify mass-transport coefficient, diffusivity, and flow velocity are easy to construct and use, but their use requires a drastic invasion into the microbial environment, including killing the biofilm microorganisms, which makes their use limited to research facilities and special circumstances. These sensors are not for routine use, but they are helpful with the general characterization of well-defined biofilms and in fundamental studies of biofilm processes.

Fiber-optic microsensors offer undeniable advantages: they do not require external reference electrodes and they are immune to electromagnetic noise, but the available technology makes it difficult to manufacture fiber-optic sensors with tip diameters small enough for probing biofilms. Microoptodes also have disadvantages. The optical signal generated at the tip of a fiber-optic sensor often depends on a relatively complex, and sometimes poorly defined, chemistry between the reactant in the solution and the reactant immobilized at the tip of the sensor. In addition, the immobilized reactants may interact with the measured reactant or affect the microbial activity in biofilms in an unknown way; this is particularly true if the immobilized reactant leaks because it is poorly immobilized. When the tip diameter of the microoptode is less than 20 μm, the performance deteriorates rapidly because of light coupling problems. Therefore, popular fiber-optic microsensors have tip diameters larger than 20 μm.

Although there are many macroscale fiber-optic sensors available, not all these sensors can be miniaturized. Many of them use large diameter fibers, often assembled in bundles, and they cannot be miniaturized to the size required by the sensors useful for probing biofilms. Fiber-optic sensors with small tips are difficult to construct and operate. As the small tip diameter limits the amount of light that reaches the sample, and even more so the amount of light that comes back from the sample, lasers are typically used as powerful light sources to deliver the light to the sample and sensitive spectrophotometers are used to analyze the spectrum of the light that comes back. Consequently, using optical microsensors requires specialized equipments, namely lasers, optical benches, spectrophotometers that can accept fiber-optic cables, lock-in amplifiers, and photomultipliers. All these make using fiber-optic microsensors more expensive than using electrochemical microsensors.

Interpreting the response of microsensors used to probe biofilms is not trivial. The most popular application of microsensors in biofilm research is in measuring the concentration profiles of dissolved substances. Using appropriate computational and experimental procedures, it is possible to quantify the kinetic parameters of microbial respiration, dynamics of nutrient transport, and the depth of nutrient penetration from these profiles. However, because of the lack of adequate mathematical models of biofilm activity, the computational procedures use modified kinetic expressions that were developed for microorganisms growing in suspension, not for attached microorganisms; the correctness of these models and the computational procedures remains to be verified.

In probing biofilms, the design of the experimental system determines the quality of the acquired information to the same extent that the availability of specific microsensors does. To get full use of the information from microsensor measurements, experimental systems should be designed to allow unambiguous interpretation of the results. This may sound trivial, but in practice so much effort is devoted to microsensor construction, computer-assisted micromanipulation, battling electromagnetic noise, and data acquisition, that this simple requirement can easily be overlooked. Microsensor response is affected by many factors, not only by the concentration of the measured substance, and the experimental system in which the microsensor measurements are conducted must be designed in such a way that the interfering factors can be isolated. For example, when microsensors are used to determine local biofilm activity it is important to control the hydrodynamics, which affect the substrate concentration profiles to the same extent that the microbial activity does. In a sense, the process of interpreting the results starts before the microelectrode measurement is taken, and it can be successfully completed only if the experimental system allows isolation of the interfering variables. For these reasons, using microsensors to probe biofilms in natural systems, where interfering factors are difficult to isolate, or cannot be isolated, requires much attention to make sure that the response of the microsensor is properly interpreted.

21.9   NOTATION

A	Sensing surface area of the electrode (m^2)
a_b	Activity of the ions in the bulk (external solution)
B	Experimental coefficient (cm^{-1})
C	Oxygen concentration within the biofilm ($mg\,l^{-1}$)
C_b	Oxygen concentration in bulk water ($mg\,l^{-1}$)
C_q	Concentration of the quencher ($mol\,l^{-1}$)
C_s	Oxygen concentration at the surface ($mg\,l^{-1}$)
D_f	Diffusion coefficient in biofilm ($cm^2\,s^{-1}$)
D_w	Diffusion coefficient in water ($cm^2\,s^{-1}$)
E	Potential across the membrane
F	Faraday's constant (96,000 C/mol)
f	Biofilm site
i	Current (ampere)
I	Light emission intensity in the presence of the quencher
I_0	Light emission intensity in the absence of the quencher
J	Flux ($mg\,cm^{-2}\,s^{-1}$)
J_{fs}	Flux into the biofilm at the biofilm surface ($mg\,cm^{-2}\,s^{-1}$)
J_{ws}	Flux to the biofilm surface from the liquid side ($mg\,cm^{-2}\,s^{-1}$)
k_q	Bimolecular rate constant for the dynamic reaction of the quencher with the fluorophore ($l\,mol^{-1}\,s^{-1}$)
K_s	Half-saturation coefficient ($mg\,l^{-1}$)

K_{SV}	Stern–Volmer quenching constant ($l\,mol^{-1}$)
L	Absorption path length in the absorption cell (m)
n	Number of electrons transferred in the balanced reaction
P_m	Permeability of the membrane
R	Gas constant
T	Absolute temperature
t	Time (s)
V_{max}	Maximum rate of reaction $= \mu_{max} * X_f/Y$ ($mg\,l^{-1}s^{-1}$)
w	Water side
x	Distance from the bottom of the biofilm (cm)
X_f	Biofilm density ($mg\,l^{-1}$)
x_s	Distance of the biofilm surface from the bottom (cm)
Y	Yield coefficient
τ	Fluorescence lifetime in the presence of the quencher (s)
τ_0	Fluorescence lifetime in the absence of the quencher (s)
α	Absorption coefficient (m^{-1})
Δ	Thickness of the membrane (m)
μ_{max}	Maximum specific microbial growth rate (s^{-1})

Acknowledgements

The work was supported by the following grants: N-00014-02-1-0567 from the US Office of Naval Research, DE-FG03-01ER63270 and DE-FG03-98ER62630/A001 from the Natural and Accelerated Bioremediation Research (NABIR) program, and by Biological and Environmental Research (BER), the US Department of Energy, and a donation from the Procter and Gamble Company.

REFERENCES

Allan, V.J.M., Macaskie, L.E. and Callow, M.E. (1999) Development of a pH gradient within a biofilm is depended upon the limiting current. *Biotechnol. Lett.* **21**, 407–413.

Andersen, K., Kjaer, T. and Revsbech, N.P. (2001) An oxygen insensitive microsensor for nitrous oxide. *Sens. Actuat. B-Chem.* **81**, 42–48.

Ardizzone, S., Carugati, A. and Trasatti, S. (1981) Properties of thermally prepared iridium dioxide electrodes. *J. Electroanal. Chem.* **126**, 287.

Augustynski, J., Koudelka, M. and Sanchez, J. (1984) ESCA study of the state of iridium and oxygen in electrochemically and thermally formed iridium oxide films. *J. Electroanal. Chem.* **160**, 233–248.

Benjamin, M.M. (2001) *Water Chemistry*, McGraw-Hill, p. 139.

Beyenal, H. and Lewandowski, Z. (2000) Combined effects of substrate concentration and flow velocity at which biofilms were grown on effective diffusivity. *Water Res.* **34**, 528–538.

Beyenal, H. and Lewandowski, Z. (2001) Mass transport dynamics, activity, and structure of sulfate reducing biofilms. *AIChE.* **47**, 1689–1697.

Beyenal, H. and Lewandowski, Z. (2002) Internal and external mass transfer in biofilms grown at various flow velocities. *Biotechnol. Prog.* **18**, 55–61.

Beyenal, H., Tanyolaç, A. and Lewandowski, Z. (1998) Measurement of local effective diffusivity in heterogeneous biofilms. *Water Sci. Technol.* **38**, 171–178.

Beyenal, H., Lewandowski, Z., Yakymyshyn, C., Lemley, B. and Wehri, J. (2000) Fiber optic microsensors to measure back scattered light intensity in biofilms. *Appl. Optics.* **39**, 3408–3412.

Bishop, P. and Yu, T. (1999) A microelectrode study of redox potential changes in biofilms. *Water Sci. Technol.* **7**, 179–186.

Brown, S.C., Grady Jr, C.P.L. and Tabak, H.H. (1990) Biodegradation kinetics of substituted phenolics: demonstration of a protocol based on electrolytic respirometry. *Water Res.* **24**, 853–861.

Buhlmann, P., Prestsch, E. and Bakker, E. (1998) Carrier-based ion-selective electrodes and bulk optodes. 2. Ionophores for potentiometric and optical sensors. *Chem. Rev.* **98**, 1593–1687.

Burke, L.D., Mulcahy, J.K. and Whelan, D.P. (1984) Preparation of an oxidized iridium electrode and the variation of its potential with pH. *J. Electroanal. Chem.* **163**, 117–128.

Cronenberg, C.C.H. and van den Heuvel, J.C. (1991) Determination of glucose diffusion coefficients in biofilms with micro-electrodes. *Biosens. Bioelectron.* **6**, 255–262.

Damgaard, L.R. and Nielsen, L.P. (1997) A microscale biosensor for methane containing methanotrophic bacteria and an internal oxygen reservoir. *Anal. Chem.* **69**, 2262–2267.

Damgaard, L.R., Nielsen, L.P., Revsbech, N.P. and Reichardt, W. (1998) Use of an oxygen-insensitive microscale biosensor for methane to measure methane concentration profiles in a rice paddy. *Appl. Environ. Microbiol.* **64**, 864–870.

Damgaard, L.R., Nielsen, L.P. and Revsbech, N.P. (2001) Methane microprofiles in a sewage biofilm determined with a microscale microsensor. *Water Res.* **35**, 1379–1386.

de Beer, D., Stoodley, P., Roe, F. and Lewandowski, Z. (1994a) Effects of biofilm structures on oxygen distribution and mass transport. *Biotechnol. Bioeng.* **43**, 1131–1138.

de Beer, D., Srinivasan, R. and Stewart, P.S. (1994b) Direct measurement of chlorine penetration into biofilms during disinfection. *Appl. Environ. Microbiol.* **60**, 4339–4344.

de Beer, D., Glud, A., Epping, E. and Kuhl, M. (1997) A fast-responding CO_2 microelectrode for profiling sediments, microbial mats and biofilms. *Limnol. Oceanogr.* **42**, 1590–1600.

Dexter, S.C. and Chandrasekaran, P. (2000) Direct measurement of pH within marine biofilms on passive metals. *Biofouling* **15**, 313–325.

Giggenbach, W.F. (1971) Optical spectra and equilibrium distribution of polysulfide ions in aqueous solution at 20°C. *Inorg. Chem.* **11**, 1201–1207.

Golla, P.S. and Overcamp, T.J. (1990) Simple solutions for steady state biofilm reactors. *J. Environ. Eng. ASCE* **116**, 829–836.

Hinke, J. (1969) Glass microelectrodes for the study of binding and compartmentalization of intracellular ions. In *Glass Microelectrodes* (ed. M. Lavalle, O.F. Schanne, N.C. Hebert), pp. 349–375.

Holst, G., Ronnie, G., Glud, R.N., Kuhl, M. and Klimant, I. (1997) A microoptodes array for fine-scale measurement of oxygen distribution. *Sens. Actuat. B.* **38–39**, 122–129.

Hseu, T. and Rechnitz, G.A. (1968) Analytical studies of a sulfide ion-selective membrane electrode in alkaline solution. *Anal. Chem.* **40**, 1054–1060.

Jeroschewski, P., Steuckart, C. and Kühl, M. (1996) An amperometric microsensor for the determination of H_2S in aquatic environments. *Anal. Chem.* **68**, 4351–4357.

Jeroschewski, P., Haase, K., Trommer, A. and Grundler, P. (1994) Galvanic sensor for determination of hydrogen sulfide. *Electroanalysis* **6**, 769–772.

Jorgensen, B.B. and Revsbech, N.P. (1988) Microsensors. *Meth. Enzymol.* **167**, 639–659.

Katsube, T., Lauks, I. and Zemel, J.N. (1982) pH sensitive sputtered iridium oxide films. *Sens. Actuat.* **2**, 399–410.

Kim, B.R. and Suidan, M.T. (1989) Approximate algebraic solution for a biofilm model with the Monod kinetic expression. *Water Res.* **23**, 1491–1498.

Kim, J.K. and Lee, Y.H. (1988) Fast response glucose microprobe. *Biotechnol. Bioeng.* **31**, 755–758.

Kjaer, T., Larsen, L.H. and Revsbech, N.P. (1999) Sensitivity control of ion-selective biosensors by electrophoretically mediated analyte transport. *Anal. Chim. Acta.* **391**, 57–63.

Klimant, I., Kuhl, M., Glud, R.N. and Holst, G. (1997) Optical measurement of oxygen and temperature in microscale: strategies and biological applications. *Sens. Actuat. B.* **38–39**, 29–37.

Klimant, I., Meyer, V. and Kuhl, M. (1995) Fiber-optic oxygen microsensor, a new tool in aquatic biology. *Limnol. Oceanogr.* **40**, 1159–1165.

Larsen, L.H., Revsbech, N.P. and Binnerup, S.J. (1996) A microsensor for nitrate based on immobilized denitrifying bacteria. *Appl. Environ. Microbiol.* **62**, 1248–1251

Lee, W., Lewandowski, Z., Morrison, M., Characklis, W.G., Avci, R. and Nielsen, P.H. (1993) Corrosion of mild steel underneath aerobic biofilms containing sulfate-reducing bacteria. Part II: At high dissolved oxygen concentration. *Biofouling* 7, 217–239

Lewandowski Z., Lee, W., Characklis, W.G. and Little, B.J. (1990) pH at polarized metal surfaces: theory, measurement and implications for MIC. *International Congress on Microbially Induced Corrosion*, October 1990, Knoxville, TN.

Lewandowski, Z. and Beyenal, H. (2001) Limiting-current-type microelectrodes for quantifying mass transport dynamics in biofilms. *Meth. Enzymol.* **331**, 337–359.

Liermann, L., Barnes, J., Kalinowski, A.S., Zhou, B.E. and Brantley, S.L. (2000) Microenvironments of pH in biofilms grown on dissolving silicate surfaces. *Chem. Geol.* **171**, 1–6.

Liu, X., Roe, F., Jesaitis, A. and Lewandowski, Z. (1998) Resistance of biofilms to the catalase inhibitor 3-amino-1,2,4-triazole. *Biotechnol. Bioeng.* **59**, 156–162.

Myers, R.J. (1986) The new low value for the second dissociation constant for H_2S. *J. Chem. Ed.* **63**, 687–690.

Okabe, S. and Watanabe, Y. (2000) Structure and function of nitrifying biofilms as determined by *in situ* hybridization and the use of microelectrodes. *Water Sci. Technol.* **42**, 21–32.

Okabe, S., Santegoeds, C. M., Watanabe, Y. and de Beer, D. (2002) Successful development of sulfate-reducing bacterial populations and their activities in an activated sludge immobilized agar gel film. *Biotechnol. Bioeng.* **78**, 119–130.

Peteu, S.F., Emerson, D. and Worden, R.M. (1996) A Clark-type oxidase enzyme-based amperometric microbiosensor for sensing glucose, galactose, or choline. *Biosens. Bioelect.* **11**, 1059–1071.

Preininger, C., Klimant, I. and Wolfbeis, S.O. (1994) Optical fiber sensors for biological oxygen demand. *Anal. Chem.* **66**, 1841–1856.

Pucacco, L.R., Corona, S.K., Jacobson, H.R. and Carter, N.W. (1986) pH microelectrode: modified Thomas recessed tip configuration. *Anal. Chem.* **153**, 251–261.

Pungor, E. and Toth, K. (1970) Ion-selective membrane electrodes. A Review. *Analyst.* **95**, 625.

Revsbech, N.P. (1989a) Microsensors: spatial gradients in biofilms. *Structure and Function of Biofilms*, Wiley, NY.

Revsbech, N.P. (1989b) An oxygen microsensor with a guard cathode. *Limnol. Oceanogr.* **34**, 474–478.

Revsbech, N.P., Jorgensen, B.B. and Blackburn, T.H. (1983) Microelectrode studies of the photosynthesis and O_2, H_2S, and pH profiles of a microbial mat. *Limnol. Oceanogr.* **28**, 1062–1074.

Revsbech, N.P., Nielsen, L.P., Christensen, P.B., and Sørensen, J. (1988) Combined oxygen and nitrous oxide microsensor for denitrification studies. *Appl. Environ. Microbiol.* **54**, 2245–2249.

Rittmann, B. and McCarty, P.L. (1980a) Model of steady-state biofilm kinetics. *Biotechnol. Bioeng.* **27**, 2343–2357.

Rittmann, B.E. and McCarty, P.L. (1980b) Evaluation of steady-state biofilm kinetics. *Biotechnol. Bioeng.* **22**, 2359–2373.

Rittmann, B., Crawford, L., Tuck, C. K. and Namkung, E. (1986) *In situ* determination of kinetic parameters for biofims: isolation and characterization of oligotrophic biofilms. *Biotechnol. Bioeng.* **27**, 1753–1760.

Saez, P.B. and Rittmann, B.E. (1988) Improved pseudo-analytical solution for steady-state biofilm kinetics. *Biotechnol. Bioeng.* **31**, 379–385.

Saez, P.B. and Rittmann, B.E. (1992) Model-parameter estimation using least squares. *Water Res.* **26**, 789–796.

Santegoeds, C.M., Schramm, A. and de Beer, D. (1998) Microsensors as a tool to determine chemical microgradients and bacterial activity in wastewater biofilms and flocks. *Biodegradation* **9**, 159–167.

Simkins, S. and Alexander, M. (1985) Nonlinear estimation of the parameters of Monod kinetics that best describe mineralization of several substrate concentrations by dissimilar bacterial densities. *Appl. Environ. Microbiol.* **50**, 816–824.

Snoeyink, V.L. and Jenkins, D. (1980) *Water Chemistry*, John Wiley and Sons, 91.

Stewart, P.S., Rayner, J., Roe, F. and Rees, W.M. (2001) Biofilm penetration and disinfection efficiency of alkaline hypochlorite and chlorosulfamates. *J. Appl. Microbiol.* **91**, 525–532.

Strand, S.E. and McDonnell, A. J. (1985) Mathematical analysis of oxygen and nitrate consumption in deep microbial films. *Water Res.* **19**, 345–352.

Suidan, M.T. and Wang, Y.T. (1985) Unified analysis of biofilm kinetics. *Environ. Eng. Div. ASCE.* **111**, 634–646.

Suidan, M.T., Wang, Y.T. and Kim, B.R. (1989) Performance evaluation of biofilm reactors using graphical techniques. *Water Res.* **23**, 837–844.

Suzuki, H., Arakawa, H., Sasaki, S. and Karube, I. (1999) Micromachined Severinghaus-type carbon dioxide electrode. *Anal. Chem.* **71**, 1737–1743.

Tankere, S.P.C., Bourne, D.G., Muller, F.L.L. and Torsvik, V. (2002) Microenvironments and microbial community structure in sediments. *Environ. Microbiol.* **4**, 97–105.

Thomas, R.C. (1978) *Ion Selective Intracellular Microelectrodes, How to Make and Use Them*, Academic Press, London.

VanHoudt, P., Lewandowski, Z. and Little, B. (1992) Iridium oxide pH microelectrode. *Biotechnol. Bioeng.* **40**, 601–608.

Verschuren, P.G., van der Baan, J.L., Blaauw, R., de Beer, D. and van den Heuvel, J.C. (1999) A nitrate selective microelectrode based on a lipophilic derivative of iodocobalt(III) salen. *Fresenius J. Anal. Chem.* **364**, 595–598.

Xia, F., Beyenal, H. and Lewandowski, Z. (1998) An electrochemical technique to measure local flow velocity in biofilms. *Water Res.* **32**, 3637–3645.

Yang, S. and Lewandowski, Z. (1995) Measurement of local mass transfer coefficient in biofilms. *Biotechnol. Bioeng.* **48**, 737–744.

Zhang, T.C. and Pang, H. (1999) Applications of microelectrode techniques to measure pH and oxidation–reduction potential in Rhizosphere soil. *Environ. Sci. Technol.* **33**, 1293–1299.

Zhao, P. and Cai, W.J. (1997) An improved potentiometric pCO_2 microelectrode. *Anal. Chem.* **69**, 5052–5058.

Zhu, X., Suidan, M.T., Alonso, C., Yu, T., Kim, B.J. and Kim, B.B. (2001) Biofilm structure and mass transfer in a gas phase trickle-bed biofilter. *Water Sci. Technol.* **43**, 285–293.

22

Use of mathematical modelling to study biofilm development and morphology

C. Picioreanu and M.C.M. van Loosdrecht

22.1 MATHEMATICAL MODELLING AND BIOFILMS

22.1.1 Why make mathematical models

Although the word 'modelling' is used for different purposes, the final result is invariably the same: models are no more than a simplified representation of reality based on hypotheses and equations used to rationalise observations. By providing a rational environment, models can lead to deeper and more general understanding. Ultimately, understanding the underlying principles becomes refined to such a state that it is possible to make accurate predictions.

In general, a quantitative representation of the studied system is superior to a merely qualitative picture. What better way is then to describe quantitatively any system than by encoding the information in sets of numerical relations? This is exactly the goal of hypotheses expressed in mathematical form, called *mathematical models*. Mathematical models consist of sets of equations containing in an abstract body the information needed to simulate a system.

© IWA Publishing. *Biofilms in Medicine, Industry and Environmental Biotechnology*. Edited by Piet Lens, Anthony P. Moran, Therese Mahony, Paul Stoodley and Vincent O'Flaherty. ISBN: 1 84339 019 1

Bailey (1998) gave reasons for justifying the use of modelling in biochemical engineering. These reasons are rephrased below in terms of biofilm modelling:

(1) *To organise disparate information into a coherent whole.* Plenty of data on biofilm structure are now available from computerised experimental techniques. The general view on biofilm structure was dramatically changed by the use of confocal scanning laser microscopy (CSLM) and computerised image analysis tools, which revealed a more complex picture of biofilm morphology (Lawrence *et al.*, 1991; Caldwell *et al.*, 1993; Costerton *et al.*, 1994). Cell clusters may be separated by interstitial voids and channels, which create a characteristic porous structure. In some particular cases, biofilms grow in the form of microbial clusters taking a 'mushroom' shape. However, biofilms may be regularly with homogeneous compact layers of microorganisms in a slime matrix. Mathematical models can answer what all these structures have in common and what leads to differences. It cannot be anymore a reason in itself acquiring impressive three-dimensional (3-D) images of biofilm structure. Constructing a mathematical model that explains the formation of biofilm morphology should complement interpretation of 3-D images of biofilm structure.

(2) *To think about and calculate logically what components and interactions are important in a complex system.* It is a common situation to have sets of experimental data from different laboratories, which lead to often opposing qualitative interpretations. When hypotheses can be expressed as mathematical models, the task of testing the theories becomes easier because experimental data from various sources can be directly and quantitatively compared. For example, the formation of finger-like biofilms has been attributed to the control of the genome of microorganisms involved (e.g. Davies *et al.*, 1998). The hypothesis of cell-to-cell signalling was initially suggested as a result of the difficulty in explaining by other physical mechanisms the pore formation in biofilms. However, using only laws of physics that are generally valid, regardless of the particular biological system studied, the formation of a variety of heterogeneous biofilm structures can be logically explained (Picioreanu *et al.*, 1998b; Picioreanu, 1999).

(3) *To make important corrections in the conventional wisdom.* An illustrative example is the belief that convective transport of nutrients through biofilm channels significantly contributes to the increased mass transfer from bulk liquid into the biofilm. Numerical simulations of flow and mass transfer with the two-dimensional (2-D) and 3-D biofilm models by Picioreanu *et al.* (2001) and Eberl *et al.* (2001a) clearly show that for the usual hydrodynamic regimes, diffusion and not convection is the dominating mass transport mechanism in biofilms. Although there is flow inside the channels, the flux of nutrient transported is minimum, and the channels cannot be seen as nutrient suppliers to deep layers of bacteria.

(4) *To understand the essential qualitative features of a complex system.* Again, the example of finger-like biofilms is suggestive. As the following sections of

this review will make clear only three processes – diffusion, reaction and microbial growth – are essential to explain, at least qualitatively, the formation of heterogeneous biofilms.

(5) *To discover new strategies.* Different types of biofilm structure are needed for different engineering applications like wastewater treatment. Whereas smooth and compact biofilms are desired for settleability in particle biofilm reactors, fluffy biofilms can be advantageous in bio-filtration units by better catching the solid impurities. Engineering biofilm structure based on model predictions could eventually lead to the formation of the desired biofilm type in a particular reactor system.

22.1.2   How to make mathematical models

22.1.2.1   Models based on first-principles

Mathematical models can be derived in two ways, somehow opposing, but nevertheless complementary. The first approach, which is often the expedient choice, is to build empirical relationships from analysis of observations. These black-box correlations are strictly restricted to the conditions in which the experimental data were obtained, and extrapolation is hazardous. The second line, which we shall approach in this review, is to represent reality based on fundamental laws of physics, chemistry and biology. These general laws are also called *first principles*. The development of models based on first principles of science leads invariably to strengthening the intuitive process and to developing orderly and rational methods for approaching a problem. As an example, biofilm models based on reaction/ transport principles (e.g. Wanner and Gujer, 1986) have proved useful not only to test soundness of different scientific concepts, but also to establish rational strategies for many design problems involving biofilms. Models based on first principles promote lateral transfer of insight between various scientific domains. Diffusion– reaction models routinely used in chemical engineering are now widely used to simulate biofilm systems (Wanner and Gujer, 1986; Wanner and Reichert, 1996). Stochastic models for crystal growth based on diffusion-limited aggregation can also explain formation of bacterial colonies (e.g. Matsushita and Fujikawa, 1990; Ben-Jacob *et al.*, 1994). Fluid mechanics methods have been applied to study biofilm rheology (Stoodley *et al.*, 1999; Dockery and Klapper, 2001) and hydrodynamic conditions in the liquid environment surrounding the biofilm matrix (Dillon *et al.*, 1996; Picioreanu *et al.*, 2000a, 2000b, 2001; Dillon and Fauci, 2000; Eberl *et al.*, 2001a; Dupin *et al.*, 2001). Laws of structural mechanics and finite element analysis, methods belonging by tradition to civil engineers, have been used to study biofilm growth and detachment (Dupin *et al.*, 2001; Picioreanu *et al.*, 2001).

The choice of an appropriate level of mathematical sophistication always depends upon the balance between how accurate an answer is needed and how many resources are available to solve the problem. For pure engineering purposes, simple models are often sufficient in order to make safe design decisions.

The initial models imagined for this goal described biofilms as uniform steady-state films containing a single type of organism, governed exclusively by one-dimensional (1-D) mass transport and biochemical reactions (e.g. Rittmann and McCarty, 1980). Later, stratified models able to represent dynamics of multi-substrate–multispecies biofilms (Wanner and Gujer, 1986) were developed. This generation of 1-D models has helped enormously in understanding complex inter-actions in a multispecies biofilm. It even reached the stage in which these models are increasingly used as educational material in engineering curricula or as design tools for biofilm reactors. Yet, 1-D models are not able to provide characteristics of biofilm morphology. Biofilm structure can be input in these models, but not the output. The more accurate picture of biofilm structural heterogeneity recently underlined through numerous experimental observations has to be mathemati-cally described by new methods. New biofilm models provide more complex 2- and 3-D descriptions of the microbial biofilm. Advances in experimental tech-niques (CLSM, microelectrodes) have been paralleled by advances in computa-tional power. We owe the present boost of multi-dimensional biofilm models not only to more powerful hardware, but equally to the highly efficient newly devel-oped numerical methods and software used to solve the model equations.

22.1.2.2   Biofilm heterogeneity

Due to 3-D variations of microbial species, biofilm porosity, substrate concentra-tion and diffusivities in biofilms have now been repeatedly reported, a definition of *biofilm heterogeneity* is needed. According to Bishop and Rittmann (1995), heterogeneity may be defined as 'spatial differences in any parameter we think is important'. Some examples of possible biofilm heterogeneity are:

(1) *Geometrical heterogeneity*: biofilm thickness, surface roughness, biofilm porosity, substratum surface coverage with microbial biofilms.
(2) *Chemical heterogeneity*: diversity of chemical solutes (nutrients and meta-bolic products), diversity of reactions (aerobic or anaerobic, etc.).
(3) *Biological heterogeneity*: diversity of microbial species and their spatial dis-tribution, differences in activity (growing, EPS producing, and dead cells).
(4) *Physical heterogeneity*: biofilm density, permeability, visco-elasticity, mechan-ical strength, solute diffusivity, presence of abiotic solids, etc.

It can be stated that geometrical heterogeneity causes in a great measure the other kinds of heterogeneity. A proper description of biofilm geometrical morph-ology is probably the basis for a good description of the other types of heterogen-eity. Biofilms are multiphase systems, comprising solid particles, a liquid phase and, in some cases, a gas phase. From the modelling point of view, we understand the 'biofilm' as being only the solid phase, which includes the extracellular poly-meric gel, microbial cells and other entrapped particles of biotic or abiotic origin. The geometrical structure of a biofilm is the spatial arrangement of particulate biofilm components. Therefore, the dynamics of solid matter (gel, cells, and other solids) produces the biofilm geometrical heterogeneity.

22.1.2.3   Essential processes in biofilm models

During biofilm development, a large number of phenomena occur over a wide range of length and time-scales. It is necessary first to assemble knowledge about the various processes occurring in the biofilm system. Modelling the structural development of a biofilm is challenging just because of this complex interaction between many processes. The biofilm development is determined on one hand by 'positive' or 'gain' processes, like cell attachment, cell growth and division, and polymer production, which lead to biofilm volume expansion. On the other hand, 'negative' or 'loss' processes, like cell detachment and cell death, may contribute to biofilm shrinking. By changing the balance between these two types of processes, biofilms with different structural properties, like porosity, compactness or surface roughness, can be formed. The main biofilm expansion is due to bacterial growth and to extracellular polymers produced. The nutrients necessary for bacterial growth are dissolved in the liquid flow and, to reach the cells, they pass first through the concentration boundary layer (CBL) (external mass transfer) and then through the biofilm matrix (internal mass transfer). The external mass transfer resistance is given by the thickness of the CBL, which is directly correlated to the hydrodynamic boundary layer (HBL) resulting from the flow pattern over the biofilm surface. Therefore, on one hand, the fluid flow drives the biofilm growth by regulating the concentration of substrates and products at the liquid–solid interface. On the other hand, the flow shears the biofilm surface, eroding the protuberances. While the flow changes the biofilm surface, the interaction is reciprocal because a new biofilm shape leads to a different boundary condition and thus different flow and concentration fields.

To summarise, when building a mathematical model with the aim of describing biofilm geometrical structure, a number of basic sub-models should be defined:

(1) A model for *biomass growth and decay* based on nutrient consumption. This is built on mass balances applied to the bacterial mass, supplemented with adequate rate equations.
(2) A model for *biomass division and spreading* to describe the increase of volume due to increased number of bacteria. Possibly, also extracellular polymer production and spreading could be considered here.
(3) A model for *substrate transport and reactions* (kinetic or equilibrium reactions) built on mass balances for all relevant chemical species.
(4) A model for *biomass detachment* based on structural mechanics.
(5) A model for *liquid flow* past the biofilm based on liquid momentum balance and liquid mass continuity equations.
(6) A model for *biomass attachment*, as it determines the colonisation of substratum and also because new species from the liquid phase can be introduced in the already existent biofilm structure.

We will review here only the current modelling approaches to the main processes that change the biofilm shape namely biomass growth, spreading and detachment. Writing and solving solute mass balances and hydrodynamic models (at least for laminar flow) are at present well documented and they will only be mentioned briefly. Incorporation of all these processes in a comprehensive model has been reviewed in Picioreanu *et al.* (2000c).

22.1.2.4   Time-scales

One major problem is how to accommodate in the same biofilm model all the fast and slow physical, chemical, and biological processes. The solution comes from a time-scale analysis. It was shown in Picioreanu *et al.* (1999, 2000b) that processes changing the biofilm volume (biomass growth, decay, and detachment) are all much slower than processes involved in substrate mass balance (diffusion, convection, and reaction). In addition, momentum transport (by convection or viscous dissipation) is much faster than the slowest step (i.e. diffusion) of substrate mass transfer. Hence, it is justified to work at three time-scales:

(1) biomass growth, in the order of hours or days,
(2) mass transport of solutes, in the order of minutes, and
(3) hydrodynamic processes, in the order of seconds.

In other words, while solving the mass balance equation, the flow pattern can be considered at pseudo-equilibrium for a given biofilm shape, and at the same time the biomass growth, decay, and detachment are in frozen state. By exploiting the natural time-scale separation in biofilms, the step by which the whole algorithm advances in time is the one necessary for the slowest process, here biomass growth.

22.2   ELEMENTS OF BIOFILM MODELS

22.2.1   Biomass representation

An essential aspect that must be considered when building any biofilm model is the foreseen spatial scale (spatial resolution). For each spatial scale, some approaches can be more suitable than the others. Current models for biofilm structure deal in two different ways with bacteria, depending mostly on the biofilm scale targeted. The first approach, *individual-based* modelling (IbM), makes an attempt to model the biofilm community by describing the actions and properties of individual bacteria. IbM allows individual variability and treats bacterial cells as the fundamental entities. Essential state variables are, e.g. the cell mass, m; cell volume, V; etc. The second line of models treats biofilms as multiphase systems and uses volume averaging to develop macroscopic equations for biomass evolution. Such models that use mass of cells per unit volume (density or concentration, C_X) as state variable will be called *biomass-based* models (BbM) in this review (Kreft *et al.*, 2001). Analysis of conditions under which biomass averaging is a valid computational tool is comprehensively made in Wood and Whitaker (1998, 1999). The representative element of volume (REV) over which the average is made should be much larger than the size of a bacterium, but also larger than the typical distance between bacteria. However, REV size must be small compared to the characteristic length scale over which biomass significantly changes, in order to have a good spatial resolution of the biomass distribution in the biofilm. Typical REV sizes are in the order of tens of microns. The BbMs can be further divided in two classes, according to the mechanism used for biomass spreading. A first type is constituted of discrete biofilm models (usually cellular automata, CA) where biomass can expand only along a finite number of directions according to a set of discrete rules. The second class of models treats biomass as a continuum, and its spreading is generally achieved by applying differential equations widely used in mechanics and transport phenomena.

22.2.2   Biomass growth

Biomass growth kinetics is dependent on substrate concentrations. It is assumed that a bacterium increases its mass by absorbing nutrient (substrates) and then when its mass reaches a critical value, it splits into two bacteria. The general rate equation governing the change in mass m of a bacterium i situated at a moment in time t at position $\mathbf{x} = [x\ y\ z]$ can be written as

$$\frac{dm_i}{dt} = r_{Xi}(m_i(t), C_S(\mathbf{x}, t), .. \tag{22.1}$$

where $\mathbf{C}_S$ is an array containing all substrate and product concentrations that might influence the bacterial growth. For example, simple equations like Monod kinetics are in many cases acceptable. In other cases, complications like: substrate/product inhibition, maintenance requirements and biomass decay can be introduced in the model rate r_X. The mass of each bacterium is usually tracked only in IbMs. For large-scale models, the space is discretised in a grid of elementary volumes (often squares in 2-D or cubes in 3-D) and only the average biomass concentration $C_{X,i}$ of a microbial species i is taken as state variable. Equation (22.1) then becomes:

$$\frac{dC_{Xi}}{dt} = r_{Xi}(C_{Xi}(\mathbf{x}, t), C_S(\mathbf{x}, t), ... \tag{22.2}$$

The result of the microbial growth process, together with production of exopolymeric substances, is an increase in the biofilm volume, called here 'biomass spreading'. Model implementation of this biofilm expansion in all space directions raises quite difficult problems, therefore, some of the current modelling approaches will be discussed below.

22.2.3   Biomass spreading

Several general requirements can be postulated for models of biomass spreading in order to reflect the experimental observations:

(1) isotropic biomass spreading (i.e. generation of round colonies when there are no growth or space limitations),

(2) generation of a sharp biofilm–liquid interface,
(3) existence of a threshold for biomass density in the biofilm,
(4) very limited biomass mixing in biofilm clusters,
(5) possibility to include the extracellular polymeric material,
(6) possibility to deal with multiple microbial species, and
(7) be quantitative, based on real and measurable parameters.

22.2.3.1   *Discrete biomass-based models*

Perhaps the easiest way to achieve some of the modelling goals for biofilm spreading listed above is by implementing a discrete dynamical model, often called cellular automaton (CA). Discrete means here that space, time, and properties of the system can have only a finite number of states. The space occupied by biomass is first discretised into boxes (grid elements). Each grid element has four first-order neighbours and another four second-order neighbours in the 2-D rectangular space discretisation (in 3-D one can consider 6, 14 or 26 neighbours). The ensemble of individual grid elements forms a lattice. Typical for CA models is that biomass can move only along the finite number of lattice directions.

The simplest models for spreading of the newly formed amount of biomass restricted bacterial activity to the biofilm surface, being similar to crystal growth. An external mass transfer process, in which the dissolved matter has to diffuse through boundary layers in order to reach a 'reactive' interface, is coupled to a surface reaction where the soluble matter (here nutrients) is transformed to solid phase (here biomass). In the pioneering structure-oriented biofilm model of Wimpenny and Colasanti (1997), growth occurred only if there is free space available in the neighbourhood of a cell that can divide. This mechanism generates growth only in the outermost cell layer, just like in crystal formation. As these spreading rules do not obey the conservation laws of the amount of substrate converted into biomass, the model cannot be quantitative. In reality, the biofilm accumulation process is unlike the crystal growth. The major difference is that the expansion of the solid–liquid biofilm interface is caused by internal pressure generated by growing biomass. The nutrients diffuse not only across an external layer of liquid, but also into the biofilm, leading to the appearance of a reaction zone in the bulk biofilm (thus, biofilms grow *in volume* and not only *at the surface*). The current difficulty in modelling the spreading of microbial colonies is that a mechanism to release the pressure generated by the growing bacteria must be implemented. Different solutions have been proposed so far, all of them still needing much improvement in order to generate a realistic picture of biomass spreading.

In order to model internal biomass spreading Hermanowicz (1998, 2001) proposed a mechanism in which the cell resulting from division pushes a whole line of cells in the direction of the nearest biofilm surface, to make place for itself. However, it can be shown that without a permanent weighting of the four or eight possible pushing directions, the colonies resulting from one inoculum cell always get either a rectangular or pyramidal shape. A similar rule for biomass propagation was proposed by Takács and Fleit (1995) for biomass growth in an activated-sludge floc.

The mechanism proposed independently by Picioreanu (1996) and Picioreanu *et al.* (1998a, 1998b) generates round cell clusters, while keeping the important feature of growth in the bulk biofilm. Once the density in one of the grid boxes becomes greater than a maximum value at the end of a time step, half of the biomass overflows into a chosen adjacent box (empty if possible, already occupied otherwise). When an occupied adjacent box is displaced, then, in its turn it has to displace another neighbour, and so on until an empty place is found either inside the biofilm or until the periphery of the aggregate is reached. It must be noted that usually the grid size (elementary box size, say 5–20 μm) is chosen to be larger than the bacterial cell size (~1 μm). Therefore, each box contains a sufficient number of bacteria to allow a reasonable averaging for the biomass. One drawback of this mechanism is that the random displacement of neighbouring biomass volumes would produce too much mixing in a multispecies biofilm (Kreft *et al.*, 2001). A newly proposed mechanism minimises biomass mixing by allowing each displaced cell to move preferentially in the direction of the nearest biofilm surface. An example of two-species nitrifying biofilm, simulated with the new mechanism, is presented in Figure 22.1 along with the spatial distribution of three chemical species: ammonia, nitrite, and nitrate.

Noguera *et al.* (1999) allows the excess biomass formed in a time step to be distributed also in volume elements non-adjacent to the source element. The newly formed cell performs a random walk starting in the 'mother's' neighbourhood and stopping when a free place is found. This biomass spreading feature makes the model non-suitable for multispecies since excess cells belonging to one species can 'jump' over a few neighbouring elements containing other species. The result is a more pronounced mixing of species in the biofilm clusters than is required by the necessary condition of keeping the cells of one type (as much as possible) together. In a newer version of CA model, Pizarro *et al.* (2001) use 'microbial particles' of certain mass. An exclusion rule forbids the existence of two microbial particles in the same lattice cell (box), which is fulfilled by redistributing particles to obtain a uniformly dense biofilm.

All these different discrete rules are easy to implement but qualitatively different and rather arbitrary. They might easily mislead the researcher to aesthetically driven rather than physically motivated model formulation. Eberl *et al.* (2001b) pointed to a series of physical drawbacks that discrete/stochastic mathematical models for biofilm spreading inherently possess:

(1) They are strongly lattice dependent. Due to the small set of possible directions for movement (e.g. 4 or 8 in 2-D and 6, 14 or 24 directions in 3-D Cartesian system), they are not invariant to changes of the coordinate system.

(2) Local symmetry cannot always be obtained under symmetric environmental conditions.

(3) When at the same time step more grid cells try to shift biomass into a shared neighbour cell, then an order of cells must be established.

(4) Many possibilities exist to formulate local biomass redistribution rules.

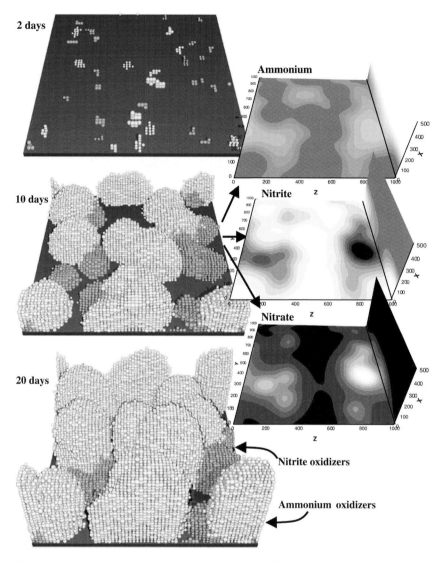

Figure 22.1   Simulation of a two-species nitrifying biofilm. Bacteria distribution at three moments (left) is visualised by light grey balls (ammonia oxidisers) and dark grey balls (nitrite oxidisers). The substrate concentration contour plots (right) correspond to day 10 and the grey scale is proportional to concentration – white is the maximum concentration and black the minimum.

22.2.3.2   *Individual-based biofilm models*

Many of the drawbacks encountered in biofilm spreading models with discrete distribution directions and discrete displacement distances, can be surmounted by allowing cells movement on a continuous set of directions and distances.

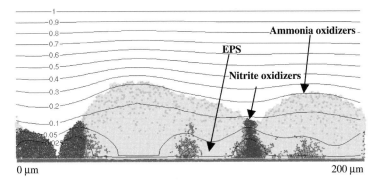

Figure 22.2   An IbM simulation of the nitrification system with ammonia oxidisers (grey dots) and nitrite oxidisers (black dots) produces more round and compact bacterial colonies than the DbM. EPS (light grey) is here produced only by one of the two species. Iso-concentration lines of oxygen (mg/l) show its limitation in the biofilm depth. The IbM is described in detail by Kreft *et al.* (2001) and the EPS model in Kreft and Wimpenny (2001).

A realistic model of this kind was proposed by Kreft *et al.* (1998) for bacterial colony growth, and it has been recently applied for simulation of a multispecies biofilm (Kreft *et al.*, 2001). Bacterial cells are represented as hard spheres, each cell having besides variable volume and mass also a set of variable growth parameters. By consuming nutrients, each cell grows and divides when a certain volume is reached. The spreading of cells occurs only when they get too close to each other. In this model, the pressure build-up due to biomass increase is, therefore, relaxed by minimising the overlap of cells. For each cell, the position is shifted by the vectorial sum of all overlap radii. As the cells are assumed spherical and the finite volumes are cubical, when this mechanism is applied together with a finite difference scheme for solution of substrate field, averaging of bacterial growth rates of cells occupying a certain grid element must be done prior to solution of substrate mass balances.

The value of the IbM approach is strengthened by the relative easiness with which a mechanism for production and spreading of extracellular polymeric substances (EPS) can be implemented. In an extension of the above model, Kreft and Wimpenny (2001) demonstrate how biofilm structure can be influenced by EPS production (Figure 22.2). The produced EPS volume was either kept in a spherical capsule surrounding the producing cell or excreted in cell-sized blobs into the environment. Unlike bacterial cells, these EPS spheres were, of course, considered inert but the effect of cell–cell, cell–EPS and EPS–EPS interactions by different types of attractive forces were also studied. This is a micro-scale and currently a rather *ad hoc* way of introducing some kind of elasticity of the biofilm matrix.

In the IbMs referred to above, microbial division and biofilm spreading can be considered to happen with the same probability in any direction. The normal consequence is that, if no substrate limitations occur and if attachment/detachment processes are insignificant, a perfectly spherical aggregate should and will develop from one inoculum cell. However, one of the most important practical aspects in activated-sludge wastewater treatment plants is the development of filamentous bacteria. An IbM approach can be useful to study conditions in which growth of filaments becomes favourable, and the activated-sludge floc settleability deteriorates. For example, an approach to model filament propagation can assume that the filamentous bacteria 'remember' the first randomly preset growth direction, thus each new cell added to a filament will be aligned with the older cells. The simulation shown in Figure 22.3 (Picioreanu, in preparation) also considered periodical attachment and detachment of both floc-formers and filament-forming bacteria along with bacterial growth and division.

To summarise IbMs *advantages*, it seems to be very appealing to microbiologists because it allows individual variability and treats bacterial cells as the fundamental units, it can explain formation of complex macroscopic structures (e.g. biofilm, activated-sludge aggregates) by describing actions and properties of individuals, it can incorporate rare species or rare events, it can make a distinction between spreading mechanisms adopted by different bacteria, and it operates at the highest spatial resolution relevant in a biofilm. However, the very detailed level of biofilm description can be also a *disadvantage* as long as one might want to model systems presenting large-scale heterogeneity. It is evident that computer resources needed to model systems at high resolution must be higher than those

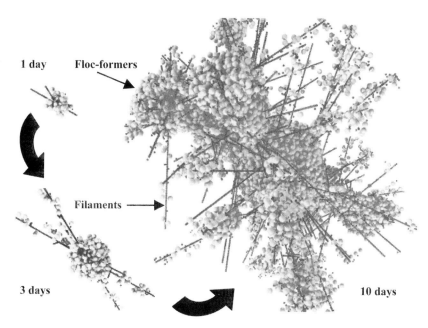

Figure 22.3   An IbM simulation of activated-sludge floc formation (Picioreanu, in preparation). The floc-formers (white balls, maximum diameter 3 μm) can grow in all directions, whereas filaments (black strings, diameter 1 μm) grow only in a preset direction. Besides bacterial growth and division, attachment and detachment were also considered in this simulation. The size of the final floc is ~400 μm.

used in BbMs (for the same spatial scale, of course). There is currently also a lack of information on the individual heterogeneity of growth parameters, on the volume fraction occupied by cells in colonies, and on actual biomass spreading mechanism adopted by different types of microorganisms.

22.2.3.3   Continuum biomass-based models (CbM)

In order to avoid the rather empirical discrete schemes for biomass spreading, recently proposed approaches look for a fully continuum biofilm description (Dockery and Klapper, 2001; Dupin *et al.*, 2001; Eberl *et al.*, 2001b). The central idea in formulation of a continuum biofilm model is that by growth bacteria generate a pressure field within the biofilm. The pressure exerted by the growing biomass (Eberl *et al.*, 2001b, call it simply 'biomass pressure') is the driving force for a very slow and viscous biofilm flow. As a result, a velocity field is created inside the biofilm. Therefore, a key notion within a model describing a sharp and moving biofilm front (biofilm–liquid interface) is the definition of convective biomass transport. An early definition of convective (also called 'advective') velocity of biomass transport was suggested in the 1-D model by Wanner and Gujer (1986). As pointed out more generally by Wood and Whitaker (1999), average biomass velocities are determined by the laws of mechanics, more precisely by the Euler equations of fluid dynamics. In the paper of Wood and Whitaker, the model was closed only for the 1-D case and no model for biomass pressure in the multi-dimensional case could be given. An important advance in this problem was made by the approaches of Dockery and Klapper (2001) and Eberl *et al.* (2001b).

Dockery and Klapper (2001) model the biofilm as a homogeneous, viscous, and incompressible fluid of constant density. When viewed together with the aqueous environment, the whole system is similar to a very slow two-phase flow with quasi-stationary aqueous phase and mobile biofilm phase. The biofilm fluid contains a field of sources and sinks, which represent the biomass growth and decay processes having the rate r_X. The state variable in the biofilm phase is pressure p, which in quasi-steady state obeys the continuity equation:

$$\lambda \nabla^2 p + r_X(C_S) = 0 \qquad (22.3)$$

The above equation represents the conservation of biofilm mass at constant biofilm density (density is here included in the constant λ). Microbial growth ($r_X > 0$) simply increases this pressure, whereas decay processes ($r_X < 0$) decrease the pressure. Any change of biofilm volume can then be represented by the normal velocity of the biofilm interface. The biofilm interface evolves quasi-statically and its velocity is proportional with the gradient of pressure at the interface. After setting a constant pressure in any static internal and boundary point, the pressure equation can be solved in the biofilm region. Simultaneously, the nutrient diffusion–reaction mass balance must be also solved, which provides the field of concentration C_S needed in Equation (22.3). Model simulations show that most growth occurs in the tips of the emerging fingers (Figure 22.4). Formation of a mushroom-like aspect is enhanced by the negative growth rate which shrinks the biofilm base exposed

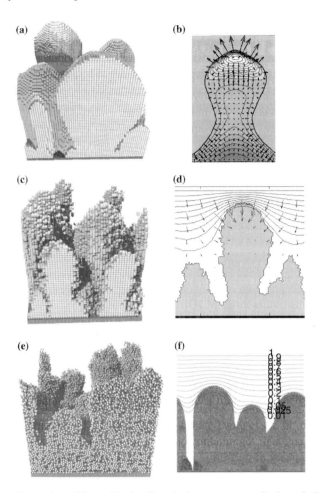

Figure 22.4   Formation of finger-like biofilms is the common prediction of all models based on diffusion and reaction of substrate even if biomass spreading is differently represented. (a) 3-D continuum biofilm model by Eberl *et al.* (2001b). (b) 2-D *continuum* biofilm model by Dockery and Klapper (2001); contour plot shows high biomass pressure (white) at the expanding biofilm tip and low pressure (dark grey) at the contracting base, together with the vector field of biofilm advective velocity. (c) 3-D and (d) 2-D *discrete* biofilm models by Picioreanu *et al.* (1998b); contour plot shows the lines of equal substrate concentration and the vector field of substrate flux much higher at the biofilm tips. (e) 3-D (Picioreanu, in preparation) and (f) 2-D *individual* biofilm models based on Kreft *et al.* (2001); the lines of substrate iso-concentration show formation of finger-like structures in case of severe substrate limitation in deep biofilm layers.

to decay due to the lack of nutrients. It is worth noting here that biofilm contraction due to biomass decay is a computational convenience, in order that the biofilm interface remains within the computational domain. Biomass growth is certainly not a reversible process, since it is most likely that cells do not completely vanish when decaying but remain as inert solids.

An alternative to a convective spreading mechanism is the biomass re-distribution by a diffusion flux proposed by Eberl *et al.* (2001b). It is evident from the differential biomass balance equation (22.2) that biomass concentration increases in any point in space where the growth rate r_X is positive. However, if a constant biomass density in time is required, a new 'diffusion' term can be introduced to obtain:

$$\frac{dC_{Xi}}{dt} = \nabla \cdot (D_{Xi}(C_{Xi}) \nabla C_{Xi}) + r_{Xi}(C_{Xi}, \mathbf{C}_S, \ldots) \qquad (22.4)$$

If the biomass 'diffusion' coefficient was taken constant, it would not be possible to guarantee neither

(1) the existence of a 'sharp front' of biomass C_X at the transition fluid/biofilm, nor
(2) biomass spreading only for C_X approaching a maximum density $C_{X,max}$, nor
(3) the threshold $C_X < C_{X,max}$.

Hence, the diffusion coefficient must be dependent on the local biomass density at a moment in time, that is $D_X(C_X(\mathbf{x}, t))$. The function:

$$D_X(C_X) = \left(\frac{\varepsilon}{C_{X,max}/C_X - 1} \right)^n \qquad (22.5)$$

proposed by Eberl *et al.* (2001b) fulfils conveniently the three above requirements when the power $n \approx 4$ and $\varepsilon \approx 10^{-4} \sim 10^{-5}$. As in the case of all the other models discussed here, Equation (22.4) is solved together with the substrate diffusion–reaction mass balance (Equation (22.6) without the convection term). Characteristic for the continuum biomass models is that no probabilistic elements are introduced in the biofilm evolution model. As a consequence, the outer surface of the finger- and mushroom-like biofilm clusters resulted from solving this model is smooth and regularly rounded (Figure 22.4).

Another deterministic approach was derived by Dupin *et al.* (2001) from material mechanics. Biofilm is modelled as a continuous, uniform, isotropic and hyper-elastic material, whose expansion and deformation are governed by material stress–strain relations. In an iterative scheme, cell growth temporarily results in an increase in biomass density without expansion. Then, the biofilm matrix is deformed to bring the density back to the required density. The biofilm deformation to a new mechanical equilibrium minimises the total potential energy of the whole biofilm aggregate. In other words, when cells divide to increase the biofilm volume, the pressure they create has to meet the resistance of the whole EPS matrix surrounding the cells. This approach is common in finite element treatment of pseudo-incompressible materials exposed to thermal stress. However, the price to be paid for rigor is a computationally intensive model. Nevertheless, more complex material rheology could be integrated in the model, such as the viscoelastic biofilm proposed by Stoodley *et al.* (1999) based on experimental data.

In conclusion, the continuum biofilm models have the *advantages* of being more rigorous, based on well-recognised laws of physics, deterministic, showing less grid-specific problems, capable of description of larger spatial domains, and being treatable within the frame of the well-developed and powerful differential calculus. Some present *disadvantages* are related to their more computationally intensive nature, necessary knowledge about more physical properties (e.g. mechanical material properties), introduction of multiple microbial species is less straightforward than in IbM, and that the continuum medium hypothesis is no longer valid when the representative volume element size approaches that of a cell (naturally, cells and EPS have different properties).

22.2.4   Biomass detachment

Detachment is essential for development of biofilm structure, because it is the main process of biomass loss. Despite its importance, biomass detachment is still scarcely included in structure-oriented biofilm models. In accordance with the biomass representations defined above, only three mechanisms for detachment have been recently proposed: discrete (Hermanowicz, 1998, 2001), individual (Dillon and Fauci, 2000) and continuum (Picioreanu *et al.*, 2001).

22.2.4.1   Discrete biofilm detachment models

The qualitative model by Hermanowicz (1998, 2001) detaches randomly cells with a probability increasing proportional with biofilm thickness and the ratio between the abstract variables 'strength of biofilm' and 'shear stress'. This mimics the observation that the biofilm region most exposed to liquid shear (i.e. outer biofilm layers) would have a larger contribution to the total amount of detached biofilm than parts sheltered from shear forces. The greatest drawback of this model is the lack of any quantitative connection with the real physical parameters characterising the biofilm.

22.2.4.2   Individual-based biofilm detachment models

Although not considering at all cell growth and division, the model reported by Dillon *et al.* (1995), Dillon *et al.* (1996), Dillon and Fauci (2000) is another fine example of individual-based biofilm modelling. Microbial cell–cell and cell–substratum adhesive interactions are represented by the creation of 'elastic springs' between the interacting entities. Following this idea, attachment occurs when the distance between two cells is less than a prescribed cohesion distance. Conversely, detachment of cells is modelled by allowing the links to break when they are stretched beyond a certain length.

22.2.4.3   Continuum-biofilm detachment models

A quantitative approach to biofilm detachment was developed in Picioreanu *et al.* (2001), based on laws of mechanics. It is founded on the hypothesis that biofilm breaks at points where the mechanical stress exceeds the biofilm mechanical strength. The mechanical stress in the biofilm builds up due to forces acting on the biofilm surface as a result of liquid flow. This model requires knowledge of mechanical properties of biofilms only scarcely measured until now, such as tensile strength (Ohashi and Harada, 1994, 1996; Ohashi *et al.*, 1999) and elasticity modulus (Stoodley *et al.*, 1999). The two known biofilm detachment mechanisms, *erosion* (loss of small biofilm parts – eventually only cells – mainly from the biofilm surface) and *sloughing* (loss of massive biofilm chunks, often broken from the substratum surface), can be modelled by considering a single breakage criterion.

In compact biofilm clusters, the highest stress develops near the biofilm surface leading to erosion. In finger-like clusters, the stress builds up near the biofilm-carrier interface, due to a high bending momentum. Cracks formed in the narrowest part of the biofilm filaments lead to sloughing of a whole biofilm patch. An avalanche effect was observed in biomass loss (Picioreanu *et al.*, 2001). Breakage of some biofilm structures left other colonies highly exposed to strong liquid shear leading to massive sloughing in a short period (Figure 22.5, between days 55 and 60).

The flow regime has a double influence on biofilm formation. As expected, at higher flow velocities the substrate flux towards the biofilm is increased because of the diminished external resistance in the boundary layer. This intensifies the biofilm growth rate. On the other hand, high flow velocity means a high shear stress at the

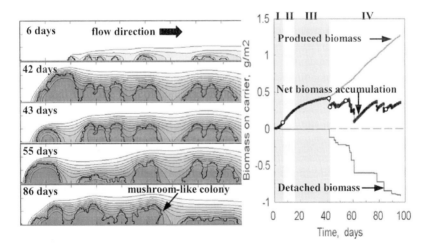

Figure 22.5   Simulation of biofilm evolution with a 2-D model including computation of flow pattern, substrate diffusion, convection and reaction, biomass growth, spreading and detachment. Model description and parameters are in Picioreanu *et al.* (2001). Net biomass accumulation (right graph) is the difference between produced and detached biomass. Biofilm evolution phases are I – exponential, II – linear, III – decay, IV – sloughing. Thick contour lines indicate the biofilm–liquid interface (left pictures). Substrate concentration field is shown both with iso-concentration lines and with patches on a grey scale. White means maximum concentration in bulk liquid. All graphs have the same scale of 1000 × 150 μm.

biofilm surface, which generates a greater detachment rate. Continuous erosion at high flow velocities shapes a more compact and thus, mechanically more stable biofilm. The biofilm grown at lower liquid velocity is more heterogeneous and mechanically weaker, thus it is more susceptible to massive biomass loss (e.g. by sloughing). Hence, the causes for sloughing must be sought not only in the biofilm strength, but also in its shape.

These results again show that the experimental observations can be directly reproduced in simulation models without a need for suggesting specific microbial interactions (e.g. chemical signalling) or other extra processes than general structural strength principles.

22.2.5   Substrate transport and conversion

22.2.5.1   Material balances

The transport of chemical compounds to the cells within the biofilm is important because the concentrations of nutrients and products determine the rates of microbial reactions. Major processes that contribute to biofilm volume increase, like microbial growth and EPS production, are driven by nutrient availability. Decay processes are also affected by the concentration level of certain compounds. Unlike with methods suggested for biofilm growth and spreading, where different solutions have been proposed because the exact mechanisms are still not fully known, the equations for substrate transport and reaction are well-established physical laws. The only difficulty we may have is to solve the, sometimes complex, system of material (moles or mass) balance equations.

Dissolved chemical species (solutes) can be transported by several mechanisms: molecular *diffusion* and *convection* (sometimes called also *advection*) are the most common. Diffusion is determined by a concentration gradient (i.e. a difference in solute concentration between two points in space). According to Fick's law, the diffusive flux of mass is proportional with the concentration gradient, and the proportionality constant is the diffusion coefficient D_i (m^2/s). These gradients of concentration occur because of substrates consumption or products generation in the biofilm. The convective flux of solute in any point in space, is proportional with the liquid velocity, $\mathbf{u}$ (m^2/s), and with the concentration, C_{Si}, of transported compound i.

To find the spatial distribution (called also 'field') of concentration of each relevant chemical species, a system of dynamic material balances must be written and then solved. That is, the rate of accumulation of an aqueous species i in an element of volume must be balanced by the rates of transport (diffusion and convection) across the volume's boundaries and the net rate of transformation (chemical reaction, R_i) in the volume. In mathematical terms, written for a small biofilm volume element the material balance for a chemical species i is

$$\frac{\partial C_{Si}}{\partial t} = D_i \nabla^2 C_{Si} - \mathbf{u} \cdot \nabla C_{Si} + R_i (C_S, C_X) \tag{22.6}$$

which must be solved together with adequate initial and boundary conditions.

A complication can arise when charged solutes (ions) are transported at different rates (e.g. different diffusion coefficients). Due to the necessity to maintain electroneutrality in each point in space, a potential function ϕ must be additionally introduced. A new contribution to the overall transport rate, called *migration*, is proportional with the gradient of electrical potential $\nabla\phi$, with the diffusion coefficient and the charge of ion, z_i (Nernst–Plank equation). This complication usually appears when pH gradients need to be calculated at low ionic strength of solution. Examples of how to deal with this situation can be found in a dental biofilm model by Dibdin (1992) or, more recently, in the biocorrosion model by Picioreanu and van Loosdrecht (2002).

Equations (22.6) can be simplified in different conditions. First, the convective term can be dropped if the material balance is written for the biofilm matrix or in a static liquid environment, where fluid velocities are near zero. This assumption can greatly shorten the calculations because the fluid velocity field **u** usually requires the solution of fluid dynamics equations. Secondly, in (quasi-)steady-state conditions (assumption explained in Section 22.1.2.4) the accumulation is zero, which again simplifies significantly the computations.

22.2.5.2   Effect of substrate gradients on biofilm morphology

The question to be addressed here is whether heterogeneous biofilm structures can occur naturally influenced by substrate gradients, but without any supposition on a deliberate influence of the bacteria directly.

In general, the slower the substrate transport processes relative to the growth process the stronger the substrate gradient will be. Simulations by Picioreanu *et al.* (1998b, 2000b) show that when the diffusion of substrate is relatively slow, strongly porous or even filamentous biofilms can be formed (Figure 22.6). Initially, the colonies of bacteria tend to grow in all directions, filling the space between them. As the biofilm thickness grows, some colonies get the chance to be closer to the substrate source than others. The tips of the biofilm will get substrate at a faster rate then the valleys, as it is shown in Figure 22.4d by the arrows representing the flux of substrate. This leads to locally higher growth rates. The process is self-enhancing: if bacteria have already the capacity for a fast growth, the tips grow out faster, and continue to grow ever faster than the bacteria in the valleys. The voids between the colonies cannot be filled with new biomass any longer. The obvious consequence is that a rough, 'finger-like' biofilm will develop on the top of a compact base layer. Conversely, compact and dense biofilms resulted at higher substrate transfer rates, when the biofilm development was limited only by the microbial metabolism. This effect is also observed in simulations including convective substrate transport (Picioreanu *et al.*, 2000b) and is independent of the biomass spreading mechanism chosen (Dockery and Klapper, 2001; Eberl *et al.*, 2001b; Kreft *et al.*, 2001; also see Figure 22.4).

Besides substrate transfer limitation, also the initial degree of substratum surface coverage with attached bacteria influences biofilm geometrical heterogeneity. A smaller fraction of the substratum surface colonised leads to a porous and

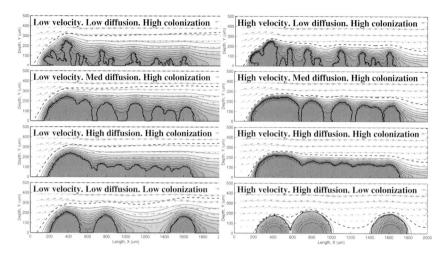

Figure 22.6   Simulations of biofilm development including computation of flow, substrate convection, diffusion and reaction, and biomass growth but no detachment. The thick continuous lines indicate the biofilm surface. Iso-concentration lines show the decrease of substrate concentration from the maximum value in the bulk liquid (white areas) to zero in the biofilm (dark-grey areas). The thick dashed contour lines indicate the limit of the CBL (98% from the bulk concentration). Model description and parameters can be found in Picioreanu et al. (2000b).

more irregular biofilm. In conclusion, these model results show that no specific microbial mechanism is needed in order to explain the incidence of biofilm structures with large geometrical differences.

22.2.5.3   Effect of biofilm channels to overall substrate conversion

By simulating the substrate conversion process at different flow rates and biofilm geometries, the contribution of pores and channels to the transport to the bacterial cells can be evaluated. This requires knowledge of liquid flow pattern past the biofilm, which affects the flux of substrate transport by convection. Due to the inherent computational complexity, accurate hydrodynamics has rarely been considered in biofilm modelling (e.g. Dillon et al., 1996; Dillon and Fauci, 2000, Picioreanu et al., 1999, 2000a, 2000b, 2001; Eberl et al., 2001a). The effect of flow velocity on the relative contribution of convective and diffusive transport mechanisms to the overall substrate transport to the same biofilm is shown in Figure 22.7 (Picioreanu et al., 2000a). It is clear that only at very high flow rates, convection will dominate the substrate transport. At such high flow rates, however, also large shear forces exist and the biofilm will adapt by becoming less porous and smoother, decreasing the pore-based convective transport. The biofilm structure adapts to the flow regime in a way that buffers mass transfer. This implies

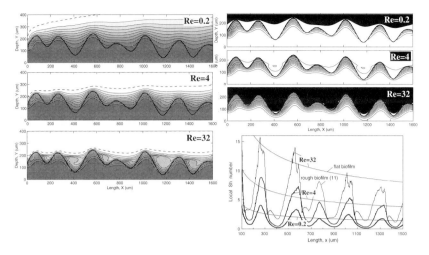

Figure 22.7   Effect of flow velocity on the relative contribution of convective and diffusive mass transport. Left: Contour lines of substrate concentration at different fluid velocities (different *Re* numbers) for a rough surface biofilm. Contour lines and grey shades have the same meaning like in Figure 22.6. Right: Convection dominates mass transfer in black areas. Inside the biofilm structure lines of equal reaction rate are drawn. The graph shows the local flux of substrate through the biofilm surface. Model description and parameters are in Picioreanu *et al.* (2000a).

that convective transport inside a biofilm might not be very important in general, unless e.g. large changes in flow rates occur in a biofilm system or the biofilm structure contains large moving parts (Stoodley *et al.*, 1998).

It can be argued whether the above simulations by a 2-D model are representative for a 3-D structure where flow can by-pass biofilm structures. Simulations by Eberl *et al.* (2001a) show that when the same biofilm is modelled in 2- or 3-D, the overall mass transfer is equivalent (Figure 22.8). This indicates that for many studies a 2-D simulation is sufficient. The effect of 3-D geometry and porosity has been further evaluated by simulating a mushroom-like biofilm. Again, it becomes apparent that in the pore region, at relatively high liquid velocities, the mass transfer is dominated by convection. That is what one would observe also by microelectrode measurements. However, the exchange of liquid between the pore region and the bulk liquid is marginal. Therefore, if the overall mass transfer from the bulk liquid to the biofilm is evaluated, the diffusive transport is dominating the convective transport even at high liquid velocities. Though, if biofilms consist of isolated colonies, the convective transport will largely contribute to the mass transfer. In conclusion, these observations show that pores probably do not contribute much to the overall conversion process (where the interest is usually from an engineering perspective) but they might have large local influences on microbial competition and selection processes (where the interest of microbiologists often is).

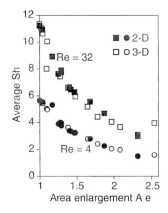

Figure 22.8   Effect of biofilm heterogeneity, expressed as area enlargement (biofilm surface area divided by substratum surface area), on mass transfer expressed as Sherwood number (total external mass transfer divided by diffusive transfer). Comparison between 2-D and 3-D simulations at low ($Re = 4$) and high ($Re = 32$) flow velocity (data from Eberl *et al.*, 2001a).

22.3   CONCLUSIONS

Despite recent successes, construction and use of mathematical models for development of biofilm structure are constrained to multiple limitations:

(1) Difficulty to express physical phenomena in mathematical form. For instance, CA and IbM for biofilm spreading were proposed because a rigorous mechanical representation for biomass was difficult to achieve.

(2) System parameters unknown. Many times the governing equations are known but an estimation of values of characteristic parameters is missing, e.g. mechanical properties of biofilms are still poorly known despite the recent efforts and the progress made.

(3) Equations too difficult to solve. There are many cases when the model equations can be easily deduced and we have also the parameters, but very costly or demanding numerical solutions are encountered. Flow and mass transfer in 3-D, turbulent flow, systems of differential mass balances coupled with equilibrium equations, especially in complex geometries, like heterogeneous biofilms.

(4) Use of a model is restricted to certain ranges of validity. Also, experiments are performed in particular conditions and extrapolation of results from one experiment to different systems must be done very carefully. Avoid designing too artificial experimental programs only with the goal to validate a certain model. As complex models also make assumptions, our intuition can be improved only to a certain point.

In conclusion, the most important result of a modelling study on biofilm structure is often not the specific numerical result or a nice graph, but rather the level of understanding reached. Mathematical models can only transform information from one form to another, hence, they help in interpretation and testing of experimental data, but they cannot create new data. The contribution of microbiologists, or experimentalists in general, is therefore essential in obtaining data needed to validate biofilm models. Anyway, one of the most important functions of mathematical models remains that of a basic 'communication tool' between scientists in the most diverse fields of research.

REFERENCES

Bailey, J.E. (1998) Mathematical modeling and analysis in biochemical engineering: past accomplishments and future opportunities. *Biotechnol. Prog.* **14**, 8–20.

Ben-Jacob, E., Schochet, O., Tenenbaum, A., Cohen, I., Czirók, A. and Vicsek, T. (1994) Generic modelling of cooperative growth patterns in bacterial colonies. *Nature* **368**, 46–49.

Bishop, P.L. and Rittmann, B.E. (1995) Modelling heterogeneity in biofilms: report of the discussion session. *Water Sci. Technol.* **32**, 263–265.

Caldwell, D.E., Korber, J.R. and Lawrence, D.R. (1993) Analysis of biofilm formation using 2D vs 3D digital imaging. *J. Appl. Bacteriol.* **74**, S52–S66.

Costerton, J.W., Lewandowski, Z., De Beer, D., Caldwell, D., Korber, D. and James, G. (1994) Minireview: biofilms, the customized microniche. *J. Bacteriol.* **176**, 2137–2142.

Davies, D.G., Parsek, M.R., Pearson, J.P., Iglewski, B.H., Costerton, J.W. and Greenberg, E.P. (1998) The involvement of cell-to-cell signals in the development of a bacterial biofilm. *Science* **280**, 295–298.

Dibdin, G. (1992) A finite-difference computer model of solute diffusion in bacterial films with simultaneous metabolism and chemical reaction. *CABIOS* **8**, 489–500.

Dillon, R. and Fauci, L. (2000) A microscale model of bacterial and biofilm dynamics in porous media. *Biotechnol. Bioeng.* **68**, 536–547.

Dillon, R., Fauci, L. and Gaver, D. (1995) A microscale model of bacterial swimming, chemotaxis and substrate transport. *J. Theoret. Biol.* **177**, 325–340.

Dillon, R., Fauci, L., Fogelson, A. and Gaver, D. (1996) Modeling biofilm processes using the immersed boundary method. *J. Comput. Phys.* **129**, 57–73.

Dockery, J. and Klapper, I. (2001) Finger formation in biofilm layers. *SIAM J. Appl. Math.* **62**, 853–869.

Dupin, H.J., Kitanidis, P.K. and McCarty, P.L. (2001) Pore-scale modeling of biological clogging due to aggregate expansion: a material mechanics approach. *Water Res.* **37**, 2965–2979.

Eberl, H.J., Picioreanu, C., Heijnen, J.J. and van Loosdrecht, M.C.M. (2001a) A three-dimensional numerical study on the correlation of spatial structure, hydrodynamic conditions, and mass transfer and conversion in biofilms. *Chem. Eng. Sci.* **55**, 6209–6222.

Eberl, H.J., Parker, D.F. and van Loosdrecht, M.C.M. (2001b) A new deterministic spatio-temporal continuum model for biofilm development. *J. Theoret. Med.* **3**, 161–175.

Hermanowicz, S.W. (1998) A model of two-dimensional biofilm morphology. *Water Sci. Technol.* **37**, 219–222.

Hermanowicz, S.W. (2001) A simple 2D biofilm model yields a variety of morphological features. *Math. Biosci.* **169**, 1–14.

Kreft, J.-U. and Wimpenny, J.W.T. (2001) Effect of EPS on biofilm structure and function as revealed by an individual-based model of biofilm growth. *Water Sci. Technol.* **43**, 135–141.

Kreft, J.-U., Booth, G. and Wimpenny, J.W.T. (1998) BacSim, a simulator for individual-based modelling of bacterial colony growth. *Microbiology* **144**, 3275–3287.

Kreft, J.-U., Picioreanu, C., Wimpenny, J.W.T. and van Loosdrecht, M.C.M. (2001) Individual-based modelling of biofilms. *Microbiology* **147**, 2897–2912.

Lawrence, J.R., Korber, D.R., Hoyle, B.D., Costerton, J.W. and Caldwell, D.E. (1991) Optical sectioning of microbial biofilms. *J. Bacteriol.* **173**, 6558–6567.

Matsushita, M. and Fujikawa, H. (1990) Diffusion-limited growth in bacterial colony formation. *Physica A* **168**, 498–506.

Noguera, D.R., Pizarro, G., Stahl, D.A. and Rittmann, B.E. (1999) Simulation of multi-species biofilm development in three dimensions. *Water Sci. Technol.* **39**, 123–130.

Ohashi, A. and Harada, H. (1994) Adhesion strength of biofilm developed in an attached growth reactor. *Water Sci. Technol.* **29**, 281–288.

Ohashi, A. and Harada, H. (1996) A novel concept for evaluation of biofilm adhesion strength by applying tensile force and shear force. *Water Sci. Technol.* **34**, 201–211.

Ohashi, A., Koyama, T., Syutsubo, K. and Harada, H. (1999) A novel method for evaluation of biofilm tensile strength resisting to erosion. *Water Sci. Technol.* **39**, 261–268.

Picioreanu, C. (1996) Modelling biofilms with cellular automata. Final report to European Environmental Research Organisation, Wageningen, The Netherlands.

Picioreanu, C. (1999) Multidimensional modeling of biofilm structure. Ph.D. thesis, Department of Bioprocess Technology, Delft University of Technology, Delft, The Netherlands.

Picioreanu, C. and van Loosdrecht, M.C.M. (2002) A mathematical model for initiation of microbiologically influenced corrosion by differential aeration. *J. Electrochem. Soc.* **149**, B211–B223.

Picioreanu, C., van Loosdrecht, M.C.M. and Heijnen, J.J. (1998a) A new combined differential-discrete cellular automaton approach for biofilm modeling: application for growth in gel beads. *Biotechnol. Bioeng.* **57**, 718–731.

Picioreanu, C., van Loosdrecht, M.C.M. and Heijnen, J.J. (1998b) Mathematical modeling of biofilm structure with a hybrid differential-discrete cellular automaton approach. *Biotechnol. Bioeng.* **58**, 101–116.

Picioreanu, C., van Loosdrecht, M.C.M. and Heijnen, J.J. (1999) Discrete-differential modelling of biofilm structure. *Water Sci. Technol.* **39**, 115–122.

Picioreanu, C., van Loosdrecht, M.C.M. and Heijnen, J.J. (2000a) A theoretical study on the effect of surface roughness on mass transport and transformation in biofilms. *Biotechnol. Bioeng.* **68**, 354–369.

Picioreanu, C., van Loosdrecht, M.C.M. and Heijnen, J.J. (2000b) Effect of diffusive and convective substrate transport on biofilm structure formation: a two-dimensional modeling study. *Biotechnol. Bioeng.* **69**, 504–515.

Picioreanu, C., van Loosdrecht, M.C.M. and Heijnen, J.J. (2000c) Modelling and predicting biofilm structure. In *Community Structure and Co-operation in Biofilms* (ed. D.G. Allison, P. Gilbert, H.M. Lappin-Scott and M. Wilson), pp. 129–166, Cambridge University Press, Cambridge, UK.

Picioreanu, C., van Loosdrecht, M.C.M. and Heijnen, J.J. (2001) Two-dimensional model of biofilm detachment caused by internal stress from liquid flow. *Biotechnol. Bioeng.* **72**, 205–218.

Pizarro, G., Griffeath, D. and Noguera, D.R. (2001) Quantitative cellular automaton model for biofilms. *J. Environ. Eng.* **127**, 782–789.

Rittmann, B.E. and McCarty, P.L. (1980) Model of steady-state-biofilm kinetics. *Biotechnol. Bioeng.* **22**, 2343–2357.

Stoodley, P., Lewandowski, Z., Boyle, J.D. and Lappin-Scott, H.M. (1998) Oscillation characteristics of biofilm streamers in turbulent flowing water as related to drag and pressure drop. *Biotechnol. Bioeng.* **57**, 536–544.

Stoodley, P., Lewandowski, Z., Boyle, J.D. and Lappin-Scott, H.M. (1999) Structural deformation of bacterial biofilms caused by short-term fluctuations in fluid shear: an *in situ* investigation of biofilm rheology. *Biotechnol. Bioeng.* **65**, 83–92.

Takács, I. and Fleit, E. (1995) Modelling of the micromorphology of the activated sludge floc: low DO, low F/M bulking. *Water Sci. Technol.* **31**, 235–243.

Wanner, O. and Gujer, W. (1986) A multispecies biofilm model. *Biotechnol. Bioeng.* **28**, 314–328.

Wanner, O. and Reichert, P. (1996) Mathematical modeling of mixed-culture biofilms. *Biotechnol. Bioeng.* **49**, 172–184.

Wimpenny, J.W.T. and Colasanti, R. (1997) A unifying hypothesis for the structure of microbial biofilms based on cellular automaton models. *FEMS Microb. Ecol.* **22**, 1–16.

Wood, B.D. and Whitaker, S. (1998) Diffusion and reaction in biofilms. *Chem. Eng. Sci.* **53**, 397–425.

Wood, B.D. and Whitaker, S. (1999) Cellular growth in biofilms. *Biotechnol. Bioeng.* **64**, 656–670.

PART THREE

Control of biofilms

Section 5: *Biofilm monitoring*   441

Section 6: *Biofilm disinfection*   471

Section 7: *Biofilm control*   535

Section 5

Biofilm monitoring

23 Biofilm monitoring by photoacoustic
    spectroscopy   443

24 Quartz crystal microbalance with dissipation monitoring
    (QCM-D): a new tool for studying
    biofilm formation in real time   450

25 Monitoring biofouling using infrared
    absorbance   461

23

Biofilm monitoring by photoacoustic spectroscopy

T. Schmid, U. Panne, C. Haisch and R. Niessner

23.1 INTRODUCTION

Biofilms are layers of microorganisms and biopolymers (extracellular polymer substances, EPSs), which occur at the interfaces of aqueous systems. The unwanted growth of biofilms in process waters is termed biofouling and is responsible for a decrease in water quality and an increase of the frictional resistance in tubes. Despite these negative effects, biofilms are used as well in beneficial applications, e.g. removal of organic and inorganic pollutants in wastewater treatment plants (Wilderer and Characklis, 1989). For improvement of anti-fouling strategies and process optimization in wastewater treatment plants, a non-destructive method for on-line monitoring of biofilms is needed. A new photoacoustic technique allows non-destructive *in situ* monitoring of growth, detachment and thickness of biofilms. Main components of the biofilm can be determined by wavelength-dependent measurements in a depth-resolved fashion.

23.2 PHOTOACOUSTIC SPECTROSCOPY

Photoacoustic spectroscopy is based on the absorption of electromagnetic radiation inside a sample, where non-radiative relaxation processes convert the absorbed energy into heat. Due to the thermal expansion of the medium, a pressure wave is generated, which can be detected by microphones or piezoelectric transducers (Rosencwaig, 1980). The amplitude p of a photoacoustic signal generated by a laser pulse inside solid or liquid samples can be generally described by

$$p \propto \frac{\beta c^2}{C_p} E_0 \mu_a \qquad (23.1)$$

where C_p is the heat capacity, β, the thermal expansion coefficient, c, the speed of sound in the medium under study, E_0, the laser-pulse energy, and μ_a, the optical absorption coefficient of the sample (Tam, 1986). After normalization of the signal with regard to laser-pulse energy and temperature, the absorption coefficient of the sample can be determined, which depends linearly on the concentration of the absorbing compound.

If short laser pulses are used for excitation, a time-resolved recording of the photoacoustic signal allows a depth-resolved investigation of the light absorption inside the irradiated part of the sample (Karabutov et al., 1995). The distance between an absorbing object inside the sample and the sample surface can be calculated as

$$z = c \cdot t \qquad (23.2)$$

where t is the time delay between the laser pulse and the arrival of the pressure wave at the sample surface. Thus, changes in the optical absorption properties of a sample can be investigated depth resolved, if the sound velocity inside the sample is known.

23.3   PHOTOACOUSTIC BIOFILM MONITORING

23.3.1   Experimental setup

Photoacoustic sensor heads for biofilm monitoring consist of a piezoelectric detector, which is coupled to a transparent prism (Figure 23.1). The biofilm grows directly on the surface of the prism and is illuminated by short laser pulses. The electromagnetic radiation is absorbed inside the biofilm, where a pressure wave is generated by the photoacoustic effect. The laser-induced pressure waves are detected by the piezoelectric transducer consisting of a 25 μm thick poly(vinylidene fluoride) (PVDF) film. The transducer has a diameter of 5 mm allowing a representative area of 20 mm². The detection limit for the optical absorption coefficient is $0.02\,\text{cm}^{-1}$ (Schmid et al., 2002a) and the depth resolution in aqueous samples (e.g. aqueous solutions, hydrogels, biological matrices) is approximately 10 μm (Kopp and Niessner, 1999).

The laser pulses are guided via optical fibers (550 μm diameter, HCG-MO550T-10, Laser Components, Santa Rosa, USA) from the laser to the sensor head. In this study, a frequency-doubled Nd:YAG laser ($\lambda = 532\,\text{nm}$, Surelite 10-I,

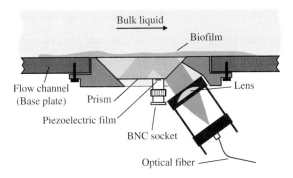

Figure 23.1   Photoacoustic sensor head for biofilm monitoring.

Continuum, Santa Clara, USA) and a tunable optical parametric oscillator (OPO, Panther, Continuum, Santa Clara, USA) were used for excitation of photoacoustic signals. The OPO allows the generation of laser pulses with wavelengths from 410 nm in the visible (Vis) to 2550 nm in the near-infrared range (NIR). The pre-amplified photoacoustic signal is recorded by a digital storage oscilloscope (TDS 540, Tektronix, Beaverton, USA). The synchronization of the oscilloscope with the laser pulse is performed with a trigger signal. All components of the sensor system are controlled by an in-house developed LabVIEW software (LABVIEW 5.1, National Instruments, Austin, USA): flash lamp and shutter of the laser via a serial interface and the digital storage oscilloscope via the IEEE 488 bus.

For extensive characterization of the new biofilm monitoring technique, an 18-l tube reactor for growth of microorganisms was set up. The reactor contained a mixture of microorganisms taken from an aerobic sequencing batch biofilm reactor and fed by a nutrient solution consisting of 690 mg l^{-1} sodium acetate, 60 mg l^{-1} potassium dihydrogenphosphate, 252 mg l^{-1} ammonium sulfate, 19 mg l^{-1} potassium chloride and 4 mg l^{-1} yeast extract. The reactor was aerated with compressed air with a volume flow of 11 min^{-1}. To generate biofilms on solid surfaces, the content of the tube reactor was pumped through a flow channel (100 ml volume; 260 mm long) by a peristaltic pump. For investigation of biofilms at different positions, three photoacoustic sensor heads were integrated into the base plate of the channel. A more detailed description of the experimental setup can be found in Schmid *et al.* (2002).

23.3.2   Photoacoustic signals of biofilms

Figure 23.2 shows a typical photoacoustic signal profile of a biofilm measured at a wavelength of 532 nm. According to Equation (23.2), the time scale of the oscilloscope was converted into the corresponding depth scale. The conversion factor was the speed of sound inside the biofilm system. Water is the predominant component of both the biofilm and the bulk liquid phase. Therefore, the sound

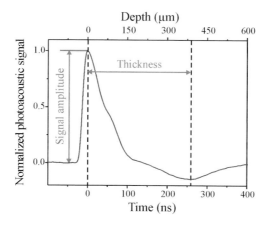

Figure 23.2   Typical photoacoustic signal profile of a biofilm ($\lambda = 532$ nm).

velocity of water, i.e. $c = 1.5 \times 10^3 \, \mathrm{m \, s^{-1}}$ can be assumed for the biofilm system. The origin of the abscissa at $0 \, \mu\mathrm{m}$ corresponds to the surface of the prism.

As exhibited in Equation (23.1), the amplitude of the photoacoustic signal is a function of the optical absorption coefficient, which depends linearly on the concentration of the absorbing compound. Most of the photoacoustic biofilm measurements presented in this chapter were performed at 532 nm. In the Vis spectral range, the incident light is absorbed by various pigments inside the cells and the EPS matrix. Therefore, the photoacoustic signal amplitude depends on the density of the immobilized biomass.

In most cases, the maximum of the signal corresponding to the highest biofilm density is at $0 \, \mu\mathrm{m}$ and, therefore, directly on the surface of the sensor head. For the whole depth range inside the absorbing and scattering biofilm, the signal has a positive value. At the interface between biofilm and water, the signal becomes negative and reaches its minimum. Thus, the biofilm thickness can be determined by measuring the distance between the maximum and minimum of the photoacoustic signal profile.

This possibility for biofilm thickness measurements was discovered by measurements of biofilm models. The biofilm models consisted of agar–agar hydrogel layers with well-defined thickness containing enclosed iron(III) oxide particles. To simulate an aqueous bulk phase, water was added to the surface of the layers. In this way, the acoustic properties of biofilms could be modeled by agar–agar and the optical properties by iron(III) oxide particles, which absorb and scatter laser radiation at 532 nm.

The thickness measurements were validated by confocal laser-scanning microscopy (CLSM) using real biofilms grown on glass slides. The result of the photoacoustic thickness measurement was in good agreement with the independent microscopic-imaging technique (Schmid *et al.*, 2002).

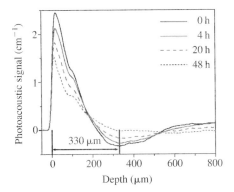

Figure 23.3   Biofilm detachment caused by a pH shift from 8.4 to 11.1. Depth-resolved photoacoustic measurements at 532 nm.

23.3.3   Photoacoustic biofilm measurements

In order to examine the new photoacoustic biofilm monitoring technique, biofilms were generated on the surfaces of the three sensor heads inside the flow channel. Subsequently, one of the process parameters (e.g. flow conditions, pH) was changed and modifications in biofilm density and thickness were monitored on-line by photoacoustic measurements.

Biofilm growth could be monitored by plotting the photoacoustic signal amplitudes versus time. The increase in signal intensity reflects the attachment of microbial cells and EPS molecules to the sensor surfaces and the corresponding increase of the density of the immobilized biomass (Schmid et al., 2001). Due to different flow conditions at the three positions inside the flow channel, biofilms with different density and, in particular, different structure were formed on the sensor surfaces. The influence of flow conditions on the biofilm structure could be verified by generation of biofilms at different flow velocities. The experiments revealed that an increase in flow velocity leads to a shift of the biomass towards the solid surface and, therefore, to the formation of thinner biofilms (Schmid et al., 2002).

Biofilm detachment caused by dissolved as well as colloidal substances was monitored by depth-resolved measurements. The pH influences electrostatic interactions between EPS molecules and has, therefore, an indirect influence on the stability of the biofilm. After a pH shift from 8.4 to 11.1, biofilm detachment could be observed over several hours (Figure 23.3). The minimum of the photoacoustic signal at 330 μm reflects the interface region between biofilm and bulk liquid. The measured biofilm thickness was approximately constant during the detachment process. It could be concluded that biofilm aggregates, which are smaller than the representative area of the sensor (20 mm^2) were detached from the biofilm surface. Thus, biofilm detachment caused by pH shifts can be interpreted as erosion of microscopic aggregates (Schmid et al., 2003b)

The photoacoustic monitoring technique was used for an extensive investigation of various anti-fouling strategies. The effect of hydrogen peroxide on biofilms

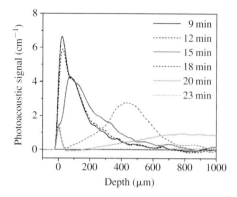

Figure 23.4   Biofilm detachment by 200 ppm hydrogen peroxide ($\lambda = 532$ nm).

was compared with diverse isothiazolinone biocides. Biofilm removal by the oxidant hydrogen peroxide was the most effective anti-fouling strategy in this context. Most of the adsorbed biomass was removed within a few minutes. The changes in signal shape and biofilm thickness indicate that sloughing off of relatively large areas of the biofilm lead to the fast biofilm detachment (Figure 23.4).

Even the interaction of biofilms with particulate matter can be investigated by photoacoustic measurements. The sorption of iron(III) oxide particles ($<5\,\mu$m) could be observed on-line. Sorbed particles can influence the stability of the biofilm and, therefore, lead to a partial detachment. This was observed by depth-resolved measurements. Predominantly, the part of the biofilm with the highest particle concentration was removed from the sensor surface. The particles could not reach the base biofilm. Therefore, detachment effects in this part of the film were relatively weak (Schmid *et al.*, 2002).

By use of a tunable OPO, photoacoustic absorption spectra of biofilms were determined. Absorption bands of main components, such as pigments, water and carbohydrates could be identified in the Vis and NIR spectra. Depth-resolved measurements revealed the possibility to investigate the distribution of chemical compounds inside the biofilm system (Schmid *et al.*, 2003).

23.4   CONCLUSIONS

Photoacoustic spectroscopy is an interesting approach to biofilm monitoring. The new technique allows *in situ* monitoring of growth, detachment and thickness of biofilms. Detachment mechanisms can be elucidated by depth-resolved measurements. By photoacoustic spectroscopy, the influence of various physical and chemical factors, such as flow conditions or pH-value was investigated. Additionally, diverse anti-fouling strategies could be compared to each other.

Wavelength-dependent measurements allow the depth-resolved determination of main components of the biofilm system due to their characteristic absorptions. Within further studies, specific staining reagents will be used to distinguish, e.g between living and dead cells. Stained cells and EPS components will be detected by photoacoustic measurements at specific wavelengths in a depth-resolved fashion.

Acknowledgements

The authors acknowledge the financial support by Deutsche Forschungs-gemeinschaft (DFG) and a grant awarded to Thomas Schmid by Max-Buchner-Forschungsstiftung (MBFSt).

REFERENCES

Karabutov, A.A., Podymova, N.B. and Letokhov, V.S. (1995) Time-resolved optoacoustic detection of absorbing particles in scattering media. *J. Mod. Opt.* **42**, 7–11.

Kopp, C. and Niessner, R. (1999) Depth-resolved determination of the absorption coefficient by photoacoustic spectroscopy within a hydrogel. *Anal. Chem.* **71**, 4663–4668.

Rosencwaig, A. (1980) *Photoacoustics and Photoacoustic Spectroscopy*, Wiley, New York.

Schmid, T., Kazarian, L., Panne, U. and Niessner, R. (2001) Depth-resolved analysis of biofilms by photoacoustic spectroscopy. *Anal. Sci.* **17**, 574–577.

Schmid, T., Panne, U., Haisch, C., Hausner, M. and Niessner, R. (2002) A photoacoustic technique for *in situ* monitoring of biofilms. *Environ. Sci. Technol.* **36**, 4135–4141.

Schmid, T., Panne, U., Haisch, C. and Niessner, R. (2002a) Photoacoustic absorption spectra of biofilms. *Rev. Sci. Instrum.* **74**, 755–757.

Schmid, T., Panne, U., Haisch, C. and Niessner, R. (2003b) Biofilm monitoring by photoacoustic spectroscopy (PAS). *Water Sci. Technol.* **47**, in press.

Tam, A.C. (1986) Applications of photoacoustic sensing techniques. *Rev. Mod. Phys.* **58**, 381–431.

Wilderer, P.A. and Characklis, W.G. (1989) *Structure and Function of Biofilms*, Wiley-Interscience, New York.

24

Quartz crystal microbalance with dissipation monitoring: a new tool for studying biofilm formation in real time

M. Rudh

24.1 INTRODUCTION

The surface-sensitive quartz crystal microbalance with dissipation monitoring (QCM-D) technique has been widely used as a real-time tool to characterize molecule–surfaces interaction ranging in size from small molecules to large proteins or polymers. In the field of bacterial attachment and biofilm formation, the use of QCM-D is not yet widely spread and few reports exist. However, there is reason to believe that the QCM-D technique offers strong advantages in real-time biofilm formation characterization. In this chapter, the QCM-D principle is explained in detail and a brief overview of the earlier studies involving protein–surfaces interactions are covered as well as the existing reports on biofilm formation. In the work presented in this chapter the QCM-D instrument from Q-Sense AB (Göteborg, Sweden) has been used (for further information see www.qsense.com).

24.2   THE QCM-D MEASUREMENT PRINCIPLE

The QCM-D technique is an electro-mechanical, surface-sensitive method that measures changes in mass and viscoelastic properties in real time. The QCM-D technique is based on a piezoelectric quartz crystal (SiO_2), which is sandwiched between two electrodes. One of the electrodes is in contact with the sample and acts as the "sensing" surface where interactions are measured. The other electrode is kept in contact with air. The interactions studied can vary in strength from inter-molecular interactions to covalent binding, and from monolayers of atoms or small molecules (less than a nanometer) to micrometers of polymers in thickness.

The sensing principle is based on the piezoelectric property of quartz, which allows the crystal to deform laterally, i.e. along the surface, if an electric field is applied over the two electrodes. If an AC voltage is applied, the crystal can be made to oscillate at its resonance frequency. In 1964, Sauerbrey showed that a change in the resonance frequency of the crystal is proportional to a change of the oscillating mass, i.e. mass adsorbed on the crystal, allowing the crystal connected to an oscillator circuit to work as a very sensitive microbalance at the sub-nanogram scale. Note that the detection limit in an aqueous solution is below 5 ng/cm^2. The QCM-D technique also makes use of the fact that systems oscillating in mechanical resonance are subjected to energy losses. The oscillation of the crystal will inevitably stop when the driving AC voltage is switched off, but it will not stop right away. The energy stored in the oscillation will gradually dissipate away to the surroundings with a rate that is determined by the sum of all energy losses. Imagine a person ringing a church bell, when the free bell rings the tone is firm and will die away slowly, but when the church bell is damped by for instance the ringer's hand the tone is lowered and it rings away almost immediately. For the quartz crystal, these losses have several sources, such as, e.g inherent imperfections in the crystal, resistive losses in the contacts to the electrical circuit and losses induced from the surrounding on the crystal. Considering the latter, it is clear that when the crystal oscillates with an attached rigid mass (e.g. a crystalized salt) the energy is dissipated away slowly (the ring-down is long) compared to when the crystal is damped (e.g. a layer of extracellular polymeric substance (EPS)-producing bacteria attached to the crystal) the energy is dissipated much faster (the ring-down is short). Figure 24.1 summarizes the QCM-D measurement principle.

The measurement technique utilizes, in a cyclic repetition, a short AC voltage pulse (with a frequency that corresponds to the resonant frequency of the crystal) applied over the crystal. Thus, a stable oscillation is produced before the AC voltage is switched off and the ring-down (an exponentially decaying sinusoidal curve) is recorded and transferred to a computer for mathematical processing. From this, the absolute frequency and the decay time constant (being inversely proportional to the energy dissipation) are derived.

The repetition rate is approximately 1 Hz, i.e. one recording of the frequency, f, and the energy dissipation, D, each second. Changes in f represent changes in mass, Δm, and changes in D represent changes in energy losses, which in aqueous solution are dominated by changes in viscous losses (viscoelastic properties)

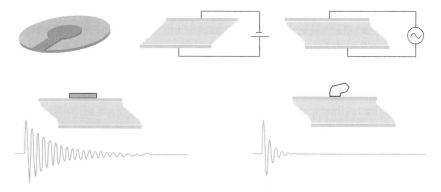

Figures 24.1   The measurement principle of the QCM-D technique with the physical deformation induced in the crystal upon applying a static and varying electric field (above), and the difference in ring-down between rigid and viscous matter (below).

either in the bulk solution or in adsorbed films. Thus, the two parameters Δf and ΔD give complementary information: changes in frequency are generally converted to mass (with the equation known as the Sauerbrey equation, see Equation 24.3) and used as mass information parameter giving information about changes in adsorbed mass versus time. The dissipation is used to extract structural information about the material attached. If the material attached on the crystal is rigid (e.g. crystalized salt) the dissipation will be low, while if it is non-rigid (e.g. bacterium) the dissipation will be high. For instance, processes or phenomena with internal structural changes (cross linking of polymer chains or structural changes of a cell membrane) are particularly interesting to evaluate.

24.2.1   QCM-D data

24.2.1.1   Mass sensitivity

Sauerbrey proposed a simple physical model for the observed proportionality between added mass and induced frequency shift (Sauerbrey, 1959). By defining the area density of quartz as $m_q = t_q \rho_q$, where ρ_q and t_q are the density and thickness of the quartz plate, respectively, he obtained

$$f = \frac{n v_q \rho_q}{2 m_q} \tag{24.1}$$

where n is the number of waves and v_q is the wave velocity in the crystal plate. By differentiating Equation (24.1) he obtained

$$df = -\frac{f}{m_q} dm_q \tag{24.2}$$

Sauerbrey made the assumption that for small mass changes, the added film can be treated as an equivalent mass change of the quartz crystal itself. By making the substitution $d \to \Delta$, the equation can, therefore, be written as

$$\Delta f = -\frac{f}{m_q}\Delta m = -\frac{f}{t_q \rho_q}\Delta m = -\frac{n}{C}\Delta m \qquad (24.3)$$

where $\Delta m = \rho_f t_f$, and ρ_f and t_f are the density and thickness of the added film, respectively, and C is the so-called mass sensitivity of the QCM. Equation (24.3) has become known as the Sauerbrey equation. For a typical quartz crystal used in QCM-D, where $\rho_q = 2650\,\text{kg/m}^3$ and $v_q = 3340\,\text{m/s}$, the mass sensitivity for a 5 MHz resonator oscillating in its fundamental mode is $17.7\,\text{ng}\,\text{cm}^{-2}\,\text{Hz}^{-1}$ (observe that $n = 3$ for the third overtone (15 MHz)). This means that the addition of one monolayer of water, which has an area density of approximately $0.25\,\text{mg/m}^2$, causes a frequency shift of $-1.4\,\text{Hz}$. Since the resonant frequency of a 5 MHz crystal resonating in vacuum can easily be measured with a precision of 0.01 Hz, very small masses can be measured.

Equation (24.3) is valid, if the added mass is

- small compared to the weight of the crystal,
- rigid,
- evenly distributed over the active area of the crystal.

24.2.1.2   The dissipation factor

The dissipation factor, D, is the inverse of the more known Q factor, defined by

$$D = \frac{1}{Q} = \frac{E_{\text{dissipated}}}{2\pi E_{\text{stored}}} \qquad (24.4)$$

where $E_{\text{dissipated}}$ is the energy dissipated during one period of oscillation, and E_{stored} is the energy stored in the oscillating system (Rodahl, 1995). Consequently, D is the sum of all mechanisms that dissipate energy from the oscillating system, such as friction and viscous losses. The dissipation factor of a quartz oscillator can be measured by recording how the oscillation decays after the oscillator has been excited into oscillation. In Figure 24.1, typical decay curves for rigid and viscoelastic matter is shown (curve that describes the decay of, e.g. electric signals or acoustic waves).

24.3   FROM PROTEIN ADSORPTION TO BIOFILM FORMATION

Although used in bio-surface research for some years, bacteria and biofilm research are so far not areas where QCM-D has been widely used. Starting with protein adsorption, this section covers some of the most relevant areas where QCM-D is used. Since the development of the QCM-D technique in the late 1990s (Rodahl, 1995; Höök, 1997) the technique has been used in many different areas of research. Studies of protein adsorption (Höök *et al.*, 1998a), was among the first areas to be explored followed by studies of antigen–antibody reactions, formation of lipid bi-layers (Keller and Kasemo, 1998), structural changes in proteins (Fant *et al.*, 2000), several topics in polymer research involving UV degradation, cross-linking and migration (Dahlqvist and Bjöörn, 1999); bacterial attachment (Otto *et al.*, 1999); biofilm formation (Green, 2001); and cell processes, such as spreading (Fredriksson *et al.*, 1998) and exocytosis (Cans *et al.*, 2001).

Since the QCM-D technique is electro-mechanical there is no need for either the sample or the surface to be transparent which in turn gives rise to a large flexibility in the choice of surface. In particular, this is useful in areas ranging from protein lipid vesicle adsorption studies (Keller and Kasemo, 1998), where the surface properties (hydrophilic, hydrophobic, etc.) are known to be critical for the outcome of the process and the formation of lipid bi-layers which occurs rapidly on very hydrophilic SiO_2 surfaces. It should be pointed out that the combination of having mass information (frequency) and structural information (dissipation) enables a time-independent analysis of $\Delta D/\Delta f$ by simply displaying the information in a ΔD versus Δf graph. Such a plot gives instructive information about changes in rigidity per coupled mass, which has been proven useful as a fingerprint of certain biomolecule and surface interactions (Höök *et al.*, 1998).

While the QCM-D technique today can be considered well documented and widely used in the fields of proteins and smaller biological molecules, its use in biofilm research is still in its infancy. Some studies of the interaction of *E. coli* with surfaces under different saline concentrations (Otto *et al.*, 1999) can be seen as a starting point for the use of QCM-D in biofilm research. Another step in this direction was the study by Green (2001) and Rudh *et al.* (2002) who investigated the attachment, growth and the production of EPSs of a *Leuconostoc mesenteroides* strain. The initial attachment of a condition layer, followed by bacterial attachment and EPS production has been reported. *L. mesenteroides* has two distinct metabolic pathways: when fructose is used as a carbon source no EPS production takes place, but when sucrose is used as a carbon source excessive production of EPS (dextran) takes place. In that study, a growth medium consisting of yeast extract, combined with either fructose or sucrose or both was used to follow the initial QCM-D signatures. In a stepwise measurement, subsequent addition of all components was performed, allowing the formation of a layer of the yeast extract on the surface to be monitored. This was followed by exposure of the yeast extract covered surface to a carbon source (in this case fructose). There was an apparent exchange reaction taking place on the surface where the previously adsorbed yeast extract was replaced by the fructose molecules. When the *L. mesenteroides* was inserted into the chamber, there was again an exchange reaction, during which a conditioning layer formed by the bacteria removed some of the previous yeast extract/fructose layer before bacterial attachment started to increase dramatically approximately 15 h after the addition of bacteria in the chamber (see Figure 24.2).

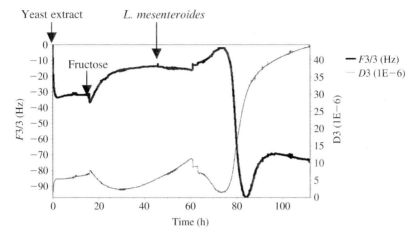

Figure 24.2   QCM-D data of biofilm formation of *L. mesenteroides* with fructose as carbon source as reported by Green (2001). Thick line frequency change; thin line dissipation change.

There were some unexplained observations in the study, especially the decrease–increase–decrease of frequency phase after 80 h. Whether this is caused by a structural change in the exoskeleton of the bacteria or rearrangement on the surface or actual detachment of bacteria remains to be determined. The dissipation factor proved to be very useful in determining the presence of EPS in the biofilm. A set of relative experiments were performed to study the difference between the different metabolic pathways. When fructose was used as a carbon source, the response was different compared to that of sucrose. With sucrose there was extensive EPS production and a considerably higher increase in D (three times higher than fructose) indicating that the structure of the biofilm was different from the fructose case. The much higher dissipation values indicate that the biofilm is more viscoelastic and non-rigid in its structure when there is a lot of EPS present in comparison to when the biofilm consists mainly of bacteria. It is not hard to imagine that the presence of a lot of EPS in the biofilm makes the structure more loose and "slime like" compared to the pure bacterial biofilm. As with the stepwise addition of the different components, there still remain unexplained results with the decrease–increase–decrease of the frequency (see Figure 24.3).

Further experiments were also performed on different types of surfaces. The biofilm growth was monitored on gold, stainless steel, SiO_2 and methyl-terminated thiolated gold surfaces (hydrophobic). The results varied biofilm formation tending to form faster on gold and stainless steel than on the hydrophobic thiol surface. Combination of the QCM-D data with optical techniques, such as epifluorescence microscopy or confocal laser scanning microscopy (CLSM), can be an even more powerful tool in biofilm research. In the study by Green (2001), the use of epifluorescence microscopy was widely used but combination with CLSM remains to be performed.

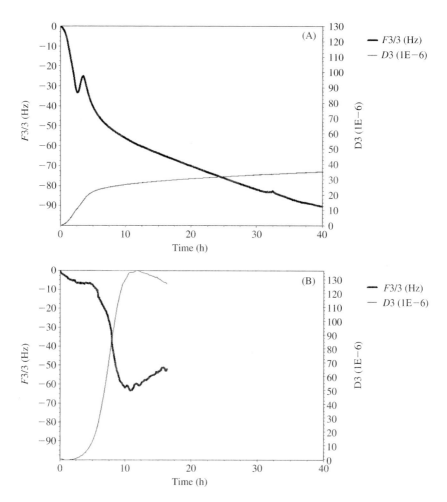

Figure 24.3   QCM-D data of biofilm formation of *L. mesenteroides* with fructose (A) and sucrose (B) as carbon source as reported by Green (2001). Thick line frequency change; thin line dissipation change.

24.4   OVERTONES – ADDITIONAL INFORMATION

The QCM-D technique does not only use one frequency, there is also the possibility to measure overtones. This section focuses on describing the additional information provided by overtone measurements and how this information can be used. For example, a guitar string has one fundamental resonance tone but also several overtones. The same holds true also for QCM crystals. The cyclic measurement principle of the QCM-D technique allows rapid switching between different overtone frequencies. Instead of applying AC pulses with 5 MHz only, i.e. the fundamental frequency, alternating pulses corresponding to the overtone

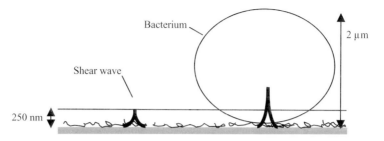

Figure 24.4   Penetration depth of the oscillary shear wave in a single spherical bacterium (of 2 μm diameter).

frequencies 15, 25, 35, ... MHz can also be used. Operating the QCM-D equipment not only at the fundamental frequency but also at several harmonics is a standard operating mode performed automatically. Operating at different overtones also improves the interpretation of the data, basically because additional experimental information is achieved.

For instance, the penetration depth of the oscillatory shear motion extending from the sensor surface through any adsorbed film and further out in solution decreases as the frequency increase: the higher the frequency, the smaller the penetration depth (or sensitivity range) above the crystal. This means that the data from the different overtones also, to some extent, contain geometrical (along the z-axis) information. For instance, the sensitivity range of a 5 MHz crystal placed in pure water is 250 nm (see Figure 24.4) which is approximately a quarter or less of one bacterium. Even if the penetration depth depends critically on the viscosity (and elasticity for a viscoelastic film), a bacterium can, to a first approximation, be considered not distinctively different from water. The bacterium consists to a large extent of water and it is reasonable to assume that the density is only slightly higher than that of water. It is thus realistic to believe that the penetration range in a single bacterium is not significantly different from that of water. But when several bacteria form colonies, undergo stress induced by shear forces and 'anchor' themselves to the surface. The biofilm properties will change in such a way that their rigidity, and thus ability to transfer mechanical motion, increases (Stoodley et al., 2001). This, in turn, is likely to increase the penetration depth significantly, up to several layers of bacteria ($\sim$μm). It is, however, not possible to generally state the penetration depth or sensitivity range. Nevertheless, recent advancement in theoretical modeling of the QCM-D response on simpler systems (see below), which are in the process of being transferred to the more complex systems as bacteria, has the potential to theoretically predict the oscillatory shear motion through bacterial films in different states of development. It is, thus, not unrealistic to believe that the operation at several harmonics will pretty soon allow us to use lower frequencies to probe one or several layers of bacteria, and higher frequencies to probe fractions of whole bacterium, i.e. the bacterial wall or the interfacial region closest to the crystal surface.

24.5   MODELING OF THE QCM-D DATA – ENHANCED UNDERSTANDING OF BIOLOGICAL PROCESSES

Changes in f and D, as described so far, allows for a qualitative interpretation of data. However, this section presents how theoretical modeling of the results can be used to improve and quantify the interpretation. A viscoelastic model based on a mechanical Voight-based representation (Voinova *et al.*, 1999), of an attached ad-layer consisting of a spring and dashpot in parallel has been used. Under the assumptions that the adsorbed material is homogenous, evenly spread and attached without slip. The model can be used to fit the physical representation of the film: density, viscosity, thickness and shear modulus to the f and D data. The results from such modeling have been shown to improve the interpretation base significantly. This has been illustrated by the recent work on adsorption and cross linking of a mussel-adhesive protein (Höök *et al.*, 2001). The mussel-adhesive protein is the mussel's mechanism to attach to marine surfaces, which is unwanted on ships hulls and in industrial applications. The mussel-adhesive proteins involve a cross-linking-mediated conformational change from an elongated structure to a collapsed cross-linked structure, which in QCM-D measurements was manifested as a transformation from a state inducing high D to a state inducing low D. Furthermore, theoretical modeling of the QCM-D response was shown to accurately determine the thickness of the protein both in its elongated and collapsed state. This was correlated with ellipsometry measurements. In addition, the changes obtained in the viscoelastic components of the film were in good agreement with the obtained changes in refractive index from ellipsometry.

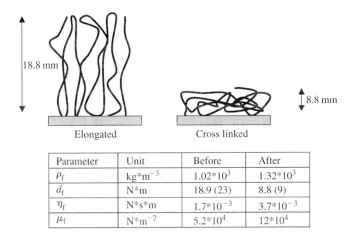

Parameter	Unit	Before	After
ρ_f	$kg*m^{-3}$	$1.02*10^3$	$1.32*10^3$
d_f	$N*m$	$18.9\ (23)$	$8.8\ (9)$
η_f	$N*s*m$	$1.7*10^{-3}$	$3.7*10^{-3}$
μ_f	$N*m^{-2}$	$5.2*10^4$	$12*10^4$

Figure 24.5   Modeling of Mefp-1 mussel-adhesive protein, as reported by Höök *et al.* (2001).In the table, the results from the modeling of the frequency and dissipation data are given. Corresponding thickness measurements (in parenthesis) where performed with elliposometry.

The use of this type of theoretical modeling of the QCM-D response in biofilm research is very likely to be a useful tool to understand better some of the processes occurring in biofilm formation, especially since knowledge about the viscoelastic components of the film can be used to estimate the sensitivity depth. As reported by Sauer (2001), bacteria undergo tremendous changes on the genetic level upon irreversible attachment to surfaces and up-regulating and switching on numerous genes coding for production of different adhesive proteins, EPS, etc. It is tempting to believe that a detailed understanding of the QCM-D response by operating at extensive harmonics and by utilizing theoretical modeling will one day allow us to correlate the response to such types of phenomena.

24.6 CONCLUSIONS

To date, the QCM-D technique has only recently been introduced to the field of biofilm research. Based on the few studies that have been conducted, combined with the type of information that has been proven possible to achieve from simpler systems, it is clear that there are numerous biofilm-related applications where the QCM-D technique is likely to contribute significantly. Such examples include, the design of bacteria-resistant surfaces (contact lenses), the study of biofilm formation in drinking water systems, the evaluation of effective biocide treatments for biofilm removal or design of protective coatings (pulp and paper industry amongst others). Another interesting approach, although until today only explored in the electronics industry, is the use of quartz crystals with multiple layer of electrode materials (for instance copper on top of gold, or stainless steel on top of gold) for studies of microbial induced corrosion (MIC), where the upper layer of the electrodes would simply be corroded by the bacteria. The real-time data recorded of both mass- and structure-related changes in the biofilm combined with the easiness and flexibility in experimental design, the QCM-D technique can, in the future, become one of the standard methods for the evaluation of biofilm formation.

REFERENCES

Cans, A.S., Höök, F., Shupliakov, O., Ewing, A.G., Eriksson, P.S., Brodin, L. and Orwar, O. (2001) Measurement of the dynamics of exocytosis and vesicle retrieval at cell populations using a quartz crystal microbalance. *Anal. Chem.* **73**, 5805–5811.

Dahlqvist, P. and Bjöörn, P. (1999) Studies of polymer film systems with the QCM-D technique, Masters thesis, Department of Applied Physics, Chalmers University of Technology.

Fant, C., Sott, K., Elwing, H. and Höök, F. (2000) Adsorption behavior and enzymatically or chemically induced cross-linking of a mussel adhesive protein. *Biofouling* **16**, 119–132.

Fredriksson, C., Kihlman, S., Steel, D.M. and Kasemo, B. (1998) *In vitro* real-time characterization of cell attachment and spreading. *Mat. Med.* **9**, 785–788.

Green, H. (2001) Detection of biofilm formation using the quartz crystal microbalance with dissipation monitoring, MSc. thesis, Department of Physics and Measurement Technology, Linköping University, Sweden.

Höök, F. (1997) Development of a novel QCM technique for protein adsorption studies, Ph.D. thesis, Department of Biochemistry and Biophysics and the Department of Applied Physics, Chalmers University of Technology.

Höök, F., Rodahl, M., Kasemo, B. and Brezezinski, P. (1998a) Structural changes in hemo-globin during adsorption to solid surfaces: effects of pH, ionic strength, and ligand binding. *Proc. Natl. Acad. Sci. USA Biophys.* **95**, 12271–12276.

Höök, F., Rodahl, M., Brezinski and Kasemo, B. (1998b) Measurements using the quartz crystal microbalance technique of ferritin monolayers on methyl-thiolated gold: depend-ence of energy dissipation and saturation coverage on salt concentration. *J. Colloid Interf. Sci.* **208**, 63–67.

Höök, F., Kasemo, B., Nylander, T., Fant, C., Sott, K. and Elwing, H. (2001) Variations in coupled water, viscoelastic properties and film thickness of a Mefp-1 protein film dur-ing adsorption and cross-linking: a QCM-D, ellipsometry and SPR study. *Anal. Chem.* **73**, 5796–5804.

Keller, C.A. and Kasemo, B. (1998) Surface specific kinetics of lipid vesicle adsorption measured with a quartz crystal microbalance. *Biophys. J.* **75**, 1397.

Otto, K., Elwing, H. and Hermansson, M. (1999) Effect of ionic strength on initial inter-actions of *Escherichia coli* with surfaces, studied on-line by a novel quartz crystal microbalance technique. *J. Bacteriol.* **181**, 5210–5218.

Rodahl, M. (1995) On the frequency and Q factor response of the quartz crystal microbalance to liquid overlayers, Ph.D. thesis, Department of Applied Physics, Chalmers University of Technology.

Rodahl, M., Höök, F., Krozer, A., Brezinski, P. and Kasemo, B. (1995) Quartz crystal microbalance setup for frequency and Q-factor measurements in gaseous and liquid environments. *Rev. Sci. Intrum.* **66**, 3924–3920.

Rudh, M., Green, H., Lie, E. and Sjöström, L. (2002) Measuring biofilm formation in real-time by quartz crystal microbalance with dissipation montoring (QCM-D). *Oral Presentation at the International Specialised Conference on Biofilm Monitoring*, Porto, March 2002.

Sauer, K. (2001) Characterization of phenotypic changes in *Pseudomonas putida* in response to surface-associated growth. *Poster and Presentation at the Center for Biofilm Engineering, Montana, ASM Meeting*, May 2001.

Sauerbrey, G. (1959) Verwendung von schwingquarzen zur wägung dünner schichten und zur mikrovägung. *Zeits. für Phys.* **155**, 206–222.

Sauerbrey, G. (1964) Einfluß der elektrodenmasse auf die schwingunsfiguren dünner schwingquartzplatten. *Archiv. Elektr. Übertrag.* **18**(10), 617–624.

Stoodley, P., Jacobsen, A., Dunsmore, B.C., Purevdorj, B., Wilson, S., Lappin-Scott, H.M. and Costerton, J.W. (2001) The influence of fluid shear and AlCl$_3$ on the material properties of *Pseudomonas aeruginosa* PA01 and *Desulfovibrio* sp. EX265 biofilms. *Water Sci. Technol.* **43**, 113–120.

Voinova, M.V., Rodahl, M., Jonsson, M. and Kasemo, B. (1999) Viscoelastic acoustic response of layered polymer films at fluid–solid interfaces. *Biophys. J.* **107**, 1397–1402.

25

Monitoring biofouling using infrared absorbance

T.R. Bott

25.1 INTRODUCTION

The assessment of the extent of biofilm formation in flowing aqueous systems in industrial operations is essential to maintain efficiency. It is also necessary in research and development programmes to provide information on the effect of changing system variables on biofilm formation. Continuous assessment is preferable so that trends in biofilm development can be identified at an early stage. A further requirement is that the technique employed should not be intrusive, yet provide an electrical output that is capable of being continuously recorded.

All these criteria are fulfilled by the use of absorption of infrared radiation with a wave length of 950 nm, in a specially designed monitor. In laboratory and pilot-plant experimentation, the total water flow passes through the monitor, whereas in full-scale operations it is necessary to use the infrared monitor in conjunction with a sidestream. A suitable monitor has been under development at Birmingham University for a number of years.

In order for the monitor to function, it is necessary to pass the infrared radiation through a transparent (usually glass) tubular section. The amount of infrared radiation absorbed is a measure of the biofilm residing on the glass surface. The use of glass that is different from the material of construction of the plant surface on which the biofilm accumulates, may be regarded as a drawback. In general, however, the results from the monitor provide a satisfactory empirical estimate of deposit accumulation, from which remedial action may be taken as required, or the effects of operating variables observed. In laboratory and pilot-plant studies, where careful control of the variables can be achieved, it is possible to correlate the infrared absorbance against biofilm accumulation in terms of thickness.

The current design of the monitor is robust, reliable and operates in conjunction with continuous recording of other system variables, such as pH. It is also capable of modifying the injection of biocide for control purposes, in response to its assessment of biofilm residing on the test section. The contaminated water under test is passed through the test section, usually mounted vertically to avoid gravitational effects. At either end of a diameter of the tubular test section, an infrared emitter and receiver are located, suitably held in a specially designed mounting. The layout is schematically shown in Figure 25.1.

The radiation from the source passes in sequence, through the glass wall, the biofilm on the glass surface, the flowing water, the biofilm and the glass wall at the other end of the diameter. The difference between the infrared radiation emitted and received indicates the extent of the microbial deposit. By running a 'blank', when no biofilm is present and adjusting the reading to zero, any subsequent change in absorption is due to the biofilm present on the two inside surfaces of the test section. Absorbance may be defined as follows:

$$\text{absorbance} = \log\left(\frac{I_e}{I_e - I_g}\right) \tag{25.1}$$

where I_e is the intensity of the emitted radiation and I_g is the difference, i.e. the reading display on the monitor.

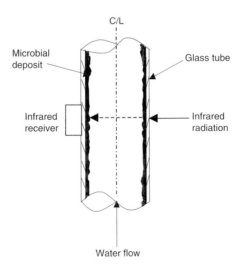

Figure 25.1   The concept of the biofouling monitor based on infrared radiation.

Further background and technical details can be found from previous publications (Bartlett *et al.*, 1999; Tinham and Bott, 2002).

25.2   PRACTICAL ASPECTS

The tubular glass section of the monitor must be carefully mounted in the circuit so that there is no disturbance to the water flow, otherwise inaccuracies in the measurements would result. An alternative would be to employ a long glass tube and to make measurements well downstream from the inlet, where flow conditions are stabilised.

It is well known that the surface of a biofilm is 'rough' (Bott, 1995). Since the monitor essentially records a point measurement, for accurate work it would be necessary to traverse a length of the test section, to obtain a mean value of biofilm accumulation. The design of the emitter and receiver housing allows for this requirement, although in the present design the traverse has to be manually carried out, it would be a relatively straightforward modification to motorise the traversing procedure.

25.3   EXAMPLES OF THE APPLICATION OF
THE MONITOR

The monitor has been used extensively at the University of Birmingham to investigate the effects of different variables on the development of biofilms in flowing water. The majority of the work has been directed towards the development of mitigation technologies to combat microbial growth, principally in cooling water systems and in paper manufacture. Physical as well as chemical methods of control have been investigated. Some limited investigations have also been carried out on industrial cooling water systems. Examples that illustrate the usefulness and responsiveness of the monitor are given in Figures 25.2–25.8.

In this series of figures, the data refer to the development of biofilms consisting of *Pseudomonas fluorescens*. The *Pseudomonas* species is a known slime former in industrial aqueous systems and is, therefore, a useful indicator of the effectiveness of any applied mitigation treatment. A monoculture is employed since it has been found difficult to maintain a mixed culture at fixed specification, for the long periods that are required for assessment tests. If a mixed culture is employed, it would still be possible, however, to assess the accumulation of the biofilm using the infrared monitor. Prior to a laboratory experiment, the equipment and the nutrient solutions to be added are effectively sterilised. The monoculture is grown in a suitable fermenter and pumped into the equipment at a known rate. In the experiments to which the figures refer, the concentration of cells in the circulating water was approximately 10^7 cells/ml. Velocities around 1 m/s were used in the experiments since this velocity represents the level of cooling water velocity generally encountered in industrial coolers.

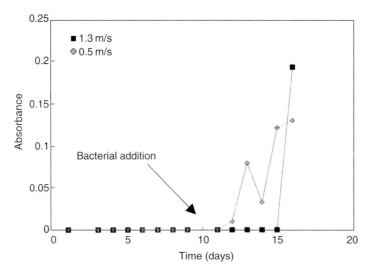

Figure 25.2   The initiation of biofouling on the tubular glass section by the introduction of bacteria to the system (Grant, 1999).

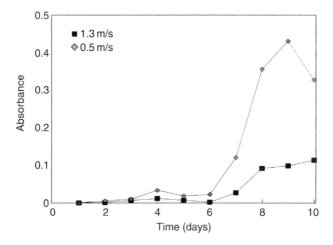

Figure 25.3   The effect of fluid velocity on biofilm accumulation at a constant Reynolds number of 11,700 (Grant, 1999).

Figure 25.2 shows how responsive the infrared monitor is to the development of biofilm. For the first 10 days of the experiment there were no bacteria present in the system. At day 10 bacteria were admitted into the equipment. The figure indicates the effect of velocity (shear forces) on biofilm development. At the lower velocity of 0.5 m/s there is a measurable accumulation after 2 days, with a rapid development of biofilm thereafter. At the higher velocity of 1.3 m/s the biofilm took much longer to begin to grow, i.e. about a further 6 days, to be fol-

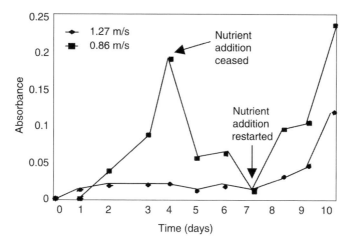

Figure 25.4   Effect of the termination of nutrient addition on biofilm development with metallic matrix inserts fitted, at two fluid velocities (Wills, 1999).

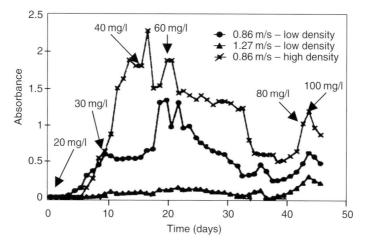

Figure 25.5   Effect of biocide 'A' dosing on biofilm development in tubes with metallic matrix inserts fitted, at two different fluid velocities (Wills, 1999).

lowed by a rapid increase in biofilm accumulation. These particular tests were carried out to demonstrate the effectiveness of the sterilisation procedure prior to the commencement of an experiment. The equipment was clearly sterile till the bacteria were added.

The development of biofilm at the same two velocities, with microbial-laden water right from the beginning of the experiment is shown on Figure 25.3. The effect of fluid velocity is made clear. Low velocity favours extensive biofilm development, whereas a higher velocity produces less accumulation due to the relatively high

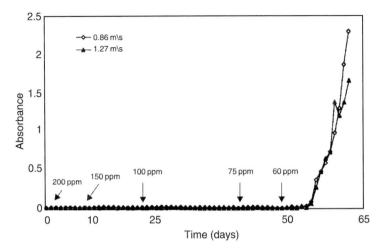

Figure 25.6   Effect of 35% hydrogen peroxide concentration on biofilm development (Wills, 1999).

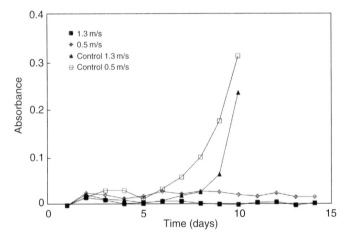

Figure 25.7   Effect of dosing biocide 'B' four times a day for 15 min at a peak concentration of 33.6 mg/l on biofilm growth (Grant, 1999).

shear forces at the higher velocity. Figure 25.3 demonstrates that some of the accumulated biofilm tends to slough away (at 0.5 m/s on Figure 25.3) due to structural weaknesses combined with the shear forces acting on the biofilm. Experience suggests that biofilms grown at low velocity tend to be less structurally robust and are, therefore, more susceptible to sloughing. The tests recorded on Figure 25.3 were carried out at a Reynolds number of 11,700. The Reynolds number was chosen, as it has been found that maximum biofilm growth occurred at a Reynolds number of

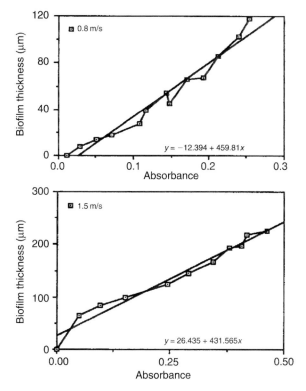

Figure 25.8   Correlation between biofilm thickness and absorbance at the fluid velocities of 0.8 and 1.5 m/s (Pujo, 1993).

about 12,000, assumed to be due to the combined effects of mass transfer and removal forces (Pujo, 1993).

Figure 25.3 also demonstrates that an initiation period is required before the biofilm begins to develop. It is thought that during the first few days the virgin glass surface is 'conditioned' by the adsorption of organic macromolecules that facilitate the adhesion of the bacteria to the surface (Bott, 1995).

Figure 25.4 clearly shows the effect of eliminating the nutrient supply from the water stream flowing at 0.86 m/s, i.e. a relatively rapid fall in microbial deposit. After a period of 3 days the nutrient feed was restored resulting in a rapid redevelopment of the biofilm. Figure 25.4 presents data on tests with wire-wound turbulence promotors inserted in the test section, to investigate whether or not the enhanced turbulence by the inserts affected the effectiveness of the biocide. Wire-wound inserts are used in heat exchangers to improve heat transfer. It is possible that increased turbulence could increase the removal forces acting on the biofilm, and improve the mass transfer of biocide to the biofilm, which would facilitate the action of the biocide in killing the bacteria and removing them from the heat exchanger surface.

Figure 25.5 records data obtained using a proprietry biocide (designated 'A' for commercial reasons), in conjunction with the wire-wound inserts used to obtain the data for Figure 25.4. 'High density' refers to an insert with a relatively large number of wires compared with 'low density'. The former would exert a greater disturbance of the flow, i.e. greater turbulence and at the same time imposing a higher pressure loss.

Biocide 'A' was fed continuously to the test circuit to give an initial concentration of 20 mg/l. Figure 25.5 shows that the biofilm growth was inhibited at a water velocity of 1.27 m/s with a low density insert fitted, whereas after an induction period of approximately 3 days, growth entered the exponential phase when the water velocity was 0.86 m/s. With a high density insert in place at the same water velocity biofilm growth did not enter the exponential phase for approximately 5 days.

On Figure 25.5 the changes in biocide concentration that were made are recorded. An increase of biocide concentration to 60 mg/l had an immediate effect on biofilm development at the lower velocity. After a further period of about 15 days, the biofilm began to develop again and this development was only arrested when the biocide concentration was raised to 100 mg/l. Although the accumulation of biofilm is relatively small at the higher velocity (1.27 m/s), the associated curve roughly mirrors the curves obtained at the lower velocity (0.86 m/s). It confirms the effects of velocity in relation to Figure 25.4. The reduction in biocide efficacy may have been due to physiological changes within the maturing biofilm (Christensen et al., 1990; Korber et al., 1994). The development of resistance to the biocide through mutation could be another explanation.

Hydrogen peroxide is a biocide that is used industrially to combat biofouling. Figure 25.6 demonstrates its effectiveness at two velocities. It can be seen that biocide concentrations of 200, 150, 100 and 75 mg/l of 35% hydrogen peroxide all completely inhibited biofilm growth at the experimental water velocities of 0.86 and 1.27 m/s. When the concentration was lowered to 60 mg/l, control was lost and after an induction period of 4 days, a rapid growth in biofilm occurred.

It is not usual to 'dose' continuously cooling water with biocide to control microbial growth on industrial surfaces; as this is not necessary and the cost would be much higher than it need be. It is more usual to employ intermittent dosing based on experience with a particular cooling water system.

Figure 25.7 indicates how a particular dosing system pattern of a biocide (designated 'B' for commercial reasons) can be effective in preventing biofilm development. The dosing regime involved just 15-min doses four times in 24 h. The peak concentration of biocide during the dosing period in this experiment was 33.6 mg/l. Under these conditions, a very low level of biofouling was observed compared with no biocide present.

Earlier it was stated that a correlation could be obtained between infrared absorbance and biofilm thickness. The method involves taking measurements of absorbance at different times of biofilm development, along the length of a 1-m tube to obtain a mean value. By accurately weighing the tube at the same time, the mass accumulated is determined. On the reasonable assumption that the biofilm consists largely of water, it is possible to estimate the volume of the biofilm present. Then by knowing the internal diameter of the test section, the mean thickness of the biofilm can be determined:

$$\text{biofilm thickness} = \frac{M_B}{dl\pi\rho_w} \tag{25.2}$$

where M_B is the mass of biofilm; d, the internal diameter of test section; l, the length of test section; and ρ_w, the density of water.

It has to be emphasised that the correlation obtained is only applicable to the conditions under which the data were obtained, i.e. fluid flow velocity and quality of the water in terms of nutrient levels and microbial content. Figure 25.8 shows such correlations for a given set of conditions including two velocities. It is interesting to note that the data collected at a velocity of 1.5 m/s show less scatter than those gathered at 0.8 m/s. This observation is consistent with the earlier statement that biofilms grown under relatively low velocities tend to have a more 'open' structure and, therefore, less structurally stable than those obtained at higher velocities.

25.4 CONCLUSION

Infrared absorbance in a suitably designed monitor, is a useful tool for the study of the trends in biofilm formation in flowing water system under different operating conditions. It is a valuable technique in the operation of pilot plants to investigate the control of biofilm formation prior to full-scale tests. It is also possible to employ such infrared absorbance on full-scale plant, using a sidestream, in order to monitor the development of biofouling. The information so obtained can be used to adjust the treatment programme being used to improve microbial control. Automatic response is possible, based on the electronic signal from the monitor.

Acknowledgements

The author would like to acknowledge the input to the design of the monitor by Geoff Bartlett and Peter Tinham with the researchers involved with the use of the infrared monitor. Some of their names are given in the list of references.

REFERENCES

Bartlett, H.G., Santos, R., Bott, T.R. and Grant, D. (1999) Measurement of biofilm development within flowing water using infrared absorbance. In *Understanding Heat Exchanger Fouling and its Mitigation* (ed. T.R. Bott, *et al.*), United Engineering Foundation and Begell House Inc., New York.
Bott, T.R. (1995) *Fouling of Heat Exchangers*, Elsevier, Amsterdam.

Christensen, B.E., Tronnes, M.N., Vollan, K., Smidsrod, O. and Bakke, R. (1990) Biofilm removal by low concentrations of hydrogen peroxide. *Biofouling* **2**, 165–175.

Grant, D.M. (1999) Biofilm control through optimised biocide dosing, Ph.D. thesis, University of Birmingham.

Korber, D.R., Lawrence, J.R. and Caldwell, D.E. (1994) Effect of motility on surface colonisation and reproductive success of *Pseudomonas fluorescens* in dual-dilution continuous culture and batch culture systems. *Appl. Env. Microbial.* **60**, 1421–1429.

Tinham, P. and Bott, T.R. (2002) Biofouling assessment using an infrared monitor. *Proceedings of the International Specialised Conference on Biofouling Monitoring*, Porto.

Pujo, M. (1993) Effects of hydrodynamic conditions and biocides on biofilm control, Ph.D. thesis, University of Birmingham.

Wills, A.J. (1999) Mitigation of biofouling using tube inserts in conjunction with biocides, Ph.D. thesis, University of Birmingham.

Section 6

Biofilm disinfection

26 Factors that affect disinfection of
    pathogenic biofilms   473

27 Device-associated infection: the
    biofilm-related problem in health care   503

28 Bacterial resistance to biocides:
    current knowledge and future problems   512

26

Factors that affect disinfection of pathogenic biofilms

S.B.I. Luppens, M.W. Reij, F.M. Rombouts and T. Abee

26.1 INTRODUCTION

Every year food-borne diseases cause millions of illnesses worldwide (Cooke, 1990; Todd, 1992; Mead *et al.*, 1999). The symptoms include diarrhoea, nausea, vomiting and fever. In a limited number of cases, there may be severe complications and sequelae (such as meningitis, septicaemia, arthritis, kidney disfunctioning and even death). It is, therefore, necessary to guard the microbiological quality of our food very strictly. Bacteria are always present on and in raw foods. Historically, numerous processing techniques have been developed to kill, inactivate or prevent further growth of pathogenic and spoilage bacteria. These techniques include cooking, pasteurisation, salting and fermentation. New techniques, such as pulsed electric field (PEF), high-pressure treatment and modified atmosphere packaging, are currently being developed and introduced (Jeyamkondan *et al.*, 1999; Amanatidou *et al.*, 2000; Amanatidou, 2001; Pol *et al.*, 2001). However, when the bacteria from the raw food have been killed or inactivated food can still be recontaminated with bacteria. Thus, everything that comes into contact with food should be free of bacteria or have a level of contamination that is not harmful. This applies to all food contact surfaces and processing equipment in the food industry. Therefore, these surfaces have to be cleaned and disinfected on a regular basis to prevent accumulation of pathogenic and spoilage organisms. Significant problems are posed by bacteria that can form biofilms on these surfaces, since biofilm cells are difficult to remove and are more resistant to disinfection than free-living bacteria (Brown and Gilbert, 1993; Costerton *et al.*, 1995; Kumar and Anand, 1998; Whiteley *et al.*, 2001). Disinfectants are traditionally tested on cells in suspension (free-floating or planktonic cells) (Anonymous, 1997a, b). Thus, disinfectant testing would greatly profit from the inclusion of a biofilm disinfectant test.

Furthermore, the disinfectant tests that are currently used employ plate counting for viability assessment of the survivors. This is a time-consuming method (2 days) and rapid alternatives would further improve the method. One such alternative is the use of fluorescent probes, which allows reduction of the time for viability assessment to approximately 2 h.

This chapter gives an introduction to two human pathogenic bacteria that we focus on in our studies, namely *Staphylococcus aureus* and *Listeria monocytogenes* (both are known as biofilm formers) and an overview of disinfectants and how they are tested. Furthermore, the chapter discusses several topics that are of importance for disinfectant testing and biofilm control.

26.1.1   *Listeria monocytogenes*

L. monocytogenes is a Gram-positive non-spore-forming facultatively anaerobic rod, motile by means of flagella. *Listeria* was named after Lord Lister, an English surgeon. 'Monocytogenes' means monocyte (blood cell) producing. This refers to the ability of *L. monocytogenes* to infect blood cells (Seeliger and Jones, 1986). Some studies suggest that 1–10% of humans may be intestinal carriers of *L. monocytogenes*. The bacterium has been found in mammals, birds and in some species of fish and shellfish. It can be isolated from soil, silage and other environmental sources and in food-processing plants (Norwood and Gilmour, 1999, 2001; Leriche and Carpentier, 2000). *L. monocytogenes* is quite hardy and resists the harmful effects of freezing, drying and heat remarkably well for a non-spore-forming bacterium. *L. monocytogenes* has been associated with raw milk, supposedly pasteurised milk, cheeses, ice creams, raw vegetables, fermented raw-meat sausages, raw and cooked poultry, raw meats, and raw and smoked fish. It is able to grow at temperatures as low as 0°C, which permits multiplication in refrigerated foods (Seeliger and Jones, 1986; Farber and Peterkin, 1991; Anonymous, 1992a; Farber, 1993; Jay, 1996).

L. monocytogenes causes listeriosis, which manifests itself in septicaemia, meningitis, encephalitis and intrauterine or cervical infections in pregnant women, which may result in spontaneous abortion or stillbirth. The onset of these disorders is usually preceded by influenza-like symptoms including persistent fever. Gastrointestinal symptoms may precede the more serious forms or may be the only symptoms expressed. Overall mortality for listeriosis may be as high as

23% (Sleator *et al.*, 1999). The main target populations for listeriosis are pregnant women, immunocompromised people, the elderly and sometimes healthy individuals, particularly when food is heavily contaminated with the organism. The infective dose of *L. monocytogenes* is believed to vary with the strain and susceptibility of the victim. From cases contracted through milk, it has been concluded that in susceptible persons fewer than 1000 total organisms may cause disease (Seeliger and Jones, 1986; Farber and Peterkin, 1991; Anonymous, 1992a; Farber, 1993; Jay, 1996). *L. monocytogenes* can form biofilms on surfaces of materials that are present in the food industry, such as stainless steel, glass, cast iron and plastic (Frank and Koffi, 1990; Spurlock and Zottola, 1991; Krysinski *et al.*, 1992; Sasahara and Zottola, 1993; Blackman and Frank, 1996; Lou and Yousef, 1999; Norwood and Gilmour, 1999; Maukonen *et al.*, 2000; Kalmokoff *et al.*, 2001). Its biofilm formation has been mostly studied in batch systems (Frank and Koffi, 1990; Spurlock and Zottola, 1991; Krysinski *et al.*, 1992; Sasahara and Zottola, 1993; Blackman and Frank, 1996; Lou and Yousef, 1999; Norwood and Gilmour, 1999; Maukonen *et al.*, 2000; Kalmokoff *et al.*, 2001). Inhibition or enhancement of *L. monocytogenes* growth in mixed species biofilms has also been studied (Leriche *et al.*, 1999; Leriche and Carpentier, 2000; Norwood and Gilmour, 2001). *L. monocytogenes* biofilm cells are more resistant to quaternary ammonium compounds (quats), acid anionics, hypochlorite, iodine, peroxyacetic acid and phenol than planktonic cells (Gibson *et al.*, 1999; Lou and Yousef, 1999). Despite the multitude of studies on *L. monocytogenes* biofilms, only few have focused on the genetic basis of attachment and biofilm formation by this food-borne pathogen. With the help of a flagellin mutant it was shown that flagella facilitate the early stage of attachment of *L. monocytogenes* to stainless steel (Vatanyoopaisarn *et al.*, 2000). Two genes involved in stringent response *relA* and *hpt* were shown to be important for surface attached growth of *L. monocytogenes* (Taylor *et al.*, 2002). Proteomic analysis showed that several proteins are more highly expressed in biofilms than in planktonic cells (Tremoulet *et al.*, 2002). These are proteins that are involved in global carbon metabolism, (oxidative) stress response, DNA repair and protection, and cellular division.

26.1.2   *Staphylococcus aureus*

S. aureus is a Gram-positive, non-spore-forming facultatively anaerobic spherical bacterium. The word 'staphylococcus' is derived from 'staphyle' (which is Greek for bunch of grapes) and 'coccus' (which means grain or berry). The name refers to the grape bunch-like clusters that are formed by *S. aureus* during growth and the spherical form of the bacterium. 'Aureus' means golden which refers to the yellow colour of *S. aureus* colonies on plates (Kloos and Schleifer, 1986).

Staphylococcal food poisoning or staphyloenterotoxicosis is caused by the heat-stable enterotoxins that some *S. aureus* strains produce under favourable growth conditions (Jay, 1996). The most common symptoms are nausea, vomiting, retching, abdominal cramping and prostration. In more severe cases, headache, muscle cramping and transient changes in blood pressure and pulse rate may occur. A toxin dose of less than 1.0 μg in contaminated food will produce symptoms of staphylococcal intoxication. This toxin level is reached when the *S. aureus* population exceeds 100,000 g^{-1}. *S. aureus* is also involved in toxic shock syndrome and is a common cause of community-acquired infections including endocarditis, osteomylitis, septic arthritis, pneumonia and abscesses (Anonymous, 1992b; Jay, 1996).

Foods that are frequently incriminated in staphylococcal food poisoning include meat products, poultry and egg products, salads, bakery products, sandwich fillings, and milk and dairy products. Foods that require considerable handling during preparation and that are kept at slightly elevated temperatures after preparation are frequently involved in staphylococcal food poisoning (Anonymous, 1992b; Jay, 1996).

Staphylococci exist in air, dust, sewage, water, milk and food or on food equipment, environmental surfaces, humans and animals. Humans and animals are the primary reservoirs. Staphylococci are present in the nasal passages and throats and on the hair and skin of 50% or more of healthy individuals. Although food handlers are usually the main source of food contamination in food poisoning outbreaks, equipment and environmental surfaces can also be sources of contamination with *S. aureus* (Anonymous, 1992b; Jay, 1996; Becker *et al.*, 2001).

Most studies on biofilm formation by *S. aureus* have been done with surfaces that are used in the medical field. This is because *S. aureus* is one of the main bacteria that forms biofilms on prosthetic devices and in this way causes infections that are very difficult to treat with antibiotics (Becker *et al.*, 2001). *S. aureus* forms biofilms on a variety of surfaces, namely hydroxyapatite, stainless steel, glass and on all kinds of plastics like polystyrene, polypropylene and silicone (Anwar *et al.*, 1992; Oie *et al.*, 1996; Johansen *et al.*, 1997; Cramton *et al.*, 1999; Gracia *et al.*, 1999; Pratten *et al.*, 2001; Luppens *et al.*, 2002b). On most of these materials biofilms have been grown statically in a batch system, with sometimes one to three medium refreshments (Oie *et al.*, 1996; Williams *et al.*, 1997; Akiyama *et al.*, 1998; Gracia *et al.*, 1999; Cucarella *et al.*, 2001). Especially for genetic studies, batch biofilm formation in microtiter plate wells is widely used (Anwar *et al.*, 1992; Fournier and Hooper, 2000; Rachid *et al.*, 2000; Cucarella *et al.*, 2001). In the batch systems, biofilm formation varies from 5×10^6 to 8×10^7 CFU cm^{-2} after 24–48 h. *S. aureus* biofilms grown in chemostat vary from 3×10^6 CFU cm^{-2} after 4 days to 2×10^8 CFU cm^{-2} after 13 days (Anwar *et al.*, 1992). *S. aureus* biofilm cells are more resistant to antibiotics (Anwar *et al.*, 1992; Williams *et al.*, 1997; Akiyama *et al.*, 2000), disinfectants (Oie *et al.*, 1996) and enzymes (Johansen *et al.*, 1997). It is speculated that the resistance to enzymes may depend on the thickness of the biofilm (Johansen *et al.*, 1997) and resistance to antibiotics may be due to alteration of the physiological status of the cells, such as altered membrane permeability for antibiotics, alteration of molecular targets of antibiotics, induction of antibiotic-degrading enzymes, exopolysaccharide production and slow growth (Anwar *et al.*, 1992; Williams *et al.*, 1997; Campanac *et al.*, 2002).

Recently, several studies have focused on the genes that are involved in *S. aureus* biofilm formation on abiotic surfaces. *S. aureus* biofilm formation is believed to consists of two steps: adhesion of bacteria to a surface followed by cell–cell adhesion, which results in a multiple layer biofilm (Cramton *et al.*, 1999). Disruption of the *arlR/arlS* locus which codes for a two-component regulatory system or disruption of *sar* (staphylococcal accessory regulator) results in increased adherence of *S. aureus* (first step) (Fournier and Hooper, 2000; Pratten *et al.*, 2001). Gross *et al.* (2001) suggest a role for teichoic acids, which are cell wall polymers, in the first step of biofilm formation. Bap (biofilm-associated protein), an *S. aureus* surface protein, is involved in both steps (Cucarella *et al.*, 2001). Repression or absence of the *agr* quorum-sensing system enhances both steps of biofilm formation (Vuong *et al.*, 2000; Otto, 2001). The *ica* locus that encodes for the polysaccharide intercellular adhesin (PIA) which is composed of linear β-1,6-linked glucosaminylglycans is involved in the second step (Cramton *et al.*, 1999). Furthermore, several studies show that stress is involved in *S. aureus* biofilm formation. Osmotic stress induces biofilm formation and *ica* transcription (Rachid *et al.*, 2000) and anaerobic growth stimulates PIA production (Cramton *et al.*, 2001). Deletion of the alternative σ-factor σ^B, which is involved in regulation of stress response, results in a biofilm-negative phenotype (Rachid *et al.*, 2000). Induction of σ^B expression leads to dose-dependent *S. aureus* attachment and micro-colony formation on surfaces (Bateman *et al.*, 2001). Additionally, Becker *et al.* (2001) showed that homologues to genes encoding glycolytic enzymes and other metabolic enzymes (phospoglycerate mutase, triosephosphate isomerase and alcohol dehydrogenase) are upregulated in biofilms. This may be correlated to oxygen limitation under these conditions. Of the two other upregulated genes in *S. aureus* biofilms, one showed homology to *ClpC*, a general stress protein, and the other to threonyl-tRNA synthetase that is indicative of threonine starvation (Becker *et al.*, 2001).

26.1.3 Disinfectants

To prevent accumulation of pathogenic and spoilage organisms, food contact surfaces have to be cleaned and disinfected on a regular basis. Cleaning agents are used to facilitate the removal of adhering organic material and soil. In dictionary definitions of what a disinfectant is there are five recurring elements: it is an agent that removes infectious agents, kills microorganisms, may not kill spores, may be chemical or physical and is used on inanimate objects (Block, 1991). The British Standard Institution defines disinfection as the destruction of vegetative microbial cells; bacterial spores are generally not affected. A disinfectant does not necessarily kill all microorganisms, but reduces their numbers to a level acceptable for a defined purpose. For example, a level which is neither harmful to health nor to the quality of perishable foods (Bloomfield *et al.*, 1994).

Disinfectant efficiency is increased when it is applied to a surface that first has been cleaned. A wide range of disinfectants is used in the food industry. These disinfectants can be divided in different groups:

- oxidising agents, such as chlorine-based compounds, hydrogen peroxide, ozone and peracetic acid

- surface-active compounds, such as quats and acid anionics;
- iodophores
- enzymes (Wirtanen, 1995).

The efficiency of disinfection is influenced by interfering organic substances, pH, temperature, concentration and contact time (Wirtanen, 1995). Desired characteristics of disinfectants are that they must be effective, safe and easy to use and easily rinsed off surfaces leaving no toxic residues or residues that affect the sensory properties of the product (Wirtanen, 1995).

In our studies four disinfectants were used, namely sodium hypochlorite, hydrogen peroxide, benzalkonium chloride (BAC) and dodecylbenzene sulphonic acid (DSA). Sodium hypochlorite (NaOCl) is a chlorine-based oxidising agent. Its mechanism of action is reaction with and alteration of proteins and DNA, especially enzymes with thiol or amino groups (Edelmeyer, 1982; Dychdala, 1991; Russell et al., 1992). Its recommended concentrations in the food industry range from $2 \, mg \, l^{-1}$ for rinse and cooling water to $5000 \, mg \, l^{-1}$ for concrete surfaces. NaOCl is cheap, effective and easy to use, it detaches the biofilm matrix and it has a broad spectrum of activity. Disadvantages are its poor stability, its toxicity, corrosiveness and its lack of initial adhesion control, discolouration of the product and occurrence of rapid aftergrowth (Dychdala, 1991; Wirtanen, 1995).

Hydrogen peroxide is an oxidising agent. Its mechanism of action is oxidation or formation of free radicals (Acworth and Bailey, 1997; Henle and Linn, 1997) which affect enzymes and proteins, DNA, membranes and lipids resulting in damage of transport systems and receptors, difficulty in maintaining ionic gradients over the cytoplasmic membrane, impairment of replication and (in)activation of enzyme systems (Russell et al., 1992; Acworth and Bailey, 1997; Henle and Linn, 1997). Hydrogen peroxide decomposes to water and oxygen, is relatively non-toxic, can easily be used in situ, weakens the biofilm and supports biofilm detachment. Disadvantages are that high concentrations are necessary and it is corrosive (Wirtanen, 1995).

BAC is a quat. Its mechanism of action involves alteration of the semipermeable properties of cell membranes, which leads to leakage of metabolites and coenzymes and disturbance in the delicate balance of metabolite concentrations within the cell (Merianos, 1991). The recommended concentrations of BAC in the food industry range from 200 to $800 \, mg \, l^{-1}$ depending on the type of surface (Wirtanen, 1995). The advantages of quats are that they are effective at non-toxic concentrations, support biofilm detachment and prevent growth, are non-corrosive, non-irritating and have no flavour or odour. The disadvantages are that they are inactivated by low pH, calcium salts and magnesium salts, are relatively ineffective against Gram-negatives and microorganisms may develop resistance against them (Mattila Sandholm and Wirtanen, 1992; Wirtanen, 1995).

DSA is an acid anionic surfactant. To DSA solutions phosphoric acid is added to increase its performance (Dychdala and Lopes, 1991) because at neutral pH DSA is negatively charged and thus repelled by the negatively charged bacterial surface (Dychdala and Lopes, 1991). The mechanism of action at low pH is disorganisation of the cell membrane, inhibition of key enzyme activities and interruption of cellular transport and denaturation of cellular proteins (Dychdala and Lopes, 1991; Russell *et al.*, 1992). Advantages of DSA use are it has cleaning/detergent properties, it removes particles from surfaces, it has a broad spectrum of activity, it is non-corrosive and does not stain equipment, gives no odour, it is stable, works rapidly and has residual bacteriostatic activity. Disadvantages are that it is only effective at low pH, it generates foam and it has a slow activity against spores (Dychdala and Lopes, 1991; Wirtanen, 1995).

26.1.4   Disinfectant tests

The procedure used for the testing of candidate disinfectants in Europe consists of three phases. In phase 1 the basic activity of the product is tested in a suspension test. Phase 2 consists of two steps. In the first step, the product is tested in a suspension test under conditions representative of different practical use. The second step consists of other laboratory tests, e.g. handwash, handrub and surface tests simulating practical conditions. Phase 3 consists of field tests under practical conditions (Anonymous, 1997a).

Until now, almost all European tests for disinfectants used in the food, industrial, domestic and institutional areas are suspension tests. In these tests, a suspension is made from colonies grown on solid medium. The main organisms that are used for these tests are *Pseudomonas aeruginosa* ATCC 15442, representative for the Gram-negative bacteria and *S. aureus* ATCC 6538 representative for the Gram-positive bacteria. The suspended bacteria are exposed for 60 min (phase 1) or 5 (phase 2) to the candidate disinfectant (Anonymous, 1997a, b). The disinfectant is then neutralised and bacterial survival determined by plate counting. If the candidate disinfectant reduces the concentration of viable cells by more than 5-log units and the concentration of cells in the suspension is within a certain limit, it is approved as a disinfectant.

There have been concerns about the suitability of the standard European disinfectant tests. Some of the concerns about these suspension tests are the lack of test reproducibility and the predictive value of laboratory-grown cultures for naturally occurring strains (Langsrud and Sundheim, 1998; Payne *et al.*, 1999). Langsrud and Sundheim (1998) showed that pre-growth in the absence of oxygen and using spread plates instead of pour plates increases the survival of *S. aureus* after exposure to BAC and grape fruit extract. On the other hand, Payne *et al.* (1999) showed that clinical isolates were not more resistant than *S. aureus* ATCC 6538 (DSM, 799), the strain that is used in suspension tests.

Apart from the nature of the strain, growth phase could influence the resistance of bacteria. In our studies, we have compared the susceptibility of suspended *S. aureus* cells grown according to the European suspension tests (Anonymous, 1997a, b) on solid medium, with cells in different growth phases grown in liquid medium (Luppens *et al.*, 2002c). Liquid medium cells in three growth phases were used: exponential-phase cells (2.5 h), stationary-phase cells (22 h) and decline-phase cells (7 days). Decline-phase cells were generally the most resistant cell type. Exponential-phase cells were less resistant than decline-phase cells and, surprisingly, stationary-phase cells were the least resistant of the three. Cells grown on solid medium as prescribed in the European tests were in none of these tests the most resistant cells. Their survival was 1–3-log units less than that of the most resistant cells. This shows that the solid-medium cells currently used are not the most resistant type of cells that can be used in a disinfectant test and thus the current suspension tests underestimate the phenotypic resistance of *S. aureus* cells to disinfectants. To improve the test, it would be advisable to use cells for suspension tests that are grown differently as compared to the currently used solid-medium cells, as was also suggested for *P. aeruginosa* (Gilbert *et al.*, 1987).

However, a good disinfectant test must be able to predict the value of the disinfectant in practice (Reybrouck, 1998) where cells can be found much more frequently on surfaces than in suspension. Thus, it is questionable whether suspension test cells are really representative of the cells found in practice. In this light a new surface test has been developed for phase 2 step 2 (Bloomfield *et al.*, 1994; Klingeren *et al.*, 1998; Anonymous, 1999b). In this surface test, a suspension of cells is put on a surface and dried for 1 h. Then the disinfectant is applied, the disinfectant is neutralised and the bacterial survival is determined with the help of plate counting. This surface tests is already a step forward compared to the suspension tests. However, there is still some concern about the suitability of this surface test. The cells in a surface test only attach to the surface and do not grow, whereas it is known that attached cells that are allowed time to grow form biofilms. Therefore, a standard biofilm test would be a very useful addition to the current tests.

Several methods have been described for antimicrobial (predominantly antibiotic) testing on biofilms. Most of these methods are medically orientated. They often use the minimal inhibitory concentration to assess antibiotic efficacy (Wright *et al.*, 1997; Das *et al.*, 1998). However, disinfectant efficacy has to be assessed by viable counting, since growth-inhibited cells can still contaminate food and resume growth after recovery. Very often batch systems are used for biofilm formation. In these systems, coupons are placed in inoculated rich medium and, optionally, the medium is replaced several times. Alternatively, inoculated medium is used for development of biofilms on the surface of microtiter plate wells or Erlenmeyer flasks (Christensen *et al.*, 1985; Becker *et al.*, 2001; Monzon *et al.*, 2001; Ramage *et al.*, 2001). A disadvantage of these batch methods is that, since little or no shear force is applied to them, the cells are very loosely attached to the surface, and thus are not representative of biofilms in practice. In some of the studies in which these kinds of systems are used shear force is applied by shaking the surface during biofilm formation (Ceri *et al.*, 1999; Brooun *et al.*, 2000). Still, biofilms grown in batch and on rich media are not representative of biofilms in the food industry. An interesting method for biofilm formation is to trap planktonic cells in a poloxamer-hydrogel that is liquid at temperatures below 15°C and solid at temperatures above 15°C (Härkönen *et al.*, 1999). With this method the viability of cells can be easily analysed. However, the cells do not have the biofilm physiology. Another method is to apply cells on a filter and place this on solid medium (Wentland *et al.*, 1996) or perfuse the filter with liquid medium (Gilbert *et al.*, 1989). The resulting biofilms are different from biofilms on inert surfaces because the cells receive their nutrients from the surface side and not from the air or bulk liquid side like in biofilms on inert surfaces. The biofilms that come closest to biofilms in food industry are the ones that are formed in special reactors that apply a certain shear force to the biofilm cells while they are growing, and that continuously provide the cells with relatively poor medium. Examples of this kind of reactors are the Robbins device (Adams and McLean, 1999), a chemostat with coupons in it (Zelver *et al.*, 2001), the concentric cylinder reactor (Willcock *et al.*, 2000), and the constant depth film fermenter (Vroom *et al.*, 1999). However, all these methods are very sophisticated and require expensive equipment. Thus, they are not very suitable for disinfectant testing. A biofilm test should be as simple as possible (Holah *et al.*, 1998). Furthermore, a standard disinfectant test for biofilm cells should resemble as closely as possible the current suspension tests in order to make comparison between the results simple. The biofilm disinfectant test we developed is discussed in Section 26.2.3.

26.2   DISINFECTANTS, TESTING AND BIOFILMS

26.2.1   Factors that influence the efficacy of a disinfectant

In our studies on the effect of disinfectants on pathogens we examined several factors that may influence the efficacy of disinfectants. One of the factors studied was the presence of nutrients during disinfection. Exposure of *L. monocytogenes* planktonic cells in the presence of glucose or BHI medium did not influence the efficacy of BAC (Luppens *et al.*, 2001) and hydrogen peroxide (data not shown) as compared to exposure in buffer. Exposure of *L. monocytogenes* cells in the absence of oxygen did not influence the efficacy of BAC (Luppens *et al.*, 2001) as compared to exposure in the presence of oxygen. It is important to know if oxygen and nutrients have an additional influence on killing, because in biofilms the distribution of these compounds is not even and not constant.

Another factor, the presence of dead cells, does influence the efficacy of disinfectants (Luppens *et al.*, 2002c). Dead cells may react with disinfectants either directly, e.g. by binding membrane-active compounds (such as BAC) or indirectly by the contribution of their enzymes (such as catalase in the degradation of hydrogen peroxide). In this way they may protect viable cells. Our results with planktonic cells show that decline-phase cells are indeed more resistant to disinfectants in the presence of dead cells, which account for more than 90% of the *S. aureus* decline-phase population (Luppens *et al.*, 2002c).

The efficacy of disinfection is influenced by growth phase (Luppens *et al.*, 2001, 2002a, c). In one of our studies, we exposed *S. aureus* exponential-phase, stationary-phase and decline-phase cells to BAC, hydrogen peroxide, DSA and NaOCl (Luppens *et al.*, 2002c). It is assumed that cells in stationary phase and decline phase are less susceptible to disinfectants than exponential-phase cells because cells in stationary phase and decline phase lack certain nutrients. Bacteria are known to respond to this starvation stress by growth rate reduction and induction of defence mechanisms (Breeuwer *et al.*, 1996; Dodd *et al.*, 1997; de Beer and Schramm, 1999). As a result, they may become more resistant to other types of stress, such as stress caused by disinfectants (Kolter, 1992; Foster and Spector, 1995; Hibma *et al.*, 1996; Denyer and Stewart, 1998; Davies, 1999). In our results, this was the case for *S. aureus* decline-phase cells, but stationary-phase cells were less resistant than exponential-phase cells. Watson *et al.* (1998) found similar results. They explain the sensitivity of stationary-phase cells with the biphasic death curves they observed for stationary-phase cells. They proposed that the stationary-phase cells that show reduced resistance to the treatments are the same cells that will die upon long-term starvation. This proposal is based on the concept that the cells become committed to survival or death early after entry into the stationary phase unless they are rescued by provision of nutrients. In our experiments, biphasic death patterns were also observed for *S. aureus* stationary-phase cells exposed to NaOCl and BAC. When exposed to other disinfectants, *S. aureus* stationary-phase cells died too rapidly for any conclusions to be drawn (Luppens *et al.*, 2002c).

Similar experiments for *L. monocytogenes* exponential-phase cells and stationary-phase cells exposed to BAC (Luppens *et al.*, 2001, 2002a), hydrogen peroxide (Luppens *et al.*, 2002a), DSA and NaOCl (data not shown) revealed that *L. monocytogenes* stationary-phase cells are less susceptible to surfactants than exponential-phase cells, but more susceptible to oxidising agents than exponential-phase cells. Decreased susceptibility of *L. monocytogenes* stationary-phase cells to environmental stresses may be explained by induction of the alternative σ-factor, σ^B, in stationary-phase cells. σ^B is involved in regulation of response to environmental stresses in *L. monocytogenes* (Becker *et al.*, 1998; Wiedmann *et al.*, 1998; Ferreira *et al.*, 2001). This factor is also involved in protection of *L. monocytogenes* stationary-phase cells to oxidative stress (Ferreira *et al.*, 2001). Since σ^B is not induced in unstressed exponential-phase cells it appears that in these cells other mechanisms are involved in protection from oxidative stress.

Catalase activity could be an important factor in the survival of cells exposed to hydrogen peroxide. However, we found that *S. aureus* exponential-phase, stationary-phase and solid-medium cells show no substantial differences in catalase activity, although exponential-phase cells die more rapidly than the other two cell types (Luppens *et al.*, 2002c). Apparently, *S. aureus* exponential-phase cells have a lower inherent resistance to hydrogen peroxide than the other two cell types. In a decline-phase cell suspension that contained the same total number of cells as the above-mentioned other cell type suspensions, the catalase broke down the hydrogen peroxide at least 2.5 times as slow as it did in the other cell types. Nevertheless, cell survival was still higher than that of the other cell types. *S. aureus* decline-phase cells apparently have a higher inherent resistance to hydrogen peroxide than other cell types (Luppens *et al.*, 2002c). Similar experiments for *L. monocytogenes* exponential-phase and stationary-phase cells revealed that catalase activity was not substantially different in these two cell types (data not shown). Thus, the difference in susceptibility to hydrogen peroxide between these cell types cannot be explained by differences in catalase activity.

Growth in a biofilm influences the efficacy of disinfectants. This may be caused by several factors. First of all, the extracellular material of the biofilm may exclude or influence the access of the disinfectant and the outer layers of the biofilm may react with and quench the disinfectant (Brown and Gilbert, 1993; Kumar and Anand, 1998). Our results show that susceptibility to NaOCl treatment does not differ significantly in biofilm cells and suspended biofilm cells (Luppens *et al.*, 2002b). Susceptibility to BAC treatment differs significantly in biofilm cells and suspended biofilm cells, but only by 1 log-unit (Luppens *et al.*, 2002b). Thus, for the *S. aureus* biofilms used in our study the biofilm matrix contributes little to disinfectant susceptibility. Additionally, susceptibility of cells in a biofilm may be altered because a cell that is attached to a surface has a different phenotype than a cell grown in suspension. This may be because of the attachment itself, nutrient stress or presence of a large number of cells on the surface (quorum sensing). For the importance of each of these three factors, evidence has been found with the help of mutants and reporter genes (Wentland *et al.*, 1996; Davies *et al.*, 1998; Dorel *et al.*, 1999). Furthermore, our results indicate that cells in biofilms formed by a single strain can differ even genotypically. A 6-day-old biofilm of *S. aureus* DSM 799 grown at 25°C contained several white (unpigmented) mutants of which two were selected. These mutants are stable and show increased slime production after 2 days of (planktonic) incubation in tryptone soya broth (TSB) (data not shown).

It is very likely that all these factors together cause a large heterogeneity in structure and phenotypes in biofilms, which increases the probability of survival of a small population of cells in the biofilm after exposure to disinfectants. In our results, this can be demonstrated by comparing the death curves of free-living cells (especially exponential-phase cells) to those of biofilm cells. Disinfectants have only a very small effect on the viability of free-living cells at low concentrations, but after a very small increase of disinfectant concentration a very sharp decline in viability occurs (Luppens *et al.*, 2001). Biofilm cells show the same response to disinfectants as free-living cells at low disinfectant concentrations, but a very large increase in disinfectant concentration is needed to cause a substantial decline in viability of biofilm cells (Luppens *et al.*, 2002b).

26.2.2   Viability assessment after exposure to a disinfectant

The traditional method to determine viability of bacteria, also in disinfectant testing, is plate counting. This method is based on the reproductive capacity of cells. However, this method has some disadvantages. The plate count technique requires long incubation times (2 days). Furthermore, for viability assessment of attached or biofilm cells, the cells have to be removed from the surface for analysis. Additionally, several studies report that cells are metabolically active while they are incapable of the cellular division required to form a colony on a plate. This is also known as a viable but non-culturable (VBNC) state (Kaprelyants and

Kell, 1993a; Braux *et al.*, 1997; Comas and Vives-Rego, 1998; Lleo *et al.*, 1998) or better, an active but non-culturable state (ANC) (Joux and Lebaron, 2000). In the case of disinfectants this may lead to overestimation of the efficacy of the disinfectant.

In our studies, we use several other methods for viability assessment. Respiration, acidification and ATP measurements (Luppens *et al.*, 2001) are very rapid, give an indication of the energy status of the cells and the immediate effect of the disinfectant can be analysed (<1 min). Unfortunately, a high cell concentration ($>10^9 ml^{-1}$) is needed to be able to detect a signal, and thus the sensitivity of these methods is very low. Furthermore, these methods analyse the response of the total population and thus they cannot be used to quantify viability. For hydrogen peroxide-exposed cells, these methods pose additional problems because of the production of large amounts of oxygen, i.e. gas bubbles, which interferes with sample taking and proper functioning of measuring instruments. This problem may be overcome by removing the disinfectant from the cell suspension by washing before the measurements, but then the immediate effect of the disinfectant cannot be analysed.

Another rapid alternative for plate counting is labelling with fluorescent probes. These probes indicate if a cell possesses other physiological characteristics required for a cell to be viable, such as membrane integrity, enzyme activity and energy production (Breeuwer and Abee, 2000; Joux and Lebaron, 2000). Fluorescent probes can be used directly to assess viability of attached cells when their use is combined with fluorescence microscopy (thin layers) (Yu and McFeters, 1994; McDowell *et al.*, 1995; Camper *et al.*, 1999; Maukonen *et al.*, 2000) or confocal laser scanning microscopy (CLSM) (thick layers) (Costerton *et al.*, 1995; McFeters *et al.*, 1995; Hartmann *et al.*, 1998; Vroom *et al.*, 1999; Zaura-Arite *et al.*, 2001). Fluorescent labelling can also be combined with flow cytometry (FCM). In this way viability assessment can be done in 0.5–2 h. FCM is a technique for individual cell analysis and it has been applied earlier to analyse heterogeneous populations, such as starved cells (Watson *et al.*, 1998) and attached cells (Williams *et al.*, 1999), and may be used for biofilm cell analysis (see below).

Several categories of fluorescent probes are used for detection and viability assessment of microorganisms (Breeuwer, 1996; Barer and Harwood, 1999; Breeuwer and Abee, 2000; Joux and Lebaron, 2000; Nebe von Caron *et al.*, 2000). *Membrane potential probes* are compounds such as rhodamine 123, oxonol and carboxycyanine dyes. Unfortunately, some of these probes can be extruded actively out of the cell and loss of membrane potential does not necessarily indicate cell death (Nebe von Caron *et al.*, 1998; Breeuwer and Abee, 2000). *pH probes*, for instance (carboxy)fluorescein, carboxyfluorescein succinimidyl ester (CFSE), BCECF, calcein, SNARF-1 and pyranine are used to determine the internal pH of microorganisms. Maintenance of a pH gradient (i.e. pH_{in} higher than pH_{out}) is indicative of cellular activity and membrane integrity. Disadvantages of some of these probes are that they are extruded actively out of the cell (Breeuwer, 1996; Breeuwer *et al.*, 1996; Luppens *et al.*, 2002a). *Redox indicators* are 5-cyano-2,3-ditolyl tetrazolium chloride (CTC), INT and Alamar

Blue. Results with these probes depend on endogenous and exogenous substrates, may be independent of the electron transport system and the probes may be toxic at higher concentrations (Smith and McFeters, 1996, 1997; Collins and Franzblau, 1997; Servais *et al.*, 2001). *Enzyme substrates*, such as (carboxy)fluorescein diacetate, fluorescein digalactoside and Chemchrome B, are used to indicate enzyme activity. These non-fluorescent substrates are converted to fluorescent products and then they can be detected. These products are often used as *dye-retention probes*. Dye retention is an indication for an intact cell membrane. Disadvantages of these probes are that the dyes can be actively extruded, enzyme activity could be absent or poor under certain physiological conditions, enzyme activity is typically energy independent and the dye can sometimes be retained in vacuoles, even when the membrane is permeable (Ueckert *et al.*, 1997; Barer and Harwood, 1999; Joux and Lebaron, 2000; Nebe von Caron *et al.*, 2000). *Dye-exclusion probes* are, e.g. propidium iodide (PI), ethidium homodimer, SYTOX Green, TOTO and TO-PRO-3. These probes are only able to penetrate cells with compromised membranes. Unfortunately, some of them are able to label intact cells under certain conditions (Nebe von Caron *et al.*, 2000). Furthermore, some live cells have permeant membranes and some dead cells have impermeant membranes (Nebe von Caron *et al.*, 2000; Luppens *et al.*, 2002a). *Nucleic acid stains*, such as Hoechst 33258/33342, DAPI, SYTO series and YOYO, are able to penetrate the cell membrane of all cells. Other nucleic acid stains, such as acridine orange, first require fixation of the cells (Wentland *et al.*, 1996). Nucleic acid stains are used to detect the total number of cells or to quantify the amount of nucleic acid. Of course, with these probes one must take into account that the amount of target nucleic acid depends on the physiological state of the cell (Barer and Harwood, 1999). *Fluorescent in situ hybridisation* (FISH) probes are for instance 16S rRNA-directed probes. These probes are used to assess ribosomal content of microorganisms and to detect different species in environmental samples (Barer and Harwood, 1999; Joux and Lebaron, 2000).

In our studies, we used several fluorescent probes for viability assessment after exposure to disinfectants: CFSE, DiSC3[5] (membrane potential probe), CTC, PI and TOTO (Luppens *et al.*, 2001, 2002a) (data not shown). At first a fluorimeter was used for analysis of fluorescent cells (Luppens *et al.*, 2001). The advantage of this method is that the immediate effect of the disinfectant can be analysed and followed during a chosen time period. The drawback of the method is that the total population is analysed and a high cell concentration is needed (10^9 ml^{-1}). In further studies, FCM was used for analysis of fluorescently labelled cells (Luppens *et al.*, 2002a). The advantages of this method are that individual cells can be rapidly analysed in great quantities (1000 cells s^{-1}). The disadvantages are that all intact cells are detected and thus large numbers of cells have to be analysed to detect possible low numbers of viable cells among a large number of dead but intact cells. Furthermore, the sensitivity is not so high, the lower detection limit is a 2–3-log reduction depending on the signal-to-noise ratio (Langsrud and Sundheim, 1996). Therefore, 4- or 5-log reductions in viability are difficult to determine.

A general advantage of the above-mentioned alternatives compared to plate counting is that they show the reaction of the cells to a disinfectant at the sub-cellular level. These methods provide information on cell processes and cell properties that are affected at low and relatively harmless disinfectant concentrations (such as loss of the proton-motive force after exposure to BAC (Luppens *et al.*, 2001) or complete loss of ATP after exposure to hydrogen peroxide (data not shown)). They also show which processes or properties are affected at a high disinfectant concentration that results in complete loss of viability according to plate counts (such as loss of respiration and membrane integrity after exposure to BAC (Luppens *et al.*, 2001, 2002a)) and which proper-ties are not altered at this high concentration and thus may indicate a VBNC state (such as membrane integrity after exposure to hydrogen peroxide (Luppens *et al.*, 2002a)). However, insight into disinfectant damage at the sub-cellular level does not allow conclusions to be drawn about cell viability as determined by plate counts. To be able to do that, it has to be known which of the damages to the cell caused by the disinfectant correspond with complete loss of viability according to plate counts, before a method can be selected as an alternative to plate counting. If the suitable killing concentration has to be determined for a disinfectant of which the precise mechanism of action is known, this choice will not be such a problem, but for screening of new disinfectants it will be. This problem could be solved in several ways.

First of all, the 'golden standard of viability', plate counting, could be abol-ished and replaced by another standard. In that case, the use of the term VBNC or ANC will be redundant, but a similar term may have to be introduced for the newly accepted standard. Several authors suggest that a good new standard would be maintenance of an intact membrane, as shown by the exclusion of dye-exclusion probes (Nebe von Caron *et al.*, 1998; Bunthof *et al.*, 2001). However, membrane integrity is not always a good viability indicator. Cells that are non-viable accord-ing to several other methods may exclude dye-exclusion probes because the membrane is not damaged or damage is not extensive enough for dye-exclusion probes to enter the cells, e.g. in hydrogen peroxide-exposed *L. monocytogenes* cells that exclude TOTO (Luppens *et al.*, 2002a) and PI (data not shown). Furthermore, cells without an intact membrane can be viable, for instance after electroporation or exposure to heat (Nebe von Caron *et al.*, 2000), or membrane integrity can be restored (Votyakova *et al.*, 1994).

Another approach would be to use multiple alternative methods simultan-eously instead of plate counting, for instance double labelling of cells with fluores-cent probes. This approach will give multiple indices of physiological activity and will provide insight in the overall effect of a disinfectant on physiological activity and on the site(s) of sub-lethal injury (Lisle *et al.*, 1999; Joux and Lebaron, 2000). In short, there are perspectives for the use of fluorescent probes as an alternative for plate counting in disinfectant testing when the proper probes are selected. They can be used in combination with FCM provided the cells can be easily removed from the surface and in combination with non-destructive methods, such as CLSM, if they are tightly bound to the surface.

26.2.3   Biofilm disinfectant test

As was mentioned in the introduction of this chapter, a biofilm disinfectant test would be a very useful addition to the currently used disinfectant tests. Therefore, we have developed a biofilm disinfectant test (Luppens *et al.*, 2002b). The development of this test consisted of several stages. First the experimental conditions were selected, such as temperature, medium, strain, inoculum and a reactor for biofilm formation. The apparatus for biofilm formation is a flow-through system that consists of a medium vessel and a waste vessel, a pump and a container that contains 23 coupons on which biofilms are formed, connected by silicon tubing (Luppens *et al.*, 2002b). For convenience and to make comparison to the previous results with suspension test cells easier, the chosen conditions are as close as possible to the current tests. Furthermore, the test is as simple as possible. A suspension of *S. aureus* DSM 799 cells, prepared according to the prescriptions of the European test (Anonymous, 1997a), is used to inoculate the coupons and biofilms are allowed to develop for 24 h. Then the coupons are rinsed, exposed to disinfectant for 5 min, neutraliser is added and viability is assessed by plate counting. Ten times diluted TSB is used as medium to prevent too much planktonic growth in the system and because it is generally thought that relatively poor media support biofilm formation (Anwar *et al.*, 1989; Brown and Gilbert, 1993; Dorel *et al.*, 1999; Hassett *et al.*, 1999). Secondly, the best method to quantify biofilm formation was selected, in this case swabbing the surface of the coupons and vortexing the swab together with the coupon. Thirdly, we assessed the reproducibility of biofilm formation $(8.1 \times 10^7 \pm 4.4 \times 10^7$ $CFU\,cm^{-2})$ which is in between the limits set by the current suspension test in which a 3.3 times difference in cell concentration is allowed. In the context of standardised biofilm formation Jackson *et al.* (2001) found a 50 times difference in biofilm cell concentration for mixed biofilms of *Pseudomonas fluorescens*, *P. aeruginosa* and *Klebsiella pneumoniae*. When they used the specific number of viable cells the difference was reduced to 5 times. Ceri *et al.* (1999) found approximately a 3 times difference in biofilm cell concentrations for *P. aeruginosa*. Lastly, we assessed the reproducibility of biofilm disinfection. The reproducibility of disinfection in this test was the same as what we found for the phase 1 suspension test (Luppens *et al.*, 2002c). Zelver *et al.* (2001) found in a literature survey of standard antimicrobial suspension and dried surface tests that the repeatability standard deviation of log reduction values varied between 0.2 and 1.2. They found a repeatability standard deviation of 0.66 for *P. aeruginosa* biofilms exposed to chlorine (Zelver *et al.*, 2001). Our results showed similar values (0.11–0.98). The following criterion was set for a candidate disinfectant to pass the test: more than a 4-log reduction in 5 min in a biofilm with a cell concentration that falls within the 3.3 times variation allowed in the current suspension tests (Luppens *et al.*, 2002b). Other authors (Wirtanen, 1995) have proposed that for a biofilm test only a 3-log reduction is necessary, but that is too small a reduction for these biofilms that can contain up to $1.3 \times 10^8\,CFU\,cm^{-2}$.

In the *S. aureus* biofilm test described above, cells were less susceptible to disinfectants than suspension test cells and one disinfectant (BAC) was relatively more effective against biofilm cells than the other (NaOCl) (Luppens *et al.*, 2002b). This difference may be explained by the fact that BAC can more easily penetrate the biofilm than NaOCl because of its surfactant properties. In addition, biofilm cells may be better adapted to oxidative damage than to membrane damaging stresses.

The described biofilm test is a general test that could be used as a replacement for the phase 1 test, provided that a candidate disinfectant will only be used on biofilms, or as an additional test for efficacy testing of disinfectants against biofilms in phase 2 steps 1 or 2. For phase 2 step 2, more specific tests could be developed and added such as the ones that already exist for the human mouth, toilet bowls, oilfield water injection systems and cooling water or water distribution systems (Drake, 2001; Pitts *et al.*, 2001). Furthermore, these tests should also include the effect of hard water, soil, sanitisers and mechanical action, in order to mimic practical conditions.

Several suggestions can be done to further develop the test. A number of studies with *L. monocytogenes*, *P. aeruginosa* and *K. pneumoniae* showed that older biofilms (1–2 weeks) are more resistant to antimicrobials than young biofilms (1–2 days) (LeChevallier *et al.*, 1988; Anwar *et al.*, 1989; Frank and Koffi, 1990; Anwar *et al.*, 1992). However, older biofilms are more difficult to remove from the surface which would make plate counting more difficult (Chae and Schraft, 2001) and extension of the current protocol by more than 1 day would mean that the test cannot be performed in one working week. Furthermore, older biofilms may not be very stable because of cycles in biofilm formation as were reported for *L. monocytogenes* biofilms (Chae and Schraft, 2001) and mixed biofilms of *P. aeruginosa*, *K. pneumoniae* and *P. fluorescens* (Jackson *et al.*, 2001) or because of selection of mutants. Finally, the most important argument is that a biofilm over 24-h old is not very likely to be present on a food contact surface, since in most factories these surfaces are cleaned and disinfected every day (Beumer, 2002).

Another suggestion for further development of the test is to use another *S. aureus* strain than *S. aureus* DSM 799. Figure 26.1A shows that several other strains of *S. aureus* are stronger biofilm formers than the currently used *S. aureus* strain. Thus, with the use of one of these strains a thicker biofilm may be developed, which will widen the detection range of the test.

Furthermore, biofilms could be grown at different temperatures that represent the conditions in the food industry more closely. As can be seen in Figure 26.1 some *S. aureus* strains are much stronger biofilm formers at 15°C than at 25°C.

It would be interesting to study whether the test developed with *S. aureus* would perform equally well with biofilms from other bacteria. For Gram-negative bacteria the obvious candidate would be *P. aeruginosa*, because it is used as a representative of the Gram-negatives in the current standard European tests (Anonymous, 1997a, b). Other Gram-negative bacteria, such as *Salmonella enteritidis* or *Escherichia coli*, may be selected on basis of their role in causing problems in specific products. For Gram-positive bacteria an obvious candidate is

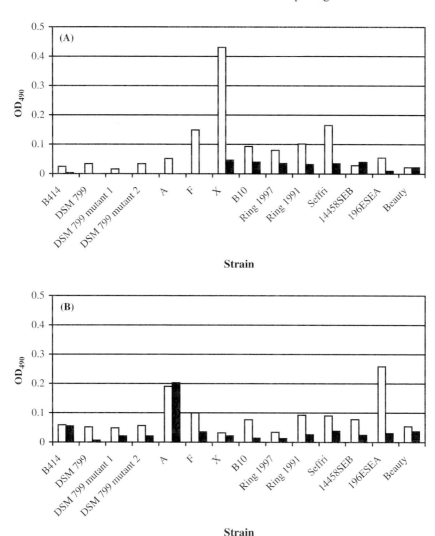

Figure 26.1   Biofilm formation by 12 *S. aureus* strains at 25°C (A) and 15°C (B) in a 96-well polystyrene microtiter plate. Wells were inoculated with 200 μl of a 100 times in TSB-diluted pre-culture (grown for 22 h, shaking, at 37°C, in TSB). After 48 (25°C) or 194 h (15°C) of incubation wells were washed twice with PBS (Luppens *et al.*, 2001), fixed with Bouin's fixative solution, stained with saffranine (5 g l^{-1}, total stain, white bars) and Congo red (1 g l^{-1}, polysaccharide stain, black bars), washed twice with PBS, after which OD$_{490}$ was measured. Values represent the average of eight wells.

L. monocytogenes. A considerable number of biofilm studies have focused on *L. monocytogenes* (Frank and Koffi, 1990; Spurlock and Zottola, 1991; Krysinski *et al.*, 1992; Sasahara and Zottola, 1993; Blackman and Frank, 1996; Lou and Yousef, 1999; Norwood and Gilmour, 1999; Maukonen *et al.*, 2000; Kalmokoff *et al.*, 2001). However, it has been questioned whether pure cultures of *L. mono-cytogenes* are able to form a biofilm. It was suggested that *L. monocytogenes* may use primary colonisers to form a biofilm (Sasahara and Zottola, 1993; Kalmokoff *et al.*, 2001). If this is the case, such a primary coloniser can be used together with *L. monocytogenes* for mixed biofilm formation. However, Chavant *et al.* (2002) demonstrate that *L. monocytogenes* can form single-species biofilms under various conditions. We found that several *L. monocytogenes* strains are able to form biofilms at 25°C and/or at 15°C (Figure 26.2) that are comparable in thickness to that of *S. aureus* as determined by the saffranine method described in Figure 26.1.

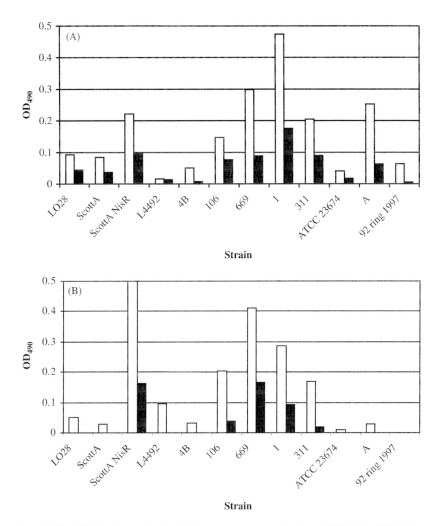

Figure 26.2   Biofilm formation by 12 *L. monocytogenes* strains at 25°C after 30 h (A) and 15°C after 148 h (B) according to the method described in Figure 26.1 (except for the pre-culture which was grown for 15.5 h, at 30°C, in BHI and diluted in BHI).

The difference in biofilm formation between Scott A and its resistant variant Scott A Nis[R] (Verheul *et al.*, 1997) clearly demonstrates the effect of *L. monocytogenes* strain diversity on biofilm formation.

Lastly, the viability assessment by plate counting in the test could be replaced with fluorescent labelling. This would shorten the protocol with 2 days. Figure 26.3 shows that it is possible to use fluorescent labelling in combination with FCM instead of plate counting for viability assessment of *S. aureus* biofilm cells

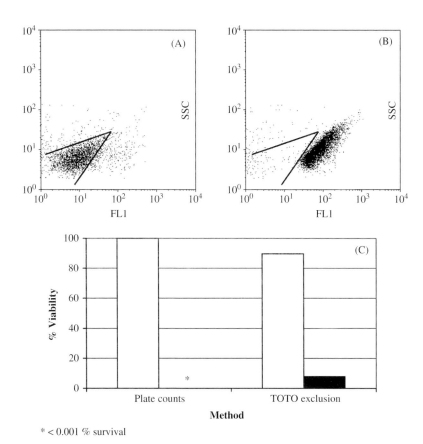

* < 0.001 % survival

Figure 26.3   Viability assessment of at 30°C grown 24-h *S. aureus* biofilm (Luppens *et al.*, 2002b) cells. Intact biofilms were exposed to 3 ml of water (untreated) or 300 mg l^{-1} of BAC. After 5 min 27 ml of PBS (Luppens *et al.*, 2001) was added. After 5 min cells were removed (Luppens *et al.*, 2002b) from the surface and either labelled with TOTO (0.3 μmol l^{-1}, 15 min) without washing and analysed with FCM or the cell suspension was diluted in PBS, plated on TSA and incubated at 30°C for 48 h. FCM results show FL1-fluorescence and SSC of untreated cells (A) and cells exposed to BAC (B) stained with TOTO (the triangle shows the region with unlabelled/viable cells). Furthermore, the percentage viability determined with plate counting and TOTO exclusion is shown (C) for untreated cells (white bars) and cells exposed to BAC (black bars). Experiments were done in four-fold

exposed to BAC. Fluorescent probes could also be used in combination with fluorescence microscopy together with phase-contrast microscopy or CLSM of intact biofilms, which will give information about the spatial distribution of disinfectant killed cells. A fluorescent probe that has been used for this purpose in several studies is the redox indicator CTC (Huang *et al.*, 1995; McDowell *et al.*, 1995; McFeters *et al.*, 1999). However, CTC is not suitable for *L. monocytogenes* viability assessment after disinfection (Luppens *et al.*, 2002a) and also in other studies its suitability has been questioned (Kaprelyants and Kell, 1993b; Servais *et al.*, 2001). An alternative fluorescent method would be measurement of the internal pH of bacteria in the biofilm with CFSE. This probe has been used for biofilm analysis in combination with CLSM and two-photon excitation microscopy (Vroom *et al.*, 1999). However, surrounding the biofilm cells with medium of a low pH, allowing for assessment of pH gradients, i.e. pH_{in} versus pH_{out}, may be a problem in thick biofilms. Another alternative may be one of the dye-exclusion probes, but these probes should only be used for cells exposed to non-oxidative disinfectants (Luppens *et al.*, 2002a). Furthermore, probes may also be used in combination to obtain more information (see above). Of course, once an alteration to the test has been made, one must keep in mind that the suitability of the biofilm quantification method and the reproducibility of biofilm formation and biofilm disinfection have to be validated again, because the alteration may influence biofilm formation.

Finally, standardised biofilms may be valuable tools to study the genes that are induced or repressed in biofilm cells compared to planktonic cells. Such studies may lead to the discovery of genes that are important for resistance of biofilms cells (Whiteley *et al.*, 2001) and genes that code for targets for disinfectants in biofilms. The flow-through system used in our test offers a good alternative for the batch biofilms that are mostly used for such studies.

26.2.4   Biofilm control in practice

There are several ways to control biofilm formation in practice. The best way of control is of course prevention of biofilm formation. Surfaces can be treated in a way that they repel bacteria, they can be coated with antimicrobials (Kumar and Anand, 1998) or with substances that interfere with their quorum-sensing system (Givskov *et al.*, 1996), they can be made out of material that leaches biocides or antimicrobials can be attached covalently to the surface to prevent exhaustion of the antimicrobial (Tiller *et al.*, 2001). Another method is to place a current over the surface or to allow a biofilm of harmless microorganisms to develop that will competitively exclude pathogenic microorganisms (Kumar and Anand, 1998; Leriche *et al.*, 1999; Leriche and Carpentier, 2000). Most of these methods are expensive or developed for other purposes than the food industry, e.g. to prevent colonisation of medical devices and implants. Thus, more research is needed before they can be applied in the food industry. A simpler and more effective method for biofilm prevention is to follow the guidelines of hygienic design of the European Hygiene Engineering and Design Group (EHEDG) (Anonymous,

2000). Examples of these guidelines for hygienic design are use of smooth surfaces, avoidance of dead ends, corners and crevices in the used materials.

Despite all methods and precautions, biofilm formation can never be completely prevented (Elvers *et al.*, 1999). Thus, biofilm formation should be monitored so that cleaning and disinfection of surfaces can be done at the proper time. For the monitoring of biofilm formation a large variety of techniques is available. The presence of a biofilm can be monitored by removal of biofilm cells from the surface by swabbing, vortexing with glass beads, scraping or by sonication and further analysis by plating or possibly FCM (Carpentier and Cerf, 1993; Holah, 1996; Luppens *et al.*, 2002b). Intact biofilms can be analysed with phase-contrast microscopy, electron microscopy, CLSM, fluorescence microscopy (autofluorescent cells, green fluorescent protein or fluorescently labelled cells), direct viable count, radioactive labelling of cells (Holah *et al.*, 1988; Hussain *et al.*, 1992; Yu and McFeters, 1994; Costerton *et al.*, 1995; Wirtanen, 1995; Hibma *et al.*, 1996; Holah, 1996; McFeters *et al.*, 1999; Sternberg *et al.*, 1999), with a Malthus microbiological growth analyser (based on changes in conductance) (Foster, 1999) or by detection of cell metabolites or cell constituents, such as ATP, proteins or exopolysaccharides (Wirtanen, 1995; Holah, 1996; Gracia *et al.*, 1999). In the future, new viability techniques based on the detection of mRNA (Klein and Juneja, 1997; Norton and Batt, 1999) may be used. Biofilm formation in closed pipe systems can be monitored without opening them by measurement of pressure change or current change in a small device that contains a large surface area for biofilm formation and receives the same liquid flow as the pipe system. These systems are already commercially available (Dickinson, 1997; Anonymous, 1999a). Which of these techniques is preferable depends on the situation. With the data obtained from these measurements, biofilm build-up can be monitored in time. Furthermore, with these data the efficiency of cleaning and disinfection can be assessed. With the help of the acquired data a model can be developed to estimate the appropriate time for cleaning and disinfection and a cleaning and disinfection regime can be developed accordingly (den Aantrekker *et al.*, 1999).

For efficient cleaning and disinfection, it is not only important that the cells in the biofilm are killed but also the biofilm matrix has to be broken down, so that the biofilm can be completely removed from the surface. Any left over organic material provides nutrients that facilitate the rapid formation of a new biofilm (Wirtanen, 1995). Biofilm matrix breakage, which also makes the biofilm cells better accessible for disinfection, can be done by a cleaning agent, an additional agent, such as EDTA or by the disinfectant itself (Wirtanen, 1995).

26.3   CONCLUSION

In this chapter, we discussed several factors that are important in disinfectant testing on bacteria in general and on biofilms in particular. Our conclusion is that there are several ways to further improve the currently used standard disinfectant tests. The most obvious way for improvement is to use a biofilm disinfectant test in addition to the tests that are currently used, for instance the biofilm test described in this chapter. With the help of this test, known and candidate disinfectants, such as for instance electrolysed water (Kim *et al.*, 2000), can be evaluated for their biofilm killing efficacy. This evaluation will teach us which disinfectants are most effective against biofilms, which in the end will contribute to food safety and food quality and the control of cleaning costs in the food industry.

Acknowledgement

The authors thank Lisette Elsinga for her technical assistance.

REFERENCES

Acworth, I.N. and Bailey, B. (1997) *The Handbook of Oxidative Metabolism*, Esa Inc., Chelmsford, Maryland, USA.

Adams, J.L. and McLean, R.J.C. (1999) Impact of rpoS deletion on *Escherichia coli* biofilms. *Appl. Environ. Microbiol.* **65**, 4285–4287.

Akiyama, H., Yamasaki, O., Kanzaki, H., Tada, J. and Arata, J. (1998) Effects of sucrose and silver on *Staphylococcus aureus* biofilms. *J. Antimicrob. Chemoth.* **42**, 629–634.

Akiyama, H., Yamasaki, O., Tada, J., Kubota, K. and Arata, J. (2000) Antimicrobial effects of acidic hot-spring water on *Staphylococcus aureus* strains isolated from atopic dermatitis patients. *J. Dermatol. Sci.* **24**, 112–118.

Amanatidou, A. (2001) High oxygen as an additional factor in food preservation, Ph.D. thesis, Food Hygiene and Microbiology Group, Wageningen University.

Amanatidou, A., Schlueter, O., Lemkau, K., Gorris, L.G.M., Smid, E.J. and Knorr, D. (2000) Effect of combined application of high pressure treatment and modified atmospheres on the shelf life of fresh Atlantic salmon. *Innov. Food Sci. Emerg. Technol.* **1**, 87–98.

Anonymous (1992a) *Listeria monocytogenes*. US Food and Drug Administration Center for Food Safety and Applied Nutrition. http://www.cfsan.fda.gov/~mow/chap6.html (online).

Anonymous (1992b) *Staphylococcus aureus*. US Food and Drug Administration Center for Food Safety and Applied Nutrition. http://www.cfsan.fda.gov/~mow/chap3.html (online).

Anonymous (1997a) Chemical disinfectants and antiseptics – Basic bactericidal activity – Test method and requirement (phase 1). EN-1040. Nederlands Normalistatie-instituut, Delft, The Netherlands.

Anonymous (1997b) Chemical disinfectants and antiseptics – Quantitative suspension test for the evaluation of bactericidal activity of chemical disinfectants and antiseptics used in food, industrial, domestic and institutional areas – Test method and requirement (phase 2, step 1). EN-1276. Nederlands Normalistatie-instituut, Delft, The Netherlands.

Anonymous (1999a) BIoGEORGE(tm) – Electrochemical Biofilm Activity Monitoring System. Structural Integrity associates, San Jose, USA. http://www.structint.com/tekbrefs/sim98009/ (online).

Anonymous (1999b) Chemical disinfectants and antiseptics – Quantitative surface test for the evaluation of bactericidal and/or fungicidal activity of chemical disinfectants used in food, industrial, domestic and institutional areas – Test method without mechanical action and requirements (phase 2, step 2). EN-13697. Nederlands Normalistatie-instituut, Delft, The Netherlands.

Anonymous (2000) Guidelines. European Hygienic Equipment Design Group. http://www.ehedg.org/guidelines.htm (online).

Anwar, H., van Biesen, T., Dasgupta, M., Lam, K. and Costerton, J.W. (1989) Interaction of biofilm bacteria with antibiotics in a novel *in vitro* chemostat system. *Antimicrob. Agents Chemoth.* **33**, 1824–1826.

Anwar, H., Strap, J.L. and Costerton, J.W. (1992) Eradication of biofilm cells of *Staphylococcus aureus* with tobramycin and cephalexin. *Can. J. Microbiol.* **38**, 618–625.

Barer, M.R. and Harwood, C.R. (1999) Bacterial viability and culturability. *Adv. Microb. Physiol.* **41**, 93–137.

Bateman, B.T., Donegan, N.P., Jarry, T.M., Palma, M. and Cheung, A.L. (2001) Evaluation of a tetracycline-inducible promoter in *Staphylococcus aureus in vitro* and *in vivo* and its application in demonstrating the role of *sigB* in microcolony formation. *Infect. Immun.* **69**, 7851–7857.

Becker, L.A., Sevket Çetin, M., Hutkins, R.W. and Benson, A.K. (1998) Identification of the gene encoding the alternative sigma factor sigmaB from *Listeria monocytogenes* and its role in osmotolerance. *J. Bacteriol.* **180**, 4547–4554.

Becker, P., Hufnagle, W., Peters, G. and Herrmann, M. (2001) Detection of differential gene expression in biofilm-forming versus planktonic populations of *Staphylococcus aureus* using micro-representational-difference analysis. *Appl. Environ. Microbiol.* **67**, 2958–2965.

Beumer, R. (2002) Personal communication.

Blackman, I.C. and Frank, J.F. (1996) Growth of *Listeria monocytogenes* as a biofilm on various food-processing surfaces. *J. Food Protect.* **59**, 827–831.

Block, S.S. (1991) Definition of terms. In *Disinfection Sterilization and Preservation*, 4th ed. (ed. S.S. Block), pp. 18–25, Lea & Febiger, Malvern, USA.

Bloomfield, S.F., Arthur, M., Klingeren, B.V., Pullen, W., Holah, J.T. and Elton, R. (1994) An evaluation of the repeatability and reproducibility of a surface test for the activity of disinfectants. *J. Appl. Bacteriol.* **76**, 86–94.

Braux, A.S., Minet, J., Tamanai, S.Z., Riou, G. and Cormier, M. (1997) Direct enumeration of injured *Escherichia coli* cells harvested onto membrane filters. *J. Microbiol. Meth.* **31**, 1–8.

Breeuwer, P. (1996) Assessment of viability in micro-organisms employing fluorescence techniques, Ph.D. thesis, Food hygiene and microbiology, Wageningen University.

Breeuwer, P. and Abee, T. (2000) Assessment of viability of microorganisms employing fluorescence techniques. *Int. J. Food Microbiol.* **55**, 193–200.

Breeuwer, P., Drocourt, J.L., Rombouts, F.M. and Abee, T. (1996) A novel method for continuous determination of the intracellular pH in bacteria with the internally conjugated fluorescent probe 5(and 6-)-carboxyfluorescein succinimidyl ester. *Appl. Environ. Microbiol.* **62**, 178–183.

Brooun, A., Liu, S. and Lewis, K. (2000) A dose–response study of antibiotic resistance in *Pseudomonas aeruginosa* biofilms. *Antimicrob. Agent. Chemoth.* **44**, 640–646.

Brown, M.R.W. and Gilbert, P. (1993) Sensitivity of biofilms to antimicrobial agents. *J. Appl. Bacteriol. Symp. Suppl.* **74**, 87–97.

Bunthof, C.J., Bloemen, K., Breeuwer, P., Rombouts, F.M. and Abee, T. (2001) Flow cytometric assessment of viability of lactic acid bacteria. *Appl. Environ. Microbiol.* **67**, 2326–2335.

Campanac, C., Pineau, L., Payard, A., Baziard-Mouysset, G. and Roques, C. (2002) Interactions between biocide cationic agents and bacterial biofilms. *Antimicrob. Agent. Chemoth.* **46**, 1469–1474.

Camper, A., Burr, M., Ellis, B.D., Butterfield, P. and Abernathy, C. (1999) Development and structure of drinking water biofilms and techniques for their study. *J. Appl. Microbiol. Symp. Suppl.* **85**, 1s–12s.

Carpentier, B. and Cerf, O. (1993) Review: biofilms and their consequences, with particular reference to hygiene in the food industry. *J. Appl. Bacteriol.* **75**, 499–511.

Ceri, H., Olson, M.E., Stremick, C., Read, R.R., Morck, D. and Buret, A. (1999) The calgary biofilm device: new technology for rapid determination of antibiotic susceptibilities of bacterial biofilms. *J. Clin. Microbiol.* **37**, 1771–1776.

Chae, M.S. and Schraft, H. (2001) Cell viability of *Listeria monocytogenes* biofilms. *Food Microbiol.* **18**, 103–112.

Christensen, G.D., Simpson, W.A., Younger, J.J., Baddour, L.M., Barrett, F.F., Melton, D.M. and Beachey, E.H. (1985) Adherence of coagulase-negative staphylococci to plastic tissue culture plates: a quantitative model for the adherence of staphylococci to medical devices. *J. Clin. Microbiol.* **22**, 996–1006.

Collins, L. and Franzblau, S.G. (1997) Microplate alamar blue assay versus BACTEC 460 system for high-throughput screening of compounds against *Mycobacterium tuberculosis* and *Mycobacterium avium. Antimicrob. Agent. Chemoth.* **41**, 1004–1009.

Comas, J. and Vives-Rego, J. (1998) Enumeration, viability, and heterogeneity in *Staphylococcus aureus* cultures by flow cytometry. *J. Microbiol. Meth.* **32**, 45–53.

Cooke, E.M. (1990) Epidemiology of foodborne illness: UK. *Lancet* **336**, 790–793.

Costerton, J.W., Lewandowski, Z., Caldwell, D.E., Korber, D.R. and Lappin Scott, H.M. (1995) Microbial biofilms. *Ann. Rev. Microbiol.* **45**, 711–745.

Cramton, S.E., Gerke, C., Schnell, N.F., Nichols, W.W. and Götz, F. (1999) The intercellular adhesion (ica) locus is present in *Staphylococcus aureus* and is required for biofilm formation. *Infect. Immun.* **67**, 5427–5433.

Cramton, S.E., Gerke, C. and Gotz, F. (2001) *In vitro* methods to study staphylococcal biofilm formation. *Meth. Enzymol.* **336**, 239–255.

Cucarella, C., Solano, C., Valle, J., Amorena, B., Lasa, I. and Penades, J.R. (2001) Bap, a surface protein involved in biofilm formation. *J. Bacteriol.* **183**, 2888–2896.

Das, J.R., Bhakoo, M., Jones, M.V. and Gilbert, P. (1998) Changes in the biocide susceptibility of *Staphylococcus epidermidis* and *Escherichia coli* cells associated with rapid attachment to plastic surfaces. *J. Appl. Microbiol.* **84**, 852–858.

Davies, D.G. (1999) Regulation of matrix polymer in biofilm formation and dispersion. In *Microbial Extracellular Polymeric Substances: Characterization, Structure and Function* (ed. J. Wingender, T.R. Neu and H.-C. Flemming), pp. 93–117, Springer-Verlag, Berlin, Germany.

Davies, D.G., Parsek, M.R., Pearson, J.P., Iglewski, B.H., Costerton, J.W. and Greenber g, E.P. (1998) The involvement of cell-to-cell signals in the development of a bacterial biofilm. *Science* **280**, 295–298.

de Beer, D. and Schramm, A. (1999) Micro-environments and mass transfer phenomena in biofilms studied with microsensors. *Water Sci. Technol.* **39**, 173–178.

den Aantrekker, E.D., Zwietering, M.C. and Schothorst, M. (1999) Modelling of biofilm formation. In *Biofilms: The Good, The Bad and The Ugly* (ed. J. Wimpenny, P. Gilbert, J. Walker, M. Brading and R. Bayston), pp. 251–255, Bioline, Cardiff, UK.

Denyer, S.P. and Stewart, G.S.A.B. (1998) Mechanisms of action of disinfectants. *Int. Biodeterior. Biodegrad.* **41**, 261–268.

Dickinson, W.H. (1997) Biofouling assessment using an on-line monitor. Buckman Laboratories International, Memphis, USA. http://www.buckman.com/eng/randd/whd-tappi.html (online).

Dodd, C.E.R., Sharman, R.L., Bloomfield, S.F., Booth, I.R. and Stewart, G.S.A.B. (1997) Inimical processes: bacterial self-destruction and sub-lethal injury. *Trend. Food Sci. Technol.* **8**, 238–241.

Dorel, C., Vidal, O., Prigent-Combaret, C., Vallet, I. and Lejeune, P. (1999) Involvement of the Cpx signal transduction pathway of *E. coli* in biofilm formation. *FEMS Microbiol. Lett.* **178**, 169–175.

Drake, D. (2001) Assessment of antimicrobial activity against biofilms. *Meth. Enzymol.* **337**, 385–389.

Dychdala, G.R. (1991) Chlorine and chlorine compounds. In *Disinfection, Sterilization and Preservation*, 4th ed. (ed. S.S. Block), pp. 131–151, Lea & Febiger, Malvern, USA.

Dychdala, G.R. and Lopes, J.A. (1991) Surface-active agents: acid–anionic compounds. In *Disinfection, Sterilization and Preservation*, 4th ed. (ed. S.S. Block), pp. 256–262, Lea & Febiger, Malvern, USA.

Edelmeyer, H. (1982) Über Eigenschaften, Wirkmechanismen und Wirkungen chemischer Desinfektionsmittel. *Arch. Lebensmittelhyg.* **33**, 1–32.

Elvers, K.T., Peters, A.C. and Griffith, C.J. (1999) Development and control of biofilms in the food industry. In *Biofilms: The Good, the Bad and the Ugly* (ed. J. Wimpenny, P. Gilbert, J. Walker, M. Brading and R. Bayston), pp. 139–145, Bioline, Cardiff, UK.

Farber, J.M. (1993) Current research on *Listeria monocytogenes* in foods: an overview. *J. Food Protect.* **56**, 640–643.

Farber, J.M. and Peterkin, P.I. (1991) *Listeria monocytogenes*, a food-borne pathogen. *Microbiol. Rev.* **55**, 476–511.

Ferreira, A., O'Byrne, C.P. and Boor, K.J. (2001) Role of sigma(B) in heat, ethanol, acid, and oxidative stress resistance and during carbon starvation in *Listeria monocytogenes*. *Appl. Environ. Microbiol.* **67**, 4454–4457.

Foster, J.W. (1999) When protons attack: microbial strategies of acid adaptation. *Curr. Opin. Microbiol.* **2**, 170–174.

Foster, J.W. and Spector, M.P. (1995) How *Salmonella* survive against the odds. *Ann. Rev. Microbiol.* **49**, 145–174.

Fournier, B. and Hooper, D.C. (2000) A new two-component regulatory system involved in adhesion, autolysis, and extracellular proteolytic activity of *Staphylococcus aureus*. *J. Bacteriol.* **182**, 3955–3964.

Frank, J.F. and Koffi, R.A. (1990) Surface-adherent growth of *Listeria monocytogenes* is associated with increased resistance to surfactant sanitizers and heat. *J. Food Protect.* **53**, 550–554.

Gibson, H., Taylor, J.H., Hall, K.E. and Holah, J.T. (1999) Effectiveness of cleaning techniques used in the food industry in terms of the removal of bacterial biofilms. *J. Appl. Microbiol.* **87**, 41–48.

Gilbert, P., Brown, M.R.W. and Costerton, J.W. (1987) Inocula for antimicrobial sensitivity testing: a critical review. *J. Antimicrob. Chemoth.* **20**, 147–154.

Gilbert, P., Allison, D.G., Evans, D.J., Handley, P.S. and Brown, R.W. (1989) Growth rate control of adherent bacterial populations. *Appl. Environ. Microbiol.* **55**, 1308–1311.

Givskov, M., de Nys, R., Manefield, M., Gram, L., Maximilien, R., Eberl, L., Molin, S., Steinberg, P.D. and Kjelleberg, S. (1996) Eukaryotic interference with homoserine lactone-mediated prokaryotic signalling. *J. Bacteriol.* **178**, 6618–6622.

Gracia, E., Fernández, A., Conchello, P., Alabart, J.L., Pérez, M. and Amorena, B. (1999) *In vitro* development of *Staphylococcus aureus* biofilms using slime-producing variants and ATP-bioluminescence for automated bacterial quantification. *Luminescence* **14**, 23–31.

Härkönen, P., Salo, S., Matilla-Sandholm, T., Wirtanen, G., Allison, D.G. and Gilbert, P. (1999) Development of a simple *in vitro* test system for the disinfection of bacterial biofilms. *Water Sci. Technol.* **39**, 219–225.

Hartmann, A., Lawrence, J.R., Assmus, B. and Schloter, M. (1998) Detection of microbes by laser confocal microscopy. In *Molecular Microbial Ecology Manual* (ed. A.D.L. Akkermans, J.D. van Elsas and F.J. de Bruijn), pp. 1–34, Kluwer, The Netherlands.

Hassett, D.J., Ma, J.F., Elkins, J.G., McDermott, T.R., Ochsner, U.A., West, S.E., Huang, C.T., Fredericks, J., Burnett, S., Stewart, P.S., McFeters, G., Passador, L. and Iglewski, B.H. (1999) Quorum sensing in *Pseudomonas aeruginosa* controls expression of catalase and superoxide dismutase genes and mediates biofilm susceptibility to hydrogen peroxide. *Mol. Microbiol.* **34**, 1082–1093.

Henle, E.S. and Linn, S. (1997) Formation, prevention, and repair of DNA damage by iron/hydrogen peroxide. *J. Biol. Chem.* **272**, 19095–19098.

Hibma, A.M., Jassim, S.A. and Griffiths, M.W. (1996) *In vivo* bioluminescence to detect the attachment of L-forms of *Listeria monocytogenes* to food and clinical contact surfaces. *Int. J. Food Microbiol.* **33**, 157–167.

Holah, J. (1996) Test methods development for the practical assessment of food processing equipment cleanability. In *Proceedings of the Fourth Asept International Conference on 'Food Safety '96'* (ed. A. Amgar), pp. 300–319, Laval, France.

Holah, J.T., Betts, R.T. and Thorpe, R.H. (1988) The use of direct epifluorescencemicroscopy (DEM) and the epifluorescence filter technique (DEFT) to assess microbial populations on food contact surfaces. *J. Appl. Bacteriol.* **65**, 215–221.

Holah, J.T., Lavaud, A., Peters, W. and Dye, K.A. (1998) Future techniques for disinfectant efficacy testing. *Int. Biodeterior. Biodegrad.* **41**, 273–279.

Huang, C.T., Yu, F.P., McFeters, G.A. and Stewart, P.S. (1995) Nonuniform spatial patterns of respiratory activity within biofilms during disinfection. *Appl. Environ. Microbiol.* **61**, 2252–2256.

Hussain, M., Collins, C., Hastings, J.G.M. and White, P.J. (1992) Radiochemical assay to measure the biofilm produced by coagulase-negative staphylococci on solid surfaces and its use to quantitate the effects of various antibacterial compounds on the formation of the biofilm. *J. Med. Microbiol.* **37**, 62–69.

Jackson, G., Beyenal, H., Rees, W.M. and Lewandowski, Z. (2001) Growing reproducible biofilms with respect to structure and viable cell counts. *J. Microbiol. Meth.* **47**, 1–10.

Jay, J.M. (1996) *Modern Food Microbiology*, 5th ed., Chapman & Hall, New York, USA.

Jeyamkondan, S., Jayas, D.S. and Holley, R.A. (1999) Pulsed electric field processing of foods: a review. *J. Food Protect.* **62**, 1088–1096.

Johansen, C., Falholt, P. and Gram, L. (1997) Enzymatic removal and disinfection of bacterial biofilms. *Appl. Environ. Microbiol.* **63**, 3724–3728.

Joux, F. and Lebaron, P. (2000) Use of fluorescent probes to assess physiological functions of bacteria at single-cell level. *Microbe. Infect.* **2**, 1523–1535.

Kalmokoff, M.L., Austin, J.W., Wan, X.D., Sanders, G., Banerjee, S. and Farber, J.M. (2001) Adsorption, attachment and biofilm formation among isolates of *Listeria monocytogenes* using model conditions. *J. Appl. Microbiol.* **91**, 725–734.

Kaprelyants, A.S. and Kell, D.B. (1993a) Dormancy in stationary-phase cultures of *Micrococcus luteus*: flow cytometric analysis of starvation and resuscitation. *Appl. Environ. Microbiol.* **59**, 3187–3196.

Kaprelyants, A.S. and Kell, D.B. (1993b) The use of 5-cyano-2,3-ditolyl tetrazolium chloride and flow cytometry for the visualisation of respiratory activity in individual cells of *Micrococcus luteus*. *J. Microbiol. Meth.* **17**, 115–122.

Kim, C., Hung, Y.C. and Brackett, R.E. (2000) Efficacy of electrolyzed oxidizing (EO) and chemically modified water on different types of foodborne pathogens. *Int. J. Food Microbiol.* **61**, 199–207.

Klein, P.G. and Juneja, V.K. (1997) Sensitive detection of viable *Listeria monocytogenes* by reverse transcription-PCR. *Appl. Environ. Microbiol.* **63**, 4441–4448.

Klingeren, B.V., Koller, W., Bloomfield, S.F., Böhm, R., Cremieux, A., Holah, J., Reybrouck, R. and Rödger, H.J. (1998) Assessment of the efficacy of disinfectants on surfaces. *Int. Biodeterior. Biodegrad.* **41**, 289–296.

Kloos, W.E. and Schleifer, K.H. (1986) *Staphyloccoccus*. In *Bergey's Manual of Systematic Bacteriology* (ed. J.G. Holt, N.R. Krieg, D.H. Bergey, P.H.A. Sneath, N.S. Mair, M.E. Sharpe, J.T. Staley and S.T. Williams), pp. 1013–1019, Williams, Baltimore, USA.

Kolter, R. (1992) Life and death in stationary phase. *ASM News* **58**, 75–79.

Krysinski, E.P., Brown, L.J. and Marchisello, T.J. (1992) Effect of cleaners and sanitizers on *Listeria monocytogenes* attached to product contact surfaces. *J. Food Protect.* **55**, 246–251.

Kumar, C.G. and Anand, S.K. (1998) Significance of microbial biofilms in food industry: a review. *Int. J. Food Microbiol.* **42**, 9–27.

Langsrud, S. and Sundheim, G. (1996) Flow cytometry for rapid assessment of viability after exposure to a quaternary ammonium compound. *J. Appl. Bacteriol.* **81**, 411–418.

Langsrud, S. and Sundheim, G. (1998) Factors influencing a suspension test method for antimicrobial activity of disinfectants. *J. Appl. Microbiol.* **85**, 1006–1012.

LeChevallier, M.W., Cawthon, C.D. and Lee, R.G. (1988) Factors promoting survival of bacteria in chlorinated water supplies. *Appl. Environ. Microbiol.* **54**, 649–654.

Leriche, V. and Carpentier, B. (2000) Limitation of adhesion and growth of *Listeria monocytogenes* on stainless steel surfaces by *Staphylococcus sciuri* biofilms. *J. Appl. Microbiol.* **88**, 594–605.

Leriche, V., Chassaing, D. and Carpentier, B. (1999) Behaviour of *L. monocytogenes* in an artificially made biofilm of a nisin-producing strain of *Lactococcus lactis*. *Int. J. Food Microbiol.* **51**, 169–182.

Lisle, J.T., Pyle, B.H. and McFeters, G.A. (1999) The use of multiple indices of physiological activity to access viability in chlorine disinfected *Escherichia coli* O157:H7. *Lett. Appl. Microbiol.* **29**, 42–47.

Lleo, M.D., Tafi, M.C. and Canepari, P. (1998) Nonculturable *Enterococcus faecalis* cells are metabolically active and capable of resuming active growth. *Syst. Appl. Microbiol.* **21**, 333–339.

Lou, Y. and Yousef, A.E. (1999) Characteristics of *Listeria monocytogenes* important to food processors. In *Listeria, Listeriosis and Food Safety*, 2nd ed. (ed. E.T. Ryser and E.H. Marth), pp., Marcel Dekker, New York, USA.

Luppens, S.B.I., Abee, T. and Oosterom, J. (2001) Effect of benzalkonium chloride on viability and energy metabolism in exponential- and stationary-growth-phase cells of *Listeria monocytogenes*. *J. Food Prot.* **64**, 476–482.

Luppens, S.B.I., Barbaras, B., Breeuwer, P., Rombouts, F.M. and Abee, T. (2002a) Selection of fluorescent probes for viability assessment by flow cytometry of *Listeria monocytogenes* exposed to membrane-active and oxidizing disinfectants, submitted.

Luppens, S.B.I., Reij, M.W., Rombouts, F.M. and Abee, T. (2002b) Development of a standard test to assess the resistance of *Staphylococcus aureus* biofilm cells to disinfectants. *Appl. Environ. Microbiol* **68**, 4194–4200.

Luppens, S.B.I., Rombouts, F.M. and Abee, T. (2002c) The effect of the growth phase of *Staphylococcus aureus* on resistance to disinfectants in a suspension test. *J. Food Protect.* **65**, 124–129.

Mattila Sandholm, T. and Wirtanen, G. (1992) Biofilm formation in the industry: a review. *Food Rev. Int.* **8**, 573–603.

Maukonen, J., Mattila, S.T. and Wirtanen, G. (2000) Metabolic indicators for assessing bacterial viability in hygiene sampling using cells in suspension and swabbed biofilm. *Food Sci. Technol.* **33**, 225–233.

McDowell, S.G., An, Y.H., Draugh, R.A. and Friedman, R.J. (1995) Application of a fluorescent redox dye for enumeration of metabolically active bacteria on albumin-coated titanium surfaces. *Lett. Appl. Microbiol.* **21**, 1–4.

McFeters, G.A., Yu, F.P., Pyle, B.H. and Stewart, P.S. (1995) Physiological methods to study biofilm disinfection. *J. Ind. Microbiol.* **15**, 333–338.

McFeters, G.A., Pyle, B.H., Lisle, J.T. and Broadaway, S.C. (1999) Rapid direct methods for enumeration of specific, active bacteria in water and biofilms. *J. Appl. Microbiol Symp.* Suppl. **85**, 193s–200s.

Mead, P.S., Slutsker, L., Dietz, V., McCaig, L.F., Bresee, J.S., Shapiro, C., Griffin, P.M. and Tauxe, R.V. (1999) Food-related illness and death in the United States. *Emerg. Infect. Dis.* **5**, 607–625.

Merianos, J.J. (1991) Quaternary ammonium antimicrobial compounds. In *Disinfection, Sterilization and Preservation*, 4th ed. (ed. S.S. Block), pp. 225–255, Lea & Febiger, Malvern, Pennsylvania, USA.

Monzon, M., Oteiza, C., Leiva, J. and Amorena, B. (2001) Synergy of different antibiotic combinations in biofilms of *Staphylococcus epidermidis*. *J. Antimicrob. Chemoth.* **48**, 793–801.

Nebe von Caron, G., Stephens, P. and Badley, R.A. (1998) Assessment of bacterial viability status by flow cytometry and single cell sorting. *J. Appl. Microbiol.* **84**, 988–998.

Nebe von Caron, G., Stephens, P.J., Hewitt, C.J., Powell, J.R. and Badley, R.A. (2000) Analysis of bacterial function by multi-colour fluorescence flow cytometry and single cell sorting. *J. Microbiol. Meth.* **42**, 97–114.

Norton, D.M. and Batt, C.A. (1999) Detection of viable *Listeria monocytogenes* with a 5′ nuclease PCR assay. *Appl. Environ. Microbiol.* **65**, 2122–2127.

Norwood, D.E. and Gilmour, A. (1999) Adherence of *Listeria monocytogenes* strains to stainless steel coupons. *J. Appl. Microbiol.* **86**, 576–582.

Norwood, D.E. and Gilmour, A. (2001) The differential adherence capabilities of two *Listeria monocytogenes* strains in monoculture and multispecies biofilms as a function of temperature. *Lett. Appl. Microbiol.* **33**, 320–324.

Oie, S., Huang, Y., Kamiya, A., Konishi, H. and Nakazawa, T. (1996) Efficacy of disinfectants against biofilm cells of methicillin-resistant *Staphylococcus aureus*. *Microbios* **85**, 223–230.

Otto, M. (2001) *Staphylococcus aureus* and *Staphylococcus epidermidis* peptide pheromones produced by the accessory gene regulator *agr* system. *Peptides* **22**, 1603–1608.

Payne, D.N., Babb, J.R. and Bradley, C.R. (1999) An evaluation of the suitability of the European suspension test to reflect *in vitro* activity of antiseptics against clinically significant organisms. *Lett. Appl. Microbiol.* **28**, 7–12.

Pitts, B., Willse, A., McFeters, G.A., Hamilton, M.A., Zelver, N. and Stewart, P.S. (2001) A repeatable laboratory method for testing the efficacy of biocides against toilet bowl biofilms. *J. Appl. Microbiol.* **91**, 110–117.

Pol, I.E., Mastwujk, H.C., Slump, R.A., Popa, M.E. and Smid, E.J. (2001) Influence of food matrix on inactivation of *Bacillus cereus* by combinations of nisin, pulsed electric field treatment, and carvacrol. *J. Food Prot.* **64**, 1012–1018.

Pratten, J., Foster, S.J., Chan, P.F., Wilson, M. and Nair, S.P. (2001) *Staphylococcus aureus* accessory regulators: expression within biofilms and effect on adhesion. *Microb. Infect.* **3**, 633–637.

Rachid, S., Ohlsen, K., Wallner, U., Hacker, J., Hecker, M. and Ziebuhr, W. (2000) Alternative transcription factor sigma(B) is involved in regulation of biofilm expression in a *Staphylococcus aureus* mucosal isolate. *J. Bacteriol.* **182**, 6824–6826.

Ramage, G., Vande Walle, K., Wickes, B.L. and Lopez-Ribot, J.L. (2001) Standardized method for *in vitro* antifungal susceptibility testing of *Candida albicans* biofilms. *Antimicrob. Agents Chemoth.* **45**, 2475–2479.

Reybrouck, G. (1998) The testing of disinfectants. *Int. Biodeterior. Biodegrad.* **41**, 269–272.

Russell, A.D., Hugo, W.B. and Ayliffe, G.A.J. (1992) Disinfection mechanisms. In *Principles and Practice of Disinfection, Preservation and Sterilisation*, pp. 187–210, Blackwell scientific publications, Oxford, UK.

Sasahara, K.C. and Zottola, E.A. (1993) Biofilm formation by *Listeria monocytogenes* utilizes a primary colonizing microorganism in flowing systems. *J. Food Protect.* **56**, 1022–1028.

Seeliger, H.P.R. and Jones, D. (1986) *Listeria*. In *Bergey's Manual of Systematic Bacteriology* (ed. J.G. Holt, N.R. Krieg, D.H. Bergey, P.H.A. Sneath, N.S. Mair, M.E. Sharpe, J.T. Staley and S.T. Williams), pp. 1235–1242, Williams, Baltimore, USA.

Servais, P., Agogue, H., Courties, C., Joux, F. and Lebaron, P. (2001) Are the actively respiring cells (CTC +) those responsible for bacterial production in aquatic environments? *FEMS Microbiol. Ecol.* **35**, 171–179.

Sleator, R.D., Gahan, C.G., Abee, T. and Hill, C. (1999) Identification and disruption of BetL, a secondary glycine betaine transport system linked to the salt tolerance of *Listeria monocytogenes* LO28. *Appl. Environ. Microbiol.* **65**, 2078–2083.

Smith, J.J. and McFeters, G.A. (1996) Effects of substrates and phosphate on INT (2-(4-iodophenyl)-3-(4-nitrophenyl)-5-phenyl tetrazolium chloride) and CTC (5-cyano-2,3-ditolyl tetrazolium chloride) reduction in *Escherichia coli. J. Appl. Bacteriol.* **80**, 209–215.

Smith, J.J. and McFeters, G.A. (1997) Mechanisms of INT (2-(4-iodophenyl)-3-(4-nitrophenyl)-5-phenyl tetrazolium chloride), and CTC (5-cyano-2,3-ditolyl tetrazolium chloride) reduction in *Escherichia coli* K-12. *J. Microbiol. Meth.* **29**, 161–175.

Spurlock, A.T. and Zottola, E.A. (1991) Growth and attachment of *Listeria monocytogenes* to cast iron. *J. Food Protect.* **54**, 925–929.

Sternberg, C., Christensen, B.B., Johansen, T., Nielsen, A.T., Andersen, J.B., Givskov, M. and Molin, S. (1999) Distribution of bacterial growth activity in flow-chamber biofilms. *Appl. Environ. Microbiol.* **65**, 4108–4117.

Taylor, C.M., Beresford, M., Epton, H.A., Sigee, D.C., Shama, G., Andrew, P.W. and Roberts, I.S. (2002) *Listeria monocytogenes relA* and *hpt* mutants are impaired in surface-attached growth and virulence. *J. Bacteriol.* **184**, 621–628.

Tiller, J.C., Liao, C.J., Lewis, K. and Klibanov, A.M. (2001) Designing surfaces that kill bacteria on contact. *Proc. Natl. Acad. Sci. USA* **98**, 5981–5985.

Todd, E.C.D. (1992) Foodborne disease in Canada – a 10-year summary from 1975 to 1984. *J. Food Protect.* **55**, 123–132.

Tremoulet, F., Duche, O., Namane, A., Martinie, B., The European Listeria Genome Consortium and Labadie, J.C. (2002) Comparison of protein patterns of *Listeria monocytogenes* grown in biofilm or in planktonic mode by proteomic analysis. *FEMS Microbiol. Lett.* **210**, 25–31.

Ueckert, J.E., Nebe von-Caron, G., Bos, A.P. and ter Steeg, P.F. (1997) Flow cytometric analysis of *Lactobacillus plantarum* to monitor lag times, cell division and injury. *Lett. Appl. Microbiol.* **25**, 295–299.

Vatanyoopaisarn, S., Nazli, A., Dodd, C.E., Rees, C.E. and Waites, W.M. (2000) Effect of flagella on initial attachment of *Listeria monocytogenes* to stainless steel. *Appl. Environ. Microbiol.* **66**, 860–863.

Verheul, A., Russell, N.J., van 't Hof, R., Rombouts, F.M. and Abee, T. (1997) Modifications of membrane phospholipid composition in nisin-resistant *Listeria monocytogenes* Scott A. *Appl. Environ. Microbiol.* **63**, 3451–3457.

Votyakova, T.V., Kaprelyants, A.S. and Kell, D.B. (1994) Influence of viable cells on the resuscitation of dormant cells in *Micrococcus luteus* cultures held in an extended stationary phase: the population effect. *Appl. Environ. Microbiol.* **60**, 3284–3291.

Vroom, J.M., Grauw, K.J.D., Gerritsen, H.C., Bradshaw, D.J., Marsh, P.D., Watson, G.K., Birmingham, J.J. and Allison, C. (1999) Depth penetration and detection of pH gradients in biofilms by two-photon excitation microscopy. *Appl. Environ. Microbiol.* **65**, 3502–3511.

Vuong, C., Saenz, H.L., Gotz, F. and Otto, M. (2000) Impact of the *agr* quorum-sensing system on adherence to polystyrene in *Staphylococcus aureus. J. Infect. Dis.* **182**, 1688–1693.

Watson, S.P., Clements, M.O. and Foster, S.J. (1998) Characterization of the starvation-survival response of *Staphylococcus aureus. J. Bacteriol.* **180**, 1750–1758.

Wentland, E.J., Stewart, P.S., Huang, C.T. and McFeters, G.A. (1996) Spatial variations in growth rate within *Klebsiella pneumoniae* colonies and biofilm. *Biotechnol. Prog.* **12**, 316–321.

Whiteley, M., Bangera, M.G., Bumgarner, R.E., Parsek, M.R., Teitzel, G.M., Lory, S. and Greenberg, E.P. (2001) Gene expression in *Pseudomonas aeruginosa* biofilms. *Nature* **413**, 860–864.

Wiedmann, M., Arvik, T., Hurley, R.J. and Boor, K. (1998) General stress transcription factor sigmaB and its role in acid tolerance and virulence of *Listeria monocytogenes. J. Bacteriol.* **180**, 3650–3656.

Willcock, L., Gilbert, P., Holah, J., Wirtanen, G. and Allison, D.G. (2000) A new technique for the performance evaluation of clean-in-place disinfection of biofilms. *J. Ind. Microbiol. Biotechnol.* **25**, 235–241.

Williams, I., Venables, W.A., Lloyd, D., Paul, F. and Critchley, I. (1997) The effects of adherence to silicone surfaces on antibiotic susceptibility in *Staphylococcus aureus*. *Microbiology* **143**, 2407–2413.

Williams, I., Paul, F., Lloyd, D., Jepras, R., Critchley, I., Newman, M., Warrack, J., Giokarini, T., Hayes, A.J., Randerson, P.F. and Venables, W.A. (1999) Flow cytometry and other techniques show that *Staphylococcus aureus* undergoes significant physiological changes in the early stages of surface-attached culture. *Microbiology* **145**, 1325–1333.

Wirtanen, G. (1995) Biofilm formation and its elimination from food processing equipment, Ph.D. thesis, VTT Biotechnology and Food Research, Microbiology, Helsinki University of Technology.

Wright, T.L., Ellen, R.P., Lacroix, J.M., Sinnadurai, S. and Mittelman, M.W. (1997) Effects of metronidazole on *Porphyromonas gingivalis* biofilms. *J. Periodont. Res.* **32**, 473–477.

Yu, F.P. and McFeters, A. (1994) Physiological responses of bacteria in biofilms to disinfection. *Appl. Environ. Microbiol.* **60**, 2462–2466.

Zaura-Arite, E., van Marle, J. and ten Cate, J.M. (2001) Confocal microscopy study of undisturbed and chlorhexidine-treated dental biofilm. *J. Dent. Res.* **80**, 1436–1440.

Zelver, N., Hamilton, M., Goeres, D. and Heersink, J. (2001) Development of a standardized antibiofilm test. *Meth. Enzymol.* **337**, 363–376.

27

Device-associated infection: the biofilm-related problem in health care

M. Cormican

27.1 INTRODUCTION

The use of indwelling biomedical devices is central to modern health care delivery and its importance is increasing. Biomedical devices range from the familiar peripheral venous catheter (PVC) and urinary catheter to more complex devices, such as prosthetic joints and heart valves. A limiting factor in the application of implantable biomedical devices is the associated risk of infection. For many years, it has been apparent from clinical experience that eradication of infection associated with a biomedical device using antimicrobial agents may be very difficult. In many cases, removal of the device is essential to achieve cure. It is now appreciated that difficulty in eradication of infection associated with a biomedical device is related to the formation of microbial biofilm on the surface of the device. Removal of the device, and thereby removal of the biofilm, may be essential for eradication of infection. Removal of a biomedical device may be associated with very significant problems including pain and discomfort, less satisfactory outcome for the patient, technical difficulties for the health care provider and very significant costs for the health service. Therefore, prevention and management of device-associated infection is an important area of health care research with great potential for collaboration with biofilm researchers from other disciplines.

27.2   BIOMEDICAL DEVICES AND ROUTES OF INFECTION

Biomedical devices may be considered under two headings: (a) those, which are totally implanted within the body and (b) those intended to provide access for either drainage or administration of blood or other fluids.

27.2.1   Totally implanted biomedical devices

Prosthetic joints (hips, knees and others) are the best-known examples of totally implanted devices intended to enhance or replace the function of a diseased tissue over an extended period of time (years). Placement of these devices requires major surgery. However, when healing has occurred after the procedure, the prosthetic joint is totally enclosed beneath intact skin and, therefore, is at low risk of exposure to microorganisms. Infection of such a device after wound healing is uncommon but may occur by seeding of the surface with bacteria circulating in the blood. The critical control point for prevention of infection of a totally implanted device is that the surfaces of the device are sterile when the surgical wound is closed. Production of a sterile device in a secured impermeable package by the manufacturer is an essential preliminary. A subsequent and more challenging process is the placement of the device without contamination of the surfaces. Commensal bacteria of human skin are continually shed into the air from patients and the operating team. Contamination of the surface of the device with low levels of these low-virulence bacteria at time of implantation may result in subsequent establishment of biofilm infection. As the bacteria are of low virulence, the infection may not be apparent for weeks or months after surgery, but may ultimately result in joint failure and the need for further major surgery to remove and perhaps replace the infected device. A complex set of operating room practices and elaborate controls of air quality and air flow are now routinely used in joint-replacement surgery to minimise the risk of contamination of the device at the time of surgery. With these precautions the risk of orthopaedic device-related infection is now reported to be of the order of 1–2% (Widmer, 2001). Other devices to enhance or replace the function of diseased organs are also of increasing importance including cardiac pacemakers, prosthetic hearts and left-ventricular-assist devices (to support the pumping action of failing heart muscle). In the latter case, although the device is enclosed, the risk of infection is increased through the need for wires communicating with an external power source (Malani et al., 2002).

27.2.2   Devices for access

A great deal of modern health care delivery is dependent on easy access to veins or other body compartments for administration of therapy and/or removal of fluids for drainage or sampling. Administration of therapy and drainage of blood or body fluids may be accomplished by placing a hollow tube (catheter) with one end in the relevant body compartment and the other external to the body. Familiar examples include urinary catheters and PVCs most often placed in an arm vein. Catheters represent a much greater continuing risk of infection than totally implantable devices. Catheters act as a pathway for microorganisms to gain access to nutrient rich, warm body fluids that are normally shielded from microbial exploitation by the physical barriers of intact skin or in the case of the urinary bladder by anatomical valves. The pathway may be used by the normal bacterial population of the patients skin (endogenous infection) and also by bacteria introduced to the catheter or catheter entry site on hands and equipment used to manipulate or care for the catheter (exogenous infection). Microorganisms may track along the external surface of the catheter or the internal (luminal) surface and biofilm may form on either surface.

At any time, more than 10% of patients in a hospital may have an indwelling urinary catheter (Donlan and Costerton, 2002). In some cases, this is for a short period related to surgery or other procedure, but in many cases urinary catheters may be required for extended periods. Drainage of urine into a closed drainage system (a bag) reduces the risk of infection during short periods of urinary catheterisation and is now almost universal in developed countries. Even with brief urinary tract catheterisation (less than 10 days), 26% of patients develop infection of the urinary tract (Saint, 2000). Urinary catheterisation for more than 4 weeks is associated with urinary tract infection in almost all cases (Donlan and Costerton, 2002). Catheter-associated urinary tract infection may be associated with 200 μm thick biofilm on the luminal surface of the catheter and with microbial density of 10^8–10^9/cm^2 of surface area.

PVCs are also associated with a risk of infection. With care at insertion and changing of the site every 48–72 h, the risk is modest and will not be considered further in this chapter. For patients in whom intravenous access is required over an extended period of time, frequent re-siting of PVCs is uncomfortable and it may become technically very difficult as fresh sites for placement of a PVC become more difficult to identify. A catheter placed in the large central vein inside the chest (central venous catheter or CVC) is often required to permit administration of intravenous therapy, fluids or nutrition and taking of blood samples over an extended time period. CVCs that are expected to remain *in situ* for short periods (<10 days) are generally inserted in the ward or intensive care unit through puncture of the skin. CVCs required for extended periods of months are tunnelled under the skin and require surgical insertion. There are a multiplicity of catheter materials, designs, access points and insertion methods. A useful summary of these with illustrations is available at Venousaccess.com. Perhaps the most significant distinction in respect of tunnelled CVCs is between devices in which access to the catheter is through an extension of the catheter tubing external to the skin entry site and those in which the catheter is attached to a port or reservoir that is completely enclosed under the skin. In patients with a subcutaneous port, when access is required to administer therapy a needle is pushed through the overlying skin and the silicone septum of the port. A key practical distinction between short-term and long-term tunnelled devices is the relative ease of removal of infected short-term CVCs whereas removal and replacement of infected long-term tunnelled CVCs or ports is more complex.

CVC-related infection, in particular bloodstream infection (BSI), is a major problem in modern health care. Rates of catheter-related BSI (CRBSI) range from 2.1 to 30.2 per 1000 days of CVC use (Pearson, 1996). Approximately, 75% of the 250,000 (estimated) intravascular device-related infections that occur annually in the US are related to CVCs. An estimated 30,000 cases occur annually in the UK (Raad and Hanna, 2002). CRBSI may result in death in 12–25% of cases, with added health care cost of US\$33,000–US\$35,000 (Crinch and Maki, 2002a). Due to the frequency of occurrence and the serious consequences of CRBSI, the remainder of this chapter will focus on this issue.

27.3   CVC-RELATED INFECTIONS

27.3.1   Pathogenesis of CVC-related infection

The organisms most commonly associated with CVC-related infection are coagulase-negative staphylococci (CNS), in particular *Staphylococcus epidermidis* (Pearson 1996; Raad and Hanna, 2002). *S. aureus*, *Enterococcus* spp., a variety of aerobic Gram-negative bacilli (*Pseudomonas aeruginosa*, *Acinetobacter* spp. and *Stenotrophomonas maltophilia*) and yeast-like fungi (*Candida* spp.) are also important. A CVC may be contaminated with exogenous organisms through the hands of health care workers manipulating the CVC for administration of therapy or taking of blood samples.

Bacteria may establish infection associated with a catheter by passage along the external or internal (luminal) surface of the catheter. It is considered that access via the skin exit site and extension along the external site is most important for short-term non-tunnelled CVC, whereas contamination of the device hub and extension along the luminal surface are considered more important with long-term tunnelled CVCs (Crinch and Maki, 2002b). Infection of the intravascular end of a CVC during BSI from other causes is also possible but is considered much less common.

27.3.2   Prevention of CVC-related infection

As with totally implanted devices a basic element in the prevention of CRBSI is the production, packaging and delivery of sterile CVCs. Manufacturing standards are very high and production process failures are very rarely a contributing factor in CRBSI.

The properties of the catheter are an important determinant of the risk of infection. In recent years, a great deal of research has been invested in development of catheters that resist microbial colonisation and infection. There is evidence that use of CVC coated with antimicrobial agents may help in reducing the risk of CVC-related infection (CVC-RI) in those who require short-term non-tunnelled catheters (Pearson, 1996; Crinch and Maki, 2002a; Raad and Hanna, 2002). The main alternatives for antimicrobial-coated CVCs currently available are coating with chlorhexidine–silver sulphdiazine or with a combination of minocycline and rifampicin. The minocycline–rifampicin-coated catheter is associated with lower BSI rates than the original version of the silver–sulphadiazine-coated CVC (Crinch and Maki 2002a; Raad and Hanna, 2002). Lower rates of BSI with the minocycline–rifampicin-related catheter may reflect greater antimicrobial activity of the coating and/or the coating or both external and luminal surface with these agents as compared with coating of the external surface only in the earliest chlorhexidine–sulphdiazine catheter. Development of other antimicrobial catheters, including silver-impregnated catheters is on-going. A particularly innovative approach is the development of catheters to which a low-ampereage electrical current can be applied (Crinch and Maki, 2002a).

Antimicrobial coating of long-term tunnelled catheters has not been investigated as extensively. Such evidence, as is available, suggests that this strategy is not effective for long-term tunnelled catheters. This may be because of depletion of antimicrobial activity at the catheter surfaces within days or weeks of insertion. An important design feature of tunnelled catheters is the use of a dacron cuff on the segment of the catheter that is placed in the subcutaneous tunnel. Patient tissues grow into the cuff thus helping to secure the catheter in place and close of access for microorganims along the external surface of the CVC (Crinch and Maki, 2002b). The success of this design feature explains why contamination of the CVC hub and extension of infection through the lumen is considered relatively much more important in the pathogenesis of tunnelled CVCs as compared with non-tunnelled CVC-RI. Central venous access via catheters that terminate at a subcutaneous port with no extension through the skin appears to be the method of long-term central venous access with the lowest BSI rate. As access to the subcutaneous port requires puncture of the overlying skin with a steel needle, this approach is not ideal for all patients but should be considered, in particular for those in whom intermittent central venous access is required over an extended period of time (Crinch and Maki, 2002b).

Procedures for placement of CVCs (aseptic technique) are designed to minimise the risk of contamination of the surface of the CVC with microorganisms from the skin of the patient or health care worker. Many aspects of these procedures have been studied and detailed recommendations are available (Pearson *et al.*, 1996; Crinch and Maki, 2002). Notable points are recommendations that:

(1) CVC placement by a dedicated intravascular-device team is associated with reduced rates of infection,
(2) disinfection of the patients skin with chlorhexidine-based products is associated with lower infection rates than use iodine-based disinfectants, and
(3) administration of systemic antimicrobial agents at the time of CVC placement is not beneficial.

The recommendation relating to establishment of dedicated teams for placement of CVCs and related devices has not been widely implemented even in many larger hospitals in Europe at the present time.

On-going care of the CVC after placement is critical in the prevention of infection. As with CVC placement, the delivery of continuing CVC care by a dedicated team is accepted as important in preventing CRBSI. While systemic antimicrobial prophylaxis has no role in prevention of CRBSI, the instillation of antimicrobial solution into the CVC lumen at times when therapy is not being administered through the CVC (antimicrobial lock) is of value in reducing the incidence of CRBSI (Raad and Hanna, 2002).

27.3.3   Clinical manifestations of CVC-related infection

CVC-related infection may manifest as a localised infection at the site of placement of the device. Local infection may manifest as exit-site infection (confined to the skin puncture site) or as tunnel/pocket infection when infection has extended along the CVC or the port in its subcutaneous pocket. Localised infection is generally easily recognised by pain, tenderness, inflammation and perhaps pus at the CVC site. In such cases, the association of the CVC with infection is readily apparent and removal of the CVC is generally necessary for tunnel or pocket infection.

Much more commonly, a patient has a raised temperature and/or other signs of BSI but in the absence of any manifestations specifically indicating that the CVC is the focus of infection. The patient may have other possible sources of sepsis (e.g. urinary tract or gastrointestinal tract). If the episode of infection is not CVC related, removal of the CVC may represent an unnecessary intervention with significant-associated risks and costs. On the other hand, if the infection is CVC related the patients condition may fail to improve until the CVC is removed.

27.3.4   Laboratory diagnosis of CVC-related infection

The laboratory may confirm the role of a removed CVC in an episode of BSI. For research purposes, confirmation of CRBSI usually requires (a) culture of a microorganism from blood taken from a peripheral vein and (b) culture of 'the same' microorganism in 'significant' quantity from the removed CVC (Dobbins et al., 1999). The application of this definition is less straightforward than it may seem. Culture of bacteria from peripheral blood and from the CVC that are indistinguishable on routine identification and antimicrobial susceptibility testing is generally taken as indicating that the organisms are 'the same'. However, molecular studies have shown that in 24% of cases, such pairs of CNS were distinguishable. There is also considerable uncertainty as to what represents 'significant' growth from the removed CVC, further complicated by differences in methodology for culture of the removed CVC.

Outside of the research, setting interpretation may be more difficult as quantitative or semi-quantitative culture of the removed CVC may not be performed in all cases. All culture-based methods for confirmation of CRBSI involve a delay of

1 or more days in making a diagnosis. Light microscopic and electron microscopic methods for direct visualisation of biofilm on removed CVCs have been investigated for rapid diagnosis of CRBSI, however, there are practical limitations to the methods and they are not widely used.

Confirmation that a CVC already removed was or was not a source of sepsis may be viewed as closing the stable door after the horse has bolted. Up to 75% of CVCs, removed because of a suspected role in CRBSI, are removed unnecessarily (Dobbins *et al.*, 1999). Confirmation or exclusion of CRBSI without removal of the CVC is a very desirable goal. One approach to confirmation of CRBSI without CVC removal is based on comparison of the concentration of microorganims in blood removed from a peripheral vein with the concentration in blood removed through the CVC. This is based on the assumption that blood taken through the lumen of a CVC in which there is endoluminal infection is expected to have a high concentration of microorganisms released from the infected lumen. This approach requires access to quantitative blood culture, which is not widely available in clinical laboratories for reasons of cost and time. Other problems with this approach include practical difficulties with obtaining cultures from a peripheral vein or from the CVC in some cases. Sampling of the luminal surface of the *in situ* CVC by insertion of a brush is another approach, which may be valuable. However, it has not been widely adopted at the present time.

Most clinical microbiology laboratories are limited in their ability to evaluate patients with suspected CRBSI without CVC removal. Non-quantitative blood culture from the CVC and from a peripheral vein is the most widely available diagnostic approach. Removal of CVCs based largely on clinical suspicion of CRBSI and/or result of non-quantitive cultures remains the norms in most hospitals in Europe.

27.3.5   Management of CVC-related infection

When BSI is suspected in a vulnerable patient, it is frequently necessary to commence antimicrobial therapy before laboratory results are available. Every effort should be made to collect appropriate specimens including blood cultures before administration of antimicrobials. When CRBSI is strongly suspected or confirmed, vancomycin is the drug of choice because of the importance of Gram-positive organisms (particularly, staphylococci), that are frequently resistant to other antimicrobial agents. Linezolid is a more recently introduced agent that also has activity against most Gram-positive agents and may be an alternative. Due to the wide range of possible pathogens (see above), in critically ill patients, vancomycin is used in association with other agents with activity against Gram-negative bacteria until laboratory results have identified the pathogen in the specific case. The antimicrobial regimen chose empirically in the first instance should be modified as laboratory results confirm the identity and antimicrobial susceptibility of the infecting organism.

The most difficult issue in dealing with suspected CRBSI is the decision to attempt therapy with the CVC *in situ* or to remove the CVC. The decision is made more difficult because the limitations of diagnostic methods may result in uncertainty as to the diagnosis. Raad and Hanna (2002) have proposed that the decision be supported by categorisation of the episode of infection as low, medium or high risk based an evaluation of the severity of illness and the pathogen isolated. Uncomplicated CRBSI associated with CNS represents a low-risk CRBSI and may respond to antimicrobial therapy without CVC removal in up to 80% of cases. In cases categorised as moderate risk, non-tunnelled catheters should be removed. It may be reasonable to attempt salvage of long-term tunnelled catheters through use of antimicrobial lock therapy. In high-risk CRBSI, the CVC should be removed.

When systemic antimicrobial agents are administered through an infected CVC the endoluminal biofilm is exposed to the antimicrobial agent as it passes through the CVC. However, the antimicrobial agent may be relatively dilute and the exposure may be transient if the agent is flushed out of the CVC by subsequent administration of other therapeutic agents. Antimicrobial lock therapy is intended to ensure sustained exposure of the endoluminal biofilm to high concentrations of antimicrobial agent. A volume of antimicrobial agent at sufficient to fill the internal lumen is instilled into the CVC. The device is then closed at the hub so that antimicrobial solution is not flushed out of the CVC into the vein. As the antimicrobial agent is confined within the device lumen, the concentration of antimicrobial agent that may be applied in this setting may be 50- to a 100-fold higher than concentrations that can be achieved without toxic effects in blood and tissues. The high concentration and extended exposure to antimicrobial agent may overcome the relative antimicrobial resistance that characterises biofilm. Details of proposed regimens for antimicrobial lock therapy with a variety of established antimicrobial agents have recently been reviewed (Berrington and Gould, 2002). Recent studies suggest that the new agent linezolid may also have a role in antimicrobial lock therapy (Gander et al., 2002).

27.4   CONCLUSIONS

Biomedical device-related infection has long been recognised as a key limiting factor in the application of biomedical devices in health care. Additional progress in the prevention and eradication of microbial biofilm formation on biomedical devices is essential to release the full potential of such devices to enhance human health.

REFERENCES

Berrington, A. and Gould, F.K. (2002) Use of antibiotic locks to treat colonised central venous catheters. *J. Antimicrob. Chemother.* **48**, 597–603.
Crinch, C.J. and Maki, D.G. (2002a) The promise of novel technology for the prevention of intravascular device-related bloodstream infection. I. Pathogenesis and short-term devices. *Clin. Infect. Dis.* **34**, 1232–1242.

Crinch, C.J. and Maki, D.G. (2002b) The promise of novel technology for the prevention of intravascular device-related bloodstream infection. II. Long-term devices. *Clin. Infect. Dis.* **34**, 1362–1368.

Dobbins, B.M., Kite, P. and Wilcox, M.H. (1999) Diagnosis of central venous catheter related sepsis-a critical look inside. *J. Clin. Pathol.* **52**, 165–172.

Donlan, R.M. and Costerton, J.W. (2002) Biofilms: survival mechanisms of clinically relevant microorganims. *Clin. Microbiol. Rev.* **15**, 167–193.

Gander, S., Hayward, K. and Finch, R. (2002) An investigation of the antimicrobial effects of linezolid on bacterial biofilms utilising an *in vitro* pharmacokinetic model. *J. Antimicrob. Chemother.* **49**, 301–308.

Malani, P.N., Dyke, D.B.S., Pagani, F.D. and Chenoweth, C.E. (2002) Nosocomial infections in left ventricular assist device recipients. *Clin. Infect. Dis.* **34**, 1295–1300.

Pearson, M.L. (1996) The Hospital Infection Control Practices Advisory Committee. Guideline for prevention of intravascular device-related infections. *Am. J. Infect. Control* **24**, 262–293.

Raad, I.I. and Hanna, H.A. (2002) Intravascular catheter-related infections, new horizons and recent advances. *Arch. Intern. Med.* **162**, 871–878.

Saint, S. (2000) Clinical and economic consequences of nosocomial catheter-related bacteriuria. *Am. J. Infect.Control* **28**, 68–75.

Widmer, A.F. (2001) New developments in diagnosis and treatment of infection in orthopedic implants. *Clin. Infect. Dis.* **33**(Suppl. 2), S94–106.

28

Bacterial resistance to biocides: current knowledge and future problems

A.D. Russell

28.1 INTRODUCTION

Microorganisms and microbial entities (viruses, prions) differ greatly in their response to biocides, a collective term for antiseptics, disinfectants and preservatives (Table 28.1; Russell, 1999). Prions are the least susceptible, followed by bacterial spores and some protozoa, notably *Cryptosporidium*. Of the viruses, lipid enveloped forms are inactivated much more readily than non-lipid enveloped ones.

Wide variations exist between different bacteria. As a general classification, many biocides are not sporicidal although they may possess sporistatic activity. Germinating and outgrowing cells together with the vegetative cells produced are considerably more sensitive to biocides than are dormant spores (Russell, 1990). Mycobacteria are more resistant than other non-sporulating bacteria (Russell, 1996). Gram-negative bacteria, such as *Pseudomonas aeruginosa* and *Proteus* spp. are less susceptible than staphylococci, but enterococci may present an above-average resistance (Russell and Chopra, 1996). Thus, it is clear that not only the type of organism but also the degree of differentiation within a 'life cycle' can influence the efficacy of a biocidal agent. Other factors, some of which

Table 28.1   Susceptibility of microorganisms to biocides.

Microorganism/entity	Comment
Prions	Most resistant
Bacterial spores, coccidian	Highly resistant to many biocides
Mycobacteria	Intermediate resistance
Non-enveloped viruses	Small, non-lipid viruses, e.g. poliovirus, resistant to many biocides
Yeasts and moulds	Important spoilage organisms
Gram-negative bacteria	*Pseudomonas aeruginosa*, *Proteus* spp., *Providencia* spp. have above-average resistance
Gram-positive cocci	Enterococci less susceptible than staphylococci
Lipid enveloped viruses	Most susceptible, e.g. HIV, HBV

Table 28.2   Factors influencing biocidal activity.

Factor	Comment[a]
Concentration[b]	Key factor affecting activity, action and resistance
Temperature	Activity enhanced at elevated temperatures
pH	(1) pH rise, activity enhanced: CHX, QACs
	(2) pH fall, activity enhanced: organic acids, CRAs
Organic matter	Activity reduced, e.g. CHX, QACs, CRAs
Other interfering material	Containers, closures, non-ionic surfactants
Presence of biofilm	Activity reduced

[a] CHX, chlorhexidine salts; QACs, quaternary ammonium compounds; CRAs, chlorine-releasing agents.
[b] Particularly important in inactivation of bacteria within biofilm communities.

are outside the scope of this paper, that can also affect biocidal activity include pH, temperature, biofilm formation, the presence or absence of organic or other interfering material, the biocide formulation and, above all, the biocide concentration (Table 28.2; McDonnell and Russell, 1999). Concentration gradients are of particular relevance to inactivation of bacteria within biofilms.

Microbial, and in particular bacterial, responses to biocides are now being extensively studied. There are four major reasons for the current level of interest:

(1) the possibility that bacterial resistance to biocides is increasing;
(2) even more importantly, concern that biocides might be selecting for antibiotic-resistant bacteria in hospital, industrial and domiciliary environments;
(3) the role played by biofilms in nature and disease in reducing the efficacy of biocides and, where relevant, of antibiotics; and
(4) emerging infectious, including bacterial, diseases and the possibility that the organisms responsible may be refractory to biocides and, where relevant, to antibiotics.

These aspects form the basis of this chapter. It is important at the outset to define the ways in which 'resistance' and 'resistant' are employed. They are used here to denote a strain that is either insusceptible to a concentration of biocide used in practice or is not inactivated (or sometimes not inhibited) by a biocide concentration that inactivates (or inhibits) the majority of strains of that organism. The relevance of minimum inhibitory concentrations (MICs) to denote biocidal activity must be considered. An MIC might be of value in the context of the pre-servative concentration used in a pharmaceutical, cosmetic or other formulation, but has little if any relevance in the context of a biocide used as an antiseptic or disinfectant.

Other terms used to express bacterial insusceptibility to biocides are 'tolerance' and 'reduced susceptibility'. The first term has been used to describe a situation in which a formerly effective preservative system no longer controlled bacterial growth. Unfortunately, it is also a term used to describe one particular mechanism of bacterial resistance to penicillin.

28.2   BACTERIAL RESISTANCE TO BIOCIDES

In recent years, there has been considerable interest in the responses of bacteria (and, to a lesser extent, of other microorganisms) to biocides. The outcome has been that new information has been obtained about the manner in which biocides act and the ways in which bacteria show or acquire insusceptibility. Additional data are, however, required on both counts and in particular on whether there are similarities with antibiotics. Bacteria present within biofilms both *in vivo* and out-side the body constitute an important area of study. It is also necessary to evaluate the underlying mechanisms of bacterial resistance to biocides, since all or most of these mechanisms may contribute to the refractory nature of bacteria embedded within biofilms.

The general classification of bacterial responses to biocides outlined in the preceding section (Table 28.1) actually masks wide divergences in biocidal sus-ceptibility. For example, (i) spores of *Bacillus subtilis* are less susceptible than those of *Clostridium difficile*; (ii) *Mycobacterium chelonae* strains may be highly resistant to glutaraldehyde, but not to *ortho*-phthalaldehyde (OPA); (iii) *M. avium intracellulare* is generally more resistant than *M. tuberculosis*; and (iv) antibiotic-resistant strains of staphylococci (including methicillin-resistant *Staphylococcus aureus*, MRSA) may present low-level resistance to biocides, in particular cationic types (Russell, 1999, 2002a).

Bacterial resistance mechanisms fall into two broad groups (Table 28.3; Russell, 1998).

In the first, intrinsic resistance (intrinsic insusceptibility) is a natural property of an organism, whereas in the second, acquired resistance results from genetic changes in a bacterial cell. Such a general concept has proved to be of value. It is now, however, time to re-evaluate whether a more realistic approach can be employed for examining the various resistance mechanisms.

Table 28.3 Bacterial resistance mechanisms: general aspects.

Type	Nature	Example(s)	Specific reasons for biocide resistance[a]
Intrinsic	Impermeability	Staphylococci	Fattened cells, cells with thickened cell walls
		Gram-negatives	OM permeability barrier
		Mycobacteria	Mycoylarabinogalactan
		Spores	Spore coat(s); cortex also?
	Efflux	MDR Gram negatives	proton-motive force (PMF)-mediated efflux pumps
	Phenotypic adaptation	Gram-positives and Gram-negatives	Biofilm formation
	Enzymatic	Gram-positives and Gram-negatives	Biocide degradation[b]
	Other	Bacterial spores	Presence of DNA-protecting SASPs
Acquired	Mutation	Gram-positives and Gram-negatives	Single mutation: low-level resistance to biocides
		Gram-positives and Gram-negatives	Modified enoyl reductase conferring resistance to triclosan
	Efflux	Staphylococci	Plasmid-associated *qac* genes conferring low-level resistance to cationic biocides
	Enzymatic	Some Gram-negatives	Biocide degradation[b]

[a] OM, outer membrane; MDR, multidrug-resistant.
[b] Of potential importance as a mechanism of resistance of biofilm communities as opposed to planktonic cultures.

The original classification may no longer serve to provide the most appropriate means of considering the various mechanisms. Efflux pumps, for instance, may be responsible for both an intrinsic and an acquired mechanism of bacterial resistance to biocides and to antibiotics (Table 28.3). Accordingly, individual mechanisms of resistance are discussed below.

28.2.1 Cellular impermeability

Impermeability is an innate (natural) mechanism of bacterial insusceptibility to biocides (Stickler and King, 1999). A biocide has to reach its appropriate target site(s) within (usually) a cell and the outer cell components may limit the uptake of biocide molecules. The reasons for this, however, vary from one type of cell to another.

A 'typical' bacterial spore (Foster, 1994; Setlow, 1994) consists of a central core (protoplast, germ cell) and germ cell wall surrounded by a cortex, external to which are one or two spore coats. Whereas the vegetative cell wall consists of linear glycan chains held together by cross-linked peptides, the endospore cell wall contains two layers of peptidoglycan, the inner of which is the primordial wall and the outer the cortex, about 50% of which is present as a spore-specific muramic acid lactam. During sporulation, two membranes, the inner and outer forespore membranes, surround the forespore. Some spores contain an outer exosporium. The outer spore layers, predominantly the spore coats and to some extent, possibly, the cortex, act as an impermeability barrier to the intracellular penetration of many biocides. This barrier is believed to be mainly responsible for the lack of sporicidal activity of chlorhexidine salts and QACs (Russell, 1990; Russell and Chopra, 1996). Degradative changes occur during spore germination and biosynthetic processes during spore outgrowth (Paidhungat and Setlow, 2002). Biocide uptake increases during these phases and presumably this is associated with an increase in permeability. During germination, the cells show an increased sensitivity to phenols and parabens and during outgrowth to chlorhexidine and QACs (Russell, 1990). The reasons for the selectivity of action of these biocides on a particular phase are unknown. Variations exist in spore susceptibility to biocides: *C. sporogenes* may be less sensitive than *B. subtilis*, whereas *C. difficile* spores tend to be more susceptible. The reasons for this have not been elucidated, but it is conceivable that differences in permeability arising from different spore coat composition could play a role (Russell and Chopra, 1996).

Of non-sporulating bacteria, mycobacteria are the most resistant to biocides (McDonnell and Russell, 1999). The impermeability of the complex mycobacterial cell wall, in particular its mycolic acid and arabinogalactan components, is a major factor. The uptake into mycobacteria of biocides, such as chlorhexidine and QACs is, therefore, of a low order (Russell, 1996). These biocides may, however, be mycobacteriostatic at concentrations that are similar to those inhibiting the growth of staphylococci (Broadley *et al.*, 1995). Thus, a sufficient concentration of these biocides must traverse the cell wall of mycobacteria to have an effect on, presumably, the cytoplasmic membrane. In recent years, *M. chelonae* strains with high resistance to glutaraldehyde have been isolated from endoscope washers.

This high level of resistance is maintained *in vitro*. It has been proposed that the mechanism of this resistance is associated with altered cell wall polysaccharide. These *M. chelonae* isolates are, however, inactivated by the aromatic dialdehyde, OPA, which is lipophilic and which is thus believed to readily penetrate mycobacterial cells (Simons *et al.*, 2000; Fraud *et al.*, 2001; Walsh *et al.*, 2001).

Other Gram-positive non-sporulating bacteria and Gram-negative organisms show marked differences in susceptibility to biocides, such as QACs, diamidines and hexachlorophane (hexachlorophene). Particularly resistant Gram-negatives are *Ps. aeruginosa*, *Proteus* spp., *Providencia stuartii* and *Burkholderia cepacia*. *Ps. aeruginosa* tolerates high levels of QACs and is also more resistant to chlorhexidine and triclosan than many other types of Gram-negative bacteria. The OM of cells of this organism has a high Mg^{2+} content that aids in producing strong lipopolysaccharide (LPS)-LPS links (Stickler and King, 1999). By contrast, *P. aerugunosa* is very susceptible to the chelating agent ethylenediaminetetraacetic acid (EDTA). EDTA sequesters Mg^{2+} thereby increasing the permeability of its OM to a range of chemically unrelated biocides and antibiotics (Ayres *et al.*,

1998a). In *Proteus* spp., resistance to cationic biocides may be caused by a less acidic type of LPS, so that biocide binding is reduced (Russell and Chopra, 1996). Repeated exposure of *P. aeruginosa* strains to QACs results in changes to the OM (Loughlin, Jones and Lambert, 2002) and to inner membrane fatty acids (Guerin-Mechin *et al.*, 1999; Loughlin *et al.*, 2002).

28.2.2   Efflux

Bacterial efflux is now considered to be a major mechanism for antibiotic resistance (Poole, 2000, 2002; Levy, 2002a, b). Several different types of efflux pumps are known. The best characterised efflux pump is P-glycoprotein, encoded by a human or rodent gene. Resistance to many cytotoxic drugs is mediated by ATP-dependent exporters. P-glycoprotein is a member of the ABC (ATP-binding cassette) superfamily of transporters. However, other multidrug efflux systems are secondary transporters driven by the PMF. These multidrug efflux systems are widespread among Gram-positive and -negative bacteria and belong to distinct families of proteins. These are (a) SMR, small multidrug resistance, family; (b) MFS, major facilitator superfamily; (c) RND, resistance/nodulation /cell division family; and (d) the MATE (multidrug and toxic compound extrusion) family.

The RND family exporters are unique to Gram-negative bacteria (Poole, 2000, 2002) and work in conjunction with a periplasmic membrane-fusion protein (MFP) and an OM protein. In *Ps. aeruginosa*, different types of regulatory genes (*mexR*, *nfxB*, *mexT* and *mexZ*) are involved. Antibacterial agents effluxed depend on the actual regulatory gene system, but can include β-lactams, erythromycin, tetracyclines, chloramphenicol and some biocides (Poole, 2002).

In staphylococci, various genes confer low-level resistance to cationic biocides and dyes (Table 28.4; Paulsen *et al.*, 1996).

These genes are mainly (i) *qacA* and the closely related *qacB*, the corresponding gene products (qacA and qacB) being members of the MFS family, (ii) *smr* (formerly *qacC* and *qacD* or *ebr*), where the gene product, Smr, belongs to the SMR family. The SMR family also includes the *qacG*-encoded QacG protein and the *qacH* gene product, qacG (Heir *et al.*, 1998, 1999). The *qacF* gene was

Table 28.4   *qac* genes and susceptibility of staphylococci to cationic biocides.[a]

Gene	Gene product (protein)	Protein family	Low-level resistance to
qacA	QacA	MFS	Acr, CV, QACs, Am, CHX, EB (higher than Smr)
qacB	QacB	MFS	Acr, CV, QACs, Am, EB (higher than Smr)
smr	Smr	SMR	EB (lower than QacA/QacB)
qacG	QacG	SMR	EB, QACs
qacH	QacH	SMR	EB (high level), QACs, Acr

[a] Acr, acridines; CV, crystal violet; QACs; Am, amidines; CHX, chlorhexidine salts; EB, ethidium bromide.

originally detected on a *Klebsiella aerogenes* plasmid and *qacEΔ1* is a defective version of *qacE* (Sundheim *et al.*, 1998). Rosser and Young (1999) have suggested that the carriage of *qacE* on the 3′-conserved region of an integron could result in the selection by QACs of integron-associated antibiotic resistance genes.

The overexpression of *marA*, *soxS* or *acrAB* in laboratory or clinical strains of *Escherichia coli* reduces their susceptibility to triclosan (McMurry *et al.*, 1998a; Levy, 2002b). In *Ps. aeruginosa*, resistance to triclosan is expressed via efflux pumps (Chuanchen *et al.*, 2001).

28.2.3   Biocide degradation

Enzymatic breakdown of antibiotics is a mechanism whereby bacteria can resist the action of important chemotherapeutic drugs, such as β-lactams, aminoglycosides-aminocyclitols, macrolides and chloramphenicol. By contrast, biocide molecules are generally more recalcitrant to bacterial enzymes (Beveridge, 1999). There have been claims of breakdown of QACs, but this only occurs at low concentrations and not at in-use levels. Inactivation of chlorhexidine has been observed but does not appear to have been confirmed to date. Utilisation of parabens (esters of *para*(4)-hydroxybenzoates) by *Ps. aeruginosa* in a growth medium devoid of any other carbon source has been demonstrated (Hugo, 1991, 1999). Plasmid-mediated degradation of formaldehyde by resistant strains of *Serratia marcescens* has been noted (Candal and Eagon, 1984), but again it is to be wondered whether this is likely to occur at in-use concentrations. Recent studies purport to demonstrate that some bacteria show resistance to triclosan by virtue of enzymatic inactivation (Meade *et al.*, 2001).

The most significant relationship between enzyme activity and bacterial resistance is shown in plasmid-mediated resistance to inorganic *E. coli* and organic mercury compounds. 'Narrow-spectrum' plasmids encode an enzyme, mercuric reductase, that converts an inorganic mercury salt to mercury metal which is vaporised. 'Broad-spectrum' plasmids specify resistance to organic (and inorganic) mercury compounds by encoding an enzyme, hydrolasase, that releases Hg^{2+} from organomercurials. This is then acted upon by reductase as before (Foster, 1983). Inorganic mercury compounds are of little significance as antimicrobial agents, whereas organomercurials, such as phenylmercuric nitrate (or acetate) and thiomersal are sometimes still employed as preservatives, although to a much smaller extent than hitherto.

In general, therefore, bacterial enzyme-induced degradation of biocides at concentrations used in practice is unlikely to be a major resistance mechanism.

28.2.4   Mutation and adaptation

Repeated exposure of bacterial cells to an antibiotic may result in the development of bacterial resistance. Such an effect has been known for many years (Levy, 2002a). It was also demonstrated over 50 years ago that cells could develop resistance in a similar manner to QACs (Chaplin, 1951, 1952), although it has never been clear how very high concentrations (10,000 ppm, 10000 mg/l) were dissolved in a liquid nutrient medium. However, subsequent studies have amply confirmed the ability of bacteria to acquire resistance to a variety of biocidal agents by repeated exposure to them (Russell and Chopra, 1996).

The resistance so developed may be low-level or high-level. It may be stable upon removal of the biocide or unstable, i.e. reversion to susceptibility, or the adaptive increase in resistance is lost when the biocide is removed. Cross-resistance to chemically-related compounds is usually shown, but there have been instances of decreased susceptibility to chemically unrelated compounds also, both biocides and antibiotics (Russell *et al.*, 1998; Tattawasart *et al.*, 1999a, b, 2000a, b; Lear *et al.*, 2002). One mechanism for this is believed to be a non-specific alteration in the OM of Gram-negative bacteria. It is not possible to predict the outcome, because some strains of a species rapidly develop stable resistance to a biocide whereas others do not. This appears to depend on (i) the type of organism, and (ii) the nature of the biocide in question.

Biocides are believed to have multiple actions, depending on their concentration, on bacterial cells. Single mutations would, therefore, not be expected to lead to large increases in biocide resistance. However, a reduced susceptibility of *S. aureus* and *E. coli* to triclosan might be the result of a mutation in the primary target site, enoyl reductase (McMurry *et al.*, 1998b).

28.2.5   Other mechanisms

A significant mechanism of resistance to chemical and physical agents is found in bacterial spores and involves small, acid-soluble proteins (SASPs). Large amounts of these low molecular weight basic proteins are present in the spore core. They are rapidly degraded during germination. These SASPs appear to play an important role in determining spore sensitivity/resistance to several antibacterial agents. α, β-type SASPs can coat the DNA in wild-type spores of *B. subtilis*, thereby conferring protection against enzymes and several biocidal agents. Spores ($\alpha^- \beta^-$) that lack these SASPs are much more sensitive to hydrogen peroxide and hypochlorite (Tennen *et al.*, 2000; Loshon *et al.*, 2001).

28.2.6   Bacterial stress responses

Vegetative bacterial cells respond to stress in a variety of ways (Gould, 1989). These include the activation and expression of new groups of genes. The stringent response is mediated by an accumulation of guanosine $3'$-diphosphate-$5'$-diphosphate (guanosine tetraphosphate, ppGpp), a product of the *rel* gene. In cells lacking the stringent response (*rel$^-$*) as a result of *relA* mutation, there is an accumulation of rRNA. When *E. coli* is subject to nutrient limitation (Gilbert *et al.*, 1990) or to antimicrobial agents, growth rate decreases and gene expression is markedly altered. This is essential for long-term survival of the cell and is partly mediated by alternative sigma factors, e.g. σ^s encoded by the *rpoS* gene. Positive regulating of *rpoS* expression results in the synthesis of σ^s being activated by ppGpp. In *E. coli* and other Gram-negative bacteria, cells that produce ppGpp are more resistant to biocides and antibiotics than isogenic mutants with an impaired ability to do so. Furthermore, an insertional mutant of *E. coli* that is unable to produce σ^s is considerably more sensitive than the parent strain (Greenaway and England, 1999a, b).

Tolerance to stress by hydrogen peroxide results from intrinsic defence mechanisms. The oxidative stress response involves the production of neutralising enzymes to prevent cellular damage and the SOS response is a defence system that comes into operation when DNA damage occurs (Demple and Harrison, 1994; Pomposiello and Demple, 2002; Imlay, 2002). Increased tolerance takes place when bacteria are exposed to a subinhibitory concentration of hydrogen peroxide. A series of proteins, many under the control of a sensor/regulator protein (OxyR), is induced and cross-resistance to heat, ethanol and hypochlorous acid has been described. Peroxide-induced proteins overlap with other stress proteins, such as heat-shock ones (Demple and Harrison, 1994). The role of the alternative sigma factor σ^B in the induction of genes encoding general stress proteins in *B. subtilis* and other Gram-positive bacteria has been considered by Hecker and Volker (2001).

Rowbury (2001) has discussed the presence in cultures of extracellular sensing components (ESCs) that are directly converted by stresses to extracellular induction components (EICs). These ESCs and EICs are believed to function for several stress tolerance and sensitisation responses. An early warning system against stress is thereby invoked.

28.3   EMERGING PATHOGENS: POSSIBLE RESISTANCE TO BIOCIDES

28.3.1   Bacteria of clinical relevance

Emerging pathogenic disease organisms may pose a threat not only to human beings but also to medical, containment and hygienic procedures. Emerging infections may arise

(i) as a consequence of improved treatments, as with immunocompromised patients who may possess increased chances of survival while simultaneously being more susceptible to infections,

(ii) as a result of the increasing usage of invasive procedures as part of normal hospital programme,

(iii) because of the possible enhanced ability of microorganisms to survive in new ecological niches and their acquisition of adaptive genes, in particular those associated with virulence and antibiotic resistance.

Bacterial resistance to antibiotics is a major clinical problem but there is no evidence that antibiotic-resistant isolates are significantly less susceptible to biocides. This statement cannot, at present, cover all types of emerging pathogens since

Table 28.5   Emerging pathogens of potential or actual clinical significance.

Type	Examples[a,b]
Gram-positive bacteria	MRSA (including EMRSA)
	VRSA
	VRE
	MDRTB
	Bacillus anthracis
Gram-negative bacteria	*B. cepacia*
	Stenotrophomonas maltophilia
	Acinetobacter spp.
	Alcaligenes spp.
	E. coli O157:H7
	Helicobacter pylori

[a] MRSA, methicillin-resistant *S. aureus*; EMRSA, epidemic MRSA; VRSA, vancomycin-resistant *S. aureus*; VRE, vancomycin-resistant enterococci; MDRTB, multidrug-resistant *Mycobacterium tuberculosis*.
[b] Significance within biofilm communities of many of these has yet to be addressed.

information about their sensitivity or resistance to biocides is often sparse (Weber and Rutala, 1999; Higgins *et al.*, 2001; Russell, 2002b).

Several emerging pathogenic bacteria must be discussed (Table 28.5). Important examples are MRSA, *S. aureus* strains with intermediate sensitivity to vancomycin (VISA) or to the glycopetides vancomycin and teicoplanin (GISA), vancomycin-resistant *S. aureus* (VRSA; Hiramatsu, 1998), vancomycin-resistant enterococci (VRE), multidrug-resistant *M. tuberculosis* (MDRTB), *B. anthracis*, MDR and other Gram-negative bacteria, including *E. coli* O157:H7 and *H. pylori* (Weber and Rutala, 1999). The former is a major food-borne organism that has been responsible for a large number of gastrointestinal cases with some deaths. Nosocomial outbreaks are known to occur (Weber and Rutala, 1999). *H. pylori* is a fastidious, curved to S-shaped Gram-negative organism, with coccoid forms present in older cultures. It is responsible for chronic active (type B) gastritis and *H. pylori* is a prerequisite for the large majority of duodenal ulcers.

It may be considered odd for MRSA to be included in this list because infections caused by such organisms have been known for many years. However, infections, especially those due to EMRSA, have been increasing in recent years. EMRSA-15 and EMRSA-16, in particular, have given rise to severe health and hygiene problems. It must be added that some MRSA strains do not have epidemic potential (Day and Russell, 1999).

28.3.2   Susceptibility to biocides

MRSA strains that possess *qacA* or *qacB* genes may show low-level resistance to cationic biocides, such as chlorhexidine and QACs, but this is not a problem with in-use concentrations of these biocides (Russell, 1999, 2002a). It has, however, been claimed that QAC-resistant mutants of MRSA show greatly increased resistance to some β-lactam antibiotics (Akimitsu et al., 1999), although it is unclear as to whether efflux was involved in this resistance. Fraise (2002) has shown that thickened cell walls in GISA strains may be responsible for their decreased susceptibility to some biocides. The susceptibility of VRSA to biocides has yet to be investigated.

VRE have been isolated from several hospitals throughout the world. For reasons that remain unknown, enterococci are generally less susceptible than staphylococci but VRE are not more resistant to biocides than VSE . As such, VRE do not pose any problem to inactivation by biocides (Alqurashi et al., 1996; Anderson, 1998).

As already pointed out, mycobacteria generally are less susceptible to biocides than other Gram-positive, non-sporulating bacteria (Russell, 1996). MDRTB strains comprise a fairly recent, highly unwelcome development especially in highly developed countries. Such organisms do not, however, appear to be more resistant to biocides than other mycobacteria.

Several MDR Gram-negative bacteria have been implicated in hospital-acquired infections (HAI). Non-fermenting Gram-negative bacteria (NFGNB) have historically been regarded as environmental contaminants of low pathogenicity and not of clinical significance. Several species have, however, emerged as significant causes of nosocomial bacteraemias, especially in immunocompromised patients. They include Acinetobacter spp., Alcaligenes spp. and Stenotrophomonas maltophilia. Some of these, notably Sten. maltophilia, may be antibiotic resistant but are readily inactivated by in-use concentrations of biocides (Higgins et al., 2001). Emerging, epidemic, multiple-antibiotic-resistant strains of Sten. maltophilia and B. cepacia have been described. B. cepacia is an opportunistic pathogen in immunocompromised patients and is also associated with cystic fibrosis sufferers. The organism has been found as an intrinsic contaminant of iodophor formulations and an outbreak of nosocomial B. cepacia infection in ventilated intensive care patients in one hospital was traced to an intrinsically-contaminated alcohol-free mouthwash (Russell, 2002b). Acinetobacter baumanni, a naturally-resistant Gram-negative organism that resides on human skin, could in time pose a more serious risk than MRSA. It rapidly acquires antibiotic resistance and there are few antibiotics effective against it (Roberts et al., 2001).

Although recent studies have demonstrated that in-use concentrations of biocides are lethal to the MDR Gram-negative organisms described here (Higgins et al., 2001), it is evident that considerably more research is needed about the effects of biocides on these organisms to determine whether biocide resistance is or can be of significance and whether biocides can select for antibiotic-resistant bacteria. E. coli O157:H7 is not more resistant to biocides than other strains of E. coli (Weber and Rutala, 1999). Many biocides rapidly inactivate H. pylori although it is not yet clear as to whether the coccoid forms can withstand their action and whether they can transform into the reproduction-competent, pathogenic helical forms (Gebel et al., 2001).

Due to the recent events in the United States, interest has refocused on the inactivation of B. anthracis spores by biocides. In the absence of recent publications it must be assumed that inactivation of 'surrogate' spore-forming organisms with a similar coat structure would provide a reasonable measure of sporicidal activity of appropriate biocides towards *B. anthracis*.

28.4   LINKED BACTERIAL RESISTANCE TO BIOCIDES AND ANTIBIOTICS

It is now apparent that antibiotics and biocides share at least some mechanisms of action. For example, by virtue of effects on peptidoglycan or DNA biosynthesis, several biocides (acridines, phenylethanol, phenoxyethanol and chloracetamide) and antibiotics (β-lactams, novobiocin, 4-quinolones and fluroquinolones) all induce filaments in Gram-negative bacteria (Russell and Chopra, 1996; McDonnell and Russell, 1999). The diversity of the chemical structures of these compounds, however, suggests that they do so in different ways. Thus, β-lactams, such as ampicillin and several cephalosporins that inhibit penicillin-binding protein (PBP) 3 in *E. coli* and norfloxacin and ciprofloxacin that inhibit DNA synthesis by acting on DNA gyrase (Russell, 1998) are capable of bringing about this effect. Acridines also inhibit DNA synthesis but do so as a result of intercalation and also have other effects (Russell and Chopra, 1996), whereas the aromatic alcohols and possibly chloroacetamide have a multiplicity of effects that include inhibition of DNA synthesis. The end result, therefore (filamentation; Beveridge *et al.*, 1991), cannot be ascribed to exactly the same effect in each case. In fact, preliminary studies involving *E.coli* cells pre-exposed to any one of these biocides have demonstrated that they retain sensitivity to ampicillin and norfloxacin; the converse was also true (Ng *et al.*, 2002). It has yet to be shown that cells adapted to resistance to one of these biocides will still be converted into filaments on exposure to these antibiotics (or the converse).

Another example of possible similarity of action of a biocide and antibiotic is shown by the phenylether (bisphenol), triclosan and the antitubercular drug, isoniazid (isonicotinyl acid hydrazine, INH). Triclosan inhibits an enoyl reductase in Gram-negative bacteria and in Gram-positive organisms, including mycobacteria. Thus, there is the potential for the development of resistance to triclosan in mycobacteria, including *M. tuberculosis*, to be translated into resistance to INH also. However, there do appear to be differences. INH acts essentially as a pro-drug that is converted in *M. tuberculosis* into an active form by a *katG*-encoded oxidase–peroxidase system. Absence of this enzyme is the major mechanism responsible for INH resistance. A protein target (InhA) of INH, encoded by the *inhA* gene, is involved in mycolic acid biosynthesis. Mutations in this gene in *M. smegmatis* confer resistance to both INH and triclosan (McMurry *et al.*, 1999). However, it has not been demonstrated that triclosan is responsible for selecting for MDRTB strains.

Resistance to biocides and antibiotics also share some mechanisms, notably impermeability and efflux (Levy, 2002b). Several papers and reviews have appeared that consider a possible linkage between the two types of chemical agents (Russell, 1998, 1999, 2000, 2002a; Russell *et al.*, 1998; Heir *et al.*, 1998, 1999; Sundheim *et al.*, 1998; Suller and Russell, 1999, 2000; Tattawasart *et al.*, 1999a, b, 2000a, b; Levy, 2000, 2001, 2002a, b; Thomas *et al.*, 2000; Gilbert and McBain, 2001a, b; Lambert *et al.*, 2001; Joynson *et al.*, 2002; Lear *et al.*, 2002; Loughlin *et al.*, 2002).

Antibacterial agents are selectively toxic towards bacteria and generally have a high degree of specificity in their action, although more than one target site may be involved (Russell and Chopra, 1996). Resistance may arise by mutation at a susceptible target site, enzymatic inactivation, reduced uptake (cellular impermeability and/or efflux), duplication of a target site, with the second version insusceptible, or overproduction of a target. By contrast, biocides are not selectively toxic towards bacteria and they tend to have multiple, concentration-dependent effects on a bacterial cell (Russell, 1998). Resistance may arise by enzymatic inactivation (unlikely at in-use concentrations), possible mutations producing resistance at low biocide concentrations, and reduced uptake (impermeability and/or efflux, with the latter probably being effective at low, not in-use, biocide levels). Some (cationic) antibiotics and biocides may share a common (self-promoted) mode of entry into Gram-negative bacteria (Russell and Chopra, 1996).

Three major types of biocides (triclosan, QACs and chlorhexidine) have been suggested as being implicated in antibiotic resistance (Chuanchen *et al.*, 2001; White and McDermott, 2001; Russell, 2002a). Possible reasons for this are (a) reduced cellular permeability, (b) degradation, and (c) the widespread occurrence of multidrug efflux pumps responsible for antibiotic and low-level biocide resistance (Sulavik *et al.*, 2001).

'Cross-resistance' to antibiotics and biocides has sometimes been observed, but this is usually considered in terms of MIC levels only. Whereas MICs are undoubtedly of importance with antibiotics, because they might be close to blood or tissue concentrations, they are of much less significance with biocides. Concerns have nevertheless been expressed (Levy, 2000, 2001) that biocide usage is selecting for antibiotic-resistant bacteria in clinical and domiciliary environments. The huge problems associated with antibiotic-resistant bacteria result mainly from the inadequate usage and over-usage of these drugs, including their incorporation into animal feeds (Gorbach, 2001; Russell, 2002a).

28.5   BACTERIAL RESISTANCE: RELEVANCE TO BIOFILMS

Attachment to surfaces is a vital element in bacterial infections, including those related to indwelling medical devices (Denyer, 2002). It is a well-known phenomenon that sessile bacteria on surfaces or contained within biofilms are much less readily inactivated than are planktonic cells (Donlan and Costerton, 2002; Dunne, 2002). Although the reasons for this have not been fully elucidated, several possibilities exist (Spoering and Lewis, 2001; Lewis, 2001; Gilbert *et al.*, 2002a, b).

A biofilm is envisaged as being a functional consortium of cells embedded within a matrix of extracellular polymers and concentrated products of their own environment (Gilbert et al., 2002a). Originally considered to be homogeneous structures, biofilms are now regarded as highly structured habitats with spacial heterogeneity accompanied by physiological heterogeneity that develops at a phase interface (Wimpenny et al., 2000). The close spatial arrangements of different bacterial species can prove to be advantageous to the community as a whole, as exemplified by the potential degradation of harmful chemical agents, such as antibiotics and biocides by means of enzymes, such as β-lactamases and proteases. Organisms can deposit enzymes within the matrix (Gilbert and McBain, 2001a, b). Cell–cell signalling (Davies et al., 1998; Swift et al., 2002) is an important feature of microbial communities and also in biofilm regulation. The quorum sensing systems in P. aeruginosa provide an excellent example (Wimpenny et al., 2000).

Microorganisms within biofilms can survive exposure to biocides and antibiotics that inactivate planktonic cultures. Early stress responses (Section 28.2.6) may be involved to some extent in this. Gilbert et al. (2001) have developed planktonic/ biofilm indices (PBIs) for a range of biocides, the PBI being defined as the iso-effeffective concentrations of a biocide to produce a 95% kill after 30 min. PBI values varied between 1.02 (cetrimide, little or no decrease in susceptibility of the cells) and 0.02 (polyhexamethylene biguanide, PHMB, and Staphylococcus epidermidis). Biocides with the highest activity against planktonic bacteria (PHMB, peracetic acid) were more prone to a biofilm effect (i.e. very low PBI), although they were still the most active, than those with a lower activity against planktonic cells. Diffusion of biocides into biofilms and interaction with biofilm constituents are clearly important factors. Stewart et al. (2001) described the biofilm penetration of hypochlorite and chlorosulphamates and showed that the penetration was controlled by reactivity with the biofilm components. Although penetration of the biofilm was achieved, only a low order of bacterial inactivation resulted. Thus, protective measures other than reactivity and penetration must apply. A biocide gradient is produced; in the case of a thick biofilm, there will be a use-concentration of biocide at the surface but a decreased concentration as the biocide penetrates into the community (Gilbert and McBain, 2001a, b). Degradative enzymes might thus be more effective against these reduced concentrations than in-use concentrations acting on planktonic cultures.

Apart from this chemical gradient, there will also be a physiological gradient. Nutrients and oxygen will be consumed on the periphery of the biofilm, whereas cells deeply placed within the community are starved of both (Gilbert and McBain, 2001a, b; Gilbert et al., 2002a, b). It has long been known that nutrient-limited cells expressing starvation phenotypes are more resistant to biocides than 'normal' cells (Gilbert et al., 1990). Thus, the more biocide-resistant cells will be present towards the base of the biofilm where they are exposed to the lowest concentration of the biocide. Biocide concentration was earlier mentioned (Table 28.2) as being a key factor in biocide activity and this provides an example of its importance in the biofilm context. Thus, a physiological heterogeneity is an important element, with slow-growing deeply recessed bacteria with a less sensitive phenotype being subjected to a lowered biocide concentration.

Pockets of surviving organisms may occur as small clusters in biocide-treated biofilms (Huang *et al.*, 1995), although neighbouring cells have been inactivated. It is possible that these clusters arise as a consequence of biocide concentration and physiological gradients within a biofilm. Gilbert *et al.* (2002a) have postulated that the clusters might include efflux mutants in addition to genotypes with modification in single gene products. Clonal expansion following exposure with a sublethal concentration could then result in the emergence of a population resistant to antibiotics but less likely to biocides, because the latter have multiple target sites (Russell, 2002c). In other words, a selection process might be applied to produce less sensitive clones.

Sublethal treatment with an antimicrobial agent may induce the expression of multidrug efflux pumps and efflux mutants. Efflux confers resistance to low but not high biocide concentrations (Russell, 2000, 2002a). Maira-Litran *et al.* (2000) found that *mar* expression was greatest within the depths of a biofilm where growth rates were the lowest; *mar* expression is known to be inversely related to growth rate, but neither *mar* nor *acrAB* is specifically induced within biofilms.

An interesting concept of programmed cell death (PCD) has been put forward by Lewis (2000). This envisages the potential for damaged bacterial cells to undergo PCD, i.e. inactivation of a cell does not result from direct action of a biocide (Russell and Chopra, 1996; Hugo, 1999) but from a programmed suicide mechanism and cell lysis. Persisters of such a programme are regarded as being cells defective in PCD that can then grow rapidly in the presence of exudate released from lysed community cells. Persisters were first described by Bigger (1944) in a study of penicillin action on staphylococci and have since been shown to occur with penicillin-treated Gram-negative bacteria (Hugo and Russell, 1961). Dense planktonic populations show persisters after treatment with chlorhexidine (Fitzgerald *et al.*, 1992). Persisters are not resistant in the normal sense of the word and cultures obtained from them show the same response when re-exposed to that particular agent. Likewise, the proposal that lysed or 'leaky' cells provide nutrients for surviving ones was put forward many years ago by Bean and Walters (1961) who found that survivors of phenol treatment could recover and grow in the exudate from non-surviving (inactivated) cells.

Thus, a variety of complex factors contribute to the recalcitrance towards biocides and antibiotics of bacterial communities within biofilms. On the basis of what has been discussed in this chapter and elsewhere in this volume, these can be summarised as follows:

(1) reduced access of biocide molecules to bacterial cells,
(2) chemical interaction between the biofilm and the biocide molecules,
(3) modulation of the micro-environment,
(4) reduced oxygen tensions within the biofilm,
(5) production of degradative enzymes against some antimicrobial agents,
(6) nutrient-limited and starved cells,

(7) genetic exchange between cells,
(8) quorum sensing, and
(9) presence of persisters.

Impermeability to biocides of bacterial cells, whether they exist within a biofilm or within an ordinary culture, remains a factor of importance. Ayres *et al.* (1998a, b) found that permeabilisers, such as EDTA potentiated the effects of some biocides against planktonic cultures and simple biofilms of *P. aeruginosa.* Additionally, efflux will play a role in limiting biocide uptake and biocide degradation may also be of some importance at sites where diffusible biocide concentrations are low. It is also possible that sessile cells can adapt to biocides during repeated exposure. The ways in which bacteria respond to, and seek to evade, biocide-induced stress has entered an exciting phase. Thus, several different types of approaches are necessary to understand more fully the responses to biocides and antibiotics of bacteria present within biofilm communities.

All of the likely mechanisms of bacterial resistance to biocides considered in preceding sections could, thus, contribute towards the overall level of insusceptibility demonstrated by sessile organisms. These alone, however, cannot explain the higher degree of resistance of sessile bacteria to biocides and can only be considered of importance in conjunction with the other postulated reasons (1–9) outlined above and discussed in more detail elsewhere in this volume. In this context, it is pertinent to note that cells removed from a biofilm and recultured in culture media are generally no more insusceptible to biocides than 'ordinary' planktonic cells (Stewart *et al.*, 2001).

28.6   CONCLUSIONS

Studies on bacterial resistance to biocides continue to provide useful and interesting information. It is now known that bacteria can develop resistance to these agents, although usually to a low level only, and that the potential exists for selection for antibiotic-resistant bacteria. This aspect has been widely studied, but there is as yet no convincing argument that such selection is a problem in the clinical, food or domiciliary environments (Russell, 2000, 2002a; Lambert *et al.*, 2001; McBain and Gilbert, 2001). Whereas a possible link may occur in laboratory studies, this has not been found in the general environment where complex, multispecies predominate (Gilbert *et al.*, 2002). A better understanding is needed of the mechanisms of biocidal action not only against bacteria but also against other microorganisms (Russell, 2002c).

Some biocides have only a low order of activity against bacteria, notably mycobacteria and bacterial spores that possess efficient permeability barriers. This is also one reason for the reduced susceptibility of Gram-negative organisms in comparison with Gram-positive cocci.

Another important mechanism of resistance is efflux, an additional or alternative means of limiting the uptake of sufficiently high biocide levels within bacterial cells. Adaptation and possible mutation and stress responses provide further means of alleviating, overcoming or preventing biocide action. Bacteria within biofilms continue to be widely studied, but the mechanisms responsible for the higher resistance of sessile over planktonic cells have yet to be fully elucidated. It is believed that the reduced susceptibility of cells within biofilm communities results from a complex variety of reasons.

Emerging pathogens are likely to continue to pose a severe health problem and it is important that effective procedures for controlling them, including the appropriate use of biocides, are in place. Other bacteria are likely to emerge as problem organisms in the future and it is imperative that these be identified as soon as possible with a view to understanding the nature of their responses to biocides. In all instances, it is necessary to consider the responses to biocides and antibiotics of biofilm-associated pathogens.

Ishikawa *et al.* (2002) have made the interesting observation that once a strain (*E. coli* in their studies) has mutated to resist a relatively low concentration of a QAC, it can be converted to an MDR type. Furthermore, it has been claimed that, as with antibiotics, 'true' biocide resistance will eventually emerge (Levy, 2002a, b). It is in the interests of manufacturers and users of biocidal products that biocides continue to be employed in a logical and responsible manner to ensure that this does not occur.

REFERENCES

Akimutsu, N., Hamamoto, H., Inoue, R., Shoji, M., Akamine, A., Takeemori, K., Hamasaki, N. and Saiekimizu, K. (1999) Increase in resistance of methicillin-resistant *Staphylococcus aureus* to beta-lactams caused by mutations conferring resistance to benzalkonium chloride, a disinfectant widely used in hospitals. *Antimicrob. Agents Chemother.* **43**, 3042–3043.

Alqurashi, A.M., Day, M.J. and Russell, A.D. (1996) Susceptibility of some strains of enterococci and streptococci to antibiotics and biocides. *J. Antimicrob. Chemother.* **38**, 745.

Anderson, R.L. (1998) Susceptibility of antibiotic-resistant microbes to chemical germicides. In *Disinfection, Sterilization and Antisepsis in Health Care* (ed. W.A. Rutala), pp. 241–253, APIC, Washington, DC.

Ayres, H.M., Furr, J.R. and Russell, A.D. (1998a) Use of the Malthus-AT system to assess the efficacy of permeabilizing agents on biocide activity against *Pseudomonas aeruginosa. Lett. Appl. Microbiol.* **26**, 422–426.

Ayres, H.M., Payne, D.N., Furr, J.R. and Russell, A.D. (1998b) Effect of permeabilizing agents on antibacterial activity against a simple *Pseudomonas aeruginosa* biofilm. *Lett. Appl. Microbiol.* **27**, 79–82.

Bean, H.S. and Walters, V. (1961) Studies on bacterial populations in solutions of phenols. Part II. The influence of cell exudate upon the shape of the time-survivor curve. *J. Pharm. Pharmacol.* **13**, 183T-194T.

Beveridge, E.G. (1999) Preservation of medicines and cosmetics. In *Principles and Practice of Disinfection, Preservation and Sterilization*, 3rd ed. (ed. A.D. Russell, W.B. Hugo, and G.A.J. Ayliffe) pp. 457–484, Blackwell Science, Oxford.

Beveridge, E.G., Boyd, I., Dew, I., Haswell, M. and Lowe, C.W.G. (1991) Electron and light microscopy of damaged bacteria. *Soc. Appl. Bact. Tech. Ser.* **27**, 135–153.

Biggar, J.W. (1944) Treatment of staphylococcal infection with penicillin. *Lancet* **ii**, 497–500.

Broadley, S.J., Jenkins, P.A., Furr, J.R. and Russell, A.D. (1995) Potentiation of the effects of chlorhexidine diacetate and cetylpyridinium chloride on mycobacteria by ethambutol. *J. Med. Microbiol.* **43**, 458–460.

Candal, E.J. and Eagon, R.G. (1984) Evidence for plasmid-mediated bacterial resistance to industrial biocides. *Int. Biodet.* **20**, 221–224.

Chaplin, C.E. (1951) Observations on quaternary ammonium disinfectants. *Can. J. Bot.* **29**, 373–382.

Chaplin, C.E. (1952) Bacterial resistance to quaternary ammonium disinfectants. *J. Bacteriol.* **63**, 453–458.

Chuanchen, R., Beinlick, K., Hoang, T.T., Becher, A., Karkoff-Schweizer, R.R. and Schweizer, H.P. (2001) Cross-resistance between triclosan and antibiotics in *Pseudomonas aeruginosa* is mediated by multidrug efflux pumps: exposure of a susceptible mutant strain to triclosan selects *nfxB* mutants overexpressing MexCD-OprJ. *Antimicrob. Agents Chemother.* **45**, 428–432.

Davies, D.G., Parsek, M.R., Pearson, J.P., Iglewski, B.H., Costerton, J.W. and Greenberg, E.P. (1998) The involvement of cell-to-cell signals in the development of a bacterial biofilm. *Science* **280**, 295–298.

Day, M.J and Russell, A.D (1999) Antibiotic-resistant cocci. In *Principles and Practice of Disinfection, Preservation and Sterilization*, 3rd ed. (ed. A.D. Russell, W.B. Hugo and G.A.J. Ayliffe), pp. 344–359, Blackwell Science, Oxford.

Demple, B. and Harrison, L. (1994) Repair of oxidative damage to DNA: enzymology and biology. *Ann. Rev. Biochem.* **63**, 915–948.

Denyer, S.P. (2002) Close encounters of the microbial kind. *Pharm. J.* **269**, 451–454.

Donlan, R.M. and Costerton, J.W. (2002) Biofilms: survival mechanisms of clinically relevant microorganisms. *Clin. Microbiol. Rev.* **15**, 167–193.

Dunne Jr., W.M. (2002) Bacterial adhesion: seen any good biofilms recently? *Clin. Microbiol. Rev.* **15**, 155–166.

Fitzgerald, K., Davies, A. and Russell, A.D. (1992) Sensitivity and resistance of *Escherichia coli* and *Staphylococcus aureus* to chlorhexidine. *Lett. Appl. Microbiol.* **14**, 33–36.

Foster, T.J. (1983) Plasmid-determined resistance to antimicrobial drugs and toxic metal ions in bacteria. *Microbiol. Rev.* **47**, 361–409.

Foster, S.J. (1994) The role and regulation of cell wall structural dynamics during differentiation of endospore-forming bacteria. *J. Appl. Bact.* **76**(Suppl.), 25S–39S.

Fraise, A. (2002) Susceptibility of antibiotic-resistant cocci to biocides. *J. Appl. Microbiol.* **92**(Suppl.), 158S–162S.

Fraud, S.R., Maillard, J.-Y. and Russell, A.D. (2001) Comparison of the mycobactericidal activity of *ortho*-phthalaldehyde, glutaraldehyde and other dialdehydes by a quantitative suspension test. *J. Hosp. Infect.* **48**, 214–221.

Gebel, J., Vacata, V., Sigler, K., Pietsch, H., Rechenberg, A., Exner, M. and Kistemann, T. (2001) Disinfectant activity against different morphological forms of *Helicobacter pylori:* first results. *J. Hosp. Infect.* **48**(Suppl.), S58–S63.

Gilbert, P. and McBain, A.J. (2001a) Biocide usage in the domestic setting and concern about antibacterial and antibiotic resistance. *J. Infect.* **43**, 85–91.

Gilbert, P. and McBain, A.J. (2001b) Biofilms: their impact upon health and their recalcitrance towards biocides. *Am. J. Inf. Control* **29**, 252–255.

Gilbert, P., Collier, P.J. and Brown, M.R.W. (1990) Influence of growth rate on susceptibility to antimicrobial agents: biofilms, cell cycle, dormancy and stringent response. *Antimicrob. Agents Chemother.* **34**, 1865–1868.

Gilbert, P., Das, J.R., Jones, M.V. and Allison, D.G. (2001) Assessment of resistance towards biocides following the attachment of micro-organisms to, and growth on, surfaces. *J. Appl. Microbiol.* **91**, 248–254.

Gilbert, P., Allison, D.G. and McBain, A.J. (2002a) Biofilms *in vitro* and *in vivo*: do singular mechanisms imply cross-resistance? *J. Appl. Microbiol.* **92**(Suppl.), 98S–110S.

Gilbert, P., Maira-Litran, T., McBain, A.J., Rickard, A.H. and Whyte, F. (2002b) The physiology and collective recalcitrance of microbial biofilm communities. *Adv. Microbiol. Physiol.* **46**, 205–256.

Gorbach, S.L. (2001) Antimicrobial use in animal feed – time to stop. *New Engl. J. Med.* **345**, 1202–1203.

Gould, G.W. (1989) Heat-induced injury and inactivation. In *Mechanisms of Food Preservation Procedures* (ed. G.W. Gould), pp. 11–42, Elsevier Applied Science, London.

Greenaway, D.L.A. and England, R.R. (1999a) ppGpp accumulation in *Pseudomonas aeruginosa* and *Pseudomonas fluorescens* subjected to nutrient limitation and biocide exposure. *Lett. Appl. Microbiol.* **29**, 298–302.

Greenaway, D.L.A. and England, R.R. (1999b) The intrinsic resistance of *Escherichia coli* to various antimicrobial agents requires ppGpp and σ^s. *Lett. Appl. Microbiol.* **29**, 323–326.

Guerin-Mechin, L., Dubois-Brissonnet, F., Heyd, B. and Leveau, J.Y. (1999) Specific variations of fatty acid composition of *Pseudomonas aeruginosa* ATCC 15442 induced by quaternary ammonium compounds and relation with resistance to bactericidal activity. *J. Appl. Microbiol.* **87**, 735–742.

Hecker, M. and Volker, U. (2001) General stress response of *Bacillus subtilis* and other bacteria. *Adv. Microb. Physiol.* **44**, 35–91.

Heir, E., Sundheim, G. and Holck, A.L. (1998) The *Staphylococcus qacH* gene product: a new member of the SMR family encoding multidrug resistance. *FEMS Microbiol. Lett.* **163**, 49–56.

Heir, E., Sundheim, G. and Holck, A.L. (1999) The *qacG* gene on plasmid pST94 confers resistance to quaternary ammonium compounds in staphylococci isolated from the food industry. *J. Appl. Microbiol.* **86**, 378–388.

Higgins, C.S., Murtough, S.M., Williams E., Hiom, S.J., Payne, D.J., Russell, A.D. and Walsh, T.R. (2001) Resistance to antibiotics and biocides among non-fermenting Gram-negative bacteria. *Clin. Microbiol. Infect.* **7**, 308–315.

Hiramatsu, K. (1998) Vancomycin resistance in staphylococci. *Drug Resist. Updates* **1**, 135–150.

Huang, C.T., Fu, Y.P., McFeters, G.A. and Stewart, P.S. (1995) Non-uniform spatial patterns of respiratory activity within biofilms during disinfection. *Appl. Environ. Microbiol.* **61**, 2252–2256.

Hugo, W.B. (1991) The degradation of preservatives by microorganisms. *Int. Biodet.* **27**, 185–194.

Hugo, W.B. (1999) Disinfection mechanisms. In *Principles and Practice of Disinfection, Preservation and Sterilization*, 3rd ed. (ed. A.D. Russell, W.B. Hugo and G.A.J. Ayliffe), pp. 256–283, Blackwell Science, Oxford.

Hugo, W.B. and Russell, A.D. (1961) The mode of action of penicillin. *J. Pharm. Pharmacol.* **13**, 705–721.

Imlay, J.A. (2002) How oxygen damages microbes: oxygen tolerance and obligate anaerobiosis. *Adv. Microb. Physiol.* **46**, 111–153.

Ishikawa, S., Matsumura, Y., Yoshizako, F. and Tsuchido, T. (2002) Characterization of a cationic surfactant-resistant mutant isolated spontaneously from *Escherichia coli. J. Appl. Microbiol.* **92**, 261–268.

Joynson, J.A., Forbes, B. and Lambert, R.J.W. (2002) Adaptive resistance to benzalkonium chloride, amikacin and tobramycin: the effect on susceptibility to other antimicrobials. *J. Appl. Microbiol.* **93**, 96–106.

Lambert, R.J.W., Joynson, J. and Forbes, B. (2001) The relationships and susceptibilities of some industrial, laboratory and clinical isolates of *Pseudomonas aeruginosa* to some antibiotics and biocides. *J. Appl. Microbiol.* **91**, 972–984.

Lear, J.C., Maillard, J.-Y., Dettmar, P.W., Goddard, P.A. and Russell, A.D. (2002) Chloroxylenol- and triclosan-tolerant bacteria from industrial sources. *J. Indust. Microbiol. Biotechnol.* (in press).

Levy, S.B. (2000) Antibiotic and antiseptic resistance: impact on public health. *Pediatr. Infect. Dis. J.* **19**, S120–122.

Levy, S.B. (2001) Antibacterial household products: cause for concern. *Emerg. Infect. Dis.* **7**(Suppl. 3), 512–515.

Levy, S.B. (2002a) Factors impacting on the problem of antibiotic resistance. *J. Antimicrob. Chemother.* **49**, 25–30.

Levy, S.B. (2002b) Active efflux: a common mechanism for biocide and antibiotic resistance. *J. Appl. Microbiol.* **92**(Suppl. 6), 5S–71S.

Lewis, K. (2000) Programmed cell death in bacteria. *Microbiol. Mol. Biol. Rev.* **64**, 503–514.

Lewis, K. (2001) Riddle of biofilm resistance. *Antimicrob. Agents Chemother.* **45**, 997–1007.

Loshon, C.A., Melly, E., Setlow, B. and Setlow, P. (2001) Analysis of the killing of spores of *Bacillus subtilis* by a new disinfectant, Sterilox[R]. *J. Appl. Microbiol.* **91**, 1051–1058.

Loughlin, M.F., Jones, M.V. and Lambert, P.A. (2002) *Pseudomonas aeruginosa* adapted to benzalkonium chloride show resistance to other membrane-active agents but not to clinically relevant antibiotics. *J. Antimicrob. Chemother.* **44**, 631–639.

McBain, A.J. and Gilbert, P. (2001) Biocide tolerance and the harbingers of doom. *Int. Biodet. Biodeg.* **47**, 55–61.

McDonnell, G. and Russell, A.D. (1999) Antiseptics and disinfectants: activity, action and resistance. *Clin. Microbiol. Rev.* **12**, 147–179.

McMurry, L.M., Oethinger, M. and Levy, S.B. (1998a) Overexpression of *marA*, *soxS* or *acrAB* produces resistance to triclosan in *Escherichia coli*. *FEMS Microbiol. Lett.* **166**, 305–309.

McMurry, L.M., Oethinger, M. and Levy, S.B. (1998b) Triclosan targets lipid synthesis. *Nature* **394**, 531–532.

McMurry, L.M., McDermott, P.F. and Levy, S.B. (1999) Genetic evidence that InhA of *Mycobacterium smegmatis* is a target for triclosan. *Antimicrob. Agents Chemother.* **43**, 711–713.

Maira-Litran, T., Allison, D.G. and Gilbert, P. (2000) An evaluation of the potential role of the multiple antibiotic resistance operon (*mar*) and the multidrug efflux pump *acrAB* in the resistance of *Excherichia coli* biofilms towards ciprofloxacin. *J. Antimicrob. Chemother.* **45**, 789–795.

Meade, M.J., Waddell, R.L. and Callahan, T.M. (2001) Soil bacteria *Pseudomonas putida* and *Alcaligenes xylosoxidans* subsp. *denitrificans* inactivate triclosan in liquid and solid substrates. *FEMS Microbiol. Lett.* **204**, 45–48.

Ng, E.G.L., Jones, S., Leong, S.H. and Russell, A.D. (2002) Biocides and antibiotics with apparently similar actions on bacteria: is there the potential for cross-resistance? *J. Hosp. Infect.* **51**, 147–149.

Paidhungat, M. and Setlow, P. (2002) Spore germination and outgrowth. In *Bacillus subtilis and Its Closest Relatives* (ed. A.L. Sonenshin, J. Hoch and R. Losick), pp. 537–548, ASM Press, Washington, DC.

Paulsen, I.T., Brown, M.H. and Skurray, R.A. (1996) Proton-dependent multidrug efflux systems. *Microbiol. Rev.* **60**, 575–608.

Pomposiello, P.J. and Demple, B. (2002) Global adjustment of microbial physiology during free radical stress. *Adv. Microbiol. Physiol.* **46**, 318–341.

Poole, K. (2000) Efflux-mediated resistance to fluoroquinolones in Gram-negative bacteria. *Antimicrob. Agents Chemother.* **44**, 2233–2241.

Poole, K. (2002) Mechanisms of bacterial biocide and antibiotic resistance. *J. Appl. Microbiol.* **92**(Suppl.), 55S–64S.

Roberts, S.A., Findlay, R. and Lang, S.D.R. (2001) Investigation of an outbreak of multidrug resistant *Acinetobacter baumannii* in an intensive care burns unit. *J. Hosp. Infect.* **48**, 228–232.

Rowbury, R.J. (2001) Extracellular sensing components and extracellular induction component alarmones give early warning against stress in *Escherichia coli*. *Adv. Microb. Physiol.* **44**, 215–257.

Rosser, S.J. and Young, H.K. (1999) Identification and characterization of class 1 integrons in bacteria from an aquatic environment. *J. Antimicrob. Chemother.* **44**, 11–18.

Russell, A.D. (1990) The bacterial spore and chemical sporicidal agents. *Clin. Microbiol. Rev.* **3**, 99–119.

Russell, A.D. (1996) Activity of biocides against mycobacteria. *J. Appl. Bact.* **81**(Suppl.), 87S–101S.

Russell, A.D. (1998) Mechanisms of bacterial resistance to antibiotics and biocides. *Prog. Med. Chem.* **35**, 133–197.

Russell, A.D. (1999) Bacterial resistance to disinfectants: present knowledge and future problems. *J. Hosp. Infect.* **43**(Suppl.), S57–S68.

Russell, A.D. (2000) Do biocides select for antibiotic resistance? *J. Pharm. Pharmacol.* **52**, 227–233.

Russell, A.D. (2002a) Introduction of biocides into clinical practice and the impact on antibiotic-resistant bacteria. *J. Appl. Microbiol.* **92**(Suppl.), 121S–135S.

Russell, A.D. (2002b) Emerging infectious organisms and their susceptibility to disinfectants. *Steriliz. Aust.* **20**, 12–19.

Russell, A.D. (2002c) Mechanisms of antimicrobial action of antiseptics and disinfectants: an increasingly important area of investigation. *J. Antimicrob. Chemother.* **49**, 597–599.

Russell, A.D. and Chopra, I. (1996) *Understanding Antibacterial Action and Resistance*, 2nd ed., Ellis Horwood, Chichester.

Russell, A.D., Tattawasart, U., Maillard, J.-Y., Furr, J.R and Russell, A.D. (1998) Possible link between bacterial resistance and use of antibiotics and biocides. *Antimicrob. Agents Chemother.* **42**, 2151.

Setlow, P. (1994) Mechanisms which contribute to the long-term survival of spores of *Bacillus* species. *J. Appl. Bact.* **76**(Suppl.), 49S–60S.

Simons, C., Walsh, S.E., Maillard, J.-Y. and Russell, A.D. (2000) A note: *ortho*-phthalaldehyde: mechanism of action of a new antimicrobial agent. *Lett. Appl. Microbiol.* **31**, 299–302.

Spoering, A.L. and Lewis, K. (2001) Biofilms and planktonic cells of *Pseudomonas aeruginosa* have similar resistance to killing by antimicrobials. *J. Bacteriol.* **183**, 6746–6751.

Stewart, P.S., Rayner, J., Roe, F. and Rees, W.M. (2001) Biofilm penetration and disinfection efficacy of alkaline hypochlorite and chlorosulfamates. *J. Appl. Microbiol.* **91**, 525–532.

Stickler, D.J. and King, J.B. (1999) Intrinsic resistance. In *Principles and Practice of Disinfection, Preservation and Sterilization*, 3rd ed. (ed. A.D. Russell, W.B. Hugo and G.A.J. Ayliffe), pp. 284–296, Blackwell Science, Oxford.

Sulavik, M.C., Houseweart, C., Cramer, C., Jiwani, N., Murgolo, N., Greene, J., DiDomenico, B., Shaw, K.J., Miller, G.H., Hare, R. and Shimer, G. (2001) Antibiotic susceptibility profiles of *Escherichia coli* lacking multidrug efflux pumps. *Antimicrob. Agents Chemother.* **45**, 1126–1136.

Suller, M.T.E. and Russell, A.D. (1999) Antibiotic and biocide resistance in methicillin-resistant *Staphylococcus aureus* and vancomycin-resistant enterococcus. *J. Hosp. Infect.* **43**, 281–291.

Suller, M.T.E. and Russell, A.D. (2000) Triclosan and antibiotic resistance in *Staphylococcus aureus*. *J. Antimicrob. Chemother.* **46**, 11–18.

Sundheim, G., Langsrud, S., Heir, E. and Holck, A.L. (1998) Bacterial resistance to disinfectants containing quaternary ammonium compounds. *Int. Biodet. Biodeg.* **41**, 235–239.

Swift, S., Downie, J.A., Whitehead, N.A., Barnard, A.M.L., Salmond, G.P.C. and Williams, P. (2001) Quorum sensing as a population-density determinant of bacterial physiology. *Adv. Microb. Physiol.* **45**, 199–270.

Tattawasart, U., Maillard, J.-Y., Furr, J.R. and Russell, A.D. (1999a) Development of resistance to chlorhexidine diacetate and cetylpyridinium chloride in *Pseudomonas stutzeri* and changes in antibiotic susceptibility. *J. Hosp. Infect.* **42**, 219–229.

Tattawasart, U., Maillard, J.-Y., Furr, J.R. and Russell, A.D. (1999b) Comparative responses of *Pseudomonas stutzeri* and *Pseudomonas aeruginosa* to antibacterial agents. *J. Appl. Microbiol.* **87**, 323–331.

Tattawasart, U., Maillard, J.-Y., Furr, J.R. and Russell, A.D. (2000a) Outer membrane changes in *Pseudomonas stutzeri* resistant to chlorhexidine diacetate and cetylpyridinium chloride. *Int. J. Antimicrob. Agents* **16**, 233–238.

Tattawasart, U., Hann, A.C., Maillard, J.-Y., Furr, J.R. and Russell, A.D. (2000b) Cytological changes in chlorhexidine-resistant isolates of *Pseudomonas stutzeri*. *J. Antimicrob. Chemother.* **45**, 145–152.

Tennen, R., Setlow, B., Davis, K.L., Loshon, C.A. and Setlow, P. (2000) Mechanisms of killing of spores of *Bacillus subtilis* by iodine, glutaraldehyde and nitrous acid. *J. Appl. Microbiol.* **89**, 330–338.

Walsh, S., Maillard, J.-Y. and Russell, A.D. (2001) Possible mechanisms for the relative efficacies of *ortho*-phthalaldehyde against glutaraldehyde-resistant *Mycobacterium chelonae*. *J. Appl. Microbiol.* **91**, 80–92.

Weber, D.J. and Rutala, W.A. (1999) The emerging nosocomial pathogens *Cryptosporidium, Escherichia coli* O157:H7, *Helicobacter pylori* and hepatitis C: epidemiology, environmental survival and control measures. *Infect. Cont. Hosp. Epidemiol.* **22**, 306–315.

White, D.G. and McDermott, P.F. (2001) Biocides, drug resistance and microbial evolution. *Curr. Opin. Microbiol.* **4**, 313–317.

Wimpenny, J., Manz, W. and Szewyk, U. (2000) Heterogeneity in biofilms. *FEMS Microbiol. Lett.* **24**, 661–671.

Section 7

Biofilm control

29 Resistance of medical biofilms   537

30 Control of biofilm in the food industry:
   a microbiological survey of high-risk
   processing facilities   554

31 Industrial biofilms: formation, problems
   and control   568

32 Microbial fouling control for industrial
   systems   591

29

Resistance of medical biofilms

M.R.W. Brown and A.W. Smith

29.1 INTRODUCTION

There is increasing awareness of the difficulties posed by the role of microbial biofilms in the treatment of infection. The significance of biofilms extends beyond the well-known examples of medical device-related infection, such as those associated with artificial joints, prosthetic heart valves and catheters, but also include many chronic infections not related to medical devices. These are now recognised to be due to bacteria either not growing and relatively dormant or growing slowly as biomasses or adherent biofilms on mucosal surfaces. Recent surveys indicate that catheter-associated bacteraemia, consequent from catheter-related infection, is by far the leading cause of nosocomial bloodstream infection in intensive care units (Brub-Buisson, 2001). The issue of biofilm eradication extends way beyond the infected patient, since bacteria in the environment will typically exist within complex multi-species biofilm ecosystems.

Biofilm growth almost always leads to a large increase in resistance to antimicrobial agents, including antibiotics, biocides and preservatives, compared with cultures grown in suspension (planktonic) in conventional liquid media (Gilbert *et al.*, 1990; Stewart and Costerton, 2001). However, a recent study with high-density planktonic cultures indicated similar resistance to antimicrobials as did biofilm cultures (Spoering and Lewis, 2001). Currently, there is no generally agreed mechanism to account for the broad resistance to chemical agents. We suggest that dormancy, related to the general stress response (GSR) and associated survival responses, offers an explanation of the overall general resistance of biofilm microbes (Foley *et al.*, 1999; Brown and Smith, 2001).

29.2   BIOFILM FORMATION

All surfaces, synthetic or otherwise, will be coated with constituents from the local environment: first water and salt ions, then organic material. This conditioning film exists before the arrival of the first microbe. The initial weak and reversible contact between microbe and conditioning film results from Brownian motion, gravitation, diffusion or microbial motility and involving electrostatic interactions. Detailed accounts of the physics of attachment can be found elsewhere (Gilbert *et al.*, 1995; An *et al.*, 2000). The surface interaction is a function of the cell surface (determined by the cell physiology) and the nature of the film. Recent work with a *Staphylococcus aureus* mutant bearing a stronger negative charge due to the lack of D-alanine esters in its teichoic acids could no longer colonise polystyrene or glass surfaces is highlighted by the contribution of electrostatic forces to biofilm formation, particularly on medical devices (Gross *et al.*, 2001). Charge attraction or repulsion could also contribute to interaction between bacteria and substratum (Gottenbos *et al.*, 2001). Moreover, growth in the presence of sub-inhibitory antibiotic concentrations can influence cell surface hydrophobicity (Domingue *et al.*, 1989b; Gottenbos *et al.*, 2001), as can the physiological state of the cell (Williams *et al.*, 1986; Domingue *et al.*, 1989a; Allison *et al.*, 1990a, b). It has long been known that cell surface hydrophobicity also greatly influences susceptibility to phagocytosis (van Oss, 1978). Specific interactions with bacterial surface structures can also be important in establishing a biofilm. For example, flagella and pilus-mediated twitching motility are required/important for *Escherichia coli* and *P. aeruginosa* biofilms (O'Toole and Kolter, 1998; Pratt and Kolter, 1998). There are a number of excellent reviews of the role of specific adhesin–receptor binding to host cell surfaces, with consequent complex signal transduction cascades in the host cell (Hopelman and Tuomanen, 1992; Boland *et al.*, 2000).

The penultimate step is when the adsorption becomes irreversible. This is partly due to surface appendages overcoming the repulsive forces between the two surfaces and also because of sticky exopolymers secreted by the cells. Commonly, the entire biofilm may be coated with a hydrophilic exopolymer (the glycocalyx), which is itself a complex and dynamic structure (Sutherland, 2001). When the host is unable to opsonise this hydrophilic glycocalyx, the entire biofilm is resistant to phagocytosis (Pier *et al.*, 1987). The final stage is when the biofilm population increases, typically as a result of adherent cells replicating, but including a contribution from fresh cells adhering to the biofilm (Gilbert *et al.*, 1997; Al-Bakri *et al.*, 1999; Stoodley *et al.*, 2001). In Staphylococcal species, commonly associated with device-related infections, biofilm formation requires cell–cell adhesion mediated by the *ica* locus following adhesion to the substratum

(Cramton *et al.*, 1999). Sub-inhibitory concentrations of tetracycline and the semi-synthetic streptogramin antibiotic quinupristin-dalfopristin enhanced *ica* expression and biofilm formation (Rachid *et al.*, 2000b).

The biofilm structure/phenotype depends on numerous factors, notably the organism and its physiology, the substratum, the surrounding nutrient environment and the rate of flow of any liquid over its surface (Costerton *et al.*, 1995; Karthikeyan *et al.*, 2000). The resulting biofilms may vary from sparse amorphous masses to highly structured consortia with mushroom-like cell stacks surrounded by channels with rapid aqueous movement (Costerton *et al.*, 1999). Signal molecules (see below), which influence cell physiology including virulence and the GSR (and thus dormancy) have also been shown to influence biofilm structure and susceptibility to antimicrobials.

29.3   BIOFILM AND PLANKTONIC CULTURES: COMPARING LIKE WITH LIKE?

Many biofilm models exist and are reviewed by others (Allison *et al.*, 1999; Dibdin and Wimpenny, 1999; Kharazmi *et al.*, 1999; Yasuda *et al.*, 1999; Sissons *et al.*, 2000). The majority of the literature examining the resistance of bacterial biofilms is driven by comparisons with cells grown planktonically. Although there is general acceptance that there are numerous planktonic phenotypes, many papers refer to 'the biofilm phenotype', implicitly assuming (wrongly) that there is only one. Those same parameters known to influence planktonic physiology, including growth rate and/or specific nutrient limitation (Brown, 1977; Brown *et al.*, 1990), also apply to biofilm physiology (Gilbert *et al.*, 1990). Valid comparisons between biofilm and planktonic cultures are, therefore, difficult to make, especially if such key parameters are not comparable between the two states. A model which controls biofilm growth rate in a defined, nutrient-limited medium consists of surface growth on the underside of a bacteria-proof cellulose membrane (Gilbert *et al.*, 1989). The membrane is perfused with fresh, defined medium from the sterile side and cells eluted from the biofilm are collected. Growth rate of the nutrient-limited, adherent population is controlled by rate of perfusion of fresh medium. A method of controlling density and nutrient limitation of biofilm growth consists of membrane culture on the surface of defined agar medium, growth limited by any specified major nutrient (Bühler *et al.*, 1998; Desai *et al.*, 1998). These methods enable comparison of planktonic and biofilm cultures at similar stages of growth and under similar eventual nutrient restriction (as well as temperature and pH). However, it is not possible to obtain comparable planktonic cells at a density equivalent even to that of a sparse biofilm. Such planktonic cultures would require growth in extremely high concentrations of nutrients including massive sparging with oxygen. Thus, the price of comparable density would be lack of comparison in terms of osmolarity and the damaging consequences of oxygen.

29.4   MECHANISMS OF RESISTANCE

The antibiotic susceptibility of biofilms of resistant mutants will obviously be low. However, what is not obvious is why strains susceptible in planktonic culture are typically resistant when growing as a biofilm. Eradication of infection by antibiotic treatment requires elimination of all the bacteria, typically assisted by the host defences. Otherwise infection re-occurs and chronicity is established. In other words, biofilm resistance can be determined by the susceptibility of the most resistant cell. It is not the case that all cells within a biofilm are always highly resistant (Brooun *et al.*, 2000; Lewis, 2001), but the most resistant members of a biofilm population are typically orders of magnitude more resistant than similar members of a planktonic population, and yet subculture rarely shows the existence of resistant mutants. Hence, biofilm resistance is characteristically phenotypic. Conventionally recognised mechanisms of antibiotic resistance have been reviewed elsewhere (Xu *et al.*, 2000; Lewis, 2001; Mah and O'Toole, 2001). This work will touch on these mechanisms briefly in the context of a possible enhanced role in biofilms and then focus on stress responses and dormancy.

29.4.1   Antibiotic penetration and modification

A number of reports question the possible lack of antibiotic/biocide penetration as an explanation of biofilm resistance (Gilbert *et al.*, 1995; Stewart, 1996; Xu *et al.*, 2000; Lewis, 2001; Mah and O'Toole, 2001). Given, in some cases, biofilms consisting largely of stacks of cells with flowing aqueous channels (even though coated with glycocalyx), then impenetrability seems highly unlikely (Nichols, 1991), a finding confirmed with biofilms of *Klebsiella pneumoniae* (Anderl *et al.*, 2000). If the antimicrobial agent either reacts chemically with components of the exopolymer or is significantly adsorbed by these typically anionic polymers, then the net effect is as if there is a penetration barrier. There will be a similar effect if such interactions occur with cells, perhaps dead ones, in the outer parts of the biofilm. Heterogeneity has also been given as a reason for biofilm resistance. But, in terms of eradication/sterilisation, resistance is caused by the most resistant members of the biofilm. Hence, the question still remains as to what is the mechanism. Heterogeneity *per se* is not a mechanism. Given that biofilms are indeed typically heterogeneous, these generalisations do not preclude the possibility of areas in a biofilm, where diffusion is restricted (Lewis, 2001).

Antimicrobial agents can frequently induce the production of inactivating enzymes in microbes. The relatively large amounts of antibiotic-inactivating enzymes, such as β-lactams, which accumulate within the glycocalyx produce concentration gradients of antimicrobials across it and the underlying cells have been shown to be thus protected (Giwercman *et al.*, 1991; Bagge *et al.*, 2000), although in other systems, enzyme inactivation appears not to contribute to the resistance of biofilm bacteria (Anderl *et al.*, 2000).

29.4.2 Repair

The decreased susceptibility of cells within a biofilm to antimicrobial agents could partly be due to enhanced repair systems. One example is the inducible SOS system, which increases the survival of bacteria exposed to damaging agents by increasing the capacity of error-free and error-prone repair mechanisms (Sutton *et al.*, 2000). While there are no reports of this system in biofilms *per se*, SOS induction has been reported in ageing colonies on agar plates (Taddei *et al.*, 1995). Also, the SOS system can be induced by quinolone antibiotics and trimethoprim (Kimmitt *et al.*, 2000). Enzymes involved with detoxification of reactive oxygen species, notably superoxide dismutase and catalase have been extensively studied in biofilms. Hassett and co-workers have shown that levels of the manganese- and iron-cofactored superoxide dismutases and the major catalase KatA are decreased in mutants of *P. aeruginosa* devoid of one or both quorum-sensing molecules grown planktonically with a concomitant increase in sensitivity to hydrogen peroxide and phenazine methosulphate (Hassett *et al.*, 1999). Perhaps surprisingly, biofilm-grown cells had less catalase activity and yet were more resistant to hydrogen peroxide than their planktonic counterparts. Catalase levels were even lower in quorum-sensing deficient mutants and yet they were also resistant to hydrogen peroxide. One resistance mechanism appears to be prevention of hydrogen peroxide penetration fully into the biofilm (Stewart *et al.*, 2000). Recent evidence with *P. aeruginosa* highlights the influence of iron on quorum-sensing and biofilm-specific oxidative stress response gene regulation (Bollinger *et al.*, 2001).

The susceptibility to reactive oxygen species may also be related to repair of sub-lethal injury following antimicrobial treatment. It has been proposed that sub-lethal injury by antimicrobial agents leads to an imbalance in anabolism and catabolism and a burst of damaging oxygen free-radical production (Dodd *et al.*, 1997) on attempting to recover treated cultures. Viable competitive microflora at high density can protect exponential-phase cells of another organism at lower density from the lethal effects of heat (chemical antimicrobials have not been tested). The authors propose that this addition creates an immediate reduction in the oxygen tension of the culture and oxidative metabolism is reduced (Dodd *et al.*, 1997). Although not tested, variation in oxygen tension gradients within a biofilm, together with the varying metabolic activities within a mixed microbial biofilm, could conceivably contribute to resistance.

29.4.3 Efflux

There has been much work on the contribution of efflux systems to the resistance of bacteria to antimicrobial agents (Zgurskaya and Nikaido, 2000; Lewis, 2001). Several classes of antibiotic are substrates for the pumps and include the tetracyclines, macrolides, β-lactams and fluoroquinolones (Van Bambeke *et al.*, 2000) and their role within biofilms studied. In a comprehensive study of multidrug efflux pumps in *P. aeruginosa*, none of the four systems present in the genome contributed to resistance in a biofilm. Temporal and spatial analyses using fusions to *gfp* indicated that expression of *mexAB–oprM* and *mexCD–oprJ* decreased over time in the developing biofilm with maximal expression occurring at the biofilm substratum (De Kievit *et al.*, 2001b). Also, Gilbert and co-workers have addressed the contribution of the multiple antibiotic resistance (MAR) operon and AcrAB in biofilms of *E. coli*. The Mar operon is present in a number of Gram-negative bacteria and the antibiotic resistance phenotype is mediated by upregulation of AcrAB (Moken *et al.*, 1997). Mutants deleted for MAR showed similar sensitivity to the fluoroquinolone antibiotic ciprofloxacin as wild-type cells grown in a biofilm, whereas a constitutive MAR mutant showed decreased susceptibility. Isolates in which the acrAB efflux pump was deleted also showed similar sensitivity, whereas constitutive expression of acrAB protected biofilms at low antibiotic concentrations but not high, leading the authors to conclude that ciprofloxacin resistance in *E. coli* biofilms is not mediated by upregulation of the marR and acrAB operons (Maira-Litrán *et al.*, 2000).

The contribution of efflux pumps to quorum-sensing and biofilm formation is now becoming apparent, and thus, there is potential for efflux pumps to contribute to resistance of biofilm cells through mechanisms relating to cell density, stress responses and dormancy rather than drug efflux *per se* (see below).

29.4.4   Slow or no growth

Reduced culture growth rate is associated with reduced susceptibility to numerous chemical antimicrobial agents, some of which have a requirement for replication (Brown *et al.*, 1988, 1990; Mah and O'Toole, 2001). Clearly, in any stress response by a growing culture, a reduction in growth rate and even growth cessation is associated. This makes it difficult to separate out the individual contributions to the start of a resistance cascade of growth rate *per se* or/and an enforced change in rate, cell density (and/or quorum sensing, QS) and the nature of a nutrient starvation or other stress. In the case of a biofilm, density is high at an early stage relative to the same number of cells growing in a conventional planktonic culture. It is also difficult to make a valid comparison between biofilm cells and planktonic cells when they have been cultured often in different media and harvested in different physiological states. Using growth rate-controlled cultures of *P. aeruginosa*, *E. coli* and *S. epidermidis*, in both planktonic and biofilm modes of growth, sensitivity increased with increased growth rate (Evans *et al.*, 1991; Duguid *et al.*, 1992a, b). However, increases in growth rate caused greater changes in sensitivity with planktonic cells, indicating factors operating in addition to growth rate. Using chemically and nutritionally defined cultures (Brown *et al.*, 1995; Bühler *et al.*, 1998), which ultimately entered stationary phase because of iron limitation, susceptibility to ciprofloxacin and to ceftazidime was measured along the exponential phase of batch cultures, planktonic and biofilm (Desai *et al.*, 1998). In both growth modes, there were dramatic changes in resistance throughout the exponential phase and before measurable growth reduction for stationary phase. Increases in resistance to both agents occurred in planktonic culture about three to four generations before onset of stationary phase and with biofilms about ten generations before stationary phase. Cell density may well have played a part, although effects were noted well below the commonly recorded quorum-sensing density. Events underpinning changes in growth rate are complex. The alarmone ppGpp, part of the stringent response, has been shown to play a role in regulating growth rate (Cashel *et al.*, 1996) as has the accumulation of inorganic polyphosphate (poly P) (Rao *et al.*, 1998; Kornberg *et al.*, 1999) and serves to highlight the complexity of overlapping regulatory networks that could operate within a biofilm (see below).

29.4.5   Dormancy and stress responses

Stress responses in dividing are always accompanied by a reduction in growth rate. The response of quiescent cells to stress is an open question. Even static suspensions used for susceptibility assays are from cultures that have ceased growth due to starvation or the presence of inhibitory substances. Also, handling techniques, such as centrifugation and re-suspension can contribute to stress (Gilbert *et al.*, 1991). Consequently, a reduction in growth rate is an indication of a stress and a specific slow growth rate may well maintain the cells in the initial stages of a stress response. Thus, slow growth rate is a likely contributor but not the main reason for reduced susceptibility.

There is a large literature on an aspect of the behaviour of the stationary phase of planktonic bacteria known as the GSR (Hengge-Aronis, 1999). This stress response has been implicated directly (Foley *et al.*, 1999) in chronic infection involving biofilms and could clearly occur in circumstances, where high density and QS have been reported (Stickler *et al.*, 1998; Singh *et al.*, 2000; Wu *et al.*, 2001). While most of the work is with Gram-negative bacteria, there are well-characterised stress response systems in Gram-positive bacteria. Systems under the control of alternative sigma factors are found in *Bacillus subtilis* (Scott *et al.*, 2000); *S. aureus* (Chan *et al.*, 1998; Clements and Foster, 1999), where biofilm formation is also affected (Rachid *et al.*, 2000a) and *Mycobacterium tuberculosis* (DeMaio *et al.*, 1996).

The GSR involves a late log and stationary-phase cascade during which structures are protected and the cells become quiescent. It resembles sporulation in its physiological consequences. The result is an ability to survive prolonged periods of nutrient starvation and multiple environmental stresses, such as heat, oxidising agents and hyperosmolarity. Unlike sporulation, the GSR does not involve an all or nothing switch nor an irreversible commitment. Also, some genes involved exhibit expression, which is inversely related to growth rate and are already partially induced under conditions of slow growth (Hengge-Aronis, 1996). The final, slow/non-replicating stages have been described as quiescent, resting or dormant. The expression of many proteins is regulated on entry into stationary phase, of which a core set is induced regardless of the cause of cessation of growth, e.g. the nature of the depleted nutrient (Matin, 1991). In *E. coli*, the *rpoS*-encoded sigma factor σ^s is a master regulator of the GSR (Hengge-Aronis, 1996). While some of the functions of RpoS are common to many Gram-negative bacteria, some are species specific (Suh *et al.*, 1999). There are numerous papers showing a general tendency for nutrient depletion and slow- or no growth to be associated with antibiotic and biocide resistance (Brown, 1977; Brown and Williams, 1985; Brown *et al.*, 1990). In retrospect, it seems probable that, in addition to the consequences of adaptation to the specific nutrient depletion and reduced growth rate, e.g. cell surface changes, a major role in resistance is played by the GSR. There is also evidence that specific nutrient depletion has major effects on sensitivity of microorganisms to host defences (Finch and Brown, 1978; Anwar *et al.*, 1983). Nevertheless, there is as yet little work on the effects of the GSR *per se* on susceptibility to antibiotics (McLeod and Spector, 1996). Its role in biofilm formation is emerging. For example, *E. coli* biofilm density was reduced in an *rpoS* mutant grown in a modified Robbins device (Adams and McLean, 1999). Interestingly, recent work with planktonic chemostat cultures showed that culture density played a role in *rpoS* expression (Liu *et al.*, 2000). While infections caused by Gram-negative bacteria are not treated with glycopeptide antibiotics, the gene encoding D-alanine-D-alanine dipeptidase, part of the vancomycin resistance cluster, is transcribed in stationary phase by RpoS (Lessard and Walsh, 1999). In *E. coli*, it is thought that the D-alanine could be used as an energy source for cell survival under starvation conditions. Recent work with microarrays and *P. aeruginosa* biofilms has shown reduced transcription of *rpoS*, while biofilms established with *rpoS* mutants were thicker and with larger-structured groups of bacteria than those formed by the isogenic wild type (Whiteley *et al.*, 2001). Interestingly, biofilm cells of a *P. aeruginosa rpoS*-deficient mutant were more resistant to tobramycin than wild-type cells. The authors also noted that tobramycin induced the differential expression of 20 genes.

29.4.6   Quorum sensing and antimicrobial susceptibility

It is now clear that quorum-sensing systems contribute significantly to biofilm development. QS systems comprise a transcriptional activator protein that acts in concert with a low molecular weight autoinducer (AI) signalling molecule to increase expression of target genes. As the cell population density increases, so does AI density, providing a means to monitor cell density. By definition, cell density will be high in a compact, adherent biofilm population and, consequently, relatively small biofilm populations probably demonstrate signal-driven, stationary-phase survival responses, which equivalent numbers of free-growing planktonic counterparts would not (Gilbert *et al.*, 1995). This could partially explain the general high resistance of biofilm organisms to exogenous stress. Using the example of *P. aeruginosa*, biofilm structure is dependent on QS, which may itself be dependent on accumulation of inorganic poly P (see below). LasI mutants deficient in production of the N-(3-oxododecanoyl) homoserine lactone (3OC12-HSL) AI molecule formed flat, undifferentiated biofilms that were sensitive to treatment with sodium dodecyl sulphate (Davies *et al.*, 1998). Other work has produced similar results (Shih and Huang, 2002). Evidence suggests that the

3OC12-HSL quorum-sensing molecule is a substrate for the MexAB–OprM multidrug efflux system, since mutants, which hyperexpress this system showed reduced levels of extracellular virulence factors known to be regulated by QS (Evans *et al.*, 1998). Also, a defined mutant lacking the pump accumulated more 3OC12-HSL, comparable with wild-type cells treated with cytoplasmic membrane proton gradient inhibitors (Pearson *et al.*, 1999). *P. aeruginosa* produces two AI molecules, 3OC12-HSL and N-butyryl-L-homoserine lactone (C4-HSL). In conventional planktonic culture, *P. aeruginosa* produces 3OC12-HSL at a rate between three and ten times that of C4-HSL, whereas more C4-HSL was produced by biofilm-grown cells (Singh *et al.*, 2000). Greater C4-HSL levels were noted in extracts of sputum from cystic fibrosis patients, indicating that bacteria are perhaps growing as a biofilm in the lungs of these patients (Singh *et al.*, 2000).

Signal diffusion within and from a biofilm will be affected by the nature of the biofilm matrix and of the substratum. Thus, an impermeable substratum for biofilm growth would concentrate any signal, while the degree of hydrophobicity of a boundary could influence entrapment or diffusion, depending on the chemistry of the signal. A hydrophobic signal could be trapped as aggregates/micelles within cell exopolymer and maintain an equilibrium concentration of hydrophobic monomer close to the cell, while hydrophilic molecules could diffuse away. Consistent with this hypothesis, expression of *P. aeruginosa* AI synthase genes was greatest at the interface with the impermeable substratum (De Kievit *et al.*, 2001a), permitting rapid amplification of cell-density-dependent responses, since the AI synthase genes themselves are subject to autoregulation. The N-acyl substituted HSLs vary with respect to the state of oxidation at the C-3 position and the fatty acid chain length. Relative hydrophobicity/lipophilicity for a bioactive compound can be predicted by the parameter $\log P$. Optimum permeation through a Gram-negative envelope is commonly at about $\log P$ of 4. Longer chain length AI molecules, such as 3OC12-HSL from *P. aeruginosa*, having a $\log P$ of approximately 3, would be predicted to enter or exit cells by simple diffusion (Heys *et al.*, 1997), but its amphipathic nature may well result in a tendency to form large aggregates under some conditions. Indeed, in at least the case of *P. aeruginosa* 3OC12-HSL, active efflux appears to be required (Pearson *et al.*, 1999).

Acylated homoserine lactones do not appear to operate in members of the enterobacteriacae, such as *E. coli* and *Salmonella enteritidis*, although AI activity has been reported (Surette *et al.*, 1999) and QS regulates activity of the locus of enterocyte attachment and effacement and intestinal colonisation (Sperandio *et al.*, 1999). There are, however, thus far no studies on the influence of antimicrobial agents.

A number of QS or cell density-dependent systems also operate in Gram-positive species, although here the AI molecules are typically small peptides. Examples include regulation of streptococcal competence for genetic transformation (Havarstein and Morrison, 1999); cell-density control of gene expression and sporulation in *Bacillus* spp. (Lazazzera *et al.*, 1999); the sex pheromone systems regulating conjugative plasmid transfer in *Enterococcal* spp., some of which have associated antibiotic resistance (Clewell, 1999) and the regulation of

S. aureus pathogenicity (Novick, 1999). Agents, such as chloramphenicol and tetracycline at sub-inhibitory concentrations can perturb signalling cascades through inhibition of AI peptide synthesis. Studies with the *agr* quorum-sensing system in *S. aureus* have shown that QS-deficient mutants were more able to form a biofilm on polystyrene than wild-type strains, leading the authors to question the utility of anti-QS molecules to eradicate biofilm infections (Vuong *et al.*, 2000).

Evidence for the direct contribution of AI or quorum-sensing molecules to the antimicrobial susceptibility of bacteria is lacking. Nevertheless, given their central role in virulence factor production and biofilm formation, they are themselves attractive targets for antimicrobial drug design. However, it is already clear that there are fundamental differences in the contribution of these systems to biofilm formation between species. Such differences are not surprising. It is necessary to bear in mind the plethora both of bacterial phenotypes and of potential surfaces for colonisation.

29.5   OVERLAPPING REGULATORY NETWORKS IN BIOFILMS

It is clear that there are several overlapping regulatory networks with high degrees of interaction operating in slow- or non-growing, relatively dormant bacteria that likely contribute to their susceptibility to antimicrobials. As yet, there are few studies on developing or mature biofilms. The GSR regulator RpoS is itself subject to complex transcriptional, translational and post-translational control by factors including (p)ppGpp and the stringent response (Cashel *et al.*, 1996), inorganic poly P (Shiba *et al.*, 1997), OxyR (Pomposiello and Demple, 2001), SOS (Sutton *et al.*, 2000) and cAMP-CRP (Spector, 1998), which are themselves subject to complex regulatory control. Moreover, RpoS is not solely responsible for stress responses in the many Gram-negative species in which it is found (Jørgensen *et al.*, 1999). Examples of complex overlapping networks include the rhl QS in *P. aeruginosa*, one of two QS systems, which themselves interact, which is regulated by RpoS (Whiteley *et al.*, 2000). Again in studies with *P. aeruginosa*, recent evidence indicates an important role for poly P kinase (PPK) and the accumulation of inorganic poly P in biofilm development (Rashid *et al.*, 2000). Inorganic poly P is a linear polymer of many orthophosphate residues. In bacteria, the highly conserved PPK polymerises the terminal phosphate of ATP to the poly P chain. PPK-deficient mutants of *E. coli* have been shown to be unable to adapt to nutritional stringencies and environmental stress, attributed in part to the failure to express *rpoS* (Shiba *et al.*, 1997). Now a *P. aeruginosa ppk* mutant has been shown to be unable to form thick, differentiated biofilms. Synthesis of both quorum-sensing AI molecules was reduced by 50% in the *ppk* mutant and expression of two QS target genes was reduced by more than 90%. These data led the authors to suggest that PPK and/or poly P may affect AI complex formation or perhaps interaction between RNA polymerase and promoter sequences (Rashid *et al.*, 2000). Further complexity lies in the regulation of SOS genes by PPK and poly P (Tsutsumi *et al.*, 2000). Again, no studies have been reported on the contribution of PPK to antimicrobial susceptibility; however, *P. aeruginosa ppk* mutants showed reduced virulence in a mouse model of infection (Rashid *et al.*, 2000). These data clearly imply a role in susceptibility to host defences, which as noted earlier in this chapter, will likely always make a significant contribution to 'eradication' of biofilm infections even when treated with antibiotics. Given that PPK is conserved in bacteria and not present in eukaryotes, it has been proposed as an attractive target for antimicrobial drugs (Rashid *et al.*, 2000).

29.6   CONCLUSIONS

The general resistance of biofilms is clearly phenotypic. Recent evidence showed that antibiotic-resistant phenotypic variants of *P. aeruginosa* with enhanced ability to form biofilms arose at high frequency, both *in vitro* and in the lungs of cystic fibrosis patients (Drenkard and Ausubel, 2002). A regulatory protein (PvrR) was also identified that controls the conversion between antibiotic-resistant and susceptible forms.

The well-characterised resistance mechanisms are lack of antibiotic penetration, inactivation, efflux and repair make contributions in some circumstances. However, compelling evidence that they are uniquely responsible for biofilm resistance is lacking. The suspicion that reduced growth rate has an involvement is true in that it is associated with responses to stress. Key structures are protected and cellular processes close down to a state of dormancy, leading to resistance to numerous stresses (for references see (Hengge-Aronis, 1996)). Such stress responses, linked with reduced growth rate, will be driven at least in part by the high density and QS events occurring within the biofilm. It is interesting to note that in a study of susceptibility profiles of isolates from cystic fibrosis patients receiving intensive antibiotic treatment for several years, some of the isolates remained susceptible *in vitro* (Ciofu *et al.*, 2001). This could be interpreted as phenotypic resistance, possibly due to dormancy. It may be worth remembering that dormancy is a widespread and evolutionary ancient biological survival response to stress, and we propose that exceptional vegetative cell dormancy is the basic explanation of biofilm resistance.

REFERENCES

Adams, J.L. and McLean, R.J.C. (1999) Impact of *rpoS* deletion on *Escherichia coli* biofilms. *Appl. Environ. Microbiol.* **65**, 4285–4287.

Al-Bakri, A.G., Gilbert, P. and Allison, D.G. (1999) Mixed species biofilms of *Burkholderia cepacia* and *Pseudomonas aeruginosa*. In *Biofilms: The Good, the Bad and the Ugly* (ed. J. Wimpenny, P. Gilbert, J. Walker, M. Brading and R. Bayston), pp. 327–337, BioLine, Cardiff.

Allison, D.G., Brown, M.R.W., Evans, D.J. and Gilbert, P. (1990a) Surface hydrophobicity and dispersal of *Pseudomonas aeruginosa* from biofilms. *FEMS Microbiol. Lett.* **71**, 101–104.

Allison, D.G., Evans, D.J., Brown, M.R.W. and Gilbert, P. (1990b) Possible involvement of the division cycle in dispersal of *Escherichia coli* from biofilms. *J. Bacteriol.* **172**, 1667–1669.

Allison, D., Maira-Litrán, T. and Gilbert, P. (1999) Perfused biofilm fermenters. *Meth. Enzymol.* **310**, 232–248.

An, Y.H., Dickinson, R.B. and Doyle, R.J. (2000) Mechanisms of bacterial adhesion and pathogenesis of implant and tissue infections. In *Handbook of Bacterial Adhesion: Principles, Methods, and Applications* (ed. Y.H. An and R.J. Friedman), pp. 1–27, Humana Press Inc, Totowa, NJ.

Anderl, J.N., Franklin, M.J. and Stewart, P.S. (2000) Role of antibiotic penetration limitation in *Klebsiella pneumoniae* biofilm resistance to ampicillin and ciprofloxacin. *Antimicrob. Agents Chemother.* **44**, 1818–1824.

Anwar, H., Brown, M.R.W. and Lambert, P.A. (1983) Effect of nutrient depletion on sensitivity of *Pseudomonas cepacia* to phagocytosis and serum bactericidal activity at different temperatures. *J. Gen. Microbiol.* **129**, 2021–2027.

Bagge, N., Ciofu, O., Skovgaard, L.T. and Høiby, N. (2000) Rapid development *in vitro* and *in vivo* of resistance to ceftazidime in biofilm-growing *Pseudomonas aeruginosa* due to chromosomal beta-lactamase. *APMIS* **108**, 589–600.

Boland, T., Latour, R.A. and Stutzenberger, F.J. (2000) Molecular basis of bacterial adhesion. In *Handbook of Bacterial Adhesion: Principles, Methods, and Applications* (ed. Y.H. An and R.J. Friedman), pp. 29–41, Humana Press Inc, Totowa, NJ.

Bollinger, N., Hassett, D.J., Iglewski, B.H., Costerton, J.W. and McDermott, T.R. (2001) Gene expression in *Pseudomonas aeruginosa*: evidence of iron override effects on quorum sensing and biofilm-specific gene regulation. *J. Bacteriol.* **183**, 1990–1996.

Brooun, A., Liu, S. and Lewis, K. (2000) A dose-response study of antibiotic resistance in *Pseudomonas aeruginosa* biofilms. *Antimicrob. Agents Chemother.* **44**, 640–646.

Brown, M.R.W. (1977) Nutrient depletion and antibiotic susceptibility. *J. Antimicrob. Chemother.* **3**, 198–201.

Brown, M.R.W. and Smith, A.W. (2001) Dormancy and persistence in chronic infection: role of the general stress response in resistance to chemotherapy. *J. Antimicrob. Chemother.* **48**, 141–142.

Brown, M.R.W. and Williams, P. (1985) The influence of environment on envelope properties affecting survival of bacteria in infections. *Ann. Rev. Microbiol.* **39**, 527–556.

Brown, M.R.W., Allison, D. and Gilbert, P. (1988) Resistance of bacterial biofilms to antibiotics: a growth-rate related effect? *J. Antimicrob. Chemother.* **22**, 777–780.

Brown, M.R.W., Collier, P.J. and Gilbert, P. (1990) Influence of growth rate on susceptibility to antimicrobial agents: Modification of the cell envelope and batch and continuous culture studies. *Antimicrob. Agent. Chemother.* **34**, 1623–1628.

Brown, M.R.W., Collier, P.J., Courcol, R.J. and Gilbert, P. (1995) Definition of phenotype in batch culture. In *Microbiological Quality Assurance. A Guide Towards Relevance and Reproducibility of Inocula* (ed. M.R.W. Brown and P. Gilbert), pp. 13–20, CRC Press, Boca Raton, FL.

Brub-Buisson, C. (2001) New technologies and infection control practices to prevent intravascular catheter-related infections. *Am. J. Resp. Crit. Care Med.* **164**, 1557–1558.

Bühler, T., Ballestero, S., Desai, M. and Brown, M.R.W. (1998) Generation of a reproducible nutrient-depleted biofilm of *Escherichia coli* and *Burkholderia cepacia*. *J. Appl. Bact.* **85**, 457–462.

Cashel, M., Gentry, D.R., Hernandez, V.J. and Vinella, D. (1996) The stringent response. In *Escherichia coli and Salmonella* (ed. F.C. Neidhardt, R. Curtiss III, J.L. Ingraham, E.C.C. Lin, K.B. Low, B. Magasanik, W.S. Reznikoff, M. Riley, M. Schaechter and H.E. Umbarger),pp. 1458–1496, ASM Press, Washington DC.

Chan, P.F., Foster, S.J., Ingham, E. and Clements, M.O. (1998) The *Staphylococcus c* alternative sigma factor sigma B controls the environmental stress response but not starvation survival or pathogenicity in a mouse abscess model. *J. Bacteriol.* **180**, 6082–6089.

Ciofu, O., Fussing, V., Bagge, N., Koch, C. and Hoiby, N. (2001) Characterization of paired mucoid/non-mucoid *Pseudomonas aeruginosa* isolates from Danish cystic fibrosis patients: antibiotic resistance, β-lactamase activity and RiboPrinting. *J. Antimicrob. Chemother.* **48**, 391–396.

Clements, M.O. and Foster, S.J. (1999) Stress resistance in *Staphylococcus aureus. Trend. Microbiol.* **7**, 458–462.

Clewell, D.B. (1999) Sex pheromone systems in Enterococci. In *Cell–Cell Signaling in Bacteria* (ed. G.M. Dunny and S.C. Winans), pp. 47–66, ASM Press, Washington DC.

Costerton, J.W., Lewandowski, Z., Caldwell, D.E., Korber, D.R. and Lappin-Scott, H.M. (1995) Microbial biofilms. *Ann. Rev. Microbiol.* **49**, 711–745.

Costerton, J.W., Stewart, P.S. and Greenberg, E.P. (1999) Bacterial biofilms: a common cause of persistent infections. *Science* **284**, 1318–1322.

Cramton, S.E., Gerke, C., Nichols, W.W. and Gotz, F. (1999) The intercellular adhesion (*ica*) locus is present in *Staphylococcus aureus* and is required for biofilm. *Infect. Immun.* **67**, 5427–5433.

Davies, D.G., Parsek, M.R., Pearson, J.P., Iglewski, B.H., Costerton, J.W. and Greenberg, E.P. (1998) The involvement of cell-to-cell signals in the development of a bacterial biofilm. *Science* **280**, 295–298.

De Kievit, T.R., Gillis, R., Marx, S., Brown, C. and Iglewski, B.H. (2001a) Quorum-sensing genes in *Pseudomonas aeruginosa* biofilms: their role and expression patterns. *Appl. Environ. Microbiol.* **67**, 1865–1873.

De Kievit, T.R., Parkins, M.D., Gillis, R.J., Srikumar, R., Ceri, H., Poole, K., Iglewski, B.H. and Storey, D.G. (2001b) Multidrug efflux pumps: expression patterns and contribution to antibiotic resistance in *Pseudomonas aeruginosa* biofilms. *Antimicrob. Agent. Chemother.* **45**, 1761–1770.

DeMaio, J., Zhang, Y., Ko, C., Young, D.B. and Bishai, W.R. (1996) A stationary-phase stress-response sigma factor from *Mycobacterium tuberculosis. Proc. Natl. Acad. Sci. USA* **93**, 2790–2794.

Desai, M., Bühler, T., Weller, P.H. and Brown, M.R.W. (1998) Increasing resistance of planktonic and biofilm cultures of *Burkholderia cepacia* to ciprofloxacin and ceftazidime during exponential growth. *J. Antimicrob. Chemother.* **42**, 153–160.

Dibdin, G. and Wimpenny, J. (1999) Steady-state biofilm: practical and theoretical models. *Meth. Enzymol.* **310**, 296–322.

Dodd, C.E.R., Sharman, R.L., Bloomfield, S.F., Booth, I.R. and Stewart, G.S.A.B. (1997) Inimical processes: bacterial self destruction and sub-lethal injury. *Trend. Food Sci. Technol.* **8**, 238–241.

Domingue, P.A.G., Lambert, P.A. and Brown, M.R.W. (1989a) Iron depletion alters surface-associated properties of *Staphylococcus aureus* and its association to human neutrophils in chemiluminescence. *FEMS Microbiol. Lett.* **59**, 265–268.

Domingue, P.A.G., Schwarzinger, E. and Brown, M.R.W. (1989b) Growth rate, iron depletion, and a sub-minimal inhibitory concentration of pencillin G affect the surface hydrophobicity of *Staphylococcus aureus*. In *The Influence of Antibiotics on the Host–Parasite Relationship III* (ed. G. Gillissen, W. Opferkuch, G. Peters and G. Pulverer), pp. 50–62, Springer-Verlag, Berlin, Germany.

Drenkard, E. and Ausubel, F.M. (2002) *Pseudomonas* biofilm formation and antibiotic resistance are to linked to phenotypic variation. *Nature* **416**, 740–743.

Duguid, I.G., Evans, E., Brown, M.R.W. and Gilbert, P. (1992a) Effect of biofilm culture upon the susceptibility of *Staphylococcus epidermidis* to tobramycin. *J. Antimicrob. Chemother.* **30**, 803–810.

Duguid, I.G., Evans, E., Brown, M.R.W. and Gilbert, P. (1992b) Growth-rate-independent killing by ciprofloxacin of biofilm-derived *Staphylococcus epidermidis*; evidence for cell-cycle dependency. *J. Antimicrob. Chemother.* **30**, 791–802.

Evans, D.J., Allison, D.G., Brown, M.R.W. and Gilbert, P. (1991) Susceptibility of *Pseudomonas aeruginosa* and *Escherichia coli* biofilms towards ciprofloxacin: effect of specific growth rate. *J. Antimicrob. Chemother.* **27**, 177–184.

Evans, K., Passador, L., Srikumar, R., Tsang, E., Nezezon, J. and Poole, K. (1998) Influence of the MexAB-OprM multidrug efflux system on quorum sensing in *Pseudomonas aeruginosa. J. Bacteriol.* **180**, 5443–5447.

Finch, J.E. and Brown, M.R.W. (1978) Effect of growth environment on *Pseudomonas aeruginosa* killing by rabbit polymorphonuclear leukocytes and cationic proteins. *Infect. Immun.* **20**, 340–346.

Foley, I., Marsh, P., Wellington, E.M.H., Smith, A.W. and Brown, M.R.W. (1999) General stress response master regulator *rpoS* is expressed in human infection: a possible role in chronicity. *J. Antimicrob. Chemother.* **43**, 164–165.

Gilbert, P., Allison, D.G., Evans, D.J., Handley, P.S. and Brown, M.R.W. (1989) Growth rate control of adherent bacterial populations. *Appl. Environ. Microbiol.* **55**, 1308–1311.

Gilbert, P., Collier, P.J. and Brown, M.R.W. (1990) Influence of growth rate on susceptibility to antimicrobial agents: biofilms, cell cycle, dormancy, and stringent response. *Antimicrob. Agent. Chemother.* **34**, 1865–1868.

Gilbert, P., Coplan, F. and Brown, M.R.W. (1991) Centrifugation injury of Gram-negative bacteria. *J. Antimicrob. Chemother.* **27**, 550–551.

Gilbert, P., Hodgson, A.E. and Brown, M.R.W. (1995) Influence of the environment on the properties of microorganisms grown in association with surfaces. In *Microbiological Quality Assurance: A Guide Towards Relevance and Reproducibility of Inocula* (ed. M.R.W. Brown and P. Gilbert), pp. 61–82, CRC Press Inc, Boca Raton, FL.

Gilbert, P., Allison, D.G., Jacob, A., Korner, D., Wolfaa, G. and Foley, I. (1997) Immigration of planktonic *Enterococcus faecalis* cells into mature *E. faecalis* biofilms. In *Biofilms: Community Interactions and Control* (ed. J.T. Wimpenny, P. Handley, P. Gilbert, and H.M. Lappin-Scott), pp. 133–142, Bioline, Cardiff.

Giwercman, B., Jensen, E.T., Høiby, N., Kharazmi, A. and Costerton, J.W. (1991) Induction of beta-lactamase production in *Pseudomonas aeruginosa* biofilm. *Antimicrob. Agent. Chemother.* **35**, 1008–1010.

Gottenbos, B., Grijpma, D.W., van der Mei, H.C., Feijen, J. and Busscher, H.J. (2001) Antimicrobial effects of positively charged surfaces on adhering Gram-positive and Gram-negative bacteria. *J. Antimicrob. Chemother.* **48**, 7–13.

Gross, M., Cramton, S.E., Gotz, F. and Peschel, A. (2001) Key role of teichoic acid net charge in *Staphylococcus aureus* colonization of artificial surfaces. *Infect. Immun.* **69**, 3423–3426.

Hassett, D.J., Ma, J.F., Elkins, J.G., McDermott, T.R., Ochsner, U.A., West, S.E.H., Huang, C.T., Fredericks, J., Burnett, S., Stewart, P.S., McFeters, G., Passador, L. and Iglewski, B.H. (1999) Quorum sensing in *Pseudomonas aeruginosa* controls expression of catalase and superoxide dismutase genes and mediates biofilm susceptibility to hydrogen peroxide. *Mol. Microbiol.* **34**, 1082–1093.

Havarstein, L.S. and Morrison, D.A. (1999) Quorum sensing and peptide pheromones in Streptococcal competence for genetic transformation. In *Cell–Cell Signaling in Bacteria* (ed. G.M. Dunny and S.C. Winans), pp. 9–26, ASM Press, Washington DC.

Hengge-Aronis, R. (1996) Regulation of gene expression during entry into stationary phase. In *Escherichia coli and Salmonella. Cellular and Molecular Biology* (ed. F.C. Neidhardt, R. Curtiss III, J.K. Ingraham, E.C.C. Lin, K.B. Low, B. Magasanik, W.S. Reznikoff, M. Riley, M. Schaechter and H.E. Umbarger), Vol. 1, pp. 1497–1512, ASM Press, Washington DC.

Hengge-Aronis, R. (1999) Interplay of global regulators and cell physiology in the general stress response of *Escherichia coli. Curr. Opin. Microbiol.* **2**, 148–152.

Heys, S.J.D., Gilbert, P., Eberhard, A. and Allison, D.G. (1997) Homoserine lactones and bacterial biofilms. In *Biofilms: Community Interactions and Control* (ed. J. Wimpenny, P. Handley, P. Gilbert, H.M. Lappin-Scott and M. Jones), pp. 103–112, Bioline, Cardiff.

Hopelman, A.I.M. and Tuomanen, E. (1992) Consequences of microbial attachment: directing host cell functions with adhesins. *Infect. Immun.* **60**, 1729–1733.

Jørgensen, F., Bally, M., Chapon-Herve, V., Stewart, G.S.A.B., Michel, G., Lazdunski, A. and Williams, P. (1999) RpoS-dependent stress tolerance in *Pseudomonas aeruginosa*. *Microbiology* **145**, 835–844.

Karthikeyan, S., Korber, D.R., Wolfaardt, G.M. and Caldwell, D.E. (2000) Monitoring the organization of microbial biofilm communities. In *Handbook of Bacterial Adhesion: Principles, Methods, and Applications* (ed. Y.H. An and R.J. Friedman), pp. 171–188, Humana Press, Totowa NJ.

Kharazmi, A., Giwercman, B. and Høiby, N. (1999) Robbins device in biofilm research. *Meth. Enzymol.* **310**, 207–215.

Kimmitt, P.T., Harwood, C.R. and Barer, M.R. (2000) Toxin gene expression by shiga toxin-producing *Escherichia coli*: the role of antibiotics and the bacterial SOS response. *Emerg. Infect. Dis.* **6**, 458–465.

Kornberg, A., Rao, N.N. and Ault-Richie, D. (1999) Inorganic polyphosphate: a molecule of many functions. *Ann. Rev. Biochem.* **68**, 89–125.

Lazazzera, B.A., Palmer, T., Quisle, J. and Grossman, A.D. (1999) Cell-density control of gene expression and development in *Bacillus subtilis*. In *Cell–Cell Signaling in Bacteria* (ed. G.M. Dunny and S.C. Winans), pp. 27–46, ASM Press, Washington DC.

Lessard, I.A. and Walsh, C.T. (1999) VanX, a bacterial D-alanyl-D-alanine dipeptidase: resistance, immunity, or survival function? *Proc. Natl. Acad. Sci. USA* **96**, 11028–11032.

Lewis, K. (2001) Riddle of biofilm resistance. *Antimicrob. Agent. Chemother.* **45**, 999–1007.

Liu, X., Ng, C. and Ferenci, T. (2000) Global adaptations resulting from high population densities in *Escherichia coli* cultures. *J. Bacteriol.* **182**, 4158–4164.

Mah, T.-F.C. and O'Toole, G.A. (2001) Mechanisms of biofilm resistance to antimicrobial agents. *Trend. Microbiol.* **9**, 34–39.

Maira-Litrán, T., Allison, D.G. and Gilbert, P. (2000) Expression of the multiple antibiotic resistance operon(*mar*) during growth of *Escherichia coli* as a biofilm. *J. Appl. Microbiol.* **88**, 243–247.

Matin, A. (1991) The molecular basis of carbon-starvation-induced general resistance in *Escherichia coli*. *Mol. Microbiol.* **5**, 3–10.

McLeod, G.I. and Spector, M.P. (1996) Starvation- and stationary-phase-induced resistance to the antimicrobial peptide polymyxin B in *Salmonella typhimurium* is RpoS (sigma(S)) independ ent and occurs through both phoP-dependent and -independent pathways. *J. Bacteriol.* **178**, 3683–3688.

Moken, M.C., McMurray, L.M. and Levy, S.B. (1997) Selection of multiple-antibiotic-resistant (Mar) mutants of *Escherichia coli* by using the disinfectant pine oil: roles of the *mar* and *acrAB* loci. *Antimicrob. Agent. Chemother.* **41**, 2770–2772.

Nichols, W.W. (1991) Biofilms, antibiotics and penetration. *Rev. Med. Microbiol.* **2**, 177–181.

Novick, R.P. (1999) Regulation of pathogenicity in *Staphylococcus aureus* by a peptide-based density-sensing system . In *Cell–Cell Signaling in Bacteria* (ed. G.M. Dunny and S.C. Winans), pp. 129–146, ASM Press, Washington DC.

O'Toole, G.A. and Kolter, R. (1998) Flagella and twitching motility are necessary for *Pseudomonas aeruginosa* biofilm development. *Mol. Microbiol.* **30**, 295–304.

Pearson, J.P., Van Delden, C. and Iglewski, B.H. (1999) Active efflux and diffusion are involved in transport of *Pseudomonas aeruginosa* cell-to-cell signals. *J. Bacteriol.* **181**, 1203–1210.

Pier, G.B., Saunders, J.M., Ames, P., Edwards, M.S., Auerbach, H., Speert, D.P. and Hurwitch, S. (1987) Opsonophagocytic killing antibody to *Pseudomonas aeruginosa* mucoid exopolysaccharide in older noncolonized patients with cystic fibrosis. *N. Engl. J. Med.* **317**, 793–798.

Pomposiello, P.J. and Demple, B. (2001) Redox-operated genetic switches: the SoxR and OxyR transcription factors. *Trend, Biotechnol.* **19**, 109–114.

Pratt, L.A. and Kolter, R. (1998) Genetic analysis of *Escherichia coli* biofilm formation: roles of flagella, motility, chemotaxis and type I pili. *Mol. Microbiol.* **30**, 285–293.

Rachid, S., Ohlsen, K., Wallner, U., Hacker, J., Hecker, M. and Ziebuhr, W. (2000a) Alternative transcription factor sigma B is involved in regulation of biofilm expression in a *Staphylococcus aureus* mucosal isolate. *J. Bacteriol.* **182**, 6824–6826.

Rachid, S., Ohlsen, K., Witte, W., Hacker, J. and Ziebuhr, W. (2000b) Effect of subinhibitory antibiotic concentrations on polysaccharide intercellular adhesin expression in biofilm-forming *Staphylococcus epidermidis*. *Antimicrob. Agent. Chemother.* **44**, 3357–3363.

Rao, N.N., Liu, S.J. and Kornberg, A. (1998) Inorganic polyphosphate in *Escherichia coli*: the phosphate regulon and the stringent response. *J. Bacteriol.* **180**, 2186–2193.

Rashid, M.H., Rumbaugh, K., Passador, L., Davies, D.G., Hamood, A.N., Iglewski, B.H. and Kornberg, A. (2000) Polyphosphate kinase is essential for biofilm development, quorum sensing, and virulence of *Pseudomonas aeruginosa*. *Proc. Natl. Acad. Sci. USA* **97**, 9636–9641.

Scott, J.M., Mitchell, T. and Haldenwang, W.G. (2000) Stress triggers a process that limits activation of the *Bacillus subtilis* stress transcription factor sigma B. *J. Bacteriol.* **182**, 1452–1456.

Shiba, T., Tsutsumi, K., Yano, H., Iahara, Y., Yameda, A., Tanaka, T., Takahashia, M., Munekata, M., Rao, N.N. and Kornberg, A. (1997) Inorganic polyphosphate and the induction of *rpoS* expression. *Proc. Natl. Acad. Sci. USA* **94**, 11210–11215.

Shih, P.-C. and Huang, C.T. (2002) Effect of quorum-sensing deficiency on *Pseudomonas aeruginosa* biofilm formation and antibiotic resistance. *J. Antimicrob. Chemother.* **49**, 309–314.

Singh, P.K., Schaefer, A.L., Parsek, M.R., Moninger, T.O., Welsh, M.J. and Greenberg, E.P. (2000) Quorum-sensing signals indicate that cystic fibrosis lungs are infected with bacterial biofilms. *Nature* **407**, 762–764.

Sissons, C.H., Wong, L. and An, Y.H. (2000) Laboratory culture and analysis of biofilms. In *Handbook of Microbial Adhesion: Principles, Methods and Applications* (ed. Y.H. An and R.J. Friedman), pp. 133–169, Humana Press, Totowa, NJ.

Spector, M.P. (1998) The starvation–stress response (SSR) of *Salmonella*. *Adv. Microb. Physiol.* **40**, 233–279.

Sperandio, V., Mellies, J.L., Nguyen, W., Shin, S. and Kaper, J.B. (1999) Quorum sensing controls expression of the type III secretion gene transcription and protein secretion in enterohemorrhagic and enteropathogenic *Escherichia coli*. *Proc. Natl. Acad. Sci. USA* **96**, 15196–15201.

Spoering, A.L. and Lewis, K. (2001) Biofilms and planktonic cells of *Pseudomonas aeruginosa* have similar resistance to killing by antimicrobials. *J. Bacteriol.* **182**, 6746–6751.

Stewart, P.S. (1996) Theoretical aspects of antibiotic diffusion into microbial biofilms. *Antimicrob. Agent. Chemother.* **38**, 2125–2133.

Stewart, P.S. and Costerton, J.W. (2001) Antibiotic resistance of bacteria in biofilms. *Lancet* **358**, 135–138.

Stewart, P.S., Roe, F., Rayner, J., Elkins, J.G., Lewandowski, Z., Ochsner, U.A. and Hassett, D.J. (2000) Effect of catalase on hydrogen peroxide penetration into *Pseudomonas aeruginosa* biofilms. *Appl. Environ. Microbiol.* **66**, 836–838.

Stickler, D.J., Morris, N.S., McLean, R.J. and Fuqua, C. (1998) Biofilms on indwelling urethral catheters produce quorum-sensing signal molecules *in situ* and *in vitro*. *Appl. Environ. Microbiol.* **64**, 3486–3490.

Stoodley, P., Wilson, S., Hall-Stoodley, L., Lappin-Scott, H.M. and Costerton, J.W. (2001) Growth and detachment of cell clusters from mature mixed-species biofilms. *Appl. Environ. Microbiol.* **67**, 5608–5613.

Suh, S.-J., Silo-Suh, L., Woods, D.E., Hassett, D.J., West, S.E.H. and Ohman, D.E. (1999) Effect of *rpoS* mutation on the stress response and expression of virulence factors in *Pseudomonas aeruginosa. J. Bacteriol.* **181**, 3890–3897.

Surette, M.G., Miller, M.B. and Bassler, B.L. (1999) Quorum sensing in *Escherichia coli, Salmonella typhimurium*, and *Vibrio harveyi*: a new family of genes responsible for autoinducer production. *Proc. Natl. Acad. Sci. USA* **96**, 1639–1644.

Sutherland, I.W. (2001) The biofilm matrix – an immobilised but dynamic environment. *Trend. Microbiol.* **9**, 222–227.

Sutton, M.D., Smith, B.T., Godoy, V.G. and Walker, G.C. (2000) The SOS response: recent insights into umuDC-dependent mutagenesis and DNA damage tolerance. *Ann. Rev. Genet.* **34**, 479–497.

Taddei, F., Matic, I. and Radman, M. (1995) cAMP-dependent SOS induction and mutagenesis in resting bacterial populations. *Proc. Natl. Acad. Sci. USA* **92**, 11736–11740.

Tsutsumi, K., Munekata, M. and Shiba, T. (2000) Involvement of inorganic polyphosphate in expression of SOS genes. *Biochim. Biophys. Acta* **1493**, 73–81.

Van Bambeke, F., Balzi, E. and Tulkens, P.M. (2000) Antibiotic efflux pumps. *Biochem. Pharmacol.* **60**, 457–470.

van Oss, C.J. (1978) Phagocytosis as a surface phenomenon. *Ann. Rev. Microbiol.* **32**, 19–39.

Vuong, C., Saenz, H.L., Gotz, F. and Otto, M. (2000) Impact of the *agr* quorum-sensing system on adherence to polystyrene in *Staphylococcus aureus. J. Infect. Dis.* **182**, 1688–1693.

Whiteley, M., Parsek, M.R. and Greenberg, E.P. (2000) Regulation of quorum sensing by RpoS in *Pseudomonas aeruginosa. J. Bacteriol.* **182**, 4356–4360.

Whiteley, M., Bangera, M.G., Bumgarner, R.E., Parsek, M.R., Teitzel, G.M., Lory, S. and Greenberg, E.P. (2001) Gene expression in *Pseudomonas aeruginosa* biofilms. *Nature* **413**, 860–864.

Williams, P., Lambert, P.A., Haigh, C.G. and Brown, M.R.W. (1986) The influence of the O and K antigens of *Klebsiella aerogenes* on surface hydrophobicity and susceptibility to phagocytosis and antimicrobial agents. *J. Med. Microbiol.* **21**, 125–132.

Wu, H., Song, Z., Givskov, M., Doring, G., Worlitzsch, D., Rygaard, J. and Høiby, N. (2001) *Pseudomonas aeruginosa* mutations in *lasI* and *rhlI* quorum sensing systems result in milder chronic lung infection. *Microbiology* **147**, 1105–1113.

Xu, K.D., McFeters, G. and Stewart, P.S. (2000) Biofilm resistance to antimicrobial agents. *Microbiology* **146**, 547–549.

Yasuda, H., Koga, T. and Fukoka, T. (1999) *In vitro* and *in vivo* models of bacterial biofilms. *Meth. Enzymol.* **310**, 577–595.

Zgurskaya, H.I. and Nikaido, H. (2000) Multidrug resistance mechanisms: drug efflux across two membranes. *Mol. Microbiol.* **37**, 219–225.

30

Control of biofilm in the food industry: a microbiological survey of high-risk processing facilities

A. Peters

30.1 INTRODUCTION

Bacteria attach to available surfaces in many natural, industrial and medical environments and can develop into extensive biofilm. Biofilm consist of a complex consortium of microorganisms enmeshed within an extracellular matrix, which is largely composed of water and various polymers, commonly polysaccharides and glycoproteins (Christensen and Characklis, 1990). Biofilms are considered to have a heterogeneous structure consisting of microcolonies and exhibit microenvironments different from the bulk phase (Costerton *et al.*, 1995). The biofilm mode of growth confers a number of advantages on the bacteria, including an increased resistance against antimicrobial agents.

Food-processing environments provide a variety of conditions, which might favour the formation of biofilm, i.e. presence of moisture, attachment surfaces, nutrients and inocula of microorganisms from the raw materials or the environment (Holah and Kearney, 1992; Holah *et al.*, 1994). Such biofilm may result in contamination of food with spoilage and pathogenic microorganisms. A number of reviews have addressed the significance of biofilm in food processing (Notermans *et al.*, 1991; Hood and Zottola, 1995; Kumar and Anand, 1998), although relatively few studies have been published on *in situ* observations of biofilm in food-processing industry (Austin and Bergeron, 1995; Lindsay *et al.*, 1996; Hood and Zottola, 1997). Most studies describe bacterial attachment to food-contact surfaces under laboratory conditions (e.g. Zoltai *et al.*, 1981) or the isolation of food spoilage and pathogenic organisms from food environments (e.g. Cox *et al.*, 1989; te Giffel *et al.*, 1996). The time available for biofilm development is relatively short due to frequent cleaning and disinfection of high-risk processing areas. Cleaning, that is undertaken correctly, should reduce the risk of biofilm formation, and in turn reduce the contamination of food by bacteria associated with biofilm.

This chapter does not attempt to review biofilm development and its control in the food industry; a number of published reviews, including those cited above, have examined these issues in some detail. Instead, it examines the potential for biofilm in high-risk food manufacturing premises, reporting the results of case studies in food plants in South Wales, most of which had small and medium enterprise (SME) status. The findings are discussed in relation to the control of biofilm and the potential food-safety risk posed by the development of biofilm in high-risk food-processing environments.

30.2   SELECTION OF PREMISES

Eight SME and two large businesses producing high-risk foods were selected for this study. The selection process was a convenience sample drawn from a database of food producers in South Wales held by the Food Industry Centre at University of Wales Institute (Cardiff, UK). All companies were contacted with an initial letter stating the purpose of the study, and pre-meetings were arranged to present the rationale for the study and assure the companies of confidentiality. Data on the size and nature of the businesses were also collected at this meeting.

High-risk foods were defined as food products that could be eaten without further heat processing, where there was a potential for the survival and growth of foodborne pathogens. Such foods are normally produced under carefully controlled conditions to minimise the risk of contamination by pathogens. Measures commonly taken include physical separation of raw and processed products, careful control of personnel, the use of the hazard analysis critical control point (HACCP) system to identify and prevent specific hazards and the use of codes of practice to ensure general hazards are under control. The nature of each of the businesses involved in the study is shown in Table 30.1.

30.3   MICROBIOLOGICAL SURVEY

Each of the premises was visited a number of times during a normal production run and following cleaning. An initial screening identified sites with the potential

Table 30.1   Nature and size of food businesses in survey.

Product	Number of food handlers
Cooked, chilled or frozen chicken halves and pieces	50
Cooked meats	28
Ready-to-eat curry, catering packs, and cook–chill–freeze	9
Pasteurised milk	119
Filled sandwiches and rolls	35
Sauces, spare ribs	125
Cheeses	60–70
Cheese and butter	37
Hot dogs, black puddings, saveloy	400
Ready-to-eat foods, chilled and frozen	600

for biofilm formation. These were then studied in more detail. Only limited details of the microbiological methods are given here.

30.3.1   Intial survey

An initial study examined high-risk areas for environmental and food-contact sites that might support the growth of biofilm. Criteria for selection of sites included presence of moisture, evidence of nutrient source and ease of accessibility for cleaning. Environmental and food-contact sites were considered, but sites, such as drains and those where technical support would be required to gain access were not sampled. Information was gathered on the duration of the production runs and the cleaning schedules in place including frequency of cleaning and which detergents and disinfectants were used.

Sites with the potential for biofilm development were screened by taking samples using sterile cotton tipped wood-stick swabs (Technical Service Consultants Ltd. Lancashire, UK) avoiding any food debris on the surface. Areas of approximately $100\,cm^2$ were swabbed wherever possible. Swabs were pre-moistened with maximum recovery diluent (MRD; Oxoid Ltd., Basingstoke, UK). Samples were analysed for total aerobic plate counts (APCs). Sites from the initial survey with counts $>1 \times 10^3$ colony forming units (CFU)/$100\,cm^2$ were selected for more detailed investigation.

30.3.2   Main survey

For the main microbiological survey, each plant was visited at least twice to allow sampling during a normal production run and after cleaning had been carried out. In some cases, it was impossible to sample immediately after cleaning and so samples were taken immediately before a production run. Environmental and food-contact surfaces were sampled according to the sampling plan based on the results of the initial survey. The aim of the main survey was to determine the nature of microbial populations and samples were analysed for total aerobic heterotrophs at 30°C and 4°C, *Pseudomonas* spp., Enterobacteriaceae, *Salmonella* spp., *Listeria* spp., *Escherichia coli* O157, *Staphylococcus aureus* and *Bacillus cereus*. In plants where poultry was a main component, samples were additionally examined for *Campylobacter* spp.

30.3.2.1   Swabbing protocol

Sterile cotton tipped wood-stick swabs (TSC) were pre-moistened using sterile MRD (Oxoid). Where possible, 100 cm² areas were swabbed using a bi-directional swabbing technique (Roberts *et al.*, 1995). Swabs were rotated during the swabbing and a consistent pressure was applied by always handling the swab by the plastic cap fitting. Although food-contact areas did not always allow consistent swabbing of 100 cm² between sites, care was taken to ensure swabbing consistency at each site. The areas swabbed were estimated so that results could be presented as CFU/100 cm². Multiple swabs were taken of adjacent areas of the same sample to allow microbial enumeration and pathogen detection. Prior to swabbing, sterile phosphate buffered saline was used to rinse surfaces to remove food debris and any non-adhered bacteria. After swabbing, swabs were returned to their plastic carrying tube and returned to the laboratory in a cold box for analysis within 4 h of sampling.

The sampling was carried out before and after normal cleaning schedules used in each food industry. In some instances, scheduled cleaning had not been carried out prior to the second sampling indicating the persistence of microbial populations in the absence of cleaning. These sites have been excluded from the data presented on cleaning efficacy, as no attempt to control populations was evident.

30.3.2.2   Criteria for selection of biofilm sites

The presence of bacterial populations exceeding 1×10^4 CFU/100 cm² both during processing and after cleaning were taken as representing a significant attached load and identified as presumptive biofilm or sites with a potential to develop biofilm populations.

A number of preliminary laboratory studies on the attachment and development of biofilm of *Staphylococcus aureus* ATCC 29213 to stainless steel coupons indicated that the attachment and development of an attached population as a biofilm was evident at recovery levels below 1×10^4 CFU/100 cm² suggesting that this was a reasonable lower limit for the purposes of this study. It was also established that approximately 46% of the population was recovered through swabbing using the standard protocol and that between 92% and 100% of the swab load was released after 30 s of vortexing in sterile MRD.

30.3.2.3   ATP bioluminescence analysis

Swab samples were taken of each area and analysed for microbial ATP using a method devised by the research group at University of Wales Institute, Cardiff

(UWIC) and described by Davidson *et al.* (1999). Swabs were released into 0.6 ml MRD (Oxoid). Two 0.2 ml portions were then analysed for total ATP and free/ somatic ATP, respectively. The former utilised an extractant based on chlorhexidine (Biotrace, Bridgend, UK) that released ATP from all cells including microbial cells. The latter measured free ATP and used a mild detergent to extract ATP from non-microbial cells. ATP was estimated using commercially available enzyme/reagent test kits (Biotrace). Measurements were made in a handheld luminometer (Xcel, Biotrace) and recorded as relative light units (RLU)

Subtracting the free/somatic ATP estimate from the total ATP gave an estimate of the ATP that was associated with the microbial populations on the swab. The microbial ATP data were then correlated with the heterotrophic plate count data.

30.3.2.4   Biocide activity

The biocidal activity of a specific biocide employed on one of the premises was tested against a number of bacteria isolated from presumptive biofilm sites on environmental surfaces. The test method employed was based on the quantitative suspension test for the evaluation of bactericidal activity of chemical disinfectants as defined by BS EN 1276 (1997).

Test suspensions of a *Listeria* sp. isolate and two different *Pseudomonas* spp. isolates were tested against a commercial terminal disinfectant with a quaternary ammonium compound as active agent. The disinfectant had been used regularly on the surface from which the isolates were obtained.

The tests were performed at three different concentrations for the manufacturer's recommended contact time. Reduction in viability was calculated by the standard method and expressed as mean-log reduction.

30.4   SURVEY RESULTS

A total of 156 sites from ten processing plants with the potential for biofilm formation were screened, and 79 of these sites were selected for detailed analysis. Thirty-nine sites were food contact (e.g. conveyors) and 40 sites were environmental (e.g. floors, walls). Table 30.2 shows the total number of sites selected using the following criteria:

(1) APCs $\geqslant 1 \times 10^4$ CFU/100 cm^2 and/or
(2) *Pseudomonas* count $\geqslant 1 \times 10^4$ CFU/100 cm^2 before and after cleaning.

Table 30.2   Percentage of sites with APC and/or *Pseudomonas* count = 10^4 CFU/100 cm^2.

	Aerobic plate count		*Pseudomonas* spp.	
	Pre-clean	Post-clean	Pre-clean	Post-clean
Food contact	28	31	15	15
Environmental	58	45	45	35

Providing one or both criteria were fulfilled, data for the other species were included.

30.4.1   Bacterial isolations

The number of food contact and environmental presumptive biofilm sites positive for *Listeria* spp. showed a decline after cleaning from 18% to 8% and 30% to 20% of sites, respectively. Thirty-seven isolates belonging to the genus *Listeria* were isolated from the ten food premises. *L. innocua* was the most frequently isolated species 89% (33/37) followed by *L. welshimeri* 8% (3/37) and *L. grayi* 3% (1/37). *L. monocytogenes* was not isolated. *Bacillus cereus* was isolated from the floor of a dairy. *Staphylococcus aureus* was isolated from the following sites: floor and conveyor in a plant producing cooked poultry; floor in a dairy; floor and hand washing sink in a sandwich producer and a conveyor in a factory that produced ready-to-eat meals. No *Salmonella* spp., *E. coli* 0157 or *Campylobacter* spp. were isolated from any site.

30.4.2   Biofilm contamination case studies

A more detailed examination of two food-contact sites where *Listeria* spp. were isolated during processing and after cleaning illustrates the role biofilm may have in contributing to cross-contamination of foods by this organism. Both sites were fabric-based conveyor belts coming into direct contact with food product immediately prior to packaging.

In one, cooked chicken for the catering industry left a nitrogen freezer tunnel and passed along a conveyor. The frozen chicken caused local chilling of the line resulting in condensation forming as it passed through the high-risk area. These conditions supported the growth of a biofilm. *Listeria* spp. were consistently isolated from the belt associated with high aerobic heterotroph and *Pseudomonas* counts ($>1 \times 10^4$ CFU/100 cm^2) and the layer of moisture would have increased the opportunity for product contamination. The belt was not easily removed for cleaning and it is doubtful that this was done, although schedules indicated that cleaning had been completed.

In the second case, creamed cheese portions that had been processed to incorporate a variety of flavours including herbs and mustard seeds passed along a conveyor prior to packing. The surface of the conveyor was wet, partly as a result of cleaning and partly from whey expressed from the cheese. *Listeria* spp. were isolated from the conveyor during processing and following cleaning. They were associated with aerobic heterotroph and *Pseudomonas* counts above 1×10^4 CFU/100 cm^2, indicative of the presence of biofilm.

Water-distribution systems in food plants also provide sites for biofilm formation and a possible route for contamination of food-contact sites and perhaps food. In one milk-processing dairy, the cold, potable supply was distributed to required points in the high-risk bottling plant via 15 mm, nylon reinforced plastic tubing. The water was in constant use to flush floors and equipment of milk spilt during the bottling process. During one site visit, operatives were observed using the water to rinse an aseptic plastic-bottle filler in operation, compromising the aseptic shield. A series of swabs from the bore of the tubing indicated the presence of a biofilm and *Listeria* spp. were isolated at each sampling time. The business did not attempt to clean the tubing or to check the microbiological quality of the water.

30.4.3   Biofilm control

Floors and conveyors were highlighted as the sites most poorly cleaned. For example, Table 30.3 shows the results for a floor in a utensil washing area and Table 30.4 for a food-contact conveyor in a dairy.

These results are representative of flooring in areas where there is frequent pooling of water. Other problem floors are those in holding chill-stores that are not regularly cleaned as they were infrequently emptied of all product.

Although cleaning reduced the microbial load of most sites, this reduction was often small ($<$1 log order reduction) or if greater reduction was achieved, low levels of microorganisms remained on the surface. The mean-log reduction $\pm$ standard deviation (S.D.) for APC following cleaning was 1.1 $\pm$ 1.9 and -0.2 ± 1.7 for food-contact and environmental sites, respectively. If only sites with initial APC $\geqslant 1 \times 10^4$ CFU/100 cm^2 were included, the mean-log reduction $\pm$ S.D.

Table 30.3   Bacterial levels before and after cleaning a floor in a utensil washing area (numbers expressed as CFU/100 cm^2).

	Before cleaning	After cleaning
APC	1.7×10^7	1.1×10^8
Enterobacteriaceae	1.8×10^7	1.3×10^7
Staphylococcus spp.	2.0×10^6	2.4×10^6
Pseudomonas spp.	3.2×10^5	1.2×10^7
Psychrotrophic plate count	6.2×10^6	1.3×10^8
Listeria spp.	positive	positive

Table 30.4   Bacterial levels before and after cleaning for a food-contact conveyor in a dairy packaging process (numbers expressed as CFU/100 cm^2).

	Before cleaning	After cleaning
APC	3.0×10^6	3.0×10^5
Enterobacteriaceae	2.6×10^4	2.2×10^6
Staphylococcus spp.	1.2×10^5	7.7×10^5
Pseudomonas spp.	3.2×10^5	1.1×10^6
Listeria sp.	positive	positive

Table 30.5   Mean-log reduction in APCs for food-contact and environmental sites.

	Food contact mean ± S.D.	Environmental mean ± S.D.
All sites†	1.1 ± 1.9	−0.2 ± 1.7
Sites where initial APC $\geqslant 1 \times 10^4\,CFU/100\,cm^2$	2.1 ± 2.3	0.2 ± 1.7

†Assuming (a) cleaning was carried out and (b) the population before or after cleaning was above the detection limit for the assay ($>1 \times 10^1\,CFU/100\,cm^2$).
APC: aerobic plate count

was 2.1 ± 2.3 and 0.2 ± 1.7 for food-contact and environmental sites, respectively (Table 30.5).

These data indicate that the food industry has a control problem. Cleaning was often not effective and interviews with management and cleaning operatives indicated inadequate understanding of the cleaning process. Most businesses used combined detergent disinfectants that were applied for the recommended period and then rinsed. This approach was relatively inexpensive, quick and removed food debris leaving a wet surface that appeared clean.

More effort was used on food-contact surfaces, but again surfaces were routinely left wet and would still be wet when production restarted. The smaller businesses did not assess the microbiological status of surfaces regularly and none used trend analysis to examine cleaning performance over a period of time.

30.4.4   ATP bioluminescence

Figure 30.1 presents a scatter plot indicating the correlation between microbial ATP (recorded as RLU) and bacterial counts (recorded as $CFU/100\,cm^2$). Both axes are shown on a log scale. Results are shown for mid-shift and post-cleaning samples.

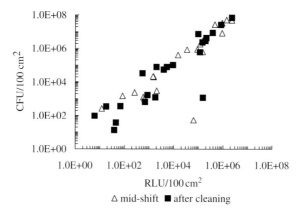

Figure 30.1   Correlation between ATP ($RLU/100\,cm^2$) and bacterial count ($CFU/100\,cm^2$) from all surface types sampling mid-shift and post-cleaning.

These results show that there is no significant difference between mid-shift and post-cleaning samples, supporting the data presented in Table 30.5. There is correlation between the microbial ATP and CFU data with just two 'fail-safe' outliers, where high ATP results were recorded against low CFU counts. The data suggest that measuring microbial ATP is a useful indicator of the microbial load of surfaces. Commercial test kits are available to allow manufacturers to measure microbial ATP and this could be a valuable tool in validating the efficacy of cleaning processes in reducing surface contamination to acceptable levels. Hygiene assessment tests based on estimating total ATP or protein are also widely available and can be of use in trend analysis of cleaning performance.

30.4.5   Biocide activity

Biocide activity against environmental isolates of a *Listeria* sp. and two *Pseudomonas* spp. are shown in Figure 30.2. Tests were carried out on newly acquired isolates (following sub-culturing necessary to ensure a pure culture and storage on tryptic soy agar (TSA, Oxoid) slopes) and on isolates that had been sub-cultured seven times on tryptic soy broth (TSB, Oxoid) at 30° for 24 h. The biocide had been tested for compliance with BS EN 1276 (1997) and control tests with a defined *Pseudomonas aeruginosa* culture showed biocidal activity in excess of a 5 log reduction.

The results for *Listeria* sp. (Figure 30.2) indicate that there was a small increase in biocidal activity following sub-culturing and that the disinfectant was active against this organism, achieving around a 5 log reduction in viability at all concentrations.

The results for the two *Pseudomonas* spp. show very different responses to the biocide. Figure 30.3 shows a set of data, where the biocidal activity was very low, never exceeding a 1 log reduction in viability. Subculturing and biocide concentration did not have a significant effect.

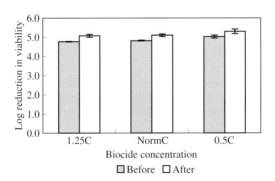

Figure 30.2   Effect of biocide at three concentrations (1.25 × normal, normal and 0.5 × normal, where normal was manufacturer's recommended concentration) on the viability of *Listeria* sp. isolated from a presumptive biofilm site. Error bars indicate 95% confidence intervals. 'Before' refers to tests conducted on a newly isolated culture, and 'after' following seven periods of sub-culture.

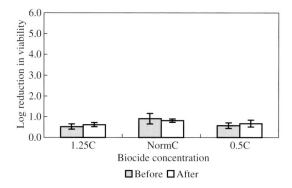

Figure 30.3   Effect of biocide at three concentrations (1.25 × normal, normal and 0.5 × normal, where normal was manufacturer's recommended concentration) on the viability of *Pseudomonas* sp. isolated from a presumptive biofilm site. Error bars indicate 95% confidence intervals. 'Before' refers to tests conducted on a newly isolated culture, and 'after' following seven periods of sub-culture.

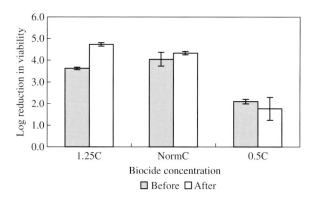

Figure 30.4   Effect of biocide at three concentrations (1.25 × normal, normal and 0.5 × normal, where normal was manufacturer's recommended concentration) on the viability of *Pseudomonas* sp. isolated from a presumptive biofilm site. Error bars indicate 95% confidence intervals. 'Before' refers to tests conducted on a newly isolated culture, and 'after' following seven periods of sub-culture.

Figure 30.4 shows a second *Pseudomonas* isolate against which there was moderate biocidal activity at normal and 1.25 × normal concentration and that this biocidal activity increased following the sub-culturing regime. At the lower concentration, however, a significantly lower level of activity was recorded.

These results are interesting, suggesting that populations contain individual bacteria with very different responses to the disinfectants in use.

It is recognised that biofilm populations often exhibit greater resistance to disinfectants and this is due to a number of reasons including bulk transfer and concentration gradients protecting cells, an extracellular polymeric substances providing physical protection. However, these biocide efficacy tests were conducted using a standard suspension method and the populations had been sub-cultured seven times on a nutrient rich broth under ideal conditions.

The data presented suggests that one of the *Pseudomonas* sp. was resistant to the biocide and would persist in that plant. Rotation of biocide use may help prevent this occurring.

30.5   DISCUSSION

Biofilm may develop on surfaces in food-processing environments. This study has attempted to identify the potential for biofilm in high-risk food-processing environments, using threshold numbers of bacteria on surfaces as a crude indicator of the presence of biofilm. This approach has the benefit of allowing a relatively large-scale microbiological survey though it is by no means a wholly satisfactory means of defining biofilm. It also allowed the evaluation of control measures, as sampling of representative areas was possible before and after cleaning. Particular problem areas identified in all plants included floors and conveyors. Isolation of specific bacteria from large mixed populations also presents problems; it is often impossible to select a small population from a large background count and impossible to know that the target organism was involved in the biofilm community. Precautions were taken during sampling to reduce the potential for isolating pathogens associated with food debris. In common with many reports (e.g. Cox *et al.*, 1989; Slade, 1992; Pritchard *et al.*, 1995) *Listeria* spp. were frequently isolated in this study, although not always associated with a large microbial population indicative of biofilm formation. *B. cereus*, isolated in this study from the floor of a dairy, has been isolated at all stages of processing of pasteurised milk and from processing plant pipelines and heat exchangers (Rönner and Husmark, 1992; te Giffel *et al.*, 1996). The attachment of *B. cereus* in process lines may act as a continual source of post-pasteurisation contamination. *S. aureus* was isolated from some food-contact and environmental sites in this study. Notermans *et al.* (1982) showed that contamination of poultry by *S. aureus* was due to a strain indigenous to the processing plant particularly at the plucking and evisceration stages. The source of *S. aureus* is almost always from humans (e.g. food handlers) or by cross-contamination from another source (such as utensils) previously contaminated by humans (Eley, 1996).

All industries involved in the study identified cleaning as an important control measure in the prevention of food contamination. All expended a considerable amount of time and money on cleaning and yet for the majority of sites cleaning was not effective. Many of the sites were difficult to clean due to poor accessibility and there was little or no recognition that bacteria might attach to surfaces and the population increase during normal production. Cleaning and disinfection is the major control measure for hygiene of surfaces and, if implemented correctly, can control biofilm growth. Cleaning is an essential prerequisite of food-safety management systems, such as HACCP; and in high-risk environments, it is important that cleaning efficacy is assessed. None of the manufacturers had identified surface cleanliness as a critical control point, although in some cases foods were in direct contact with surfaces that had large bacterial populations and were positive for *Listeria* spp. In these situations, it may be useful to employ ATP bioluminescence as a rapid assessment tool to 'monitor' control. The data presented here shows the potential for this. The dairy whose food-contact conveyor features in Table 30.4 took action following the study. The plant invested in a quick release belt that was removed for cleaning each day and left to dry overnight before being refitted. The cleaning process was revised and environmental microbiological and ATP bioluminescence swabs are taken regularly to assess cleaning. Production does not start unless an acceptable ATP reading is obtained.

The efficiency of cleaning schedules in this study was shown to be variable, indicating a need for validation of cleaning. Kinetic energy, such as the physical scrubbing of a surface, may also be needed in cleaning, yet was absent from some of the schedules examined. All equipment for food handling must, by law, be designed and constructed to minimise harbourage of soil and microorganisms and to enable thorough cleaning and disinfection. Cleaned equipment should also be allowed to dry, as residual pools of water and trapped layers of moisture between fabric conveyors and stainless steel surfaces always resulted in unacceptable surface contamination.

Biofilm may result in surface contamination that is difficult to clean effectively. There is, however, no evidence from this study that significant pathogen populations are associated with biofilms. *Listeria* spp. were the only organisms of concern isolated as they may be indicative of the presence of *Listeria monocytogenes*, although this was not isolated. No attempt was made to determine the relative importance of these organisms to the biofilm population, though this would have been conducted had initial isolation results indicated the need for it. Two plants that had *Listeria* spp. present on conveyor surfaces prior to packing reported a *Listeria* problem in their product.

It is very difficult to determine the source of bacteria isolated from foods and this is usually only attempted retrospectively, after an outbreak of food poisoning. Although a clear route of contamination can be identified in the above cases, it is not the only one possible. More detailed studies using genome typing would allow elucidation of the nature of the *Listeria* strains associated with the surface. This would help to establish whether they were transient visitors or an established 'house' population, and determine whether they are the source of contamination in any product.

30.6   CONCLUSIONS AND RECOMMENDATIONS

Low levels of microorganisms remaining after cleaning may develop into biofilms with time; acting as a continual reservoir of microorganisms that can be picked up as food product is moved across a surface. Environmental sites are also a potential source of contamination through cross-contamination by personnel, pests, utensils, cleaning processes and air currents.

Although few specific food hazards were identified in this work, a number of recommendations can be made:

(1) Food industries need to construct cleaning schedules adequate for each site and to clearly identify areas of concern, where ineffective cleaning would result in unacceptable risk of contamination and those cleaning schedules should be validated on site.

(2) There is a need for applied research into cleaning and, particularly, the control of biofilm through effective cleaning. A number of Teaching Company Schemes between UWIC and the food industry have proved very effective in ensuring SME businesses have stringent, validated cleaning processes.

(3) Cleaning efficacy needs to be assessed regularly, particularly for food-contact sites, where cleaning may represent an important control point. ATP bioluminescence may provide a useful real time assessment tool.

(4) There is evidence that some bacteria become resistant to certain chemical disinfectants when exposed to sub-lethal concentrations of the agent. Work should be undertaken to elucidate this and to determine whether biofilm presents bacterial populations with an environment that may assist in the selection of resistant strains.

The food industry needs to be aware that biofilms are dynamic ecosystems and present a control problem that can be reduced through the development of properly evaluated, effective cleaning. The design of some equipment, particularly conveyors, is not conducive to cleaning, although modern, quick release belts can be disassembled and cleaned more easily.

Biofilm has the potential to form in high-risk food-processing environments. Without proper control through cleaning, there is an increased risk of contamination, threatening the microbiological quality, if not the safety, of the product.

Acknowledgements

This research was supported by the UK Food Standards Agency (Project B01016). The microbiological survey, laboratory work and data analysis was undertaken by Dr. Karen T. Elvers during her tenure as Postdoctoral Research Assistant on the project.

REFERENCES

Austin, J.W. and Bergeron, G. (1995) Development of bacterial biofilms in dairy processing lines. *J. Dairy Res.* **62**, 509–519.
BS EN 1276 (1997) Chemical disinfectants and antiseptics. Quantitative suspension test for the evaluation of bactericidal activity of chemical disinfectants and antiseptics used

in food, industrial, domestic, and institutional areas. Test method and requirements (phase 2, step 1), British Standards Institute, London.

Christensen, B.E. and Characklis, W.G. (1990) Physical and chemical properties of biofilms. In *Biofilms* (ed. W.G. Characklis and K.C. Marshall), pp. 93–130, John Wiley and Sons, Inc., New York.

Costerton, J.W., Lewandowski, Z., Caldwell, D.E., Korber, D.R. and Lappin-Scott, H.M. (1995) Microbial biofilms. *Ann. Rev. Microbiol.* **49**, 711–745.

Cox, L.J., Kleiss, T., Cordier, J.L., Cordellana, C., Konkel, P., Pedrazzini, C., Beumer, R. and Siebenga, A. (1989) *Listeria* spp. in food processing, non-food and domestic environments. *Food Microbiol.* **6**, 49–61.

Davidson, C.A., Griffith, C.J., Peters, A.C. and Fielding, L.M. (1999) Evaluation of two methods for monitoring surface cleanliness – ATP bioluminescence and traditional hygiene swabbing. *Luminescence* **14**, 33–38.

Eley, A.R. (1996) Toxic bacterial food poisoning. In *Microbial Food Poisoning*, 2nd ed. (ed. A.R. Eley), pp. 37–55, Chapman and Hall, London.

Holah, J.T. and Kearney, L.R. (1992) Introduction to biofilms in the food industry. In *Biofilms: Science and Technology* (ed. L.F. Melo, T.R. Bott, M. Fletcher and B. Capdeville), pp. 35–41, Kluwer Academic Publishers, Netherlands.

Holah, J.T., Bloomfield, S.F., Walker, A.J. and Spenceley, H. (1994) Control of biofilms in the food industry. In *Bacterial Biofilms and Their Control in Medicine and Industry* (ed. J.W.T. Wimpenny, W. Nichols, D. Stickler and H.M. Lappin-Scott), pp. 163–168, Cardiff, Bioline.

Hood, S.K. and Zottola, E.A. (1995) Biofilms in food processing. *Food Cont.* **6**, 9–18.

Hood, S.K. and Zottola, E.A. (1997) Isolation and identification of adherent Gram-negative microorganisms from four meat-processing facilities. *J. Food Prot.* **60**, 1135–1138.

Kumar, C.G. and Anand, S.K. (1998) Significance of microbial biofilms in the food industry: a review. *Int. J. Food Microbiol.* **42**, 9–27.

Lindsay, D., Geornaras, I. and von Holy, A. (1996) Biofilms associated with poultry processing equipment. *Microbios* **86**, 105–116.

Notermans, S., Dufrenne, J. and van Leeuwen, W.J. (1982) Contamination of broiler chickens by *Staphylococcus aureus* during processing: incidence and origin. *J. Appl. Bacteriol.* **52**, 275–280.

Notermans, S., Dormans, J.A.M.A. and Mead, G.C. (1991) Contribution of surface attachment to the establishment of microorganisms in food processing plants: a review. *Biofouling* **5**, 21–36.

Pritchard, T.J., Flanders, K.J. and Donnelly, C.W. (1995) Comparison of the incidence of *Listeria* on equipment versus environmental sites within dairy processing plants. *Int. J. Food Microbiol.* **26**, 375–384.

Roberts, D., Hooper, W. and Greenwood, M. (ed.) (1995) *Practical Food Microbiology*, 2nd ed., Public Health Laboratory Service, London.

Rönner, U. and Husmark, U. (1992) Adhesion of *Bacillus cereus* spores: a hazard to the dairy industry. In *Biofilms: Science and Technology* (ed. L.F. Melo, T.R. Bott, M. Fletcher and B. Capdeville), pp. 403–406, Kluwer Academic Publishers, The Netherlands.

Slade, P.J. (1992) Monitoring *Listeria* in the food production environment. I. Detection of *Listeria* in processing plants and isolation methodology. *Food Res. Int.* **25**, 45–56.

te Giffel, M.C., Beumer, R.R., Bonestroo, M.H. and Rombouts, F.M. (1996) Incidence and characterization of *Bacillus cereus* in two dairy processing plants. *Neth. Milk Dairy J.* **50**, 479–492.

Zoltai, P.T., Zottola, E.A. and McKay, L.L. (1981) Scanning electron microscopy of microbial attachment to milk contact surfaces. *J. Food Prot.* **44**, 204–208.

31

Industrial biofilms: formation, problems and control

J.W. Patching and G.T.A. Fleming

31.1 INTRODUCTION

With the exception of processes operating under sterile conditions, any industrial system where water contacts a surface will be vulnerable to biofilm formation. The development of a biofilm is summarised in Figure 31.1. Organisms will be attracted to the surface and adhere, initially by weak physicochemical interactions. Adherence is consolidated by the formation of a polysaccharide matrix. Once established, the biofilm provides an ideal niche for its component microorganisms, enabling them to grow in flowing systems containing minimal levels of nutrients. Biofilms have been reported in high-purity water systems (Mittelman, 1991). Biofilm communities are also more resistant to biocides than planktonic (free-floating) communities. The film thus enters a growth phase where biofilm accumulation is primarily due to its growth rather than further adhesion of planktonic cells. Eventually, growth will be balanced by senescence and biofilm detachment. The biofilm thickness at which this plateau phase occurs will be the result of factors influencing growth and senescence (nutrient supply, temperature, etc.) and removal (film stability and shear forces).

In practice, the mere presence of biofilms is not a problem, but rather the effects that they may have on a process. These may include decreased efficiency, increased running costs, system damage or breakdown and deterioration in product quality.

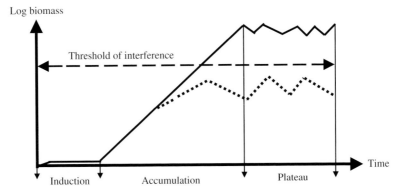

Figure 31.1   The development of a biofilm (based on Flemming and Griebe, 2000).

The objective of control strategies is to limit biofilm development to levels below a threshold of interference (Flemming, 1997), which will vary from process to process and may be defined as the level at which the biofilm causes significant problems and may be more correctly referred to as biofouling. Limitation may be achieved by delaying primary colonisation, inhibiting growth and increasing removal. This chapter deals with strategies to control of biofouling, using cooling water systems and reverse osmosis (RO)/nanofiltration systems as examples. Morton (2000) may be referred to for a recent review of the problems caused by biofilms in industry and Percival (2000) for methods used for monitoring them. Other recent reviews on industrial biofilms and their control include those by Flemming and Griebe (2000) and Donlon (2000).

31.1.1   Water-based heat transfer systems

Water is widely used for heat transfer in industry to cool processes and to maximise the efficiency of power generation. By their nature, heat exchangers have a large solid–liquid interface relative to the volume of water passing through them, thus making them vulnerable to problems caused by biofilms and biofouling. Cutting fluids used in metal working processes have both cooling and lubricating functions. These fluids may be neat oil, oil-in-water emulsions, 'semi-synthetic' oil-in-water emulsions or chemical solutions (American Society for Testing and Materials, 1978). With the exception of neat oils (uncontaminated with water) all these fluids are susceptible to bacterial or fungal attack. Biofilms are frequently found in cutting fluid screen filters and swarf accumulations. Microbial contamination of metal working fluids may result in their de-emulsification, pipe and filter blockage and the presence of pathogens (Morton, 2000).

   The design and operation of a cooling system has a significant effect on the establishment of biofilms, their effects and the ease with which they can be controlled. Unfortunately, financial and practical considerations usually limit the use of construction and operating variables as a way of dealing with biofilm problems. Cooling water may be utilised in three ways (Bott, 1998):

(1) Closed systems where the water circulates, transferring heat through heat exchangers, but does not come into contact with the environment. Due to the closed nature of the system, chemical methods of biofilm control can be used which might otherwise be ruled out on considerations of cost per litre and the possibility of environmental pollution. Unfortunately, these systems typically use air as the final heat sink and this results in high costs, both capital (air-blown heat exchangers) and operational (the movement of large volumes of air).

(2) Open recirculating systems in which the water circulates between a heat exchanger and a cooling tower or spray pond, where heat is dissipated by evaporation of some of the water. This results in an increase in the concentration of dissolved solids in the circulating water, which would eventually result in precipitation and clogging. This is avoided by a periodic 'blow-down' or discharge of the circulating water, usually to the original water source. Water is abstracted from this source to make up for losses due to evaporation, windage and 'blowdown'. 'Blowdown' may present problems of pollution, both thermal and chemical. The latter include the concentrated salts and any chemical that has been used for biofilm control.

(3) 'Once through' systems are relatively simple. Cold water is taken from a natural source (of high-thermal capacity), passed through a heat exchanger and returned to its origin. Due to the large volumes of water used and pollution concerns, physical, rather than chemical methods of biofilm control are preferred.

31.1.2   Reverse osmosis

RO technology was first developed in the USA during the late 1950s for the desalination of seawater. The technology utilises semi-permeable membranes commonly composed of cellulose acetate (Voros et al., 1996), cellulose diacetates/triacetates (Kosutic and Kunst, 2001) or composite materials including polyamides (Roh, 2002). Asymmetric membranes, which are composed of two layers of two different polymers are becoming more prevalent in industrial use. The membranes are semi-permeable (pore size ca. 0.5 nm) which allows the passage of water but prevents the passage of bacteria, dissolved salts and particulates.

The concept of RO can be described using a simple model consisting of a high- and low-salt solution separated by a semi-permeable membrane (Figure 31.2). In a simple osmotic system consisting of brine and water, water will flow from low to higher concentrations. Equilibrium is reached in relation to differential osmotic pressures between the two solutions. When a head pressure somewhat greater than this osmotic pressure is applied to the brine side of the membrane, the flow of water is reversed and water permeates yielding pure product water.

RO systems are constructed as hollow-fibre or spiral-wound membrane modules. The latter consist of sheets of semi-permeable membranes separated by feed and permeate spacers (Figure 31.3). These are wound around a hollow tube in a

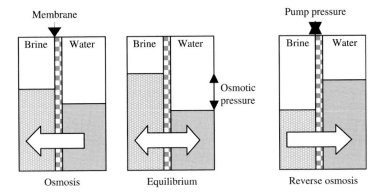

Figure 31.2   Principle of RO.

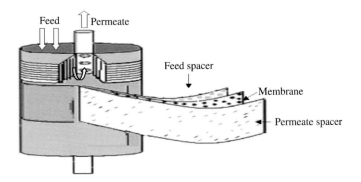

Figure 31.3   Spiral-wound membrane.

spiral fashion. The spacers separate the membranes and allow free flow of liquid. Spacer fibres enhance tangential cross flow and create turbulence between membrane sheets. The spiral design facilitates optimum permeate flow rate and maximises surface area. The arrangement is positioned within carbon steel pressure vessels, many of which can withstand pressures exceeding 40 atm. Industrial modules typically process 6000–60,000 l/day. The design of RO filter systems will depend on both the application for which they are employed and plant design. They can be combined with carbon filter absorption and ion exchange modules to polish and produce ultra-clean water. The RO units themselves can consist of a single module or a bank of modules (in series or in parallel) depending on the application.

RO processes are used for the desalination of brackish/seawater to provide freshwater for domestic, agricultural and industrial use (Al-Mutaz, 2001). They also find extensive use in the food, electronics, paper and fine-chemical industries (Leeper and Tsao, 1987; Franken, 1999; Koyuncu et al., 1999; Girard and Fukumoto, 2000). The pharmaceutical sector's requirement for parenteral water necessitates the stringent control of dissolved salts and bacterial endotoxin concentrations in product water

(Minuth, 1997; Weyandt, 2001). This is achieved using RO systems in conjunction with pre-treatment processes. RO is used for element recovery in heavy metal processing (Karakulski and Morawski, 2000), the recovery of water-soluble fractions in the paint sector (Anderson *et al.*, 1981) and the recovery of water during oil fractionation (Gioli *et al.*, 1987). The worldwide market for ultra-pure water is currently (2002) worth $2.3 billion and is expected to grow to $6.2 billion over the next 4 years (Anon, 2002). Membrane fouling control is crucial if this growth is to be sustained.

31.2   PROBLEMS CAUSED BY BIOFILMS

It can be argued that the raison d'être for the existence of bacterial biofilms is that the biofilm, rather than the planktonic phase, is the natural bacterial form in the environment. Biofilms are intrinsic to liquid–surface interfaces – all that is required are favourable conditions for microoganisms to grow. Industrial waters (feed or process) are rarely devoid of sufficient nutrients, or are at such restrictive temperatures or pH to prevent incurred bacteria forming biofilms. For example, heat exchange systems are susceptible to bacterial fouling since operating temperatures are amenable for bacterial growth (Bott, 1994). Practically, every industrial sector has accrued costs as a direct result of biofouling. Overviews dealing with the occurrence and detrimental effects of biofilms in industry are provided elsewhere (Mattila-Sandholm and Wirtanen, 1992; Morton, 2000). Specific problems resulting from biofouling in heat transfer and RO systems are dealt with here.

31.2.1   Cooling water and heat exchange systems

The deposits that form on heat exchanger surfaces and cooling towers are complex and consist of both organic and inorganic materials. Biofouling influences the efficiency of processes in a number of ways:

(1) Water flow through the system is restricted because of occlusion of pipes and channels and increased frictional resistance (McCoy *et al.*, 1981; Brankevich *et al.*, 1990). This results in increased backpressure and additional pumping and replacement costs.
(2) Heat transfer is diminished because the biofilm effectively acts as an insulator on heat exchange surfaces (Melo and Bott, 1997). Energy losses associated with biofouling of heat exchangers have been estimated to be in the region of 30% over a 30-day operation period (Percival, 2000).
(3) Biofilms will encourage biocorrosion and biodeterioration. The anaerobic environment generated within thick biofilms can lead to the proliferation of sulphate-reducing bacteria and consequent biocorrosion of metals (Morton, 2000). Concete and wood in cooling towers may also be susceptible to microbial biodeterioration.
(4) Biofilms can act as reservoirs for pathogens including *Legionella* spp. (Anon, 1991).

31.2.2   RO filtration

Process difficulties caused by biofilm growth in RO systems are well documented (Flemming, 1997; Baker and Dudley, 1998; Morton, 2000; Murphy *et al.*, 2001). The effects of fouling in spiral-wound membranes may be summarised as follows:

(1) Biofilm attachment to the membrane surface elicits an increased hydraulic resistance. This results in an elevated differential pressure between feed and permeate lines (ΔP membrane) with associated elevated pumping costs, lower permeate productivity, increased cleaning intervals and a very real risk of membrane damage. Damage caused by the build-up of biofilm may be so extreme in some instances that the module 'telescopes' (Saad, 1992) necessitating the replacement of the RO cartridge.

(2) Biofouling of feed channel spacers (Figure 31.2) results in a pressure drop between the feed and brine lines (ΔP feed/brine). Since the efficiency of salt rejection relates to the differences in osmolarity of feed and permeate flow, fouling in these regions brings about concentration polarisation of salts on the membrane. This in turn leads to increased salt passage, yielding impure product. Flemming (1997) has also demonstrated that in some instances bacteria may transverse the membrane. Such water is inherently unacceptable for parenteral purposes.

(3) Cellulose acetate membranes are prone to biofilm-related degradation during periods of shutdown or storage. Such biodeterioration appears to be primarily caused by fungal species (Murphy *et al.*, 2001). This can cause irreversible damage necessitating complete replacement of improperly sanitised units.

31.3   BIOFILM MONITORING IN INDUSTRY

Any medium- to large-scale industrial facility will typically have three key departments, namely production, engineering and QC/QA. Production managers strive to meet or exceed product quotas. Process engineers endeavour to maximise process capability whereas quality personnel are required to carry out appropriate testing to ensure that standards laid down by regulatory bodies are adhered to. The production manager is all, too frequently the motivating force within this structure. 'Downtime' for corrective action is not a term of endearment. Very often, the process is stretched to limits of capability (or beyond) and difficulties associated with biofouling are recorded as process failure or product non-conformance. This is perhaps the most costly form of monitoring. Such a scenario can be avoided by putting *realistic* biofilm monitoring protocols in place. While the following is proposed to provide a short overview of monitoring practices, the subject is addressed in detail by others (Flemming 1997; Flemming *et al.*, 1998; Donlon 2000; Klahre and Flemming, 2000).

31.3.1   Sensory/system evaluation

Sensory evaluation (by a trained and experienced operator) is often the first step in recognising the presence of biofilm. For example, regular inspection of pre-treatment piping, dosing tanks, media filters cooling towers, etc. may reveal the presence of microbial deposition. Filter colour and odour may also act as indicators of biofouling of RO membranes. While visual inspection is always recommended, practical difficulties associated with examination of sealed units can make this procedure impossible. Performance characteristics, such as thermal efficiency and flow characteristics, should be routinely monitored for heat exchange units. Low efficiencies may point to fouling. It is important that both chemical and microbiological analysis be carried out in this instance to assess if fouling is of a biotic or scaling nature.

For RO systems, permeate purity (by conductivity) and salt rejection rates should be analysed. Feed, brine and permeate flows should also be monitored for organic carbon as total organic carbon (TOC) and chemical oxygen demand (COD). These parameters often act as important indicators of microbial fouling in the process (Al-Ahmad et al., 2000). Pressure differences along RO manifolds also act as good indicators of biofouling. It is a generally accepted 'rule of thumb' that if the pressure drop between feed and reject lines has increased by >15%, then fouling has occurred within the feed path resulting in restricted flow across the membrane. Conversely, a similar pressure difference between the feed and permeate lines would indicate membrane fouling.

31.3.2   Microbiological testing

Microbiological analyses (plate counts, direct counts, etc.) are essential to monitor variations in feed water quality and the efficiency of pre-treatment. When elevated microbial counts are detected in feed water, remedial measures can be quickly put in place to counteract potential biofouling in downstream systems. Testing for planktonic microorganisms alone can grossly underestimate biofouling levels (Costerton et al., 1987) and the type of microorganisms that predominate may differ significantly from those in the biofilm (Percival, 2000). Furthermore, this type of analysis is of little use in determining the location of biofilm growth in sealed systems (Flemming et al., 1996). Microbiological testing, however, is a valuable tool for identifying and tracking offending species, especially when rapid identification methods are used. Microbiological techniques are essential for the development of effective clean-in-process/disinfection protocols for the control of biofilms. Data obtained from autopsy is often useful for process re-evaluation and development of model systems for prediction of biofouling potential (Gagnon et al., 1997).

31.3.3   Side-stream monitoring

Side-stream monitors placed in parallel to the main process and share the same water supply. They usually take the form of Robbins-type devices where coupons are exposed within the stream flow (McCoy *et al.*, 1981). Analysis of biofilm depth on coupons can provide a quantitative profile of biofilms build-up, but only from a historical perspective. Coupons are small compared with pipe diameters, and may not be subject to the same fluid hydrodynamics as the main stream.

31.3.4   On-line monitoring

Flemming and co-authors (Flemming *et al.*, 1998) have proposed a number of principles for an optimal (automated) monitoring regime. Biofouling information should be available rapidly and from representative locations, and should allow us to follow the kinetics of deposition. Monitoring devices themselves should be cheap, robust, and give accurate and reproducible readings in real time.

Many on-line instruments are currently only of scientific value but some are commercially available. The most common type of instrumentation detects light scattering/reflectance caused by the build-up of biofilm. Such devices (e.g. fibre-optic or laser turbidimetry) cannot differentiate between biological fouling and inorganic scaling (Flemming *et al.*, 1998). Differentiation is possible with FTIR cells, but these are too expensive to be used in small-scale facilities. Impedence-based methods show promise, especially for monitoring biofilm formation in membrane-based systems (Licina *et al.*, 1994; Chilcott *et al.*, 2002). The development of new techniques for on-line monitoring is exciting and to be welcomed, but they will not be adopted for routine use unless proven to meet industrial standards for ruggedness, reliability and cost.

31.4   BIOFILM CONTROL: PREVENTION AND CURE

A universal panacea for solving biofouling problems in water-based technical systems does not yet exist. A traditional (and still popular) approach to biofilm control is to use biocides to slow or halt established biofilm development, supported by removal by chemical or physical methods. There are limitations to this approach. The use of chemicals is becoming increasingly limited due to environmental concerns. 'New generation' biodegradable biocides are being introduced but with a cost penalty. The use of physical removal by scraping can be effective but its on-line use is presently limited to certain system geometries (the inside of tubes).

An alternative, but complementary strategy is to design and operate systems in an attempt to prevent the establishment of biofilms. Operating parameters, such as the nature of fluid flow through the system, operating temperatures, the quality of the raw water used, etc., may be manipulated. Common-sense approaches to system design include avoiding of excess use of gaskets, bends and dead joints (Mattila-Sandholm and Wirtanen, 1992). Saad (1992) has provided an excellent treatise of this approach to the practical control of biofilms in RO membrane systems and Bott (2001b) has summarised aspects of design and operation, which control fouling of heat exchangers. As part of this preventative strategy, surface treatments, ultrasound and changes in surface and particle charge may be used to discourage settlement. Implementation of such pre-emptive strategies may not eliminate the formation of biofilms, but may to some extent delay the onset of system interference. The adage 'prevention is often better than cure' can be readily applied to microbial fouling.

While the following sections discuss various individual strategies and treatments in more detail, the reader is reminded that effective antifouling control is best achieved by applying them in concert: physical and chemical, and prevention and cure.

31.4.1   Chemical treatments

The use of chemicals is a common, reflex response to microbial problems in industry, but an incorrect or inappropriate use of chemicals may not only fail to tackle biofouling problems but may even exacerbate them. For example, chemical treatments may act as a source of nutrition for biofilm growth and biocides may react with humic acids to increase their availability as a source of nutrition for the developing biofilm (LeChevallier, 1991). This may partly explain why biofouling of RO membranes at a desalination plant decreased when continuous chlorination of seawater was discontinued (Hamida and Moch, 1996).

A treatment which reduces or removes the planktonic bioburden will, at best delay the initiation of biofilm formation. Once a film is established and is in the growth or plateau phases of development (Figure 32.1) it has no further requirement for seeding from the planktonic population.

It is now well established that microorganisms in a biofilm are relatively resistant to antimicrobial chemicals. The reasons for this are complex, involving both the physiological state of the biofilm community and the properties of the extracellular polymer matrix (the glycocalyx) in which they are embedded (Allison et al., 2000). Effective antimicrobial treatment of biofilms may require the use of more aggressive chemicals or the use of higher levels than would be applied to control free-living microorganisms, with potential problems of cost, plant damage and environmental pollution. One solution to this problem is to apply physical treatments which enhance the effectiveness of biocides (see Section 31.4.2). Chemical treatments have also been used to destabilise biofilms (see Section 31.4.1.2), either rendering the attached microorganisms more susceptible to biocides or releasing them as planktonic organisms where they may be dealt with as such.

Flemming and Griebe (2000) have pointed out the dangers in regarding biofouling as a 'disease', which may be cured by killing or inhibiting the causative organism. If all biological activity within biofouling could be halted, it would cease to develop, but would retain many of its undesirable properties (decreased heat transfer efficiency, increased resistance to flow, membrane fouling, etc.). Dead biofilm will also act as an ideal environment for rapid re-establishment of biofouling if not effectively removed. Other treatments (chemical or physical) will be needed to destabilise or remove the fouling.

The possibility that biofilms will eventually develop resistance to an antifouling measure is a factor, which must be considered when devising a control strategy (Allison *et al.*, 2000). MacDonald *et al.* (2000) applied a biodispersant to a simulated recirculating cooling system for a period of 2 years. The dispersant initially controlled biofouling, but became ineffective in the long term, even when its concentration was increased to double the recommended dose. Changes in the composition of the biofilm community suggested that this was due to the selection of strains which could remain attached in the presence of the dispersant. The long-term use of non-oxidising biocides in RO systems may result in the development of resistant biofilms, especially if continuous, low-dose rates are used (Baker and Dudley, 1998). Shock dosing and rotation of biocides are recommended as strategies to combat this. While problems of resistance selection are generally considered in connection with chemical treatments, they may also arise as the result of physical biofouling control. Brush cleaning of heat exchange surfaces has been found to stimulate the development of a more tenacious biofilm (Nickels *et al.*, 1981).

31.4.1.1   *Biocides*

When selecting a biocide for controlling biofouling, the industrialist is faced with a wide choice (Payne, 1988). Biocides are broadly assigned to two groups, oxidising and non-oxidising. Choosing a biocide and its mode of application (concentration and dosing schedule) should be firmly linked to the specific industrial process and biofouling problem.

The following are some aspects of a biocide treatment which should be considered:

- mode of action and spectrum of activity,
- effects of environment on activity,
- development of resistance,
- effects on plant,
- effects on product,
- effects on environment,
- ease of handling,
- cost.

The effectiveness of biocide against biofouling must be considered *in situ*, remembering that the biofilm environment confers additional resistance and that environmental factors may influence biocide activity. For example, oxidising biocides tend to be neutralised by organic material, which will reduce their effectiveness in systems with a high organic load, but may help to ameliorate adverse environmental effects on discharge.

The potential for a biocide to damage the plant or contaminate the product are of particular relevance to RO systems. Polyamide-based RO membranes are damaged by prolonged contact with oxidising biocides, though some may be used to treat composite membranes if pH, temperature and contact time are strictly controlled. If chlorine is used as a pre-treatment, it must be neutralised with sodium bisulphite before RO is attempted. Pre-treatment of seawater with chlorine or ozone results in the formation of hypobromous acid, which will kill bacteria, but also damage RO membranes (Baker and Dudley, 1998). High-molecular-weight biocides will be removed from water by RO systems and discharged to waste, but low-molecular-weight compounds may still be found in water that has been processed by the system. Non-oxidising biocides are not approved for potable water supplies (Baker and Dudley, 1998).

Growing environmental awareness has resulted in increasingly stringent regulations controlling discharges. Environmentally friendly biocides or neutralisation before discharge may be considered, but both may result in increased costs. Effectiveness as well as product and environmental concerns will have conflicting requirements for biocide persistence. Low persistence in recirculating systems will result in an excessive demand for biocide but high persistence may result in unacceptable pollution on discharge or product contamination.

Oxidising biocides of proven or potential use for the control of biofilms in cooling systems include chlorine (or chlorine compounds, such as bleach), bromine-containing compounds and ozone. Traditionally, chlorine has been viewed as a cheap, effective choice for biological control in industrial water systems. It has several disadvantages, however. Use of chlorine gas creates problems of storage and safety. The pH of water in recirculating cooling systems is typically >pH 8. Under these conditions, chlorine is not found as hypochlorous acid, but the less effective hypochlorite ion. Lastly, chlorine does not deal effectively with heavy biofouling (Donlon, 2000). These problems have encouraged a movement away from reliance on chlorine to other biocides and methods of fouling control. Bromine-containing compounds (listed in Donlon, 2000) are finding increasing use in cooling systems as alternatives to chlorine. Unlike chlorine, their efficiency is not curtailed by pH values >8.

Ozone has not, as yet, found widespread application as an antifouling treatment, though its use as a means of stripping biofilm from heat exchanger walls has been patented (Watanabe et al., 1996) and its potential use in cooling water systems has been investigated in experimental simulations (Kaur et al., 1992; Videla et al., 1995; Viera et al., 1999a, b, 2000). Since ozone is produced on site by corona discharge, there are no storage or handling problems and it has a short half-life. This would minimise disposal problems, but might adversely effect the cost of its use in recirculating systems. If used as a cooling system biocide, it would have no adverse effects on stainless steel or titanium, but would corrode copper alloys (Viera et al., 2000) and it is considerably less effective against bacteria when they are incorporated in a biofilm (Viera et al., 1999a).

There are a plethora of non-oxidising biocides of actual or potential use in biofouling control and the range is increasing as manufactures strive to meet the requirements of industry. Thus, proprietary non-oxidising biocides have been developed for use with RO systems that are effective, membrane compatible and suitable for discharge. According to Baker and Dudley (1998), cleaners and biocides used for off-line defouling of RO membranes should be

- non-oxidising,
- stable at pH 2–12 and 20–35°C,
- non-ionic or anionic,
- non-filming,
- compatible with other cleaning products.

Typical non-oxidising biocides used in cooling water systems include (Mattila-Sandholm and Wirtanen, 1992; Cloete *et al.*, 1998):

- aldehydes,
- halogenated bisphenols,
- isothiazolones,
- organosulphurs,
- quaternary ammonium compounds,
- biguanides.

Since choosing the most effective biocide will be highly dependent on the process to be treated, the nature of the problem and the dosing schedule envisaged, it is not surprising that there is no consensus as to which of the non-oxidising biocides are most effective. Some are, however, more suited for tackling biofilms than others. For example, quaternary ammonium compounds and other surface-active biocides may encourage biofilm detachment (Mattila-Sandholm and Wirtanen, 1992), whereas formaldehyde can react with the biofilm matrix to make it more resistant to removal (Exner *et al.*, 1987).

31.4.1.2   Other chemical treatments

Other chemical treatments (often applied with biocides) may be used as part of anti-biofouling protocols. They serve one or both of two purposes, to loosen or disrupt the biofilm so that it may be removed by the flow of water or by mechanical cleaning and to enable a biocide to act more efficiently on biofilm microorganisms.

Surfactants (including detergents and biodispersants) are (after biocides) the most widely used chemicals for the control of fouling. By altering surface tension, they may facilitate biocide penetration and effective biofilm removal. Surfactant/non-oxidising biocide formulations are widely used for the off-line defouling and sanitizing of RO systems (Baker and Dudley, 1998). While most of the published studies on the on-line use of surfactants and surfactant/biocide treatments (cited in Donlon, 2000) are laboratory studies, surfactant/biocide treatments have been shown to function effectively in field trials with industrial cooling systems. Cooling tower field trials showed that a metronisadole/biodispersant combination penetrated the biofilm and killed obligate anaerobic bacteria, but was non-toxic to aerobic bacteria and fungi (Wiatr and Fedyniak, 1991). A combination of conventional chlorine bleach and a biodetergent caused a significant sloughing of biofilm from cooling tower fill (Yu *et al.*, 1999), though the authors state that other, unspecified biocide/biodispersant combinations were less successful.

Chelating agents (EDTA/EGTA) may be used to destabilise biofilms. Chen and Stewart (2000) reported a 26% removal of a laboratory biofilm after 1 h treatment with 0.01 M EDTA. EDTA did not kill the biofilm bacteria, but since it will chelate divalent cations associated with cell membranes, it could act synergistically with biocides. Chelating agents have been used to break down biofilm accumulations in cutting fluid systems (Mattila-Sandholm and Wirtanen, 1992) but their use would not appear to be practical in cooling water systems due to the high levels of divalent cations present and environmental problems associated with their discharge (Donlon, 2000).

The use of enzymes to attack the biofilm matrix has also been proposed. Enzyme/detergent formulations marketed for the in-place cleaning of RO and ultrafiltration membranes contain proteases (Kumar and Takagi, 1999), but the protein fouling which they are intended to solubilise is typically a result of the material being processed rather than biofilm formation. Johansen *et al.* (1997) were able to remove model biofilms from steel and polypropylene substrata by means of a complex mixture of polysaccharide-hydrolysing enzymes and Wiatr (1991) has patented the use of a mixture of cellulase, protease and alpha-amylase to control slime formation in cooling towers, but there is no evidence of the significant use of enzymes to control biofouling in cooling systems. Barriers to its acceptance would seem to be cost and the problem of formulating an enzyme mixture capable of dealing with the wide diversity of polymers found in biofilms.

31.4.2   Physical biofilm removal

Physical methods may be used to loosen and strip off biofilms. Taking equipment or pipe runs off-line and cleaning them manually or with high-pressure water jets is undesirable from an economic point of view. On-line physical methods are typically used in heat exchangers, where they can replace or reduce the use of chemicals.

The use of circulating sponge rubber balls has been found to be effective for the prevention of fouling on the inside of the tubes through which cooling water circulates in the shell-and-tube heat exchangers typically used as condensers in power stations. Figure 31.4 shows the system manufactured by Taprogge GMBH. The balls, which are slightly greater in diameter than the tubes, are introduced into the incoming cooling water which forces them through the tubes, wiping the tube walls in the process.

Balls will be introduced to individual tubes at random, and sufficient must be employed to ensure that each tube receives balls at an acceptable frequency. The balls are recovered downstream by means of a strainer and recirculated.

In a case study (cited in Bott, 1994) carried out at a 500 MW power station, 12-month trials showed that the use of the Taprogge system resulted in an overall heat transfer coefficient which was 5–10% better than that found when film control was carried out by chlorination of the circulating water. Bott (1998) has stated that, when compared with the cost of chemical antifouling agents, the Taprogge

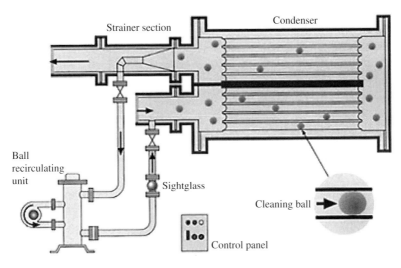

Figure 31.4   The Taprogge system.

system could justify its capital cost within a year. One problem associated with this system is the occasional blocking of tubes by balls.

Brush and cage systems may be employed in smaller tube heat exchangers. Water flow is used to shuttle brushes from one end of the tubes to the other. The periodic reversal of water flow, which is needed for this, may not be acceptable in some systems. Installation costs are higher than the ball system, but maintenance costs may be lower (Bott, 1994).

The introduction of polymer fibres (200 μm × 12 μm) into the cooling water stream has been shown to slow the development of biofilm (Bott, 2001a). The mechanism involved was not investigated, but it was suggested that the fibres could act as scrubbers, or possibly become incorporated in the biofilm and weaken it. Unlike systems employing balls or shuttling brushes, the use of fibres would not be restricted to the inside of tubes, but problems of fibre recovery after use and their effects on pumps and cooling towers need to be investigated.

Ultrasonic pulses propagated axially along the tubes of heat exchangers have been suggested as a method of biofilm removal (Mott *et al.*, 1998). Experimental studies on *Pseudomonas fluorescens* biofilms in simulated heat exchanger tubes, showed that 3 × 30 s doses of high-intensity ultrasound (20 kHz) administered axially ten times per day were sufficient to produce an 88% reduction in biofilm development when compared with controls (Bott, 2000). It was unclear whether this was achieved by loosening and removal of the film or inhibition of initial attachment (see Section 31.4.5.2).

A major advantage of physical methods is the absence of the environmental problems associated with the use of chemicals, but (with the exception of the Taprogge system) they may be relatively expensive. At present, no suitable physical method has been developed for the control of biofilms in open cooling towers, where electrochemical or chemical methods must be used.

While the search for physical methods which might substitute for expensive and potentially polluting chemicals continues (Bott, 2001a), exploiting the synergism between physical and chemical methods of biofilm removal would seem to offer a rewarding approach. Continuous ultrasound in the 2–200 m W cm^{-2} power range has been shown to enhance the killing of biofilm microorganisms by antibiotic, though these levels of ultrasound alone were not antimicrobial (Huang *et al.*, 1996; Qian *et al.*, 1996, 1997, 1999; Peterson and Pitt, 2000). It should be noted that, unlike treatments proposed for industrial use, the ultrasonic treatment did not disrupt or remove the biofilm. Such effects were deemed undesirable since these studies were carried out in connection with problem bacterial biofilms on implanted medical devices. Low-strength (± 12 V cm^{-1}), low-current (± 2.1 m A cm^{-2}) electrical fields have been shown to enhance the efficacy of biocides against biofilm bacteria to levels greater than those shown towards planktonic forms (Blenkinsopp *et al.*, 1992). Subsequent studies on this 'bioelectric effect' (Costerton *et al.*, 1994) have concentrated on the use of antibiotics (e.g. Jass *et al.*, 1995; Wattanakaroon and Stewart, 2000), and the use of this phenomenon in connection with chemicals used for the treatment of industrial biofilms does not seem to have been researched. Chemical treatments which loosen or disrupt films (discussed elsewhere in this chapter) could also make them more susceptible to physical removal.

31.4.3   Manipulating operating parameters

Increasing fluid velocity within a hydraulic process will enhance shear stress (Melo and Bott, 1997). While Bott (1994) has argued that high fluid velocity should be used as a measure to reduce the potential for biofilm growth in such systems, a recent study carried out by Soini and co-workers (Soini *et al.*, 2002) cautions against this strategy. They showed that the introduction of high shear stresses in hydraulic systems had little effect in the control of microbial attachment and biofilm growth on surfaces. A doubling of velocity will also quadruple pumping costs, a factor that must be weighed against downtime and cleaning outlay. Furthermore, increasing fluid velocity as a method of biofilm control is not always feasible for RO systems. Excessive fluid velocities can result in membrane rupture. Increasing the flux rate will also significantly increase the drag due to filtration, resulting in fouling particles being carried to the membrane surface.

Introducing turbulence (cavitation) within shell-and-tube heat exchange systems is known to discourage fouling (Gough and Rogers, 1987). Wire-wound tube inserts can act as turbulence promoters. It has been shown that a reduction of biofilm accumulation on tube walls approaching 50% can be achieved by the use of inserts under certain circumstances (Wills *et al.*, 2000). Fouling may occur on the inserts themselves. While this should not affect heat transfer, the inserts and their associated bioburden may add to the pressure drop through the exchanger, with a consequent increase in pumping costs.

Temperature is another process variable, which influences biofilm development. The thickness of an *Escherichia coli* biofilm has been shown to increase by about

80% when temperature was increased from 30°C to 35°C (Bott and Pinheiro, 1977). Based on this, Bott (1994) has suggested that lower operating temperatures could be used to inhibit biofouling, but since industrial biofilms do not develop in pure culture conditions, such a dramatic influence is unlikely. Changing to a different process temperature will select for organisms best able to function at that temperature, providing it is within the extremes for growth. For example, thermophiles were found to predominate in biofilms formed in heat exchangers operating at 41–60°C (Almeida and de Franca, 1999). Heat exchangers in temperate climates generally operate at a water temperature of 20–50°C and typically there will be little opportunity to control biofilm formation by changes in the operating temperature.

The levels of nutrients and planktonic microorganisms in feed water will influence the establishment of industrial biofilms (Flemming, 1997; Al-Ahmad et al., 2000). The development of a biofilm requires seeding microorganisms and nutrients. Water used for cooling purposes need not meet the criteria required for potable water and will often be obtained from open waters or bore holes. It may, therefore, not only carry a microbial population capable of initiating biofilms, but also low concentrations of nutrients (organic and inorganic) favouring film formation. Water in open re-circulating systems (i.e. spray ponds and cooling towers) is particularly prone to contamination with nutrients (Melo and Bott, 1997). Paradoxically, another source of organic nutrients may be antifouling treatments. Organic anti-scalants have been used to prevent the build-up of abiological fouling on RO and nanofiltration membranes, but many of these have been found to promote biofilm growth (Vrouwenvelder et al., 2000).

Installing sand filters to treat incoming or recirculated water have been suggested as a strategy for lessening its biofouling potential (Bott, 1998). Assuming that biofilm formation is inevitable they encourage its formation, effectively trapping bacteria and nutrients away from critical parts of the process (Flemming and Griebe, 2000). The performance of continuous sand filters installed as side-stream biofilters in an open recirculating cooling system has been assessed (Daamen et al., 2000). The installation of an ASTRASAND® biofilter resulted in diminished nutrient concentrations. The degree of biofouling was considerably lessened, as was the usage of oxidising biocide. This strategy further reduced the reduced requirements for 'blow-down', and lessened the corrosive potential of the cooling water. A pre-treatment strategy that used a combination of sand filtration and RO cartridges was also shown to be effective for the control of biofouling in heat exchangers (Griebe and Flemming, 1998).

Pre-treatment approaches combining sand filtration, multimedia filtration and granular-activated carbon filtration have also been suggested as strategies for the protection of RO elements (Al-Ahmad et al., 2000). Treatment of surface feed waters with ferric oxide floc resulted in a stable filtration process (Galjaard et al., 2001). The use of pre-treatment reduced the degree of membrane biofouling compared with conventional treatment systems. It must be emphasised that the use of media filters typically generates a waste volume of some 4% of their intake. Since this waste will normally contain coagulant, it is not advisable to return it to the feed source and its disposal will add to process costs. Developments in the past 7 years have witnessed the introduction of polypropylene or alumina-fixed pore-size filters positioned directly before RO elements. These also contribute to lessening of biofouling (Hernandez *et al.*, 1995; Kang and Shah, 1997). The use of biocides to lower the bioburden of cooling water or RO feeds is discussed in Section 31.4.1.

31.4.4   Discouraging film attachment

31.4.4.1   Surface properties

The composition and finish of surfaces will influence the process of film establishment. Montero and Pintado (1994) compared materials commonly used to construct heat exchangers in thermoelectric plants. They found that admiralty brass was more resistant to biofouling than stainless steel but was more prone to corrosion. The finish applied to stainless steel is of paramount importance in controlling the initial attachment of bacteria. In the paper and pulp industry, electro-polished stainless steel, rather than lower-grade steel products, is often the material of choice due to its low biofouling potential (Mattila *et al.*, 2002). Electroplating involves immersion of the steel in an acid bath through which a current is passed. This is believed to give an overall negative charge to the surface mitigating against the attachment potential of planktonic bacteria. Others have shown that streptococcal attachment to stainless steel is governed by cell surface proteins, rather than surface charge (Flint *et al.*, 2002).

The relationship between surface energy and microbial adhesion is well understood. Generally, those surfaces that possess poor surface energies (e.g. silicone polymers), are less prone to bacterial attachment (Thorpe *et al.*, 1999). Such materials are of limited use for heat exchangers because of their poor thermal conductivity. Surfaces that are hydrophobic, smooth and uncharged are less likely to form biofilms (Pasmore *et al.*, 2001). Typically, the walls of water handling systems upstream of RO systems (well castings, risers and submersible pumps) are sealed with epoxy resins to limit biofouling and consequent downstream problems.

RO membrane properties play an important role in biofouling potential. Biofilm initiation by *Pseudomonas aeruginosa* is known to increase with surface roughness (Pasmore *et al.*, 2001). Membrane roughness (as a result of deposition and repeated cleaning) is endemic to aged membranes. Neutrally charged thin film composite membrane configurations, such as those composed of polyamide (Filmtec®: Dow liquid separations, US), have reduced binding energy (Van der Waal) and electrostatic (zeta potential) forces. RO membranes composed of aromatic polyamide beneath nano-sized titanium oxide also show potential for the control of biofouling (Kwak *et al.*, 2001). These membranes have photo-catalytic bactericidal activity against planktonic *Echerichia coli* when irradiated with UV light. Their effectiveness in controlling biofouling in an industrial setting has yet to be established.

31.4.4.2   Other methods

Ultrasonic (>20 kHz) waves create cavitation and generate microbubbles at surfaces (Melo and Bott, 1997). While this may deter bacterial attachment, there is as yet no firm evidence as to the relative contribution of this (as opposed to loosening and removal of established films) to the effectiveness of high-intensity, pulsed ultrasound treatments for biofilm control in heat exchangers (Bott, 2000). Infrasonic pulsing of polymeric membranes has met with considerable success in controlling colloidal fouling during microfiltration of wine and beer (Czekaj et al., 2002). High-frequency pulses of permeate are periodically returned in the direction of the membrane, causing the element to vibrate rapidly (ca.7 Hz) thus shedding fouling particles at the intake side. This principle might find application for the control of biofilm-type fouling in membrane systems.

Another approach to discouraging initial attachment is to alter the charge properties of particles (including microorganisms) and surfaces by the application of electrical fields. This is the function of the Zeta Rod® deposit control system (Zeta Corporation, USA) which is available commercially. A ceramic-coated rod and the vessel or pipe in which it is installed together form a capacitor which is charged by a high-voltage DC supply. The electrostatic field so produced induces a high zeta potential (positive charge) on suspended particles in the system, counteracting Van der Waals attraction, which would cause flocculation and attraction of particles including microorganisms to surfaces (Pitts, 1995). The efficiency of the system when applied to RO systems in a soft drinks bottling plant has been reported (Romo and Pitts, 1999). Control RO membranes normally required cleaning at 3–4 month intervals. With the introduction of zeta rods upstream of RO modules, membranes were shown to be relatively clear of biofilm after 2 months operation and the interval between cleaning/disinfection regimes could be extended to 9 months. The technology was applied to the paper mill industry and shown to be effective in limiting the growth of biofouling (Wiltshire et al., 2000). The manufacturers recommend its use for the control of biofouling in cooling towers and other aspects of heat exchange systems. Others have proposed protocols for biofilm control where strong electric currents are applied across silver membrane surfaces (Webster et al., 2000). These non-polluting technologies should find increasing use in the future.

31.5   CONCLUSIONS

Biofilm formation is unavoidable in most industrial situations where water or aqueous solutions contact surfaces. While specific strategies for controlling establishment and growth of biofilms in heat-exchange and filtration systems have been discussed in this chapter, their *modus operandi* is also applicable to other industrial processes. The challenge to the plant operator is to delay biofilm establishment and control its development so that a threshold of interferences is not exceeded. How this challenge is met will vary from process to process, but the shrewd operator should consider the following approaches:

- The use of effective real-time monitoring of liquid and surface phases to define the biofouling problem and guide the treatment regime.
- The judicious use of biocides to inhibit both planktonic and film-associated microorganisms.
- The use of other chemical treatments (surfactants and chelating agents) that disrupt the biofilm matrix, facilitating biofilm loss by natural sloughing or mechanical removal, and enhancing biocide effectiveness.
- The use of physical methods to strip biofilms from surfaces.
- The pragmatic control of system variables to minimize the potential effect of biofouling at critical stages of the process.
- Discouraging the attachment of bacteria to surfaces from the onset of operation by careful choice of surface materials and use of technologies, such as ultrasonic waves and zeta rod systems.

Established biofilms are both a process liability and costly to remove. The prudent operator will attempt to minimise the potential for biofouling rather than contend with a chronic problem. Effective control will usually require a combination of several approaches. In general, prevention is better than cure.

REFERENCES

Al-Ahmad, M., Aleem, F.A.A., Mutiri, A. and Ubaisy, A. (2000) Biofouling in RO membrane systems. Part 1: Fundamentals and control. *Desalination* **132**, 173–179.

Allison, D.G., McBain, A.J. and Gilbert, P. (2000) Biofilms: problems of control. In *Community Structure and Co-operation in Biofilms* (ed. D.G. Allison, P. Gilbert, H. Lappin-Scott and M. Wilson), Vol. 59, SGM Symposia, Cambridge University Press, Cambridge.

Almeida, M.A.N. and de Franca, F.P. (1999) Thermophilic and mesophilic bacteria in biofilms associated with corrosion in a heat exchanger. *World J. Microbiol. Biotech.* **15**, 439–442.

Al-Mutaz, I.S. (2001) The continued challenge of capacity building in desalination. *Desalination* **141**, 145–156.

American Society for Testing and Materials. (1978) Standard classification of metal working fluids and related materials. ANSI/ASTM Designation D 2881-73 (reapproved 1978). In *Annual Book of ASTM Standards*, Vol. 24, pp. 750–751, ASTM, Philadelphia, PA.

Anderson, J.E., Springer, W.S. and Strosberg, G.G. (1981) Application of reverse osmosis to automotive electrocoat paint wastewater recycling. *Desalination* **36**, 179–188.

Anon (1991) The control of legionellosis including legionnaires disease. Report HSpG70, HMSO, UK.

Anon (2002) Ultrapure water: world markets. Report No. 29, The Mcilvaine Company, Northbrook, IL.

Baker, J.S. and Dudley, L.Y. (1998) Biofouling in membrane systems. *Desalination* **118**, 81–90.

Blenkinsopp, S.A., Khoury, A.E. and Costerton, J.W. (1992) Electrical enhancement of biocide efficacy against *Pseudomonas aeruginosa* biofilms. *Appl. Environ. Microbiol.* **58**, 3770–3773.

Bott, T.R. (1994) The control of biofilms in industrial water cooling systems. In *Bacterial Biofilms and Their Control in Medicine and Industry* (ed. J. Wimpenny, W. Nichols, D.J. Stickler and H. Lappin-Scott), pp. 173–180, BioLine, Cardiff, UK.

Bott, T.R. (1998) Techniques for reducing the amount of biocide necessary to counter-act the effects of biofilm growth in cooling water systems. *Appl. Therm. Eng.* **18**, 1059–1066.

Bott, T.R. (2000) Biofouling control with ultrasound. *Heat Trans. Eng.* **21**, 43–49.

Bott, T.R. (2001a) Potential physical methods for the control of biofouling in water systems. *Chem. Eng. Res. Des.* **79**, 484–490.

Bott, T.R. (2001b) To foul or not to foul – that is the question. *Chem. Eng. Prog.* **97**, 30–37.

Bott, T.R. and Pinheiro, M.M.V.P.S. (1977) Biological fouling: velocity and temperature effects. *Can. J. Chem. Eng.* **55**, 473–474.

Brankevich, G.J., Demele, M.L.F. and Videla, H.A. (1990) Biofouling and corrosion in coastal power-plant cooling water systems. *Mar. Technol. Soc. J.* **24**, 18–28.

Chen, X. and Stewart, P.S. (2000) Biofilm removal caused by chesmical treatments. *Water Res.* **34**, 4229–4233.

Chilcott, T.C., Gaedt, L., Nantawisarakul, T., Fane, A.G. and Coster, H.G.L. (2002) Electrical impedance spectroscopy characterisation of conducting membranes 1. Theory. *J. Membr. Sci.* **195**, 153–167.

Cloete, T.E., Jacobs, L. and Brozel, V.S. (1998) The chemical control of biofouling in industrial water systems. *Biodegradation* **9**, 23–37.

Costerton, J.W., Keng, K.J., Geesey, G.G., Ladd, T.I., Nickel, J.C., Dasyupa, M. and Marrie, T.J. (1987) Bacterial biofilms in nature and disease. *Appl. Rev. Microbiol.* **41**, 435–464.

Costerton, J.W., Ellis, B., Lam, K., Johnson, F. and Khoury, A.E. (1994) Mechanism of electrical enhancement of efficacy of antibiotics in killing biofilm bacteria. *Antimicrob. Agent. Chemother.* **38**, 2803–2809.

Czekaj, P., Lopez, F. and Guell, C. (2002) Membrane fouling by turbidity constituents of beer and wine: characterization and prevention by means of infrasonic pulsing. *J. Food Eng.* **49**, 25–36.

Daamen, E.J., Wouters, J.W. and Savelkoul, J.T.G. (2000) Side stream biofiltration for improved biofouling control in cooling water systems. *Water Sci. Technol.* **41**, 445–451.

Donlon, R.M. (2000) Biofilm control in industrial water systems: approaching an old prob-lem in new ways. In *Biofilms: Recent Advances in Their Study and Control* (ed. L.V. Evans), pp. 333–360, Harwood Academic Publishers, Amsterdam.

Exner, M., Tuschewitzki, G.J. and Scharnagel, J. (1987) Influence of biofilms by chemical disinfectants and mechanical cleaning. *Zbl. Bakt. Mik. Hyg. B.* **183**, 549–563.

Flemming, H.C. (1997) Reverse osmosis membrane biofouling. *Exp. Ther. Fluid Sci.* **14**, 382–391.

Flemming, H.C. and Griebe, T. (2000) Control of biofilms in industrial waters and processes. In *Industrial Biofouling* (ed. J. Walker, S. Surman and J. Jass), pp. 125–141, John Wiley & Sons Ltd., Chichester, UK.

Flemming, H.C., Griebe, T. and Schaule, G. (1996) Antifouling strategies in technical systems – a short review. *Water Sci. Technol.* **34**, 517–524.

Flemming, H.C., Tamachkiarowa, A., Klahre, J. and Schmitt, J. (1998) Monitoring of foul-ing and biofouling in technical systems. *Water Sci. Technol.* **38**, 291–298.

Flint, S., Brooks, J., Bremer, P., Walker, K. and Hausman, E. (2002) The resistance to heat of thermo-resistant streptococci attached to stainless steel in the presence of milk. *J. Ind. Microbiol. Biotechnol.* **28**, 134–136.

Franken, T. (1999) Ultrapure water: more than membrane technology alone. *Membr. Technol.* **1999**, 9–12.

Gagnon, G.A., Ollos, P.J. and Huck, P.M. (1997) Modelling BOM utilisation and biofilm growth in distribution systems: review and identification of research needs. *J. Water SRT Aqua.* **46**, 165–180.

Galjaard, G., van Paassen, J., Buijs, P. and Schoonenberg, F. (2001) Enhanced pre-coat engineering (EPEC) for micro- and ultra-filtration: the solution for fouling? *Water Suppl.* **1**, 1251–1256.

Gioli, P., Silingardi, G.E. and Ghiglio, G. (1987) High quality water from refinery waste. *Desalination* **67**, 271–282.

Girard, B. and Fukumoto, L.R. (2000) Membrane processing of fruit juices and beverages: a review. *Crit. Rev. Food Sci. Nut.* **40**, 91–157.

Gough, M. and Rogers, J.V. (1987) Reduced fouling by enhanced heat transfer using wire matrix radical mixing elements. *AIChe. Symp. Ser.* **83**, 16–21.

Griebe, T. and Flemming, H.C. (1998) Biocide-free antifouling strategy to protect RO membranes from biofouling. *Desalination* **118**, 153–156.

Hamida, A.B. and Moch, I. (1996) Controlling biological fouling in open sea intake RO plants without continuous chlorination. *Desalin. Water Reuse* **6**, 40–45.

Hernandez, A., Martinez, F., Martin, A. and Pradanos, P. (1995) Porous structure and surface-charge density on the walls of microporous alumina membranes. *J. Coll. Interf. Sci.* **173**, 284–296.

Huang, C.T., James, G., Pitt, W.G. and Stewart, P.S. (1996) Effects of ultrasonic treatment on the efficacy of gentamicin against established *Pseudomonas aeruginosa* biofilms. *Coll. Surf. B Biointerf.* **6**, 235–242.

Jass, J., Costerton, J.W. and Lappin-Scott, H.M. (1995) The effect of electrical currents and tobramycin on *Pseudomonas aeruginosa* biofilms. *J. Ind. Microbiol.* **15**, 234–242.

Johansen, C., Falholt, P. and Gram, L. (1997) Enzymatic removal and disinfection of bacterial biofilms. *Appl. Environ. Microbiol.* **63**, 3724–3728.

Kang, P.K. and Shah, D.O. (1997) Filtration of nanoparticles with dimethyldioctadecylammonium bromide treated microporous polypropylene filters. *Langmuir* **13**, 1820–1826.

Karakulski, K. and Morawski, W.A. (2000) Purification of copper wire drawing emulsion by application of UF and RO. *Desalination* **131**, 87–95.

Kaur, K., Bott, T.R. and Leadbeater, B.S.C. (1992) Effects of ozone as a biocide in an experimental cooling water system. *Ozone Sci. Eng.* **14**, 517–530.

Klahre, J. and Flemming, H.C. (2000) Monitoring of biofouling in papermill process waters. *Water Res.* **34**, 3657–3665.

Kosutic, K. and Kunst, B. (2001) Effect of hydrolysis on porosity of cellulose acetate reverse osmosis membranes. *J. Appl. Polym. Sci.* **81**, 1768–1775.

Koyuncu, I., Yalcin, F. and Ozturk, I. (1999) Color removal of high strength paper and fermentation industry effluents with membrane technology. *Water. Sci. Technol.* **40**, 241–248.

Kumar, C.G. and Takagi, H. (1999) Microbial alkaline proteases: from a bioindustrial viewpoint. *Biotech. Adv.* **17**, 561–594.

Kwak, S.Y., Kim, S.H. and Kim, S.S. (2001) Hybrid organic/inorganic reverse osmosis (RO) membrane for bactericidal anti-fouling. 1. Preparation and characterization of TiO_2 nanoparticle self-assembled aromatic polyamide thin-film-composite (TFC) membrane. *Environ. Sci. Technol.* **35**, 2388–2394.

LeChevallier, M.W. (1991). Biocides and the current status of biofouling in water systems. In *Biofouling and Biocorrosion in Industrial Water Systems* (ed. H.C. Flemming and G.G. Geesey), pp. 113–132, Springer, Heidelberg.

Leeper, S.A. and Tsao, G.T. (1987) Membrane separations in ethanol recovery: an analysis of two applications of hyperfiltration. *J. Membr. Sci.* **30**, 289–312.

Licina, G.W., Nekoksa, G. and Howard, R.L. (1994) An electrochemical method for online monitoring of biofilm activity in cooling water using the BioGeorge probe. Report No. 1232, American Society for Testing Materials.

MacDonald, R., Santa, M. and Brozel, V.S. (2000) The response of a bacterial biofilm community in a simulated industrial cooling water system to treatment with an anionic dispersant. *J. Appl. Microbiol.* **89**, 225–235.

Mattila, K., Weber, A. and Salkinoja-Salonen, M.S. (2002) Structure and on-site formation of biofilms in paper machine water flow. *J. Ind. Microbiol. Biotechnol.* **28**, 268–279.

Mattila-Sandholm, T. and Wirtanen, G. (1992) Biofilm formation in the food industry – a review. *Food Rev. Int.* **8**, 573–603.

McCoy, W.F., Bryers, J.D., Robbins, J. and Costerton, J.W. (1981) Observations of fouling biofilm formation. *Can. J. Microbiol.* **27**, 910–917.

Melo, L.F. and Bott, T.R. (1997) Biofouling in water systems. *Exp. Therm. Fluid Sci.* **14**, 375–381.

Minuth, W. (1997) Recent developments in the technologies for the production of water for injection. *Pharma. Ind.* **59**, 899–902.

Mittelman, M.C. (1991) Bacterial growth and biofouling control in purified water systems. In *Biofouling and Biocorrosion in Industrial Water Systems* (ed. H.C. Flemming and G.G. Geesey), pp. 113–134, Springer, Heidelberg.

Montero, F. and Pintado, J.L. (1994) Bioensuciamiento de tubos en cambiiadores de calor. *Microbiología: Publicación de la Sociedad Española de Microbiología* **10**, 93–102.

Morton, G. (2000) Problems of biofilms in industrial waters and processes. In *Industrial Biofouling: Detection, Prevention and Control* (ed. J. Walker, S. Surman and J. Jass), pp. 79–102, John Wiley & Sons Ltd., Chichester, UK.

Mott, I.E.C., Stickler, D.J., Coakley, W.T. and Bott, T.R. (1998) The removal of bacterial biofilm from water-filled tubes using axially propagated ultrasound. *J. Appl. Microbiol.* **84**, 509–514.

Murphy, A.P., Moody, C.D., Reily, R.L., Lin, S.W., Murugaverl, B. and Rusin, P. (2001) Microbial damage of cellulose acetate membranes. *J. Membr. Sci.* **193**, 111–121.

Nickels, J., Bobbie, R.J., Lott, D.F., Maritz, R.F., Benson, P.H., and White, D.C. (1981) Effect of manual brush cleaning on biomass and community structure of microfouling film formed on aluminium and titanium surfaces exposed to rapidly flowing seawater. *Appl. Environ. Microbiol.* **42**, 1422–1453.

Pasmore, M., Todd, P., Smith, S., Baker, D., Silverstein, J., Coons, D. and Bowman, C.N. (2001) Effects of ultrafiltration membrane surface properties on *Pseudomonas aeruginosa* biofilm initiation for the purpose of reducing biofouling. *J. Membr. Sci* **194**, 15–32.

Payne, K.R. (1988) *Industrial Biocides*, John Wiley and Sons Ltd., London.

Percival, S.L. (2000) Detection of biofilms in industrial waters and processes. In *Industrial Biofouling* (ed. J. Walker, S. Surman and J. Jass), pp. 103–124, John Wiley and Sons Ltd., London.

Peterson, R.V. and Pitt, W.G. (2000) The effect of frequency and power density on the ultrasonically-enhanced killing of biofilm-sequestered *Escherichia coli*. *Coll. Surf. B Biointerf.* **17**, 219–227.

Pitts, M.M. (1995) Fouling mitigation in aqueous systems using electrochemical water treatment. In *Fouling Mitigation of Industrial Heat Exchangers*, pp. 18–23, Shell Beach, California.

Qian, Z., Stoodley, P. and Pitt, W.G. (1996) Effect of low-intensity ultrasound upon biofilm structure from confocal scanning laser microscopy observation. *Biomaterials* **17**, 1975–1980.

Qian, Z., Sagers, R.D. and Pitt, W.G. (1997) The role of insonation intensity in acoustic-enhanced antibiotic treatment of bacterial biofilms. *Coll. Surf. B Biointerf.* **9**, 239–245.

Qian, Z., Sagers, R.D. and Pitt, W.G. (1999) Investigation of the mechanism of the bio-acoustic effect. *J. Biomed. Mater. Res.* **44**, 198–205.

Roh, I.J. (2002) Influence of rupture strength of interfacially polymerized thin-film structure on the performance of polyamide composite membranes. *J. Membrane Sci.* **198**, 63–74.

Romo, R.F.V. and Pitts, M.M. (1999) Application of electrotechnology for removal and prevention of reverse osmosis biofouling. *Environ. Prog.* **18**, 107–112.

Saad, M.A. (1992) Biofouling prevention in RO polymeric membrane systems. *Desalination* **88**, 85–105.

Soini, S.M., Koskinen, K.T., Vilenius, M.J. and Puhakka, J.A. (2002) Effects of fluid-flow velocity and water quality on planktonic and sessile microbial growth in water hydraulic system. *Water Res.* **36**, 3812–3820.

Thorpe, A.A., Nevell, T.G. and Tsibouklis, J. (1999) Surface energy characteristics of poly(methylpropenoxyalkylsiloxane) film structures. *Appl. Surf. Sci.* **137**, 1–10.

Videla, H.A., Viera, M.R., Guiamet, P.S. and Alais, J.C.S. (1995) Using ozone to control biofilms. *Mater. Performance* **34**, 40–44.

Viera, M.R., Guiamet, P.S., de Mele, M.F.L. and Videla, H.A. (1999a). Biocidal action of ozone against planktonic and sessile *Pseudomonas fluorescens. Biofouling* **14**, 131–141.

Viera, M.R., Guiamet, P.S., de Mele, M.F.L. and Videla, H.A. (1999b) Use of dissolved ozone for controlling planktonic and sessile bacteria in industrial cooling systems. *Int. Biodeterior. Biodegrad.* **44**, 201–207.

Viera, M.R., Guiamet, P.S., de Mele, M.F.L. and Videla, H.A. (2000) The effect of dissolved ozone on the corrosion behaviour of heat exchanger structural materials, biocidal efficacy on bacterial films. *Corros. Rev.* **18**, 205–220.

Voros, N.G., Maroulis, Z.B. and MarinosKouris, D. (1996) Salt and water permeability in reverse osmosis membranes. *Desalination* **104**, 141–154.

Vrouwenvelder, J.S., Manolarakis, S.A., Venendaal, H.R. and van der Kooij, D. (2000) Biofouling potential of chemicals used for scale control in RO and NF membranes. *Desalination* **132**, 1–10.

Watanabe, S., Onda, K., Matsuda, T. and Hayashi, A. (1996) Ozonised water for peeling off slime in the wall of heat exchangers, Patent no. JP08166197 A2, Japan.

Wattanakaroon, W. and Stewart, P.S. (2000) Electrical enhancement of *Streptococcus gordonii* biofilm killing by gentamicin. *Arch. Oral Biol.* **45**, 167–171.

Webster, R.D., Chilukuri, S.V.V., Levesley, J.A. and Webster, B.J. (2000) Electrochemical cleaning of microporous metallic filters fouled with bovine serum albumin and phosphate under low cross-flow velocities. *J. Appl. Electrochem.* **30**, 915–924.

Weyandt, R.G. (2001) Microbiological aspects of ultra pure water units in the pharmaceutical industry – an up-to-date literature review. *Pharm. Ind.* **63**, 1295-+.

Wiatr, C.L. (1991) Enzyme blend containing cellulase to control industrial slime. US Patent no. 4,994,390.

Wiatr, C.L. and Fedyniak, O.X. (1991) Development of an obligate anaerobe specific biocide. *J. Ind. Microbiol.* **7**, 7–13.

Wills, A., Bott, T.R. and Gibbard, I.J. (2000) The control of biofilms in tubes using wire-wound inserts. *Can. J. Chem. Eng.* **78**, 61–64.

Wiltshire, K., Verreault, M. and Pitts, M.M. (2000) Slime control in paper machine shower water systems with unique chemical-free electro-technology – it is a viable alternative to current chemical methods. *Pulp Pap.-Can.* **101**, 103–106.

Yu, F.P., Ginn, L.D. and McCoy, W.F. (1999) Cooling tower fill fouling control in a geothermal power plant. *Corros. Rev.* **17**, 205–217.

32

Microbial fouling control
for industrial systems

W.F. McCoy

32.1 THE MICROBIAL-FOULING PROCESS

Technical advances in the control of microbial fouling have occurred because astute descriptions of the fouling process have provided innovators with better targets at which to direct their developments. With a clear view of the fouling process in mind, practitioners apply newly developed techniques more efficiently, more safely, and to much better effect.

32.1.1 Fouling factors

Microbial fouling in industrial systems occurs because of viable microorganisms, nutrients, and surfaces. The influence of these factors on microbial fouling is interdependent.

While specifics of the fouling process vary with system type and conditions, several general facts are nearly always relevant (Table 32.1). Regarding viable microorganisms, diversity of microbial species directly relates to stability because highly diverse microbial communities more effectively resist control and adapt to stress. Regarding nutrients, microbial growth in most natural aquatic systems (lakes, rivers, ocean) is phosphate limited, but phosphate is often enriched

Table 32.1   Key microbial-fouling factors with limits and effects.

Microbial-fouling factor	Microbial-fouling limits	Effect of limits
Viable microorganisms	Microbial diversity and metabolic activity	Directly proportional to stability of microbial-fouling deposits and the difficulty in controlling them
	Temperature, pH, ionic strength, water activity	May limit microbial activity
Nutrients	Bioavailability of phosphate, nitrogen, carbon	Phosphate concentration nearly always limits growth in natural systems, but is often enriched in industrial water systems
Surfaces	Surface area Surface roughness	Directly proportional to microbial activity and to difficulty in achieving control

in engineered water systems resulting in carbon or nitrogen limited growth. Regarding surfaces, they always enhance the fouling process, since biofilms often prolifically develop upon them.

Even in systems that may not seem laden with surfaces to which micro-organisms can adhere (such as in the pelagic ocean), most microbiological activity depends on activity in biofilms (Donlan, 2002). Suspended particulate matter provides vast surface area upon which microbial biofilms proliferate. For example, biofilms on these surfaces account for most of the microbial activity in the sea.

Temperature, pH, ionic strength, and water activity have not been indicated as fouling factors in Table 32.1, but rather, they are given as limits on viable microorganisms. In so far as they effect metabolism of viable microorganisms, these parameters greatly influence microbial fouling. However, due to the remarkable versatility in the microbial world, fouling problems occur even in relatively extreme conditions. This fact leads us not to quickly dismiss biological fouling in very hot, very cold, very acid, very caustic, very salty, or very dry systems (within biotic limits, of course) because microbial-fouling control in such systems may still be necessary.

32.1.2   Operational performance effects

The value of managing microbial-fouling processes can be measured in terms of industrial operational performance effects. There are principally three operational problem categories:

(1) product contamination,
(2) corrosion,
(3) energy losses.

Table 32.2   Operational effects by microbial fouling in industrial systems.

Problem category	Effect	Examples
Product contamination	Inventory loss	In-can product spoilage, breakage in paper spooling, metalworking fluid spoilage
Accelerated corrosion	System failure, aesthetic losses	Under-deposit pitting corrosion, corrosion at welds, accelerated general corrosion
Energy losses	Increased energy consumption	Heat transfer resistance, fluid frictional resistance

Table 32.3   Biofouling in heating, ventilation, and air-conditioning systems.

Biofilm thickness (mm) on heat exchanger tubes	Percent increase in energy consumption (%)	Added energy cost due (USD annually, $)*
0.15	5.3	13,500
0.30	10.8	27,000
0.61	21.5	59,000
0.91	32.2	83,000

*Calculated for a 1000-ton chiller operated 350 days/year, 16 h a day @ $0.07/KWH.

Product contamination from microbial fouling can cause significant inventory loss and destroy brand loyalty. Microbiologically accelerated corrosion on critical surfaces can cause catastrophic system failure. Increased energy consumption due to microbial fouling of surfaces can cause remarkably higher operating cost. Table 32.2 lists effects of these problems on operational performance.

The cost of failing to control industrial microbial fouling can be directly measured. For example, biofilm fouling in heat exchanger tubes causes increased heat transfer and fluid frictional resistances. Biofilm-induced energy losses in heat exchangers are the result of thermal insulation and the viscoelasticity of the deposit. Such losses can be very expensive (Table 32.3).

32.1.3   Health-related risk

If the only value of industrial microbial-fouling control is given in terms of improving operational performance, then its real value may be severely underestimated. This is because health-related risk for industrial water systems is often tangibly linked to microbial fouling. The value of reducing health-related risk by controlling microbial fouling can be vastly greater than the value derived merely from reducing expenses through improved operational performance.

Effective management of microbial fouling to reduce health-related risk from industrial water systems has become a business priority. This is due to increasing public awareness of potentially significant health-related risk from these systems. The most important industrial water health-related issue today is the risk of legionellosis.

Legionellosis, which refers to infections caused by bacteria in the genus *Legionella*, can result in the pneumonia referred to as Legionnaires' disease and/or the flu-like symptoms known as Pontiac Fever. *Legionella* proliferation is always associated with its protozoan hosts and in biofilms (Murga *et al.*, 2001; Donlan and Costerton, 2002). Legionellosis is the second or third most frequent cause of sporadic community-acquired pneumonia in the US. Worldwide, many tens of thousands of people are affected by legionellosis each year. In susceptible people, the disease can be devastating (Atlas, 1999).

Recently, nontuberculosis mycobacterial (NTM) pneumonia has been increasingly associated with aerosol-generating water systems. The causative bacterial agents are found ubiquitously in the environment, especially associated with biofilms in aquatic and soil habitats. Sixteen species of bacteria in the genus *Mycobacterium* are known to be potential human pathogens. Disease caused by NTM is not currently a reportable incident in the US. The only published national summary information comes from occasional surveys conducted by the Centres for Disease Control and Prevention (CDC). Incidence of the disease and public concern appears to be increasing (Gorman, 2002).

Also of growing concern, especially in tropical regions, is disease caused by pathogenic amoebae, such as *Naegleria fowleri*, and also in the genus *Acanthamoeba*. Primary amoebic meningoencephalitis (PAM), granulomatious amoebic encephalitis (GAE), and acanthamoebic keratitis (amoebic infections of the eye) are all potentially serious afflictions. Although cases are rare, mortality rates among the ill can be high.

Individuals with weakened immune systems, such as transplant patients and people suffering from autoimmune diseases, can be at risk from opportunistic pathogenic microorganisms in industrial water systems. Health-related risk minimisation recommendations for industrial systems have now been specified in recent guidance from industry and government (OSHA, 1998; CTI, 2000; HSE, 2000; ASHRAE, 2001; DHS, 2001; ASTM, 2002).

32.2   MICROBIAL-FOULING CONTROL STRATEGIES

Biofilms and the nutrients that fuel their development are at the root-cause of essentially all microbial-fouling problems in industry. It is not practical to entirely eliminate them in most industrial situations. This is also observed to be true in natural systems. Control of microbial fouling, therefore, requires a process management strategy.

32.2.1   Establishing a fouling management programme

Most industrial systems are open to the environment. Usually, microbial inoculation occurs continuously. Therefore, the goal of microbial-fouling control in industrial systems is almost never to sterilise the system, but rather to manage the fouling process.

Table 32.4   Categories of practice in microbial-fouling control.

Category of practice	Purpose and process	Payoff
Recognition of problems worth solving	System surveys to identify problem root-causes using diagnostic and monitoring tools	Discovery of problems worth solving and establishment of criteria for success
Remedy	Selection and application of antimicrobials, dispersants, and other remedies	Establish microbial control
Regulation of remedy	Proactive variance of applied stresses	Maintain microbial control long term

Managing the microbial-fouling process, not sterilising the system, is the predominant strategy we observe in nature. In fact, natural strategies look like what the water-treatment industrialist would call 'programmes'. Programmes are comprised of activities to recognise problems worth solving, apply remedies to mitigate the problems, and then to carefully regulate the applied remedies. Practising in these three general categories is necessary to effectively control industrial microbial fouling (Table 32.4).

32.2.1.1   Recognition of the problems

Natural strategies to control microbial fouling in animals and plants always depend upon recognition of problems that are actually worth solving. Natural recognition systems are far more sophisticated than the current state-of-the-art in industry. For example, white blood cells in animals capture and destroy bacteria that have been marked with compounds, such as specific antibodies from the immune response. These substances, called opsonins, stimulate white blood cells to capture the invading microbes, activate release of the antimicrobial proteins, and then, trigger the oxidative respiratory burst. In this way, one part of the system (such as the circulatory system) can be cleared of infection, while another part of the system (such as the digestive system) can be teaming with microbial life. Natural microbial-fouling control strategies consist of much more than just applying antimicrobials in an effort to kill off as many bacteria as possible.

Effective industrial programmes begin with recognition of problems that are worth solving, discovering root-causes of the problems, and then establishing specific criteria for success in mitigating the problems. This discovery process is the result of system survey, diagnostics, and monitoring. Many spectacular failures in water-treatment programmes could have been avoided if the root-causes of serious microbial-fouling problems had been recognised earlier.

32.2.1.2   Remedy

Tools in nature and analogous examples from industry used to remedy critical microbial-fouling problems are given in Table 32.5.

Table 32.5   Natural microbial control tools and a few industrial examples.

Natural microbial control tool	Natural examples	Industrial examples
Stabilised halogenating agents	*In situ* halide oxidation and stabilisation by mammalian white blood cells	Stabilised bromine
Antimicrobials and antibiotics	Specific metabolic inhibitors, electrophilic organics, cell signalling inhibitors	Non-oxidising biocides
Surfactants, detergents, and enzymes	Bioactive peptides, lytic enzymes, anionic surfactants	Biodetergents targeted at specific deposits

32.2.1.3   Regulation of remedy

An obvious feature of microbial control in nature is that applied antimicrobial stresses are tightly controlled and proactively varied. This occurs as the result of concentration changes and/or use of more than one antimicrobial agent. Proactive variance of antimicrobial stresses is necessary, since complex microbial-fouling communities are diverse and can rapidly adapt.

An example in nature of remedy regulation is the feedback inhibition of stabilised halogenating agents in the mammalian immune system. The concentration of antimicrobial is limited by regulating production of reactants and the enzyme necessary for synthesis *in situ*. This regulation of remedy protects the animal against damage from excessive oxidant concentrations (Mayeno, 1989).

The goal in nature is always to maintain control of a system, not to let it foul until cleanup is the last resort. In nature, maintenance of a 'clean' system is necessary for survival. Clean in this context rarely, if ever, means sterile. Natural systems are almost always comprised of vast microbial flora in close proximity or actually a part of the protected system.

Similarly, in water treatment, management of the microbial-fouling process is necessary. Sterilisation is seldom the goal in water treatment. Management of microbial fouling in the entire system is certainly the best means to reduce the risk of legionellosis, e.g. ASHRAE (2000) and Murga *et al.* (2001).

Microbial diversity is an important indicator of the extent to which fouling control has been achieved because highly diverse microbial communities are stable and more difficult to manipulate (Cloete and Brozel, 1992). Reviews of molecular techniques to measure microbial diversity (Theron and Cloete, 2000) and integrated control techniques are available (Cloete *et al.*, 1998).

Innovation in practical ways to regulate the applied remedies needed to control microbial fouling is an active field of research in industry and academe. Recent innovations include on-line performance monitoring, dosage control, and on-site biocide active analysis (Wetegrove *et al.*, 1996; Chattoraj, 2002). This new technology promises to improve efficiency in managing the microbial-fouling process.

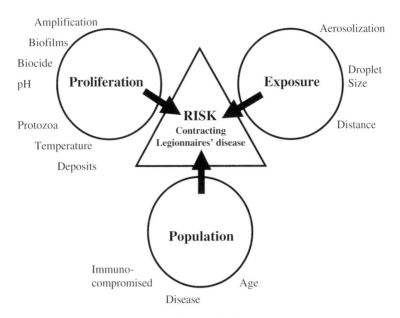

Figure 32.1   Model to assess the risk of legionellosis.

32.2.2   Assessing risk factors

Risk can be quantitatively assessed (Percival *et al.*, 2000). Assessing health-related risk in industrial and domestic water supply systems is the essential first step in minimising risk. The result of a formal risk assessment is a document that links discovery (recognition) of problem root-causes with specific corrective action recommendations (remedy) and assigns responsible personnel to action (regulation of remedy). The risk assessment document helps establish value and provides motive to get the risk minimisation job done properly. It has become an important part of the business in microbial-fouling control.

The risk assessment process is becoming more integrated into legionellosis risk management. Approved codes of practice and guidance for legionellosis risk management were first established in the UK (HSE, 2000). New legislation puts emphasis on *Legionella* risk assessment and management (DHS, 2002). In Victoria, Australian regulations require owners of land upon which there is a cooling tower system to

(1) register their systems;
(2) develop, implement, and review risk management plans;
(3) implement improved maintenance programmes;
(4) keep records of maintenance and test results for audit/inspection purposes.

The maximum penalty for failing to complete a risk management plan in Victoria, Australia is AU$6,000. The maximum penalty for failing to register a cooling tower system is AU$12,000.

Legionellosis risk is the product of three factors: pathogen proliferation potential, exposure potential, and population susceptibility (Figure 32.1).

Pathogen proliferation potential exists in nearly all water systems. Many factors are involved, but most important are the presence of microbial biofilms and the degree of microbial diversity, especially in warm water systems. Managing the microbial-fouling process to reduce the risk of legionellosis consists principally of controlling biofilms and limiting microbial diversity within the entire system.

Exposure to *Legionella* occurs by aspiration (deep inhalation into the lungs) of microscopic ($<10\,\mu$m diameter) aerosols. Certain industrial water systems produce aerosols as an inherent function of their operation. Minimising aerosol dissemination from such systems can directly reduce health-related risk. Assessment of exposure potential focuses on the proximity of aerosol-generating systems to human populations.

The susceptible population risk is the third factor assessed. The most susceptible individuals are the elderly, heavy smokers, patients with chronic pulmonary disease, and immunosuppressed people. Systems in close proximity to susceptible populations may present high risk of legionellosis.

32.3   ADVANCES IN MICROBIAL-FOULING CONTROL

Technical advances in microbial-fouling control can be described within the context of the three critical factors given in Table 32.1 (viable microorganisms, available nutrients, surfaces) and the programme approach to managing the fouling process in Table 32.4 (recognition, remedy, regulation of remedy). Recent advances include better diagnostic/monitoring tools, superior antimicrobials, practical ways to limit available nutrients, more effective surface-active agents to remove or penetrate deposits, and a practical means to automate microbial-fouling control.

32.3.1   Diagnostics and monitoring

Modern advances in biofilm monitoring have led to better ways of discovering problem root-causes. Reviews of biofilm analytical technique and applications are available (Cloete *et al.*, 1998; Schaule *et al.*, 2000; Donlan, 2002).

Biofilm-monitoring devices have been usefully categorised into Level 1, 2, 3, or 4 (Flemming, 2002). These devices, respectively, can

(1) detect the kinetics of deposition, but cannot differentiate microorganisms from abiotic deposit components;
(2) distinguish between biotic and abiotic components in a deposit;
(3) provide detailed information about the chemical composition of a deposit;
(4) discriminate living from dead microorganisms on a surface.

Microbial fouling in industrial systems can be monitored with side-stream devices or with whole-system probes. Most of the devices in use today are side-stream biofilm monitors. Among the many choices now available, the most practical appear to be optical systems that measure deposit transmittance or absorption; these biofouling monitors are sensitive, robust, and widely used (Borchardt, et al., 1997; Klahre and Flemming, 2000).

The principal advantage of a whole-system monitor is that the general microbial activity in the entire system can be measured. In complex systems, this may be preferred to local measurements of fouling from a side-stream construct that simulates just one aspect of the actual system. Recently, a whole-system metabolic activity monitor has been successfully applied for industrial systems (Chattoraj et al., 2002). In this technique, a soluble metabolic activity probe is dosed to the system and measured with in-line fluorometry. This method has now been used to automate microbial-fouling control (see Section 32.3.5).

32.3.2 Novel superior antimicrobials

Superior antimicrobials for industrial applications are designed to fit into programmes for microbial-fouling control. The trend is for antimicrobials that are more effective in biofilm control, less toxic, and easier to use compared to chlorine, which is still by far the most frequently used industrial antimicrobial. The most significant chlorine alternatives in use today are chlorine dioxide, peracetic acid, ozone, bromine, and stabilised bromine.

Stabilised bromine is an innovation designed to imitate natural microbial-fouling control (McCoy, 1998; McCoy et al., 1998). Certain white blood cells in the immune system make N-bromo-aminoethanesulphonic acid to control complex microbial infections of tissue and blood. This natural antimicrobial is superior to chlorine and other oxidants because of better performance and lower cytotoxicity (Weiss et al., 1986; Thomas et al., 1995).

Analogously designed for uses in industrial water treatment, stabilised bromine antimicrobials perform better than chlorine and are inherently less toxic to aquatic wildlife. This often results in at least 25% less oxidant required for microbial-fouling control in field applications. Table 32.6 summarises a comparison of technical attributes and the resultant benefits.

Stabilised bromine treatments provide superior microbial control with reduced environmental impact. An example is given in the adsorbable organic halide (AOX) discharge data from a cooling water system at a refinery in Germany (Figure 32.2).

This cooling system was periodically contaminated by hydrocarbon leaks. Therefore, an especially effective microbial control programme was necessary. Effective heat transfer rate standards were established at the plant and also, operating standards specified aerobic bacterial counts in the recirculating cooling water at $\leqslant 10^4$ CFU/ml. The local authority in Germany imposed an AOX discharge limit of 0.5 mg/l for discharge from the cooling system. Chlorine alternatives that generate no AOX were applied, but none could meet the microbial control requirements. The AOX discharge limit was exceeded when the system was treated with unstabilised bromine to meet microbial fouling and heat transfer rate standards. Violation of the AOX limit resulted in significant fines.

Table 32.6   Technical attributes and the resultant benefits of stabilised bromine compared to unstabilised chlorine or bromine.

Technical attributes of stabilised bromine	Improvements compared to unstabilised halogen	Benefits derived from the technical attributes
Less aquatic toxicity	3–22× less toxic	Reduced environmental toxicity
Less AOX generation	50% less produced	Reduced environmental toxicity
No bromate[a]	100× less produced	Reduced environmental toxicity
Less volatility	10× less volatile	Reduced chemical waste
Less reactivity	50% less reactive	Less waste and materials damage
Better biofilm control	At least 5× better	Less oxidant required
Better storage stability	50% less degradation	Reduced chemical waste
Ease of use	Significantly easier	Reduced potential for accidents

[a] Not detectable at use concentrations referring to the essentially complete inhibition of the bromate formation process.

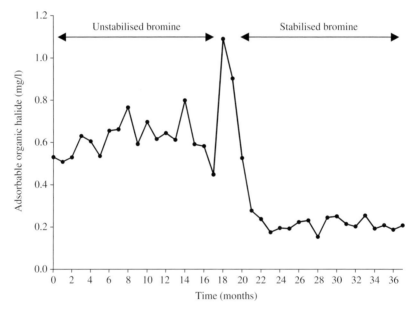

Figure 32.2   AOX discharged from a cooling water system treated to strict microbial control standards with unstabilised bromine and then after conversion to stabilised bromine.

The cooling water microbial-fouling control programme for the refinery in Germany was converted to stabilised bromine at about month 19 (Figure 32.2). Lower oxidant concentrations of stabilised bromine (25–33% lower, data not shown) were required to achieve the operational performance standards. Stabilised bromine generates 50% less AOX compared to equivalent concentrations of unstabilised halogen (McCoy, 2002). Lower applied oxidant doses and less AOX generation resulted in achieving the AOX discharge limit consistently. This cooling water system has been in compliance with operational performance standards and AOX discharge limits, since implementation of the stabilised bromine antimicrobial-treatment programme.

32.3.3   Novel superior biofilm dispersants

Dispersants have been used for many years in water treatment. Most of the products for microbial control have been formulations of ethylene oxide and/or propylene oxide non-ionic co-polymers. These polymers provide fairly good dispersancy (the attribute of a surfactant to help keep suspended materials from accumulating onto surfaces), but relatively poor detergency (the attribute of a surfactant to remove deposits from surfaces). In recent years, surfactants with much better detergency, such as hard-water tolerant halogen-resistant linear alkylbenzene sulphonates, have been successfully used in microbial-fouling control and have been called biodetergents (Yu *et al.*, 1998). Combining the best of surfactants to achieve better dispersancy and detergency has improved biofilm control.

Modern cooling towers are usually equipped with high-efficiency packing (fill) to facilitate liquid–gas contact and improve cooling by evaporation. The fill is highly susceptible to microbial fouling. Fill fouling is accelerated by microbial biofilms that accumulate suspended particulate matter from the cooling water in a process that resembles sedimentation in a natural habitat. Fill fouling can rapidly reduce thermal efficiency in a cooling tower. Fouled cooling tower fill may also provide conditions in which *Legionella* proliferation may lead to increased health-related risk.

A successful biodetergent application to control microbial fouling in high-efficiency cooling tower fill is given in Figure 32.3. The rapid recovery of thermal efficiency in the cooling tower correlated with vast amounts of deposit removal from the fill (data not shown) due to the detergency of the treatment.

32.3.4   Removal of biofouling potential by pre-treatment

Biological reactors have been successfully used to limit microbial activity, improve performance of disinfectants, and minimise disinfection by-products. The concept is simple: limit microbial fouling by removing nutrients (potential biomass) that drive the fouling process. So far, most commercial applications have been in municipal drinking water treatment.

Recently, in-line biological reactors have been successfully used to limit industrial microbial fouling (Griebe and Flemming, 1996, 1998; Flemming and Griebe, 2000; Chandy and Angles, 2001). In certain applications, biological filtration may improve microbial-fouling control substantially. This can result in less need for chemical antimicrobial treatments. In some cases, the need for antimicro-

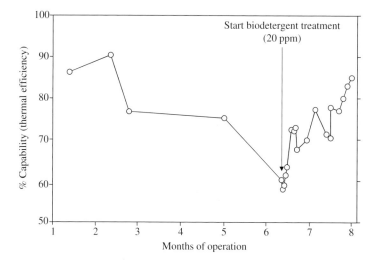

Figure 32.3   Power plant cooling tower thermal efficiency (% capability) before and after treatment with biodetergent. Microbial fouling of the high-efficiency cooling tower fill caused the decrease in % capability. The recovery of thermal efficiency resulted from continuous application of the biodetergent. Thermal efficiency rapidly improved and correlated with removal of biomass from the surface of cooling tower fill (data not shown). Redrawn from Yu *et al.* (1998).

bial-treatment chemicals could be entirely eliminated by depleting essential nutrients in the system with biological treatment.

32.3.5   Automating microbial-fouling control

Until very recently, there has been no practical means to automate the microbial-fouling control process. Techniques in diagnostics and monitoring have now advanced sufficiently to allow the first practical automation system for microbial-fouling control of industrial water treatment (Chattoraj *et al.*, 2002). An example application in an industrial system is given in Figure 32.4.

Note that in Figure 32.4, the antimicrobial feed pump run time per 600 s time interval tracks with changes in the microbial activity measurements. Pump run time depends on the microbial activity relative to predetermined set points (horizontal dashed lines) and the rate of change in microbial metabolic activity over the previous 600 s time interval. Thus, the remedy is proactively regulated. Regulation of the applied remedy results in improved antimicrobial efficiency by reducing the quantity of toxicant necessary for control of the microbial-fouling process. In this system, consumption of antimicrobial was substantially reduced (data not shown) with no loss in microbial-fouling control.

The key technical feature in automated microbial-fouling control is real-time proactive variance of antimicrobial stresses based on performance monitoring.

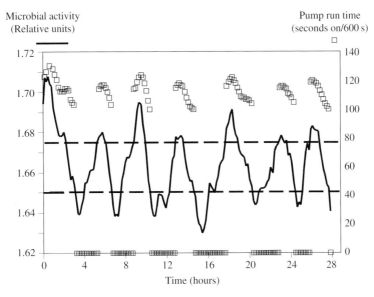

Figure 32.4   Real-time automated control of microbial fouling in an industrial water cooling system. Microbial activity was measured by in-line fluorometric analysis of a metabolic activity probe in the recirculating cooling water (solid line, expressed in relative units). Antimicrobial was fed to the system and controlled by computer algorithm that decided pump run time (seconds) during each 600 s (10 min) interval (squares). For example, at $t = 10$ min, the algorithm called for about 120 s of antimicrobial pump run time for the next 600 s interval. The pump run time decision was calculated by the algorithm based on the measurement of microbial activity relative to set points (dashed lines) and the rate of change in microbial activity during the previous 10 min interval. Substantially, less antimicrobial was consumed compared to continuous-dosing treatment programmes (data not shown). The data are redrawn from Chattoraj *et al.* (2002).

Compared to manual or timed on/off feed of antimicrobial, the benefits of automating microbial-fouling control are

(1) more efficient use of chemical,
(2) immediate response to upset conditions,
(3) less on-site expertise required.

   These benefits are valuable.

32.4   CONCLUSIONS

Microbial-fouling control can reduce industrial losses in operations and it can reduce health-related risk. Value can be measured in terms of product quality, energy conservation, and more importantly, in terms of reducing human suffering. The value of industrial microbial-fouling control is, therefore, very high.

Natural strategies to manage the microbial-fouling process can be models for industrial programmes. Sterilisation of natural and industrial systems is rarely ever practical. Root-causes of industrial-fouling problems are essentially always related to microbial diversity in biofilms.

Risk management has become an important aspect of the business in microbial-fouling control. This is the result of published guidance in government-approved codes of practice and from industry associations. Assessing risk is the essential first step in reducing it. The result of a formal risk assessment is a document that links discovery (recognition) of problem root-causes with specific corrective action recommendations (remedy) and assigns responsible personnel to action (regulation of remedy). The risk assessment document helps establish value and provides motive to get the risk management job done properly.

Significant improvements in diagnostics, monitoring, antimicrobials, biodetergents, nutrient limitation, and automation in microbial-fouling control technology for industry have been recently introduced.

REFERENCES

ASHRAE (American Society of Heating, Refrigeration and Air-Conditioning Engineers) (2000) Guideline 12-2000: Minimising the risk of Legionellosis associated with building water systems. 1791 Tullie Circle, N.E., Atlanta, GA 30329. www.ashrae.org

ASTM (American Society of Testing and Materials) D5952-96 (2002) Standard guide for inspecting water systems for *Legionellae* and investigating possible outbreaks of Legionellosis (Legionnaires' disease or pontiac fever). PO Box C700, Conshohocken, PA 19428–2959. www.astm.org

Atlas, R.M. (1999) *Legionella*: from environmental habitats to disease pathology, detection and control. *Environ. Microbiol.* **1**(4), 283–293.

Borchardt, S.A., Wetegrove, R.L. and Martens, J.D. (1997) New approaches to biocide effectiveness monitoring using on-site biocide active analysis, ATP analysis, and on-line dosage/monitoring control. *NACE Intl.* (www.nace.org) paper no. 466.

Chandy, J.P. and Angles, M.L. (2001) Determination of nutrients limiting biofilm formation and the subsequent impact on disinfectant decay. *Water Res.* **35**, 2677–2682.

Chattoraj, M., Fehr, M., Hatch, S.R. and Allain, E.J (2002) On-line measurement and control of microbiological activity in industrial water systems. *Matl. Perf.* **41**, 40–45.

Cloete, T.E. and Brozel, V.S. (1992) Practical aspects of biofouling control in industrial water systems. *Intl. Biodeterior Biodegrad.* **29**, 299–341.

Cloete, T.E., Jacobs, L. and Brozel, V.S. (1998) The chemical control of biofouling in industrial water systems. *Biodegradation* **9**, 23–37.

CTI (Cooling Technology Institute) (2000) Legionellosis. Guideline: Best practices for control of *Legionella*. Cooling Technology Institute. Houston, TX 77273. www.cti.org

DHS (Department of Human Services, Public Health Division) (2001) Supplementary notes for hospitals. Public Health Group, Victorian Government Department of Human Services, Melbourne, Victoria. www.legionella.vic.gov.au

DHS (Department of Human Services, Public Health Division) (2002) Code of practice for water treatment service providers (cooling tower systems). Public Health Group, Victorian Government Department of Human Services, Melbourne, Victoria. www.legionella.vic.gov.au

Donlan, R.M. (2002) Biofilms: microbial life on surfaces. *Emer. Inf. Dis.* **8**, 881–890.

Donlan, R.M and Costerton, J.W. (2002) Biofilms: survival mecahnaisms of clinically relevant microorganisms. *Clin. Microbiol. Rev.* **15**(2), 167–193.

Fields, B.S., Benson, R.F. and Besser, R.E. (2002) *Legionella* and Legionnaires' disease: 25 years of investigation. *Clin. Microbiol. Rev.* **15**(3), 506–526.

Flemming, H.-C. (2002) Role and levels of real time monitoring for successful antifouling strategies – An overview. In: *Proceedings of the International Specialised Conference on Biofilm Monitoring,* Porto, Portugal, March 17–20, 2002. pp. 8–10, www.deb. uminho.pt/biofilm2002

Flemming, H.-C. and Griebe, T. (2000) Control of biofilms in industrial waters and processes. In: *Biofouling of Industrial Waters and Processes*, (ed. J.T. Walker, J.J Jass and S.Surman), pp. 125–141, John Wiley & Sons Ltd.

Gorman, C. (2002) What's in your pipes? *Time Mag.* **160**(1), 56.

Griebe, T.U. and Flemming, H.-C. (1996) Vermeidung von Bioziden in Wasseraufbereitungs-Systemen durch Nährstoffentnahme (Avoiding biocides in water treatment systems by nutrient limitation). *Vom Wasser* **86**, 217–230.

Griebe, T. and Flemming, H.-C. (1998) Biocide-free antifouling strategy to protect RO membranes from biofouling. *Desalination* **118**, 153–156.

HSE (Health and Safety Commission and Executive) (2000) Legionnaires' disease. The control of legionella bacteria in water systems. Approved Code of Practice and Guidance, L8. PO Box 1999, Sudbury, UK www.hse.gov.uk

Klahre, J. and Flemming, H.-C. (2000) Monitoring of biofouling in papermill process waters. *Water Res.* **34**, 3657–3665.

Mayeno, A.B. (1989) Investigation into halogenating agents generated by human neutrophils and eosinophils. Ph.D. thesis. University of California, Los Angeles.

McCoy, W.F. (1998) Imitating natural microbial fouling control. *Matl. Perf.* **37**, 45–48.

McCoy, W.F. (2002) A new environmentally sensible chlorine alternative. In *Industrial Biocides* (ed. D.R. Karsa and D. Ashworth), pp. 52–62, Royal Society of Chemistry, Cambridge, UK.

McCoy, W.F, Allain, E.J., Yang, S. and Dallmier, A.W. (1998) Strategies used in nature for microbial fouling control: applications for industrial water treatment. *NACE Intl.* (www.nace.org) paper No. 520.

Murga, R., Forster, T.S., Brown, E., Pruckler, J.M, Fields, B.S. and Donlan, R.M. (2001) Role of biofilms in the survival of *Legionella pneumophila* in a model potable-water system. *Microbiology* **147**, 3121–3126.

OSHA (Occupational, Safety and Health Administration) (1998) Technical Manual, Section II, Chapter 7, Legionnaires' Disease. www.osha-slc.gov/TechMan_data/II_7.html

Percival, S.T., Walker, J.T. and Hunter, P.R (2000) Risk assessment. In: *Microbiological Aspects of Biofilms and Drinking Water*. Chapter 4, pp. 41–48, CRC Press, London.

Schaule, G., Griebe, T. and Flemming, H.-C. (2000) Steps in biofilm sampling and characterization of biofouling cases. In *Biofilms: Investigative Methods and Applications* (ed. H.-C. Flemming, U. Szewzyk and T. Griebe), pp. 1–21, Technomic Publishing Co., Inc. Lancaster, Pennsylvania, USA 17604.

Theron, J. and Cloete, T.E. (2000) Molecular techniques for determining microbial diverstiy and community structure in natural environments. *Crit. Rev. Microbiol.* **26**, 37–57.

Thomas, E.L., Bozeman, P.M, Jefferson, M.M. and King, C.C. (1995) Oxidation of bromide by the human leukocyte enzymes myeloperoxidase and eosinophil peroxidase. *J. Biol. Chem.* **270**(7), 2906–2913.

Weiss, S.J., Test, S.T., Eckmann, C.M., Roos, D. and Regiani, S. (1986) Brominating oxidants generated by human eosinophils. *Science* **234**, 200–203.

Wetegrove, R.L., Banks, R.H. and Hermiller M.R. (1996) Optical monitor for improved fouling control in cooling systems. Cooling Technical Institute Paper No. TP96-09. www.cti.org

Yu, F.P., Ginn, L.D. and McCoy, W.F. (1998) Cooling tower fill fouling control in a geothermal power plant. *J. Cooling Tech. Inst.* **19**(2), 34–36.

Index

Absorbance
  infrared 461
Acanthamoeba 183
Adhesion 92, 259
  initial 17, 48
Adhesin 26, 34
Adsorption
  protein 453
Aerobic 133
Aerobiology 160
Aeromonas 58, 66, 353
Aerosol
  particles 161
  generation 168
Aggregation 26, 91, 102, 146, 259, 295
Algenate 25, 57, 96, 293
Amoebae 181
Anaerobic digestion 144
Antibiotic
  application 537
  inactivation 101
  modification 540
  resistance 523
Antimicrobials 480, 537, 599
Antimicrobial therapy, *see* disinfection
Atomic force microscope (AFM) 11, 119, 270
ATP bioluminescence 216, 484, 557
Azospirillum brasilense 269

Bacillus
  cereus 557
  subtilis 266, 543
Bacteria
  non-culturable 484
Bioaerosol
  characteristics 163

formation 167
relevance 165
sampling 170
sampling devices 171
Biocides 467, 512, 537, 562, 577
  efficacy testing
Biocorrosion 334, 572
Biofilm
  activity
  channels 432
  cleaning 575
  construct 231, 293
  control 492, 554, 575
  detachment 55, 218
  development 48, 63, 321, 569
  disinfectant test 231, 487
  formation 215, 538
  fermenter
  heterogeneity 48, 416
  matrix 81, 289
  mechanical properties 197
  models 413, 539
  monitoring 222, 443, 450, 573
  mushroom 66, 426
  structure 67, 420
  substrate transport 286, 398, 430
  proteome analysis 105
  thickness 469
Biofouling
  factors 591
  health risk 593
  management 594
  monitoring 461, 598
  processes 572
Biological aerated filters (BAF) 140

Biomass
    detachment 218, 428
    growth 215, 419
    spreading 419
Biomedical device 537
Boundery layer 402, 417
Brine 574
Bromine 578, 599

Campylobacter spp. 557
Catheter 66, 503, 537
Cathodic depolarisation 122
Cellular automaton 420
Cell signalling 56, 63
Chemical gradient 100, 240, 398, 431,
    513
Chemostat 206, 237
Chlorine 389, 578
Coadhesion 32
Coaggregation 34
Collagen 277
Completely stirred tank reactor 206, 220,
    237
COMSTAT 71
Conditioning layers
Constant depth film fermenter (CDFF)
    247
Contact angle 17
Cooling water 572, 600
Copper foils 337
Corrosion
    processes 115
    products 116
Cryoporometry 301
Cystic fibrosis 65, 547

Denaturing gradient gel electrophoresis
    (DGGE) 353
Desulfovibrio sp. 123
Diffusion 285
Diffusion barrier 100, 115
Disinfectant testing 219, 230, 479, 562
Disinfection 477
DLVO-theory 5, 19, 152, 261
Dormancy 543
Double layer 260
Droplets 163

Efflux 517, 541
Electrostatic interactions 260
Energy dispersive X-ray (EDS) analysis
    332
Entamoeba histolytica 180
Enterobacteriaceae 557

Environmental
    cells 340
    scanning electron microscopy 332
    signals 48
    transmission electron microscopy 339
Enzymes 84, 120
Escherichia coli 12, 49, 97, 182, 270, 522,
    545, 557
Exopolymers 21, 81, 119
Exopolysaccharides 92
Extracellular polymeric substances (EPS)
    16, 55, 197, 308, 447, 451
    composition 85, 94
    function 93
    matrix 65, 119
    modelling 423
    physical properties 87, 95

Flagella 53
Flow cells 197, 222
    flat plate 201
    square glass capillary 202
Flow velocity 66, 464
Fuidized bed reactor 141
Fluorescence in situ hybridisation (FISH)
    37, 363, 376, 485
Food
    hygiene 564
    industry 554
Force
    distance curve 12
    measurements 273
Fouling, *see* biofouling

Gene expression 42, 197
Giardia lamblia 180
Glycoconjugates 309
Glycoproteins 83
Gradient plate 245
Gradostat 238
Granulation 146
Granular sludge 289, 145
Green fluorescent protein (GFP) 69, 85
Growth rate 215, 419
Growth systems 197, 214, 230, 236

Hart valves 504, 537
Heat exchangers 569, 593
Heterogeneity 236
Heterogeneous microenvironments 245
Homoserine lactones (HSL) 65, 544
    Acylated (AHSL) 63, 103
Hospital acquired infection 504, 522
Hybridisation 354

Hydrodynamics 203
Hydrogels 231
Hydrogen peroxide 123
Hydrophobicity
  cell surface 6, 17, 103
Hygiene 566

Image analysis 71
Imaging
  AMF 272
  ETEM 344
  NMR 294, 298
  molecular 299
  temperature 300
Infection 503
Infrared 461
Ion selective glass 378

Klebsiella pneumoniae 23, 183, 540

Laser 444
Laser scanning microscopy (LSM) 37,
  311, 375, 414, 446, 455, 493
Lectin
  binding analysis 310
  properties 24, 85, 309
  screening 311
Legionella pneumophila 162, 182, 572,
  594
Legionellosis 593
Lipopolysaccharide 53
Listeria 557
  monocytogenes 474, 231
Liquid ion exchangers (LIX) 381
London - van der Waales forces 260

Matrix
  architecture 85
  compostion 83
Membrane
  types 571
Metal
  binding 125
  ions 274, 289, 335
  reduction 124
Methanogenic bacteria 144
Methanosaeta 147
Microautoradiography (MAR)
  368
Microbiological
  influenced corrosion (MIC)
  117
  survey 556
  testing 574

Microbiosensors
  glucose 392
  methane 393
  nitrate 394
  nitrous oxide 394
Microelectrodes
  amperometric 385
  cardon dioxide 384
  chlorine 389
  dissolved oxygen (DO) 85, 385
  hydrogen peroxide 389
  ion-selective 381
  mass transport 285, 390
  pH 85, 382
  potentiometric 380
  redox 381
  sulfide 380, 387
Microscopy 331
Microoptodes 375
Microsensors
  fiber-optic 396
Microstat 245
Minimum inhibitory concentration (MIC)
  514
Modelling 412
Model distribution system (MDS) 223
Monitoring, *see* biofilm monitoring
  side stream 574
  on-line 575, 598
Multispecies 48, 416
Mutagenesis 48
Mutant 49, 70
Mycobacterium 53, 186, 353, 514, 543

Nuclear magnetic resonance (NMR)
  imaging 294
  pulsed field gradiet (PFG) 291
  spectroscopy 288
Non-sporulating bacteria

Oceanospirillum 343
Oligonucleotide probe 354
On-line monitoring 575
Oral bacteria 32
Overtones 456
Ozone 578

Pathogens 474, 594
Percolating filter 135
Perfusion 285
Piezoelectric
  transducer 444
  quartz crystal 451
Phospholipids 83, 95

Photoacoustic spectroscopy 301, 444
1-Photon excitation 325
2-Photon excitation 105, 325
Phylogenetic tree 354
Physiological gradient 100
Pili 56, 66
Planctonic cells 47, 230, 539, 568
Preservative 537
Propella reactor 222
Protozoa 179
Poloxamer 231
Polysaccharide
    chains 84, 95
    intercellular adhesin (PIA) 26, 103
Probes
    dye exclusion 485
    peptide nucleic acid (PNA) 355
    16S rRNA 353
Protein 96
Porosity 293
Pseudomonas 123, 136, 557
    aeruginosa 22, 49, 66, 96, 183, 233,
        506, 513, 545
    fluorescens 22, 233, 268, 463
    putida 344

Quartz Crystal Microbalance 450
Quorum sensing 58, 63, 544
    interspecies 65

Receptor 35
Reaction sink 101
Repair 541
Reporter genes 68
Resistance
    mechanisms 514, 540
Reverse osmosis 570
Reynolds number 464
Ribosomal RNA
    16S rRNA 352
    probes 354
Risk assessment 597
Robbins device 222, 574
Rotating
    annular reactor 205, 222, 310
    biological contactor 138
RotoTorque 222

Salmonella 66, 234, 557
    enteritidis 545
SOS system 541
Spectroscopy
    electron energy loss (EELS) 342

infrared 263, 309, 575
    nuclear magnetic resonance (NMR)
        287
    photoacoustic 443
    secondary-ion mass (SIMS) 10
    X-ray photoelectron 264
Sphingomonas paucimobilis 22
SspA/SspB 39
Stainless steels 116
Staphylococcus
    aureus 49, 475, 506, 514, 538, 557
    epidermis 23, 49, 97, 233
Streptococci 24, 66
Stress responses 184
Sulfate reducing bacteria 117, 333
Surface
    active compounds (SAC) 22
    analysis 262
    charge 260
    composition 266
    forces 273
    hydrophobicity 6
    imaging 272
    proteins 56, 277
    tention 19

Thermodynamics 152, 261
Tomography 287
Trickling filter 135
Turbidimetry 575

Ultrasound 585
Upflow anaerobic sludge blanket (UASB)
        142

Viscosity 99

Water
    drinking 181, 214
    ultrapure 571
Wastewater treatment 132

XDLVO theory 23
xPS 92
    cohesiveness 98
    constituents 94
    physical properties 95
X-ray photoelectron spectroscopy (XPS)
        10, 264

Zeta potential 7